T0367734

Introduction to Coding Theory

Error-correcting codes constitute one of the key ingredients in achieving the high degree of reliability required in modern data transmission and storage systems. This book introduces the reader to the theoretical foundations of error-correcting codes, with an emphasis on Reed–Solomon codes and their derivative codes.

After reviewing linear codes and finite fields, the author describes Reed–Solomon codes and various decoding algorithms. Cyclic codes are presented, as are MDS codes, graph codes, and codes in the Lee metric. Concatenated, trellis, and convolutional codes are also discussed in detail. Homework exercises introduce additional concepts such as Reed–Muller codes, and burst error correction. The end-of-chapter notes often deal with algorithmic issues, such as the time complexity of computational problems.

While mathematical rigor is maintained, the text is designed to be accessible to a broad readership, including students of computer science, electrical engineering, and mathematics, from senior-undergraduate to graduate level.

This book contains over 100 worked examples and over 340 exercises—many with hints.

RON M. ROTH joined the faculty of Technion—Israel Institute of Technology (Haifa, Israel) in 1988, where he is a Professor of Computer Science and holds the General Yaakov Dori Chair in Engineering. He also held visiting positions at IBM Research Division (San Jose, California) and, since 1993, at Hewlett–Packard Laboratories (Palo Alto, California). He is a Fellow of the Institute of Electrical and Electronics Engineers (IEEE).

Introduction to Coding Theory

Ron M. Roth

Technion—Israel Institute of Technology
Haifa, Israel

CAMBRIDGE UNIVERSITY PRESS
Cambridge, New York, Melbourne, Madrid, Cape Town,
Singapore, São Paulo, Delhi, Mexico City

Cambridge University Press
The Edinburgh Building, Cambridge CB2 8RU, UK

Published in the United States of America by Cambridge University Press, New York

www.cambridge.org
Information on this title: www.cambridge.org/9780521845045

First published 2006
Reprinted with corrections 2007

A catalogue record for this publication is available from the British Library

ISBN 978-0-521-84504-5 Hardback

Contents

Preface page ix

1 Introduction 1
- 1.1 Communication systems 1
- 1.2 Channel coding 3
- 1.3 Block codes 5
- 1.4 Decoding 7
- 1.5 Levels of error handling 11
- Problems 17
- Notes 22

2 Linear Codes 26
- 2.1 Definition 26
- 2.2 Encoding of linear codes 28
- 2.3 Parity-check matrix 29
- 2.4 Decoding of linear codes 32
- Problems 36
- Notes 47

3 Introduction to Finite Fields 50
- 3.1 Prime fields 50
- 3.2 Polynomials 51
- 3.3 Extension fields 56
- 3.4 Roots of polynomials 59
- 3.5 Primitive elements 60
- 3.6 Field characteristic 62
- 3.7 Splitting field 64
- 3.8 Application: double error-correcting codes 66
- Problems 70
- Notes 90

4 Bounds on the Parameters of Codes 93
4.1 The Singleton bound 94
4.2 The sphere-packing bound 95
4.3 The Gilbert–Varshamov bound 97
4.4 MacWilliams' identities 99
4.5 Asymptotic bounds 104
4.6 Converse Coding Theorem 110
4.7 Coding Theorem 115
Problems 119
Notes 136

5 Reed–Solomon and Related Codes 147
5.1 Generalized Reed–Solomon codes 148
5.2 Conventional Reed–Solomon codes 151
5.3 Encoding of RS codes 152
5.4 Concatenated codes 154
5.5 Alternant codes 157
5.6 BCH codes 162
Problems 163
Notes 177

6 Decoding of Reed–Solomon Codes 183
6.1 Introduction 183
6.2 Syndrome computation 184
6.3 Key equation of GRS decoding 185
6.4 Solving the key equation by Euclid's algorithm 191
6.5 Finding the error values 194
6.6 Summary of the GRS decoding algorithm 195
6.7 The Berlekamp–Massey algorithm 197
Problems 204
Notes 215

7 Structure of Finite Fields 218
7.1 Minimal polynomials 218
7.2 Enumeration of irreducible polynomials 224
7.3 Isomorphism of finite fields 227
7.4 Primitive polynomials 227
7.5 Cyclotomic cosets 229
Problems 232
Notes 240

8 Cyclic Codes 242
- 8.1 Definition 242
- 8.2 Generator polynomial and check polynomial 244
- 8.3 Roots of a cyclic code 247
- 8.4 BCH codes as cyclic codes 250
- 8.5 The BCH bound 253
- Problems 256
- Notes 265

9 List Decoding of Reed–Solomon Codes 266
- 9.1 List decoding 267
- 9.2 Bivariate polynomials 268
- 9.3 GRS decoding through bivariate polynomials 269
- 9.4 Sudan's algorithm 271
- 9.5 The Guruswami–Sudan algorithm 276
- 9.6 List decoding of alternant codes 280
- 9.7 Finding linear bivariate factors 284
- 9.8 Bounds on the decoding radius 289
- Problems 291
- Notes 295

10 Codes in the Lee Metric 298
- 10.1 Lee weight and Lee distance 298
- 10.2 Newton's identities 300
- 10.3 Lee-metric alternant codes and GRS codes 302
- 10.4 Decoding alternant codes in the Lee metric 306
- 10.5 Decoding GRS codes in the Lee metric 312
- 10.6 Berlekamp codes 314
- 10.7 Bounds for codes in the Lee metric 316
- Problems 321
- Notes 327

11 MDS Codes 333
- 11.1 Definition revisited 333
- 11.2 GRS codes and their extensions 335
- 11.3 Bounds on the length of linear MDS codes 338
- 11.4 GRS codes and the MDS conjecture 342
- 11.5 Uniqueness of certain MDS codes 347
- Problems 351
- Notes 361

12 Concatenated Codes 365
12.1 Definition revisited 366
12.2 Decoding of concatenated codes 367
12.3 The Zyablov bound 371
12.4 Justesen codes 374
12.5 Concatenated codes that attain capacity 378
Problems 381
Notes 392

13 Graph Codes 395
13.1 Basic concepts from graph theory 396
13.2 Regular graphs 401
13.3 Graph expansion 402
13.4 Expanders from codes 406
13.5 Ramanujan graphs 409
13.6 Codes from expanders 411
13.7 Iterative decoding of graph codes 414
13.8 Graph codes in concatenated schemes 420
Problems 426
Notes 445

14 Trellis and Convolutional Codes 452
14.1 Labeled directed graphs 453
14.2 Trellis codes 460
14.3 Decoding of trellis codes 466
14.4 Linear finite-state machines 471
14.5 Convolutional codes 477
14.6 Encoding of convolutional codes 479
14.7 Decoding of convolutional codes 485
14.8 Non-catastrophic generator matrices 495
Problems 501
Notes 518

Appendix: Basics in Modern Algebra 521
Problems 522

Bibliography 527

List of Symbols 553

Index 559

Preface

Do ye imagine to reprove words?
Job 6:26

This book has evolved from lecture notes that I have been using for an introductory course on coding theory in the Computer Science Department at Technion. The course deals with the basics of the theory of error-correcting codes, and is intended for students in the graduate and upper-undergraduate levels from Computer Science, Electrical Engineering, and Mathematics. The material of this course is covered by the first eight chapters of this book, excluding Sections 4.4–4.7 and 6.7. Prior knowledge in probability, linear algebra, modern algebra, and discrete mathematics is assumed. On the other hand, all the required material on finite fields is an integral part of the course. The remaining parts of this book can form the basis of a second, advanced-level course.

There are many textbooks on the subject of error-correcting codes, some of which are listed next: Berlekamp [36], Blahut [46], Blake and Mullin [49], Lin and Costello [230], MacWilliams and Sloane [249], McEliece [259], Peterson and Weldon [278], and Pless [280]. These are excellent sources, which served as very useful references when compiling this book. The two volumes of the *Handbook of Coding Theory* [281] form an extensive encyclopedic collection of what is known in the area of coding theory.

One feature that probably distinguishes this book from most other classical textbooks on coding theory is that generalized Reed–Solomon (GRS) codes are treated *before* BCH codes—and even before cyclic codes. The purpose of this was to bring the reader to see, as early as possible, families of codes that cover a wide range of minimum distances. In fact, the cyclic properties of (conventional) Reed–Solomon codes are immaterial for their distance properties and may only obscure the underlying principles of the decoding algorithms of these codes. Furthermore, bit-error-correcting codes, such as binary BCH codes, are found primarily in spatial communication applications, while readers are now increasingly exposed to temporal communication platforms, such as magnetic and optical storage media. And in those applications—including domestic CD and DVD—the use of GRS codes prevails.

Therefore, the treatment of finite fields in this book is split, where the first batch of properties (in Chapter 3) is aimed at laying the basic background on finite fields that is sufficient to define GRS codes and understand their decoding algorithm. A second batch of properties of finite fields is provided in Chapter 7, prior to discussing cyclic codes, and only then is the reader presented with the notions of minimal polynomials and cyclotomic cosets.

Combinatorial bounds on the parameters of codes are treated mainly in Chapter 4. In an introductory course, it would suffice to include only the Singleton and sphere-packing bounds (and possibly the non-asymptotic version of the Gilbert–Varshamov bound). The remaining parts of this chapter contain the asymptotic versions of the combinatorial bounds, yet also cover the information-theoretic bounds, namely, the Shannon Coding Theorem and Converse Coding Theorem for the q-ary symmetric channel. The latter topics may be deferred to an advanced-level course.

GRS codes and alternant codes constitute the center pillar of this book, and a great portion of the text is devoted to their study. These codes are formally introduced in Chapter 5, following brief previews in Sections 3.8 and 4.1. Classical methods for GRS decoding are described in Chapter 6, whereas Chapter 9 is devoted to the list decoding of GRS codes and alternant codes. The performance of these codes as Lee-metric codes is then the main topic of Chapter 10. GRS codes play a significant role also in Chapter 11, which deals with MDS codes.

The last three chapters of the book focus on compound constructions of codes. Concatenated codes and expander-based codes (which are, in a way, two related topics) are presented in Chapters 12 and 13, and an introduction to trellis codes and convolutional codes is given in Chapter 14. This last chapter was included in this book for the sake of an attempt for completeness: knowing that the scope of the book could not possibly allow it to touch all the aspects of trellis codes and convolutional codes, the model of state-dependent coding, which these codes represent, was still too important to be omitted.

Each chapter ends with problems and notes, which occupy on average a significant portion of the chapter. Many of the problems introduce additional concepts that are not covered in text; these include Reed–Muller codes, product codes and array codes, burst error correction, interleaving, the implementation of arithmetic in finite fields, or certain bounds—e.g., the Griesmer and Plotkin bounds. The notes provide pointers to references and further reading. Since the text is intended also for readers who are computer scientists, the notes often contain algorithmic issues, such as the time complexity of certain computational problems that are related to the discussion in the text.

Finally, the Appendix (including the problems therein) contains a short summary of several terms from modern algebra and discrete mathematics, as these terms are frequently used in the book. This appendix is meant merely to recapitulate material, which the reader is assumed to be rather familiar with from prior studies.

I would like to thank the many students and colleagues, whose input on earlier versions of this book greatly helped in improving the presentation. Special thanks are due to Shirley Halevy, Ronny Lempel, Gitit Ruckenstein, and Ido Tal, who taught the course with me at Technion and offered a wide variety of useful ideas while the book was being written. Ido was particularly helpful in detecting and correcting many of the errors in earlier drafts of the text (obviously, the responsibility for all remaining errors is totally mine). I owe thanks to Brian Marcus and Gadiel Seroussi for the good advice that they provided along the way, and to Gadiel, Vitaly Skachek, and the anonymous reviewers for the constructive comments and suggestions. Part of the book was written while I was visiting the Information Theory Research Group at Hewlett–Packard Laboratories in Palo Alto, California. I wish to thank the Labs for their kind hospitality, and the group members in particular for offering a very encouraging and stimulating environment.

Chapter 1

Introduction

In this chapter, we introduce the model of a communication system, as originally proposed by Claude E. Shannon in 1948. We will then focus on the channel portion of the system and define the concept of a probabilistic channel, along with models of an encoder and a decoder for the channel. As our primary example of a probabilistic channel—here, as well as in subsequent chapters—we will introduce the memoryless q-ary symmetric channel, with the binary case as the prevailing instance used in many practical applications. For $q = 2$ (the binary case), we quote two key results in information theory. The first result is a coding theorem, which states that information through the channel can be transmitted with an arbitrarily small probability of decoding error, as long as the transmission rate is below a quantity referred to as the capacity of the channel. The second result is a converse coding theorem, which states that operating at rates above the capacity necessarily implies unreliable transmission.

In the remaining part of the chapter, we shift to a combinatorial setting and characterize error events that can occur in channels such as the q-ary symmetric channel, and can always be corrected by suitably selected encoders and decoders. We exhibit the trade-off between error correction and error detection: while an error-detecting decoder provides less information to the receiver, it allows us to handle twice as many errors. In this context, we will become acquainted with the erasure channel, in which the decoder has access to partial information about the error events, namely, the location of the symbols that might be in error. We demonstrate that—here as well—such information allows us to double the number of correctable errors.

1.1 Communication systems

Figure 1.1 shows a communication system for transmitting information from a *source* to a *destination* through a *channel*. The communication can be

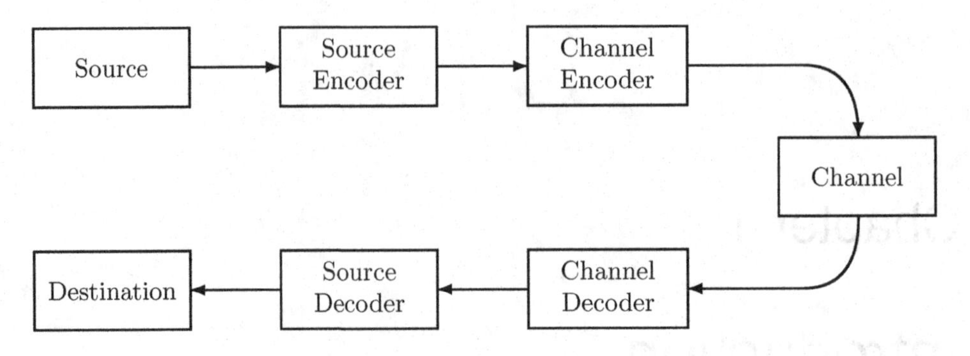

Figure 1.1. Communication system.

either in the space domain (i.e., from one location to another) or in the time domain (i.e., by storing data at one point in time and retrieving it some time later).

The role of source coding is twofold. First, it serves as a translator between the output of the source and the input to the channel. For example, the information that is transmitted from the source to the destination may consist of analog signals, while the channel may expect to receive digital input; in such a case, an analog-to-digital conversion will be required at the encoding stage, and then a back conversion is required at the decoding stage. Secondly, the source encoder may *compress* the output of the source for the purpose of economizing on the length of the transmission; at the other end, the source decoder decompresses the received signal or sequence. Some applications require that the decoder restore the data so that it is identical to the original, in which case we say that the compression is *lossless*. Other applications, such as most audio and image transmissions, allow some (controlled) difference—or distortion—between the original and the restored data, and this flexibility is exploited to achieve higher compression; the compression is then called *lossy*.

Due to physical and engineering limitations, channels are not perfect: their output may differ from their input because of noise or manufacturing defects. Furthermore, sometimes the design requires that the format of the data at the output of the channel (e.g., the set of signals that can be read at the output) should differ from the input format. In addition, there are applications, such as magnetic and optical mass storage media, where certain patterns are not allowed to appear in the recorded (i.e., transmitted) bit stream. The main role of channel coding is to overcome such limitations and to make the channel as transparent as possible from the source and destination points of view. The task of signal translation, which was mentioned earlier in the context of source coding, may be undertaken partially (or wholly) also by the channel encoder and decoder.

1.2 Channel coding

We will concentrate on the channel coding part of Figure 1.1, as shown in Figure 1.2.

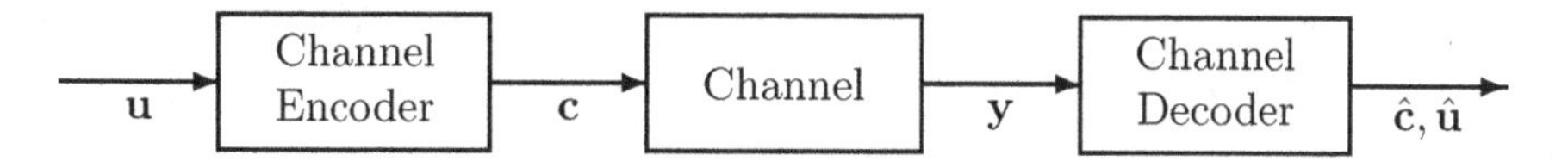

Figure 1.2. Channel coding.

Our model of the channel will be that of the *(discrete) probabilistic channel*: a probabilistic channel S is defined as a triple (F, Φ, Prob), where F is a finite *input alphabet*, Φ is a finite *output alphabet*, and Prob is a conditional probability distribution

$$\mathsf{Prob}\{\,\mathbf{y} \text{ received} \mid \mathbf{x} \text{ transmitted}\,\}$$

defined for every pair $(\mathbf{x}, \mathbf{y}) \in F^m \times \Phi^m$, where m ranges over all positive integers and F^m (respectively, Φ^m) denotes the set of all words of length m over F (respectively, over Φ). (We assume here that the channel neither deletes nor inserts symbols; that is, the length of an output word $\mathbf{y}$ always equals the length of the respective input word $\mathbf{x}$.)

The input to the channel encoder is an *information word* (or *message*) $\mathbf{u}$ out of M possible information words (see Figure 1.2). The channel encoder generates a *codeword* $\mathbf{c} \in F^n$ that is input to the channel. The resulting output of the channel is a *received word* $\mathbf{y} \in \Phi^n$, which is fed into the channel decoder. The decoder, in turn, produces a *decoded codeword* $\hat{\mathbf{c}}$ and a *decoded information word* $\hat{\mathbf{u}}$, with the aim of having $\mathbf{c} = \hat{\mathbf{c}}$ and $\mathbf{u} = \hat{\mathbf{u}}$. This implies that the channel encoder needs to be such that the mapping $\mathbf{u} \mapsto \mathbf{c}$ is one-to-one.

The *rate* of the channel encoder is defined as

$$R = \frac{\log_{|F|} M}{n} .$$

If all information words have the same length over F, then this length is given by the numerator, $\log_{|F|} M$, in the expression for R (strictly speaking, we need to round up the numerator in order to obtain that length; however, this integer effect phases out once we aggregate over a sequence of $\ell \to \infty$ transmissions, in which case the number of possible information words becomes M^ℓ and the codeword length is $\ell \cdot n$). Since the mapping of the encoder is one-to-one, we have $R \le 1$.

The encoder and decoder parts in Figure 1.2 will be the subject of Sections 1.3 and 1.4, respectively. We next present two (related) examples of probabilistic channels, which are very frequently found in practical applications.

Example 1.1 The memoryless *binary symmetric channel* (in short, BSC) is defined as follows. The input and output alphabets are $F = \Phi = \{0, 1\}$, and for every two binary words $\mathbf{x} = x_1 x_2 \ldots x_m$ and $\mathbf{y} = y_1 y_2 \ldots y_m$ of a given length m,

$$\begin{aligned} &\mathsf{Prob}\{\, \mathbf{y} \text{ received} \mid \mathbf{x} \text{ transmitted} \,\} \\ &\qquad = \prod_{j=1}^{m} \mathsf{Prob}\{\, y_j \text{ received} \mid x_j \text{ transmitted} \,\}\,, \qquad (1.1) \end{aligned}$$

where, for every $x, y \in F$,

$$\mathsf{Prob}\{\, y \text{ was received} \mid x \text{ was transmitted} \,\} = \begin{cases} 1-p & \text{if } y = x \\ p & \text{if } y \neq x \end{cases} .$$

The parameter p is a real number in the range $0 \le p \le 1$ and is called the *crossover probability* of the channel.

The action of the BSC can be described as flipping each input bit with probability p, independently of the past or the future (the adjective "memoryless" reflects this independence). The channel is called "symmetric" since the probability of the flip is the same regardless of whether the input is 0 or 1. The BSC is commonly represented by a diagram as shown in Figure 1.3. The possible input values appear to the left and the possible output values are shown to the right. The label of a given edge from input x to output y is the conditional probability of receiving the output y given that the input is x.

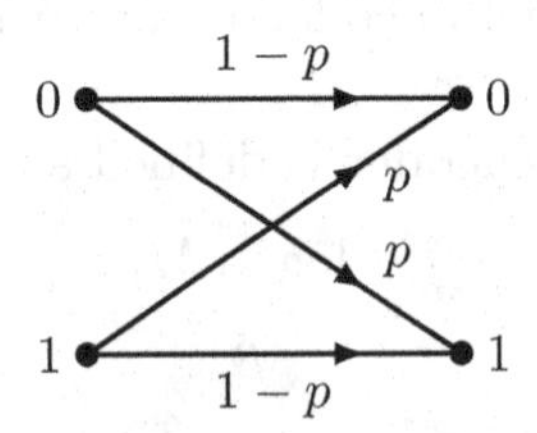

Figure 1.3. Binary symmetric channel.

The cases $p = 0$ and $p = 1$ correspond to reliable communication, whereas $p = \frac{1}{2}$ stands for the case where the output of the channel is statistically independent of its input. □

Example 1.2 The memoryless *q-ary symmetric channel* with crossover probability p is a generalization of the BSC to alphabets $F = \Phi$ of size q. The conditional probability (1.1) now holds for every two words $\mathbf{x} = x_1 x_2 \ldots x_m$ and $\mathbf{y} = y_1 y_2 \ldots y_m$ over F, where

$$\mathsf{Prob}\{\, y \text{ was received} \mid x \text{ was transmitted} \,\} = \begin{cases} 1-p & \text{if } y = x \\ p/(q-1) & \text{if } y \neq x \end{cases} .$$

(While the term "crossover" is fully justified only in the binary case, we will nevertheless use it for the general q-ary case as well.) □

In the case where the input alphabet F has the same (finite) size as the output alphabet Φ, it will be convenient to assume that $F = \Phi$ and that the elements of F form a finite Abelian group (indeed, for every positive integer q there is an Abelian group of size q, e.g., the ring $\mathbb{Z}_q$ of integer residues modulo q; see Problem A.21 in the Appendix). We then say that the channel is an *additive channel*. Given an additive channel, let $\mathbf{x}$ and $\mathbf{y}$ be input and output words, respectively, both in F^m. The *error word* is defined as the difference $\mathbf{y}-\mathbf{x}$, where the subtraction is taken component by component. The action of the channel can be described as adding (component by component) an error word $\mathbf{e} \in F^m$ to the input word $\mathbf{x}$ to produce the output word $\mathbf{y} = \mathbf{x} + \mathbf{e}$, as shown in Figure 1.4. In general, the distribution of the error word $\mathbf{e}$ may depend on the input $\mathbf{x}$. The q-ary symmetric channel is an example of a channel where $\mathbf{e}$ is statistically independent of $\mathbf{x}$ (in such cases, the term *additive noise* is sometimes used for the error word $\mathbf{e}$).

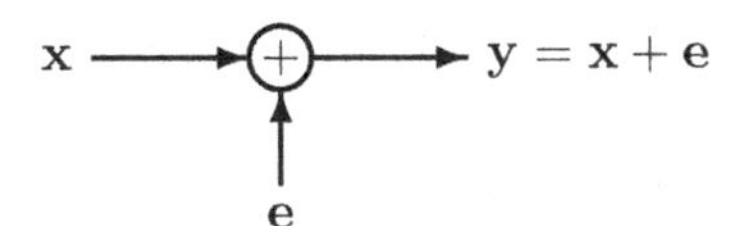

Figure 1.4. Additive channel.

When F is an Abelian group, it contains the zero (or unit) element. The *error locations* are the indexes of the nonzero entries in the error word $\mathbf{e}$. Those entries are referred to as the *error values*.

1.3 Block codes

An (n, M) *(block) code* over a finite alphabet F is a nonempty subset $\mathcal{C}$ of size M of F^n. The parameter n is called the *code length* and M is the *code size*. The *dimension* (or *information length*) of $\mathcal{C}$ is defined by $k = \log_{|F|} M$, and the *rate* of $\mathcal{C}$ is $R = k/n$. The range of the mapping defined by the channel encoder in Figure 1.2 forms an (n, M) code, and this is the context in which the term (n, M) code will be used. The elements of a code are called *codewords*.

In addition to the length and the size of a code, we will be interested in the sequel also in quantifying how much the codewords in the code differ from one another. To this end, we will make use of the following definitions.

Let F be an alphabet. The *Hamming distance* between two words $\mathbf{x}, \mathbf{y} \in F^n$ is the number of coordinates on which $\mathbf{x}$ and $\mathbf{y}$ differ. We denote the Hamming distance by $\mathsf{d}(\mathbf{x}, \mathbf{y})$.

It is easy to verify that the Hamming distance satisfies the following properties of a metric for every three words $\mathbf{x}, \mathbf{y}, \mathbf{z} \in F^n$:

- $\mathsf{d}(\mathbf{x}, \mathbf{y}) \geq 0$, with equality if and only if $\mathbf{x} = \mathbf{y}$.
- Symmetry: $\mathsf{d}(\mathbf{x}, \mathbf{y}) = \mathsf{d}(\mathbf{y}, \mathbf{x})$.
- The triangle inequality: $\mathsf{d}(\mathbf{x}, \mathbf{y}) \leq \mathsf{d}(\mathbf{x}, \mathbf{z}) + \mathsf{d}(\mathbf{z}, \mathbf{y})$.

Let F be an Abelian group. The *Hamming weight* of $\mathbf{e} \in F^n$ is the number of nonzero entries in $\mathbf{e}$. We denote the Hamming weight by $\mathsf{w}(\mathbf{e})$. Notice that for every two words $\mathbf{x}, \mathbf{y} \in F^n$,

$$\mathsf{d}(\mathbf{x}, \mathbf{y}) = \mathsf{w}(\mathbf{y} - \mathbf{x}) .$$

Turning now back to block codes, let $\mathcal{C}$ be an (n, M) code over F with $M > 1$. The *minimum distance* of $\mathcal{C}$ is the minimum Hamming distance between any two distinct codewords of $\mathcal{C}$; that is, the minimum distance d is given by

$$d = \min_{\mathbf{c}_1, \mathbf{c}_2 \in \mathcal{C} \,:\, \mathbf{c}_1 \neq \mathbf{c}_2} \mathsf{d}(\mathbf{c}_1, \mathbf{c}_2) .$$

An (n, M) code with minimum distance d is called an (n, M, d) *code* (when we specify the minimum distance d of an (n, M) code, we implicitly indicate that $M > 1$). We will sometimes use the notation $\mathsf{d}(\mathcal{C})$ for the minimum distance of a given code $\mathcal{C}$.

Example 1.3 The binary $(3, 2, 3)$ *repetition code* is the code

$$\{000, 111\}$$

over $F = \{0, 1\}$. The dimension of the code is $\log_2 2 = 1$ and its rate is $1/3$. □

Example 1.4 The binary $(3, 4, 2)$ *parity code* is the code

$$\{000, 011, 101, 110\}$$

over $F = \{0, 1\}$. The dimension is $\log_2 4 = 2$ and the code rate is $2/3$. □

1.4 Decoding

1.4.1 Definition of decoders

Let $\mathcal{C}$ be an (n, M, d) code over an alphabet F and let S be a channel defined by the triple (F, Φ, Prob). A *decoder* for the code $\mathcal{C}$ with respect to the channel S is a function

$$\mathcal{D} : \Phi^n \to \mathcal{C} .$$

The *decoding error probability* $\mathrm{P_{err}}$ of $\mathcal{D}$ is defined by

$$\mathrm{P_{err}} = \max_{\mathbf{c} \in \mathcal{C}} \mathrm{P_{err}}(\mathbf{c}) ,$$

where

$$\mathrm{P_{err}}(\mathbf{c}) = \sum_{\mathbf{y} : \mathcal{D}(\mathbf{y}) \neq \mathbf{c}} \mathsf{Prob}\{ \mathbf{y} \text{ received} \mid \mathbf{c} \text{ transmitted} \} .$$

Note that $\mathrm{P_{err}}(\mathbf{c})$ is the probability that the codeword $\mathbf{c}$ will be decoded erroneously, given that $\mathbf{c}$ was transmitted.

Our goal is to have decoders with small $\mathrm{P_{err}}$.

Example 1.5 Let $\mathcal{C}$ be the binary $(3, 2, 3)$ repetition code and let S be the BSC with crossover probability p.

Define a decoder $\mathcal{D} : \{0, 1\}^3 \to \mathcal{C}$ as follows:

$$\mathcal{D}(000) = \mathcal{D}(001) = \mathcal{D}(010) = \mathcal{D}(100) = 000$$

and

$$\mathcal{D}(011) = \mathcal{D}(101) = \mathcal{D}(110) = \mathcal{D}(111) = 111 .$$

The probability $\mathrm{P_{err}}$ equals the probability of having two or more errors:

$$\begin{aligned} \mathrm{P_{err}} = \mathrm{P_{err}}(000) = \mathrm{P_{err}}(111) &= \tbinom{3}{2} p^2 (1-p) + \tbinom{3}{3} p^3 \\ &= 3p^2 - 3p^3 + p^3 \\ &= p(2p-1)(1-p) + p . \end{aligned}$$

So, $\mathrm{P_{err}}$ is smaller than p when $p < 1/2$, which means that coding has improved the probability of error per message, compared to uncoded transmission. The price, however, is reflected in the rate: three bits are transmitted for every information bit (a rate of $(\log_2 M)/n = 1/3$). □

1.4.2 Maximum-likelihood decoding

We next consider particular decoding strategies for codes and channels. Given an (n, M, d) code $\mathcal{C}$ over F and a channel $S = (F, \Phi, \mathsf{Prob})$, a *maximum-likelihood decoder* (MLD) for $\mathcal{C}$ with respect to S is the function $\mathcal{D}_{\mathrm{MLD}} : \Phi^n \to \mathcal{C}$ defined as follows: for every $\mathbf{y} \in \Phi^n$, the value $\mathcal{D}_{\mathrm{MLD}}(\mathbf{y})$ equals the codeword $\mathbf{c} \in \mathcal{C}$ that maximizes the probability

$$\mathsf{Prob}\{\, \mathbf{y} \text{ received} \mid \mathbf{c} \text{ transmitted} \,\} .$$

In the case of a tie between two (or more) codewords, we choose one of the tying codewords arbitrarily (say, the first according to some lexicographic ordering on $\mathcal{C}$). Hence, $\mathcal{D}_{\mathrm{MLD}}$ is well-defined for the code $\mathcal{C}$ and the channel S.

A *maximum a posteriori decoder* for $\mathcal{C}$ with respect to a channel $S = (F, \Phi, \mathsf{Prob})$ is defined similarly, except that now the codeword $\mathbf{c}$ maximizes the probability

$$\mathsf{Prob}\{\, \mathbf{c} \text{ transmitted} \mid \mathbf{y} \text{ received} \,\} .$$

In order to compute such a probability, however, we also need to know the *a priori* probability of transmitting $\mathbf{c}$. So, unlike an MLD, a maximum *a posteriori* decoder assumes some distribution on the codewords of $\mathcal{C}$. Since

$$\begin{aligned} &\mathsf{Prob}\{\, \mathbf{c} \text{ transmitted} \mid \mathbf{y} \text{ received} \,\} \\ &\quad = \mathsf{Prob}\{\, \mathbf{y} \text{ received} \mid \mathbf{c} \text{ transmitted} \,\} \cdot \frac{\mathsf{Prob}\{\, \mathbf{c} \text{ transmitted} \,\}}{\mathsf{Prob}\{\, \mathbf{y} \text{ received} \,\}} , \end{aligned}$$

the terms maximum *a posteriori* decoder and MLD coincide when the *a priori* probabilities $\mathsf{Prob}\{\, \mathbf{c} \text{ transmitted} \,\}$ are the same for all $\mathbf{c} \in \mathcal{C}$; namely, they are all equal to $1/M$.

Example 1.6 We compute an MLD for an (n, M, d) code $\mathcal{C}$ with respect to the BSC with crossover probability $p < 1$. Let $\mathbf{c} = c_1 c_2 \ldots c_n$ be a codeword in $\mathcal{C}$ and $\mathbf{y} = y_1 y_2 \ldots y_n$ be a word in $\{0, 1\}^n$. Then

$$\begin{aligned} &\mathsf{Prob}\{\, \mathbf{y} \text{ received} \mid \mathbf{c} \text{ transmitted} \,\} \\ &\qquad = \prod_{j=1}^{n} \mathsf{Prob}\{\, y_j \text{ received} \mid c_j \text{ transmitted} \,\} , \end{aligned}$$

where

$$\mathsf{Prob}\{\, y_j \text{ received} \mid c_j \text{ transmitted} \,\} = \begin{cases} 1-p & \text{if } y_j = c_j \\ p & \text{otherwise} \end{cases} .$$

Therefore,

$$\begin{aligned} \mathsf{Prob}\{\, \mathbf{y} \text{ received} \mid \mathbf{c} \text{ transmitted} \,\} &= p^{\mathsf{d}(\mathbf{y},\mathbf{c})} (1-p)^{n-\mathsf{d}(\mathbf{y},\mathbf{c})} \\ &= (1-p)^n \cdot \left(\frac{p}{1-p} \right)^{\mathsf{d}(\mathbf{y},\mathbf{c})} , \end{aligned}$$

where $\mathsf{d}(\mathbf{y}, \mathbf{c})$ is the Hamming distance between $\mathbf{y}$ and $\mathbf{c}$. Observing that $p/(1-p) < 1$ when $p < 1/2$, it follows that—with respect to the BSC with crossover probability $p < 1/2$—for every (n, M, d) code $\mathcal{C}$ and every word $\mathbf{y} \in \{0,1\}^n$, the value $\mathcal{D}_{\mathrm{MLD}}(\mathbf{y})$ is a closest codeword in $\mathcal{C}$ to $\mathbf{y}$. In fact, this holds also for the q-ary symmetric channel whenever the crossover probability is less than $1-(1/q)$ (Problem 1.7). □

A *nearest-codeword decoder* for an (n, M) code $\mathcal{C}$ over F is a function $F^n \to \mathcal{C}$ whose value for every word $\mathbf{y} \in F^n$ is a closest codeword in $\mathcal{C}$ to $\mathbf{y}$, where the term "closest" is with respect to the Hamming distance. A nearest-codeword decoder for $\mathcal{C}$ is a decoder for $\mathcal{C}$ with respect to any additive channel whose input and output alphabets are F. From Example 1.6 we get that with respect to the BSC with crossover probability $p < 1/2$, the terms MLD and nearest-codeword decoder coincide.

1.4.3 Capacity of the binary symmetric channel

We have seen in Example 1.5 that coding allows us to reduce the decoding error probability $\mathrm{P}_{\mathrm{err}}$, at the expense of transmitting at lower rates. We next see that we can, in fact, achieve arbitrarily small values of $\mathrm{P}_{\mathrm{err}}$, while still transmitting at rates that are bounded away from 0.

Define the *binary entropy function* $\mathsf{H} : [0,1] \to [0,1]$ by

$$\mathsf{H}(x) = -x \log_2 x - (1-x) \log_2 (1-x) ,$$

where $\mathsf{H}(0) = \mathsf{H}(1) = 0$. The binary entropy function is shown in Figure 1.5. It is symmetric with respect to $x = 1/2$ and takes its maximum at that point ($\mathsf{H}(1/2) = 1$). It is $\cap$-concave and has an infinite derivative at $x = 0$ and $x = 1$ (a real function f is $\cap$-concave over a given interval if for every two points x_1 and x_2 in that interval, the line segment that connects the points $(x_1, f(x_1))$ and $(x_2, f(x_2))$ lies entirely on or below the function curve in the real plane; the function f is called $\cup$-convex if $-f$ is $\cap$-concave).

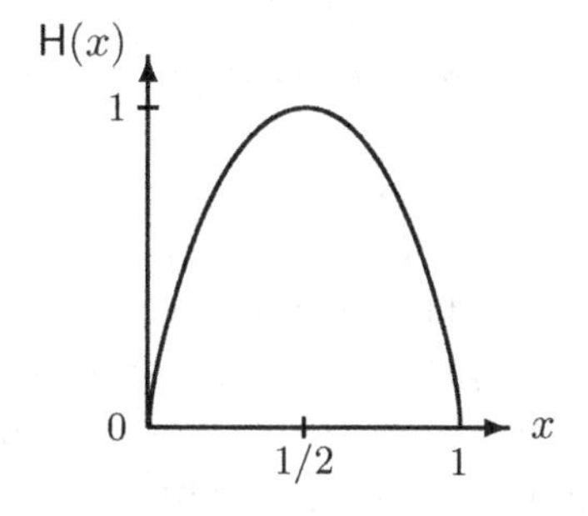

Figure 1.5. Binary entropy function.

Let S be the BSC with crossover probability p. The *capacity* of S is given by

$$\mathsf{cap}(S) = 1 - \mathsf{H}(p) .$$

The capacity is shown in Figure 1.6 as a function of p. Notice that $\mathsf{cap}(S) = 1$ when $p \in \{0, 1\}$ and $\mathsf{cap}(S) = 0$ when $p = 1/2$.

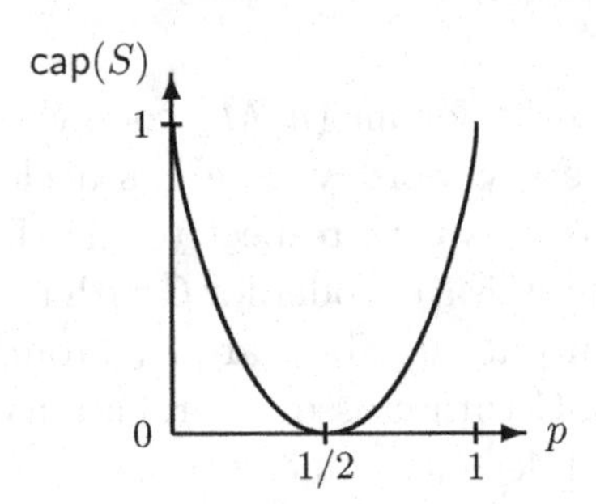

Figure 1.6. Capacity of the BSC.

The next two theorems are special cases of fundamental results in information theory. These results state that the capacity of a channel is the largest rate at which information can be transmitted reliably through that channel.

Theorem 1.1 (Shannon Coding Theorem for the BSC) *Let S be the memoryless binary symmetric channel with crossover probability p and let R be a real in the range $0 \le R < \mathsf{cap}(S)$. There exists an infinite sequence of (n_i, M_i) block codes over $F = \{0, 1\}$, $i = 1, 2, 3, \cdots$, such that $(\log_2 M_i)/n_i \ge R$ and, for maximum-likelihood decoding for those codes (with respect to S), the decoding error probability $\mathrm{P}_{\mathrm{err}}$ approaches 0 as $i \to \infty$.*

Theorem 1.2 (Shannon Converse Coding Theorem for the BSC) *Let S be the memoryless binary symmetric channel with crossover probability p and let R be a real greater than $\mathsf{cap}(S)$. Consider any infinite sequence of (n_i, M_i) block codes over $F = \{0, 1\}$, $i = 1, 2, 3, \cdots$, such that $(\log_2 M_i)/n_i \ge R$ and $n_1 < n_2 < \cdots < n_i < \cdots$. Then, for any decoding scheme for those codes (with respect to S), the decoding error probability $\mathrm{P}_{\mathrm{err}}$ approaches 1 as $i \to \infty$.*

The proofs of these theorems will be given in Chapter 4. In particular, we will show there that $\mathrm{P}_{\mathrm{err}}$ in Theorem 1.1 can be guaranteed to decrease exponentially with the code length n_i. On the other hand, our proof in that chapter will only establish the *existence* of codes with the property that is stated in the theorem, without exhibiting an efficient algorithm for producing them. The constructive part will be filled in later on in Section 12.5. At this point, we will just provide an intuition for why the rate cannot exceed $\mathsf{cap}(S)$ if $P_{\mathrm{err}} \to 0$.

Upon correct decoding of a received word $\mathbf{y} \in F^n$, the receiver reconstructs the following two pieces of information:

- The correct transmitted codeword $\mathbf{c}$ out of M possible codewords; this is equivalent to $\log_2 M$ information bits.

- The error word $\mathbf{e} = \mathbf{y} - \mathbf{c}$ (the subtraction here is taken modulo 2, component by component).

By the definition of the BSC, those two pieces of information are statistically independent. Note that the information conveyed by the error word is "forced" on the receiver: even though that word was not part of the information transmitted by the source, correct decoding implies that, in addition to the correct codeword, the receiver will know the (correct) error word as well. Such forced knowledge has a price, which will be reflected in the rate.

We now estimate the amount of information conveyed by the error word. Although there are 2^n possible error words, most of them are unlikely to occur. By the Law of Large Numbers, the error words will most likely have Hamming weight within the range $n(p \pm \delta)$ for small δ. A word whose Hamming weight lies in that range will be called *typical.* If $\mathbf{e}$ is a typical word, then the probability of having $\mathbf{e}$ as an error word is

$$p^{\mathsf{w}(\mathbf{e})}(1-p)^{n-\mathsf{w}(\mathbf{e})} = 2^{-n(\mathsf{H}(p)+\epsilon)} ,$$

where, by continuity, $\epsilon \to 0$ when $\delta \to 0$. So, each of the typical error words has probability "approximately" $2^{-n\mathsf{H}(p)}$ to occur (we neglect multiplying factors that grow or decay more slowly than exponential terms in n). This means that there are approximately $2^{n\mathsf{H}(p)}$ typical error words, all with approximately the same probability. So, the information conveyed by the error word is equivalent to $n\mathsf{H}(p)$ bits.

We conclude that upon correct decoding, the receiver reconstructs altogether $(\log_2 M) + n\mathsf{H}(p)$ bits. On the other hand, the received word $\mathbf{y}$ contains n bits. So, we must have

$$(\log_2 M) + n\mathsf{H}(p) \le n$$

or

$$\frac{\log_2 M}{n} \le 1 - \mathsf{H}(p) = \mathsf{cap}(S) .$$

1.5 Levels of error handling

While the setting in the previous sections was probabilistic, we turn now to identifying error words that are generated by an additive channel and are always recoverable, as long as the transmitted codewords are taken from a block code whose minimum distance is sufficiently large. Our results will be combinatorial in the sense that they do not depend on the particular conditional probability of the channel.

In our discussion herein, we will distinguish between three levels of handling errors: error correction, error detection, and erasure correction. The difference between the first two terms lies in the model of the decoder used, whereas the third level introduces a new family of channels.

1.5.1 Error correction

We consider channels $S = (F, \Phi, \mathsf{Prob})$ with $\Phi = F$.

Given an (n, M, d) code $\mathcal{C}$ over F, let $\mathbf{c} \in \mathcal{C}$ be the transmitted codeword and $\mathbf{y} \in F^n$ be the received word. By an *error* we mean the event of changing an entry in the codeword $\mathbf{c}$. The number of errors equals $\mathsf{d}(\mathbf{y}, \mathbf{c})$, and the error locations are the indexes of the entries in which $\mathbf{c}$ and $\mathbf{y}$ differ.

The task of error correction is recovering the error locations and the error values. In the next proposition we show that errors are always recoverable, as long as their number does not exceed a certain threshold (which depends on the code $\mathcal{C}$).

Proposition 1.3 *Let $\mathcal{C}$ be an (n, M, d) code over F. There is a decoder $\mathcal{D} : F^n \to \mathcal{C}$ that recovers correctly every pattern of up to $\lfloor (d-1)/2 \rfloor$ errors for every channel $S = (F, F, \mathsf{Prob})$.*

Proof. Let $\mathcal{D}$ be a nearest-codeword decoder, namely, $\mathcal{D}(\mathbf{y})$ is a closest (with respect to the Hamming distance) codeword in $\mathcal{C}$ to $\mathbf{y}$. Let $\mathbf{c}$ and $\mathbf{y}$ be the transmitted codeword and the received word, respectively, where $\mathsf{d}(\mathbf{y}, \mathbf{c}) \le (d-1)/2$. Suppose to the contrary that $\mathbf{c}' = \mathcal{D}(\mathbf{y}) \ne \mathbf{c}$. By the way $\mathcal{D}$ is defined,

$$\mathsf{d}(\mathbf{y}, \mathbf{c}') \le \mathsf{d}(\mathbf{y}, \mathbf{c}) \le (d-1)/2 .$$

So, by the triangle inequality,

$$d \le \mathsf{d}(\mathbf{c}, \mathbf{c}') \le \mathsf{d}(\mathbf{y}, \mathbf{c}) + \mathsf{d}(\mathbf{y}, \mathbf{c}') \le d-1 ,$$

which is a contradiction. □

We next provide a geometric interpretation of the proof of Proposition 1.3. Let F be an alphabet and let t be a nonnegative integer. The set of all words in F^n at Hamming distance t or less from a given word $\mathbf{x}$ in F^n is called a *(Hamming) sphere of radius t in F^n centered at* $\mathbf{x}$. Given an (n, M, d) code $\mathcal{C}$ over F, it follows from the proof of Proposition 1.3 that spheres of radius $\tau = \lfloor (d-1)/2 \rfloor$ that are centered at distinct codewords of $\mathcal{C}$ must be disjoint (i.e., their intersection is the empty set). Figure 1.7 depicts two such spheres (represented as circles) that are centered at codewords $\mathbf{c}$ and $\mathbf{c}'$. Let $\mathbf{y} \in F^n$ be a word that is contained in a sphere of radius $\tau = \lfloor (d-1)/2 \rfloor$ centered at a codeword $\mathbf{c}$. A nearest-codeword decoder, when applied to $\mathbf{y}$, will return the center $\mathbf{c}$ as the decoded codeword. That center is the transmitted codeword if the number of errors is τ or less.

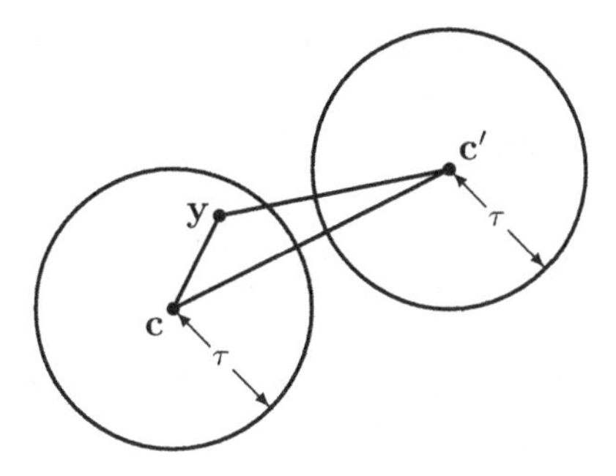

Figure 1.7. Spheres of radius $\tau = \lfloor (d-1)/2 \rfloor$ centered at distinct codewords $\mathbf{c}$ and $\mathbf{c}'$.

Example 1.7 The binary $(n, 2, n)$ repetition code consists of the codewords $00\ldots0$ and $11\ldots1$. A nearest-codeword decoder for the repetition code corrects any pattern of up to $\lfloor (n-1)/2 \rfloor$ errors. □

Proposition 1.3 is tight in the following sense: for every (n, M, d) code $\mathcal{C}$ over F and for every decoder $\mathcal{D} : F^n \to \mathcal{C}$ there is a codeword $\mathbf{c} \in \mathcal{C}$ and a word $\mathbf{y} \in F^n$ such that $\mathsf{d}(\mathbf{y}, \mathbf{c}) \leq \lfloor (d+1)/2 \rfloor$ and $\mathcal{D}(\mathbf{y}) \neq \mathbf{c}$ (Problem 1.10).

Example 1.8 Consider the binary $(3, 4, 2)$ parity code. Suppose that the received word is 001 and that one error has occurred. The correct codeword could be either 000, 011, or 101. □

Observe that while the result of Proposition 1.3 does not depend on the conditional probability distribution Prob of the channel S, an application of this proposition to the design of an encoding–decoding scheme for S does typically take that distribution into account. Specifically, based on Prob, the designer first computes an integer parameter τ such that the probability of having more than τ errors does not exceed a prescribed requirement of the transmission. The (n, M, d) code $\mathcal{C}$ is then selected so that $d \geq 2\tau + 1$.

1.5.2 Error detection

Error detection means an indication by the decoder that errors have occurred, without attempting to correct them. To this end, we generalize the definition of a decoder for a code $\mathcal{C}$ so that its range is $\mathcal{C} \cup \{\text{"e"}\}$, where "e" is the indication that errors have been detected.

As shown in the next proposition, limiting the decoder to only detecting errors (rather than attempting to correct them) allows us to handle more errors than guaranteed by Proposition 1.3.

Proposition 1.4 *Let $\mathcal{C}$ be an (n, M, d) code over F. There is a decoder $\mathcal{D} : F^n \to \mathcal{C} \cup \{\text{"e"}\}$ that detects (correctly) every pattern of up to $d-1$ errors.*

Proof. Let $\mathcal{D}$ be defined by

$$\mathcal{D}(\mathbf{y}) = \begin{cases} \mathbf{y} & \text{if } \mathbf{y} \in \mathcal{C} \\ \text{"e"} & \text{otherwise} \end{cases} .$$

Detection will fail if and only if the received word $\mathbf{y}$ is a codeword other than the transmitted codeword. This occurs only if the number of errors is at least d. □

Example 1.9 The binary parity code of length n consists of all words in $\{0, 1\}^n$ with even number of 1's. This is an $(n, 2^{n-1}, 2)$ code with which we can detect one error. □

The next result combines Propositions 1.3 and 1.4.

Proposition 1.5 *Let $\mathcal{C}$ be an (n, M, d) code over F and let τ and σ be nonnegative integers such that*

$$2\tau + \sigma \leq d-1 .$$

There is a decoder $\mathcal{D} : F^n \to \mathcal{C} \cup \{\text{"e"}\}$ with the following properties:

- *If the number of errors is τ or less, then the errors will be recovered correctly.*
- *Otherwise, if the number of errors is $\tau + \sigma$ or less, then they will be detected.*

Proof. Consider the following decoder $\mathcal{D} : F^n \to \mathcal{C} \cup \{\text{"e"}\}$:

$$\mathcal{D}(\mathbf{y}) = \begin{cases} \mathbf{c} & \text{if there is } \mathbf{c} \in \mathcal{C} \text{ such that } \mathsf{d}(\mathbf{y}, \mathbf{c}) \leq \tau \\ \text{"e"} & \text{otherwise} \end{cases} .$$

(Referring to Figure 1.7, the value of τ may now be smaller than $\lfloor (d-1)/2 \rfloor$. If $\mathbf{y}$ is contained in a sphere of radius τ centered at a codeword $\mathbf{c}$, then $\mathbf{c}$ is the return value of $\mathcal{D}$; otherwise, $\mathbf{y}$ is in the space between spheres and the return value is "e".)

Suppose that $\mathbf{c}$ is the transmitted codeword and $\mathbf{y}$ is the received word and $\mathsf{d}(\mathbf{y}, \mathbf{c}) \leq \sigma + \tau$. Decoding will fail if $\mathbf{y}$ is contained in a sphere of radius τ that is centered at a codeword $\mathbf{c}' \neq \mathbf{c}$. However, this would mean that

$$d \leq \mathsf{d}(\mathbf{c}, \mathbf{c}') \leq \mathsf{d}(\mathbf{y}, \mathbf{c}) + \mathsf{d}(\mathbf{y}, \mathbf{c}') \leq (\tau + \sigma) + \tau \leq d-1 ,$$

which is a contradiction. □

1.5.3 Erasure correction

An erasure is a concealment of an entry in a codeword; as such, an erasure can be viewed as an error event whose location is known (while the correct entry at that location still needs to be recovered).

Example 1.10 The diagram in Figure 1.8 represents the memoryless *binary erasure channel* where the input alphabet is $\{0, 1\}$ and the output alphabet is $\{0, 1, ?\}$, with "?" standing for an erasure. An input symbol is erased with probability p, independently of past or future input symbols.

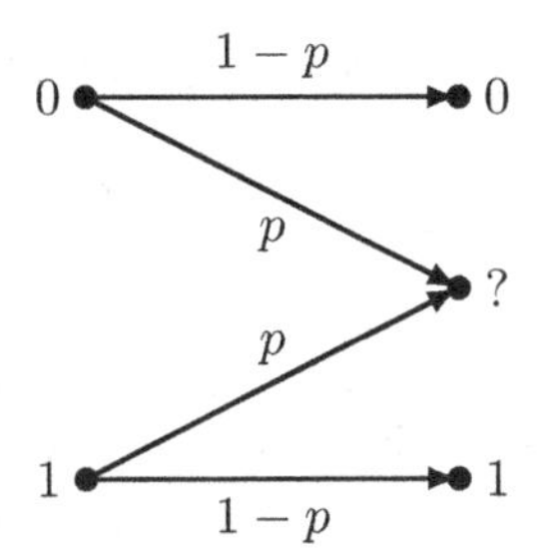

Figure 1.8. Binary erasure channel.

Similarly, we define the memoryless *q-ary erasure channel* with erasure probability p as a triple (F, Φ, Prob), where F is an input alphabet of size q and Φ is the output alphabet $F \cup \{?\}$ (of size $q+1$). The conditional probability distribution Prob satisfies for every two words, $\mathbf{x} = x_1 x_2 \ldots x_m \in F^m$ and $\mathbf{y} = y_1 y_2 \ldots y_m \in \Phi^m$, of the same length m, the independence condition

$$\begin{aligned} &\mathsf{Prob}\{\, \mathbf{y} \text{ received} \mid \mathbf{x} \text{ transmitted} \,\} \\ &\qquad = \prod_{j=1}^{m} \mathsf{Prob}\{\, y_j \text{ received} \mid x_j \text{ transmitted} \,\} , \end{aligned}$$

and for every $x \in F$ and $y \in \Phi$,

$$\mathsf{Prob}\{\, y \text{ was received} \mid x \text{ was transmitted} \,\} = \begin{cases} 1-p & \text{if } y = x \\ p & \text{if } y = \text{?} \\ 0 & \text{otherwise} \end{cases} .$$

It can be shown that the capacity of this channel is $1-p$; namely, this is the highest rate with which information can be transmitted through the channel with a decoding error probability that goes to zero as the code length goes to infinity. □

In general, an *erasure channel* is a triple $S = (F, \Phi, \mathsf{Prob})$, where $\Phi = F \cup \{?\}$ for an erasure symbol "?" not contained in F, and the following property holds for every two words, $\mathbf{x} = x_1 x_2 \ldots x_m \in F^m$ and $\mathbf{y} = y_1 y_2 \ldots y_m \in \Phi^m$, of the same length m:

$$\mathsf{Prob}\{\, \mathbf{y} \text{ was received} \mid \mathbf{x} \text{ was transmitted} \,\} > 0$$

only if $y_j \in \{x_j, ?\}$ for $j = 1, 2, \ldots, m$. For such channels, we have the next result.

Proposition 1.6 *Let $\mathcal{C}$ be an (n, M, d) code over F and let $\Phi = F \cup \{?\}$. There is a decoder $\mathcal{D} : \Phi^n \to \mathcal{C} \cup \{\text{"e"}\}$ that recovers every pattern of up to $d-1$ erasures.*

Proof. Consider a decoder $\mathcal{D} : \Phi^n \to \mathcal{C} \cup \{\text{"e"}\}$ defined as follows:

$$\mathcal{D}(\mathbf{y}) = \begin{cases} \mathbf{c} & \text{if } \mathbf{y} \text{ agrees with } \textit{exactly one } \mathbf{c} \in \mathcal{C} \text{ on the entries in } F \\ \text{"e"} & \text{otherwise} \end{cases} .$$

Suppose that the number of erasures (i.e., the number of occurrences of the symbol "?") in a word $\mathbf{y} \in \Phi^n$ does not exceed $d-1$. Since every two distinct codewords in $\mathcal{C}$ differ on at least d locations, there can be at most one codeword in $\mathcal{C}$ that agrees with $\mathbf{y}$ on its non-erased locations. And if $\mathbf{y}$ is received through an erasure channel, then the transmitted codeword agrees with $\mathbf{y}$ on those locations. □

The following theorem combines Propositions 1.3, 1.4, and 1.6: it covers the case where a channel $S = (F, F \cup \{?\}, \mathsf{Prob})$ inserts either erasures or errors (or both), i.e., the channel S may change any input symbol in F to any of the symbols in $F \cup \{?\}$.

Theorem 1.7 *Let $\mathcal{C}$ be an (n, M, d) code over F and let $S = (F, \Phi, \mathsf{Prob})$ be a channel with $\Phi = F \cup \{?\}$. For each number ρ of erasures in the range $0 \le \rho \le d-1$, let $\tau = \tau_\rho$ and $\sigma = \sigma_\rho$ be nonnegative integers such that*

$$2\tau + \sigma + \rho \le d-1 .$$

There is a decoder $\mathcal{D} : \Phi^n \rightarrow \mathcal{C} \cup \{\text{"e"}\}$ *with the following properties:*

- *If the number of errors (excluding erasures) is* τ *or less, then all the errors and erasures will be recovered correctly.*
- *Otherwise, if the number of errors is* $\tau + \sigma$ *or less, then the decoder will return "e".*

Proof. Let $\mathbf{y}$ be the received word and let J be the set of indexes of all entries of $\mathbf{y}$ that are in F; that is, J points at the non-erased entries of $\mathbf{y}$. For a word $\mathbf{x} \in \Phi^n$, we denote by $\mathbf{x}_J$ the sub-word of $\mathbf{x}$ indexed by J. Consider the code

$$\mathcal{C}_J = \{ \mathbf{c}_J \, : \, \mathbf{c} \in \mathcal{C} \} .$$

Clearly, the minimum distance of $\mathcal{C}_J$ is at least $d - \rho$. Now apply the decoder of Proposition 1.5 to the code $\mathcal{C}_J$ to decode the word $\mathbf{y}_J \in F^{n-\rho}$. If $\mathsf{d}(\mathbf{y}_J, \mathbf{c}_J) \leq \tau$ for some codeword $\mathbf{c}_J \in \mathcal{C}_J$, then the decoder will return this (unique and correct) codeword $\mathbf{c}_J$, which corresponds to a unique codeword $\mathbf{c} \in \mathcal{C}$; otherwise, the decoder will return "e". □

Theorem 1.7 exhibits the "exchange rate" that exists between error correction on the one hand, and error detection or erasure correction on the other hand: error correction costs twice as much as any of the other two levels of handling errors. Indeed, each corrected error requires increasing the minimum distance of the code by 2, while for each detected error or corrected erasure we need to increase the minimum distance only by 1.

Problems

[Section 1.2]

Problem 1.1 (Additive white Gaussian noise channel) This problem presents an example of a channel, called the *additive white Gaussian noise* (in short, AWGN) *channel*, whose input and output alphabets are the real field $\mathbb{R}$ (and are thus infinite and continuous).

Given a positive integer m and an input word $x_1 x_2 \ldots x_m \in \mathbb{R}^m$, the respective output of the AWGN channel is a random word $y_1 y_2 \ldots y_m \in \mathbb{R}^m$, whose cumulative conditional probability distribution satisfies for every real word $z_1 z_2 \ldots z_m \in \mathbb{R}^m$ the independence condition

$$\mathsf{Prob}\{ y_1 \leq z_1, y_2 \leq z_2, \ldots, y_m \leq z_m \mid x_1 x_2 \ldots x_m \} = \prod_{j=1}^{m} \mathsf{Prob}\{ y_j \leq z_j \mid x_j \} ,$$

and for every $x, z \in \mathbb{R}$,

$$\frac{d}{dz} \mathsf{Prob}\{ y_j \leq z \mid x \text{ transmitted} \} = \frac{1}{\sqrt{2\pi\sigma^2}} \, \mathsf{e}^{-(z-x)^2/(2\sigma^2)} ,$$

where $\mathsf{e} = 2.71828\cdots$ is the base of natural logarithms, $\pi = 3.14159\cdots$, and σ is a positive real. Thus, each output value y_j can be written as a sum $x_j + \nu_j$, where $\nu_1, \nu_2, \ldots, \nu_m$ are mutually independent Gaussian random variables taking values in $\mathbb{R}$, with each ν_j having expected value 0 and variance σ^2 (see Figure 1.9; the term "white" captures the fact that the random variables ν_j are uncorrelated).

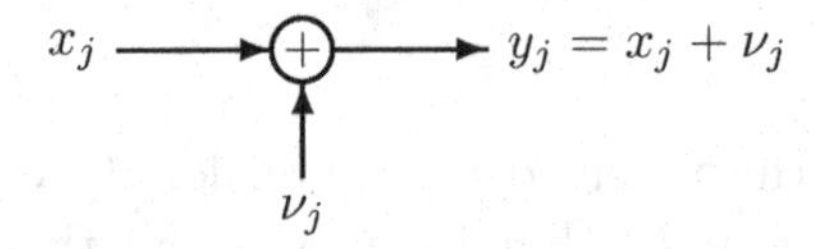

Figure 1.9. AWGN channel.

A binary word is transmitted through the AWGN channel by feeding each entry of the word into a *modulator*, which is a mapping $\Lambda : \{0, 1\} \rightarrow \mathbb{R}$ defined by

$$\Lambda(0) = \alpha \quad \text{and} \quad \Lambda(1) = -\alpha ,$$

for some positive real α. A respective *demodulator* $\Delta : \mathbb{R} \rightarrow \{0, 1\}$ is placed at the output of the channel, where

$$\Delta(y) = \begin{cases} 0 & \text{if } y \geq 0 \\ 1 & \text{otherwise} \end{cases} .$$

Show that the sequence of modulator, AWGN channel, and demodulator, as shown in Figure 1.10, behaves as a memoryless binary symmetric channel with crossover probability

$$p = \tfrac{1}{2} - \mathrm{erf}(\alpha/\sigma) ,$$

where $\mathrm{erf} : \mathbb{R} \rightarrow [0, \frac{1}{2}]$ is the *error function*, which is defined by

$$\mathrm{erf}(z) = \frac{1}{\sqrt{2\pi}} \int_{t=0}^{z} \mathsf{e}^{-t^2/2} \, dt .$$

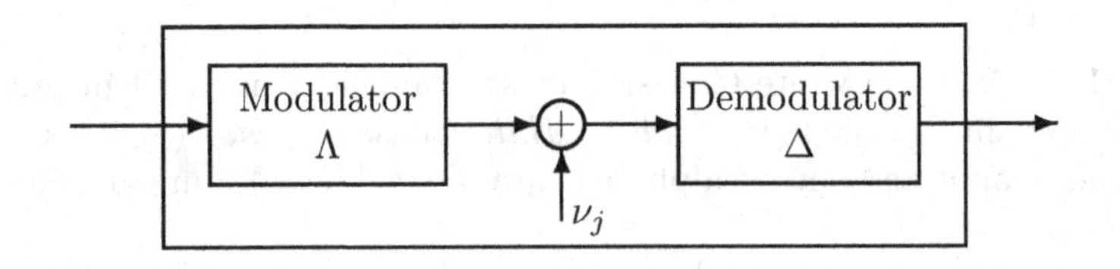

Figure 1.10. AWGN channel with modulator and demodulator.

(The quantity $(\alpha/\sigma)^2$ is commonly referred to as the *signal-to-noise ratio* or, in short, SNR, with α^2 and σ^2 standing for the "energy per symbol" of the signal x_j and the noise ν_j, respectively. The SNR is usually measured in decibel units (dB), i.e., it is expressed as the value $20 \log_{10}(\alpha/\sigma)$. Clearly, any transmitted binary information word can undergo encoding before being fed into the system in Figure 1.10, in which case a respective decoder will be placed after the demodulator. Such a scheme is called *hard-decision* decoding, as opposed to *soft-decision* decoding, where the decoder has direct access to the output of the AWGN channel.)

[Section 1.3]

Problem 1.2 Let $F = \{0,1\}$ and let $\mathbf{x}$, $\mathbf{y}$, and $\mathbf{z}$ be words in F^n that form an "equilateral triangle," that is,

$$\mathsf{d}(\mathbf{x},\mathbf{y}) = \mathsf{d}(\mathbf{y},\mathbf{z}) = \mathsf{d}(\mathbf{z},\mathbf{x}) = 2t \ .$$

Show that there is exactly one word $\mathbf{v}$ in F^n such that

$$\mathsf{d}(\mathbf{x},\mathbf{v}) = \mathsf{d}(\mathbf{y},\mathbf{v}) = \mathsf{d}(\mathbf{z},\mathbf{v}) = t \ .$$

Problem 1.3 (Rank distance) Let A and B be $m \times n$ matrices over a field F. Define the *rank distance* between A and B by $\mathrm{rank}(A-B)$. Show that the rank distance is a metric over the set of all $m \times n$ matrices over F.

Hint: Recall that $\mathrm{rank}(A+B) \le \mathrm{rank}(A) + \mathrm{rank}(B)$.

Problem 1.4 Show that when one adds an overall parity bit to each codeword of an (n, M, d) code over $\{0,1\}$ where d is odd, an $(n+1, M, d+1)$ code is obtained. (The overall parity bit of a word over $\{0,1\}$ is 0 if the Hamming weight of the word is even, and is 1 otherwise.)

[Section 1.4]

Problem 1.5 A codeword of the code $\{01010, 10101\}$ is transmitted through a BSC with crossover probability $p = 0.1$, and a nearest-codeword decoder $\mathcal{D}$ is applied to the received word. Compute the decoding error probability $\mathrm{P_{err}}$ of $\mathcal{D}$.

Problem 1.6 Let $\mathcal{C}$ be a $(7,16)$ code over $F = \{0,1\}$ such that every word in F^7 is at Hamming distance at most 1 from exactly one codeword of $\mathcal{C}$. A codeword of $\mathcal{C}$ is transmitted through a BSC with crossover probability $p = 10^{-2}$.

1. Compute the rate of $\mathcal{C}$.
2. Show that the minimum distance of $\mathcal{C}$ equals 3.
3. What is the probability of having more than one error in the received word?
4. A nearest-codeword decoder $\mathcal{D}$ is applied to the received word. Compute the decoding error probability $\mathrm{P_{err}}$ of $\mathcal{D}$.
5. Compare the value of $\mathrm{P_{err}}$ to the error probability when no coding is used: compute the probability of having at least one bit in error when an (uncoded) word of four bits is transmitted through the given BSC.

Problem 1.7 Show that for every code $\mathcal{C}$ over an alphabet of size q, a nearest-codeword decoder is an MLD with respect to the memoryless q-ary symmetric channel with crossover probability $p < 1 - (1/q)$.

Problem 1.8 A channel $S = (F, \Phi, \mathsf{Prob})$ is called *memoryless* if for every positive integer m and every two words $x_1 x_2 \ldots x_m \in F^m$ and $y_1 y_2 \ldots y_m \in \Phi^m$,

$$\mathsf{Prob}\{\, y_1 y_2 \ldots y_m \text{ received} \mid x_1 x_2 \ldots x_m \text{ transmitted} \,\} = \prod_{j=1}^{m} \mathsf{Prob}\{\, y_j \text{ received} \mid x_j \text{ transmitted} \,\} \ .$$

The channel is called *binary* if $F = \{0,1\}$. Thus, a memoryless binary channel is characterized by the output alphabet Φ and the values of the function

$$\mathsf{Prob}(y|x) = \mathsf{Prob}\{\, y \text{ received} \mid x \text{ transmitted} \,\}\ , \quad x \in F\ , \quad y \in \Phi\ .$$

(According to this definition, the conditional probability of a memoryless channel at any given time index j depends only on the input symbol and the output symbol at that time index; in particular, the behavior of the channel does not depend on the past or the future, neither does it depend on the index j itself. A memoryless channel is therefore also *time-invariant.*)

Let $S = (F = \{0,1\}, \Phi, \mathsf{Prob})$ be a memoryless binary channel and assume that $\mathsf{Prob}(y|x) > 0$ for every $x \in F$ and $y \in \Phi$. For every element $y \in \Phi$, define the *log likelihood ratio* of y by

$$\mu(y) = \log_2 \left(\frac{\mathsf{Prob}(y|0)}{\mathsf{Prob}(y|1)} \right) .$$

Let $\mathcal{C}$ be an (n, M) code over F, and define $\mathcal{D} : \Phi^n \to \mathcal{C}$ to be the decoder that maps every word $y_1 y_2 \ldots y_n \in \Phi^n$ to the codeword $c_1 c_2 \ldots c_n \in \mathcal{C}$ that maximizes the expression

$$\sum_{j=1}^{n} (-1)^{c_j} \cdot \mu(y_j)$$

(with the entries c_j acting here as integers, i.e., $(-1)^0 = 1$ and $(-1)^1 = -1$). Show that $\mathcal{D}$ is a maximum-likelihood decoder for $\mathcal{C}$ with respect to the channel S.

Problem 1.9 Let $S = (F, \Phi, \mathsf{Prob})$ be a memoryless channel as defined in Problem 1.8. Let $\mathcal{C}$ be an (n, M, d) code over F and $\mathcal{D}_{\mathrm{MLD}} : \Phi^n \to \mathcal{C}$ be a maximum-likelihood decoder for $\mathcal{C}$ with respect to S. For a codeword $\mathbf{c} \in \mathcal{C}$, denote by $Y(\mathbf{c})$ the set of pre-images of $\mathbf{c}$ under $\mathcal{D}_{\mathrm{MLD}}$; namely,

$$Y(\mathbf{c}) = \{\, \mathbf{y} \in \Phi^n \ : \ \mathcal{D}_{\mathrm{MLD}}(\mathbf{y}) = \mathbf{c} \,\} .$$

1. Show that for every $\mathbf{c} \in \mathcal{C}$,

$$\mathrm{P}_{\mathrm{err}}(\mathbf{c}) = \sum_{\mathbf{c}' \in \mathcal{C} \setminus \{\mathbf{c}\}} \ \sum_{\mathbf{y} \in Y(\mathbf{c}')} \mathsf{Prob}(\mathbf{y}|\mathbf{c})\ ,$$

where

$$\mathsf{Prob}(\mathbf{y}|\mathbf{c}) = \mathsf{Prob}\{\, \mathbf{y} \text{ received} \mid \mathbf{c} \text{ transmitted} \,\} .$$

2. Show that for every $\mathbf{c} \in \mathcal{C}$,

$$\mathrm{P}_{\mathrm{err}}(\mathbf{c}) \le \sum_{\mathbf{c}' \in \mathcal{C} \setminus \{\mathbf{c}\}} \ \sum_{\mathbf{y} \in Y(\mathbf{c}')} \sqrt{\mathsf{Prob}(\mathbf{y}|\mathbf{c}) \cdot \mathsf{Prob}(\mathbf{y}|\mathbf{c}')}$$

and, so,

$$\mathrm{P}_{\mathrm{err}}(\mathbf{c}) \le \sum_{\mathbf{c}' \in \mathcal{C} \setminus \{\mathbf{c}\}} \ \sum_{\mathbf{y} \in \Phi^n} \sqrt{\mathsf{Prob}(\mathbf{y}|\mathbf{c}) \cdot \mathsf{Prob}(\mathbf{y}|\mathbf{c}')}\ .$$

3. (The Bhattacharyya bound) Show that for every codeword $\mathbf{c} = c_1 c_2 \ldots c_n$ in $\mathcal{C}$,

$$\mathrm{P}_{\mathrm{err}}(\mathbf{c}) \leq \sum_{c'_1 c'_2 \ldots c'_n \in \mathcal{C} \setminus \{\mathbf{c}\}} \prod_{j=1}^{n} \sum_{y \in \Phi} \sqrt{\mathsf{Prob}(y|c_j) \cdot \mathsf{Prob}(y|c'_j)}$$

and, so,

$$\mathrm{P}_{\mathrm{err}}(\mathbf{c}) \leq \sum_{c'_1 c'_2 \ldots c'_n \in \mathcal{C} \setminus \{\mathbf{c}\}} \prod_{j \,:\, c_j \neq c'_j} \sum_{y \in \Phi} \sqrt{\mathsf{Prob}(y|c_j) \cdot \mathsf{Prob}(y|c'_j)} \, .$$

4. Show that for every $c, c' \in F$,

$$\sum_{y \in \Phi} \sqrt{\mathsf{Prob}(y|c) \cdot \mathsf{Prob}(y|c')} \leq 1 \, .$$

When does equality hold?

Hint: Use the Cauchy–Schwartz inequality.

5. Show that if S is a memoryless binary channel (as defined in Problem 1.8) then for every $\mathbf{c} \in \mathcal{C}$,

$$\mathrm{P}_{\mathrm{err}}(\mathbf{c}) \leq \sum_{\mathbf{c}' \in \mathcal{C} \setminus \{\mathbf{c}\}} \gamma^{\mathsf{d}(\mathbf{c}, \mathbf{c}')} \, ,$$

where

$$\gamma = \sum_{y \in \Phi} \sqrt{\mathsf{Prob}(y|0) \cdot \mathsf{Prob}(y|1)} \, .$$

In particular, if S is the BSC with crossover probability p, then $\gamma = 2\sqrt{p(1-p)}$.

6. Suppose now that S is the memoryless q-ary symmetric channel with crossover probability p. Show that for every $\mathbf{c} \in \mathcal{C}$,

$$\mathrm{P}_{\mathrm{err}}(\mathbf{c}) \leq \sum_{\mathbf{c}' \in \mathcal{C} \setminus \{\mathbf{c}\}} \gamma^{\mathsf{d}(\mathbf{c}, \mathbf{c}')} \, ,$$

where

$$\gamma = 2\sqrt{\frac{p(1-p)}{q-1}} + \frac{p(q-2)}{q-1} \, .$$

Hint: For two distinct elements $c, c' \in F$, compute the expression

$$\sqrt{\mathsf{Prob}(y|c) \cdot \mathsf{Prob}(y|c')} \, ,$$

assuming first that $y \in \{c, c'\}$ and then that y takes any other value of F.

7. Let S and γ be either as in part 5 or part 6. Show that the decoding error probability of $\mathcal{D}_{\mathrm{MLD}}$ can be bounded from above by

$$\mathrm{P}_{\mathrm{err}} \leq (M-1) \cdot \gamma^{d} \, .$$

[Section 1.5]

Problem 1.10 Show that for every (n, M, d) code $\mathcal{C}$ over F and for every decoder $\mathcal{D} : F^n \rightarrow \mathcal{C}$ there is a codeword $\mathbf{c} \in \mathcal{C}$ and a word $\mathbf{y} \in F^n$ such that $\mathsf{d}(\mathbf{y}, \mathbf{c}) \leq \lfloor (d+1)/2 \rfloor$ and $\mathcal{D}(\mathbf{y}) \neq \mathbf{c}$.

Problem 1.11 Let $\mathcal{C}$ be an $(8, 16, 4)$ code over $F = \{0, 1\}$. A codeword of $\mathcal{C}$ is transmitted through a BSC with crossover probability $p = 10^{-2}$.

1. Compute the rate of $\mathcal{C}$.
2. Given a word $\mathbf{y} \in F^8$, show that if there is a codeword $\mathbf{c} \in \mathcal{C}$ such that $\mathsf{d}(\mathbf{y}, \mathbf{c}) \leq 1$, then every other codeword $\mathbf{c}' \in \mathcal{C} \setminus \{\mathbf{c}\}$ must satisfy $\mathsf{d}(\mathbf{y}, \mathbf{c}') \geq 3$.
3. Compute the probability of having exactly two errors in the received word.
4. Compute the probability of having three or more errors in the received word.
5. The following decoder $\mathcal{D} : F^8 \rightarrow \mathcal{C} \cup \{\text{“e”}\}$ is applied to the received word:

$$\mathcal{D}(\mathbf{y}) = \begin{cases} \mathbf{c} & \text{if there is } \mathbf{c} \in \mathcal{C} \text{ such that } \mathsf{d}(\mathbf{y}, \mathbf{c}) \leq 1 \\ \text{“e”} & \text{otherwise} \end{cases} .$$

   Compute the decoding error probability of $\mathcal{D}$; namely, compute the probability that $\mathcal{D}$ produces either “e” or a wrong codeword.
6. Show that the value computed in part 4 bounds from above the probability that the decoder $\mathcal{D}$ in part 5 produces a wrong codeword (the latter probability is called the *decoding misdetection probability* of $\mathcal{D}$: this probability does not count the event that the decoder produces “e”).

Problem 1.12 Let S denote the memoryless binary erasure channel with input alphabet $F = \{0, 1\}$, output alphabet $\Phi = F \cup \{?\}$, and erasure probability $p = 0.1$. A codeword of the binary $(4, 8, 2)$ parity code is transmitted through S and the following decoder $\mathcal{D} : \Phi^4 \rightarrow \mathcal{C} \cup \{\text{“e”}\}$ is applied to the received word:

$$\mathcal{D}(\mathbf{y}) = \begin{cases} \mathbf{c} & \text{if } \mathbf{y} \text{ agrees with \textit{exactly one} } \mathbf{c} \in \mathcal{C} \text{ on the entries in } F \\ \text{“e”} & \text{otherwise} \end{cases} .$$

Compute the probability that $\mathcal{D}$ produces “e”. Does this probability depend on which codeword is transmitted?

Problem 1.13 Repeat Problem 1.12 for the code $\mathcal{C} = \{0000, 0111, 1011, 1101\}$.

Notes

[Section 1.1]

Figure 1.1 is taken from Shannon's seminal paper [330], which laid the foundations of what is known today as information theory. This area is treated in several textbooks, such as Cover and Thomas [87], Csiszár and Körner [88], Gallager [140], and McEliece [259].

[Section 1.4]

The expression for the capacity of the BSC is a special case of the notion of capacity which can be defined for a wider family of channels, and Theorems 1.1 and 1.2 can then be shown to hold for these channels. We next introduce the definition of capacity for the family of *(discrete) memoryless channels* (in short, DMC), which were defined in Problem 1.8.

Consider a probability distribution $Q : F \times \Phi \to [0,1]$; i.e., $Q(x,y)$ is the probability of the pair $(x,y) \in F \times \Phi$. Define the marginal distributions

$$P(x) = \sum_{y \in \Phi} Q(x,y) \quad \text{and} \quad \Psi(y) = \sum_{x \in F} Q(x,y) .$$

The *mutual information* of Q is defined by

$$I(Q) = \sum_{(x,y) \in F \times \Phi} Q(x,y) \cdot \log_{|F|} \left(\frac{Q(x,y)}{P(x)\Psi(y)} \right) ,$$

where we assume that the summation skips pairs $(x,y) \in F \times \Phi$ for which either $P(x) = 0$ or $\Psi(y) = 0$. Denoting by $\mathsf{E}_Q\{f(x,y)\}$ the expected value of a function $f : F \times \Phi \to \mathbb{R}$ with respect to the probability distribution Q, we have

$$I(Q) = \mathsf{E}_Q \left\{ \log_{|F|} \left(\frac{Q(x,y)}{P(x)\Psi(y)} \right) \right\} .$$

The following result provides bounds on $I(Q)$.

Proposition 1.8 *For* $Q : F \times \Phi \to [0,1]$,

$$0 \le I(Q) \le 1 .$$

Proof. Starting with the lower bound and letting $q = |F|$,

$$I(Q) = -\mathsf{E}_Q \left\{ \log_q \left(\frac{P(x)\Psi(y)}{Q(x,y)} \right) \right\} \ge -\log_q \mathsf{E}_Q \left\{ \frac{P(x)\Psi(y)}{Q(x,y)} \right\} , \qquad (1.2)$$

where the (last) inequality follows from the concavity of the logarithmic function (in general, every $\cap$-concave real function $f : x \to f(x)$ satisfies the inequality $\mathsf{E}\{f(X)\} \le f(\mathsf{E}\{X\})$, which is known as *Jensen's inequality*). Now,

$$\begin{aligned} \mathsf{E}_Q \left\{ \frac{P(x)\Psi(y)}{Q(x,y)} \right\} &= \sum_{(x,y) \in F \times \Phi} Q(x,y) \cdot \frac{P(x)\Psi(y)}{Q(x,y)} \\ &= \sum_{x \in F} P(x) \sum_{y \in \Phi} \Psi(y) = 1 , \end{aligned}$$

and, so, from (1.2) we get that $I(Q) \ge -\log_q 1 = 0$.

As for the upper bound, since $Q(x,y) \le \Psi(y)$ we have

$$\begin{aligned} I(Q) &= \mathsf{E}_Q \left\{ \log_q \left(\frac{Q(x,y)}{P(x)\Psi(y)} \right) \right\} \\ &\le \mathsf{E}_Q \left\{ \log_q \left(\frac{1}{P(x)} \right) \right\} = \mathsf{E}_P \left\{ \log_q \left(\frac{1}{P(x)} \right) \right\} , \end{aligned}$$

and by concavity we obtain

$$\begin{aligned} \mathsf{E}_P\left\{\log_q\left(\frac{1}{P(x)}\right)\right\} &\le \log_q \mathsf{E}_P\left\{\frac{1}{P(x)}\right\} \\ &= \log_q \sum_{x\in F} P(x)\cdot\frac{1}{P(x)} = \log_q q = 1 . \end{aligned}$$

Thus,

$$I(Q) \le \mathsf{E}_P\left\{\log_q\left(\frac{1}{P(x)}\right)\right\} \le 1 ,$$

as desired. □

The lower bound in Proposition 1.8 is attained when $Q(x,y) = P(x)\Psi(y)$; this corresponds to the case where a random symbol taken from F is statistically independent of the symbol taken from Φ. As for the other extreme case, one can verify that the upper bound is attained when $Q(x,y)$ defines a deterministic mapping from Φ onto F and the symbols in F are uniformly distributed.

The *capacity* of a DMC $S = (F, \Phi, \mathsf{Prob})$ is defined by

$$\mathsf{cap}(S) = \max_P I(Q) ,$$

where the maximum is taken over all probability distributions $P : F \to [0,1]$ and Q is related to P by

$$Q(x,y) = P(x)\cdot \mathsf{Prob}(y|x) ,$$

with $\mathsf{Prob}(y|x)$ standing hereafter for $\mathsf{Prob}\{\, y \text{ received} \mid x \text{ transmitted}\,\}$. Note that $P(x)$ equals the marginal distribution $\sum_{y\in\Phi} Q(x,y)$.

Example 1.11 Let $S = (F, F, \mathsf{Prob})$ be the memoryless q-ary symmetric channel with crossover probability p. We show that

$$\mathsf{cap}(S) = 1 - \mathsf{H}_q(p) ,$$

where $\mathsf{H}_q : [0,1] \to [0,1]$ is the *q-ary entropy function*

$$\mathsf{H}_q(p) = -p\log_q p - (1-p)\log_q(1-p) + p\log_q(q-1) .$$

Notice that for $q = 2$ we get $\mathsf{H}_2(p) = \mathsf{H}(p)$.

Let $Q(x,y) = P(x)\cdot\mathsf{Prob}(y|x)$ for some probability distribution $P : F \to [0,1]$ and denote the marginal distribution $\sum_{x\in F} Q(x,y)$ by $\Psi(y)$. We have,

$$\begin{aligned} I(Q) &= \mathsf{E}_Q\left\{\log_q\left(\frac{\mathsf{Prob}(y|x)}{\Psi(y)}\right)\right\} \\ &= \mathsf{E}_Q\Bigl\{\log_q \mathsf{Prob}(y|x)\Bigr\} + \mathsf{E}_Q\Bigl\{\log_q\left(1/\Psi(y)\right)\Bigr\} . \end{aligned}$$

Now,

$$\begin{aligned} \mathsf{E}_Q\Bigl\{\log_q \mathsf{Prob}(y|x)\Bigr\} &= \sum_{x\in F} P(x) \sum_{y\in F} \mathsf{Prob}(y|x)\cdot\log_q \mathsf{Prob}(y|x) \\ &= \sum_{x\in F} P(x)\Bigl((1-p)\log_q(1-p) + p\log_q(p/(q-1))\Bigr) \\ &= (1-p)\log_q(1-p) + p\log_q(p/(q-1)) \\ &= -\mathsf{H}_q(p) \end{aligned}$$

(regardless of P) and

$$\mathsf{E}_Q\Big\{\log_q(1/\Psi(y))\Big\} \le \log_q \mathsf{E}_Q\Big\{1/\Psi(y)\Big\} = \log_q q = 1 , \tag{1.3}$$

where the inequality in (1.3) follows from the concavity of the logarithmic function; furthermore, this inequality holds with equality when $\Psi(y) = 1/q$ for every $y \in F$. Such a marginal distribution Ψ can be realized by taking $P(x) = 1/q$ for every $x \in F$. Hence, we obtain

$$\begin{aligned} \mathsf{cap}(S) = \max_P I(Q) &= \mathsf{E}_Q\Big\{\log_q \mathsf{Prob}(y|x)\Big\} + \max_P \mathsf{E}_Q\Big\{\log_q(1/\Psi(y))\Big\} \\ &= 1 - \mathsf{H}_q(p) . \end{aligned}$$

In particular, for $q = 2$ (i.e., the BSC) the capacity equals $1 - \mathsf{H}(p)$. □

The proofs of Theorems 1.1 and 1.2 for the case of the memoryless q-ary symmetric channel will be given in Sections 4.7 and 4.6, respectively.

Example 1.12 Let $S = (F, \Phi, \mathsf{Prob})$ be the memoryless q-ary erasure channel with erasure probability p. We show that

$$\mathsf{cap}(S) = 1 - p .$$

Fix a probability distribution $P : F \to [0,1]$ and let $Q : F \times \Phi \to [0,1]$ be given by $Q(x,y) = P(x) \cdot \mathsf{Prob}(y|x)$. The marginal distribution $\Psi(y) = \sum_{x \in F} Q(x,y)$ is related to P as follows:

$$\Psi(y) = \begin{cases} (1-p) \cdot P(y) & \text{if } y \in F \\ p & \text{if } y = ? \end{cases} .$$

Therefore,

$$\begin{aligned} I(Q) &= \mathsf{E}_Q\left\{\log_q\left(\frac{\mathsf{Prob}(y|x)}{\Psi(y)}\right)\right\} \\ &= \sum_{x \in F} P(x) \sum_{y \in \Phi} \mathsf{Prob}(y|x) \cdot \log_q\left(\frac{\mathsf{Prob}(y|x)}{\Psi(y)}\right) \\ &= \sum_{x \in F} P(x) \cdot (1-p) \cdot \log_q\left(\frac{1-p}{\Psi(x)}\right) \\ &= (1-p) \cdot \sum_{x \in F} P(x) \cdot \log_q\left(\frac{1}{P(x)}\right) \\ &= (1-p) \cdot \mathsf{E}_P\Big\{\log_q(1/P(x))\Big\} \\ &\le (1-p) \cdot \log_q \mathsf{E}_P\Big\{1/P(x)\Big\} \\ &= 1 - p , \end{aligned}$$

where the inequality follows from concavity; furthermore, the inequality holds with equality when $P(x) = 1/q$ for every $x \in F$. Hence, $\mathsf{cap}(S) = \max_P I(Q) = 1 - p$. □

Problem 1.9 demonstrates a useful technique for bounding the decoding error probability from above, based on the Bhattacharyya bound; see [43].

Chapter 2

Linear Codes

In this chapter, we consider block codes with a certain structure, which are defined over alphabets that are fields. Specifically, these codes, which we call linear codes, form linear spaces over their alphabets. We associate two objects with these codes: a generator matrix and a parity-check matrix. The first matrix is used as a compact representation of the code and also as a means for efficient encoding. The parity-check matrix will be used as a tool for analyzing the code (e.g., for computing its minimum distance) and will also be part of the general framework that we develop for the decoding of linear codes.

As examples of linear codes, we will mention the repetition code, the parity code, and the Hamming code with its extensions. Owing to their structure, linear codes are by far the predominant block codes in practical usage, and virtually all codes that will be considered in subsequent chapters are linear.

2.1 Definition

Denote by $\mathrm{GF}(q)$ a finite (*Galois*) field of size q. For example, if q is a prime, the field $\mathrm{GF}(q)$ coincides with the ring of integer residues modulo q, also denoted by $\mathbb{Z}_q$. We will see more constructions of finite fields in Chapter 3.

An (n, M, d) code $\mathcal{C}$ over a field $F = \mathrm{GF}(q)$ is called *linear* if $\mathcal{C}$ is a linear subspace of F^n over F; namely, for every two codewords $\mathbf{c}_1, \mathbf{c}_2 \in \mathcal{C}$ and two scalars $a_1, a_2 \in F$ we have $a_1\mathbf{c}_1 + a_2\mathbf{c}_2 \in \mathcal{C}$.

The *dimension* of a linear (n, M, d) code $\mathcal{C}$ over F is the dimension of $\mathcal{C}$ as a linear subspace of F^n over F. If k is the dimension of $\mathcal{C}$, then we say that $\mathcal{C}$ is a linear $[n, k, d]$ code over F (depending on the context, we may sometimes omit the specification of the minimum distance and use the abbreviated notation $[n, k]$ instead). The difference $n-k$ is called the *redundancy* of $\mathcal{C}$. The case $k = 0$ corresponds to the trivial linear code, which consists only of the all-zero word $\mathbf{0}$ (i.e., the word whose entries are all zero).

Every basis of a linear $[n, k, d]$ code $\mathcal{C}$ over $F = \mathrm{GF}(q)$ contains k codewords, the linear combinations of which are distinct and generate the whole set $\mathcal{C}$. Therefore, $|\mathcal{C}| = M = q^k$ and the code rate is $R = (\log_q M)/n = k/n$.

Words $\mathbf{y} = y_1 y_2 \ldots y_n$ over a field F—in particular, codewords of a linear $[n, k, d]$ code over F—will usually be denoted by $(y_1\ y_2\ \ldots\ y_n)$, to emphasize that they are elements of a vector space F^n.

Example 2.1 The $(3, 4, 2)$ parity code over $\mathrm{GF}(2)$ is a linear $[3, 2, 2]$ code since it is spanned by $(1\ 0\ 1)$ and $(0\ 1\ 1)$. □

A *generator matrix* of a linear $[n, k, d]$ code over F is a $k \times n$ matrix whose rows form a basis of the code. In most cases, a generator matrix of a given linear code is not unique (see Problem 2.2). We will typically denote a generator matrix by G. Obviously, the rank of a generator matrix G of a linear code $\mathcal{C}$ over F equals the dimension of $\mathcal{C}$.

Example 2.2 The matrix

$$G = \begin{pmatrix} 1 & 0 & 1 \\ 0 & 1 & 1 \end{pmatrix}$$

is a generator matrix of the $[3, 2, 2]$ parity code over $\mathrm{GF}(2)$, and so is the matrix

$$\hat{G} = \begin{pmatrix} 0 & 1 & 1 \\ 1 & 1 & 0 \end{pmatrix} .$$

In general, the $[n, n-1, 2]$ *parity code* over a field F is defined as the code with a generator matrix

$$G = \left(\begin{array}{c|c} & -1 \\ I & -1 \\ & \vdots \\ & -1 \end{array} \right) ,$$

where I is the $(n-1) \times (n-1)$ identity matrix and -1 is the additive inverse of the (multiplicative) unity element 1 of F. The entries along each row of G sum to zero, and so do the entries in every linear combination of these rows. Hence, the $[n, n-1, 2]$ parity code over F can be equivalently defined as the $(n-1)$-dimensional linear subspace over F that consists of all vectors in F^n whose entries sum to zero. From this characterization of the code we easily see that its minimum distance is indeed 2.

The definition of the parity code applies to any finite field F, even though it is the binary case where the term "parity" really carries its ordinary meaning: only in GF(2) is a zero sum equivalent to having an even number of 1's. □

Example 2.3 The $(3, 2, 3)$ repetition code over GF(2) is a linear $[3, 1, 3]$ code generated by

$$G = (\ 1 \ \ 1 \ \ 1\ ) .$$

In general, the $[n, 1, n]$ *repetition code* over a field F is defined as the code with a generator matrix

$$G = (\ 1 \ \ 1 \ \ \ldots \ \ 1\ ) .$$

□

Every linear code, being a linear space, contains the all-zero vector $\mathbf{0}$ as one of its codewords. In the next proposition, we show that the minimum distance of a linear code equals the minimum Hamming weight of any nonzero codeword in the code.

Proposition 2.1 *Let $\mathcal{C}$ be a linear $[n, k, d]$ code over F. Then*

$$d = \min_{\mathbf{c} \in \mathcal{C} \setminus \{\mathbf{0}\}} \mathsf{w}(\mathbf{c}) .$$

Proof. Since $\mathcal{C}$ is linear,

$$\mathbf{c}_1, \mathbf{c}_2 \in \mathcal{C} \quad \Longrightarrow \quad \mathbf{c}_1 - \mathbf{c}_2 \in \mathcal{C} .$$

Now, $\mathsf{d}(\mathbf{c}_1, \mathbf{c}_2) = \mathsf{w}(\mathbf{c}_1 - \mathbf{c}_2)$ and, so,

$$d = \min_{\mathbf{c}_1, \mathbf{c}_2 \in \mathcal{C} \,:\, \mathbf{c}_1 \neq \mathbf{c}_2} \mathsf{d}(\mathbf{c}_1, \mathbf{c}_2) = \min_{\mathbf{c}_1, \mathbf{c}_2 \in \mathcal{C} \,:\, \mathbf{c}_1 \neq \mathbf{c}_2} \mathsf{w}(\mathbf{c}_1 - \mathbf{c}_2) = \min_{\mathbf{c} \in \mathcal{C} \setminus \{\mathbf{0}\}} \mathsf{w}(\mathbf{c}) .$$

□

2.2 Encoding of linear codes

Let $\mathcal{C}$ be a linear $[n, k, d]$ code over F and G be a generator matrix of $\mathcal{C}$. We can encode information words to codewords of $\mathcal{C}$ by regarding the former as vectors $\mathbf{u} \in F^k$ and using a mapping $F^k \to \mathcal{C}$ defined by

$$\mathbf{u} \mapsto \mathbf{u}G .$$

Since $\text{rank}(G) = k$, this mapping is one-to-one. Also, we can apply elementary operations to the rows of G to obtain another generator matrix that contains a $k \times k$ identity matrix as a sub-matrix.

A $k \times n$ generator matrix is called *systematic* if it has the form

$$(I \,|\, A) ,$$

where I is a $k \times k$ identity matrix and A is a $k \times (n-k)$ matrix. A code $\mathcal{C}$ has a systematic generator matrix if and only if the first k columns of any generator matrix of $\mathcal{C}$ are linearly independent. Now, there are codes for which this condition does not hold; however, if $\mathcal{C}$ is such a code, we can always permute the coordinates of $\mathcal{C}$ to obtain an *equivalent* (although different) code $\hat{\mathcal{C}}$ for which the condition does hold. The code $\hat{\mathcal{C}}$ has the same length, dimension, and minimum distance as the original code $\mathcal{C}$.

When using a systematic generator matrix $G = (I \,|\, A)$ for encoding, the mapping $\mathbf{u} \mapsto \mathbf{u}G$ takes the form $\mathbf{u} \mapsto (\mathbf{u} \,|\, \mathbf{u}A)$; that is, the first k entries in the encoded codeword form the information word.

2.3 Parity-check matrix

Let $\mathcal{C}$ be a linear $[n, k, d]$ code over F. A *parity-check matrix* of $\mathcal{C}$ is an $r \times n$ matrix H over F such that for every $\mathbf{c} \in F^n$,

$$\mathbf{c} \in \mathcal{C} \qquad \Longleftrightarrow \qquad H\mathbf{c}^{\mathrm{T}} = \mathbf{0} .$$

In other words, the code $\mathcal{C}$ is the (right) kernel, $\ker(H)$, of H in F^n. From the well-known relationship between the rank of a matrix and the dimension of its kernel we get

$$\mathrm{rank}(H) = n - \dim \ker(H) = n - k .$$

So, in (the most common) case where the rows of H are linearly independent we have $r = n-k$.

Let G be a $k \times n$ generator matrix of $\mathcal{C}$. The rows of G span $\ker(H)$ and, in particular,

$$HG^T = 0 \qquad \Longrightarrow \qquad GH^T = 0$$

(where $(\cdot)^T$ stands for transposition). Also,

$$\dim \ker(G) = n - \mathrm{rank}(G) = n - k .$$

Hence, the rows of H span $\ker(G)$. So, a parity-check matrix of a linear code can be computed by finding a basis of the kernel of a generator matrix of the code.

In the special case where G is a systematic matrix $(I \,|\, A)$, we can take the $(n-k) \times n$ matrix $H = (-A^T \,|\, I)$ as a parity-check matrix.

Example 2.4 The matrix

$$(\; 1 \;\; 1 \;\; \ldots \;\; 1 \;)$$

is a parity-check matrix of the $[n, n-1, 2]$ parity code over a field F, and

$$\left(\begin{array}{c|c} & -1 \\ I & -1 \\ & \vdots \\ & -1 \end{array} \right)$$

is a parity-check matrix of the $[n, 1, n]$ repetition code over F. □

Let $\mathcal{C}$ be a linear $[n, k, d]$ code over F. The *dual code* of $\mathcal{C}$, denoted by $\mathcal{C}^\perp$, consists of all vectors $\mathbf{x} \in F^n$ such that $\mathbf{x} \cdot \mathbf{c}^T = 0$ for all $\mathbf{c} \in \mathcal{C}$. That is, the codewords of $\mathcal{C}^\perp$ are "orthogonal" to $\mathcal{C}$ (yet the notion of orthogonality over finite fields should be used with care, since a vector can be orthogonal to itself). An equivalent definition of a dual code is given by

$$\mathcal{C}^\perp = \{ \mathbf{x} \in F^n \; : \; \mathbf{x} G^T = \mathbf{0} \} ,$$

where G is a generator matrix of $\mathcal{C}$. It follows that the dual code $\mathcal{C}^\perp$ is a linear $[n, n-k, d^\perp]$ code over F having G as a parity-check matrix. Conversely, a generator matrix of $\mathcal{C}^\perp$ is a parity-check matrix of $\mathcal{C}$ (see Problem 2.15); so, $(\mathcal{C}^\perp)^\perp = \mathcal{C}$, and we refer to $(\mathcal{C}, \mathcal{C}^\perp)$ as a *dual pair*.

The $[n, 1, n]$ repetition code and the $[n, n-1, 2]$ parity code form a dual pair.

Example 2.5 The linear $[7, 4, 3]$ *Hamming code* over $F = \mathrm{GF}(2)$ is defined by the parity-check matrix

$$H = \begin{pmatrix} 0 & 0 & 0 & 1 & 1 & 1 & 1 \\ 0 & 1 & 1 & 0 & 0 & 1 & 1 \\ 1 & 0 & 1 & 0 & 1 & 0 & 1 \end{pmatrix} ,$$

whose columns range over all the nonzero vectors in F^3. A respective generator matrix is given by

$$G = \begin{pmatrix} 1 & 1 & 1 & 1 & 1 & 1 & 1 \\ 0 & 0 & 0 & 1 & 1 & 1 & 1 \\ 0 & 1 & 1 & 0 & 0 & 1 & 1 \\ 1 & 0 & 1 & 0 & 1 & 0 & 1 \end{pmatrix} .$$

Indeed, $HG^T = 0$ and $\dim \ker(H) = 7 - \mathrm{rank}(H) = 4 = \mathrm{rank}(G)$. One can check exhaustively that the minimum Hamming weight of any nonzero codeword in the code is 3. □

Example 2.6 The linear $[8, 4, 4]$ *extended Hamming code* over GF(2) is obtained from the $[7, 4, 3]$ Hamming code by preceding each codeword with an overall parity bit. Based on this definition, a parity-check matrix H_{e} of the code can be obtained from the matrix H in Example 2.5 by adding an all-zero column and then an all-one row, i.e.,

$$H_{\mathrm{e}} = \left(\begin{array}{c|ccccccc} 1 & 1 & 1 & 1 & 1 & 1 & 1 & 1 \\ \hline 0 & 0 & 0 & 0 & 1 & 1 & 1 & 1 \\ 0 & 0 & 1 & 1 & 0 & 0 & 1 & 1 \\ 0 & 1 & 0 & 1 & 0 & 1 & 0 & 1 \end{array} \right) .$$

Due to the additional overall parity bit, codewords of Hamming weight 3 in the $[7, 4, 3]$ Hamming code become codewords of Hamming weight 4 in the extended code.

For the extended code, the matrix H_{e} is also a generator matrix. So, the $[8, 4, 4]$ extended Hamming code over GF(2) is a *self-dual code*; namely, $\mathcal{C} = \mathcal{C}^{\perp}$. □

The following theorem provides a characterization of the minimum distance of a linear code through any parity-check matrix of the code.

Theorem 2.2 *Let H be a parity-check matrix of a linear code $\mathcal{C} \neq \{\mathbf{0}\}$. The minimum distance of $\mathcal{C}$ is the largest integer d such that every set of $d-1$ columns in H is linearly independent.*

Proof. Write $H = (\mathbf{h}_1\ \mathbf{h}_2\ \ldots\ \mathbf{h}_n)$ and let $\mathbf{c} = (c_1\ c_2\ \ldots\ c_n)$ be a codeword in $\mathcal{C}$ with Hamming weight $t > 0$. Let $J \subseteq \{1, 2, \ldots, n\}$ be the support of $\mathbf{c}$, i.e., J is the set of indexes of the t nonzero entries in $\mathbf{c}$. From $H\mathbf{c}^T = \mathbf{0}$ we have

$$\sum_{j \in J} c_j \mathbf{h}_j = \mathbf{0} ,$$

namely, the t columns of H that are indexed by J are linearly dependent.

Conversely, every set of t linearly dependent columns in H defines at least one vanishing nontrivial linear combination of the columns of H, with at most t nonzero coefficients in that combination. The coefficients in such a combination, in turn, form a nonzero codeword $\mathbf{c} \in \mathcal{C}$ with $\mathsf{w}(\mathbf{c}) \leq t$.

Given d as defined in the theorem, it follows that no nonzero codeword in $\mathcal{C}$ has Hamming weight less than d, but there is at least one codeword in $\mathcal{C}$ whose Hamming weight is d. □

The next two examples generalize the constructions in Examples 2.5 and 2.6.

Example 2.7 For an integer $m > 1$, the $[2^m-1, 2^m-1-m, 3]$ *Hamming code* over $F = \mathrm{GF}(2)$ is defined by an $m \times (2^m-1)$ parity-check matrix H whose columns range over all the nonzero elements of F^m. Every two columns in H are linearly independent and, so, the minimum distance of the code is at least 3. In fact, the minimum distance is exactly 3, since there are three dependent columns, e.g., $(0 \ldots 0\,0\,1)^T$, $(0 \ldots 0\,1\,0)^T$, and $(0 \ldots 0\,1\,1)^T$. □

Example 2.8 The $[2^m, 2^m-1-m, 4]$ *extended Hamming code* over $F = \mathrm{GF}(2)$ is derived from the $[2^m-1, 2^m-1-m, 3]$ Hamming code by preceding each codeword of the latter code with an overall parity bit. An $(m+1) \times 2^m$ parity-check matrix of the extended code can be obtained by taking as columns all the elements of F^{m+1} whose first entry equals 1. It can be verified that every three columns in this matrix are linearly independent. □

Hamming codes are defined also over non-binary fields, as demonstrated in the next example.

Example 2.9 Let $F = \mathrm{GF}(q)$ and for an integer $m > 1$ let n be given by $(q^m - 1)/(q - 1)$. The $[n, n-m, 3]$ *Hamming code* over F is defined by an $m \times n$ parity-check matrix H whose columns range over all the nonzero elements of F^m whose leading nonzero entry is 1. Again, every two columns in H are linearly independent, yet H contains three dependent columns. So, the minimum distance of the code is 3.

An extended code can be obtained by preceding each codeword with an entry whose value is set so that the (weighted) sum of entries in the codeword is zero (by a weighted sum we mean that each coordinate is assigned a constant of the field which multiplies the entry in that coordinate before taking the sum). However, for $q > 2$, the extended code may still have minimum distance 3. □

2.4 Decoding of linear codes

Let $\mathcal{C}$ be a linear $[n, k, d]$ code over $F = \mathrm{GF}(q)$. Recall from Example 1.6 and Problem 1.7 that maximum-likelihood decoding for $\mathcal{C}$ with respect to a memoryless q-ary symmetric channel with crossover probability $p < 1-(1/q)$ is the same as nearest-codeword decoding; namely:

- Given a received word $\mathbf{y} \in F^n$, find a codeword $\mathbf{c} \in \mathcal{C}$ that minimizes the value $\mathsf{d}(\mathbf{y}, \mathbf{c})$.

Equivalently:

- Given a received word $\mathbf{y} \in F^n$, find a word $\mathbf{e} \in F^n$ of minimum Hamming weight such that $\mathbf{y} - \mathbf{e} \in \mathcal{C}$.

Below are two methods for implementing nearest-codeword decoding. The first method, called *standard array decoding*, is rather impractical, but it demonstrates how the linearity of the code is incorporated into the decoding. The second method, called *syndrome decoding*, is in effect a more efficient way of implementing standard array decoding.

2.4.1 Standard array decoding of linear codes

Let $\mathcal{C}$ be a linear $[n, k, d]$ code over $F = \mathrm{GF}(q)$. A *standard array for* $\mathcal{C}$ is a $q^{n-k} \times q^k$ array of elements of F^n defined as follows.

- The first row in the array consists of the codewords of $\mathcal{C}$, starting with the all-zero codeword.
- Each subsequent row starts with a word $\mathbf{e} \in F^n$ of a smallest Hamming weight that has not yet appeared in previous rows, followed by the words $\mathbf{e} + \mathbf{c}$, where $\mathbf{c}$ ranges over all the nonzero codewords in $\mathcal{C}$ in their order of appearance in the first row.

Example 2.10 Let $\mathcal{C}$ be a linear $[5, 2, 3]$ code over GF(2) with a generator matrix

$$G = \begin{pmatrix} 1 & 0 & 1 & 1 & 0 \\ 0 & 1 & 0 & 1 & 1 \end{pmatrix} .$$

A standard array of this code is shown in Table 2.1. Clearly, the standard

Table 2.1. Standard array.

00000	10110	01011	11101
00001	10111	01010	11100
00010	10100	01001	11111
00100	10010	01111	11001
01000	11110	00011	10101
10000	00110	11011	01101
00101	10011	01110	11000
10001	00111	11010	01100

array is not unique. For example, we can permute the five rows that start with the words of Hamming weight 1. Furthermore, the penultimate row could start with any of the words 00101, 11000, 10001, or 01100. □

Each row in the standard array is a *coset* of $\mathcal{C}$ in F^n. Indeed, two words $\mathbf{y}_1, \mathbf{y}_2 \in F^n$ are in the same row if and only if $\mathbf{y}_1 - \mathbf{y}_2 \in \mathcal{C}$. The cosets of $\mathcal{C}$ form a partition of F^n into q^{n-k} subsets, each of size $|\mathcal{C}| = q^k$.

The first word in each row is called a *coset leader*. By construction, a coset leader is always a minimum-weight word in its coset. However, as the last two rows in the example show, a minimum-weight word in a coset is not necessarily unique.

Let $\mathbf{y}$ be a received word. Regardless of the decoding strategy, the error word $\hat{\mathbf{e}}$ found by the decoder must be such that $\mathbf{y} - \hat{\mathbf{e}} \in \mathcal{C}$. Hence, the decoded error word must be in the same coset as $\mathbf{y}$. Nearest-codeword decoding means that $\hat{\mathbf{e}}$ is a minimum-weight word in its coset. Our decoding strategy will therefore be as follows.

- Given a received word $\mathbf{y} \in F^n$, find the row (coset) that contains $\mathbf{y}$, and let the decoded error word be the coset leader $\mathbf{e}$ of that row.

The decoded codeword is $\mathbf{c} = \mathbf{y} - \mathbf{e}$. By construction, $\mathbf{c}$ is the first entry in the *column* containing $\mathbf{y}$.

Referring to the example, suppose that the received word is $\mathbf{y} = 01111$. This word appears in the fourth row and the third column of the standard array in Table 2.1. The coset leader of the fourth row is 00100, and the decoded codeword is 01011, which is the first entry in the third column.

Notice that when the Hamming weight of the coset leader does not exceed $(d-1)/2$, then it is the unique minimum-weight word in its coset. This must be so in view of Proposition 1.3, where we have shown that nearest-codeword decoding yields the correct codeword when the number of errors does not exceed $(d-1)/2$.

In the example, the coset leaders of Hamming weight 1 are the unique minimum-weight words in their cosets.

2.4.2 Syndrome decoding of linear codes

Let $\mathcal{C}$ be a linear $[n, k, d]$ code over $F = \mathrm{GF}(q)$ and fix H to be an $(n-k) \times n$ parity-check matrix of $\mathcal{C}$; that is, we assume here that the rows of H are linearly independent.

The *syndrome* of a word $\mathbf{y} \in F^n$ (with respect to H) is defined by

$$\mathbf{s} = H\mathbf{y}^T .$$

Recall that for every vector $\mathbf{c} \in F^n$,

$$\mathbf{c} \in \mathcal{C} \qquad \Longleftrightarrow \qquad H\mathbf{c}^T = \mathbf{0} .$$

That is, the codewords of $\mathcal{C}$ are precisely the vectors of F^n whose syndromes are $\mathbf{0}$. Now, if $\mathbf{y}_1$ and $\mathbf{y}_2$ are vectors in F^n, then

$$\mathbf{y}_1 - \mathbf{y}_2 \in \mathcal{C} \qquad \Longleftrightarrow \qquad H\mathbf{y}_1^T = H\mathbf{y}_2^T ,$$

which means that $\mathbf{y}_1$ and $\mathbf{y}_2$ are in the same coset of $\mathcal{C}$ in F^n if and only if their syndromes are equal. So, given an $(n-k) \times n$ parity-check matrix H of $\mathcal{C}$, there is a one-to-one correspondence between the q^{n-k} cosets of $\mathcal{C}$ in F^n and the q^{n-k} possible values of the syndromes, with the trivial coset $\mathcal{C}$ corresponding to the syndrome $\mathbf{0}$.

Nearest-codeword decoding can thus be performed by the following two steps:

1. *Finding the syndrome of (the coset of) the received word:* given a received word $\mathbf{y} \in F^n$, compute $\mathbf{s} = H\mathbf{y}^T$.

2. *Finding a coset leader in the coset of the received word:* find a minimum-weight word $\mathbf{e} \in F^n$ such that
$$\mathbf{s} = H\mathbf{e}^T .$$

Step 1 is simply a matrix-by-vector multiplication. As for Step 2, it is equivalent to finding a smallest set of columns in H whose linear span contains the vector $\mathbf{s}$. This problem is known to be computationally difficult (NP-complete) for general matrices H and vectors $\mathbf{s}$. However, when the redundancy $n-k$ is small, we can implement Step 2 through a look-up table of size q^{n-k} that lists for every syndrome its respective coset leader. Step 2 can sometimes be tractable also when $n-k$ is large: in fact, most codes that we treat in upcoming chapters have parity-check matrices with a special structure that allows efficient decoding when the number of decoded errors does not exceed $(d-1)/2$ (and sometimes even when it does). Thus, with such codes and decoding algorithms, the coset leader in Step 2 can be computed efficiently, and no look-up table will be necessary.

The definition of syndrome can be extended to $r \times n$ parity-check matrices H that have dependent rows. In this case, the syndrome of $\mathbf{y} \in F^n$ will be a vector $H\mathbf{y}^T \in F^r$ where $r > n-k$. Still, there will be q^{n-k} possible values for the syndrome, which correspond to the q^{n-k} cosets of the code in F^n.

Example 2.11 Let $\mathcal{C}$ be the $[2^m-1, 2^m-1-m, 3]$ Hamming code over GF(2) and let $H = (\, \mathbf{h}_1 \; \mathbf{h}_2 \; \ldots \; \mathbf{h}_{2^m-1} \,)$ be an $m \times (2^m-1)$ parity-check matrix of the code, where $\mathbf{h}_j$ is the m-bit binary representation of the integer j.

Suppose that $\mathbf{y} \in F^n$ is the received word and that at most one error has occurred. Then $\mathbf{y} = \mathbf{c} + \mathbf{e}$ where $\mathbf{c} \in \mathcal{C}$ and $\mathbf{e}$ is either $\mathbf{0}$ or a unit vector.

The syndrome of $\mathbf{y}$ is the same as that of $\mathbf{e}$, i.e.,
$$\mathbf{s} = H\mathbf{y}^T = H\mathbf{e}^T .$$

Now, if $\mathbf{e} = \mathbf{0}$ then $\mathbf{s} = \mathbf{0}$ and $\mathbf{y}$ is an error-free codeword. Otherwise, suppose that the nonzero entry of $\mathbf{e}$ is at location j. In this case,
$$\mathbf{s} = H\mathbf{y}^T = H\mathbf{e}^T = \mathbf{h}_j .$$

That is, the syndrome, when regarded as an m-bit binary representation of an integer, is equal to the location index j of the error. □

Problems

[Section 2.1]

Problem 2.1 Let $\mathcal{C}_1$ and $\mathcal{C}_2$ be linear codes of the same length n over $F = \mathrm{GF}(q)$ and let G_1 and G_2 be generator matrices of $\mathcal{C}_1$ and $\mathcal{C}_2$, respectively. Define the following codes:

- $\mathcal{C}_3 = \mathcal{C}_1 \cup \mathcal{C}_2$
- $\mathcal{C}_4 = \mathcal{C}_1 \cap \mathcal{C}_2$
- $\mathcal{C}_5 = \mathcal{C}_1 + \mathcal{C}_2 = \{ \mathbf{c}_1 + \mathbf{c}_2 \ : \ \mathbf{c}_1 \in \mathcal{C}_1 \text{ and } \mathbf{c}_2 \in \mathcal{C}_2 \}$
- $\mathcal{C}_6 = \{ (\mathbf{c}_1 \,|\, \mathbf{c}_2) \ : \ \mathbf{c}_1 \in \mathcal{C}_1 \text{ and } \mathbf{c}_2 \in \mathcal{C}_2 \}$

(here $(\cdot|\cdot)$ stands for concatenation of words). For $i = 1, 2, \ldots, 6$, denote by k_i the dimension $\log_q |\mathcal{C}_i|$ and by d_i the minimum distance of $\mathcal{C}_i$. Assume that both k_1 and k_2 are greater than zero.

1. Show that $\mathcal{C}_3$ is linear if and only if either $\mathcal{C}_1 \subseteq \mathcal{C}_2$ or $\mathcal{C}_2 \subseteq \mathcal{C}_1$.
2. Show that the codes $\mathcal{C}_4$, $\mathcal{C}_5$, and $\mathcal{C}_6$ are linear.
3. Show that if $k_4 > 0$ then $d_4 \geq \max\{d_1, d_2\}$.
4. Show that $k_5 \leq k_1 + k_2$ and that equality holds if and only if $k_4 = 0$.
5. Show that $d_5 \leq \min\{d_1, d_2\}$.
6. Show that
$$\left(\begin{array}{c|c} G_1 & 0 \\ \hline 0 & G_2 \end{array} \right)$$
is a generator matrix of $\mathcal{C}_6$ and, so, $k_6 = k_1 + k_2$.
7. Show that $d_6 = \min\{d_1, d_2\}$.

Problem 2.2 Show that the number of distinct generator matrices of a linear $[n, k, d]$ code over $F = \mathrm{GF}(q)$ is $\prod_{i=0}^{k-1}(q^k - q^i)$.

Hint: First show that the sought number equals the number of $k \times k$ nonsingular matrices over F. Next, count the latter matrices: show that given a set U of $i < k$ linearly independent vectors in F^k (with U standing for the first i rows in a $k \times k$ nonsingular matrix over F), there are $q^k - q^i$ ways to extend U by one vector to form a set of $i+1$ linearly independent vectors in F^k.

Problem 2.3 Let $\mathcal{C}$ be a linear $[n, k, d]$ code over F where $n > k$. For $i \in \{1, 2, \ldots, n\}$, denote by $\mathcal{C}_i$ the code

$$\mathcal{C}_i = \{(c_1\, c_2\, \ldots\, c_{i-1}\, c_{i+1}\, \ldots\, c_n) \ : \ (c_1\, c_2\, \ldots\, c_n) \in \mathcal{C}\} \ .$$

The code $\mathcal{C}_i$ is said to be obtained by *puncturing* $\mathcal{C}$ at the ith coordinate.

1. Show that $\mathcal{C}_i$ is a linear $[n-1, k_i, d_i]$ code over F where $k_i \geq k-1$ and $d_i \geq d-1$.

   Hint: Show that $\mathcal{C}_i$ is spanned by the rows of a matrix obtained by deleting the ith column from a generator matrix of $\mathcal{C}$.

2. Show that there are at least $n-k$ indexes i for which $k_i = k$.

Problem 2.4 Let $(a_1\ a_2\ \ldots\ a_k)$ be a nonzero vector over $F = \mathrm{GF}(q)$ and consider the mapping $f : F^k \to F$ defined by $f(x_1, x_2, \ldots, x_k) = \sum_{i=1}^{k} a_i x_i$. Show that each element of F is the image under f of exactly q^{k-1} vectors in F^k.

Problem 2.5 Show that in every linear code over $F = \mathrm{GF}(2)$, either all codewords have even Hamming weight or exactly half of the codewords have even Hamming weight.

Hint: Let G be a generator matrix of the code; see when Problem 2.4 can be applied to the mapping $F^k \to F$ that is defined by $\mathbf{u} \mapsto \mathbf{u}G(1\ 1\ \ldots\ 1)^T$.

Problem 2.6 Let $\mathcal{C}$ be a linear $[n, k, d]$ code over $F = \mathrm{GF}(q)$ and let T be a $q^k \times n$ array whose rows are the codewords of $\mathcal{C}$. Show that each element of F appears in every nonzero column in T exactly q^{k-1} times.

Hint: Use Problem 2.4.

Problem 2.7 (The Plotkin bound for linear codes) Show that every linear $[n, k, d]$ code over $F = \mathrm{GF}(q)$ satisfies the inequality

$$d \leq \frac{n \cdot (q-1) \cdot q^{k-1}}{q^k - 1} .$$

Hint: Using Problem 2.6, show that the average Hamming weight of the $q^k - 1$ nonzero codewords in the code is at most $n \cdot (q-1) \cdot q^{k-1}/(q^k - 1)$. Then argue that the minimum distance of the code is bounded from above by that average.

Problem 2.8 Let G be a $k \times n$ generator matrix of a linear code $\mathcal{C} \neq \{\mathbf{0}\}$ over a field F. Show that the minimum distance of $\mathcal{C}$ is the largest integer d such that every $k \times (n-d+1)$ sub-matrix of G has rank k.

Hint: Show that if J is the set of indexes of the zero entries in a nonzero codeword, then the columns in G that are indexed by J form a $k \times |J|$ sub-matrix whose rows are linearly dependent; conversely, show that if the columns indexed by a set J form a sub-matrix whose rows are linearly dependent, then there is a nonzero codeword in which the entries that are indexed by J equal zero.

Problem 2.9 (Group codes) Let F be an Abelian group. An (n, M) code over F is called a *group code* over F if it is a subgroup of F^n under the addition in F^n, where the addition of two words in F^n is defined as their sum, component by component. Show that for every (n, M) group code $\mathcal{C}$ over F of size $M > 1$,

$$\mathsf{d}(\mathcal{C}) = \min_{\mathbf{c} \in \mathcal{C} \setminus \{\mathbf{0}\}} \mathsf{w}(\mathbf{c}) .$$

[Section 2.2]

Problem 2.10 Let F be the ternary field GF(3) and let $\mathcal{C}$ be a linear code over F that is generated by

$$G = \begin{pmatrix} 2 & 1 & 2 & 1 \\ 1 & 1 & 1 & 0 \end{pmatrix}.$$

1. List all the codewords of $\mathcal{C}$.
2. Find a systematic generator matrix of $\mathcal{C}$.
3. Compute the minimum distance of $\mathcal{C}$.
4. A codeword of $\mathcal{C}$ is transmitted through an additive channel (F, F, Prob) and the word $\mathbf{y} = (1\ 1\ 1\ 1)$ is received. Find all the codewords of $\mathcal{C}$ that can be produced as an output by a nearest-codeword decoder for $\mathcal{C}$ when applied to $\mathbf{y}$.
5. Suppose that the encoding is carried out by the mapping $\mathbf{u} \mapsto \mathbf{u}G$. A nearest-codeword decoder for $\mathcal{C}$ is applied to the word $\mathbf{y}$ in part 4 to produce a codeword $\hat{\mathbf{c}}$. Denote by $\hat{\mathbf{u}}$ the information word that is associated with $\hat{\mathbf{c}}$. Can any of the entries of $\hat{\mathbf{u}}$ be uniquely determined—regardless of which nearest-codeword decoder is used?
6. Repeat part 5 for the case where a systematic generator matrix replaces G in the encoding.

[Section 2.3]

Problem 2.11 Let $\mathcal{C}$ be a linear code over $F = \mathrm{GF}(2)$ with a parity-check matrix

$$H = \left(\begin{array}{ccc|ccc|ccc} 1 & 1 & 0 & 0 & 0 & 0 & 0 & 0 & 0 \\ 1 & 0 & 1 & 0 & 0 & 0 & 0 & 0 & 0 \\ 0 & 0 & 0 & 1 & 1 & 0 & 0 & 0 & 0 \\ 0 & 0 & 0 & 1 & 0 & 1 & 0 & 0 & 0 \\ 0 & 0 & 0 & 0 & 0 & 0 & 1 & 1 & 0 \\ 0 & 0 & 0 & 0 & 0 & 0 & 1 & 0 & 1 \end{array}\right).$$

1. What are the parameters n, k, and d of $\mathcal{C}$?
2. Write a generator matrix of $\mathcal{C}$.
3. Find the largest integer t such that every pattern of up to t errors will be decoded correctly by a nearest-codeword decoder for $\mathcal{C}$.
4. A codeword of $\mathcal{C}$ is transmitted through an additive channel (F, F, Prob) and the word $\mathbf{y} = (1\ 0\ 1\ 0\ 1\ 0\ 1\ 0\ 1)$ is received. What will be the respective output of a nearest-codeword decoder for $\mathcal{C}$ when applied to $\mathbf{y}$?
5. Given that the answer to part 4 is the correct codeword, how many errors does a nearest-codeword decoder correct in this case? And how is this number consistent with the value t in part 3?

Problem 2.12 Let $\mathcal{C}_1$ be a linear $[n, k_1, d_1]$ code and $\mathcal{C}_2$ be a linear $[n, k_2, d_2]$ code (of the same length)—both over F. Define the code $\mathcal{C}$ by

$$\mathcal{C} = \{ (\mathbf{c}_1 \,|\, \mathbf{c}_1 + \mathbf{c}_2) \;:\; \mathbf{c}_1 \in \mathcal{C}_1 \text{ and } \mathbf{c}_2 \in \mathcal{C}_2 \} \;.$$

1. Let G_1 and G_2 be generator matrices of $\mathcal{C}_1$ and $\mathcal{C}_2$, respectively. Show that $\mathcal{C}$ is a linear code over F generated by

$$\left(\begin{array}{c|c} G_1 & G_1 \\ \hline 0 & G_2 \end{array} \right) .$$

2. Let H_1 and H_2 be parity-check matrices of $\mathcal{C}_1$ and $\mathcal{C}_2$, respectively. Show that

$$\left(\begin{array}{c|c} H_1 & 0 \\ \hline H_2 & -H_2 \end{array} \right)$$

is a parity-check matrix of $\mathcal{C}$.

3. Show that $\mathcal{C}$ is a linear $[2n, k, d]$ code over F where $k = k_1 + k_2$ and $d = \min\{2d_1, d_2\}$.

4. Write $\Phi = F \cup \{?\}$, where ? stands for the erasure symbol, and let $\mathcal{D}_1 : \Phi^n \rightarrow \mathcal{C}_1$ be a decoder for $\mathcal{C}_1$ that recovers correctly any pattern of τ errors and ρ erasures, whenever $2\tau + \rho < d_1$ (see Theorem 1.7). Also, let $\mathcal{D}_2 : F^n \rightarrow \mathcal{C}_2$ be a decoder for $\mathcal{C}_2$ that recovers correctly any pattern of up to $\lfloor (d_2-1)/2 \rfloor$ errors.

   Consider the decoder $\mathcal{D} : F^{2n} \rightarrow \mathcal{C}$ that maps every vector $\mathbf{y} = (y_1 \; y_2 \; \ldots \; y_{2n})$ in F^{2n} to a codeword $(\mathbf{c}_1 \,|\, \mathbf{c}_1 + \mathbf{c}_2) \in \mathcal{C}$ as follows. Write $\mathbf{y} = (\mathbf{y}_1 \,|\, \mathbf{y}_2)$, where $\mathbf{y}_1$ (respectively, $\mathbf{y}_2$) consists of the first (respectively, last) n entries of $\mathbf{y}$. The codeword $\mathbf{c}_2 \in \mathcal{C}_2$ is given by

$$\mathbf{c}_2 = \mathcal{D}_2(\mathbf{y}_2 - \mathbf{y}_1) \;.$$

   As for the codeword $\mathbf{c}_1 \in \mathcal{C}_1$, denote by J the support of $\mathbf{y}_2 - \mathbf{y}_1 - \mathbf{c}_2$ and let the vector $\mathbf{z} = (z_1 \; z_2 \; \ldots \; z_n)$ be defined by

$$z_j = \begin{cases} y_j & \text{if } j \notin J \\ ? & \text{otherwise} \end{cases} , \quad 1 \le j \le n \;.$$

   Then,

$$\mathbf{c}_1 = \mathcal{D}_1(\mathbf{z}) \;.$$

   Show that $\mathcal{D}$ recovers correctly any pattern of up to $\lfloor (d-1)/2 \rfloor$ errors, where d is as in part 3.

Problem 2.13 Let $\mathcal{C}_1$ be a linear $[n_1, k_1, d_1]$ code over F and let $\mathcal{C}_2$ be a linear $[n_2, k_2, d_2]$ code over F where $k_2 \ge k_1$. Let G_1 be a generator matrix of $\mathcal{C}_1$ and

$$G_2 = \left(\frac{G_{2,1}}{G_{2,2}} \right)$$

be a generator matrix of $\mathcal{C}_2$, where $G_{2,1}$ consists of the first k_1 rows of G_2. Consider the linear $[n_1+n_2, k, d]$ code $\mathcal{C}$ over F with a generator matrix

$$G = \left(\begin{array}{c|c} G_1 & G_{2,1} \\ \hline 0 & G_{2,2} \end{array} \right) .$$

1. Show that $k = k_2$.

2. Show that if $k_1 = k_2$ then $d \geq d_1 + d_2$. Provide examples of codes $\mathcal{C}_1$ and $\mathcal{C}_2$, where $k_1 < k_2$ and $d < d_1 + d_2$.

3. Show that there is a $k_1 \times n_1$ matrix Q_1 over F such that $Q_1 G_1^T = I$, where I is the $k_1 \times k_1$ identity matrix.

4. Let Q_1 be as in part 4 and let H_1 be parity-check matrix of $\mathcal{C}_1$. Show that the matrix

$$\left(\begin{array}{c} H_1 \\ \hline Q_1 \end{array} \right)$$

has rank n_1.

5. Show that there is a $k_1 \times n_2$ matrix Q_2 over F such that $Q_2 G_{2,1}^T = I$ and $Q_2 G_{2,2}^T = 0$.

6. Let Q_1, H_1, and Q_2 be as in the previous parts and let H_2 be a parity-check matrix of $\mathcal{C}_2$. Show that

$$\left(\begin{array}{c|c} H_1 & 0 \\ \hline 0 & H_2 \\ \hline Q_1 & -Q_2 \end{array} \right)$$

is a parity-check matrix of $\mathcal{C}$.

Problem 2.14 Let $\mathcal{C}$ be a linear $[n, k, d]$ code over F where $k > 1$. For $i \in \{1, 2, \ldots, n\}$, denote by $\mathcal{C}^{(i)}$ the code

$$\mathcal{C}^{(i)} = \{(c_1\ c_2\ \ldots\ c_{i-1}\ c_{i+1}\ \ldots\ c_n) \ : \ (c_1\ c_2\ \ldots\ c_{i-1}\ 0\ c_{i+1}\ \ldots\ c_n) \in \mathcal{C}\} .$$

The code $\mathcal{C}^{(i)}$ is said to be obtained by *shortening* $\mathcal{C}$ at the ith coordinate.

1. Show that $\mathcal{C}^{(i)}$ is a linear $[n-1, k^{(i)}, d^{(i)}]$ code over F where $k^{(i)} \geq k-1$ and $d^{(i)} \geq d$.

2. Show that a parity-check matrix of $\mathcal{C}^{(i)}$ is obtained by deleting the ith column from a parity-check matrix of $\mathcal{C}$.

Problem 2.15 Let $\mathcal{C}$ be a linear code over F. Show that every generator matrix of the dual code $\mathcal{C}^\perp$ is a parity-check matrix of $\mathcal{C}$. Conclude that $(\mathcal{C}^\perp)^\perp = \mathcal{C}$.

Problem 2.16 For an integer $m > 1$ let $\mathcal{C}$ be the $[n, n-m, 3]$ Hamming code over GF(2) where $n = 2^m - 1$.

1. Show that for every two distinct columns $\mathbf{h}_1$ and $\mathbf{h}_2$ in a parity-check matrix of $\mathcal{C}$ there is a unique third column in that matrix that equals the sum $\mathbf{h}_1 + \mathbf{h}_2$.

2. Using part 1, show that the number of codewords of Hamming weight 3 in $\mathcal{C}$ is $n(n-1)/6$.
3. Show that $\mathcal{C}$ contains a codeword of Hamming weight n (namely, $\mathcal{C}$ contains the all-one codeword).
4. How many codewords are there in $\mathcal{C}$ of Hamming weight $n-1$? $n-2$? and $n-3$?

Problem 2.17 (First-order Reed–Muller codes) Let $F = \mathrm{GF}(q)$ and for a positive integer m let $n = q^m$. The *first-order Reed–Muller code* over F is defined as the linear $[n, m+1]$ code $\mathcal{C}$ over F with an $(m+1) \times n$ generator matrix whose columns range over all the vectors in F^{m+1} with a first entry equaling 1. (Observe that for $q = 2$, the first-order Reed–Muller code is the dual code of the extended Hamming code defined in Example 2.8.)

1. Show that the minimum distance of $\mathcal{C}$ equals $q^{m-1}(q-1)$ and that this number is the Hamming weight of $q(q^m - 1)$ codewords in $\mathcal{C}$. What are the Hamming weights of the remaining q codewords in $\mathcal{C}$?

   Hint: Use Problem 2.4.
2. Show that no linear $[n, m+1]$ code over F can have minimum distance greater than $q^{m-1}(q-1)$.

   Hint: Use the Plotkin bound in Problem 2.7.
3. The *shortened* first-order Reed–Muller code over F is defined as the linear $[n-1, m]$ code $\mathcal{C}'$ over F with an $m \times (n-1)$ generator matrix whose columns range over all the nonzero vectors in F^m. Show that $\mathcal{C}'$ can be obtained from $\mathcal{C}$ by the shortening operation defined in Problem 2.14.
4. Show that every nonzero codeword in the shortened code $\mathcal{C}'$ has Hamming weight $q^{m-1}(q-1)$.
5. Verify that the shortened code $\mathcal{C}'$ attains the Plotkin bound in Problem 2.7.

Problem 2.18 (Simplex codes, or dual codes of Hamming codes) Let $F = \mathrm{GF}(q)$ and for an integer $m > 1$ let $n = (q^m - 1)/(q - 1)$. Consider the dual code $\mathcal{C}$ of the $[n, n-m]$ Hamming code over F.

1. Show that the Hamming weight of each nonzero codeword in $\mathcal{C}$ is q^{m-1}.

   Hint: Use Problem 2.17.
2. Verify that $\mathcal{C}$ attains the Plotkin bound in Problem 2.7.

Problem 2.19 (Binary Reed–Muller codes) Let $F = \mathrm{GF}(2)$ and let m and r be integers such that $m > 0$ and $0 \le r \le m$. Denote by $\mathcal{S}(m, r)$ the Hamming sphere of radius r in F^m centered at $\mathbf{0}$, that is

$$\mathcal{S}(m, r) = \{\mathbf{e} \in F^m \ : \ \mathsf{w}(\mathbf{e}) \le r\} \ ;$$

note that

$$|\mathcal{S}(m, r)| = \sum_{j=0}^{r} \binom{m}{j} \ .$$

For two elements $a, e \in F$, the expression a^e will be interpreted as if e is an integer, namely,

$$a^e = \begin{cases} 0 & \text{if } a = 0 \text{ and } e = 1 \\ 1 & \text{if } a = 1 \text{ or } e = 0 \end{cases} .$$

Given vectors $\mathbf{a} = (a_0 \; a_1 \; \ldots \; a_{m-1})$ and $\mathbf{e} = (e_0 \; e_1 \; \ldots \; e_{m-1})$ in F^m, the notation $\mathbf{a}^{\mathbf{e}}$ will stand hereafter as a shorthand for the product $a_0^{e_0} a_1^{e_1} \cdots a_{m-1}^{e_{m-1}}$.

Let $G_{\mathrm{RM}}(m, r)$ be the $|\mathcal{S}(m, r)| \times 2^m$ matrix over F whose rows and columns are indexed by the elements of $\mathcal{S}(m, r)$ and F^m, respectively, and for every $\mathbf{e} \in \mathcal{S}(m, r)$ and $\mathbf{a} \in F^m$, the entry of $G_{\mathrm{RM}}(m, r)$ that is indexed by $(\mathbf{e}, \mathbf{a})$ equals $\mathbf{a}^{\mathbf{e}}$. Denote by $\mathcal{C}_{\mathrm{RM}}(m, r)$ the linear code over F that is spanned by the rows of $G_{\mathrm{RM}}(m, r)$. The code $\mathcal{C}_{\mathrm{RM}}(m, r)$ is called the *rth order Reed–Muller code* of length 2^m over F.

1. Identify the codes $\mathcal{C}_{\mathrm{RM}}(m, 0)$ and $\mathcal{C}_{\mathrm{RM}}(m, 1)$.

2. Show that the rows of $G_{\mathrm{RM}}(m, r)$ are linearly independent.

   Hint: It suffices to show that $\mathrm{rank}(G_{\mathrm{RM}}(m, m)) = 2^m$. Show that when the rows and columns of $G_{\mathrm{RM}}(m, m)$ are arranged according to the ordinary lexicographic ordering of their indexes, then $G_{\mathrm{RM}}(m, m)$ becomes upper-triangular, with 1's along the main diagonal.

   (The following alternate proof uses the correspondence between the rows of $G_{\mathrm{RM}}(m, m)$ and the functions $\mathbf{x} \mapsto \mathbf{x}^{\mathbf{e}}$ for $\mathbf{e} \in F^m$. For every vector $\mathbf{u} = (u_{\mathbf{e}})_{\mathbf{e} \in F^m}$ in F^{2^m} whose entries are indexed by the elements of F^m, the entry in $\mathbf{u} G_{\mathrm{RM}}(m, m)$ that is indexed by $\mathbf{a} \in F^m$ equals the value of the function $f_{\mathbf{u}} : \mathbf{x} \mapsto \sum_{\mathbf{e} \in F^m} u_{\mathbf{e}} \mathbf{x}^{\mathbf{e}}$ at $\mathbf{x} = \mathbf{a}$. Now, it is known that Boolean "and" and "exclusive-or"—namely, multiplication and addition in F—and the constant "1" form a *functionally complete* set of Boolean operations, in that they generate every Boolean function. Thus, when $\mathbf{u}$ ranges over all the elements of F^{2^m}, the functions $f_{\mathbf{u}}$ range over all the 2^{2^m} Boolean functions with m variables. Deduce that the rows of $G_{\mathrm{RM}}(m, m)$ are linearly independent.)

3. Show that for $0 < r < m$,

$$G_{\mathrm{RM}}(m, r) = \left(\begin{array}{c|c} G_{\mathrm{RM}}(m{-}1, r) & G_{\mathrm{RM}}(m{-}1, r) \\ \hline 0 & G_{\mathrm{RM}}(m{-}1, r{-}1) \end{array} \right) .$$

   Hint:

$$\mathcal{S}(m, r) = \{(0 \; \mathbf{e}) \; : \; \mathbf{e} \in \mathcal{S}(m{-}1, r)\} \cup \{(1 \; \mathbf{e}) \; : \; \mathbf{e} \in \mathcal{S}(m{-}1, r{-}1)\} .$$

4. Show that the minimum distance of $\mathcal{C}_{\mathrm{RM}}(m, r)$ is 2^{m-r}.

   Hint: Apply Problem 2.12 to part 3 inductively.

5. For $\mathbf{e} \in \mathcal{S}(m, r)$, denote by $(G_{\mathrm{RM}}(m, r))_{\mathbf{e}}$ the row of $G_{\mathrm{RM}}(m, r)$ that is indexed by $\mathbf{e}$. Show that for elements $\mathbf{e}$ and $\mathbf{e}'$ in F^m,

$$(G_{\mathrm{RM}}(m, m))_{\mathbf{e}} \cdot (G_{\mathrm{RM}}(m, m))_{\mathbf{e}'}^T = 0$$

   if and only if there is at least one location in which both $\mathbf{e}$ and $\mathbf{e}'$ have a zero entry.

6. Show that for $0 \le r < m$,

$$(\mathcal{C}_{\mathrm{RM}}(m,r))^{\perp} = \mathcal{C}_{\mathrm{RM}}(m, m-r-1) .$$

Hint: Use part 5 to show that $G_{\mathrm{RM}}(m,r)(G_{\mathrm{RM}}(m,m-r-1))^T = 0$, and verify that

$$\mathrm{rank}(G_{\mathrm{RM}}(m,r)) + \mathrm{rank}(G_{\mathrm{RM}}(m,m-r-1)) = \sum_{j=0}^{m} \binom{m}{j} = 2^m .$$

Problem 2.20 (Linear codes over rings) Let F be a commutative ring of finite size and let the addition in F^n be defined as in Problem 2.9. Using the vector notation $(c_1\ c_2\ \ldots c_n)$ for words in F^n, the product of a word $\mathbf{c} = (c_1\ c_2\ \ldots c_n)$ in F^n by a scalar $a \in F$ is defined—similarly to vectors over fields—by

$$a \cdot \mathbf{c} = (ac_1\ ac_2\ \ldots\ ac_n) .$$

A set of words $\mathbf{u}_1, \mathbf{u}_2, \ldots, \mathbf{u}_m \in F^r$ is *linearly independent* over F if and only if for every m elements $a_1, a_2, \ldots, a_m \in F$,

$$\sum_{i=1}^{m} a_i \mathbf{u}_i = \mathbf{0} \quad \iff \quad a_1 = a_2 = \ldots = a_m = 0 .$$

An (n, M) code $\mathcal{C}$ over F is called *linear* over F if $\mathcal{C}$ is a group code over F (as defined in Problem 2.9) and for every $a \in F$,

$$\mathbf{c} \in \mathcal{C} \quad \Longrightarrow \quad a \cdot \mathbf{c} \in \mathcal{C} .$$

Let H be an $r \times n$ matrix over F whose columns are given by $\mathbf{h}_1, \mathbf{h}_2, \ldots, \mathbf{h}_n \in F^r$. Extend the definition of matrix-by-vector multiplication from fields to the ring F: for a word $\mathbf{c} = (c_1\ c_2\ \ldots\ c_n)$, let $H\mathbf{c}^T$ be $\sum_{j=1}^{n} c_j \mathbf{h}_j$. Consider the (n, M) code $\mathcal{C}$ over F that is defined by

$$\mathcal{C} = \left\{ \mathbf{c} \in F^n \ :\ H\mathbf{c}^T = \mathbf{0} \right\} .$$

1. Show that $\mathcal{C}$ is a linear code over F.
2. Show that if $M > 1$, then $\mathsf{d}(\mathcal{C})$ equals the largest integer d such that every set of $d-1$ columns in H is linearly independent.
3. Suppose that every set of $d-1$ columns in H contains a $(d-1) \times (d-1)$ sub-matrix whose determinant has a multiplicative inverse in F (such a sub-matrix is then invertible over F). Show that $\mathsf{d}(\mathcal{C}) \ge d$.

   (Observe that the converse is not true; for example, if F is the ring $\mathbb{Z}_6$ and

$$H = \begin{pmatrix} 1 & 0 & 3 & 3 \\ 1 & 0 & 2 & 2 \\ 0 & 1 & 1 & 0 \end{pmatrix} ,$$

   then $\mathsf{d}(\mathcal{C}) = 3$, yet none of the determinants of the 2×2 sub-matrices in the last three columns has a multiplicative inverse in $\mathbb{Z}_6$.)

Problem 2.21 (Product codes and interleavers) Let $\mathcal{C}_1$ be a linear $[n_1, k_1, d_1]$ code over $F = \mathrm{GF}(q)$ and let $\mathcal{C}_2$ be a linear $[n_2, k_2, d_2]$ code over the same field F. Assume that both codes have systematic generator matrices.

Define a mapping $\mathcal{E} : F^{k_1 k_2} \to F^{n_1 n_2}$ through the algorithm shown in Figure 2.1. The algorithm makes use of an array Γ over F with n_1 columns and n_2 rows; the array Γ is divided into three sub-arrays, U, V, and W, as shown in Figure 2.2.

Input: word $\mathbf{u} \in F^{k_1 k_2}$.
Output: word $\mathbf{c} \in F^{n_1 n_2}$.

(a) Fill in the sub-array U, row by row, by the contents of $\mathbf{u}$.

(b) For $i = 1, 2, \ldots, k_2$, determine the contents of the ith row in the sub-array V so that the ith row in Γ is a codeword of $\mathcal{C}_1$.

(c) For $j = 1, 2, \ldots, n_1$, determine the contents of the jth column in the sub-array W so that the jth column in Γ is a codeword of $\mathcal{C}_2$.

(d) Read the contents of Γ, row by row, into a word $\mathbf{c}$ of length $n_1 n_2$, and output $\mathbf{c}$ as the value $\mathcal{E}(\mathbf{u})$.

Figure 2.1. Algorithm that defines the mapping $\mathbf{u} \mapsto \mathcal{E}(\mathbf{u})$.

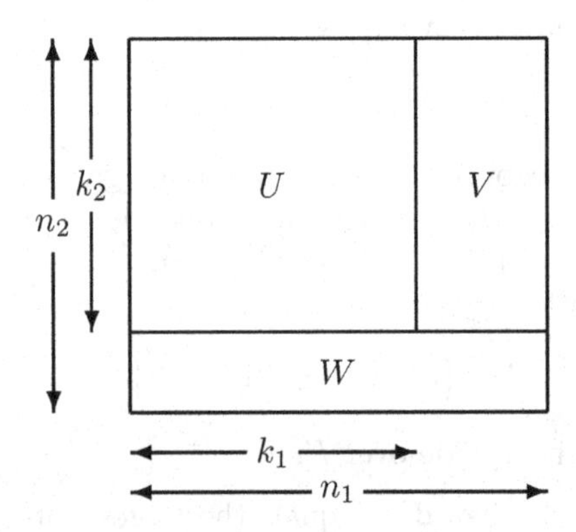

Figure 2.2. Array Γ of a product code.

The *product code* $\mathcal{C}_1 * \mathcal{C}_2$ is defined by

$$\mathcal{C}_1 * \mathcal{C}_2 = \left\{ \mathbf{c} \in F^{n_1 n_2} \ : \ \mathbf{c} = \mathcal{E}(\mathbf{u}) \text{ for some } \mathbf{u} \in F^{k_1 k_2} \right\} .$$

1. Why is it always possible to determine the contents of the sub-array V so that it satisfies the condition in Step (b) in Figure 2.1?

2. Is the contents of V uniquely determined by the contents of U?

3. Show that the *rows* of W form codewords of $\mathcal{C}_1$.

4. Show that $\mathcal{C}_1 * \mathcal{C}_2$ is a linear $[n_1 n_2, k_1 k_2, d_1 d_2]$ code over F.

5. Given two matrices $A = (A_{i,j})_{i=1}^{m}{}_{j=1}^{n}$ and $B = (B_{i,j})_{i=1}^{r}{}_{j=1}^{s}$ of orders $m \times n$ and $r \times s$, respectively, over a field F, define their *Kronecker product* (or *direct product*) as the $mr \times ns$ matrix $A \otimes B$ whose entries are given by

$$(A \otimes B)_{r(i-1)+i', s(j-1)+j'} = A_{i,j} B_{i',j'} \,, \qquad 1 \le i \le m,\ 1 \le j \le n,\ 1 \le i' \le r,\ 1 \le j' \le s \,.$$

Let H_1 and H_2 be parity-check matrices of $\mathcal{C}_1$ and $\mathcal{C}_2$, respectively, and denote by I_n the $n \times n$ identity matrix over F. Show that

$$\left(\frac{I_{n_2} \otimes H_1}{H_2 \otimes I_{n_1}} \right)$$

is a parity-check matrix of $\mathcal{C}_1 * \mathcal{C}_2$.

6. An *interleaver* is a special case of a product code where $\mathcal{C}_1$ is the $[n_1, n_1, 1]$ code F^{n_1}. A *burst of length* ℓ is the event of having errors in a codeword such that the locations i and j of the first (leftmost) and last (rightmost) errors, respectively, satisfy $j - i = \ell - 1$.

Suggest an encoding–decoding scheme that is based on interleavers and allows one to correct every burst of length up to $\lfloor (d_2 - 1)/2 \rfloor \cdot n_1$.

[Section 2.4]

Problem 2.22 Let H be a parity-check matrix of a linear code over F and let t be a positive integer. Recall the definition of a burst as in part 6 of Problem 2.21.

1. Find a necessary and sufficient condition on the columns of H so that every burst of length up to t that occurs in a codeword can be detected.
2. What are the respective conditions on the columns of H so that every burst of length up to t can be corrected?
3. A *burst erasure* of length ℓ is the event where erasures occur in ℓ consecutive locations within a codeword. Find a necessary and sufficient condition on the columns of H so that every burst erasure of length up to t can be recovered.

Problem 2.23 (Sylvester-type Hadamard matrices) Let k be a positive integer and F be the field GF(2). The $2^k \times 2^k$ *Sylvester-type Hadamard matrix*, denoted by $\mathcal{H}_k$, is the real matrix whose rows (respectively, columns) are indexed by the row (respectively, column) vectors in F^k, and

$$(\mathcal{H}_k)_{\mathbf{u},\mathbf{v}} = \begin{cases} 1 & \text{if } \mathbf{u} \cdot \mathbf{v} = 0 \\ -1 & \text{if } \mathbf{u} \cdot \mathbf{v} = 1 \end{cases} , \qquad \mathbf{u}, \mathbf{v} \in F^k \,.$$

For $k = 0$, define $\mathcal{H}_0$ to be the 1×1 matrix $(\,1\,)$.

1. Assuming that the rows and columns of $\mathcal{H}_k$ are written according to the standard lexicographic ordering on the elements of F^k, show that for every $k \ge 0$,

$$\mathcal{H}_{k+1} = \left(\begin{array}{c|c} \mathcal{H}_k & \mathcal{H}_k \\ \hline \mathcal{H}_k & -\mathcal{H}_k \end{array} \right) .$$

2. Suggest a recursive algorithm for multiplying the matrix $\mathcal{H}_k$ by a vector in $\mathbb{R}^{2^k}$ while requiring only $k \cdot 2^k$ real additions and subtractions.

   Hint: For $\mathbf{x}_0, \mathbf{x}_1 \in \mathbb{R}^{2^k}$,

$$\mathcal{H}_{k+1} \begin{pmatrix} \mathbf{x}_0 \\ \mathbf{x}_1 \end{pmatrix} = \begin{pmatrix} \mathcal{H}_k \mathbf{x}_0 + \mathcal{H}_k \mathbf{x}_1 \\ \mathcal{H}_k \mathbf{x}_0 - \mathcal{H}_k \mathbf{x}_1 \end{pmatrix} .$$

3. Show that the rows of the $2^{k+1} \times 2^k$ matrix

$$\begin{pmatrix} \mathcal{H}_k \\ -\mathcal{H}_k \end{pmatrix}$$

   form the codewords of the $[2^k, k+1]$ first-order Reed–Muller code over F, which was defined in Problem 2.17, with 1 (respectively, -1) representing the element 0 (respectively, 1) of F.

4. Show that

$$\mathcal{H}_k^2 = \mathcal{H}_k \mathcal{H}_k^T = 2^k \cdot I .$$

Problem 2.24 Let $F = \mathrm{GF}(2)$ and let $S = (F, \Phi, \mathsf{Prob})$ be a memoryless binary channel as defined in Problem 1.8; namely, for every positive integer m and every two words $x_1 x_2 \ldots x_m \in F^m$ and $y_1 y_2 \ldots y_m \in \Phi^m$, the conditional probability distribution takes the form

$$\mathsf{Prob}\{\, y_1 y_2 \ldots y_m \text{ received} \mid x_1 x_2 \ldots x_m \text{ transmitted} \,\} = \prod_{j=1}^{m} \mathsf{Prob}(y_j | x_j) .$$

Assume hereafter that $\mathsf{Prob}(y|x) > 0$ for every $x \in F$ and $y \in \Phi$.

Let $\mathcal{C}$ be a linear $[n, k, d]$ code over F, and let the function $\mathcal{D} : \Phi^n \to \mathcal{C}$ map every word $y_1 y_2 \ldots y_n \in \Phi^n$ to the codeword $(c_1\ c_2\ \ldots\ c_n) \in \mathcal{C}$ that maximizes the expression

$$\sum_{j=1}^{n} (-1)^{c_j} \cdot \mu(y_j) ,$$

where

$$\mu(y) = \log_2 \left(\frac{\mathsf{Prob}(y|0)}{\mathsf{Prob}(y|1)} \right) , \quad y \in \Phi$$

(the elements c_j act in the expression $(-1)^{c_j}$ as if they were integers). Recall from Problem 1.8 that $\mathcal{D}$ is a maximum-likelihood decoder for $\mathcal{C}$ with respect to the channel S.

Fix $G = (\mathbf{g}_1\ \mathbf{g}_2\ \ldots\ \mathbf{g}_n)$ to be a $k \times n$ generator matrix of $\mathcal{C}$. For every word $\mathbf{y} = y_1 y_2 \ldots y_n$ in Φ^n, associate a vector $\boldsymbol{\mu}(\mathbf{y}) \in F^{2^k}$ whose entries are indexed by the column vectors in F^k, and the entry indexed by $\mathbf{v} \in F^k$ is given by

$$(\boldsymbol{\mu}(\mathbf{y}))_{\mathbf{v}} = \sum_{j \,:\, \mathbf{g}_j = \mathbf{v}} \mu(y_j)$$

(a sum over an empty set is defined as zero).

An information word $\mathbf{u} \in F^k$ is transmitted through the channel S by encoding it into the codeword $\mathbf{c} = \mathbf{u}G$.

1. Show that for every $\mathbf{y} \in \Phi^n$, the value $\mathcal{D}(\mathbf{y})$ is a codeword $\hat{\mathbf{u}}G$ that corresponds to an index $\hat{\mathbf{u}} \in F^k$ of a largest entry in the vector $\mathcal{H}_k \boldsymbol{\mu}(\mathbf{y})$, where $\mathcal{H}_k$ is the $2^k \times 2^k$ Sylvester-type Hadamard matrix defined in Problem 2.23.

2. Assume that the values $\mu(y)$ are pre-computed for all $y \in \Phi$. Show that the decoder $\mathcal{D}$ can then be implemented with less than $(k+1) \cdot 2^k + n$ real additions, subtractions, and comparisons (as opposed to $n \cdot 2^k - 1$ operations, which are required in a brute force search over the codewords of $\mathcal{C}$).

Notes

[Section 2.2]

Two (n, M) codes $\mathcal{C}$ and $\hat{\mathcal{C}}$ over F are said to be equivalent if the sets $\mathcal{C}$ and $\hat{\mathcal{C}}$ are the same, up to a fixed permutation on the coordinates of the codewords of $\mathcal{C}$. For example, the code in Example 2.10 is equivalent to the linear code over $F = \mathrm{GF}(2)$ with a generator matrix

$$\hat{G} = \begin{pmatrix} 1 & 1 & 0 & 0 & 1 \\ 0 & 0 & 1 & 1 & 1 \end{pmatrix} .$$

Given two $k \times n$ matrices G and $\hat{G}$ over a field F, the complexity of deciding whether these matrices generate equivalent linear codes over F is still not known. On the one hand, this problem is unlikely to be too easy, since it is at least as difficult as the *Graph Isomorphism* problem (see Babai *et al.* [26'] and Petrank and Roth [279]). On the other hand, as shown in [279], the equivalence problem of linear codes is unlikely to be intractable (NP-complete) either; see also Sendrier [327]. For more on the theory of NP-completeness, refer to the book by Garey and Johnson [143].

[Section 2.3]

Hamming codes were discovered by Golay in [149] and [150] and by Hamming in [169]. Unlike the binary case, an attempt to extend Hamming codes over an arbitrary field $\mathrm{GF}(q)$ generally results in a code whose minimum distance is still 3. The problem of determining the length, $\Lambda_q(r)$, of the longest linear code with minimum distance 4 and redundancy r over $\mathrm{GF}(q)$ has been treated primarily in the context of projective geometries over finite fields.

Specifically, let $\mathrm{PG}(r{-}1, q)$ denote the $(r{-}1)$-dimensional projective geometry over $F = \mathrm{GF}(q)$ (the elements of $\mathrm{PG}(r{-}1, q)$ are all the nonzero vectors of F^r, with linearly dependent vectors being identified as one element of $\mathrm{PG}(r{-}1, q)$; thus, $|\mathrm{PG}(r{-}1, q)| = (q^r - 1)/(q - 1)$). An *$n$-cap* in $\mathrm{PG}(r{-}1, q)$ is a set of n points in $\mathrm{PG}(r{-}1, q)$ such that no three points in the set are collinear. The elements of an n-cap over $F = \mathrm{GF}(q)$ thus form an $r \times n$ parity-check matrix of a linear $[n, \geq n{-}r, \geq 4]$ code over F, and $\Lambda_q(r)$ is the size of the largest n-cap in $\mathrm{PG}(r{-}1, q)$.

The case $q = 2$ is fully settled: every linear $[n, n{-}r, 4]$ code over F implies (by puncturing) the existence of a linear $[n{-}1, n{-}r{-}1, 3]$ code $\mathcal{C}$ over F, and, since no two columns in an $(r{-}1) \times (n{-}1)$ parity-check matrix of $\mathcal{C}$ are linearly dependent, we have the upper bound $n{-}1 \leq 2^{r-1} - 1$. This bound is tight as it is attained by

$[2^{r-1}, 2^{r-1}-r, 4]$ extended Hamming codes; therefore,

$$\Lambda_2(r) = 2^{r-1} .$$

For $r = 3$, the values of $\Lambda_q(r)$ are known to be

$$\Lambda_q(3) = \begin{cases} q+1 & \text{when } q \text{ is odd} \\ q+2 & \text{when } q \text{ is even} \end{cases}$$

(these values follow from a result that will be proved in Section 11.3: see Proposition 11.15 therein). It is also known that when $r = 4$ and $q > 2$,

$$\Lambda_q(4) = q^2 + 1$$

(an attaining code construction will be presented in Problem 3.44). For $r \geq 5$ and sufficiently large q we have the upper bound

$$\Lambda_q(r) \leq \begin{cases} q^{r-2} - \frac{1}{4}q^{r-(5/2)} + 2q^{r-3} & \text{when } q \text{ is odd} \\ q^{r-2} - \frac{1}{2}q^{r-3} & \text{when } q \text{ is even} \end{cases} .$$

See Hirschfeld and Thas [186, Chapter 27] and Thas [361] for details.

Reed–Muller codes (Problems 2.17 and 2.19) are named after Reed [288] and Muller [265]. For more on Reed–Muller codes, including their generalization to non-binary alphabets, see Assmus and Key [18, Chapter 5], Berlekamp [36, Sections 15.3 and 15.4], MacWilliams and Sloane [249, Chapters 13–15], and Peterson and Weldon [278, Section 5.5]. Refer also to the discussion on Reed–Muller codes in the notes on Sections 5.1 and 5.2.

[Section 2.4]

The complexity of nearest-codeword decoding of linear codes was first treated by Berlekamp *et al.* [38]. Specifically, they showed that given an $r \times n$ matrix H over $F = \mathrm{GF}(2)$, a vector $\mathbf{s} \in F^r$, and an integer m, the problem of deciding whether there is a solution to $H\mathbf{e}^T = \mathbf{s}$ for a vector $\mathbf{e} \in F^n$ such that $\mathsf{w}(\mathbf{e}) \leq m$, is NP-complete. Furthermore, Arora *et al.* [17] showed that the problem remains NP-hard even if it is relaxed only to approximating, within a constant factor, the Hamming weight of the lightest vector $\mathbf{e}$ that satisfies $\mathbf{s} = H\mathbf{e}^T$. The approximation within certain constant factors turns out to be hard also if the decoder—which depends on H, but not on $\mathbf{s}$—can be preprocessed, i.e., the time required to design the decoder is not counted towards the complexity; see Bruck and Naor [68], Feige and Micciancio [122], Lobstein [236], and Regev [291].

The problem of deciding whether the minimum distance. of a given linear code is at most m is, in a way, a special case of the nearest-codeword decoding problem and can be stated as follows. Given an $r \times n$ (parity-check) matrix H over F and an integer m, decide whether there is a solution to $H\mathbf{e}^T = \mathbf{0}$ for a *nonzero* vector $\mathbf{e} \in F^n$ such that $\mathsf{w}(\mathbf{e}) \leq m$. This problem was shown to be NP-complete over the rational field by Khachiyan [212] and over finite fields by Vardy [370]. Dumer *et al.* [110] showed that the approximation—within a constant factor—of the minimum distance of linear codes is unlikely to be polynomial-time.

Sylvester-type Hadamard matrices are used quite often in signal and image processing, and the algorithm sought in Problem 2.23 is known as the Fast Hadamard Transform (in short, FHT). For more on these matrices, see for example, Agaian [3], Wallis *et al.* [378], and Yarlagadda and Hershey [391].

Chapter 3

Introduction to Finite Fields

For most of this chapter, we deviate from our study of codes to become acquainted with the algebraic concept of finite fields. These objects will serve as our primary tool for constructing codes in upcoming chapters. As a motivating example, we present at the end of this chapter a construction of a double-error-correcting binary code, whose description and analysis make use of finite fields. This construction will turn out to be a special case of a more general family of codes, to be discussed in Section 5.5.

Among the properties of finite fields that we cover in this chapter, we show that the multiplicative group of a finite field is cyclic; this property, in turn, suggests a method for implementing the arithmetic operations in finite fields of moderate sizes through look-up tables, akin to logarithm tables. We also prove that the size of any finite field must be a power of a prime and that this necessary condition is also sufficient, that is, every power of a prime is a size of some finite field. The practical significance of the latter property is manifested particularly through the special case of the prime 2, since in most coding applications, the data is sub-divided into symbols—e.g., bytes—that belong to alphabets whose sizes are powers of 2.

3.1 Prime fields

For a prime p, we let $\mathrm{GF}(p)$ (*Galois field of size* p) denote the ring of integer residues modulo p (this ring is also denoted by $\mathbb{Z}_p$).

By Euclid's algorithm for integers (Problem A.3), for every integer $a \in \{1, 2, \ldots, p-1\}$ there exist integers s and t such that

$$s \cdot a + t \cdot p = 1 .$$

The integer s, taken modulo p, is the multiplicative inverse a^{-1} of a in $\mathrm{GF}(p)$. Therefore, $\mathrm{GF}(p)$ is indeed a field.

Example 3.1 In GF(7) we have

$$2 \cdot 4 = 3 \cdot 5 = 6 \cdot 6 = 1 \cdot 1 = 1 .$$

Also, $a^6 = 1$ for every nonzero $a \in \mathrm{GF}(7)$.

The multiplicative group of GF(7) is cyclic, and the elements 3 and 5 generate all the nonzero elements of the field:

$$\begin{array}{lllll} 3^0 &=& 1 &=& 5^0 \\ 3^1 &=& 3 &=& 5^5 \\ 3^2 &=& 2 &=& 5^4 \end{array} \qquad \begin{array}{lllll} 3^3 &=& 6 &=& 5^3 \\ 3^4 &=& 4 &=& 5^2 \\ 3^5 &=& 5 &=& 5^1 \end{array}$$

□

Let F be a field. The symbols 0 and 1 will stand for the additive and multiplicative unity elements, respectively, of F. The multiplicative group of F will be denoted by F^*, and the multiplicative order of an element $a \in F^*$ (whenever such an order exists) will be denoted by $\mathcal{O}(a)$. In particular, this order exists whenever F is finite, in which case we get the following result.

Proposition 3.1 *Let F be a finite field. For every $a \in F$,*

$$a^{|F|} = a .$$

Proof. Clearly, $0^{|F|} = 0$. As for $a \in F^*$, by Lagrange's Theorem it follows that $\mathcal{O}(a)$ divides $|F^*|$ (Problem A.16); therefore, $a^{|F|-1} = 1$. □

We show in Section 3.5 below that the multiplicative group of every finite field is cyclic. A generator of F^* is called a *primitive element* in F.

3.2 Polynomials

Let F be a field. A *polynomial* over F (in the indeterminate x) is an expression of the form

$$a(x) = a_0 + a_1 x + \ldots + a_n x^n ,$$

where n is a nonnegative integer and $a_0, a_1, \ldots, a_n$ are elements of F: these elements are called the *coefficients* of the polynomial $a(x)$. We regard two polynomials as equal if they agree on their nonzero coefficients. The set of all polynomials over F in the indeterminate x is denoted by $F[x]$.

We next mention several terms and notations that are associated with polynomials.

The *zero polynomial* over F is the (unique) polynomial whose coefficients are all zero.

The *degree* of a nonzero polynomial $a(x) = \sum_{i=0}^{n} a_i x^i$ over F is the largest index i for which $a_i \neq 0$. The degree of the zero polynomial is defined as $-\infty$.

The degree of $a(x)$ will be denoted by $\deg a(x)$ or $\deg a$, and the set of all polynomials over F of degree less than n will be denoted by $F_n[x]$.

A nonzero polynomial $a(x)$ is called *monic* if the coefficient of $x^{\deg a}$ equals 1.

Let $a(x) = \sum_{i=0}^{n} a_i x^i$ and $b(x) = \sum_{i=0}^{n} b_i x^i$ be two polynomials over F (of possibly different degrees). Their sum and difference are defined by

$$a(x) \pm b(x) = \sum_{i=0}^{n} (a_i \pm b_i) x^i ,$$

and their product is the polynomial $c(x) = \sum_{i=0}^{2n} c_i x^i$ whose coefficients are

$$c_i = \sum_{j=0}^{i} a_j b_{i-j} , \quad 0 \le i \le 2n$$

(where $a_i = b_i = 0$ for $i > n$). Under these operations, $F[x]$ forms an integral domain (see the Appendix and Problem 3.2).

Let $a(x)$ and $b(x)$ be polynomials over F such that $a(x) \neq 0$. As with integers, we can apply a "long division" on $a(x)$ and $b(x)$ and compute unique polynomials—a *quotient* $q(x)$ and a *remainder* $r(x)$—such that

$$b(x) = a(x)q(x) + r(x) ,$$

where $\deg r < \deg a$. In particular, $a(x) \,|\, b(x)$ ("$a(x)$ divides $b(x)$") means that dividing $b(x)$ by $a(x)$ yields a zero remainder; in this case we also say that $a(x)$ is a *factor* of $b(x)$. The notation

$$b(x) \equiv c(x) \pmod{a(x)}$$

("$b(x)$ is congruent to $c(x)$ modulo $a(x)$") is the same as saying that $a(x)$ divides $b(x) - c(x)$.

Euclid's algorithm (see Problem 3.3) can be applied to polynomials to compute their greatest common divisor (gcd). Such a divisor is defined only up to a multiple by a nonzero scalar in F; so, $\gcd(a(x), b(x)) = 1$ is the same as writing $\deg \gcd(a(x), b(x)) = 0$.

Example 3.2 Let $F = \mathrm{GF}(2)$ and let $a(x) = x^4 + x^2 + x + 1$ and $b(x) = x^3 + 1$. To find $\gcd(a(x), b(x))$ we compute a sequence of remainders $r_i(x)$ as follows, starting with $r_{-1}(x) = a(x)$ and $r_0(x) = b(x)$.

1. Divide $r_{-1}(x) = x^4 + x^2 + x + 1$ by $r_0(x) = x^3 + 1$:

$$\begin{array}{ccccccc|l}
x & & & & & & & \\
\hline
x^4 & + & x^2 & + & x & + & 1 & x^3 + 1 \\
x^4 & + & & & x & & & \\
\cline{3-7}
 & & x^2 & + & & & 1 &
\end{array}$$

That is, the quotient is x and the remainder is $r_1(x) = x^2 + 1$:

$$\underbrace{x^4 + x^2 + x + 1}_{r_{-1}(x)} = x(\underbrace{x^3 + 1}_{r_0(x)}) + \underbrace{x^2 + 1}_{r_1(x)} . \tag{3.1}$$

2. Divide $r_0(x) = x^3 + 1$ by the new remainder $r_1(x) = x^2 + 1$:

$$\begin{array}{llll|l} x & & & & \\ \hline x^3 & + & & 1 & x^2+1 \\ x^3 & + & x & & \\ \cline{3-4} & & x & + \; 1 & \end{array}$$

(Recall that $-x = x$ over GF(2).) Thus,

$$\underbrace{x^3 + 1}_{r_0(x)} = x(\underbrace{x^2 + 1}_{r_1(x)}) + \underbrace{x + 1}_{r_2(x)} . \tag{3.2}$$

3. Divide $r_1(x) = x^2 + 1$ by the new remainder $r_2(x) = x + 1$:

$$\begin{array}{llll|l} x & + & 1 & & \\ \hline x^2 & + & & 1 & x+1 \\ x^2 & + & x & & \\ \cline{3-4} & & x & + \; 1 & \end{array}$$

So,

$$\underbrace{x^2 + 1}_{r_1(x)} = (x + 1)(\underbrace{x + 1}_{r_2(x)})$$

and the new remainder, $r_3(x)$, is zero. Hence,

$$\gcd(x^4 + x^2 + x + 1, x^3 + 1) = r_2(x) = x + 1 .$$

□

For two polynomials $a(x), b(x) \in F[x]$, not both zero, there always exist polynomials $s(x), t(x) \in F[x]$ such that

$$s(x) \cdot a(x) + t(x) \cdot b(x) = \gcd(a(x), b(x))$$

(substitute $i = \nu$ in part 2 of Problem 3.3). In the previous example we have

$$\begin{aligned} \gcd(a(x), b(x)) = r_2(x) &= x + 1 \\ &\overset{(3.2)}{=} \underbrace{x^3+1}_{r_0(x)} + x(\underbrace{x^2+1}_{r_1(x)}) \end{aligned}$$

$$
\begin{aligned}
&\stackrel{(3.1)}{=} \underbrace{x^3+1}_{r_0(x)} + x[\underbrace{x^4+x^2+x+1}_{r_{-1}(x)} + x(\underbrace{x^3+1}_{r_0(x)})] \\
&= x(\underbrace{x^4+x^2+x+1}_{r_{-1}(x)}) + (x^2+1)(\underbrace{x^3+1}_{r_0(x)}) \\
&= x \cdot a(x) + (x^2+1) \cdot b(x) .
\end{aligned}
$$

So here $s(x) = x$ and $t(x) = x^2 + 1$.

A polynomial $P(x) \in F[x]$ is called *irreducible* over F if (i) $\deg P(x) > 0$ and (ii) for any $a(x), b(x) \in F[x]$ such that $P(x) = a(x) \cdot b(x)$, either $\deg a(x) = 0$ or $\deg b(x) = 0$. That is, the only non-scalar divisors of $P(x)$ are its multiples by elements of F^*. A polynomial that is not irreducible is called *reducible.*

Example 3.3 The polynomial $x^2 + 1$ is irreducible over the real field $\mathbb{R}$. Indeed, if it were reducible, then we would have $x^2 + 1 = (x - \alpha)(x - \beta)$ for some real polynomials $x - \alpha$ and $x - \beta$. Yet this would imply that $\alpha^2 + 1 = (\alpha - \alpha)(\alpha - \beta) = 0$, which is impossible for every real α. □

Example 3.4 We construct the irreducible polynomials over $F = \mathrm{GF}(2)$ for small degrees.

Clearly, x and $x+1$ are all the irreducible polynomials of degree 1 over F.

The *reducible* polynomials of degree 2 over F are

$$x^2 , \quad x(x+1) = x^2 + x , \quad \text{and} \quad (x+1)^2 = x^2 + 1 .$$

This leaves one irreducible polynomial of degree 2, namely, $x^2 + x + 1$.

Turning to degree 3, consider the polynomial

$$a(x) = x^3 + a_2x^2 + a_1x + a_0$$

over F. If $a(x)$ is reducible, then it must be divisible by some polynomial of degree 1. Now, it is easy to see that $x \,|\, a(x)$ if and only if $a_0 = 0$. In addition, it will follow from Proposition 3.5 below that $x+1 \,|\, a(x)$ if and only if $a(x)$ takes the zero value at $x = 1$, i.e.,

$$a(1) = 1 + a_2 + a_1 + a_0 = 0 .$$

We thus conclude that $a(x)$ is irreducible if and only if $a_0 = a_1 + a_2 = 1$, resulting in the two polynomials

$$x^3 + x + 1 \quad \text{and} \quad x^3 + x^2 + 1 .$$

Finally, we consider the polynomial

$$a(x) = x^4 + a_3x^3 + a_2x^2 + a_1x + a_0$$

of degree 4 over F. From similar arguments to those used earlier, we deduce that neither x nor $x+1$ divides $a(x)$ if and only if $a_0 = a_1 + a_2 + a_3 = 1$, thereby resulting in the following four polynomials:

$$x^4 + x + 1 \, , \quad x^4 + x^2 + 1 \, , \quad x^4 + x^3 + 1 \, , \quad \text{and} \quad x^4 + x^3 + x^2 + x + 1 \, .$$

Yet, these polynomials are not necessarily irreducible: we still need to check the case where $a(x)$ is a product of two (irreducible) polynomials of degree 2. Indeed, the polynomial

$$(x^2 + x + 1)^2 = x^4 + x^2 + 1$$

is reducible.

Table 3.1 summarizes our discussion in this example. □

Table 3.1. List of irreducible polynomials over GF(2).

n	Irreducible polynomials of degree n over GF(2)
1	x, $x+1$
2	x^2+x+1
3	x^3+x+1, x^3+x^2+1
4	x^4+x+1, x^4+x^3+1, $x^4+x^3+x^2+x+1$

Irreducible polynomials play a role in $F[x]$ which is similar to that of prime numbers in the integer ring $\mathbb{Z}$. Specifically, we show in Theorem 3.4 below that $F[x]$ is a unique factorization domain: every polynomial in $F[x]$ can be expressed in an essentially unique way as a product of irreducible polynomials over F. We precede the theorem by a lemma and a proposition.

Lemma 3.2 *Let $a(x)$, $b(x)$, and $c(x)$ be polynomials over F such that $c(x) \neq 0$ and $\gcd(a(x), c(x)) = 1$. Then*

$$c(x) \,|\, a(x) \cdot b(x) \quad \Longrightarrow \quad c(x) \,|\, b(x) \, .$$

Proof. Since $\gcd(a(x), c(x)) = 1$, there are polynomials $s(x)$ and $t(x)$ over F such that

$$s(x) \cdot a(x) + t(x) \cdot c(x) = \gcd(a(x), c(x)) = 1 \, ,$$

i.e.,

$$s(x) \cdot a(x) \equiv 1 \pmod{c(x)} \, .$$

Multiplying both sides by $b(x)$ yields

$$s(x) \cdot a(x) \cdot b(x) \equiv b(x) \pmod{c(x)} .$$

However, $s(x) \cdot a(x) \cdot b(x) \equiv 0 \pmod{c(x)}$. So, $b(x) \equiv 0 \pmod{c(x)}$. □

Proposition 3.3 *Let $P(x)$ be an irreducible polynomial over F and let $a(x)$ and $b(x)$ be polynomials over F. Then,*

$$P(x) \,|\, a(x) \cdot b(x) \quad \Longrightarrow \quad P(x) \,|\, a(x) \quad \text{or} \quad P(x) \,|\, b(x) .$$

Proof. This is an immediate corollary of Lemma 3.2. □

Theorem 3.4 (The unique factorization theorem) *The factorization of a nonzero polynomial into irreducible polynomials over F is unique (up to permutation and scalar multiples).*

Proof. Let $a(x)$ be a monic polynomial over F. The proof is by induction on the (smallest) number of irreducible factors of $a(x)$ (the factors are not necessarily distinct).

When the number of factors is zero, $a(x) = 1$ and the result is obvious.

Now, suppose that $a(x) = \prod_{i=1}^{n} P_i(x) = \prod_{i=1}^{m} Q_i(x)$, where $P_i(x)$ and $Q_i(x)$ are monic irreducible polynomials. By Proposition 3.3, $P_1(x)$ divides one of the $Q_i(x)$'s, say $Q_1(x)$. Hence, $P_1(x) = Q_1(x)$ and we can apply the induction hypothesis to $a(x)/P_1(x)$. □

3.3 Extension fields

Let F be a field and $P(x)$ be an irreducible polynomial of degree h over F.

Consider the ring of residues of the polynomials in $F[x]$ modulo $P(x)$. This ring is denoted by $F[x]/P(x)$, and the residues can be regarded as elements of $F_h[x]$ with polynomial arithmetic modulo $P(x)$. We show that this ring is a field. To this end, we verify that every nonzero element in that ring has a multiplicative inverse with respect to the constant polynomial 1.

Recall that $\gcd(a(x), P(x)) = 1$ for every $a(x) \in F_h[x] \setminus \{0\}$. Hence, there exist polynomials $s(x)$ and $t(x)$ over F such that

$$s(x) \cdot a(x) + t(x) \cdot P(x) = \gcd(a(x), P(x)) = 1$$

or

$$s(x) \cdot a(x) \equiv 1 \pmod{P(x)} .$$

So, $s(x)$, when reduced modulo $P(x)$, is a multiplicative inverse of $a(x)$ in $F[x]/P(x)$ (one can effectively compute $s(x)$ by substituting $b(x) = P(x)$ in Euclid's algorithm in Problem 3.3).

Let F and Φ be fields. We say that Φ is an *extension field* of F if F is a *subfield* of Φ; that is, $F \subseteq \Phi$, and the addition and multiplication operations of Φ, when acting on the elements of F, coincide with the respective operations of F.

An extension field Φ of F is a vector space over F. The *extension degree* of Φ over F, denoted by $[\Phi : F]$, is the dimension of Φ as a vector space over F.

If $P(x)$ is an irreducible polynomial of degree h over F, then the field $F[x]/P(x)$ is an extension field of F with extension degree h.

Example 3.5 Let F be the real field $\mathbb{R}$ and $P(x)$ be the irreducible polynomial $x^2 + 1$ over $\mathbb{R}$. The field $\mathbb{R}[x]/(x^2 + 1)$ is an extension field of $\mathbb{R}$ with extension degree 2. In this field, the sum of two elements, $a + bx$ and $c + dx$, is given by

$$(a + bx) + (c + dx) = (a + c) + (b + d)x ,$$

while their product is

$$\begin{aligned}(a + bx)(c + dx) &= ac + (ad + bc)x + bdx^2 \\ &\equiv (ac - bd) + (ad + bc)x \pmod{(x^2 + 1)} .\end{aligned}$$

One can easily identify these operations as addition and multiplication in the complex field $\mathbb{C}$, with x substituting for $\sqrt{-1}$. Indeed, the fields $\mathbb{R}[x]/(x^2+1)$ and $\mathbb{C}$ are isomorphic. □

When $F = \mathrm{GF}(q)$, the field $F[x]/P(x)$ has size q^h and is denoted by $\mathrm{GF}(q^h)$. This notational convention, which does not specify the polynomial $P(x)$, will be justified in Chapter 7: we will show in Theorem 7.13 that all finite fields of the same size are isomorphic. Hence, the field $\mathrm{GF}(q^h)$ is essentially unique, even though it may have several (seemingly) different representations.

Example 3.6 Let $F = \mathrm{GF}(2)$ and $P(x) = x^3 + x + 1$. We construct the field $\Phi = \mathrm{GF}(2^3)$ as a ring of residues of the polynomials over F modulo $P(x)$. The second column in Table 3.2 contains the elements of the field, which are written as polynomials in the indeterminate ξ.

The third column in the table expresses each nonzero element in the field as a power of the element $\xi = 0 \cdot 1 + 1 \cdot \xi + 0 \cdot \xi^2$. Indeed,

$$\xi^3 \equiv 1 + \xi \pmod{P(\xi)} ,$$

Table 3.2. Representation of $\mathrm{GF}(2^3)$ as $F[\xi]/(\xi^3+\xi+1)$.

Coefficients	Field element	Power of ξ
000	0	0
100	1	1
010	ξ	ξ
110	$1+\xi$	ξ^3
001	ξ^2	ξ^2
101	$1+\xi^2$	ξ^6
011	$\xi+\xi^2$	ξ^4
111	$1+\xi+\xi^2$	ξ^5

$$\begin{aligned}
\xi^4 &\equiv \xi\cdot\xi^3 \equiv \xi(1+\xi) \equiv \xi+\xi^2 \pmod{P(\xi)}\,,\\
\xi^5 &\equiv \xi\cdot\xi^4 \equiv \xi(\xi+\xi^2) \equiv \xi^2+\xi^3 \equiv 1+\xi+\xi^2 \pmod{P(\xi)}\,,\\
\xi^6 &\equiv \xi\cdot\xi^5 \equiv \xi(1+\xi+\xi^2) \equiv \xi+\xi^2+\xi^3 \equiv 1+\xi^2 \pmod{P(\xi)}\,,
\end{aligned}$$

and

$$\xi^7 \equiv \xi\cdot\xi^6 \equiv \xi(1+\xi^2) \equiv \xi+\xi^3 \equiv 1 \pmod{P(\xi)}\,.$$

□

Several remarks are worth making about Example 3.6. First, while the elements of $\Phi = \mathrm{GF}(2^3)$ are represented in the example as polynomials in $F_3[\xi]$, we can also define polynomials over the field Φ itself (in which case we will typically use the indeterminate x); for example, $(\xi+\xi^2)+(1+\xi)x^2+x^3$ is a polynomial in $\Phi[x]$.

Secondly, we can obtain an alternate representation of the field $\mathrm{GF}(2^3)$ by using the other irreducible polynomial of degree 3 over F, namely, x^3+x^2+1. The third column of Table 3.2 will look different under such a representation (see Problem 3.6).

Thirdly, notice the difference between $\mathrm{GF}(2^3)$ and $\mathbb{Z}_8$: the latter is the ring of integer residues modulo 8 and, since it has zero divisors (e.g., $4\cdot 2 \equiv 0 \pmod 8$), it is certainly not a field.

Let F be a field and let Φ be an extension field of F with extension degree $[\Phi : F] = h$. We can represent Φ as a vector space over F using any basis

$$\Omega = (\omega_1\ \omega_2\ \ldots\ \omega_h)$$

of Φ over F. The representation of an element $u \in \Phi$ will be a column vector

$\mathbf{u} = (u_1 \ u_2 \ \ldots \ u_h)^T$ in F^h such that

$$u = \Omega \mathbf{u} = (\omega_1 \ \omega_2 \ \ldots \ \omega_h) \begin{pmatrix} u_1 \\ u_2 \\ \vdots \\ u_h \end{pmatrix} = \sum_{i=1}^{h} u_i \omega_i \ .$$

When $\Phi = F[\xi]/P(\xi)$ for an irreducible polynomial $P(x)$ of degree h over F, it is convenient to select the basis

$$\Omega = (1 \ \xi \ \xi^2 \ \ldots \ \xi^{h-1}) \ ,$$

in which case each element $u = u_1 + u_2\xi + \ldots + u_h\xi^{h-1}$ in Φ is represented by the vector of coefficients of u when the latter is regarded as a polynomial in $F_h[\xi]$.

Example 3.7 Continuing Example 3.6, we can represent the elements of GF(2^3) as vectors in F^3 (see the first column in Table 3.2), e.g., the vectors (1 0 1) and (0 1 1) stand for the elements $1+\xi^2$ and $\xi+\xi^2$, respectively. Addition in the field then becomes the conventional component-by-component addition of vectors over F. In order to perform multiplication, we switch back to the polynomial representation; for example, the product of (1 0 1) and (0 1 1) equals (1 1 0) since

$$(1+\xi^2)(\xi+\xi^2) = \xi + \xi^2 + \xi^3 + \xi^4 \equiv 1 + \xi \pmod{P(\xi)} \ .$$

Inversion can be implemented using Euclid's algorithm.

Alternatively, we can use the third column of Table 3.2 as a "logarithm table" for multiplication as follows. We first see in the table that the elements (1 0 1) and (0 1 1) equal ξ^6 and ξ^4, respectively. The product of these two powers of ξ can then be easily computed by adding the exponents modulo the multiplicative order, 7, of ξ; i.e., $\xi^6 \cdot \xi^4 = \xi^{10} \equiv \xi^3 \pmod{P(\xi)}$. Finally, we look again at Table 3.2 and see that ξ^3 is represented by the vector (1 1 0). This table approach, which allows us to compute inversions as well, is commonly used when implementing arithmetic operations in fields of moderate sizes. □

3.4 Roots of polynomials

Let F be a field and Φ be an extension field of F. An element $\beta \in \Phi$ is a *root* of a polynomial $a(x) \in F[x]$ if the equality $a(\beta) = 0$ holds in Φ.

In the representation of the field GF(2^3) in Example 3.6, the element ξ is a root of the polynomial $P(x) = x^3 + x + 1$. In fact, the elements ξ^2 and $\xi^4 = \xi^2 + \xi$ are also roots of this polynomial in GF(2^3).

Proposition 3.5 *Let $a(x)$ be a polynomial over F and let β be an element in an extension field Φ of F. Then $a(\beta) = 0$ if and only if $x - \beta \,|\, a(x)$, where the latter division is in $\Phi[x]$.*

Proof. Write $a(x) = b(x)(x - \beta) + c$, where $\deg c < 1$. Then $a(\beta) = 0$ if and only if $c = 0$. □

Assuming the representation of $\mathrm{GF}(2^3)$ as in Example 3.6, we have

$$x^3 + x + 1 = (x - \xi)(x - \xi^2)(x - \xi^4) .$$

Thus, while $x^3 + x + 1$ is irreducible over $\mathrm{GF}(2)$, it is reducible over $\mathrm{GF}(2^3)$, since it factors into three linear terms over that field.

Proposition 3.6 *Let F be a finite field. Then*

$$\prod_{\beta \in F} (x - \beta) = x^{|F|} - x . \tag{3.3}$$

Proof. By Proposition 3.1, every element of F is a root of the polynomial $x^{|F|} - x$. Therefore, by Proposition 3.5, the polynomial $x - \beta$ divides $x^{|F|} - x$ for every $\beta \in F$. The claim follows by observing that both sides of (3.3) have the same degree and both are monic. □

Let Φ be an extension field of F and let $\beta \in \Phi$ be a root of a nonzero polynomial $a(x) \in F[x]$. The *multiplicity* of β in $a(x)$ is the largest integer m such that $(x - \beta)^m \,|\, a(x)$. A root is called *simple* if its multiplicity is 1.

Theorem 3.7 *A polynomial of degree $n \geq 0$ over a field F has at most n roots (counting multiplicity) in every extension field of F.*

Proof. Let $\beta_1, \beta_2, \ldots$ be the roots of $a(x)$ in an extension field Φ of F and let m_i be the multiplicity of the root β_i. The product $\prod_i (x - \beta_i)^{m_i}$, which is a polynomial of degree $m = \sum_i m_i$ over Φ, necessarily divides $a(x)$ in $\Phi[x]$; so, $m \leq n$. □

3.5 Primitive elements

In this section, we will demonstrate a key property of the multiplicative group of a finite field, namely, that this group is cyclic.

Example 3.8 We construct the field $\mathrm{GF}(2^4)$ as a ring of residues of the polynomials over $F = \mathrm{GF}(2)$ modulo the polynomial $P_1(x) = x^4 + x + 1$. The elements of the field are listed in Table 3.3 as polynomials in $F_4[\xi]$ and sorted

Table 3.3. Representation of $\mathrm{GF}(2^4)$ as $F[\xi]/(\xi^4+\xi+1)$.

Power of ξ	Field element	Coefficients
0	0	0000
ξ^0	1	1000
ξ^1	ξ	0100
ξ^2	ξ^2	0010
ξ^3	ξ^3	0001
ξ^4	$1+\xi$	1100
ξ^5	$\xi+\xi^2$	0110
ξ^6	$\xi^2+\xi^3$	0011
ξ^7	$1+\xi+\xi^3$	1101
ξ^8	$1+\xi^2$	1010
ξ^9	$\xi+\xi^3$	0101
ξ^{10}	$1+\xi+\xi^2$	1110
ξ^{11}	$\xi+\xi^2+\xi^3$	0111
ξ^{12}	$1+\xi+\xi^2+\xi^3$	1111
ξ^{13}	$1+\xi^2+\xi^3$	1011
ξ^{14}	$1+\xi^3$	1001

according to increasing powers of the element $\xi = 0\cdot 1 + 1\cdot\xi + 0\cdot\xi^2 + 0\cdot\xi^3$. Thus, the element ξ is a primitive element if we represent the field $\mathrm{GF}(2^4)$ as $F[\xi]/P_1(\xi)$.

We can represent the field $\mathrm{GF}(2^4)$ also as $F[\xi]/P_2(\xi)$, where

$$P_2(x) = x^4 + x^3 + x^2 + x + 1 \,.$$

Yet, in this representation, the element ξ has multiplicative order 5 and is therefore non-primitive: the polynomial $P_2(x)$ divides x^5-1 and, so, $\xi^5 \equiv 1 \pmod{P_2(\xi)}$ (refer also to Problems 3.17 and 3.18). □

Theorem 3.8 *Every finite field contains a primitive element; that is, the multiplicative group of a finite field is cyclic.*

Proof. Let F be a finite field and let $\alpha \in F^*$ be of maximal multiplicative order $\mathcal{O}(\alpha)$ in F^*. We first show that $\mathcal{O}(\beta)$ divides $\mathcal{O}(\alpha)$ for every $\beta \in F^*$.

For an element $\beta \in F^*$, let r be a prime divisor of $\mathcal{O}(\beta)$ and write

$$\mathcal{O}(\beta) = r^m \cdot n \quad \text{and} \quad \mathcal{O}(\alpha) = r^s \cdot t \,,$$

where $\gcd(r,n) = \gcd(r,t) = 1$. We have

$$\mathcal{O}(\beta^n) = \frac{\mathcal{O}(\beta)}{\gcd(\mathcal{O}(\beta), n)} = \frac{\mathcal{O}(\beta)}{n} = r^m$$

and

$$\mathcal{O}(\alpha^{r^s}) = \frac{\mathcal{O}(\alpha)}{\gcd(\mathcal{O}(\alpha), r^s)} = \frac{\mathcal{O}(\alpha)}{r^s} = t$$

(see Problem A.9). Now, $\gcd(r^m, t) = 1$ and, so,

$$\mathcal{O}(\beta^n \cdot \alpha^{r^s}) = \mathcal{O}(\beta^n) \cdot \mathcal{O}(\alpha^{r^s}) = r^m \cdot t$$

(Problem A.10). But $\mathcal{O}(\beta^n \cdot \alpha^{r^s}) \leq \mathcal{O}(\alpha)$, since α has a maximal multiplicative order in F^*; therefore, $r^m \cdot t \leq r^s \cdot t$, i.e., $m \leq s$. Hence, r^m divides $\mathcal{O}(\alpha)$.

By applying the previous argument to every prime divisor r of $\mathcal{O}(\beta)$ we obtain that $\mathcal{O}(\beta)$ divides $\mathcal{O}(\alpha)$.

We thus conclude that every $\beta \in F^*$ is a root of the polynomial $x^{\mathcal{O}(\alpha)} - 1$. By Theorem 3.7 it follows that

$$\mathcal{O}(\alpha) = \deg\left(x^{\mathcal{O}(\alpha)} - 1\right) \geq |F^*| = |F| - 1 .$$

In fact, the inequality must hold with equality, since $\mathcal{O}(\alpha)$ divides $|F^*|$. $\square$

If α is a primitive element in F, then α^i is primitive if and only if $\gcd(i, |F^*|) = 1$ (Problem A.11). Therefore, the number of primitive elements in $F = \mathrm{GF}(q)$ is $\phi(q-1)$, where $\phi(\cdot)$ is the Euler function defined in Problem A.1.

3.6 Field characteristic

We now turn to demonstrating several properties of the *additive* groups of finite fields.

Let F be a field. The *characteristic* of F, denoted by $\mathsf{c}(F)$, is the order of the element 1 in the additive group of F, provided that this order is finite. If 1 does not have a finite additive order, then $\mathsf{c}(F)$ is defined to be zero.

For example, $\mathsf{c}(\mathrm{GF}(7)) = 7$, $\mathsf{c}(\mathrm{GF}(2^4)) = 2$, and $\mathsf{c}(\mathbb{R}) = 0$.

Given a positive integer $m \in \mathbb{Z}$, the element

$$\underbrace{1 + 1 + \ldots + 1}_{m \text{ times}}$$

in a field F will be denoted by $\overline{m}$; in most cases, we omit the bar if no confusion arises. Similarly, for an element $\beta \in F$,

$$m\beta = (\underbrace{1 + 1 + \ldots + 1}_{m \text{ times}}) \cdot \beta = \underbrace{\beta + \beta + \ldots + \beta}_{m \text{ times}} .$$

The distinct elements of F among

$$0,\ \pm 1,\ \pm 2,\ \ldots,\ \pm m,\ \ldots$$

are called the *integers* of F. When $\mathsf{c}(F) > 0$, the (ordinary) integers $m, n \in \mathbb{Z}$ represent the same integer of F if and only if $m \equiv n \pmod{\mathsf{c}(F)}$; thus, when $\mathsf{c}(F) > 0$, we may assume without loss of generality that the integers of F are

$$0,\ 1,\ 2,\ \ldots,\ \mathsf{c}(F){-}1\ .$$

The next proposition imposes a restriction on the values that $\mathsf{c}(F)$ may take.

Proposition 3.9 *If F is a field with $\mathsf{c}(F) > 0$, then $\mathsf{c}(F)$ is a prime.*

Proof. Suppose that $\mathsf{c}(F) = mn$ for some positive integers m and n. Then, by the distributive law,

$$0 = \sum_{i=1}^{mn} 1 = \Bigl(\sum_{i=1}^{m} 1\Bigr)\Bigl(\sum_{i=1}^{n} 1\Bigr)\ .$$

Therefore, either $\sum_{i=1}^{m} 1 = 0$ or $\sum_{i=1}^{n} 1 = 0$, which implies that either m or n is at least $\mathsf{c}(F)$ $(= mn)$. However, this is possible only when either n or m equals 1. □

Suppose that F is a field with $\mathsf{c}(F) > 0$. By the last proposition, $\mathsf{c}(F) = p$ for a prime p, and—without loss of generality—the p distinct integers of F are $0, 1, \ldots, p{-}1$. Since p is the additive order of 1 in F, the integers of F, with the addition and multiplication operations of F, form a field which is isomorphic to $\mathrm{GF}(p)$. We verify this next: given two nonnegative integers $m, n \in \mathbb{Z}$, let c be the remainder obtained when the (ordinary) integer $m{+}n$ is divided by the characteristic p. Then,

$$\overline{m} + \overline{n} = \Bigl(\sum_{i=1}^{m} 1\Bigr) + \Bigl(\sum_{i=1}^{n} 1\Bigr) = \sum_{i=1}^{m+n} 1 = \sum_{i=1}^{c} 1 = \overline{m+n}\ .$$

Likewise, if d is the remainder of mn modulo p then, by the distributive law,

$$\overline{m} \cdot \overline{n} = \Bigl(\sum_{i=1}^{m} 1\Bigr) \cdot \Bigl(\sum_{i=1}^{n} 1\Bigr) = \sum_{i=1}^{mn} (1 \cdot 1) = \sum_{i=1}^{mn} 1 = \sum_{i=1}^{d} 1 = \overline{mn}\ .$$

We conclude that $\mathrm{GF}(p)$ is a subfield of F. This, in turn, leads to the following property of the size of finite fields.

Theorem 3.10 *Let F be a finite field with $\mathsf{c}(F) = p$. Then $|F| = p^n$ for some integer n.*

Proof. The field F is an extension field of $\mathrm{GF}(p)$ and, as such, it is a vector space over $\mathrm{GF}(p)$ of a finite dimension n. The size of F must therefore be p^n. □

Next, we provide several useful facts about the characteristic.

Proposition 3.11 *Let F be a field with $\mathsf{c}(F) = p > 0$. For every $\alpha, \beta \in F$,*

$$(\alpha \pm \beta)^p = \alpha^p \pm \beta^p .$$

Proof. For $0 < i < p$, the binomial coefficient

$$\binom{p}{i} = \frac{p(p-1)(p-2)\cdots(p-i+1)}{i!} \tag{3.4}$$

is an integer multiple of p: the numerator in (3.4) is divisible by p whereas the denominator is not. □

Corollary 3.12 *Let F be a field with $\mathsf{c}(F) = p > 0$. For every $\alpha, \beta \in F$ and integer $m \geq 0$,*

$$(\alpha \pm \beta)^{p^m} = \alpha^{p^m} \pm \beta^{p^m} .$$

Proof. Iterate Proposition 3.11 m times. □

Corollary 3.13 *Let F be a finite field with $\mathsf{c}(F) = p$. For every integer $m \geq 0$, the mapping $f_m : F \to F$ defined by $f_m : x \mapsto x^{p^m}$ is an automorphism of F (i.e., f_m is an isomorphism of F into itself).*

Proof. Clearly, $f_m(\alpha \cdot \beta) = f_m(\alpha) \cdot f_m(\beta)$ for every $\alpha, \beta \in F$, and by Corollary 3.12 we also have $f_m(\alpha + \beta) = f_m(\alpha) + f_m(\beta)$. It remains to show that f_m has an inverse mapping. Observe that if $|F| = p^n$ then $f_{m+n} = f_m$. Hence, we can assume that $m < n$, and the inverse of f_m is then f_{n-m}. □

A mapping $x \mapsto x^{p^m}$ over a finite field of characteristic p is called a *Frobenius mapping*.

3.7 Splitting field

The following result complements Theorem 3.7.

Proposition 3.14 *Let $a(x)$ be a polynomial of degree $n \geq 0$ over a field F. Then there exists an extension field Φ of F with $[\Phi : F] < \infty$ in which $a(x)$ has n roots (counting multiplicity).*

Proof. The proof is by induction on n for every field, and the result is obvious for $n = 0$. As for the induction step, suppose that $P(x)$ is an irreducible factor of degree m of $a(x)$, and consider the extension field $K = F[\xi]/P(\xi)$ of F, where ξ is an indeterminate. The element ξ is a root of $P(x)$ in K and, as such, it is also a root of $a(x)$ in K. Write $a(x) = (x - \xi)b(x)$, where $b(x)$ is a polynomial of degree $n-1$ over K. By the induction hypothesis, there is an extension field Φ of K with $[\Phi : K] < \infty$ in which $b(x)$ has $n-1$ roots. Clearly, the field Φ is also an extension field of F and $[\Phi : F] = [\Phi : K][K : F]$. □

Given a field F and a polynomial $a(x) \in F[x]$, a field Φ which satisfies the conditions of the last proposition with the smallest possible extension degree $[\Phi : F]$ is called a *splitting field* of $a(x)$ over F. (As mentioned earlier, we will show in Chapter 7 that all finite fields of the same size are isomorphic. This result, in turn, will imply that given a finite field F and a polynomial $a(x) \in F[x]$, the splitting field of $a(x)$ over F is unique. Until then, however, we will use the indefinite article when referring to splitting fields.)

Let $a(x) = \sum_{i=0}^{n} a_i x^i$ be a polynomial over F. The *formal derivative* of $a(x)$, denoted by $a'(x)$, is defined as

$$a'(x) = \sum_{i=1}^{n} i a_i x^{i-1} .$$

We have

$$(a(x)b(x))' = a'(x)b(x) + a(x)b'(x)$$

and

$$[a(b(x))]' = a'(b(x))b'(x)$$

(Problem 3.38).

Through the formal derivative, we can determine whether the roots of a given polynomial over F are simple in all the extension fields of F.

Lemma 3.15 *Let $a(x)$ be a nonzero polynomial over a field F. The roots of $a(x)$ in every extension field of F are simple if and only if $\gcd(a(x), a'(x)) = 1$.*

Proof. Let β be a root of $a(x)$ with multiplicity m in an extension field Φ of F and write $a(x) = (x - \beta)^m b(x)$ (where $b(\beta) \neq 0$). We have

$$a'(x) = m(x - \beta)^{m-1} b(x) + (x - \beta)^m b'(x) . \tag{3.5}$$

Hence, $x - \beta$ divides $r(x) = \gcd(a(x), a'(x))$ in $\Phi[x]$ if and only if $m > 1$. In particular, $\deg r(x) = 0$ implies $m = 1$. Conversely, if $\deg r(x) > 0$, then $r(x)$ has a root β in an extension field of F, in which case $x - \beta$ divides both

$a(x)$ and $a'(x)$. By (3.5), the multiplicity of β as a root of $a(x)$ must be greater than 1, or else we would have $a'(\beta) = b(\beta) \neq 0$. □

In Section 3.3, we described how extension fields can be constructed based on irreducible polynomials. Through that construction method, we effectively provide an explicit representation of the extension field, as in Examples 3.6 or 3.8. The next proposition, on the other hand, shows the existence of such fields in a more implicit manner.

Proposition 3.16 *Let $F = \mathrm{GF}(q)$ and consider the polynomial*

$$Q(x) = x^{q^n} - x$$

over F where n is a positive integer. The roots of $Q(x)$ in a splitting field of $Q(x)$ over F form an extension field K of F with extension degree $[K : F] = n$.

Proof. Denote by Φ a splitting field of $Q(x)$ over F. Since q is a power of $\mathsf{c}(F)$, we have

$$Q'(x) = q^n x^{q^n - 1} - 1 = -1 .$$

Hence, $\gcd(Q(x), Q'(x)) = 1$ and, so, by Lemma 3.15, the polynomial $Q(x)$ has q^n *distinct* roots in Φ. We denote by K the set of roots of $Q(x)$ in Φ.

For any two elements $\alpha, \beta \in K$ we have

$$(\alpha + \beta)^{q^n} = \alpha^{q^n} + \beta^{q^n} = \alpha + \beta$$

and

$$(\alpha\beta)^{q^n} = \alpha^{q^n} \beta^{q^n} = \alpha\beta .$$

This implies that K is closed under addition and multiplication in Φ. Since K is finite, it follows from Problem A.13 that K is a subfield of Φ. Noting that $F \subseteq K$ and $|K| = q^n$, we thus conclude that K is an extension field of F with $[K : F] = n$. (Observe that K itself is in fact a splitting field of $Q(x)$ over F, which readily implies that $[\Phi : F] = [K : F] = n$.) □

The previous proposition implies the following converse of Theorem 3.10.

Theorem 3.17 *For every prime p and integer $n > 0$ there is an extension field of $\mathrm{GF}(p)$ with extension degree n.*

3.8 Application: double error-correcting codes

In this section, we exhibit how finite fields can be instrumental in designing codes: we present a construction of a linear code over $F = \mathrm{GF}(2)$ with minimum distance at least 5. We will start off with a parity-check matrix of the Hamming code over $F = \mathrm{GF}(2)$—thereby guaranteeing a minimum distance of at least 3—and then increase the minimum distance by adding (carefully-selected) rows to that matrix.

We can represent a parity-check matrix of a $[2^m-1, 2^m-1-m, 3]$ Hamming code $\mathcal{C}$ over $F = \mathrm{GF}(2)$ as

$$H = \left(\alpha_1 \; \alpha_2 \; \ldots \; \alpha_{2^m-1} \right) ,$$

where α_j ranges over all the nonzero elements of $\mathrm{GF}(2^m)$.

Example 3.9 Construct the field $\mathrm{GF}(2^4)$ as in Example 3.8, and take $\alpha_j = \xi^{j-1}$ for $1 \le j \le 15$. In this case,

$$H = \begin{pmatrix} 1&0&0&0&1&0&0&1&1&0&1&0&1&1&1 \\ 0&1&0&0&1&1&0&1&0&1&1&1&1&0&0 \\ 0&0&1&0&0&1&1&0&1&0&1&1&1&1&0 \\ 0&0&0&1&0&0&1&1&0&1&0&1&1&1&1 \end{pmatrix} .$$

□

Let $\mathbf{c} = (c_1 \; c_2 \; \ldots \; c_n)$ be a vector of length $n = 2^m-1$ over $F = \mathrm{GF}(2)$. The vector $\mathbf{c}$ is a codeword of the $[n, n-m, 3]$ Hamming code $\mathcal{C}$ if and only if $H\mathbf{c}^T = \mathbf{0}$ or, equivalently,

$$\sum_{j=1}^{n} c_j \alpha_j = 0 ,$$

where the equality is over $\mathrm{GF}(2^m)$.

Taking squares of both sides and recalling that $c_j^2 = c_j$, we obtain

$$\sum_{j=1}^{n} c_j \alpha_j^2 = 0 .$$

Hence, the matrix

$$H_2 = \begin{pmatrix} \alpha_1 & \alpha_2 & \ldots & \alpha_{2^m-1} \\ \alpha_1^2 & \alpha_2^2 & \ldots & \alpha_{2^m-1}^2 \end{pmatrix} ,$$

where each element of $\mathrm{GF}(2^m)$ is represented as a column vector in F^m, is again a parity-check matrix of $\mathcal{C}$, yet of order $2m \times n$. So, when adding rows to H to form H_2, we did *not* change the code; in particular, neither the minimum distance nor the rate has changed.

Example 3.10 For the parameters of Example 3.9 we get

$$H_2 = \begin{pmatrix} 1 & \xi & \xi^2 & \xi^3 & \xi^4 & \xi^5 & \xi^6 & \xi^7 & \xi^8 & \xi^9 & \xi^{10} & \xi^{11} & \xi^{12} & \xi^{13} & \xi^{14} \\ 1 & \xi^2 & \xi^4 & \xi^6 & \xi^8 & \xi^{10} & \xi^{12} & \xi^{14} & \xi & \xi^3 & \xi^5 & \xi^7 & \xi^9 & \xi^{11} & \xi^{13} \end{pmatrix},$$

and, in binary form,

$$H_2 = \left(\begin{array}{ccccccccccccccc} 1 & 0 & 0 & 0 & 1 & 0 & 0 & 1 & 1 & 0 & 1 & 0 & 1 & 1 & 1 \\ 0 & 1 & 0 & 0 & 1 & 1 & 0 & 1 & 0 & 1 & 1 & 1 & 1 & 0 & 0 \\ 0 & 0 & 1 & 0 & 0 & 1 & 1 & 0 & 1 & 0 & 1 & 1 & 1 & 1 & 0 \\ 0 & 0 & 0 & 1 & 0 & 0 & 1 & 1 & 0 & 1 & 0 & 1 & 1 & 1 & 1 \\ \hline 1 & 0 & 1 & 0 & 1 & 1 & 1 & 1 & 0 & 0 & 0 & 1 & 0 & 0 & 1 \\ 0 & 0 & 1 & 0 & 0 & 1 & 1 & 0 & 1 & 0 & 1 & 1 & 1 & 1 & 0 \\ 0 & 1 & 0 & 1 & 1 & 1 & 1 & 0 & 0 & 0 & 1 & 0 & 0 & 1 & 1 \\ 0 & 0 & 0 & 1 & 0 & 0 & 1 & 1 & 0 & 1 & 0 & 1 & 1 & 1 & 1 \end{array}\right).$$

The latter matrix has order 8×15, yet its rank (as a matrix over GF(2)) is only 4: each of the last four rows can be obtained as a linear combination of the first four rows (specifically, rows 6 and 8 are copies of rows 3 and 4, respectively; row 5 equals the sum of rows 1 and 3; and row 7 is the sum of rows 2 and 4). □

Next, we try to replace the second powers with the third powers of the nonzero elements of $\mathrm{GF}(2^m)$, resulting in

$$H_3 = \begin{pmatrix} \alpha_1 & \alpha_2 & \ldots & \alpha_{2^m-1} \\ \alpha_1^3 & \alpha_2^3 & \ldots & \alpha_{2^m-1}^3 \end{pmatrix}.$$

This matrix, when represented over $F = \mathrm{GF}(2)$, is a $2m \times n$ parity-check matrix of some linear $[n, k, d]$ code over F with $k \geq n - 2m$. We refer to this code in the present section as $\mathcal{C}_3$.

Example 3.11 For the parameters of Example 3.9 we get

$$H_3 = \begin{pmatrix} 1 & \xi & \xi^2 & \xi^3 & \xi^4 & \xi^5 & \xi^6 & \xi^7 & \xi^8 & \xi^9 & \xi^{10} & \xi^{11} & \xi^{12} & \xi^{13} & \xi^{14} \\ 1 & \xi^3 & \xi^6 & \xi^9 & \xi^{12} & 1 & \xi^3 & \xi^6 & \xi^9 & \xi^{12} & 1 & \xi^3 & \xi^6 & \xi^9 & \xi^{12} \end{pmatrix}.$$

The respective binary matrix takes the form

$$H_3 = \left(\begin{array}{ccccccccccccccc} 1 & 0 & 0 & 0 & 1 & 0 & 0 & 1 & 1 & 0 & 1 & 0 & 1 & 1 & 1 \\ 0 & 1 & 0 & 0 & 1 & 1 & 0 & 1 & 0 & 1 & 1 & 1 & 1 & 0 & 0 \\ 0 & 0 & 1 & 0 & 0 & 1 & 1 & 0 & 1 & 0 & 1 & 1 & 1 & 1 & 0 \\ 0 & 0 & 0 & 1 & 0 & 0 & 1 & 1 & 0 & 1 & 0 & 1 & 1 & 1 & 1 \\ \hline 1 & 0 & 0 & 0 & 1 & 1 & 0 & 0 & 0 & 1 & 1 & 0 & 0 & 0 & 1 \\ 0 & 0 & 0 & 1 & 1 & 0 & 0 & 0 & 1 & 1 & 0 & 0 & 0 & 1 & 1 \\ 0 & 0 & 1 & 0 & 1 & 0 & 0 & 1 & 0 & 1 & 0 & 0 & 1 & 0 & 1 \\ 0 & 1 & 1 & 1 & 1 & 0 & 1 & 1 & 1 & 1 & 0 & 1 & 1 & 1 & 1 \end{array}\right),$$

and one can verify that the eight rows in this matrix are linearly independent over F. Hence, in this case, the code $\mathcal{C}_3$ turns out to be a linear $[15, 7]$ code over F. □

We next show that the minimum distance of $\mathcal{C}_3$ is at least 5. We do this by presenting an algorithm for decoding correctly every pattern of up to two errors.

Let $\mathbf{c} \in \mathcal{C}_3$ be the transmitted codeword and $\mathbf{y}$ be the received word. The error word is given by $\mathbf{e} = \mathbf{y} - \mathbf{c}$, and we compute the syndrome of $\mathbf{y}$ with respect to the matrix H_3 (see Section 2.4.2). For the analysis in the sequel, we find it convenient to write the syndrome as a vector of length 2 over $\mathrm{GF}(2^m)$:

$$\begin{pmatrix} s_1 \\ s_3 \end{pmatrix} = H_3\mathbf{y}^T = H_3\mathbf{e}^T .$$

Suppose that at most two errors have occurred, i.e., $\mathsf{w}(\mathbf{e}) \le 2$. We distinguish between the following cases.

Case 1: $\mathbf{e} = \mathbf{0}$. Here

$$\begin{pmatrix} s_1 \\ s_3 \end{pmatrix} = H_3\mathbf{e}^T = \mathbf{0} ,$$

namely, $s_1 = s_3 = 0$.

Case 2: $\mathsf{w}(\mathbf{e}) = 1$. Let i be the location of the (only) entry equaling 1 in $\mathbf{e}$. Then

$$\begin{pmatrix} s_1 \\ s_3 \end{pmatrix} = H_3\mathbf{e}^T = \begin{pmatrix} \alpha_i \\ \alpha_i^3 \end{pmatrix} ;$$

namely, $s_3 = s_1^3 \neq 0$, and the error location is the index i such that $\alpha_i = s_1$.

Case 3: $\mathsf{w}(\mathbf{e}) = 2$. Let i and j be the distinct locations of the two entries in $\mathbf{e}$ equaling 1. We have

$$\begin{pmatrix} s_1 \\ s_3 \end{pmatrix} = H_3\mathbf{e}^T = \begin{pmatrix} \alpha_i + \alpha_j \\ \alpha_i^3 + \alpha_j^3 \end{pmatrix} .$$

Since $s_1 = \alpha_i + \alpha_j \neq 0$, we can write

$$\frac{s_3}{s_1} = \frac{\alpha_i^3 + \alpha_j^3}{\alpha_i + \alpha_j} = \alpha_i^2 + \alpha_i\alpha_j + \alpha_j^2 .$$

Also,

$$s_1^2 = \alpha_i^2 + \alpha_j^2 .$$

By adding the left-hand sides and, respectively, the right-hand sides of the last two equations we obtain

$$\frac{s_3}{s_1} + s_1^2 = \alpha_i\alpha_j ;$$

in particular, $\alpha_i \alpha_j \neq 0$ implies $s_3 \neq s_1^3$, thereby making Case 3 distinguishable from Cases 1 and 2. Recalling that

$$s_1 = \alpha_i + \alpha_j \ ,$$

it follows that α_i and α_j are the solutions to the following quadratic equation over $\mathrm{GF}(2^m)$ in the unknown x:

$$x^2 + s_1 x + s_3 s_1^{-1} + s_1^2 = 0 \ . \tag{3.6}$$

Now, the problem of finding roots of a quadratic polynomial over $\mathrm{GF}(2^m)$ can be translated into a set of m linear equations over GF(2) (see Problem 3.42). Alternatively, we can also find the solutions to (3.6) simply by substituting the elements $\alpha_1, \alpha_2, \ldots, \alpha_{2^m-1}$ into the left-hand side of (3.6) and checking the result; such a procedure requires a number of operations in $\mathrm{GF}(2^m)$ that is at most linear in the code length.

Summing up Cases 1–3, we have ended up with a decoding algorithm for correcting up to two errors.

The *double-error-correcting [narrow-sense] alternant code* over $F = \mathrm{GF}(2)$ is a linear $[n, k, d]$ code of length $n \leq 2^m - 1$ with a $(2m) \times n$ parity-check matrix

$$H_3 = \begin{pmatrix} \alpha_1 & \alpha_2 & \ldots & \alpha_n \\ \alpha_1^3 & \alpha_2^3 & \ldots & \alpha_n^3 \end{pmatrix} , \tag{3.7}$$

for distinct nonzero elements $\alpha_j \in \mathrm{GF}(2^m)$. Such codes are obtained from $\mathcal{C}_3$ by shortening (see Problem 2.14) and, so, we have $k \geq n - 2m$ and $d \geq 5$. Obviously, the decoding algorithm that we have described for $\mathcal{C}_3$ applies to these shortened codes as well.

Figure 3.1 summarizes the decoding algorithm for correcting up to two errors while using a double-error-correcting $[n, k{\geq}n{-}2m]$ alternant code over $F = \mathrm{GF}(2)$. We will see in Section 5.5 that these codes are instances of a more general family of codes.

Problems

[Section 3.1]

Problem 3.1 Verify that the multiplicative group of GF(11) is cyclic, and find all the generators of this group (i.e., all the primitive elements in GF(11)).

[Section 3.2]

Problem 3.2 Show that $F[x]$ is an integral domain for any field F.

Input: received word $\mathbf{y} \in F^n$.
Output: error word $\mathbf{e} \in F^n$ or an error-detection indicator "e".

1. Compute the syndrome
$$\begin{pmatrix} s_1 \\ s_3 \end{pmatrix} = H_3 \mathbf{y}^T \in (\mathrm{GF}(2^m))^2 ,$$
where H_3 is given by (3.7);
2. if $s_1 = s_3 = 0$ then return $\mathbf{e} = \mathbf{0}$;
3. else if $s_1^3 = s_3$ then return an error word $\mathbf{e}$ with only one nonzero entry at location i, if any, such that $\alpha_i = s_1$;
4. else if $s_1 \neq 0$ then return an error word $\mathbf{e}$ with two nonzero entries at locations i and j, where i and j correspond to two distinct solutions α_i and α_j, if any, to the quadratic equation
$$x^2 + s_1 x + s_3 s_1^{-1} + s_1^2 = 0$$
in the unknown x over $\mathrm{GF}(2^m)$;
5. otherwise, or if any of the previous steps fails (by not finding the elements α_i or α_j), return an error-detection indicator "e" to signal that at least three errors have occurred.

Figure 3.1. Decoding algorithm for a double-error-correcting binary alternant code.

Problem 3.3 (Extended Euclid's algorithm for polynomials) Let $a(x)$ and $b(x)$ be polynomials over a field F such that $a(x) \neq 0$ and $\deg a > \deg b$ and consider the algorithm in Figure 3.2 for computing the polynomials $r_i(x)$, $q_i(x)$, $s_i(x)$, and $t_i(x)$ over F (the algorithm is written in the style of the C programming language, and the notation "$r_{i-2}(x)$ div $r_{i-1}(x)$" stands for the quotient obtained when $r_{i-2}(x)$ is divided by $r_{i-1}(x)$).

Let ν be the largest index i for which $r_i(x) \neq 0$. Show the following properties by induction on i:

1. $s_i(x)t_{i-1}(x) - s_{i-1}(x)t_i(x) = (-1)^{i+1}$ for $i = 0, 1, \ldots, \nu{+}1$.
2. $s_i(x)a(x) + t_i(x)b(x) = r_i(x)$ for $i = -1, 0, \ldots, \nu{+}1$.
3. $\deg t_i + \deg r_{i-1} = \deg a$ for $i = 0, 1, \ldots, \nu{+}1$.
4. $\deg s_i + \deg r_{i-1} = \deg b$ for $i = 1, 2, \ldots, \nu{+}1$.
5. If a polynomial $g(x)$ divides both $a(x)$ and $b(x)$ then $g(x)$ divides $r_i(x)$ for $i = -1, 0, \ldots, \nu{+}1$.
6. $r_\nu(x)$ divides $r_i(x)$ for $i = \nu{-}1, \nu{-}2, \ldots, -1$.

(From parts 5 and 6 it follows that $r_\nu(x) = \gcd(a(x), b(x))$.)

```
r_{-1}(x) ← a(x); r_0(x) ← b(x);
s_{-1}(x) ← 1; s_0(x) ← 0;
t_{-1}(x) ← 0; t_0(x) ← 1;
for (i ← 1; r_{i-1}(x) ≠ 0; i++) {
    q_i(x) ← r_{i-2}(x) div r_{i-1}(x);
    r_i(x) ← r_{i-2}(x) − q_i(x)r_{i-1}(x);
    s_i(x) ← s_{i-2}(x) − q_i(x)s_{i-1}(x);
    t_i(x) ← t_{i-2}(x) − q_i(x)t_{i-1}(x);
}
```

Figure 3.2. Euclid's algorithm.

Problem 3.4 Let r and s be positive integers. Show that over every field, the polynomial $x^r - 1$ divides $x^s - 1$ if and only if $r \mid s$.

Problem 3.5 Let r and s be positive integers and let $t = \gcd(r, s)$. The purpose of this problem is to generalize Problem 3.4 and show that over every field, $\gcd(x^r - 1, x^s - 1) = x^t - 1$.

1. Show that over every field, the polynomial $x^t - 1$ divides both $x^r - 1$ and $x^s - 1$.

Let $g(x)$ be a polynomial that divides both $x^r - 1$ and $x^s - 1$ over a field F.

2. Show that the polynomial x has a finite *multiplicative* order in the *ring* $F[x]/g(x)$.
3. Denote by e the multiplicative order of x in $F[x]/g(x)$. Show that e divides both r and s and, therefore, it divides t.

   Hint: See part 1 of Problem A.9.
4. Show that $g(x)$ divides $x^e - 1$ and, therefore, it divides $x^t - 1$.
5. Deduce that $\gcd(x^r - 1, x^s - 1) = x^t - 1$.

[Section 3.3]

Problem 3.6 Let $F = \mathrm{GF}(2)$ and suppose that the field $\Phi = \mathrm{GF}(2^3)$ is represented as $F[\xi]/(\xi^3 + \xi^2 + 1)$. Express each nonzero element of Φ as a power of ξ.

Problem 3.7 Let $F = \mathrm{GF}(2)$ and represent the field $\Phi = \mathrm{GF}(2^3)$ as $F[\xi]/(\xi^3 + \xi + 1)$. In particular, each element $u \in \Phi$ is associated with a vector $\mathbf{u} = (u_0\ u_1\ u_2)^T$ in F^3 such that $u = u_0 + u_1\xi + u_2\xi^2$; that is,

$$u = (1\ \xi\ \xi^2) \begin{pmatrix} u_0 \\ u_1 \\ u_2 \end{pmatrix} = (1\ \xi\ \xi^2)\mathbf{u}\,.$$

Consider the mapping $f : \Phi \to \Phi$ defined by $f : u \mapsto \xi^2 u$. Express f as a linear transformation by finding a 3×3 matrix A over F such that

$$f : (1\ \xi\ \xi^2)\mathbf{u} \mapsto (1\ \xi\ \xi^2)A\mathbf{u}\,.$$

Problem 3.8 Let $F = \mathrm{GF}(2)$ and let the field $\Phi = \mathrm{GF}(2^2)$ be represented as $F[\xi]/(\xi^2+\xi+1)$.

1. Let u_0, u_1, v_0, and v_1 denote elements in F. Show that the product of the elements $u_0 + u_1\xi$ and $v_0 + v_1\xi$ in Φ is given by

$$(u_0v_0 + u_1v_1) + (u_0v_1 + u_1v_0 + u_1v_1)\xi$$

   and, so, it can be computed using three additions and four multiplications in F.

2. Show that two elements in Φ can be multiplied using four additions and three multiplications in F.

   Hint: Consider the terms u_0v_0, u_1v_1, and $(u_0+u_1)(v_0+v_1)$.

3. Show that the square of an element in Φ can be computed using one addition in F.

4. Show that the mapping $\Phi \to \Phi$ defined by $x \mapsto \xi x^2$ can be computed only by re-positioning of elements of F, without any arithmetic operations in F.

Problem 3.9 Let $a(x) = \sum_{i=0}^{n} a_i x^i$ be a monic polynomial of degree n over a field F. The *companion matrix* of $a(x)$, denoted by C_a, is an $n \times n$ matrix over F defined by

$$C_a = \begin{pmatrix} 0 & 0 & \ldots & 0 & -a_0 \\ 1 & 0 & \ldots & 0 & -a_1 \\ 0 & 1 & \ldots & 0 & -a_2 \\ \vdots & \ddots & \ddots & 0 & \vdots \\ 0 & \ldots & 0 & 1 & -a_{n-1} \end{pmatrix} .$$

1. Show that $a(x)$ is the characteristic polynomial of C_a; i.e., $a(x)$ equals the determinant $\det(xI - C_a)$.

2. Let

$$u(x) = \sum_{i=0}^{n-1} u_i x^i \quad \text{and} \quad v(x) = \sum_{i=0}^{n-1} v_i x^i$$

   be two polynomials in $F_n[x]$, and define the associated vectors

$$\mathbf{u} = (u_0\ u_1\ u_2\ \ldots u_{n-1})^T \quad \text{and} \quad \mathbf{v} = (v_0\ v_1\ v_2\ \ldots v_{n-1})^T$$

   in F^n. Show that

$$v(x) \equiv x \cdot u(x) \pmod{a(x)}$$

   if and only if

$$\mathbf{v} = C_a \mathbf{u} .$$

3. Consider the circuit shown in Figure 3.3. Each of the n boxes represents a delay unit, which can store an element of F. The delay units are synchronous through the control of a clock. A circle labeled $-a_i$ represents a multiplication by the constant $-a_i$, and the circled "+" represents addition in F. Let $\mathbf{u} = (u_0\ u_1\ u_2\ \ldots u_{n-1})^T$ be the initial contents of the delay units. Show that right after the ℓth clock tick, the contents of the delay units equal $C_a^\ell \mathbf{u}$; hence, the circuit computes the remainder in $F_n[x]$ obtained when dividing $x^\ell \cdot u(x)$ by $a(x)$.

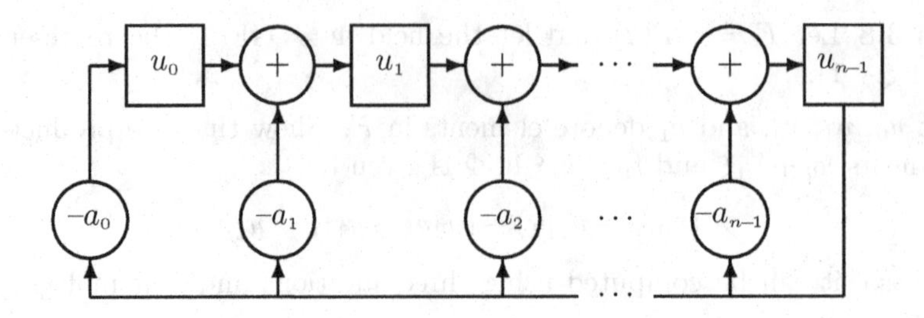

Figure 3.3. Multiplication by powers of x modulo $a(x)$.

4. (Multiplication by a fixed polynomial modulo $a(x)$) Let $b(x) = \sum_i b_i x^i$ be a polynomial over F. Consider the mapping $\psi_b : F_n[x] \to F_n[x]$ defined by $\psi_b : u(x) \mapsto v(x)$, where $v(x)$ is the remainder in $F_n[x]$ obtained when dividing $b(x) \cdot u(x)$ by $a(x)$; namely,

$$v(x) \equiv b(x) \cdot u(x) \pmod{a(x)} .$$

Show that ψ_b can be expressed as a linear transformation $F^n \to F^n$ defined by $\mathbf{u} \mapsto b(C_a)\mathbf{u}$, where

$$b(C_a) = \sum_i b_i C_a^i .$$

5. Show that the set of matrices

$$\{ b(C_a) \; : \; b(x) \in F_n[x] \} ,$$

with ordinary matrix addition and multiplication, forms a ring isomorphic to $F[x]/a(x)$; in particular, if $a(x)$ is irreducible over F then the resulting ring is a field.

Hint: Recall that $a(C_a) = 0$ (the Cayley–Hamilton Theorem).

6. Show that $a(x)$ is irreducible over F if and only if $b(C_a)$ is nonsingular for every nonzero polynomial $b(x) \in F_n[x]$.

Problem 3.10 Let $P(x) = x^2 + x + \beta$ be an irreducible polynomial over $F = \mathrm{GF}(q)$ and consider the field $\Phi = F[\xi]/P(\xi)$. Show that the multiplicative inverse of a nonzero element $u_0 + u_1\xi \in \Phi$ is given by

$$\frac{u_0 - u_1 - u_1\xi}{u_0^2 - u_0 u_1 + u_1^2 \beta}$$

(notice that the denominator of the latter expression is a nonzero element of F).

[Section 3.4]

Problem 3.11 Let Φ be an extension field of $F = \mathrm{GF}(q)$ and consider the polynomial $Q(x) = x^q - x$ over Φ. Show that the roots of $Q(x)$ in Φ are the elements of F.

Problem 3.12 Let F be a field and Φ be an extension field of F with extension degree $[\Phi : F] = h < \infty$. Let β be an element in Φ and denote by m the smallest positive integer such that the elements $1, \beta, \beta^2, \ldots, \beta^m$ are linearly dependent over F.

1. Verify that $m \leq h$.
2. Show that if $a(x)$ is a nonzero polynomial in $F[x]$ such that $a(\beta) = 0$, then $\deg a(x) \geq m$.
3. Show that there exists a unique monic polynomial $M_\beta(x)$ of degree exactly m over F such that $M_\beta(\beta) = 0$.
4. Show that the polynomial $M_\beta(x)$ is irreducible over F.
5. Show that if $a(x)$ is a nonzero polynomial in $F[x]$ such that $a(\beta) = 0$, then $M_\beta(x)$ divides $a(x)$.

   Hint: Show that β is a root of the remainder polynomial obtained when $a(x)$ is divided by $M_\beta(x)$.
6. Let $F = \mathrm{GF}(2)$ and $\Phi = F[\xi]/(\xi^3 + \xi + 1)$. Compute the polynomial $M_{\xi^3}(x)$.

   Hint: Using Table 3.2, identify the representations in F^3 of ξ^{3i} for $i = 0, 1, 2, 3$ according to the basis $\Omega = (1 \ \xi \ \xi^2)$. Then check the linear dependence of those representations.

Problem 3.13 (Vandermonde matrix) Let $\beta_1, \beta_2, \ldots, \beta_r$ be r distinct elements of a field F and let X be the $r \times r$ matrix that is given by

$$X = \begin{pmatrix} 1 & 1 & \ldots & 1 \\ \beta_1 & \beta_2 & \ldots & \beta_r \\ \beta_1^2 & \beta_2^2 & \ldots & \beta_r^2 \\ \vdots & \vdots & \vdots & \vdots \\ \beta_1^{r-1} & \beta_2^{r-1} & \ldots & \beta_r^{r-1} \end{pmatrix} .$$

Show that

$$\det(X) = \prod_{\substack{(i,j): \\ 1 \leq i < j \leq r}} (\beta_j - \beta_i) .$$

Hint: Let z be an indeterminate and consider the parametric matrix

$$X(z) = \begin{pmatrix} 1 & 1 & \ldots & 1 & 1 \\ \beta_1 & \beta_2 & \ldots & \beta_{r-1} & z \\ \beta_1^2 & \beta_2^2 & \ldots & \beta_{r-1}^2 & z^2 \\ \vdots & \vdots & \vdots & \vdots & \vdots \\ \beta_1^{r-2} & \beta_2^{r-2} & \ldots & \beta_{r-1}^{r-2} & z^{r-2} \\ \beta_1^{r-1} & \beta_2^{r-1} & \ldots & \beta_{r-1}^{r-1} & z^{r-1} \end{pmatrix} .$$

Show that the polynomial

$$b(z) = \det(X(z))$$

is divisible by $z - \beta_i$ for every $1 \leq i < r$ and that the coefficient of z^{r-1} in $b(z)$ is given by $\det(\tilde{X})$, where $\tilde{X}$ denotes the $(r-1) \times (r-1)$ upper-left sub-matrix of X. Deduce by induction on r that

$$b(z) = \left(\prod_{i=1}^{r-1}(z - \beta_i)\right) \prod_{\substack{(i,j): \\ 1 \leq i < j < r}} (\beta_j - \beta_i) .$$

Complete the proof by arguing that $\det(X) = b(\beta_r)$.

Problem 3.14 (Unique interpolation) Let $\beta_1, \beta_2, \ldots, \beta_r$ be r distinct elements of a field F.

1. Given a vector $(v_1 \; v_2 \; \ldots \; v_r)$ in F^r, show that there exists a unique polynomial $u(x) \in F_r[x]$ that interpolates through the points $\{(\beta_j, v_j)\}_{j=1}^r$; namely, $u(x)$ satisfies

$$u(\beta_j) = v_j \, , \quad 1 \leq j \leq r \, .$$

   Hint: Use Problem 3.13 to show that the set of linear equations

$$\sum_{i=0}^{r-1} u_i \beta_j^i = v_j \, , \quad 1 \leq j \leq r \, ,$$

   has a unique solution for $(u_0 \; u_1 \; \ldots \; u_{r-1})$.

2. Show that the polynomial $u(x)$ in part 1 is given by

$$u(x) = \sum_{\ell=1}^{n} v_\ell \prod_{\substack{1 \leq m \leq r: \\ m \neq \ell}} \frac{x - \beta_m}{\beta_\ell - \beta_m} \, .$$

   Hint: Verify that $u(\beta_j) = v_j$ for all j.

Problem 3.15 Let F be the finite field $\mathrm{GF}(q)$. Show that every function $f : F \to F$ can be realized uniquely by a polynomial $u(x) \in F_q[x]$; that is, there exists a unique polynomial $u(x) \in F_q[x]$ that satisfies

$$f(\beta) = u(\beta) \quad \text{for every } \beta \in F \, .$$

Hint: See Problem 3.14.

Problem 3.16 Let $a(x)$ be a polynomial over F and β be an element in an extension field of F and define $b(x) = a(x + \beta)$.

1. Show that β is a root of multiplicity m of $a(x)$ if and only if 0 is a root of multiplicity m of $b(x)$.

2. Write $b(x) = \sum_i b_i x^i$. Show that β is a root of multiplicity m of $a(x)$ if and only if $b_0 = b_1 = \ldots = b_{m-1} = 0$ and $b_m \neq 0$.

[Section 3.5]

Problem 3.17 Let $F = \mathrm{GF}(2)$ and suppose that the field $\Phi = \mathrm{GF}(2^4)$ is represented as $F[\xi]/(\xi^4 + \xi^3 + 1)$. Express each nonzero element of Φ as a power of ξ.

Problem 3.18 Let $F = \mathrm{GF}(2)$ and suppose that the field $\Phi = \mathrm{GF}(2^4)$ is represented as $F[\xi]/(\xi^4 + \xi^3 + \xi^2 + \xi + 1)$. Show that in this representation, $\xi + 1$ is a primitive element in Φ.

Hint: Verify that for $m \in \{3, 5\}$, the powers $(\xi + 1)^m$ (when computed in Φ) do not equal 1.

Problem 3.19 Let m be an integer greater than 1. Show that all the elements in $\mathrm{GF}(2^m) \setminus \{0, 1\}$ are primitive if and only if $2^m - 1$ is a prime (such primes are called *Mersenne primes*).

Problem 3.20 Let $F = \mathrm{GF}(q)$ where q is a power of an odd prime and let α be a primitive element in F. Show that $\alpha^{(q-1)/2} = -1$.

Hint: Identify $\alpha^{(q-1)/2}$ as one of the roots in F of the polynomial $x^2 - 1$.

Problem 3.21 Let $\alpha_1, \alpha_2, \ldots, \alpha_{q-1}$ be the nonzero elements of $F = \mathrm{GF}(q)$. Show that

$$\prod_{j=1}^{q-1} \alpha_j = -1 .$$

Hint: Let α be a primitive element in F. First, argue that $\prod_{j=1}^{q-1} \alpha_j = \alpha^{(q-1)(q-2)/2}$. Then distinguish between odd and even values of q: for the former values apply Problem 3.20, and for the latter values justify the equalities $(\alpha^{q-1})^{(q-2)/2} = 1 = (-1)^{q-1} = -1$.

Problem 3.22 Let $\alpha_1, \alpha_2, \ldots, \alpha_q$ be the elements of $F = \mathrm{GF}(q)$ and let r be an integer in the range $0 \le r < q-1$. Show that

$$\sum_{j=1}^{q} \alpha_j^r = 0$$

(where 0^0 is defined as 1).

Hint: The case $r = 0$ follows by applying Problem A.16 to the additive group of F. Next, assume that $0 < r < q-1$ and let α be a primitive element in F. First, argue that $\sum_{j=1}^{q} \alpha_j^r = \sum_{i=0}^{q-2} \alpha^{ir}$. Then, using the known formula for the sum of a geometric progression, show that the latter sum is equal to $(\alpha^{(q-1)r} - 1)/(\alpha^r - 1)$.

[Section 3.6]

Problem 3.23 (Quadratic residues) Let $F = \mathrm{GF}(q)$, where q is a power of an odd prime. An element $\beta \in F$ is called a *quadratic residue* in F if it is a square of a nonzero element of F. Equivalently, β is a quadratic residue if the polynomial $x^2 - \beta$ has nonzero roots in F. A nonzero element that is not a quadratic residue is called a *quadratic non-residue*.

1. List all the quadratic residues in GF(7) and in GF(13).
2. Show that there are exactly $(q-1)/2$ quadratic residues in F.

   Hint: An element is a quadratic residue if and only if it is an even power of a primitive element in F.
3. Show that the quadratic residues in F are the roots of the polynomial $x^{(q-1)/2} - 1$ in F, and the quadratic non-residues are the roots of $x^{(q-1)/2}+1$.
4. Show that -1 is a quadratic residue in F if and only if $q \equiv 1 \pmod 4$.

   Hint: Use part 3 (or see Problem 3.20).
5. Suppose that $q \equiv 1 \pmod 4$ and let α be a quadratic non-residue in F. Show that the set

$$\{\alpha + \beta \,:\, \beta \text{ is a quadratic residue in } F\}$$

   contains exactly $(q-1)/4$ quadratic residues in F.

   Hint: The problem is equivalent to finding the number of solutions $(x, y) \in F^* \times F^*$ of the equation $\alpha + x^2 = y^2$. This equation can also be written as $\alpha = (y - x)(y + x)$. Denote $z = y + x$, in which case

$$x = \frac{z - \alpha/z}{2} \quad \text{and} \quad y = \frac{z + \alpha/z}{2}$$

   (where the division by 2 stands for multiplication by the inverse of the integer $\overline{2}$ of F). Next, consider the values $y \in F^*$ obtained when z ranges over the elements of F^*, and show that each such y is obtained for two distinct values of z, which are the two distinct roots of the polynomial $z^2 - 2yz + \alpha$.
6. Show that when $q \equiv 1 \pmod 4$ and α is a quadratic residue in F, the set defined in part 5 contains exactly $(q-5)/4$ quadratic residues.
7. Show that when $q \equiv 3 \pmod 4$ and α is a nonzero element in F, the set defined in part 5 contains exactly $(q-3)/4$ quadratic residues.

Problem 3.24 (Quadratic polynomials over fields with odd characteristic) Let $F = \mathrm{GF}(q)$, where q is a power of an odd prime, and let $P(x) = x^2 + \alpha x + \beta$ be a polynomial over F.

1. Find a necessary and sufficient condition on α and β for $P(x)$ to be irreducible over F.

   Hint: The condition involves the discriminant of $P(x)$ (which is defined similarly to the real field); see Problem 3.23.
2. Fix α and let β range over all the elements of F. Find the number of polynomials $P(x)$ that are irreducible over F.
3. Fix β and let α range over all the elements of F. Find the number of polynomials $P(x)$ that are irreducible over F (this number may depend on β).

Problem 3.25 (Multiplicative characters) Let Φ be the field GF(q) and $\mathbb{C}$ be the complex field. A *multiplicative character* of Φ is a nonzero mapping $\psi : \Phi^* \to \mathbb{C}$ that satisfies $\psi(\beta \cdot \gamma) = \psi(\beta)\psi(\gamma)$ for every $\beta, \gamma \in \Phi^*$.

1. Let ω be a root of order $q-1$ of unity in $\mathbb{C}$; that is, $\omega = \mathrm{e}^{2\pi\imath/(q-1)}$, where $\mathrm{e} = 2.71828\cdots$ is the base of natural logarithms, $\pi = 3.14159\cdots$, and $\imath = \sqrt{-1}$. Let α be a primitive element in Φ, and consider the mapping $\psi_\ell : \Phi^* \to \mathbb{C}$ that is defined for some integer $\ell \in \{0, 1, \ldots, q-2\}$ by

$$\psi_\ell(\alpha^i) = \omega^{\ell i}, \quad 0 \le i \le q-2 .$$

   Show that ψ_ℓ is a multiplicative character of Φ.

2. Show that $\psi(1) = 1$ for every multiplicative character ψ of Φ.

3. Show that $(\psi(\beta))^{q-1} = 1$ for every multiplicative character ψ of Φ and every $\beta \in \Phi^*$.

4. Show that a multiplicative character of Φ is completely determined by its value at a primitive element in Φ.

5. Show that the mappings ψ_ℓ considered in part 1 range over all the multiplicative characters of Φ.

6. Show that the set of multiplicative characters of Φ is orthogonal: for every two multiplicative characters $\psi_\ell, \psi_{\ell'}$ of Φ,

$$\sum_{\beta\in\Phi^*} \psi_\ell(\beta)\psi_{\ell'}^*(\beta) = \begin{cases} q-1 & \text{if } \ell = \ell' \\ 0 & \text{otherwise} \end{cases},$$

   where a^* denotes the complex conjugate of an element $a \in \mathbb{C}$. In particular, the *trivial character* ψ_0, which takes the value 1 for every nonzero field element, is orthogonal to any other multiplicative character; therefore, for every nontrivial multiplicative character ψ of Φ,

$$\sum_{\beta\in\Phi^*} \psi(\beta) = 0 .$$

7. Let ψ be a nontrivial multiplicative character of Φ and extend its domain to include the zero value by defining $\psi(0) = 0$. Show that for every $\gamma \in \Phi$,

$$\sum_{\beta\in\Phi^*} \psi(\beta)\psi^*(\beta+\gamma) = \begin{cases} q-1 & \text{if } \gamma = 0 \\ -1 & \text{otherwise} \end{cases}.$$

   Hint: Verify the chain of equalities

$$\begin{aligned} \psi(\beta)\psi^*(\beta+\gamma) &= \psi(\beta) \cdot \psi^*\left(\beta(\gamma\beta^{-1}+1)\right) \\ &= |\psi(\beta)|^2\, \psi^*\left(\gamma\beta^{-1}+1\right) \\ &= \psi^*\left(\gamma\beta^{-1}+1\right) . \end{aligned}$$

   Then compute the sum $\sum_{\beta\in\Phi^*} \psi^*(\gamma\beta^{-1}+1)$.

Problem 3.26 (Legendre symbol) Let p be an odd prime. An integer $a \in \mathbb{Z}$ is called a quadratic residue (respectively, non-residue) modulo p if a is not a multiple of p and the integer $\bar{a}$ of the field $F = \mathrm{GF}(p)$ is a quadratic residue (respectively, non-residue) in F (see Problem 3.23).

Given an integer $a \in \mathbb{Z}$, define the *Legendre symbol* of a modulo p by the following value in $\mathbb{Z}$:

$$\left(\frac{a}{p}\right) = \begin{cases} 1 & \text{if } a \text{ is a quadratic residue modulo } p \\ 0 & \text{if } p \,|\, a \\ -1 & \text{otherwise} \end{cases} .$$

Thus, the mapping $\psi : F^* \to \{-1, 1\}$ that is given by

$$\psi(\bar{a}) = \left(\frac{a}{p}\right) , \quad a = 1, 2, \ldots, p-1 ,$$

is the multiplicative character $\psi_{(p-1)/2}$ of F, as defined in part 1 of Problem 3.25; this character is also referred to as the *quadratic character* of F.

1. Show that for every two integers $a, b \in \mathbb{Z}$,

$$\left(\frac{ab}{p}\right) = \left(\frac{a}{p}\right) \cdot \left(\frac{b}{p}\right) .$$

2. (Euler's criterion) Show that if a is an integer that is not a multiple of p, then

$$\left(\frac{a}{p}\right) \equiv a^{(p-1)/2} \pmod{p} .$$

Hint: Use part 3 of Problem 3.23.

3. (Legendre sequence) Let the sequence $\mathbf{x} = (x_0 \; x_1 \; x_2 \; \cdots)$ over $\mathbb{Z}$ be defined by

$$x_i = \begin{cases} 1 & \text{if } \left(\frac{i}{p}\right) = 1 \\ -1 & \text{otherwise} \end{cases} , \quad i \geq 0 .$$

Show that

$$\sum_{i=0}^{p-1} x_i = -1 .$$

4. Let $\mathbf{x}$ be as in part 3, and define the *autocorrelation function* of $\mathbf{x}$ by

$$R_{\mathbf{x}}(\tau) = \sum_{i=0}^{p-1} x_i x_{i+\tau} , \quad \tau = 0, 1, \ldots, p-1 .$$

Show that when $p \equiv 3 \pmod{4}$,

$$R_{\mathbf{x}}(\tau) = \begin{cases} p & \text{if } \tau = 0 \\ -1 & \text{otherwise} \end{cases} ,$$

and when $p \equiv 1 \pmod{4}$,

$$R_{\mathbf{x}}(\tau) = \begin{cases} p & \text{if } \tau = 0 \\ -3 & \text{if } \tau \text{ is a quadratic residue modulo } p \\ 1 & \text{otherwise} \end{cases} .$$

Hint: Use part 4 of Problem 3.23 and part 7 of Problem 3.25.

Problem 3.27 (Fourier transform) Let $F = \mathrm{GF}(q)$ and let n be a positive integer that divides $q-1$. Fix an element $\alpha \in F$ of multiplicative order n (why does such an element exist?), and denote by X the $n \times n$ Vandermonde matrix

$$\left(\alpha^{ij}\right)_{i,j=0}^{n-1} = \begin{pmatrix} 1 & 1 & 1 & \ldots & 1 \\ 1 & \alpha & \alpha^2 & \ldots & \alpha^{n-1} \\ 1 & \alpha^2 & \alpha^4 & \ldots & \alpha^{2(n-1)} \\ \vdots & \vdots & \vdots & \vdots & \vdots \\ 1 & \alpha^{n-1} & \alpha^{2(n-1)} & \ldots & \alpha^{(n-1)(n-1)} \end{pmatrix},$$

over F.

Given a column vector $\mathbf{y} = (y_j)_{j=0}^{n-1}$ in F^n, the *Fourier transform* of $\mathbf{y}$ is defined by

$$\begin{pmatrix} Y_0 \\ Y_1 \\ \vdots \\ Y_{n-1} \end{pmatrix} = X\mathbf{y} .$$

Also, associate with $(y_j)_{j=0}^{n-1}$ and $(Y_i)_{i=0}^{n-1}$ the polynomial representations

$$y(x) = y_0 + y_1 x + \ldots + y_{n-1}x^{n-1} \quad \text{and} \quad Y(x) = Y_0 + Y_1 x + \ldots + Y_{n-1}x^{n-1} ,$$

both in $F_n[x]$.

1. Show that the inverse of X is given by

$$X^{-1} = (1/n) \cdot \left(\alpha^{-ij}\right)_{i,j=0}^{n-1} ,$$

where $(1/n)$ stands here for the multiplicative inverse of the integer $\overline{n}$ of F.

2. Show that $y_j = 0$ if and only if α^{-j} is a root of the polynomial $Y(x)$.

3. Let $\mathbf{y} = (y_j)_{j=0}^{n-1}$ and $\mathbf{z} = (z_j)_{j=0}^{n-1}$ be two column vectors in F^n with the associated polynomial representations $y(x) = \sum_{j=0}^{n-1} y_j x^j$ and $z(x) = \sum_{j=0}^{n-1} z_j x^j$ in $F_n[x]$, and let $(Y_i)_{i=0}^{n-1}$ and $(Z_i)_{i=0}^{n-1}$ be the respective Fourier transforms of $\mathbf{y}$ and $\mathbf{z}$. Denote by $c(x) = \sum_{j=0}^{n-1} c_j x^j$ the remainder of dividing $y(x)z(x)$ by $x^n - 1$; that is, $c(x)$ is the unique polynomial in $F_n[x]$ that satisfies

$$c(x) \equiv y(x)z(x) \pmod{(x^n - 1)} ;$$

the associated vector $\mathbf{c} = (c_j)_{j=0}^{n-1}$ is commonly referred to as the *cyclic convolution* of $\mathbf{y}$ and $\mathbf{z}$. Show that the Fourier transform of the vector $\mathbf{c}$ is given by $(Y_i Z_i)_{i=0}^{n-1}$.

Problem 3.28 (Discrete logarithms in prime fields) Let $F = \mathrm{GF}(p)$ where p is an odd prime, and let α be a primitive element in F. Define the function

$$\log_\alpha : F^* \to \{0, 1, \ldots, p-2\}$$

by

$$\log_\alpha(\alpha^i) = i , \quad i = 0, 1, \ldots, p-2 .$$

The function $\log_\alpha(x)$ is referred to as the *discrete logarithm in F with the basis α*.

Show that when the images of $\log_\alpha(x)$ are regarded as integers of the field F, then $\log_\alpha(x)$ is realized by the polynomial

$$u(x) = -1 - \sum_{j=1}^{p-2} \frac{x^j}{1-\alpha^{-j}}$$

in $F_{p-1}[x]$; namely,

$$\log_\alpha(\beta) = u(\beta) \quad \text{for every } \beta \in F^* .$$

Hint: Writing $u(x) = \sum_{j=0}^{p-2} u_j x^j$ and letting X be a $(p-1) \times (p-1)$ Vandermonde matrix as in Problem 3.27, first verify that the polynomial sought satisfies

$$\begin{pmatrix} 0 \\ 1 \\ 2 \\ \vdots \\ p-2 \end{pmatrix} = X \begin{pmatrix} u_0 \\ u_1 \\ u_2 \\ \vdots \\ u_{p-2} \end{pmatrix} .$$

Deduce from part 1 of Problem 3.27 that

$$u_j = -\sum_{i=0}^{p-2} i \cdot \alpha^{-ij} , \quad 0 \le j \le p-2 .$$

Proceed by proving the following identity, which holds for every positive integer n and every element $\beta \in F$:

$$\sum_{i=0}^{n-1} i \cdot \beta^i = \begin{cases} \dfrac{n\beta^n}{\beta-1} - \dfrac{\beta(\beta^n-1)}{(\beta-1)^2} & \text{if } \beta \ne 1 \\ n(n-1)/2 & \text{if } \beta = 1 \end{cases} .$$

Finally, apply this identity to $n = p-1$ and $\beta = \alpha^{-j}$. (Why is $(p-1)(p-2)/2$ equal to 1?)

(The images of $\log_\alpha(x)$ are interpreted in this problem as elements of the field F $(= \mathbb{Z}_p)$ only for the purpose of an exercise. In view of the fact that $\alpha^{i+p-1} = \alpha^i$ for every integer i, it would be more natural in practice to regard those images as elements of the ring $\mathbb{Z}_{p-1}$.)

Problem 3.29 Let n and a be positive integers such that $\gcd(n, a) = 1$ and consider the polynomial $(x-a)^n$ over the ring of integers. Show that n is a prime if and only for every $i = 1, 2, \ldots, n-1$, the coefficient of x^i in $(x-a)^n$ is divisible by n.

Hint: Obtain the "only if" part from Proposition 3.11. As for the "if" part, let p be a prime divisor of n and let k be the multiplicity of p in the prime factorization of n. Show that p^k does not divide $\binom{n}{p} a^{n-p}$.

Problem 3.30 Let Φ be an extension field of $F = \mathrm{GF}(q)$. Show that if an element $\alpha \in \Phi$ is a root of a polynomial $a(x)$ over F, then so are the elements α^{q^r} for every $r \ge 0$.

Problem 3.31 (Trace of an element) Let $F = \mathrm{GF}(q)$ and $\Phi = \mathrm{GF}(q^m)$. Define the *trace polynomial* over Φ with respect to F by

$$\mathrm{T}_{\Phi:F}(x) = x + x^q + x^{q^2} + \ldots + x^{q^{m-1}} .$$

The *trace* with respect to F of an element $\beta \in \Phi$ is defined as $\mathrm{T}_{\Phi:F}(\beta)$.

1. Show that for every $\beta \in \Phi$,

$$(\mathrm{T}_{\Phi:F}(\beta))^q = \mathrm{T}_{\Phi:F}(\beta^q) = \mathrm{T}_{\Phi:F}(\beta) .$$

2. Show that $\mathrm{T}_{\Phi:F}(\beta) \in F$ for every $\beta \in \Phi$.

   Hint: See Problem 3.11.

3. Show that the mapping $\Phi \to F$ defined by $x \mapsto \mathrm{T}_{\Phi:F}(x)$ is a linear transformation over F; namely, for every $b, c \in F$ and $\beta, \gamma \in \Phi$,

$$\mathrm{T}_{\Phi:F}(b \cdot \beta + c \cdot \gamma) = b \cdot \mathrm{T}_{\Phi:F}(\beta) + c \cdot \mathrm{T}_{\Phi:F}(\gamma) .$$

4. Show by a counting argument that every linear transformation $\Phi \to F$ (over F) can be written in the form $x \mapsto \mathrm{T}_{\Phi:F}(\mu x)$ for some $\mu \in \Phi$.

5. Show that when β ranges over all the elements of Φ, the trace $\mathrm{T}_{\Phi:F}(\beta)$ takes each value of F exactly q^{m-1} times.

   Hint: Use Problem 2.4.

6. Show that for every $b \in F$,

$$\mathrm{T}_{\Phi:F}(x) - b = \prod_{\beta \in \Phi \,:\, \mathrm{T}_{\Phi:F}(\beta) = b} (x - \beta) .$$

7. Show that

$$x^{q^m} - x = \prod_{b \in F} (\mathrm{T}_{\Phi:F}(x) - b) .$$

8. Find a necessary and sufficient condition on m for having $\mathrm{T}_{\Phi:F}(b) = 0$ for every $b \in F$.

9. Let F be the field $\mathrm{GF}(2)$ and let the extension field $\Phi = \mathrm{GF}(2^4)$ be represented as $F[\xi]/(\xi^4 + \xi + 1)$ (see Table 3.3). Identify the elements $\beta \in \Phi$ for which $\mathrm{T}_{\Phi:F}(\beta) = 1$.

Problem 3.32 (Linearized polynomials) Let $F = \mathrm{GF}(q)$ and $\Phi = \mathrm{GF}(q^n)$. A *linearized polynomial* over Φ with respect to F is a polynomial over Φ of the form

$$a(x) = \textstyle\sum_{i=0}^{t} a_i x^{q^i}$$

for some $t \geq 0$. Hereafter in this problem, all the linearized polynomials are assumed to be over Φ with respect to F.

1. Let $a(x)$ be a linearized polynomial. Show that the mapping $\Phi \to \Phi$ defined by $x \mapsto a(x)$ is a linear transformation over F.

2. Show by a counting argument that every linear transformation $\Phi \to \Phi$ over F can be represented as a mapping $x \mapsto a(x)$ for some linearized polynomial of degree at most q^{n-1}.

3. Let $a(x)$ be a linearized polynomial of degree q^t where $0 \le t < n$. Fix a basis $\Omega = (\alpha_1 \; \alpha_2 \; \ldots \; \alpha_n)$ of Φ over F, and let A be an $n \times n$ matrix representation of the mapping $\Phi \to \Phi$ defined by $x \mapsto a(x)$; that is, if $\mathbf{u}$ is a column vector in F^n that represents an element $u \in \Phi$ as $u = \Omega\mathbf{u}$, then $A\mathbf{u}$ is a vector representation of $a(u)$, i.e., $a(u) = \Omega A\mathbf{u}$. Show that $\mathrm{rank}(A) \ge n-t$.

   Hint: Find an upper bound on the size of the right kernel of A.

Problem 3.33 Let $F = \mathrm{GF}(q)$ and $\Phi = \mathrm{GF}(q^n)$ and let $\beta_1, \beta_2, \ldots, \beta_m$ be elements of Φ. Show that the $m \times m$ matrix

$$B = \begin{pmatrix} \beta_1 & \beta_2 & \ldots & \beta_m \\ \beta_1^q & \beta_2^q & \ldots & \beta_m^q \\ \beta_1^{q^2} & \beta_2^{q^2} & \ldots & \beta_m^{q^2} \\ \vdots & \vdots & \vdots & \vdots \\ \beta_1^{q^{m-1}} & \beta_2^{q^{m-1}} & \ldots & \beta_m^{q^{m-1}} \end{pmatrix}$$

is nonsingular if and only if $\beta_1, \beta_2, \ldots, \beta_m$, as elements belonging to a linear space Φ over F, are linearly independent over F.

Hint: To show the "if" part, observe that $(a_0 \; a_1 \; \ldots \; a_{m-1})B = \mathbf{0}$ if and only if $\beta_1, \beta_2, \ldots, \beta_m$ are roots of the polynomial $a(x) = \sum_{i=0}^{m-1} a_i x^{q^i}$. Then use Problem 3.32 to argue that if $\beta_1, \beta_2, \ldots, \beta_m$ are roots of $a(x)$, then so are all their linear combinations over F. Conclude that $a(x)$ must be the zero polynomial whenever $\beta_1, \beta_2, \ldots, \beta_m$ are linearly independent over F.
To show the "only if" part, observe that for every $c_1, c_2, \ldots, c_m \in F$,

$$\sum_{j=1}^{m} c_j \beta_j = 0 \quad \Longrightarrow \quad \sum_{j=1}^{m} c_j \beta_j^{q^i} = 0 \quad \text{for every } i \ge 0 \, .$$

Problem 3.34 Let $F = \mathrm{GF}(q)$ and let U be a linear subspace of $\Phi = \mathrm{GF}(q^n)$ over F. Show that

$$a(x) = \prod_{\gamma \in U} (x - \gamma)$$

is a linearized polynomial over Φ with respect to F (see Problem 3.32 for the definition of linearized polynomials).

Hint: Let $\beta_1, \beta_2, \ldots, \beta_m$ be a basis of U over F. Use Problem 3.33 to show that the set of m linear equations

$$\beta_j^{q^m} + \sum_{i=0}^{m-1} a_i \beta_j^{q^i} = 0 \, , \quad 1 \le j \le m \, ,$$

has a unique solution for $(a_0 \; a_1 \; \ldots \; a_{m-1}) \in \Phi^m$. Then verify that $a(x)$ equals $x^{q^m} + \sum_{i=0}^{m-1} a_i x^{q^i}$.

Problem 3.35 (Dual basis) Let $F = \mathrm{GF}(q)$ and let $\Omega = (\beta_1\ \beta_2\ \ldots\ \beta_n)$ be a basis of $\Phi = \mathrm{GF}(q^n)$ over F. The *dual basis* (or *complementary basis*) of Ω over F is a basis $(\lambda_1\ \lambda_2\ \ldots\ \lambda_n)$ of Φ over F for which

$$\mathrm{T}_{\Phi:F}(\lambda_i\beta_j) = \begin{cases} 1 & \text{if } i = j \\ 0 & \text{if } i \neq j \end{cases}, \quad 1 \leq i, j \leq n ,$$

where $\mathrm{T}_{\Phi:F}(\cdot)$ denotes the trace with respect to F, as defined in Problem 3.31.

1. Show that a dual basis always exists and that it is unique.

   Hint: Using Problem 3.33, show that the set of n linear equations

   $$\sum_{j=1}^{n} \lambda_j \beta_j^{q^i} = \begin{cases} 1 & \text{if } i = 0 \\ 0 & \text{if } i \neq 0 \end{cases}, \quad 0 \leq i < n ,$$

   has a unique solution for $(\lambda_1\ \lambda_2\ \ldots\ \lambda_n) \in \Phi^n$. Given that solution, show that the $n \times n$ matrix

   $$\begin{pmatrix} \lambda_1 & \lambda_1^q & \lambda_1^{q^2} & \ldots & \lambda_1^{q^{n-1}} \\ \lambda_2 & \lambda_2^q & \lambda_2^{q^2} & \ldots & \lambda_2^{q^{n-1}} \\ \vdots & \vdots & \vdots & \vdots & \vdots \\ \lambda_n & \lambda_n^q & \lambda_n^{q^2} & \ldots & \lambda_n^{q^{n-1}} \end{pmatrix}$$

   is a right inverse of the matrix $B = (\beta_j^{q^i})_{i=0}^{n-1}{}_{j=1}^{n}$ and, as such, it is also a left inverse.

2. Let γ be an element of Φ and let $\mathbf{u} = (u_1\ u_2\ \ldots\ u_n)^T$ be the vector in F^n that represents γ with respect to the basis Ω, that is, $\gamma = \Omega\mathbf{u}$. Show that

   $$u_i = \mathrm{T}_{\Phi:F}(\gamma\lambda_i) , \quad 1 \leq i \leq n .$$

Problem 3.36 (Additive characters) Let $F = \mathrm{GF}(p)$ and $\Phi = \mathrm{GF}(p^m)$ where p is a prime. An *additive character* of Φ is a nonzero mapping $\chi : \Phi \to \mathbb{C}$ that satisfies $\chi(\beta + \gamma) = \chi(\beta)\chi(\gamma)$ for every $\beta, \gamma \in \Phi$.

1. Let ω be a root of order p of unity in $\mathbb{C}$ (see Problem 3.25), and consider the mapping $\chi_\mu : \Phi \to \mathbb{C}$ that is defined for some $\mu \in \Phi$ by

   $$\chi_\mu(\beta) = \omega^{\mathrm{T}_{\Phi:F}(\mu\beta)} , \quad \beta \in \Phi ,$$

   where $x \mapsto \mathrm{T}_{\Phi:F}(x)$ is the trace with respect to F as defined in Problem 3.31 (raising ω to a power $b \in F$ is the same as writing ω^t for some positive integer t such that $\bar{t} = b$). Show that χ_μ is an additive character of Φ.

2. Show that $\chi(0) = 1$ for every additive character χ of Φ.

3. Show that $(\chi(\beta))^p = 1$ for every additive character χ of Φ and every $\beta \in \Phi$.

4. Let $\Omega = (\alpha_1\ \alpha_2\ \ldots\ \alpha_m)$ be a basis of Φ over F. Show that for every additive character χ of Φ and every column vector $\mathbf{u} = (u_1\ u_2\ \ldots\ u_m)^T \in F^m$,

$$\chi(\Omega\mathbf{u}) = \prod_{j=1}^{m} (\chi(\alpha_j))^{u_j} .$$

That is, an additive character of Φ is completely determined by its values at a basis of Φ over F.

5. Let ω be a root of order p of unity in $\mathbb{C}$. Show that for every additive character χ of Φ there exists a row vector $\mathbf{v} \in F^m$ such that

$$\chi(\Omega\mathbf{u}) = \omega^{\mathbf{v}\cdot\mathbf{u}} , \quad \mathbf{u} \in F^m .$$

6. Show that the mappings χ_μ considered in part 1 range over all the additive characters of Φ.

   Hint: See parts 3 and 4 of Problem 3.31.

7. Show that the set of additive characters of Φ is orthogonal, i.e., for every two additive characters $\chi_\mu, \chi_{\mu'}$ over Φ,

$$\sum_{\beta\in\Phi} \chi_\mu(\beta)\chi_{\mu'}^*(\beta) = \begin{cases} p^m & \text{if } \mu = \mu' \\ 0 & \text{otherwise} \end{cases} ,$$

   where $(\cdot)^*$ denotes a complex conjugate. Deduce that with the exception of the trivial character χ_0 (which takes the value 1 for every field element), every additive character χ of Φ satisfies

$$\sum_{\beta\in\Phi} \chi(\beta) = 0$$

   (compare with part 6 of Problem 3.25).

8. Let χ be a nontrivial additive character of Φ. Show that for every $\gamma \in \Phi$,

$$\sum_{\beta\in\Phi} \chi(\beta)\chi^*(\gamma\beta) = \begin{cases} p^m & \text{if } \gamma = 1 \\ 0 & \text{otherwise} \end{cases} .$$

Problem 3.37 Let $F = \mathrm{GF}(p)$ and $\Phi = \mathrm{GF}(p^m)$ where p is a prime, and let $\mathrm{T}(x) = \mathrm{T}_{\Phi:F}(x)$ be the trace polynomial over Φ with respect to F, as defined in Problem 3.31.

Fix α to be a primitive element in Φ and ω to be a root of order p of unity in the complex field $\mathbb{C}$, and consider the sequence $\mathbf{x} = (x_0\ x_1\ x_2\ \cdots)$ over $\mathbb{C}$ that is defined by

$$x_i = \omega^{\mathrm{T}(\alpha^i)} , \quad i \ge 0 .$$

1. Show that

$$\sum_{i=0}^{p^m-2} x_i = -1 .$$

   Hint: Use part 7 of Problem 3.36.

2. Let $\mathbf{x}$ be as in part 1, and define the autocorrelation function of $\mathbf{x}$ by

$$R_{\mathbf{x}}(\tau) = \sum_{i=0}^{p^m-2} x_i x_{i+\tau}^* , \qquad \tau = 0, 1, \ldots, p^m-2$$

(where $x_{i+\tau}^*$ denotes the complex conjugate of $x_{i+\tau}$). Show that

$$R_{\mathbf{x}}(\tau) = \begin{cases} p^m-1 & \text{if } \tau = 0 \\ -1 & \text{otherwise} \end{cases} .$$

Hint: Use again part 7 of Problem 3.36.

(Compare the properties of $\mathbf{x}$ herein with those in parts 3 and 4 of Problem 3.26.)

[Section 3.7]

Problem 3.38 Let $a(x)$ and $b(x)$ be polynomials over F and let c be an element of F. Prove the following properties of the formal derivative:

1. $(a(x) + b(x))' = a'(x) + b'(x)$.
2. $(c \cdot a(x))' = c \cdot a'(x)$.
3. $(a(x)b(x))' = a'(x)b(x) + a(x)b'(x)$.
4. $[a(b(x))]' = a'(b(x))b'(x)$.

Problem 3.39 Let $a(x)$ be a polynomial over a field F and let β be an element in an extension field of F.

1. Show that β is a multiple root of $a(x)$ if and only if $a(\beta) = a'(\beta) = 0$.
2. Show by example that there are cases where β is a root of multiplicity less than 3 of $a(x)$, yet $a(\beta) = a'(\beta) = a''(\beta) = 0$.

   Hint: Consider fields with characteristic 2.

Problem 3.40 Let $a(x) = \sum_{i=0}^{n} a_i x^i$ be a polynomial over F. The *ℓth Hasse derivative* (also known as the *ℓth hyper-derivative*) of $a(x)$, denoted by $a^{[\ell]}(x)$, is defined as

$$a^{[\ell]}(x) = \sum_{i=\ell}^{n} \binom{i}{\ell} a_i x^{i-\ell} .$$

1. Verify that $a'(x) = a^{[1]}(x)$.
2. Show that the following properties hold for every two polynomials $a(x)$ and $b(x)$ over F and every element c in F:

   (a) $(a(x) + b(x))^{[\ell]} = a^{[\ell]}(x) + b^{[\ell]}(x)$.

   (b) $(c \cdot a(x))^{[\ell]} = c \cdot a^{[\ell]}(x)$.

   (c) $(a(x)b(x))^{[\ell]} = \sum_{m=0}^{\ell} a^{[m]}(x) b^{[\ell-m]}(x)$.

3. Show that for every field element β and every nonnegative integer i,
$$((x-\beta)^i)^{[\ell]} = \binom{i}{\ell}(x-\beta)^{i-\ell}$$
($\binom{i}{\ell}$ is defined as 0 when $i < \ell$).

4. Show that for every polynomial $b(x) = \sum_{i=0}^{n} b_i x^i$ over F and every element β in an extension field of F,
$$(b(x-\beta))^{[\ell]} = \sum_{i=\ell}^{n} \binom{i}{\ell} b_i \cdot (x-\beta)^{i-\ell} = b^{[\ell]}(\xi)|_{\xi=x-\beta} \,.$$

5. Let β be an element in an extension field of F. Show that β is a root of $a(x)$ of multiplicity exactly m if and only if m is the smallest ℓ such that $a^{[\ell]}(x)|_{x=\beta} \neq 0$.

   Hint: Define $b(x) = a(x+\beta)$ and use part 4 and Problem 3.16.

Problem 3.41 Let α be a nonzero element with multiplicative order n in a field F.

1. Show that F is a splitting field of the polynomial $x^n - 1$ over F, and find all the roots of this polynomial in F. What is the multiplicity of each root?

2. Show that
$$\prod_{i=1}^{n-1}(x-\alpha^i) = 1 + x + x^2 + \ldots + x^{n-1}$$
and, so,
$$\prod_{i=1}^{n-1}(1-\alpha^i) = \overline{n} \,.$$
Hint:
$$(x-1)(1+x+x^2+\ldots+x^{n-1}) = x^n - 1 \,.$$

[Section 3.8]

Problem 3.42 (Quadratic polynomials over fields with characteristic 2) Let $F = \mathrm{GF}(2)$ and $\Phi = \mathrm{GF}(2^m)$ and for a fixed element $\beta \in \Phi$ define the polynomial $P_\beta(x)$ over Φ by
$$P_\beta(x) = x^2 + x + \beta \,.$$
Let $\mathrm{T}_{\Phi:F}(x)$ be the trace polynomial over Φ with respect to F, as defined in Problem 3.31.

1. Can $P_\beta(x)$ have multiple roots in Φ?

2. Show that $P_\beta(x)$ has roots in Φ only if $\mathrm{T}_{\Phi:F}(\beta) = 0$.

   Hint: Given an element $\gamma \in \Phi$, compute the value $\mathrm{T}_{\Phi:F}(P_\beta(\gamma))$.

3. The polynomial $P_0(x)$ (i.e., $P_\beta(x)$ for $\beta = 0$) is evaluated at all the elements of Φ. Show that $P_0(x)$ takes each value $\gamma \in \Phi$ at most twice. What can be said about the trace $\mathrm{T}_{\Phi:F}(\gamma)$ of each value γ taken by $P_0(x)$?

4. Based on part 3, show that if $\mathrm{T}_{\Phi:F}(\beta) = 0$ then $P_\beta(x)$ has two distinct roots in Φ.
5. Let α be a nonzero element in Φ. Show that the polynomial $x^2 + \alpha x + \beta$ is irreducible over Φ if and only if $\mathrm{T}_{\Phi:F}(\beta/\alpha^2) = 1$. How many roots does this polynomial have in Φ when $\alpha = 0$?

   Hint: $x^2 + \alpha x + \beta = \alpha^2 \cdot P_{\beta/\alpha^2}(x/\alpha)$.

(Recall from Corollary 3.13 that taking the square of an element in Φ is an automorphism of Φ and, so, it is a linear transformation over F. Hence, the equation $x^2 + \alpha x = \beta$ over Φ can be translated into a set of m linear equations over F, whose solutions, if any, are the roots of $x^2 + \alpha x + \beta$ in Φ.)

Problem 3.43 Verify that the algorithm in Figure 3.1 will produce the error-detection indicator "e" if the computed syndrome is such that $s_1 = 0$ and $s_3 \neq 0$.

Problem 3.44 (Linear $[q^2{+}1, q^2{-}3, 4]$ codes over $\mathrm{GF}(q)$) Let $F = \mathrm{GF}(q)$ and for $b, c \in F$ let $f : F \times F \to F$ be the function

$$f(x, y) = x^2 + bxy + cy^2 .$$

Consider the 4×3 matrix over F which is given by

$$B = \begin{pmatrix} 1 & 1 & 1 \\ \alpha_1 & \alpha_2 & \alpha_3 \\ \beta_1 & \beta_2 & \beta_3 \\ f(\alpha_1, \beta_1) & f(\alpha_2, \beta_2) & f(\alpha_3, \beta_3) \end{pmatrix},$$

where $(\alpha_1\ \beta_1)$, $(\alpha_2\ \beta_2)$, and $(\alpha_3\ \beta_3)$ are distinct elements of F^2.

1. Show that the columns of B are linearly dependent over F only if

   $$f(\alpha_1{-}\alpha_2, \beta_1{-}\beta_2) = f(\alpha_2{-}\alpha_3, \beta_2{-}\beta_3) = f(\alpha_3{-}\alpha_1, \beta_3{-}\beta_1) = 0 .$$

   Hint: Start by showing that if the columns of B are linearly dependent, then either $\alpha_1 = \alpha_2 = \alpha_3$ or

   $$\frac{\beta_1{-}\beta_2}{\alpha_1{-}\alpha_2} = \frac{\beta_2{-}\beta_3}{\alpha_2{-}\alpha_3} = \frac{\beta_3{-}\beta_1}{\alpha_3{-}\alpha_1} .$$

2. Show that if $f(x, 1) = x^2 + bx + c$ is an irreducible polynomial over F, then the columns of B are linearly independent over F.
3. Suppose that $f(x, 1) = x^2 + bx + c$ is an irreducible polynomial over F, and let $\mathcal{C}$ be the linear code of length q^2 over F whose parity-check matrix consists of all distinct column vectors of the form

   $$\begin{pmatrix} 1 \\ \alpha \\ \beta \\ f(\alpha, \beta) \end{pmatrix},$$

   where $(\alpha\ \beta)$ ranges over all the elements of F^2. Show that the minimum distance of $\mathcal{C}$ is at least 4.
4. Show that the minimum distance remains at least 4 when one adds to the parity-check matrix in part 3 the column vector $(0\,0\,0\,1)^T$.

Notes

An extensive treatment of finite fields can be found in the book by Lidl and Niederreiter [229]. More material can be found in books on coding theory, e.g., in Blahut [46, Chapter 4] and MacWilliams and Sloane [249, Chapter 4].

[Section 3.2]

Euclid's algorithm and its complexity analysis are covered in many sources. See, for example, Aho *et al.* [6, Section 8.8], von zur Gathen and Gerhard [144, Section 11.1], Knuth [215, Sections 4.5 and 4.6], and Zippel [403, Chapters 1 and 8]. There are known methods for accelerating Euclid's algorithm: the algorithm in [144, Section 11.1] computes the gcd of two polynomials in $F_n[x]$ using $O(n \log^2 n \log\log n)$ arithmetic operations in F (hereafter, the notation $O(f)$ stands for an expression that grows at most linearly with f).

There are known efficient algorithms for finding the irreducible factors—in particular, finding the linear factors and hence the roots—of polynomials over $\mathrm{GF}(p^m)$. Some of these algorithms are deterministic and are applicable to cases where p is small (e.g., fixed): see Berlekamp [34], Berlekamp *et al.* [39], and Lidl and Niederreiter [229, Chapter 4]. Other algorithms are probabilistic and their time complexity grows polynomially with $\log p$ (rather than with p), thereby making them applicable to large fields as well: see Ben-Or [32], Berlekamp [34], Lidl and Niederreiter [229, Sections 4.2 and 4.3], Rabin [286], and Shoup [336].

Factorization algorithms can also serve for testing irreduciblity. In Section 7.2 we show that irreducible polynomials are rather dense among all polynomials of the same degree; hence, any algorithm for testing irreduciblity implies a probabilistic algorithm for finding irreducible polynomials (refer also to Shoup [334], [335]). As for explicit constructions of irreducible polynomials over finite fields, the polynomial

$$x^{2\cdot 3^n} + x^{3^n} + 1$$

is irreducible over GF(2) for every positive integer n (see Golomb [152, p. 96] and Problem 7.11).

[Section 3.3]

The implementation of arithmetic in an extension field Φ of a field $F = \mathrm{GF}(q)$ depends on the particular representation of the elements of Φ. When $\Phi = F[\xi]/P(\xi)$ for an irreducible polynomial $P(x)$ of degree n over F, the elements of Φ are commonly represented as polynomials in $F_n[x]$. Addition in Φ is then implemented by n additions in F. Multiplication can be performed by first computing the product in $F[x]$ and then reducing the result modulo $P(x)$. The (asymptotically) fastest known algorithm for multiplying polynomials of degree less than n over F is due to Schönhage and Strassen [322], [323], requiring $O(n \log n \log\log n)$ additions and multiplications in F. This is also the complexity of computing the remainder modulo $P(x)$ ([6, Section 8.3], [144, Section 9.1]), and is thus the complexity of multiplying elements in Φ. A multiplicative inverse of an element in Φ can be obtained by fast methods of computing the gcd, requiring $O(n \log^2 n \log\log n)$ arithmetic operations in F [144, Section 11.1].

If only multiplications in F are to be counted, then the multiplication of two elements in $\Phi = F[\xi]/P(\xi)$ can be carried out in $O(n)$ multiplications in F. See Chudnovsky and Chudnovsky [80] and Winograd [386]. Reducing the number of multiplications at the expenses of additions in F can lead to savings in complexity if recursion is used through an intermediate field K, where $F \subseteq K \subseteq \Phi$: multiplication in Φ is implemented using operations in K, and then each multiplication in K is realized by operations in F.

[Section 3.5]

There is no general efficient algorithm known for testing whether a given element α in $\mathrm{GF}(q)$ is primitive. When the prime factorization of $q-1$ is known, then one can perform the test by checking if $\alpha^{(q-1)/\ell} \neq 1$ for every prime divisor ℓ of $q-1$. Yet, the best known algorithms for integer factorization have super-polynomial time complexity.

The *discrete logarithm* problem over finite fields is defined as follows. Let $F = \mathrm{GF}(p^n)$ where p is a prime and let α be a primitive element in F; given $\beta \in F^*$, compute the (unique) integer $i \in \{0, 1, \ldots, |F|-2\}$ such that $\beta = \alpha^i$ (Problem 3.28 deals with the case where $n = 1$). The fastest known algorithms for computing the discrete logarithm have a time complexity that is still super-polynomial in $\log |F|$: see Adleman [1], Coppersmith [85], Coppersmith *et al.* [86], Gordon [159], Odlyzko [273], and Schirokauer [320], [321].

[Section 3.6]

The notion of characters (Problems 3.25 and 3.36) can be generalized to every finite Abelian group G. Specifically, a nonzero mapping $\chi : G \to \mathbb{C}$ is a character of G if

$$\chi(\beta + \gamma) = \chi(\beta)\chi(\gamma) \qquad \text{for every } \beta, \gamma \in G$$

(see, for example, Lidl and Niederreiter [229, Section 5.1]). Denoting by 0 the unity of G, every character χ of G satisfies $\chi(0) = 1$. For the special case of the ring $\mathbb{Z}_n$ we thus have

$$(\chi(1))^n = \chi(\underbrace{1 + 1 + \ldots + 1}_{n \text{ times}}) = \chi(0) = 1$$

for every character χ of $\mathbb{Z}_n$. This, in turn, implies that the characters of $\mathbb{Z}_n$ are given by

$$\chi_\mu(\beta) = \omega_n^{\mu\beta} , \quad \beta \in \mathbb{Z}_n ,$$

where ω_n is a fixed root of order n of unity in $\mathbb{C}$ and μ ranges over the elements of $\mathbb{Z}_n$. It follows that the characters of $\mathbb{Z}_n$ form a group, where χ_0 serves as the unity element and the product of two characters χ_μ and $\chi_{\mu'}$ is defined by

$$(\chi_\mu \cdot \chi_{\mu'})(\beta) = \chi_\mu(\beta)\chi_{\mu'}(\beta) , \quad \beta \in \mathbb{Z}_n ;$$

furthermore, this group is isomorphic to $\mathbb{Z}_n$. Now, every finite Abelian group G is isomorphic to the direct sum

$$\mathbb{Z}_{n_1} \oplus \mathbb{Z}_{n_2} \oplus \ldots \oplus \mathbb{Z}_{n_m}$$

for some integers $n_1, n_2, \ldots, n_m$ (see, for example, the book by MacLane and Birkhoff [246, p. 387]). Thus, one can conclude that the characters of G take the form

$$\chi(\beta_1, \beta_2, \ldots, \beta_m) = \prod_{i=1}^{m} \omega_{n_i}^{\mu_i \beta_i}, \quad (\beta_1, \beta_2, \ldots, \beta_m) \in \mathbb{Z}_{n_1} \oplus \mathbb{Z}_{n_2} \oplus \ldots \oplus \mathbb{Z}_{n_m},$$

for some elements $\mu_i \in \mathbb{Z}_{n_i}$, and these characters form a group that is isomorphic to G (see [246, pp. 416–417]). In Problem 3.25, G is the multiplicative group of GF(q) (in which case $m = 1$ and $n_1 = q-1$), while in Problem 3.36, G is the additive group of GF(p^m) (in which case $n_1 = n_2 = \ldots = n_m = p$).

Additional properties of the Legendre symbol (Problem 3.26) will be presented in the notes on Section 13.5.

For a treatment of the Fourier transform over finite fields (Problem 3.27), see Blahut [46, Chapter 8], [48] and MacWilliams and Sloane [249, Section 8.6].

Problem 3.29 is taken from Agrawal and Biswas [4]; it exhibits one of the characterizations of prime numbers that has eventually led to the deterministic polynomial-time algorithm for primality testing due to Agrawal *et al.* [5].

For more on linearized polynomials (Problem 3.32), see Berlekamp [36, Chapter 11], Lidl and Niederreiter [229, Section 3.4], and MacWilliams and Sloane [249, Section 4.9].

Chapter 4

Bounds on the Parameters of Codes

In this chapter, we establish conditions on the parameters of codes. In the first part of the chapter, we present bounds that relate between the length n, size M, minimum distance d, and the alphabet size q of a code. Two of these bounds—the Singleton bound and the sphere-packing bound—imply necessary conditions on the values of n, M, d, and q, so that a code with the respective parameters indeed exists. We also exhibit families of codes that attain each of these bounds. The third bound which we present—the Gilbert–Varshamov bound—is an existence result: it states that there exists a linear $[n, k, d]$ code over $\mathrm{GF}(q)$ whenever n, k, d, and q satisfy a certain inequality. Additional bounds are included in the problems at the end of this chapter. We end this part of the chapter by introducing another example of necessary conditions on codes—now in the form of MacWilliams' identities, which relate the distribution of the Hamming weights of the codewords in a linear code with the respective distribution in the dual code.

The second part of this chapter deals with asymptotic bounds, which relate the rate of a code to its relative minimum distance $\delta = d/n$ and its alphabet size, as the code length n tends to infinity.

In the third part of the chapter, we shift from the combinatorial setting of (n, M, d) codes to the probabilistic framework of the memoryless q-ary symmetric channel. We first prove the Shannon Converse Coding Theorem, which states that at code rates above the capacity of the channel, the decoding error probability of any decoder for any code approaches 1 as the code length goes to infinity. We then present and prove the respective Coding Theorem, which applies to rates below the channel capacity. Specifically, we show that in this range, the expected decoding error probability of a nearest-codeword decoder for a random linear code decreases exponentially with the code length.

4.1 The Singleton bound

The next theorem is our first bound on the parameters of a code.

Theorem 4.1 (The Singleton bound) *For any (n, M, d) code over an alphabet of size q,*

$$d \leq n - (\log_q M) + 1 .$$

Proof. Let $\ell = \lceil \log_q M \rceil - 1$. Since $q^\ell < M$, there must be at least two codewords that agree on their first ℓ coordinates. Hence, $d \leq n - \ell$. □

For a linear $[n, k, d]$ code over $\mathrm{GF}(q)$, the Singleton bound becomes

$$d \leq n - k + 1 .$$

This inequality can be derived also from Theorem 2.2: since the rank of a parity-check matrix of the code is $n-k$, such a matrix contains a set of $n-k+1$ linearly dependent columns (in fact, every set of $n-k+1$ columns is linearly dependent). Thus, the minimum distance is at most $n-k+1$. Alternatively, the Singleton bound for linear codes can also be obtained by looking at a systematic generator matrix of the code: the Hamming weight of each row in that matrix is at most $n-k+1$.

An (n, M, d) code over an alphabet of size q is called *maximum distance separable* (in short, MDS) if it attains the Singleton bound, namely, it satisfies the equality $d = n - (\log_q M) + 1$.

The following codes are examples of MDS codes over $F = \mathrm{GF}(q)$:

- The whole space F^n, which is a linear $[n, n, 1]$ code over F.
- The $[n, n-1, 2]$ parity code over F.
- The $[n, 1, n]$ repetition code over F.

We next present another important family of MDS codes.

Let $\alpha_1, \alpha_2, \ldots, \alpha_n$ be distinct elements of $F = \mathrm{GF}(q)$. A *[normalized generalized] Reed–Solomon code* over F is a linear $[n, k, d]$ code with the parity-check matrix

$$H_{\mathrm{RS}} = \begin{pmatrix} 1 & 1 & \ldots & 1 \\ \alpha_1 & \alpha_2 & \ldots & \alpha_n \\ \alpha_1^2 & \alpha_2^2 & \ldots & \alpha_n^2 \\ \vdots & \vdots & \vdots & \vdots \\ \alpha_1^{n-k-1} & \alpha_2^{n-k-1} & \ldots & \alpha_n^{n-k-1} \end{pmatrix} .$$

This construction requires that the code length n be at most the field size q.

Proposition 4.2 *Every Reed–Solomon code is MDS.*

Proof. Every $(n-k) \times (n-k)$ sub-matrix of H_{RS} has a *Vandermonde* form

$$B = \begin{pmatrix} 1 & 1 & \ldots & 1 \\ \beta_1 & \beta_2 & \ldots & \beta_{n-k} \\ \beta_1^2 & \beta_2^2 & \ldots & \beta_{n-k}^2 \\ \vdots & \vdots & \vdots & \vdots \\ \beta_1^{n-k-1} & \beta_2^{n-k-1} & \ldots & \beta_{n-k}^{n-k-1} \end{pmatrix},$$

where $\beta_1, \beta_2, \ldots, \beta_{n-k}$ are distinct elements of the field (see Problem 3.13). Now, the determinant of B is given by

$$\det(B) = \prod_{\substack{(i,j): \\ 1 \le i < j \le n-k}} (\beta_j - \beta_i)$$

and, therefore, $\det(B) \neq 0$ and B is nonsingular. It follows that every set of $n-k$ columns in H_{RS} is linearly independent and, so, by Theorem 2.2 we have $d \geq n - k + 1$. □

We will learn more on Reed–Solomon codes in subsequent chapters.

4.2 The sphere-packing bound

Let F be an alphabet of size q. Recall that a sphere of radius t in F^n is a set of words in F^n at Hamming distance t or less from a given word in F^n. The number of words in such a sphere, or the *volume* of the sphere, is given by

$$V_q(n,t) = \sum_{i=0}^{t} \binom{n}{i} (q-1)^i$$

(see Problem 4.5). This quantity appears in our second bound, presented next.

Theorem 4.3 (The sphere-packing bound) *For any (n, M, d) code over an alphabet F of size q,*

$$M \cdot V_q(n, \lfloor (d-1)/2 \rfloor) \leq q^n . \tag{4.1}$$

Proof. The spheres in F^n of radius $t = \lfloor (d-1)/2 \rfloor$ that are centered at the codewords of an (n, M, d) code must be disjoint, as shown in Figure 1.7; indeed, if two such spheres intersected, then, by the triangle inequality, their centers would be at Hamming distance less than d from one another. It follows that the total volume of these spheres, which is given by the left-hand side of (4.1), is at most q^n. □

The sphere-packing bound is also known as the *Hamming bound.*

For a linear $[n, k, d]$ code over $\mathrm{GF}(q)$, the sphere-packing bound becomes

$$V_q(n, \lfloor (d-1)/2 \rfloor) \leq q^{n-k} ,$$

and specifically for $q = 2$ we get

$$\sum_{i=0}^{\lfloor (d-1)/2 \rfloor} \binom{n}{i} \leq 2^{n-k} .$$

A code is called *perfect* if it attains the sphere-packing bound. The minimum distance of such a code is necessarily odd (Problem 4.6).

Clearly, the whole space F^n over an alphabet F is a perfect code. Here are more examples of perfect codes.

Example 4.1 Consider the $[n, 1, n]$ repetition code over $\mathrm{GF}(2)$ where n is odd. We have

$$V_2(n, (n-1)/2) = \sum_{i=0}^{(n-1)/2} \binom{n}{i} = \frac{1}{2} \cdot \sum_{i=0}^{n} \binom{n}{i} = 2^{n-1} .$$

Hence, this code is perfect. □

Example 4.2 Consider the $[n, n-m, 3]$ Hamming code over $\mathrm{GF}(q)$ where $m > 1$ and $n = (q^m - 1)/(q - 1)$. Here

$$V_q(n, 1) = 1 + n(q - 1) = q^m = q^{n-k} ,$$

which implies that the code is perfect. □

Apart from those examples, there are only two other linear perfect codes:

- The $[23, 12, 7]$ Golay code over $\mathrm{GF}(2)$.
- The $[11, 6, 5]$ Golay code over $\mathrm{GF}(3)$.

These two codes are defined in the notes on this section at the end of the chapter and will also be discussed in Section 8.5.

Example 4.3 We consider codes of minimum distance 5 over $F = \mathrm{GF}(2)$. We have

$$V_2(n, 2) = 1 + n + \tbinom{n}{2} = \tfrac{1}{2}(n^2 + n + 2)$$

and, so, every $(n, M, 5)$ code $\mathcal{C}$ over F satisfies

$$M \le \frac{2^n}{\frac{1}{2}(n^2 + n + 2)} .$$

In particular, for $n = 2^m - 1$ we get

$$M \le \frac{2^{2^m - 1}}{2^{2m-1} - 2^{m-1} + 1} .$$

If, in addition, $\mathcal{C}$ is a linear $[n, k, 5]$ code over F then

$$\begin{aligned} k = \log_2 M &\le 2^m - 1 - \lceil \log_2 (2^{2m-1} - 2^{m-1} + 1) \rceil \\ &= 2^m - 2m \end{aligned}$$

(whenever $m > 1$). In comparison, the double-error-correcting code, $\mathcal{C}_3$, that we constructed in Section 3.8 satisfies

$$k \ge 2^m - 2m - 1 .$$

$\square$

4.3 The Gilbert–Varshamov bound

While the Singleton bound and the sphere-packing bound provide *necessary* conditions on the parameters of codes, the following theorem presents a *sufficient* condition for the existence of a linear code with given parameters.

Theorem 4.4 (The Gilbert–Varshamov bound) *Let $F = \mathrm{GF}(q)$ and let n, k, and d be positive integers such that*

$$V_q(n-1, d-2) < q^{n-k} .$$

Then there exists a linear $[n, k]$ code over F with minimum distance at least d.

Proof. We construct iteratively an $(n-k) \times n$ parity-check matrix H in which every $d-1$ columns are linearly independent, by starting with an $(n-k) \times (n-k)$ identity matrix and then adding a new column in each iteration. Assume that we have already selected the $\ell-1$ columns

$$\mathbf{h}_1, \mathbf{h}_2, \ldots, \mathbf{h}_{\ell-1} \tag{4.2}$$

to H. To maintain the property that every $d-1$ columns are linearly independent, a vector in F^{n-k} is eligible to be selected as an ℓth column to H, if and only if it cannot be expressed as a linear combination of any $d-2$

columns taken from (4.2). Equivalently, the *ineligible* columns are all the vectors in F^{n-k} that can be written as

$$(\mathbf{h}_1\, \mathbf{h}_2\, \ldots\, \mathbf{h}_{\ell-1})\mathbf{x} ,$$

for some vector $\mathbf{x} \in F^{\ell-1}$ of Hamming weight at most $d-2$. The number of such vectors $\mathbf{x}$, in turn, is $V_q(\ell-1, d-2)$, and this number is therefore an upper bound on the number of ineligible columns in F^{n-k} (an ineligible column may be associated with more than one vector $\mathbf{x}$). Hence, in order to be able to select an ℓth column to H, it is sufficient to require that

$$V_q(\ell-1, d-2) < q^{n-k} .$$

And under the conditions of the theorem, this inequality holds for every $\ell \leq n$. □

The following result shows that, in fact, most linear codes have parameters that are close to the Gilbert–Varshamov bound.

Theorem 4.5 *Let* $F = \mathrm{GF}(q)$ *and for positive integers* n*,* k*, and* d*, let* ρ *be given by*

$$\rho = \frac{q^k - 1}{q - 1} \cdot \frac{V_q(n, d-1)}{q^n} .$$

Then all but a fraction at most ρ *of the linear* $[n, k]$ *codes over* F *have minimum distance at least* d*.*

Proof. We first recall from Problem 2.2 that the number of generator matrices of a linear $[n, k]$ code over F is the same for all such codes. Therefore, it suffices to show that all but a fraction at most ρ of the $k \times n$ matrices over F generate linear $[n, k]$ codes over F with minimum distance at least d (note that the ensemble of $k \times n$ matrices over F includes also matrices whose rank is less than k, thus making the result slightly stronger than what we need).

Let $\mathcal{S}$ be the sphere of radius $d-1$ in F^n that is centered at $\mathbf{0}$ and let U denote the set of all nonzero vectors in F^k whose leading nonzero entry is 1. Clearly, $|\mathcal{S}| = V_q(n, d-1)$ and $|U| = (q^k - 1)/(q - 1)$. A $k \times n$ matrix over F is called "bad" if it does not generate a linear $[n, k]$ code over F with minimum distance at least d. Equivalently, G is bad if $\mathbf{u}G \in \mathcal{S}$ for some $\mathbf{u} \in U$. Assuming a uniform distribution over all $k \times n$ matrices over F, for every $\mathbf{u} \in U$ the random vector $\mathbf{u}G$ is uniformly distributed over F^n. Therefore,

$$\begin{aligned} \mathsf{Prob}\{ G \text{ is bad} \} &= \mathsf{Prob}\{ \mathbf{u}G \in \mathcal{S} \text{ for some } \mathbf{u} \in U \} \\ &\leq \sum_{\mathbf{u} \in U} \mathsf{Prob}\{ \mathbf{u}G \in \mathcal{S} \} = |U| \cdot \frac{|\mathcal{S}|}{q^n} = \rho , \end{aligned}$$

as claimed. □

It follows from Theorem 4.5 that if $V_q(n, d-1) \leq ((q-1)/2) \cdot q^{n-k}$, then more than half of the linear $[n, k]$ codes over $\mathrm{GF}(q)$ have minimum distance at least d. Yet, the proof of the theorem is not constructive as it does not imply an efficient algorithm for finding even one such code for given n, k, d, and q.

4.4 MacWilliams' identities

Unlike the Singleton bound or the sphere-packing bound, this section presents constraints on the parameters of codes through *equalities* rather than inequalities. Yet, as we point out in the sequel, such equalities can serve as a basis for new bounding techniques. The discussion here will be limited to linear codes, even though generalizations are known for the non-linear case as well.

Let $\mathcal{C}$ be a linear $[n, k, d]$ code over $F = \mathrm{GF}(q)$. The *(Hamming) weight distribution* of $\mathcal{C}$ is a list $(W_i)_{i=0}^n$, where W_i equals the number of codewords in $\mathcal{C}$ of Hamming weight i. Clearly, $W_0 = 1$ and $W_i = 0$ for $1 \leq i < d$. The respective generating function,

$$W_{\mathcal{C}}(z) = \sum_{i=0}^{n} W_i z^i ,$$

is called the *(Hamming) weight enumerator* of $\mathcal{C}$.

Example 4.4 Let $F = \mathrm{GF}(q)$ and let $n = (q^m - 1)/(q - 1)$ where m is a positive integer. The dual code of the $[n, n-m, 3]$ Hamming code over F is a linear $[n, m, q^{m-1}]$ code $\mathcal{C}$ (known as the *simplex code*) whose nonzero codewords all have Hamming weight q^{m-1} (see Problem 2.18). The weight enumerator of $\mathcal{C}$ is then given by

$$W_{\mathcal{C}}(z) = 1 + (q^m - 1) \cdot z^{q^{m-1}} .$$

□

In what follows, we will find it convenient to consider also the *homogeneous weight enumerator*, which is given by the bivariate polynomial

$$W_{\mathcal{C}}^{\mathrm{h}}(x, z) = x^n \cdot W_{\mathcal{C}}(z/x) = \sum_{i=0}^{n} W_i x^{n-i} z^i .$$

The next theorem relates the weight distribution of the dual code $\mathcal{C}^\perp$ to that of $\mathcal{C}$. The proof of the theorem makes use of the properties of *characters* of F. Recall from Problem 3.36 that an additive character of F is a nonzero mapping $\chi : F \to \mathbb{C}$ that satisfies

$$\chi(\beta + \gamma) = \chi(\beta)\chi(\gamma) \qquad \text{for every } \beta, \gamma \in F\,.$$

There are q distinct additive characters of F, including the trivial character, which takes the value 1 for every field element. For every nontrivial additive character χ of F we have $\chi(0) = 1$ and

$$\sum_{\beta \in F} \chi(\beta) = 1 + \sum_{\beta \in F^*} \chi(\beta) = 0\,. \tag{4.3}$$

Theorem 4.6 (MacWilliams' Theorem) *For every linear $[n, k, d]$ code $\mathcal{C}$ over $F = \mathrm{GF}(q)$,*

$$W_{\mathcal{C}^\perp}(z) = \frac{1}{|\mathcal{C}|} \cdot W^{\mathrm{h}}_{\mathcal{C}}(1 + (q-1)z, 1 - z)\,. \tag{4.4}$$

Proof. Let $\chi : F \to \mathbb{C}$ be a nontrivial additive character of F. It follows from (4.3) that for every row vector $\mathbf{u} \in F^n$,

$$\sum_{\mathbf{c} \in \mathcal{C}} \chi(\mathbf{u} \cdot \mathbf{c}^T) = \begin{cases} |\mathcal{C}| & \text{if } \mathbf{u} \in \mathcal{C}^\perp \\ 0 & \text{otherwise} \end{cases}$$

(indeed, a simple generalization of Problem 2.4 implies that when $\mathbf{u} \notin \mathcal{C}^\perp$, each element of F is an image under the mapping $\mathbf{c} \mapsto \mathbf{u} \cdot \mathbf{c}^T$ of exactly q^{k-1} codewords $\mathbf{c} \in \mathcal{C}$).

Expressing the weight enumerator of $\mathcal{C}^\perp$ as

$$W_{\mathcal{C}^\perp}(z) = \sum_{\mathbf{u} \in \mathcal{C}^\perp} z^{\mathrm{w}(\mathbf{u})}\,,$$

we obtain

$$\begin{aligned} W_{\mathcal{C}^\perp}(z) &= \frac{1}{|\mathcal{C}|} \sum_{\mathbf{u} \in F^n} \Big(\sum_{\mathbf{c} \in \mathcal{C}} \chi(\mathbf{u} \cdot \mathbf{c}^T) \Big) \cdot z^{\mathrm{w}(\mathbf{u})} \\ &= \frac{1}{|\mathcal{C}|} \sum_{\mathbf{c} \in \mathcal{C}} \Big(\sum_{\mathbf{u} \in F^n} \chi(\mathbf{u} \cdot \mathbf{c}^T) \cdot z^{\mathrm{w}(\mathbf{u})} \Big)\,. \end{aligned}$$

Writing $\mathbf{u} = (u_1\ u_2\ \ldots\ u_n)$ and $\mathbf{c} = (c_1\ c_2\ \ldots\ c_n)$, the definition of additive characters implies that

$$\chi(\mathbf{u} \cdot \mathbf{c}^T) = \chi\Big(\textstyle\sum_{j=1}^n u_j c_j \Big) = \displaystyle\prod_{j=1}^n \chi(u_j c_j)\,.$$

Hence,

$$\begin{aligned} W_{\mathcal{C}^\perp}(z) &= \frac{1}{|\mathcal{C}|} \sum_{\mathbf{c} \in \mathcal{C}} \sum_{\mathbf{u} \in F^n} \prod_{j=1}^{n} \left(\chi(u_j c_j) \cdot z^{\mathsf{w}(u_j)} \right) \\ &= \frac{1}{|\mathcal{C}|} \sum_{\mathbf{c} \in \mathcal{C}} \prod_{j=1}^{n} \Bigl(\sum_{u \in F} \chi(u \cdot c_j) \cdot z^{\mathsf{w}(u)} \Bigr) . \end{aligned} \tag{4.5}$$

Now, from (4.3) we have

$$\sum_{u \in F^*} \chi(u \cdot c_j) = \begin{cases} q-1 & \text{if } c_j = 0 \\ -1 & \text{otherwise} \end{cases} .$$

Therefore,

$$\sum_{u \in F} \chi(u \cdot c_j) \cdot z^{\mathsf{w}(u)} = \chi(0) + z \cdot \sum_{u \in F^*} \chi(u \cdot c_j) = \begin{cases} 1 + (q-1)z & \text{if } c_j = 0 \\ 1 - z & \text{otherwise} \end{cases} ;$$

so, for every codeword $\mathbf{c} = (c_1 \; c_2 \; \ldots \; c_n)$ in $\mathcal{C}$ we get

$$\prod_{j=1}^{n} \Bigl(\sum_{u \in F} \chi(u \cdot c_j) \cdot z^{\mathsf{w}(u)} \Bigr) = (1 + (q-1)z)^{n - \mathsf{w}(\mathbf{c})} (1 - z)^{\mathsf{w}(\mathbf{c})} .$$

Combining the latter equality with (4.5) thus yields

$$\begin{aligned} W_{\mathcal{C}^\perp}(z) &= \frac{1}{|\mathcal{C}|} \sum_{\mathbf{c} \in \mathcal{C}} (1 + (q-1)z)^{n - \mathsf{w}(\mathbf{c})} (1 - z)^{\mathsf{w}(\mathbf{c})} \\ &= \frac{1}{|\mathcal{C}|} \cdot W_{\mathcal{C}}^{\mathrm{h}}(1 + (q-1)z, 1 - z) , \end{aligned}$$

as claimed. □

Example 4.5 Letting $\mathcal{C}$ be the simplex code in Example 4.4, the weight enumerator of the Hamming code of length $n = (q^m - 1)/(q - 1)$ over $\mathrm{GF}(q)$ is given by

$$\begin{aligned} W_{\mathcal{C}^\perp}(z) &= (1 + (q-1)z)^{(q^{m-1}-1)/(q-1)} \\ &\quad \cdot \left(\frac{1}{q^m} \cdot (1 + (q-1)z)^{q^{m-1}} + \frac{q^m - 1}{q^m} \cdot (1 - z)^{q^{m-1}} \right) . \end{aligned}$$

When $q = 2$ this expression becomes

$$W_{\mathcal{C}^\perp}(z) = \frac{1}{n+1} \cdot (1 + z)^{(n-1)/2} \left((1 + z)^{(n+1)/2} + n \cdot (1 - z)^{(n+1)/2} \right) .$$

See also Problem 4.8. □

Let $(W_i)_{i=0}^n$ be the weight distribution of a linear $[n,k,d]$ code $\mathcal{C}$ over $F = \mathrm{GF}(q)$ and $(W_i^\perp)_{i=0}^n$ be the weight distribution of the dual code $\mathcal{C}^\perp$ of $\mathcal{C}$. We next use Theorem 4.6 to express each value $W_i^\perp$ as a (linear) function of the values $W_0, W_1, \ldots, W_n$.

For every $i = 0, 1, \ldots, n$ we can write

$$(1 + (q-1)z)^{n-i}(1-z)^i = \sum_{\ell=0}^{n} \mathcal{K}_\ell(i)\, z^\ell\,, \tag{4.6}$$

where

$$\mathcal{K}_\ell(i) = \mathcal{K}_\ell(i; n, q) = \sum_{r=0}^{\ell} \binom{i}{r}\binom{n-i}{\ell-r}(-1)^r (q-1)^{\ell-r}$$

(we define a binomial coefficient $\binom{a}{b}$ to be 0 if $a < b$). By Theorem 4.6 we have

$$\sum_{i=0}^{n} W_i^\perp z^i = \frac{1}{q^k}\sum_{i=0}^{n} W_i \cdot (1 + (q-1)z)^{n-i}(1-z)^i\,. \tag{4.7}$$

Substituting (4.6) into (4.7) yields

$$\sum_{i=0}^{n} W_i^\perp z^i = \frac{1}{q^k}\sum_{i=0}^{n} W_i \sum_{\ell=0}^{n} \mathcal{K}_\ell(i)\, z^\ell$$

or

$$\sum_{\ell=0}^{n} W_\ell^\perp z^\ell = \frac{1}{q^k}\sum_{\ell=0}^{n} z^\ell \sum_{i=0}^{n} \mathcal{K}_\ell(i)\, W_i\,.$$

For every $\ell = 0, 1, \ldots, n$, the coefficients of z^ℓ on both sides of the last equation must be equal. Hence,

$$\boxed{W_\ell^\perp = \frac{1}{q^k}\sum_{i=0}^{n} \mathcal{K}_\ell(i)\, W_i}\,, \quad 0 \le \ell \le n\,. \tag{4.8}$$

Obviously, $W_\ell^\perp$ is nonnegative for $0 \le \ell \le n$. Noting that

$$\mathcal{K}_\ell(0) = \binom{n}{\ell}(q-1)^\ell$$

and recalling that

$$W_0 = 1 \quad \text{and} \quad W_1 = W_2 = \ldots = W_{d-1} = 0\,,$$

we get from (4.8) that the values $W_d, W_{d+1}, \ldots, W_n$ must satisfy the following set of inequalities:

$$\sum_{i=d}^{n} \mathcal{K}_\ell(i)\, W_i \ge -\binom{n}{\ell}(q-1)^\ell\,, \quad 0 \le \ell \le n\,.$$

On the other hand, we have $|\mathcal{C}| = 1 + \sum_{i=d}^{n} W_i$. This, in turn, leads to an upper bound on the size of every linear code of length n and minimum distance d over $\mathrm{GF}(q)$: the size of such a code is at most

$$1 + \max \sum_{i=d}^{n} w_i \; , \tag{4.9}$$

where the maximum is taken over all integers $w_d, w_{d+1}, \ldots, w_n$ that satisfy the following set of $2n-d+1$ linear constraints:

$$\left\{ \begin{array}{rcll} w_i & \geq & 0 & , \quad d \leq i \leq n \\ \sum_{i=d}^{n} \mathcal{K}_\ell(i)\, w_i & \geq & -\binom{n}{\ell}(q-1)^\ell \, , & \quad 1 \leq \ell \leq n \end{array} \right. . \tag{4.10}$$

Finding the maximizing integers w_i in (4.9) under the constraints (4.10) is an instance of a computational problem known as *integer programming*. The general integer programming problem is known to be intractable (i.e., NP-hard). However, this impediment can be circumvented—at the expense of weakening the computed bound—by letting the variables w_i take rational values rather than restricting them to be integers. We thus obtain an instance of a bounding technique known as the *linear programming bound*.

We can expand the binomial coefficients in the definition of $\mathcal{K}_\ell(i)$, in which case we will end up with an expression that is a polynomial in the integer variable i. If we now replace i by a *real* indeterminate y, we obtain a real polynomial, $\mathcal{K}_\ell(y)$, which is known as a *Krawtchouk polynomial*. The first three Krawtchouk polynomials are

$$\begin{aligned} \mathcal{K}_0(y) &= 1 \, , \\ \mathcal{K}_1(y) &= n(q-1) - qy \, , \qquad \text{and} \\ \mathcal{K}_2(y) &= \tbinom{n}{2}(q-1)^2 - \tfrac{1}{2}((2n-1)(q-1)+1)qy + \tfrac{1}{2}q^2y^2 \, . \end{aligned}$$

We end this section by presenting another set of identities that relates the weight distribution of $\mathcal{C}$ with that of its dual code $\mathcal{C}^\perp$. Multiplying both sides of (4.7) by z^{-n} and substituting $z = 1/(\xi + 1)$ yields

$$\sum_{i=0}^{n} W_i^\perp \, (\xi+1)^{n-i} = q^{-k} \cdot \sum_{i=0}^{n} W_i \cdot (q+\xi)^{n-i} \xi^i \, .$$

We now compare coefficients of every power of ξ in both sides of the equation, thereby getting

$$\sum_{i=0}^{n-\ell} \binom{n-i}{\ell} W_i^\perp = q^{n-k-\ell} \sum_{i=0}^{\ell} \binom{n-i}{\ell-i} W_i \, , \quad 0 \leq \ell \leq n \, .$$

Reversing the roles of $\mathcal{C}$ and $\mathcal{C}^\perp$ and noting that $\binom{n-i}{\ell-i} = \binom{n-i}{n-\ell}$, we obtain the linear identities

$$\boxed{\sum_{i=0}^{n-\ell} \binom{n-i}{\ell} W_i = q^{k-\ell} \sum_{i=0}^{\ell} \binom{n-i}{n-\ell} W_i^\perp} \; , \quad 0 \le \ell \le n \, . \tag{4.11}$$

Equations (4.4), (4.7), (4.8), and (4.11) are known by the collective name *MacWilliams' identities.*

Example 4.6 Let $\mathcal{C}$ be a linear $[n, k, d]$ MDS code over $F = \mathrm{GF}(q)$. We obtain by MacWilliams' identities a complete characterization of the weight distribution of $\mathcal{C}$, as a function of n, k, and q. Clearly, $W_0 = 1$ and $W_i = 0$ for $1 \le i \le n-k$. Now, the dual code of $\mathcal{C}$ is also MDS (Problem 4.1) and, so, $W_0^\perp = 1$ and $W_i^\perp = 0$ for $1 \le i \le k$. We thus get from (4.11),

$$\binom{n}{\ell} + \sum_{i=n-k+1}^{n-\ell} \binom{n-i}{\ell} W_i = q^{k-\ell} \binom{n}{\ell} \, , \quad 0 \le \ell \le k \, ,$$

or

$$\sum_{i=d}^{n-\ell} \binom{n-i}{\ell} W_i = \binom{n}{\ell} (q^{k-\ell} - 1) \, , \quad 0 \le \ell < k \, .$$

As shown in Problem 4.18, this set of linear equations can be iteratively solved for the values $W_d, W_{d+1}, \ldots, W_n$ to yield

$$\begin{aligned} W_i &= \binom{n}{i} \sum_{s=0}^{i-d} \binom{i}{s} (-1)^s (q^{i+1-d-s} - 1) \\ &= \binom{n}{i} (q-1) \sum_{s=0}^{i-d} \binom{i-1}{s} (-1)^s q^{i-d-s} \, , \quad d \le i \le n \, . \end{aligned}$$

It is interesting to observe that for any given n, k, and $F = \mathrm{GF}(q)$, all linear $[n, k]$ MDS codes over F have the same weight distribution, regardless of the code construction. □

4.5 Asymptotic bounds

Let $\mathcal{C}$ be an (n, M, d) code over an alphabet of size q. The *relative minimum distance* of $\mathcal{C}$ is the ratio $\delta = d/n$.

In this section, we derive asymptotic bounds: we find relations between δ and the rate $R = (\log_q M)/n$ as the code length tends to infinity. Hereafter, $o(1)$ stands for an expression that goes to zero as $n \to \infty$ (yet this expression may depend on q or δ).

Starting with the Singleton bound, from

$$d \le n - (\log_q M) + 1$$

we obtain

$$\delta \le 1 - R + o(1)$$

or

$$R \le 1 - \delta + o(1) .$$

For the asymptotic versions of the sphere-packing bound and the Gilbert–Varshamov bound, we will need estimates of the volume

$$V_q(n,t) = \sum_{i=0}^{t} \binom{n}{i} (q-1)^i .$$

Those estimates will make use of the *q-ary entropy function* $\mathsf{H}_q : [0,1] \to [0,1]$, which is defined by

$$\mathsf{H}_q(x) = -x \log_q x - (1-x)\log_q(1-x) + x \log_q(q-1) ,$$

where $\mathsf{H}_q(0) = 0$ and $\mathsf{H}_q(1) = \log_q(q-1)$. One can verify that the function $x \mapsto \mathsf{H}_q(x)$ is strictly $\cap$-concave, nonnegative, and attains a maximum value of 1 at $x = 1-(1/q)$. These properties allow us, in turn, to define the inverse function $z \mapsto \mathsf{H}_q^{-1}(z)$ on the interval $[0, 1-(1/q)]$. Note that for $q = 2$, the function $\mathsf{H}_2(x)$ coincides with the binary entropy function $\mathsf{H}(x)$, which was defined in Section 1.4.3.

Lemma 4.7 *For* $0 \le t/n \le 1-(1/q)$,

$$V_q(n,t) \le q^{n\mathsf{H}_q(t/n)} .$$

Proof. The case $t = 0$ is obvious. Assume now that $t > 0$ and write $\theta = t/n$. Then,

$$\begin{aligned}
& q^{-n\mathsf{H}_q(\theta)} \cdot V_q(n,t) \\
&\quad = \theta^t(1-\theta)^{n-t}(q-1)^{-t} \cdot \sum_{i=0}^{t} \binom{n}{i} (q-1)^i \\
&\quad \overset{\theta \le 1-(1/q)}{\le} \theta^t(1-\theta)^{n-t}(q-1)^{-t} \cdot \sum_{i=0}^{n} \binom{n}{i} (q-1)^i \left(\frac{\theta}{(1-\theta)(q-1)} \right)^{i-t} \\
&\quad = \sum_{i=0}^{n} \binom{n}{i} \theta^i (1-\theta)^{n-i} \\
&\quad = (\theta + (1-\theta))^n = 1 ,
\end{aligned}$$

namely, $V_q(n,t) \le q^{n\mathsf{H}_q(\theta)}$. □

Lemma 4.8 *For integers* $0 \le t \le n$,

$$V_q(n,t) \ge \binom{n}{t}(q-1)^t \ge \frac{1}{n+1} \cdot q^{n\mathsf{H}_q(t/n)} .$$

Proof. As the cases $t = 0$ and $t = n$ are obvious, we assume hereafter in the proof that $0 < t < n$. Write $\theta = t/n$ and for $i = 0, 1, \ldots, n$, define A_i by

$$A_i = \binom{n}{i}\theta^i(1-\theta)^{n-i}$$

(this expression is the probability of having i successes among n statistically independent Bernoulli trials, each with probability θ of success).

We first show that A_i attains its maximum when $i = t$. Indeed,

$$\frac{A_{i+1}}{A_i} = \frac{\binom{n}{i+1}\theta^{i+1}(1-\theta)^{n-i-1}}{\binom{n}{i}\theta^i(1-\theta)^{n-i}} = \frac{n-i}{i+1} \cdot \frac{\theta}{1-\theta} .$$

So, $A_{i+1}/A_i < 1$ if and only if $i \ge t$.

It follows that

$$(n+1) \cdot A_t \ge \sum_{i=0}^{n} A_i = (\theta + (1-\theta))^n = 1$$

and, so,

$$A_t \ge \frac{1}{n+1} .$$

On the other hand,

$$\begin{aligned} q^{-n\mathsf{H}_q(\theta)} \cdot V_q(n,t) &\ge q^{-n\mathsf{H}_q(\theta)} \cdot \binom{n}{t}(q-1)^t \\ &= \theta^t(1-\theta)^{n-t}(q-1)^{-t} \cdot \binom{n}{t}(q-1)^t \\ &= \binom{n}{t}\theta^t(1-\theta)^{n-t} = A_t . \end{aligned}$$

Hence,

$$V_q(n,t) \ge \binom{n}{t}(q-1)^t = A_t \cdot q^{n\mathsf{H}_q(t/n)} \ge \frac{1}{n+1} \cdot q^{n\mathsf{H}_q(t/n)} ,$$

as claimed. □

We mention that by using the Stirling formula for bounding the factorial of integers, one can improve Lemma 4.8 to

$$\binom{n}{t}(q-1)^t \ge \frac{1}{\sqrt{8t(1-(t/n))}} \cdot q^{n\mathsf{H}_q(t/n)} .$$

However, for our purposes herein, Lemma 4.8 will suffice.

Theorem 4.9 (Asymptotic version of the sphere-packing bound) *For every $(n, q^{nR}, \delta n)$ code over an alphabet of q elements,*

$$R \le 1 - \mathsf{H}_q(\delta/2) + o(1) .$$

Proof. Write $t = \lfloor (\delta n - 1)/2 \rfloor$. By Theorem 4.3,

$$q^{nR} \cdot V_q(n,t) \le q^n . \tag{4.12}$$

Now, by Lemma 4.8,

$$V_q(n,t) \ge \frac{1}{n+1} \cdot q^{n\mathsf{H}_q(t/n)} \ge \frac{1}{n+1} \cdot q^{n\mathsf{H}_q(\delta/2-(1/n))} , \tag{4.13}$$

where we have used the fact that the q-ary entropy function is increasing in the range $[0, 1/2)$. From (4.12) and (4.13) we get

$$R \le 1 - \mathsf{H}_q(\delta/2 - (1/n)) + o(1) ,$$

and the theorem now follows by the continuity of the entropy function. □

Theorem 4.10 (Asymptotic version of the Gilbert–Varshamov bound) *Let $F = \mathrm{GF}(q)$, let n and nR be positive integers, and let δ be a real in $(0, 1-(1/q)]$ that satisfies*

$$R \le 1 - \mathsf{H}_q(\delta) .$$

Then there exists a linear $[n, nR, {\ge}\delta n]$ code over F.

Proof. By Theorem 4.4, such a code exists whenever

$$V_q\left(n, \lceil \delta n \rceil - 1\right) \le q^{n(1-R)} .$$

The theorem now follows from Lemma 4.7. □

Our next asymptotic bound, which will be stated in Theorem 4.12, improves on the sphere-packing bound. A key ingredient in the proof of that theorem is the following result, which presents a bound that is important on its own (a somewhat related result is given also in Problem 4.23, and we will use that problem in our next proof).

Proposition 4.11 (The Johnson bound) *Let C be an $(n, M, \delta n)$ code over an Abelian group F of size q and suppose that there is a real $\theta \in (0, 1-(1/q)]$ such that each codeword in C has Hamming weight at most θn. Then,*

$$\delta \le \frac{M}{M-1} \cdot \left(2\theta - \tfrac{q}{q-1}\theta^2\right) .$$

Proof. Let D be the average distance between the codewords in $\mathcal{C}$; that is,

$$D = \frac{1}{M(M-1)} \sum_{\mathbf{c},\mathbf{c}' \in \mathcal{C} \,:\, \mathbf{c} \neq \mathbf{c}'} \mathsf{d}(\mathbf{c}, \mathbf{c}') .$$

Construct an $M \times n$ array whose rows are the codewords of $\mathcal{C}$, and for every $a \in F$, let $x_{a,j}$ be the number of times that the element a appears in the jth column of this array. By part 3 of Problem 4.23 we get

$$\sum_{\mathbf{c},\mathbf{c}' \in \mathcal{C} \,:\, \mathbf{c} \neq \mathbf{c}'} \mathsf{d}(\mathbf{c}, \mathbf{c}') = \sum_{j=1}^{n} \Bigl(M^2 - \sum_{a \in F} x_{a,j}^2 \Bigr) .$$

Since $\sum_{a \in F} x_{a,j} = M$ for every column index j, we can eliminate the unknown values $x_{0,j}$ to obtain

$$\sum_{\mathbf{c},\mathbf{c}' \in \mathcal{C} \,:\, \mathbf{c} \neq \mathbf{c}'} \mathsf{d}(\mathbf{c}, \mathbf{c}') = \sum_{j=1}^{n} \left(2M \Bigl(\sum_{a \in F^*} x_{a,j} \Bigr) - \Bigl(\sum_{a \in F^*} x_{a,j} \Bigr)^2 - \Bigl(\sum_{a \in F^*} x_{a,j}^2 \Bigr) \right) , \tag{4.14}$$

where $F^* = F \setminus \{0\}$. On the other hand, each element of $\mathcal{C}$ has Hamming weight at most θn. Therefore,

$$\sum_{j=1}^{n} \sum_{a \in F^*} x_{a,j} \leq M \theta n . \tag{4.15}$$

Assuming that $\theta \leq 1 - (1/q)$, the maximum of the right-hand side of (4.14) over the *real* values $x_{a,j}$ that satisfy the constraint (4.15) is attained when $x_{a,j} = M\theta/(q-1)$ for every $a \in F^*$ (compare with part 2 of Problem 4.23). Substituting these maximizing values into the right-hand side of (4.14) we thus obtain

$$\sum_{\mathbf{c},\mathbf{c}' \in \mathcal{C} \,:\, \mathbf{c} \neq \mathbf{c}'} \mathsf{d}(\mathbf{c}, \mathbf{c}') \leq \sum_{j=1}^{n} \left(2M^2\theta - \tfrac{q}{q-1} M^2 \theta^2 \right) = M^2 \cdot (2\theta - \tfrac{q}{q-1}\theta^2) n .$$

Hence,

$$D = \frac{1}{M(M-1)} \sum_{\mathbf{c},\mathbf{c}' \in \mathcal{C} \,:\, \mathbf{c} \neq \mathbf{c}'} \mathsf{d}(\mathbf{c}, \mathbf{c}') \leq \frac{M}{M-1} \cdot (2\theta - \tfrac{q}{q-1}\theta^2) n .$$

The result now follows by observing that the minimum distance δn is bounded from above by the average distance D. $\square$

Theorem 4.12 (The Elias bound) *For every $(n, q^{nR}, \delta n)$ code with $\delta \leq 1 - (1/q)$ over an alphabet of size q,*

$$R \leq 1 - \mathsf{H}_q \left(\tfrac{q-1}{q} \left(1 - \sqrt{1 - \tfrac{q}{q-1}\delta} \right) \right) + o(1) .$$

Proof. Let $\mathcal{C}$ be an $(n, q^{nR}, \delta n)$ code over an alphabet F of q elements; without loss of generality we can assume that F is an Abelian group. We first show that for every $t \in \{0, 1, 2, \ldots, n\}$ there is a sphere $\mathcal{S} \subseteq F^n$ of radius t such that the size M of the intersection $\mathcal{S} \cap \mathcal{C}$ satisfies

$$M = |\mathcal{S} \cap \mathcal{C}| \geq q^{n(R-1)} \cdot V_q(n, t) . \tag{4.16}$$

There are q^n spheres of radius t in F^n, and each codeword of $\mathcal{C}$ belongs to $V_q(n, t)$ spheres. Therefore,

$$\sum_{\mathcal{S}} |\mathcal{S} \cap \mathcal{C}| = \sum_{\mathbf{c} \in \mathcal{C}} |\{\mathcal{S} : \mathbf{c} \in \mathcal{S}\}| = |\mathcal{C}| \cdot V_q(n, t) ,$$

where $\mathcal{S}$ ranges over all spheres of radius t in F^n. It follows that the average number of codewords in a sphere equals

$$\frac{1}{q^n} \sum_{\mathcal{S}} |\mathcal{S} \cap \mathcal{C}| = \frac{1}{q^n} \cdot |\mathcal{C}| \cdot V_q(n, t) = q^{n(R-1)} \cdot V_q(n, t)$$

and, so, there must be at least one sphere of radius t that satisfies (4.16).

Now, for a given $t = \theta n \leq (1 - (1/q))n$, let $\mathcal{S}$ be a sphere of radius t that satisfies (4.16). By translating the code $\mathcal{C}$, we can assume without loss of generality that $\mathcal{S}$ is centered at $\mathbf{0}$. Noting that δ is bounded from above by the relative minimum distance of the code $\mathcal{S} \cap \mathcal{C}$, we can obtain an upper bound on δ by applying Proposition 4.11 to $\mathcal{S} \cap \mathcal{C}$; that is,

$$\delta \leq \frac{M}{M-1} \cdot (2\theta - \tfrac{q}{q-1}\theta^2) . \tag{4.17}$$

We now select

$$t = \theta n = \left\lceil n \mathsf{H}_q^{-1}\Bigl(1 - R + (2 \log_q(n+1))/n\Bigr) \right\rceil ,$$

where we assume that R is at a sufficient margin from zero so that the argument of the inverse function $\mathsf{H}_q^{-1}(\cdot)$ is less than 1 and θ is less than $1-(1/q)$ (otherwise, it would mean that $R = o(1)$, in which case the theorem trivially holds). Clearly,

$$\mathsf{H}_q(\theta) = 1 - R + o(1) , \tag{4.18}$$

and by (4.16) and Lemma 4.8 we also have $M \geq n+1$. From (4.17) we readily obtain

$$\delta \leq 2\theta - \tfrac{q}{q-1}\theta^2 + o(1)$$

or

$$\theta \geq \tfrac{q-1}{q} \left(1 - \sqrt{1 - \tfrac{q}{q-1}\delta}\right) + o(1) .$$

The theorem follows by combining the latter inequality with (4.18). $\square$

For $\delta = 1 - (1/q)$ the bound of Theorem 4.12 becomes $R = o(1)$. This implies that when $\delta \geq 1-(1/q)$, the rate must go to zero as the code length increases.

The asymptotic bounds that relate the largest attainable rate R to the relative minimum distance δ can be verified to satisfy

$$1 - \mathsf{H}_q(\delta) \leq 1 - \mathsf{H}_q\left(\tfrac{q-1}{q}\left(1 - \sqrt{1 - \tfrac{q}{q-1}\delta}\right)\right) \leq 1 - \mathsf{H}_q(\delta/2)$$

whenever $\delta \leq 1 - (1/q)$; namely,

$$\text{Gilbert–Varshamov} \leq \text{Elias} \leq \text{Sphere-packing} .$$

The Singleton bound, $R \leq 1 - \delta + o(1)$, is generally weaker than the sphere-packing and Elias bounds for small values of q; on the other hand, when $q \to \infty$, it actually coincides with the Gilbert–Varshamov bound (up to an additive term $o(1)$).

Specializing now to the binary case, the bounds are plotted in Figure 4.1 for $q = 2$. The Elias bound is not the best upper bound currently known. We mention here without proof another bound, called the (first) McEliece–Rodemich–Rumsey–Welch (MRRW) bound, which takes for $F = \mathrm{GF}(2)$ the form

$$R \leq \mathsf{H}\left(\tfrac{1}{2} - \sqrt{\delta(1-\delta)}\right) + o(1) .$$

This bound is shown in Figure 4.1 as a thin curve, and it is currently the best known upper bound for values of δ that are greater than approximately 0.273. The MRRW bound is based on the linear programming technique, which was mentioned in Section 4.4. (The Elias bound is better than this bound for values of δ that are smaller than approximately 0.150; however, a second upper bound due to McEliece *et al.* supersedes both their first bound and the Elias bound for $\delta \in (0, \approx 0.273)$.)

4.6 Converse Coding Theorem

In the preceding section, we studied the relationship between the code rate and the relative minimum distance of the code. In this section, we consider instead the relationship between the code rate and the decoding error probability with respect to the q-ary symmetric channel.

Hereafter we assume that F is an Abelian group of size q and that $S = (F, F, \mathsf{Prob})$ is the memoryless q-ary symmetric channel with crossover probability $p \in (0, 1-(1/q)]$. The *capacity* of S is defined by

$$\mathsf{cap}(S) = 1 - \mathsf{H}_q(p) . \tag{4.19}$$

(To be more precise, the definition of the notion of capacity is more general, as discussed in the notes on Section 1.4; but we have also shown there that the

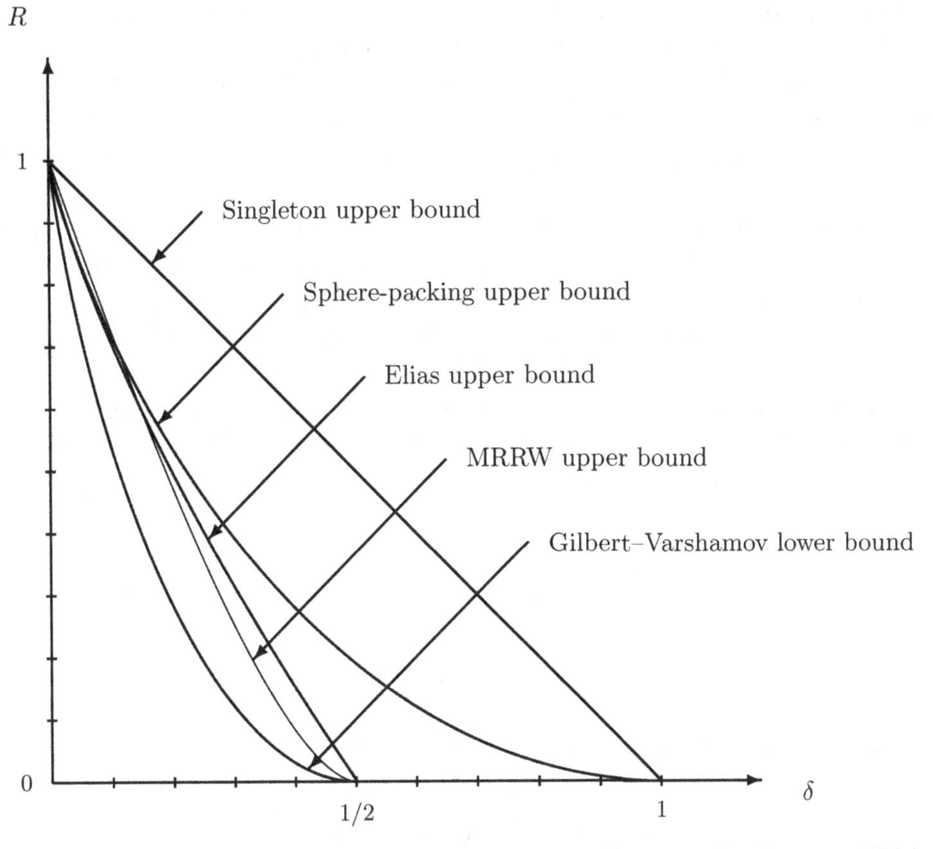

Figure 4.1. Asymptotic bounds on the largest attainable rates of codes over GF(2).

general definition becomes (4.19) for the special case of the q-ary symmetric channel S.)

The main result to be proved in this section is the Shannon Converse Coding Theorem for the q-ary symmetric channel, which states that when the code rate exceeds $\mathsf{cap}(S)$, the decoding error probability must approach 1 for sufficiently long codes. The range of rates below capacity will be the subject of Section 4.7, where we state and prove the Shannon (Direct) Coding Theorem. Both theorems were already mentioned in Section 1.4 for the special case of the binary symmetric channel.

For $\theta \in [0, 1]$ define the *(information) divergence* (or *Kullback–Leibler distance*) with respect to p by

$$\mathsf{D}_q(\theta \| p) = \theta \log_q \left(\frac{\theta}{p} \right) + (1-\theta) \log_q \left(\frac{1-\theta}{1-p} \right) ,$$

where $\mathsf{D}_q(0\|p) = -\log_q(1-p)$ and $\mathsf{D}_q(1\|p) = -\log_q p$. (We again specialize here from a broader term: information divergence is defined in general between two *distributions*. In our case, those distributions are Bernoulli trials, with probabilities θ and p, respectively, of success.)

By simple differentiation one can see that the function $\theta \rightarrow \mathsf{D}_q(\theta \| p)$ is strictly $\cup$-convex with a minimum at $\theta = p$. At that minimum, the function is zero, thereby implying that $\mathsf{D}_q(\theta \| p)$ is strictly positive for all other values of θ.

Denote by $\mathcal{S}_q(n, t)$ the set of all words in F^n of Hamming weight at most t; for convenience, here we also allow t to take non-integer real values (and the same applies to $V_q(n, t)$, which equals $|\mathcal{S}_q(n, t)|$).

The next lemma will play a key role in our subsequent analysis.

Lemma 4.13 (Large deviation estimates) *Let* $\mathbf{e}$ *denote a random error word in* F^n *that is generated by the channel* S. *Then the following conditions hold for every real* $\theta \in [0, 1]$*:*

(i) For $\theta \le p$,

$$\mathsf{Prob}\{\, \mathbf{e} \in \mathcal{S}_q(n, \theta n) \,\} \le q^{-n \mathsf{D}_q(\theta \| p)} .$$

(ii) For $\theta \ge p$,

$$\mathsf{Prob}\{\, \mathbf{e} \notin \mathcal{S}_q(n, \theta n) \,\} \le q^{-n \mathsf{D}_q(\theta \| p)} .$$

Proof. We start with part (i). For every real $z \in (0, 1]$ we have,

$$\begin{aligned}
\mathsf{Prob}\{\, \mathbf{e} \in \mathcal{S}_q(n, \theta n) \,\} &= \sum_{i=0}^{\lfloor \theta n \rfloor} \binom{n}{i} p^i (1-p)^{n-i} \\
&\le \sum_{i=0}^{n} \binom{n}{i} p^i (1-p)^{n-i} z^{i - \theta n} \\
&= z^{-\theta n} \cdot \sum_{i=0}^{n} \binom{n}{i} (pz)^i (1-p)^{n-i} \\
&= \left(z^{-\theta} (pz + 1-p) \right)^n
\end{aligned}$$

(compare with the proof of Lemma 4.7). We now select

$$z = \frac{\theta(1-p)}{p(1-\theta)} ,$$

in which case

$$\left(z^{-\theta} (pz + 1-p) \right)^n = \left(\frac{p^\theta (1-p)^{1-\theta}}{\theta^\theta (1-\theta)^{1-\theta}} \right)^n = q^{-n \mathsf{D}_q(\theta \| p)}$$

(note that this holds also for $\theta = 0$ if we define 0^0 to be 1).

As for part (ii), observe that

$$\mathsf{Prob}\{\, \mathbf{e} \notin \mathcal{S}_q(n, \theta n) \,\} \le \sum_{i=0}^{\lfloor (1-\theta)n \rfloor} \binom{n}{i} (1-p)^i p^{n-i} \;.$$

Hence, by part (i) we have

$$\mathsf{Prob}\{\, \mathbf{e} \notin \mathcal{S}_q(n, \theta n) \,\} \le q^{-n\mathsf{D}_q(1-\theta \| 1-p)} \;.$$

But $\mathsf{D}_q(\theta \| p) = \mathsf{D}_q(1-\theta \| 1-p)$. □

Lemma 4.7 can be seen as a special case of Lemma 4.13(i) obtained when $p = 1-(1/q)$, in which case the error word $\mathbf{e}$ that is generated by the channel is uniformly distributed over F^n: here,

$$q^{-n} \cdot V_q(n, \theta n) = \mathsf{Prob}\{\, \mathbf{e} \in \mathcal{S}_q(n, \theta n) \,\} \le q^{-n\mathsf{D}_q(\theta \| 1-(1/q))}$$

whenever $\theta \le 1 - (1/q)$, and one can verify that

$$\mathsf{D}_q(\theta \| 1-(1/q)) = 1 - \mathsf{H}_q(\theta) \;.$$

As before, we use the notation $o(1)$ to stand for an expression that goes to zero as $n \to \infty$ (the expression may depend on q, p, or R).

Theorem 4.14 (Shannon Converse Coding Theorem for the q-ary symmetric channel) *Let $\mathcal{C}$ be an (n, q^{nR}) code over F where n and nR are integers such that $1 - \mathsf{H}_q(p) < R \le 1$, and let $\mathcal{D} : F^n \to \mathcal{C} \cup \{\text{“e”}\}$ be a decoder for $\mathcal{C}$ with respect to the channel S. Then the decoding error probability $\mathrm{P}_{\mathrm{err}}$ of $\mathcal{D}$ satisfies*

$$\mathrm{P}_{\mathrm{err}} \ge 1 - q^{-n(\mathsf{D}_q(\theta_q(R) \| p) - o(1))} \;,$$

where $\theta_q(R) = \mathsf{H}_q^{-1}(1-R)$.

Proof. For a codeword $\mathbf{c} \in \mathcal{C}$, let $Y(\mathbf{c})$ be the set of pre-images of $\mathbf{c}$ under $\mathcal{D}$; namely,

$$Y(\mathbf{c}) = \{\, \mathbf{y} \in F^n \;:\; \mathcal{D}(\mathbf{y}) = \mathbf{c} \,\} \;.$$

Since $\sum_{\mathbf{c} \in \mathcal{C}} |Y(\mathbf{c})| = q^n$, it follows that there is a codeword $\mathbf{c}_0 \in \mathcal{C}$ such that

$$|Y(\mathbf{c}_0)| \le q^n / |\mathcal{C}| = q^{n(1-R)} \;.$$

By applying the translation $\mathbf{c} \mapsto \mathbf{c} - \mathbf{c}_0$ to each codeword and, respectively, replacing the decoder $\mathbf{y} \mapsto \mathcal{D}(\mathbf{y})$ by its translation $\mathbf{y} \mapsto \mathcal{D}(\mathbf{y} + \mathbf{c}_0)$, we can assume without loss of generality that $\mathbf{c}_0 = \mathbf{0}$. We point out that the distribution of the error words that are generated by the channel does not depend on the input to the channel; therefore, such translations will not affect the value of $\mathrm{P}_{\mathrm{err}}$.

Recall that

$$\mathrm{P}_{\mathrm{err}} = \max_{\mathbf{c} \in \mathcal{C}} \mathrm{P}_{\mathrm{err}}(\mathbf{c}) ,$$

where $\mathrm{P}_{\mathrm{err}}(\mathbf{c})$ is the probability that a codeword $\mathbf{c}$ will be decoded erroneously, given that $\mathbf{c}$ was transmitted. Thus, $\mathrm{P}_{\mathrm{err}} \geq \mathrm{P}_{\mathrm{err}}(\mathbf{0})$, where

$$\begin{aligned} 1 - \mathrm{P}_{\mathrm{err}}(\mathbf{0}) &= \sum_{\mathbf{e} \in Y} \mathsf{Prob}\{\, \mathbf{e} \text{ received} \mid \mathbf{0} \text{ transmitted}\,\} \\ &= \sum_{\mathbf{e} \in Y} (p/(q-1))^{\mathsf{w}(\mathbf{e})} (1-p)^{n-\mathsf{w}(\mathbf{e})} , \end{aligned}$$

with Y standing for $Y(\mathbf{0})$. Now, since $p \leq 1 - (1/q)$, the value

$$(p/(q-1))^{w}(1-p)^{n-w}$$

is decreasing with w. It follows that if θ is such that $V_q(n, \theta n) \geq |Y|$, then

$$\sum_{\mathbf{e} \in Y} (p/(q-1))^{\mathsf{w}(\mathbf{e})}(1-p)^{n-\mathsf{w}(\mathbf{e})} \leq \sum_{\mathbf{e} \in \mathcal{S}_q(n,\theta n)} (p/(q-1))^{\mathsf{w}(\mathbf{e})}(1-p)^{n-\mathsf{w}(\mathbf{e})} .$$

If, in addition $\theta \leq p$, then we have by Lemma 4.13(i),

$$1 - \mathrm{P}_{\mathrm{err}}(\mathbf{0}) \leq \sum_{\mathbf{e} \in \mathcal{S}_q(n,\theta n)} (p/(q-1))^{\mathsf{w}(\mathbf{e})}(1-p)^{n-\mathsf{w}(\mathbf{e})} \leq q^{-n\mathsf{D}_q(\theta \| p)} . \qquad (4.20)$$

Take n sufficiently large so that

$$1 - R + (\log_q(n+1))/n \leq \mathsf{H}_q(p - (1/n)) ,$$

and select θ to be

$$\theta = \frac{1}{n} \left\lceil n\mathsf{H}_q^{-1}\Big(1 - R + (\log_q(n+1))/n\Big) \right\rceil .$$

For this value of θ and by Lemma 4.8 we indeed have

$$V_q(n, \theta n) \geq \frac{1}{n+1} \cdot q^{n\mathsf{H}_q(\theta)} \geq q^{n(1-R)} \geq |Y| .$$

Furthermore, by the choice of n we guarantee that $\theta \leq \lceil np - 1 \rceil / n < p$. The theorem follows from (4.20) and the continuity of the function $x \mapsto \mathsf{D}_q(x \| p)$. □

Theorem 4.9 can be obtained as a corollary from Theorem 4.14 as follows. Let $\mathcal{C}$ be an $(n, q^{nR}, \delta n)$ code over an alphabet of size q and assume that

$R > 1 - \mathsf{H}_q(\delta/2)$. Write $t = \lfloor (\delta n - 1)/2 \rfloor$ and let the crossover probability p of the channel be such that

$$1 - R < \mathsf{H}_q(p) < \mathsf{H}_q(\delta/2) ;$$

in particular, $p < \delta/2$. By Lemma 4.13(ii) we have

$$\mathsf{Prob}\{ \mathbf{e} \notin \mathcal{S}_q(n,t) \} \le q^{-n\mathsf{D}_q(\delta/2\|p)} .$$

Since $\mathsf{D}_q(\delta/2\|p) > 0$, it follows that a nearest-codeword decoder for $\mathcal{C}$ will fail to return the correct codeword with probability $\mathrm{P}_{\mathrm{err}} = o(1)$. On the other hand, by Theorem 4.14 we must have $\mathrm{P}_{\mathrm{err}} = 1 - o(1)$. We conclude that the assumed inequality, $R > 1 - \mathsf{H}_q(\delta/2)$, can hold only for finitely many values of n.

4.7 Coding Theorem

Theorem 4.14 states that the code rate cannot exceed the capacity of the q-ary symmetric channel if information is to be transmitted through the channel with a decoding error probability that is bounded away from 1. The goal of the forthcoming discussion is to show that the capacity can be approached from below while attaining a decoding error probability that is arbitrarily small. In fact, we show that this can be achieved by linear codes (assuming that q is such that there is a field of size q).

Our analysis starts with two lemmas, which hold for every additive channel in which the (input and output) alphabet is a given finite field $F = \mathrm{GF}(q)$.

For a code $\mathcal{C} \subseteq F^n$ and a word $\mathbf{e} \in F^n$, let $\mathrm{P}_{\mathrm{err}}(\mathcal{C}|\mathbf{e})$ be the decoding error probability of a nearest-codeword decoder $\mathcal{D} : F^n \to \mathcal{C}$, conditioned on the error word being $\mathbf{e}$; that is, $\mathrm{P}_{\mathrm{err}}(\mathcal{C}|\mathbf{e})$ takes two possible values, as follows:

$$\mathrm{P}_{\mathrm{err}}(\mathcal{C}|\mathbf{e}) = \begin{cases} 1 & \text{if there is } \mathbf{c} \in \mathcal{C} \text{ such that } \mathcal{D}(\mathbf{c}+\mathbf{e}) \neq \mathbf{c} \\ 0 & \text{otherwise} \end{cases} .$$

Lemma 4.15 *Given n, k, and $\mathbf{e} \in \mathcal{S}_q(n,t)$, let $\overline{\mathrm{P}_{\mathrm{err}}(\mathcal{C}|\mathbf{e})}$ denote the average of $\mathrm{P}_{\mathrm{err}}(\mathcal{C}|\mathbf{e})$ over all linear $[n,k]$ codes $\mathcal{C}$ over $F = \mathrm{GF}(q)$. Then,*

$$\overline{\mathrm{P}_{\mathrm{err}}(\mathcal{C}|\mathbf{e})} < q^{k-n} \cdot V_q(n,t) .$$

Proof. Fix an error word $\mathbf{e} \in \mathcal{S}_q(n,t)$ and assume a uniform distribution over all $k \times n$ matrices G over F. Then, for every nonzero vector $\mathbf{u} \in F^k$, the random vector $\mathbf{u}G$ is uniformly distributed over F^n. Therefore,

$$\mathsf{P}_G\left\{ \mathbf{e} + \mathbf{u}G \in \mathcal{S}_q(n,t) \;\middle|\; \mathbf{e} \right\} = q^{-n} \cdot V_q(n,t) , \tag{4.21}$$

where the notation $\mathsf{P}_G\{\cdot\,|\mathbf{e}\}$ stands for the probability induced by the uniform distribution on G and conditioned on $\mathbf{e}$ being the error word generated by the channel.

A $k \times n$ matrix G is "bad" with respect to $\mathbf{e}$ if either $\mathrm{rank}(G) < k$ or G generates a linear $[n,k]$ code for which $\mathbf{e}$ is not a coset leader (in particular, the coset that contains $\mathbf{e}$ contains yet another element of $\mathcal{S}_q(n,t)$). Then,

$$\begin{aligned}\mathsf{P}_G\Big\{ G \text{ is bad w.r.t. } \mathbf{e} \;\Big|\; \mathbf{e}\Big\} & \\ &\le \mathsf{P}_G\Big\{ \mathbf{e}+\mathbf{u}G \in \mathcal{S}_q(n,t) \text{ for some } \mathbf{u} \in F^k \setminus \{\mathbf{0}\} \;\Big|\; \mathbf{e}\Big\} \\ &\le \sum_{\mathbf{u}\in F^k\setminus\{\mathbf{0}\}} \mathsf{P}_G\Big\{ \mathbf{e}+\mathbf{u}G \in \mathcal{S}_q(n,t) \;\Big|\; \mathbf{e}\Big\} \\ &< q^{k-n}\cdot V_q(n,t)\ ,\end{aligned}$$

with the last inequality following from (4.21).

Observing that $\mathrm{P}_{\mathrm{err}}(\mathcal{C}|\mathbf{e}) = 1$ only if the generator matrices of $\mathcal{C}$ are bad with respect to $\mathbf{e}$ (and recalling that all linear $[n,k]$ codes have the same number of generator matrices), it follows that

$$\overline{\mathrm{P}_{\mathrm{err}}(\mathcal{C}|\mathbf{e})} \le \mathsf{P}_G\Big\{ G \text{ is bad w.r.t. } \mathbf{e} \;\Big|\; \mathbf{e}\Big\} < q^{k-n}\cdot V_q(n,t)\ ,$$

as claimed. □

For a code $\mathcal{C} \subseteq F^n$ we denote by $\mathrm{P}_{\mathrm{err}}(\mathcal{C}|\mathcal{S}_q(n,t))$ the decoding error probability of a nearest-codeword decoder for $\mathcal{C}$ with respect to a given additive channel (F, F, Prob), conditioned on the error word $\mathbf{e}$ being in $\mathcal{S}_q(n,t)$. That is,

$$\mathrm{P}_{\mathrm{err}}(\mathcal{C}|\mathcal{S}_q(n,t)) = \max_{\mathbf{c}\in\mathcal{C}} \mathrm{P}_{\mathrm{err}}(\mathbf{c}|\mathcal{S}_q(n,t))\ ,$$

where $\mathrm{P}_{\mathrm{err}}(\mathbf{c}|\mathcal{S}_q(n,t))$ is the probability that the codeword $\mathbf{c}$ is decoded erroneously, given that $\mathbf{c}$ was transmitted and conditioned on the error word $\mathbf{e}$ being in $\mathcal{S}_q(n,t)$.

Like the previous lemma, the result of the next lemma applies to every additive channel (F, F, Prob) with alphabet $F = \mathrm{GF}(q)$.

Lemma 4.16 *Given n, k, and t, let $\overline{\mathrm{P}_{\mathrm{err}}(\mathcal{C}|\mathcal{S}_q(n,t))}$ denote the average of $\mathrm{P}_{\mathrm{err}}(\mathcal{C}|\mathcal{S}_q(n,t))$ over all linear $[n,k]$ codes $\mathcal{C}$ over $F = \mathrm{GF}(q)$. Then,*

$$\overline{\mathrm{P}_{\mathrm{err}}(\mathcal{C}|\mathcal{S}_q(n,t))} < q^{k-n}\cdot V_q(n,t)\ .$$

Proof. Let the measure μ be defined for each $\mathbf{e} \in \mathcal{S}_q(n,t)$ by the conditional probability

$$\mu(\mathbf{e}) = \mathsf{Prob}\Big\{ \mathbf{e} \text{ is the error word} \;\Big|\; \mathbf{e}\in\mathcal{S}_q(n,t)\Big\}$$

as induced by the channel. Then, for every code $\mathcal{C} \subseteq F^n$,

$$\begin{aligned}\mathrm{P}_{\mathrm{err}}(\mathcal{C}|\mathcal{S}_q(n,t)) &= \max_{\mathbf{c}\in\mathcal{C}} \mathrm{P}_{\mathrm{err}}(\mathbf{c}|\mathcal{S}_q(n,t)) \\ &= \max_{\mathbf{c}\in\mathcal{C}} \sum_{\mathbf{e}\in\mathcal{S}_q(n,t)} \mathrm{P}_{\mathrm{err}}(\mathbf{c}|\mathbf{e}) \cdot \mu(\mathbf{e}) \\ &\le \sum_{\mathbf{e}\in\mathcal{S}_q(n,t)} \max_{\mathbf{c}\in\mathcal{C}} \mathrm{P}_{\mathrm{err}}(\mathbf{c}|\mathbf{e}) \cdot \mu(\mathbf{e}) \\ &= \sum_{\mathbf{e}\in\mathcal{S}_q(n,t)} \mathrm{P}_{\mathrm{err}}(\mathcal{C}|\mathbf{e}) \cdot \mu(\mathbf{e}) ,\end{aligned}$$

where we have used the notation $\mathrm{P}_{\mathrm{err}}(\mathbf{c}|\mathbf{e})$ to indicate the decoding error probability given that the transmitted codeword was $\mathbf{c}$ and conditioned on the error word being $\mathbf{e}$. We conclude that

$$\overline{\mathrm{P}_{\mathrm{err}}(\mathcal{C}|\mathcal{S}_q(n,t))} \le \sum_{\mathbf{e}\in\mathcal{S}_q(n,t)} \overline{\mathrm{P}_{\mathrm{err}}(\mathcal{C}|\mathbf{e})} \cdot \mu(\mathbf{e}) < q^{k-n} \cdot V_q(n,t) ,$$

where the second inequality follows from Lemma 4.15. □

In the next theorem, we specialize to the memoryless q-ary symmetric channel with crossover probability $p \in (0, 1-(1/q))$ and (input and output) alphabet $F = \mathrm{GF}(q)$. The decoding error probability $\mathrm{P}_{\mathrm{err}}(\mathcal{C})$ is computed for a code $\mathcal{C}$ with respect to this channel, assuming a nearest-codeword decoder.

Theorem 4.17 (Shannon Coding Theorem for the q-ary symmetric channel) *Let n and nR be integers such that $R < 1 - \mathsf{H}_q(p)$ and let $\overline{\mathrm{P}_{\mathrm{err}}(\mathcal{C})}$ denote the average of $\mathrm{P}_{\mathrm{err}}(\mathcal{C})$ over all linear $[n, nR]$ codes $\mathcal{C}$ over $F = \mathrm{GF}(q)$. Then,*

$$\overline{\mathrm{P}_{\mathrm{err}}(\mathcal{C})} < 2q^{-n\mathrm{E}_q(p,R)} ,$$

where

$$\mathrm{E}_q(p,R) = 1 - \mathsf{H}_q(\theta_q^*(p,R)) - R$$

and

$$\theta_q^*(p,R) = \frac{\log_q(1-p) + 1 - R}{\log_q(1-p) - \log_q(p/(q-1))} .$$

Proof. Let $\mathcal{C}$ be a code of length over F and θ be a real in $[p, 1-(1/q)]$. Given a codeword $\mathbf{c}$ in $\mathcal{C}$, we can bound $\mathrm{P}_{\mathrm{err}}(\mathbf{c})$ from above by partitioning the error events into two classes, according to whether the error word $\mathbf{e}$ belongs to $\mathcal{S}_q(n, \theta n)$. Specifically,

$$\begin{aligned}\mathrm{P}_{\mathrm{err}}(\mathbf{c}) &\le \mathrm{P}_{\mathrm{err}}(\mathbf{c}|\mathcal{S}_q(n,\theta n)) \cdot \mathsf{Prob}\{\, \mathbf{e} \in \mathcal{S}_q(n,\theta n) \,\} + \mathsf{Prob}\{\, \mathbf{e} \notin \mathcal{S}_q(n,\theta n) \,\} \\ &\le \mathrm{P}_{\mathrm{err}}(\mathbf{c}|\mathcal{S}_q(n,\theta n)) + \mathsf{Prob}\{\, \mathbf{e} \notin \mathcal{S}_q(n,\theta n) \,\} \\ &\le \mathrm{P}_{\mathrm{err}}(\mathcal{C}|\mathcal{S}_q(n,\theta n)) + q^{-n\mathsf{D}_q(\theta\|p)} ,\end{aligned}$$

where we have used Lemma 4.13(ii) in the last inequality. By maximizing over $\mathbf{c} \in \mathcal{C}$ we thus obtain

$$\mathrm{P}_{\mathrm{err}}(\mathcal{C}) = \max_{\mathbf{c} \in \mathcal{C}} \mathrm{P}_{\mathrm{err}}(\mathbf{c}) \le \mathrm{P}_{\mathrm{err}}(\mathcal{C}|\mathcal{S}_q(n, \theta n)) + q^{-n\mathsf{D}_q(\theta\|p)} .$$

Next, we take the average over all linear $[n, nR]$ codes $\mathcal{C}$ over F and apply Lemma 4.16; this yields

$$\begin{aligned} \overline{\mathrm{P}_{\mathrm{err}}(\mathcal{C})} &< q^{n(R-1)} \cdot V_q(n, \theta n) + q^{-n\mathsf{D}_q(\theta\|p)} \\ &\le q^{-n(1-\mathsf{H}_q(\theta)-R)} + q^{-n\mathsf{D}_q(\theta\|p)} , \end{aligned} \tag{4.22}$$

where the last inequality follows from Lemma 4.7. Now, the function

$$x \mapsto 1 - \mathsf{H}_q(x) - R - \mathsf{D}_q(x\|p)$$

takes the positive value $1 - \mathsf{H}_q(p) - R$ when $x = p$ and the negative value $-R - \mathsf{D}_q(1-(1/q)\|p)$ when $x = 1 - (1/q)$. Therefore, there must be a value x in the interval $[p, 1-(1/q)]$ for which this function is zero. A simple computation reveals that $x = \theta_q^*(p, R)$ is (the only) such value, in which case

$$1 - \mathsf{H}_q(\theta_q^*(p, R)) - R = \mathsf{D}_q(\theta_q^*(p, R)\|p) = \mathrm{E}_q(p, R) .$$

Plugging $\theta = \theta_q^*(p, R)$ into (4.22) we obtain

$$\overline{\mathrm{P}_{\mathrm{err}}(\mathcal{C})} < q^{-n(1-\mathsf{H}_q(\theta)-R)} + q^{-n\mathsf{D}_q(\theta\|p)} = 2q^{-n\mathrm{E}_q(p,R)} .$$

Note that since the value $\theta_q^*(p, R)$ lies within the open interval $(p, 1-(1/q))$, we have $\mathrm{E}_q(p, R) = \mathsf{D}_q(\theta_q^*(p, R)\|p) > 0$. □

Corollary 4.18 *Using the notation of Theorem 4.17, for every* $\rho \in (0, 1]$, *all but a fraction less than* ρ *of the linear* $[n, nR]$ *codes* $\mathcal{C}$ *over* F *satisfy*

$$\mathrm{P}_{\mathrm{err}}(\mathcal{C}) < (1/\rho) \cdot 2q^{-n\mathrm{E}_q(p,R)} .$$

Proof. Consider the set B of codes $\mathcal{C}$ for which $\mathrm{P}_{\mathrm{err}}(\mathcal{C}) \ge (1/\rho) \cdot 2q^{-n\mathrm{E}_q(p,R)}$, and suppose to the contrary that B forms at least a fraction ρ of the linear $[n, nR]$ codes over F. Then,

$$\overline{\mathrm{P}_{\mathrm{err}}(\mathcal{C})} \ge \rho \cdot \frac{1}{|B|} \sum_{\mathcal{C} \in B} \mathrm{P}_{\mathrm{err}}(\mathcal{C}) \ge 2q^{-n\mathrm{E}_q(p,R)} ,$$

thereby contradicting Theorem 4.17. □

Corollary 4.18 states that "most" linear codes attain the Shannon Coding Theorem. Yet, the result does not suggest an efficient algorithm for finding those codes.

In many cases of linear $[n, nR, \delta n]$ codes, a nearest-codeword decoder has a known efficient implementation only when the number of errors does not exceed $\lfloor(\delta n - 1)/2\rfloor$. Given a crossover probability $p > 0$ of the channel, it follows from Lemma 4.13 that if such algorithms are to operate in their efficient range, then δ should be at least $2p + o(1)$. However, by the Elias bound (Theorem 4.12), this would force the rate to be bounded away from the capacity $1 - \mathsf{H}_q(p)$. To see this, refer to Figure 4.1 and let the abscissa stand for $2p$ instead of δ: the sphere-packing curve then coincides with the capacity curve $2p \to 1 - \mathsf{H}_q(p)$, whereas the Elias bound lies strictly below it unless $p = 0$.

Problems

[Section 4.1]

Problem 4.1 Let $\mathcal{C}$ be a linear $[n, k, d]$ code over F.

1. Show that $\mathcal{C}$ is MDS if and only if every set of k columns in its generator matrix is linearly independent.

   Hint: See Problem 2.8.

2. Show that $\mathcal{C}$ is MDS if and only if its dual code is (assuming that $k < n$).

Problem 4.2 Let $G = (\,I \,|\, A\,)$ be a systematic generator matrix of a linear $[n, k, d]$ code $\mathcal{C}$ over F. Show that $\mathcal{C}$ is MDS if and only if every square sub-matrix of A is nonsingular.

Problem 4.3 Let $\mathcal{C}$ be a linear $[n, k{>}1, d]$ over $F = \mathrm{GF}(q)$ with a generator matrix of the form

$$G = \left(\begin{array}{c|c} 0\,0\,\ldots\,0 & 1\,1\,\ldots\,1 \\ \hline G_1 & G_2 \end{array}\right)$$

where the number of 1's in the first row equals the minimum distance d of $\mathcal{C}$. Let $\mathcal{C}_1$ be the linear $[n_1{=}n{-}d, k_1, d_1]$ over F which is spanned by the rows of the $(k{-}1) \times (n{-}d)$ matrix G_1.

1. Show that $\mathrm{rank}(G_1) = k{-}1$ and, therefore, $k_1 = k{-}1$.

   Hint: Show that otherwise there would be a linear combination of the last $k{-}1$ rows of G that would result in a nonzero codeword $\mathbf{c} \in \mathcal{C}$ whose first $n{-}d$ entries are zero. Then consider linear combinations of the codeword $\mathbf{c}$ with the first row of G.

2. Let $\mathbf{c}_1$ be a codeword of $\mathcal{C}_1$. Show that there are exactly q words $\mathbf{c}_2 \in F^d$ such that the concatenation $(\,\mathbf{c}_1 \,|\, \mathbf{c}_2\,)$ is a codeword of $\mathcal{C}$.

3. Let $\mathbf{c}_1$ be a nonzero codeword of $\mathcal{C}_1$. Show that there is a word $\mathbf{c}_2 \in F^d$ of Hamming weight at most $d - \lceil d/q \rceil$ such that $(\,\mathbf{c}_1 \,|\, \mathbf{c}_2\,)$ is a codeword of $\mathcal{C}$.

4. Show that $d_1 \geq \lceil d/q \rceil$.

   Hint: Select in part 3 a codeword $\mathbf{c}_1 \in \mathcal{C}_1$ of Hamming weight d_1.

Problem 4.4 (The Griesmer bound) Denote by $N_q(k, d)$ the length of a shortest linear code of dimension k and minimum distance d over $F = \mathrm{GF}(q)$.

1. Based on Problem 4.3, show that $N_q(k, d) \geq d + N_q(k-1, \lceil d/q \rceil)$ for every $k > 1$.

2. Show by induction on k that

$$N_q(k, d) \geq \sum_{i=0}^{k-1} \left\lceil \frac{d}{q^i} \right\rceil .$$

3. Derive from part 2 the Singleton bound for linear codes.

4. Show that the following codes meet the bound in part 2:

   (a) The simplex code over $F = \mathrm{GF}(q)$, which is defined for every positive integer m and length $n = (q^m - 1)/(q - 1)$ as the $[n, m, q^{m-1}]$ dual code of the Hamming code over F (see Problem 2.18).

   (b) The first-order Reed–Muller code over F, which is defined as the linear $[q^m, m+1, q^{m-1}(q-1)]$ code over F with an $(m+1) \times q^m$ generator matrix whose columns range over all the vectors in F^{m+1} with a first entry equaling 1 (see Problem 2.17).

   (c) The shortened first-order Reed–Muller code over F, which is defined as the linear $[q^m - 1, m, q^{m-1}(q-1)]$ code over F with an $m \times (q^m - 1)$ generator matrix whose columns range over all the nonzero vectors in F^m.

[Section 4.2]

Problem 4.5 Let F be an alphabet of size q and let n be a positive integer. Consider the sphere $\mathcal{S}$ of radius t in F^n that is centered at some word $\mathbf{x} \in F^n$. Show that

$$V_q(n, t) = |\mathcal{S}| = \sum_{i=0}^{t} \binom{n}{i} (q-1)^i .$$

Hint: Given a subset $J \subseteq \{1, 2, \ldots, n\}$, how many words in F^n differ from $\mathbf{x}$ exactly on the coordinates that are indexed by J?

Problem 4.6 Show that the minimum distance of a perfect code must be odd.

Problem 4.7 Let $F = \mathrm{GF}(q)$ and let n be a prime such that $\gcd(n, q) = 1$. Denote by e the multiplicative order of (the field integer) $\overline{q}$ in $\mathrm{GF}(n)$ (see Section 3.6).

1. Show that there exists a perfect linear $[n, k]$ code over F only if e divides $n-k$.

   Hint: Show that n divides $V_q(n, t) - 1$ whenever $t < n$.

2. Find all the values of k that satisfy the necessary condition of part 1 in the following two cases:

   (a) $q = 2$ and $n = 23$.

   (b) $q = 3$ and $n = 11$.

Problem 4.8 Let $F = \mathrm{GF}(q)$ and let $\mathcal{C}$ be a Hamming code of length $n = (q^m - 1)/(q-1)$ over F. For $i = 0, 1, \ldots, n$, denote by W_i the number of codewords in $\mathcal{C}$ of Hamming weight i.

1. Let $\mathcal{D}$ be a nearest-codeword decoder for $\mathcal{C}$ and let $\mathbf{c}$ be a codeword of Hamming weight t in $\mathcal{C}$. For each of the following values of i, find the number of words of Hamming weight i in F^n that will be decoded by $\mathcal{D}$ to $\mathbf{c}$:

   (a) $i = t-1$.

   (b) $i = t+1$.

   (c) $i = t$.

   Hint: Recall that $\mathcal{C}$ is perfect with minimum distance 3.

2. Show that for $0 < i < n$,

$$(i+1) \cdot W_{i+1} + (i(q-2) + 1) \cdot W_i + (n-i+1)(q-1) \cdot W_{i-1} = \binom{n}{i}(q-1)^i ,$$

   where $W_0 = 1$ and $W_1 = 0$.

3. Show that $W_3 = \frac{1}{6} \cdot n(n-1)(q-1)^2$.

Problem 4.9 Let $\mathcal{C}$ be a perfect $(n, M, d{=}2t{+}1)$ code over $F = \mathrm{GF}(q)$ and suppose that $\mathcal{C}$ contains the all-zero codeword. Show that the number, W_{2t+1}, of codewords of Hamming weight $2t{+}1$ in $\mathcal{C}$ is given by

$$W_{2t+1} = \frac{\binom{n}{t+1}(q-1)^{t+1}}{\binom{2t+1}{t}} .$$

Hint: Given a codeword $\mathbf{c}$ of Hamming weight $2t{+}1$ in $\mathcal{C}$, show that there are exactly $\binom{2t+1}{t}$ words of Hamming weight $t{+}1$ in F^n that are decoded to $\mathbf{c}$ by a nearest-codeword decoder.

Problem 4.10 (Constant-weight codes with $d = w$) An (n, M, d) code over an Abelian group F is called an $(n, M, d; w)$ *constant-weight code* if each codeword in the code has Hamming weight w.

Let $\mathcal{C}$ be an $(n, M, d{=}2t{+}1; w{=}2t{+}1)$ constant-weight code over $F = \mathrm{GF}(q)$.

1. Show that

$$M \le \frac{\binom{n}{t+1}(q-1)^{t+1}}{\binom{2t+1}{t}} .$$

   Hint: For every codeword $\mathbf{c} \in \mathcal{C}$ there are $\binom{2t+1}{t}$ words $\mathbf{y}$ of Hamming weight $t{+}1$ in F^n such that $\mathsf{d}(\mathbf{y}, \mathbf{c}) = t$. And given a word $\mathbf{y}$ of Hamming weight $t{+}1$ in F^n, how many codewords $\mathbf{c} \in \mathcal{C}$ are there such that $\mathsf{d}(\mathbf{y}, \mathbf{c}) = t$?

2. Show that the bound in part 1 can be attained whenever there exists a perfect code of length n and minimum distance $2t+1$ over F.

   Hint: See Problem 4.9.

Problem 4.11 Let $\mathcal{C}_0$ be the $[n, n-1, 2]$ parity code over $F = \mathrm{GF}(q)$ and denote by X the complement set $F^n \setminus \mathcal{C}_0$. For a word $\mathbf{c} \in F^n$, define the set $\mathcal{S}(\mathbf{c})$ by

$$\mathcal{S}(\mathbf{c}) = \{\mathbf{y} \in X \; : \; \mathsf{d}(\mathbf{y}, \mathbf{c}) \leq 1\} .$$

1. Show that for every $\mathbf{c} \in \mathcal{C}_0$,

$$|\mathcal{S}(\mathbf{c})| = n(q-1) .$$

Suppose that $n = q^m$ and let $\mathcal{C}_1$ be the linear $[n, n-m-1]$ code over F that is defined by an $(m+1) \times n$ parity-check matrix whose columns range over all the elements of F^{m+1} whose first entry equals 1 (that is, $\mathcal{C}_1$ is the dual code of the first-order Reed–Muller code).

2. Show that $\mathcal{S}(\mathbf{c}) \cap \mathcal{S}(\mathbf{c}') = \emptyset$ for every two distinct codewords $\mathbf{c}, \mathbf{c}' \in \mathcal{C}_1$.

   Hint: Show that the minimum distance of $\mathcal{C}_1$ is at least 3.

3. Show that

$$\sum_{\mathbf{c} \in \mathcal{C}_1} |\mathcal{S}(\mathbf{c})| = (q-1) \cdot q^{n-1} = |X| ,$$

   and deduce that

$$\{\mathcal{S}(\mathbf{c}) \; : \; \mathbf{c} \in \mathcal{C}_1\}$$

   forms a partition of X into q^{n-m-1} subsets of size $n(q-1)$.

Problem 4.12 Recall from Problem 2.21 that a *burst of length* ℓ is the event of having errors in a codeword such that the locations i and j of the first (leftmost) and last (rightmost) errors, respectively, satisfy $j-i = \ell-1$.

Let $\mathcal{C}$ be a linear $[n, k>0]$ code over $F = \mathrm{GF}(q)$ and suppose that there exists a decoder for $\mathcal{C}$ that corrects every burst of length t or less.

1. Show that in every nonzero codeword $\mathbf{c}$ in $\mathcal{C}$, the locations i and j of the first and last nonzero entries in $\mathbf{c}$ must satisfy $j-i \geq 2t$.

2. (The Reiger bound: a Singleton-like bound for burst-correcting codes) Show that

$$n - k \geq 2t .$$

3. (A sphere-packing-like bound for burst-correcting codes) Show that

$$q^{n-k} \geq 1 + n(q-1) + (q-1)^2 \sum_{i=0}^{t-2} (n-i-1) q^i .$$

(The bounds in parts 2 and 3 hold, in fact, also for nonlinear $(n, M>1)$ codes over an alphabet of size q, with k taken as $\log_q M$.)

Problem 4.13 Let $\mathcal{C}$ be an (n, M, d) code over an alphabet F of size q. The Hamming distance of a word $\mathbf{y} \in F^n$ from $\mathcal{C}$, denoted by $\mathsf{d}(\mathbf{y}, \mathcal{C})$, is defined as the Hamming distance between $\mathbf{y}$ and a nearest codeword in $\mathcal{C}$ to $\mathbf{y}$; that is,

$$\mathsf{d}(\mathbf{y}, \mathcal{C}) = \min_{\mathbf{c} \in \mathcal{C}} \mathsf{d}(\mathbf{y}, \mathbf{c}) .$$

The *covering radius* of $\mathcal{C}$, denoted by r, is the largest distance from $\mathcal{C}$ of any word in F^n; namely,

$$\mathsf{r} = \max_{\mathbf{y} \in F^n} \mathsf{d}(\mathbf{y}, \mathcal{C}) .$$

1. Find the covering radii of the repetition code and of the Hamming code over $F = \mathrm{GF}(q)$.
2. (The sphere-covering bound) Show that

$$M \cdot V_q(n, \mathsf{r}) \geq q^n .$$

3. Show that $\mathsf{r} \geq (d-1)/2$ and that equality holds if and only if $\mathcal{C}$ is perfect.
4. Show that if $\mathcal{C}$ is a linear $[n, k, d]$ code over $\mathrm{GF}(q)$ then $\mathsf{r} \leq n-k$.
5. Show that if $\mathcal{C}$ is a linear $[n, k, d]$ code over $F = \mathrm{GF}(q)$ then r is the largest among the Hamming weights of the coset leaders of $\mathcal{C}$ in F^n.
6. Show that if $\mathcal{C}$ is a linear $[n, k, d]$ code over $F = \mathrm{GF}(q)$ and H is an $(n-k) \times n$ parity check of $\mathcal{C}$, then r is the smallest nonnegative integer such that every vector in F^{n-k} can be expressed as a linear combination of up to r columns in H.
7. An (n, M, d) code is called *maximal* if the addition of any new codeword to $\mathcal{C}$ reduces its minimum distance. Show that if $\mathcal{C}$ is maximal then $\mathsf{r} < d$.
8. Let $\mathcal{C}$ be a linear $[n, k, d]$ code over F such that any *lengthening* of $\mathcal{C}$ obtained by adding a column to an $(n-k) \times n$ parity-check matrix H of $\mathcal{C}$ generates $d-1$ dependent columns in H (i.e., the minimum distance drops below d). Show that $\mathsf{r} < d-1$.

Problem 4.14 A soccer betting form contains a list of 13 matches. Next to each listed match there are three fill-in boxes which correspond to the following three possible guesses: "first team wins," "second team wins," or "tied match." The bettor checks one box for each match.

Describe a strategy for filling out the *smallest* number of forms so that at least one of the forms contains at least 12 correct guesses. How many forms need to be filled out under this strategy?

Hint: Consider a perfect code of length 13 and minimum distance 3 over GF(3).

[Section 4.3]

Problem 4.15 (Variant of Theorem 4.5) Let $F = \mathrm{GF}(q)$ and let n, k, and d be positive integers where $k \leq n-d+1$. Consider the ensemble of all $(n-k) \times n$ matrices over F of the form

$$H = (\ A \mid I\ ) ,$$

and define a probability distribution on this ensemble that is induced by assuming a uniform distribution over the $(n-k) \times k$ matrices A over F.

1. Show that for every nonzero vector $\mathbf{y} \in F^n$,

$$\mathsf{Prob}\{\, H\mathbf{y}^T = \mathbf{0} \,\} = \begin{cases} 0 & \text{if the first } k \text{ entries in } \mathbf{y} \text{ are zero} \\ q^{k-n} & \text{otherwise} \end{cases} .$$

2. Show that

$$\mathsf{Prob}\{\, H \text{ contains } d{-}1 \text{ dependent columns} \,\} \le \rho \,,$$

where

$$\begin{aligned} \rho &= q^{k-n} \cdot \frac{V_q(n,d{-}1) - V_q(n{-}k,d{-}1)}{q-1} \\ &= q^{k-n} \cdot \sum_{i=1}^{d-1} \left(\binom{n}{i} - \binom{n{-}k}{i} \right) (q-1)^{i-1} . \end{aligned}$$

Deduce that all but a fraction at most ρ of the systematic linear $[n,k]$ codes over F have minimum distance at least d.

[Section 4.4]

Problem 4.16 (Alternative form of Krawtchouk polynomials) Show that

$$\mathcal{K}_\ell(i;n,q) = \sum_{r=0}^{\ell} \binom{i}{r} \binom{n{-}r}{\ell{-}r} (-q)^r (q{-}1)^{\ell-r} .$$

Hint: Write

$$\begin{aligned} (1+(q{-}1)z)^{n-i}(1-z)^i &= (1+(q{-}1)z)^n \left(1 - \frac{qz}{1+(q{-}1)z}\right)^i \\ &= \sum_{r=0}^{i} \binom{i}{r} (1+(q{-}1)z)^{n-r}(-qz)^r \end{aligned}$$

and identify the coefficient of z^ℓ.

Problem 4.17 (Alternative form of MacWilliams' identities) Let $(W_i)_{i=0}^n$ be the weight distribution of a linear $[n,k,d]$ code $\mathcal{C}$ over $F = \mathrm{GF}(q)$ and $(W_i^\perp)_{i=0}^n$ be the weight distribution of the dual code $\mathcal{C}^\perp$ of $\mathcal{C}$. Show that

$$\sum_{i=\ell}^{n} \binom{i}{\ell} W_i = q^{k-\ell} \sum_{i=0}^{\ell} \binom{n{-}i}{n{-}\ell} (q{-}1)^{\ell-i}(-1)^i \, W_i^\perp \,, \quad 0 \le \ell \le n .$$

Hint: Substitute $z = \xi + 1$ in (4.7).

Problem 4.18 Let n, k, d, and q be positive integers such that $d = n-k+1$ and consider the following set of linear equations in the k real unknown values $W_d, W_{d+1}, \ldots, W_n$:

$$\sum_{i=d}^{n-\ell} \binom{n-i}{\ell} W_i = \binom{n}{\ell} (q^{k-\ell} - 1) , \quad 0 \le \ell < k .$$

1. Show that the solution to this set of equations is unique and verify by substitution that the solution is given by

$$W_i = \binom{n}{i} \sum_{s=0}^{i-d} \binom{i}{s} (-1)^s (q^{i+1-d-s} - 1) , \quad d \le i \le n .$$

   Hint: Use the identities $\binom{n-i}{\ell}\binom{n}{i} = \binom{n}{\ell}\binom{n-\ell}{i}$ and $\binom{n-\ell}{i}\binom{i}{s} = \binom{n-\ell}{j}\binom{n-\ell-j}{s}$, where j stands for the difference $i-s$.

2. Show that the solution can also be written as

$$W_i = \binom{n}{i} (q-1) \sum_{s=0}^{i-d} \binom{i-1}{s} (-1)^s q^{i-d-s} , \quad d \le i \le n .$$

   Hint: Use the identity $\binom{i}{s} = \binom{i-1}{s} + \binom{i-1}{s-1}$.

Problem 4.19 Let $F = \mathrm{GF}(q)$ and consider transmission through a memoryless q-ary symmetric channel with crossover probability p. For a linear $[n, k, d]$ code $\mathcal{C}$ over F, let $\mathcal{D}_{\mathrm{MLD}} : F^n \to \mathcal{C}$ be a maximum-likelihood decoder for $\mathcal{C}$ with respect to this channel. Show that the decoding error probability, $\mathrm{P_{err}}$, of $\mathcal{D}_{\mathrm{MLD}}$ is bounded from above by

$$\mathrm{P_{err}} \le W_{\mathcal{C}} \left(2\sqrt{p(1-p)/(q-1)} + (p(q-2)/(q-1)) \right) - 1 .$$

Hint: See Problem 1.9.

Problem 4.20 Let $F = \mathrm{GF}(q)$ and consider transmission through a memoryless q-ary symmetric channel with crossover probability p. For a linear $[n, k, d]$ code $\mathcal{C}$ over F, let $\mathcal{D} : F^n \to \mathcal{C} \cup \{\text{“e”}\}$ be the decoder

$$\mathcal{D}(\mathbf{y}) = \begin{cases} \mathbf{y} & \text{if } \mathbf{y} \in \mathcal{C} \\ \text{“e”} & \text{otherwise} \end{cases} .$$

Define the *decoding misdetection probability* $\mathrm{P_{mis}}(\mathcal{C})$ of $\mathcal{D}$ by

$$\mathrm{P_{mis}}(\mathcal{C}) = \max_{\mathbf{c} \in \mathcal{C}} \mathrm{P_{mis}}(\mathbf{c}) ,$$

where

$$\mathrm{P_{mis}}(\mathbf{c}) = \sum_{\mathbf{y} : \mathcal{D}(\mathbf{y}) \notin \{\mathbf{c}, \text{“e”}\}} \mathsf{Prob}\{ \mathbf{y} \text{ received} \mid \mathbf{c} \text{ transmitted} \} .$$

(Note the difference between the decoding misdetection probability and the decoding error probability: while the former is the probability of only decoding to a wrong codeword, the latter includes also the probability that the decoder detects errors without correcting them.) Show that

$$\mathrm{P}_{\mathrm{mis}}(\mathcal{C}) = W_{\mathcal{C}}^{\mathrm{h}}(1-p, p/(q-1)) - (1-p)^n .$$

Problem 4.21 Let $F = \mathrm{GF}(q)$ and consider transmission through a memoryless q-ary erasure channel with input alphabet F, output alphabet $\Phi = F \cup \{?\}$, and erasure probability p (see Example 1.10). For a linear $[n, k, d]$ code $\mathcal{C}$ over F, let $\mathcal{D} : \Phi^n \to \mathcal{C} \cup \{\text{“e”}\}$ be the decoder

$$\mathcal{D}(\mathbf{y}) = \begin{cases} \mathbf{c} & \text{if } \mathbf{y} \text{ agrees with exactly one } \mathbf{c} \in \mathcal{C} \text{ on the entries in } F \\ \text{“e”} & \text{otherwise} \end{cases} .$$

The decoding error probability, $\mathrm{P}_{\mathrm{err}}$, of $\mathcal{D}$ is given by

$$\mathrm{P}_{\mathrm{err}} = \max_{\mathbf{c} \in \mathcal{C}} \sum_{\mathbf{y} : \mathcal{D}(\mathbf{y}) = \text{“e”}} \mathsf{Prob}\{\, \mathbf{y} \text{ received} \mid \mathbf{c} \text{ transmitted} \,\} .$$

1. Show that

$$\mathrm{P}_{\mathrm{err}} \le W_{\mathcal{C}}(p) - 1 .$$

2. Assuming a uniform distribution over the codewords of $\mathcal{C}$, a random codeword $\mathbf{c}$ is selected from $\mathcal{C}$ and transmitted through the erasure channel. Show that

$$\mathsf{Prob}\{\, \text{received word } \mathbf{y} \text{ is in } \{0, ?\}^n \,\} = \frac{1}{q^k} \cdot W_{\mathcal{C}}(p) .$$

Problem 4.22 Let $\mathcal{C}$ be a linear $[n, k, d]$ code over $F = \mathrm{GF}(q)$ whose generator matrix does not contain an all-zero column. Fix an integer t in the range $0 < t < (1 - (1/q))n$, and denote by Y_t the number of codewords in $\mathcal{C}$ whose Hamming weight is t or less.

1. Show that

$$W'_{\mathcal{C}}(1) = n \cdot (q-1) \cdot q^{k-1} ,$$

where $W'_{\mathcal{C}}(z)$ stands for the derivative of $W_{\mathcal{C}}(z)$ with respect to z.

Hint: See Problem 2.6.

2. Show that

$$Y_t \le \inf_{z \in (0,1]} z^{-t} W_{\mathcal{C}}(z) .$$

Hint: For every $z \in (0, 1]$, verify that Y_t is related to the weight distribution $(W_i)_{i=0}^n$ of $\mathcal{C}$ by

$$Y_t = \sum_{i=0}^{t} W_i \le \sum_{i=0}^{n} W_i z^{i-t} .$$

3. Show that the polynomial

$$Q_t(z) = zW'_{\mathcal{C}}(z) - tW_{\mathcal{C}}(z)$$

has a unique real positive root z_0 and that this root belongs to the interval $(0, 1)$.

Hint: First, verify that $Q_t(0) < 0$ and $Q_t(1) > 0$ and deduce that there exists $z_0 \in (0, 1)$ such that $Q_t(z_0) = 0$.

Next, write $Q_t(z) = \sum_{i=0}^{n} Q_{t,i} z^i$ and observe that $Q_{t,i} \leq 0$ when $i < t$ and $Q_{t,i} \geq 0$ otherwise. Use this to show that if z_0 is a positive root of $Q_t(z)$ then $Q'_t(z_0) > 0$. Finally, argue that if a polynomial has two or more positive roots, then such a polynomial cannot be increasing at all of these roots.

4. Show that

$$Y_t \leq z_0^{-t} W_{\mathcal{C}}(z_0) ,$$

where z_0 is the unique positive root of the polynomial $Q_t(z)$ in part 3.

[Section 4.5]

Problem 4.23 (The Plotkin bound) Let $\mathcal{C}$ be an (n, M, d) code over an alphabet F of size q.

1. Let $T = (T_{i,j})$ be an $M \times n$ array whose rows are the codewords of $\mathcal{C}$. For a column index $j \in \{1, 2, \ldots, n\}$ and an element $a \in F$, denote by $x_{a,j}$ the number of times the element a appears in the jth column of T. Let P_j denote the number of (ordered) pairs of row indexes $(r, s) \in \{1, 2, \ldots, M\} \times \{1, 2, \ldots, M\}$ such that $T_{r,j} = T_{s,j}$. Show that for every column index j,

$$P_j = \sum_{a \in F} x_{a,j}^2 .$$

2. Show that

$$P_j \geq M^2/q .$$

Hint: Given the constraint $\sum_{a \in F} x_{a,j} = M$, show that the minimum of $\sum_{a \in F} x_{a,j}^2$ over the reals is attained when $x_{a,j} = M/q$; or, show that

$$\sum_{a \in F} x_{a,j}^2 = M^2/q + \sum_{a \in F} (x_{a,j} - (M/q))^2 .$$

3. Show that

$$\sum_{\mathbf{c}_1, \mathbf{c}_2 \in \mathcal{C}} \mathsf{d}(\mathbf{c}_1, \mathbf{c}_2) = \sum_{j=1}^{n} (M^2 - P_j) .$$

4. Based on the previous parts, show that

$$M(M-1)d \leq \sum_{j=1}^{n} (M^2 - P_j) \leq nM^2(1 - (1/q))$$

and obtain the upper bound

$$\frac{d}{n} \leq \frac{1-(1/q)}{1-(1/M)} .$$

(Note that when $\mathcal{C}$ is linear, this bound becomes the bound in Problem 2.7.)

5. Verify that the following codes attain the bound in part 4:
   (a) The repetition code.
   (b) The simplex code over $F = \mathrm{GF}(q)$ (see part 4 of Problem 4.4).
   (c) The shortened first-order Reed–Muller code over $F = \mathrm{GF}(q)$.
6. Show that the bound in part 4 is attained only if $M-1$ divides $n(q-1)$. (In particular, this implies $M \leq 1+n(q-1)$.)

   Hint: The expression $Mn(q-1)/((M-1)q)$ must be an integer.
7. Show that a linear $[n, k, d]$ code over $F = \mathrm{GF}(q)$ attains the bound in part 4 only if $n = \ell \cdot (q^k - 1)/(q-1)$ for some positive integer ℓ.
8. Conversely, show that for every two positive integers k and ℓ there is a linear $[n, k, d]$ code over $F = \mathrm{GF}(q)$ of length $n = \ell \cdot (q^k - 1)/(q-1)$ that attains the bound in part 4.

   Hint: Consider a linear code whose generator matrix consists of ℓ copies of a generator matrix G of the simplex code.
9. A (n, M, d) code is called *equidistant* if the Hamming distance between every two distinct codewords in the code equals d. Show that a code $\mathcal{C}$ attains the bound in part 4 only if $\mathcal{C}$ is equidistant.
10. Show that the condition in part 9 is also sufficient when $\mathcal{C}$ is a linear $[n, k, d]$ code over $F = \mathrm{GF}(q)$ and no coordinate in $\mathcal{C}$ is identically zero.

    Hint: Use Problem 2.6.

Problem 4.24 Show that the Johnson bound in Proposition 4.11 is attained by the following codes:

1. The code obtained by removing the all-zero codeword from a simplex code over $\mathrm{GF}(q)$ (see part 4 of Problem 4.4).
2. The code obtained by removing the all-zero codeword from a shortened first-order Reed–Muller code over $\mathrm{GF}(q)$.

Problem 4.25 Let M and q be positive integers and θ be a rational in $(0, 1-(1/q)]$ such that $M\theta$ is an integer multiple of $q-1$. Let F be an Abelian group of size q, and consider an $M \times n$ array $T = (T_{i,j})$ over F whose columns exhaust all the distinct words $\mathbf{w}$ in F^M with the property that each nonzero element of F appears in $\mathbf{w}$ exactly $M\theta/(q-1)$ times. (Thus, the Hamming weight of each column in T is $M\theta$, and the number of columns, n, is uniquely determined by M, q, and θ.) Denote by $\mathcal{C}$ the (n, M) code over F whose codewords are given by the rows of T. Show that $\mathcal{C}$ attains the Johnson bound in Proposition 4.11.

Hint: Verify that the inequalities in the proof of Proposition 4.11 all hold with equality. In particular, use the symmetry among the rows of T to claim that $\mathcal{C}$ is equidistant (see part 9 of Problem 4.23).

Problem 4.26 (Quadratic integer expressions) For a positive integer m, an integer s, and a real number v, define

$$B(m,s,v) = \min_{z_1,z_2,\ldots,z_m} \sum_{i=1}^{m} (z_i - v)^2 ,$$

where the minimum is taken over all integers $z_1, z_2, \ldots, z_m$ such that

$$\sum_{i=1}^{m} z_i = s .$$

Write $s = mc - t$ where $c = \lceil s/m \rceil$.

1. Show that

$$\begin{aligned} B(m,s,v) &= m(c-v)^2 + 2(v-c)t + t \\ &= mv^2 - 2\left(\tbinom{c}{2}m + s(v-c+\tfrac{1}{2})\right) , \end{aligned}$$

and that this minimum is attained when exactly t of the values z_i equal $c-1$ while the remaining $m-t$ values equal c.

Hint: Show that if $z_j \geq z_\ell + 2$, then reducing z_j by 1 and increasing z_ℓ by 1 will decrease the sum $\sum_{i=1}^{m}(z_i - v)^2$.

2. Show that

$$B(m,s,v) \leq B(m,s-1,v) \quad \text{when } s \leq m\lfloor v + \tfrac{1}{2} \rfloor$$

and

$$B(m,s,v) > B(m,s-1,v) \quad \text{when } s > m\lfloor v + \tfrac{1}{2} \rfloor$$

(that is, for fixed m and v, the value $B(m,s,v)$ is smallest when $s = m\lfloor v+\frac{1}{2} \rfloor$).

Problem 4.27 (Improvements on the Johnson bound) For integers $M > 1$ and $q > 1$ and a real $\theta \in [0,1]$, define

$$\mathcal{J}(M,\theta,q) = \frac{(M-\rho-\sigma+1)M\theta + \binom{\rho}{2} + \binom{\sigma}{2}(q-1)}{\binom{M}{2}} ,$$

where $\rho = \lceil M\theta \rceil$ and $\sigma = \lceil \rho/(q-1) \rceil$.

Let $\mathcal{C}$ be an $(n, M, \delta n)$ code over an Abelian group of size q and let θn be the largest Hamming weight of any codeword in $\mathcal{C}$.

1. Show that

$$\sigma = \lceil M\theta/(q-1) \rceil$$

(even when $M\theta$ is not an integer).

2. Show that if $\theta \le 1-(1/q)$ then

$$\delta \le \mathcal{J}(M, \theta, q) .$$

Hint: Break the right-hand side of (4.14) into a sum of the following two expressions:

$$\sum_{j=1}^{n} \Bigl(2M_1 \Bigl(\sum_{a \in F^*} x_{a,j} \Bigr) - \Bigl(\sum_{a \in F^*} x_{a,j} \Bigr)^2 \Bigr)$$

and

$$\sum_{j=1}^{n} \sum_{a \in F^*} \left(2M_2 \, x_{a,j} - x_{a,j}^2 \right) ,$$

where M_1 and M_2 are reals such that

$$M_1 \ge \lceil M\theta \rceil - \tfrac{1}{2} , \quad M_2 \ge \lceil M\theta/(q-1) \rceil - \tfrac{1}{2} , \quad \text{and} \quad M_1 + M_2 = M .$$

Fix the sum

$$\sum_{j=1}^{n} \sum_{a \in F^*} x_{a,j}$$

to be equal to some integer $x \le M\theta n$, then apply part 1 of Problem 4.26 twice: once with

$$m \leftarrow n , \quad s \leftarrow x , \quad v \leftarrow M_1 , \quad \text{and} \quad z_i \leftarrow \sum_{a \in F^*} x_{a,j} ,$$

and then with

$$m \leftarrow n(q-1) , \quad s \leftarrow x , \quad v \leftarrow M_2 , \quad \text{and} \quad z_i \leftarrow x_{a,j} .$$

Deduce that

$$\delta \le \frac{nM_1^2 + n(q-1)M_2^2 - B(n, x, M_1) - B(n(q-1), x, M_2)}{M(M-1)n} .$$

Next, use part 2 of Problem 4.26 to claim that

$$\delta \le \frac{nM_1^2 + n(q-1)M_2^2 - B(n, M\theta n, M_1) - B(n(q-1), M\theta n, M_2)}{M(M-1)n}$$

and, finally, derive the result from the latter inequality.

3. Show that when $M\theta$ is an integer multiple of $q-1$, the bound $\delta \le \mathcal{J}(M, \theta, q)$ takes the form

$$\delta \le \frac{M}{M-1} \cdot \left(2\theta - \tfrac{q}{q-1}\theta^2\right)$$

(which is the expression in Proposition 4.11).

4. Given fixed positive integers M and q, consider the function

$$\theta \mapsto \mathcal{J}(M, \theta, q) ,$$

which is defined over the real interval $[0, 1]$. Show that this function is:

(a) continuous on the interval $[0, 1]$;

(b) linear on each interval $[\frac{\rho-1}{M}, \frac{\rho}{M}]$ with slope $2\left(1 - \frac{\lceil \rho q/(q-1)\rceil - 2}{M-1}\right)$ for $\rho = 1, 2, \ldots, M$;

(c) $\cap$-concave on the interval $[0, 1]$;

(d) strictly increasing for $0 < \theta < 1 - \frac{\lceil M/q \rceil}{M}$;

(e) strictly decreasing for $1 - \frac{\lfloor M/q \rfloor}{M} < \theta < 1$;

(f) flat for $1 - \frac{\lceil M/q \rceil}{M} < \theta < 1 - \frac{\lfloor M/q \rfloor}{M}$.

Problem 4.28 Let F be an Abelian group of size q, let M be an integer greater than 1, and let θ be a rational in $(0, 1-(1/q)]$. Show that for some integer n, there is an $(n, M, \delta n)$ code over F whose codewords all have Hamming weight θn and

$$\delta = \mathcal{J}(M, \theta, q) ,$$

where $\mathcal{J}(M, \theta, q)$ is as defined in Problem 4.27.

Hint: Select ℓ to be an integer such that $M\theta\ell$ is an integer multiple of $q-1$. Construct an $M \times \ell$ array U over F by filling in its entries, row by row, with $M\theta\ell/(q-1)$ copies of some nonzero element of F, followed by $M\theta\ell/(q-1)$ copies of some other nonzero element of F, and so on. The $M(1-\theta)\ell$ remaining entries of U are then filled with zeros. Next, consider the $M!$ arrays obtained by all possible permutations of the rows of U (different permutations can result in identical arrays): concatenate these arrays to produce an $M \times n$ array T over F, where $n = M!\ell$. Based on Problem 4.26, show that the values $x_{a,j}$ that are associated with T attain the minimum in (4.14) (over the integers) under the constraint (4.15). Conclude the proof by arguing that the rows of T form an equidistant code.

Problem 4.29 (Improvements on the Plotkin bound) Show that for every (n, M, d) code over an alphabet of size q,

$$\frac{d}{n} \le \frac{\binom{M-r+1}{2} + \binom{r}{2}(q-1)}{\binom{M}{2}} ,$$

where $r = \lceil M/q \rceil$.

Hint: Show that d/n is bounded from above by the maximum value of the function $\theta \mapsto \mathcal{J}(M, \theta, q)$ in Problem 4.27; then compute that maximum.

[Sections 4.6 and 4.7]

Problem 4.30 Let F be an Abelian group of size q. Fix θ to be a real in $(0, 1-(1/q)]$ and δ to be any real such that

$$0 < \delta < 2\theta - \tfrac{q}{q-1}\theta^2 .$$

The purpose of this problem is to show via probabilistic arguments that for increasing values of n, there exist $(n, M, >\delta n)$ codes over F that consist only of codewords whose Hamming weight is θn or less, while the code sizes M grow exponentially with n. (Thus, if δ is made arbitrarily close to $2\theta - (q/(q-1))\theta^2$, these codes approach the Johnson bound in Proposition 4.11; see also Problem 4.25.)

Let n be a positive integer and p be a real such that

$$p < \theta \quad \text{and} \quad \delta < 2p - \tfrac{q}{q-1} p^2$$

(why does such p exist?). Define the following probability distribution over F^n: the entries in each word $e_1 e_2 \ldots e_n$ in F^n are statistically independent, and for every $a \in F$ and $1 \le j \le n$,

$$\mathsf{Prob}\{ e_j = a \} = \begin{cases} 1-p & \text{if } a = 0 \\ p/(q-1) & \text{if } a \ne 0 \end{cases}$$

(this distribution coincides with that of the error words of length n in the q-ary symmetric channel with crossover probability p). Assume hereafter in this problem that all random selections of words in F^n are made according to this distribution.

1. Let $e_1 e_2 \ldots e_n$ and $e'_1 e'_2 \ldots e'_n$ be two randomly selected words in F^n. Show that for every $a \in F$ and $1 \le j \le n$,

$$\mathsf{Prob}\{ e_j - e'_j = a \} = \begin{cases} 1-\pi & \text{if } a = 0 \\ \pi/(q-1) & \text{if } a \ne 0 \end{cases},$$

   where

$$\pi = 2p - \tfrac{q}{q-1} p^2 .$$

2. Let $\mathbf{e}$ and $\mathbf{e}'$ be two randomly selected words in F^n. Show that

$$\mathsf{Prob}\{ \mathsf{d}(\mathbf{e}, \mathbf{e}') \le \delta n \} \le q^{-n \mathsf{D}_q(\delta \| \pi)} .$$

   Hint: Use Lemma 4.13.

3. Let M be an integer greater than 1 such that

$$M \cdot q^{-n \mathsf{D}_q(\theta \| p)} + \binom{M}{2} \cdot q^{-n \mathsf{D}_q(\delta \| \pi)} < 1 .$$

   Show that there exists an $(n, M, >\delta n)$ code over F in which each codeword has Hamming weight at most θn.

   Hint: Bound from above the probability that a randomly selected set of M words of F^n does not form such a code.

4. Show that as n increases, the integer M in part 3 can be chosen so that it grows exponentially with n.

Problem 4.31 (Shannon Coding Theorem while allowing error detection) The purpose of this problem is to show that when error detection is allowed, there is an attainable trade-off between the decoding error probability and the misdetection probability (see Problem 4.20).

Let $F = \mathrm{GF}(q)$ and consider transmission through a q-ary symmetric channel (F, F, Prob) with crossover probability p. For a linear $[n, nR]$ code $\mathcal{C}$ over F and a nonnegative real s, let $\mathcal{D}_s : F^n \to \mathcal{C} \cup \{\text{"e"}\}$ be the following decoder:

$$\mathcal{D}_s(\mathbf{y}) = \begin{cases} \mathbf{c} & \text{if } \exists \mathbf{c} \in \mathcal{C} \text{ such that } \mathsf{d}(\mathbf{y}, \mathbf{c}') > \mathsf{d}(\mathbf{y}, \mathbf{c}) + s \text{ for every } \mathbf{c}' \in \mathcal{C} \setminus \{\mathbf{c}\} \\ \text{"e"} & \text{otherwise} \end{cases} .$$

Define the decoding misdetection probability $\mathrm{P_{mis}}(\mathcal{C})$ of $\mathcal{D}_s$ by

$$\mathrm{P_{mis}}(\mathcal{C}) = \max_{\mathbf{c} \in \mathcal{C}} \mathrm{P_{mis}}(\mathbf{c}) ,$$

where

$$\mathrm{P_{mis}}(\mathbf{c}) = \sum_{\mathbf{y} : \mathcal{D}_s(\mathbf{y}) \in \mathcal{C} \setminus \{\mathbf{c}\}} \mathsf{Prob}\{ \mathbf{y} \text{ received} \mid \mathbf{c} \text{ transmitted} \}$$

(note that the decoding *error* probability, $\mathrm{P_{err}}$, is still defined as

$$\mathrm{P_{err}}(\mathcal{C}) = \max_{\mathbf{c} \in \mathcal{C}} \mathrm{P_{err}}(\mathbf{c}) ,$$

where

$$\mathrm{P_{err}}(\mathbf{c}) = \sum_{\mathbf{y} : \mathcal{D}_s(\mathbf{y}) \neq \mathbf{c}} \mathsf{Prob}\{ \mathbf{y} \text{ received} \mid \mathbf{c} \text{ transmitted} \} ;$$

in particular, the summation is taken also over values $\mathbf{y}$ for which $\mathcal{D}_s(\mathbf{y}) = \text{"e"}$).

The notations $\mathrm{P_{err}}(\mathcal{C}|\mathbf{e})$, $\mathrm{P_{err}}(\mathcal{C}|\mathcal{S}_q(n,t))$, $\mathrm{P_{err}}(\mathcal{C})$, and their averages extend from the respective definitions made, for a nearest-codeword decoder, in Section 4.7. Similar notations will now be used also for $\mathrm{P_{mis}}$.

Assume hereafter that $p \in (0, 1-(1/q))$ and $R < 1 - \mathsf{H}_q(p)$.

1. Show that for every real $t \geq 0$,

$$\overline{\mathrm{P_{err}}(\mathcal{C}|\mathcal{S}_q(n,t))} < q^{n(R-1)} \cdot V_q(n, t+s)$$

and

$$\overline{\mathrm{P_{mis}}(\mathcal{C}|\mathcal{S}_q(n,t))} \leq q^{n(R-1)} \cdot V_q(n, t-s) .$$

2. Denote by σ the ratio s/n. Show that for every $\theta_1 \in [p, 1-(1/q)-\sigma]$,

$$\overline{\mathrm{P_{err}}(\mathcal{C})} < q^{-n(1-\mathsf{H}_q(\theta_1+\sigma)-R)} + q^{-n\mathsf{D}_q(\theta_1 \| p)} ,$$

and for every $\theta_2 \in [p, 1-(1/q)+\sigma]$,

$$\overline{\mathrm{P_{mis}}(\mathcal{C})} \leq q^{-n(1-\mathsf{H}_q(\theta_2-\sigma)-R)} + q^{-n\mathsf{D}_q(\theta_2 \| p)} ,$$

where $\mathsf{H}_q(x)$ is taken as $-\infty$ when $x < 0$ and $\mathsf{D}_q(x\|p)$ is taken as $+\infty$ when $x > 1$.

3. Let $\mathsf{H}'_q(x)$ and $\mathsf{D}'_q(x\|p)$ be the derivatives of the functions $x \mapsto \mathsf{H}_q(x)$ and $x \mapsto \mathsf{D}_q(x\|p)$, respectively. Show that

$$\mathsf{H}_q(\theta + \epsilon) < \mathsf{H}_q(\theta) + \mathsf{H}'_q(\theta)\epsilon$$

and

$$\mathsf{D}_q(\theta + \epsilon \| p) > \mathsf{D}_q(\theta \| p) + \mathsf{D}'_q(\theta \| p)\epsilon$$

for every $\theta \in (0,1)$ and $|\epsilon| > 0$ such that $\theta + \epsilon \in (0,1)$.

Hint: Use convexity.

4. Let θ^* be the value $\theta_q^*(p, R)$ in Theorem 4.17. Define $A = \mathsf{H}_q'(\theta^*)$ and $B = \mathsf{D}_q'(\theta^* \| p)$. Show that both A and B are strictly positive.

5. Let $\mathrm{E}_q(p, R)$ be as in Theorem 4.17 and denote by $\gamma = \gamma_q(p, R)$ the value $AB/(A+B)$. Show that
$$\overline{\mathrm{P}_{\mathrm{err}}(\mathcal{C})} < 2q^{-n\mathrm{E}_q(p,R)} \cdot q^{\gamma s}$$
and
$$\overline{\mathrm{P}_{\mathrm{mis}}(\mathcal{C})} < 2q^{-n\mathrm{E}_q(p,R)} \cdot q^{-\gamma s} .$$
Hint: Let $\Delta = (A\sigma)/(A+B)$ and select $\theta_1 = \theta^* - \Delta$ and $\theta_2 = \theta^* + \Delta$ in part 2. Verify that $\theta_2 \in [p, 1-(1/q)+\sigma]$, and assume that $\theta_1 \in [p, 1-(1/q)-\sigma]$. Then apply the inequalities in part 3. (When $\theta_1 \notin [p, 1-(1/q)-\sigma]$, the upper bound on $\mathrm{P}_{\mathrm{err}}$ in part 2 becomes vacuous: for either boundary value, $\theta_1 = p$ or $\theta_1 = 1-(1/q)-\sigma$, that bound is greater than 1. Hence, so is the bound $2q^{-n\mathrm{E}_q(p,R)} \cdot q^{\gamma s}$.)

6. Conclude that there is a mapping $s \mapsto K_q(p, R, s)$ from the nonnegative reals onto $[1, \infty)$ such that, when ranging over all linear $[n, nR]$ codes $\mathcal{C}$ over F, the decoder $\mathcal{D}_s : F^n \to \mathcal{C}$ satisfies
$$\overline{\mathrm{P}_{\mathrm{err}}(\mathcal{C})} < 2 \cdot K_q(p, R, s) \cdot q^{-n\mathrm{E}_q(p,R)}$$
and
$$\overline{\mathrm{P}_{\mathrm{mis}}(\mathcal{C})} < 2 \cdot (K_q(p, R, s))^{-1} \cdot q^{-n\mathrm{E}_q(p,R)}$$
for every given $s \geq 0$.

Problem 4.32 (Shannon Converse Coding Theorem for the q-ary erasure channel) The purpose of this problem is to show that $1 - p$ is the largest possible rate at which information can be transmitted reliably through the memoryless q-ary erasure channel with erasure probability p.

Denote by F the input alphabet (of size q) of the channel and by $\Phi = F \cup \{?\}$ the output alphabet. Let $\mathcal{C}$ be an (n, q^{nR}) code over F where n and nR are integers such that $1 - p < R \leq 1$. Also, let $\mathcal{D} : \Phi^n \to \mathcal{C} \cup \{\text{"e"}\}$ be a decoder for $\mathcal{C}$ with respect to the q-ary erasure channel.

For a word $\mathbf{y} \in \Phi^n$, denote by $T(\mathbf{y})$ the set of indexes of the erased entries in $\mathbf{y}$. A word $\mathbf{x} \in F^n$ is said to agree with $\mathbf{y} \in \Phi^n$ if $\mathbf{x}$ and $\mathbf{y}$ agree on all their entries except those that are indexed by $T(\mathbf{y})$. Denote by $\mathcal{C}(\mathbf{y})$ the set of codewords in $\mathcal{C}$ that agree with $\mathbf{y}$.

1. Let J be a subset of $\{1, 2, \ldots, n\}$. Show that
$$\sum_{\mathbf{y} \in \Phi^n : T(\mathbf{y}) = J} \left| \left\{ \mathbf{c} \in \mathcal{C}(\mathbf{y}) \, : \, \mathcal{D}(\mathbf{y}) \neq \mathbf{c} \right\} \right| \geq \sum_{\mathbf{y} \in \Phi^n : T(\mathbf{y}) = J} (|\mathcal{C}(\mathbf{y})| - 1) = q^{nR} - q^{n-|J|} .$$

2. Show that for every subset $J \subseteq \{1, 2, \ldots, n\}$ of size $|J| \ge \theta n$,

$$\sum_{\mathbf{c} \in \mathcal{C}} \sum_{\mathbf{y} : \mathcal{D}(\mathbf{y}) \ne \mathbf{c}} \mathsf{Prob}\left\{ \mathbf{y} \text{ received} \;\middle|\; \mathbf{c} \text{ transmitted and } T(\mathbf{y}) = J \right\} \ge q^{nR} - q^{n(1-\theta)} .$$

3. Show that for every $\theta \in [0, 1]$, there is at least one codeword $\mathbf{c} \in \mathcal{C}$ for which

$$\sum_{\mathbf{y} : \mathcal{D}(\mathbf{y}) \ne \mathbf{c}} \mathsf{Prob}\left\{ \mathbf{y} \text{ received} \;\middle|\; \mathbf{c} \text{ transmitted and } |T(\mathbf{y})| \ge \theta n \right\} \ge 1 - q^{n(1-\theta-R)} .$$

4. Show that for every $\theta \in [0, p]$, the decoding error probability $\mathrm{P}_{\mathrm{err}}$ of $\mathcal{D}$ satisfies

$$\mathrm{P}_{\mathrm{err}} \ge (1 - q^{n(1-\theta-R)})(1 - q^{-n\mathsf{D}_q(\theta \| p)}) .$$

5. Show that

$$\mathrm{P}_{\mathrm{err}} > 1 - 2q^{-n\mathsf{D}_q(\theta \| p)} ,$$

where $\theta = \theta_q(p, R)$ is a solution to the equation

$$\theta - \mathsf{D}_q(\theta \| p) = 1 - R$$

in the open interval $(1-R, p)$. Verify that such a solution indeed exists and that it is unique.

Problem 4.33 (Shannon Coding Theorem for the q-ary erasure channel) The purpose of this problem is to show that when $F = \mathrm{GF}(q)$, the bound in Problem 4.32 can be approached by linear codes over F. The error probabilities herein are all defined with respect to the memoryless q-ary erasure channel with input alphabet $F = \mathrm{GF}(q)$, output alphabet $\Phi = F \cup \{?\}$, and erasure probability p.

1. Let k and r be positive integers such that $r \ge k$. Show that all but a fraction less than q^{k-r} of the $k \times r$ matrices over F have rank k.

   Hint: Assume a uniform distribution over all $k \times r$ matrices G and consider the probability that $\mathbf{u}G = \mathbf{0}$ for at least one $\mathbf{u} \in F^k \setminus \{\mathbf{0}\}$. Alternatively, show that there are

$$\prod_{i=0}^{k-1} (q^r - q^i)$$

   $k \times r$ matrices over F that have rank k.

2. For a linear $[n, k]$ code $\mathcal{C}$ over F, let $\mathcal{D} : \Phi^n \to \mathcal{C} \cup \{\text{“e”}\}$ be the following decoder:

$$\mathcal{D}(\mathbf{y}) = \begin{cases} \mathbf{c} & \text{if } \mathbf{y} \text{ agrees with exactly one } \mathbf{c} \in \mathcal{C} \text{ on the entries in } F \\ \text{“e”} & \text{otherwise} \end{cases} .$$

   For a set $J \subseteq \{1, 2, \ldots, n\}$, let $\mathrm{P}_{\mathrm{err}}(\mathcal{C}|J)$ be the decoding error probability of $\mathcal{D}$ conditioned on the erasures being indexed by J. Show that the average, $\overline{\mathrm{P}_{\mathrm{err}}(\mathcal{C}|J)}$, of $\mathrm{P}_{\mathrm{err}}(\mathcal{C}|J)$ over all linear $[n, k]$ codes $\mathcal{C}$ over F satisfies

$$\overline{\mathrm{P}_{\mathrm{err}}(\mathcal{C}|J)} < q^{k-n+|J|} .$$

Hint: Decoding will be successful if the columns of the generator matrix of $\mathcal{C}$ that are not indexed by J form a matrix whose rank is k.

3. For a linear $[n, nR]$ code $\mathcal{C}$, let $\mathrm{P}_{\mathrm{err}}(\mathcal{C})$ be the decoding error probability of the decoder $\mathcal{D}$ defined in part 2. Show that when $R < 1 - p$, the average, $\overline{\mathrm{P}_{\mathrm{err}}(\mathcal{C})}$, of $\mathrm{P}_{\mathrm{err}}(\mathcal{C})$ over all linear $[n, nR]$ codes $\mathcal{C}$ over F satisfies

$$\overline{\mathrm{P}_{\mathrm{err}}(\mathcal{C})} < 2q^{-n\mathsf{D}_q(\theta\|p)} ,$$

where $\theta = \tilde{\theta}_q(p, R)$ is a solution to

$$\theta + \mathsf{D}_q(\theta\|p) = 1 - R$$

in the open interval $(p, 1-R)$. Verify that such a solution indeed exists and that it is unique.

Notes

[Section 4.1]

The Singleton bound (Theorem 4.1) is named after Singleton's paper [339], where the linear case was stated. Yet, the bound had been known already in the early 1950s. MDS codes will be discussed in more detail in Chapter 11.

The Griesmer bound (part 2 of Problem 4.4) was obtained in [163].

[Section 4.2]

The sphere-packing bound (Theorem 4.3) was obtained by Hamming in [169].

The binary *Golay code* is a linear $[23, 12, 7]$ code over GF(2) whose 12×23 generator matrix has the echelon form

$$G = \begin{pmatrix} g_0 & g_1 & \cdots & g_r & & & & \\ & g_0 & g_1 & \cdots & g_r & & \mathbf{0} & \\ \mathbf{0} & & \ddots & \ddots & \cdots & \ddots & \\ & & & g_0 & g_1 & \cdots & g_r \end{pmatrix} , \tag{4.23}$$

where $r = 11$ and

$$(g_0 \, g_1 \, \ldots \, g_{11}) = (1\,0\,1\,0\,1\,1\,1\,0\,0\,0\,1\,1) .$$

The ternary Golay code is a linear $[11, 6, 5]$ code over GF(3) whose 6×11 generator matrix has the form (4.23) where now $r = 5$ and

$$(g_0 \, g_1 \, \ldots \, g_5) = (2\,0\,1\,2\,1\,1) .$$

These two codes were introduced by Golay in [149]. Golay codes have been among the most-studied codes in coding theory, as they possess rather unique combinatorial properties. See MacWilliams and Sloane [249, Chapter 20] and Pless [280, Chapter 10].

It is known that the Golay codes and the odd-length binary repetition code in Example 4.1 are the only perfect codes with minimum distance $d > 3$ over finite fields; see van Lint [233], Tietäväinen [364], and Zinov'ev and Leont'ev [399]. As for perfect codes with minimum distance $d = 3$, the Hamming codes are the only such codes that are *linear*, yet there are nonlinear constructions of perfect codes for $d = 3$ which are inequivalent to Hamming codes; see Etzion and Vardy [117] and the references therein.

Burst errors (Problem 4.12, and see also Problems 2.21 and 2.22) are a common model for describing the error patterns in many communication systems. For more on burst errors, see the books by Lin and Costello [230, Chapter 9] and Peterson and Weldon [278, Chapter 14]. The Reiger bound was obtained in [292].

There is an extensive literature on the covering radius of codes (Problem 4.13). The book by Cohen *et al.* [81] contains a thorough treatment of this subject. See also Cohen *et al.* [82], Cohen *et al.* [83], and Graham and Sloane [161]. It was shown by Goblick in [148] that (the asymptotic version of) the sphere-covering bound is attained by linear codes. Blinovskii then showed in [55] that all but a small fraction of these code attain this bound. See also Delsarte and Piret [100].

[Section 4.3]

The Gilbert–Varshamov bound (Theorem 4.4) was obtained by Gilbert [146], Varshamov [372], and Sacks [311]. Theorem 4.5 is an example of a *random coding* result, in that it states a property that holds for all but a (small) fraction of codes in a given code ensemble; yet, it does not provide an efficient algorithm for finding even one instance that satisfies the stated property. The proof of Theorem 4.5 makes use of an inequality of the form

$$\mathsf{Prob}\Big\{\cup_{\mathbf{u}\in U}\mathcal{A}_{\mathbf{u}}\Big\} \le \sum_{\mathbf{u}\in U} \mathsf{Prob}\left\{\mathcal{A}_{\mathbf{u}}\right\} ,$$

where $\mathcal{A}_{\mathbf{u}}$ denotes an event that is indexed by $\mathbf{u} \in U$ (specifically in that proof, $\mathcal{A}_{\mathbf{u}}$ stands for the event "$\mathbf{u}G \in \mathcal{S}$"). This inequality is commonly referred to as the *union bound*.

[Section 4.4]

Theorem 4.6 was obtained by MacWilliams [247], [248]. The exposition here of MacWilliams' Theorem makes use of dual codes and, as such, it applies to linear codes. However, the concept of a dual weight distribution can be generalized to nonlinear codes, and MacWilliams' Theorem can be extended accordingly.

Krawtchouk polynomials were defined in [219], and a summary of their properties can be found in MacWilliams and Sloane [249, Section 5.7], Nikiforov *et al.* [270], and Szegő [351, pp. 35–37]. One obvious property is that $\deg \mathcal{K}_\ell(y) = \ell$ and, so, every polynomial $\lambda(y) \in \mathbb{R}_{n+1}[y]$ can be expressed uniquely in the form $\lambda(y) = \sum_{\ell=0}^{n} \lambda_\ell \mathcal{K}_\ell(y)$ for some reals λ_ℓ.

On integer and linear programming see Luenberger [242], Nemhauser and Wolsey [268], and Schrijver [324]. It follows from the *duality theorem* of linear programming [242, Chapter 4] that solving the rational version of (4.9) and (4.10) is equivalent to solving the *dual* linear programming problem

$$1 + \min \sum_{\ell=1}^{n} \binom{n}{\ell} (q-1)^{\ell} \lambda_{\ell} , \tag{4.24}$$

where the minimum is taken over all rationals $\lambda_1, \lambda_2, \ldots, \lambda_n$ such that

$$\left\{ \begin{array}{rcll} \lambda_{\ell} & \geq & 0 , & 1 \leq \ell \leq n \\ \sum_{\ell=1}^{n} \mathcal{K}_{\ell}(i) \, \lambda_{\ell} & \leq & -1 , & d \leq i \leq n \end{array} \right. . \tag{4.25}$$

The application of linear programming to bounding code parameters was introduced by Delsarte [93]–[95]. In particular, as shown by Delsarte, the dual linear programming problem (4.24)–(4.25) can be recast in the following manner.

Theorem 4.19 (Delsarte's linear programming bound) *Given $F = \mathrm{GF}(q)$ and positive integers n and d, let the polynomial $\lambda(y) \in \mathbb{R}_{n+1}[y]$ have a Krawtchouk expansion*

$$\lambda(y) = 1 + \sum_{\ell=1}^{n} \lambda_{\ell} \, \mathcal{K}_{\ell}(y; n, q)$$

such that $\lambda_{\ell} \geq 0$ for $1 \leq \ell \leq n$. Suppose in addition that the values

$$\lambda(d), \lambda(d{+}1), \ldots, \lambda(n)$$

are all nonpositive. Then the size of every (linear) code of length n and minimum distance d over F is bounded from above by $\lambda(0)$.

[Section 4.5]

As seen from Problem 4.15, Theorem 4.10 holds even if we restrict it to linear codes that have *systematic* $(nR) \times n$ generator matrices over $\mathrm{GF}(q)$; the size of this ensemble of codes is $q^{n^2 R(1-R)}$. Clearly, the theorem only becomes stronger if we can reduce the ensemble size even further. Indeed, it will follow from Problem 12.10 that Theorem 4.10 holds also when the linear codes are assumed to be generated by $(nR) \times n$ matrices of the form (4.23), where $r = n(1-R)$ and $g_0 + g_1 x + \ldots + g_r x^r$ is a monic irreducible polynomial over $\mathrm{GF}(q)$. The size of this ensemble is less than $q^{n(1-R)}$. Other ensembles of comparable sizes will be presented in Section 12.4 and Problem 12.11.

Using tools from algebraic geometry, Tsfasman *et al.* [365] obtained a construction of an infinite family of linear $[n, nR, \delta n]$ codes over fields $\mathrm{GF}(q)$ with q square, such that

$$R \geq 1 - \frac{1}{\sqrt{q} - 1} - \delta - o(1) .$$

These codes exceed the bound of Theorem 4.10 for field sizes $q \geq 49$. The time complexity of the fastest algorithm currently known for computing generator matrices of such codes is proportional to $(n \log_q n)^3$. See Brigand [66], Katsman *et al.* [208], and Shum *et al.* [338].

The Plotkin bound (part 4 of Problem 4.23) appeared in [282]. It follows from the Plotkin bound that given an alphabet F of size q and a fixed $\delta \in (1-(1/q), 1]$, all $(n, M, \geq\delta n)$ codes over F must satisfy

$$M \leq \left(1 - \frac{1-(1/q)}{\delta}\right)^{-1} .$$

That is, the code size is bounded from above by a constant that is independent of n. Equivalently, the code rate is bounded from above by c/n for a constant c that depends only on q and δ. On the other hand, when $\delta = 1 - (1/q)$, one can obtain codes whose rate is already $(1 + \log_q n)/n$. The first-order Reed–Muller code (Problem 2.17) is an example of such a code.

The Johnson bound (Proposition 4.11) was obtained in [198]. This bound, which is one of several bounds in coding theory that are named after Johnson, is commonly stated for constant-weight codes (see Problem 4.10), where each codeword has Hamming weight *exactly* θn. Problem 4.25 demonstrates that the bound in Proposition 4.11 is attainable for a fairly dense set of triples (M, θ, q) (see also Problem 4.30, which is taken from Goldreich *et al.* [151, Section 4.3]). The improvement on Proposition 4.11 in Problem 4.27 is achieved by optimizing over integer values rather than over the reals. We will present an application of the improved bound in Section 9.8.

The Elias bound (Theorem 4.12) can be found in Shannon *et al.* [331]. For the McEliece–Rodemich–Rumsey–Welch (MRRW) bound, see [260].

[Sections 4.6 and 4.7]

The Channel Coding Theorem and its converse were first proved by Shannon in [330], and Theorems 4.17 and 4.14 are special cases of his theorems for the q-ary symmetric channel. Shannon's theorems can be found in any textbook on information theory. See, for instance, Cover and Thomas [87], Csiszár and Körner [88], Gallager [140], McEliece [259], and Viterbi and Omura [374]. We mention that Theorem 4.17 does not provide the best—namely, fastest—exponential decay of the decoding error probability; see Gallager [140] (and the exposition below) and Viterbi and Omura [374, Section 3.10].

In Lemma 4.13 (and also in Lemma 4.7 and Problem 4.22) we used a bounding technique known as the *Chernoff bound*; see, for example, Gallager [140, Section 5.4]. One variant of this bound is given next.

Proposition 4.20 (Chernoff bound) *Let $X_1, X_2, \ldots, X_n$ be real random variables that are statistically independent and identically distributed. Then,*

$$\mathsf{Prob}\{\, X_1 + X_2 + \ldots + X_n \leq \theta n \,\} \leq \alpha_\theta^n ,$$

where

$$\alpha_\theta = \inf_{z \in (0,1]} z^{-\theta} \mathsf{E}_X\{z^X\} ,$$

and $\mathsf{E}_X\{\cdot\} = \mathsf{E}_{X_j}\{\cdot\}$ denotes expected value with respect to the distribution of X_j.

Proof. For every $z \in (0,1]$ we have,

$$\begin{aligned}
\mathsf{Prob}\{ X_1 + X_2 + \ldots + X_n \le \theta n \} &\le \mathsf{E}_{X_1,X_2,\ldots,X_n}\left\{ z^{X_1+X_2+\ldots+X_n-\theta n} \right\} \\
&= z^{-\theta n} \cdot \mathsf{E}_{X_1,X_2,\ldots,X_n}\left\{ \prod_{j=1}^{n} z^{X_j} \right\} \\
&= z^{-\theta n} \cdot \prod_{j=1}^{n} \mathsf{E}_{X_j}\left\{ z^{X_j} \right\} \\
&= \left(z^{-\theta} \mathsf{E}_X \{ z^X \} \right)^n .
\end{aligned}$$

The result is obtained by taking the infimum over z. □

Lemma 4.13(i) can be thought of as a special case of Proposition 4.20 where the random variables $\{X_j\}_j$ are independent Bernoulli trials taking values in $\{0,1\}$, and $X_j = 1$ if and only if the jth coordinate of the error word $\mathbf{e}$ is nonzero.

The trade-off between the decoding error probability and the decoding misdetection probability (Problem 4.31) can be found in Forney [129, Appendix A].

In the remaining part of the notes on this chapter, we present the Channel Coding Theorem for an arbitrary discrete memoryless channel. Our exposition is based on the work of Gallager [140, Chapter 5].

Let $S = (F, \Phi, \mathsf{Prob})$ be a channel and let n and M be positive integers. We consider the following model of random codes of length n and size M over F. Fix a probability distribution $P_n : F^n \to [0,1]$ and a *messages set* U of size M. A *random* (n, M) *encoder over* F *(with respect to* P_n*)* is a random mapping $\mathcal{E} : U \to F^n$ whose value at any message $\mathbf{u} \in U$ has the distribution P_n, independently of the values at all other messages. In other words, we assume a distribution $\mathsf{P}_{\mathcal{E}}$ on the random encoders where for every mapping $\varepsilon : U \to F^n$,

$$\mathsf{P}_{\mathcal{E}} \{ \mathcal{E} = \varepsilon \} = \prod_{\mathbf{u} \in U} P_n(\varepsilon(\mathbf{u})) .$$

We then say that $\mathsf{P}_{\mathcal{E}}$ is induced by P_n. While $\mathsf{P}_{\mathcal{E}}$ denotes hereafter the distribution of $\mathcal{E}$, we will also make use of the conditional distribution Prob of the channel S, with the shorthand notation

$$\mathsf{Prob}(\mathbf{y}|\mathbf{x}) = \mathsf{Prob}\{ \mathbf{y} \text{ received} \mid \mathbf{x} \text{ transmitted} \} .$$

The set of images of $\mathcal{E}$ forms a (random) (n, M) code $\mathcal{C}$ over F, with the additional twist that $\mathcal{C}$ is allowed to be a multi-set: distinct messages may be mapped by $\mathcal{E}$ to the same codeword. Thus, it will be more appropriate to define a decoder (for an encoder $\mathcal{E}$ with respect to S) as a mapping $\mathcal{D} : \Phi^n \to U$, where the images of $\mathcal{D}$ are now messages rather than codewords. Consequently, an MLD for $\mathcal{E}$ with respect to S is a function $\mathcal{D}_{\mathrm{MLD}} : \Phi^n \to U$ such that for every $\mathbf{y} \in \Phi^n$, the value $\mathcal{D}_{\mathrm{MLD}}(\mathbf{y})$ equals the message $\mathbf{u} \in U$ that maximizes the conditional probability

$$\mathsf{Prob}\{ \mathbf{y} \text{ received} \mid \mathcal{E} \text{ was selected and } \mathcal{E}(\mathbf{u}) \text{ transmitted} \}$$

(this probability does not depend on the probability measure $\mathsf{P}_{\mathcal{E}}$ assumed on $\mathcal{E}$).

The next theorem provides an upper bound on the average decoding error probability (per transmitted message) of an MLD for a random encoder with respect to an arbitrary channel $S = (F, \Phi, \mathsf{Prob})$.

Theorem 4.21 *Given a channel* $S = (F, \Phi, \mathsf{Prob})$, *a code length* n, *a message set* U *of size* M, *a message* $\mathbf{u} \in U$, *and a distribution* $P_n : F^n \to [0, 1]$, *let* $\mathrm{P_{err}}(\mathbf{u}|\varepsilon)$ *be the decoding error probability of an MLD for a random* (n, M) *encoder* $\mathcal{E}$, *conditioned on* $\mathbf{u}$ *being the transmitted message and on using a prescribed instance* $\varepsilon : U \to F^n$ *of* $\mathcal{E}$ *for the encoding. Denote by* $\overline{\mathrm{P_{err}}(\mathbf{u})}$ *the expected value of the random variable* $\mathrm{P_{err}}(\mathbf{u}|\mathcal{E})$, *namely,*

$$\overline{\mathrm{P_{err}}(\mathbf{u})} = \sum_{\varepsilon} \mathsf{P}_{\mathcal{E}}\{\mathcal{E}{=}\varepsilon\} \cdot \mathrm{P_{err}}(\mathbf{u}|\varepsilon) ,$$

where $\mathsf{P}_{\mathcal{E}}$ *is the distribution of* $\mathcal{E}$ *induced by* P_n. *Then for every real* $\varrho \in [0, 1]$,

$$\overline{\mathrm{P_{err}}(\mathbf{u})} \le (M{-}1)^{\varrho} \sum_{\mathbf{y} \in \Phi^n} \Bigl(\sum_{\mathbf{x} \in F^n} P_n(\mathbf{x}) \cdot (\mathsf{Prob}(\mathbf{y}|\mathbf{x}))^{1/(\varrho+1)} \Bigr)^{\varrho+1} .$$

Proof. For $\mathbf{c} \in F^n$ and $\mathbf{y} \in \Phi^n$, denote by $\overline{\mathrm{P_{err}}(\mathbf{u} \,|\, \mathcal{E}(\mathbf{u}){=}\mathbf{c}, \mathbf{y})}$ the expected decoding error probability of an MLD for a random (n, M) encoder $\mathcal{E}$, conditioned on all the following three events: (i) having $\mathbf{u}$ as the transmitted message, (ii) having $\mathcal{E}(\mathbf{u}) = \mathbf{c}$, and (iii) receiving $\mathbf{y}$ at the output of the channel. Clearly,

$$\begin{aligned} \overline{\mathrm{P_{err}}(\mathbf{u})} &= \sum_{\mathbf{y} \in \Phi^n} \sum_{\mathbf{c} \in F^n} \mathsf{P}_{\mathcal{E}}\{\mathcal{E}(\mathbf{u}) = \mathbf{c}\} \cdot \mathsf{Prob}(\mathbf{y}|\mathbf{c}) \cdot \overline{\mathrm{P_{err}}(\mathbf{u} \,|\, \mathcal{E}(\mathbf{u}){=}\mathbf{c}, \mathbf{y})} \\ &= \sum_{\mathbf{y} \in \Phi^n} \sum_{\mathbf{c} \in F^n} P_n(\mathbf{c}) \cdot \mathsf{Prob}(\mathbf{y}|\mathbf{c}) \cdot \overline{\mathrm{P_{err}}(\mathbf{u} \,|\, \mathcal{E}(\mathbf{u}){=}\mathbf{c}, \mathbf{y})} . \qquad (4.26) \end{aligned}$$

Now,

$$\begin{aligned} &\overline{\mathrm{P_{err}}(\mathbf{u} \,|\, \mathcal{E}(\mathbf{u}){=}\mathbf{c}, \mathbf{y})} \\ &\quad \le \mathsf{P}_{\mathcal{E}} \Bigl\{ \mathsf{Prob}(\mathbf{y}|\mathcal{E}(\mathbf{u}')) \ge \mathsf{Prob}(\mathbf{y}|\mathbf{c}) \text{ for some } \mathbf{u}' \in U \setminus \{\mathbf{u}\} \Bigm| \mathcal{E}(\mathbf{u}){=}\mathbf{c}, \mathbf{y} \Bigr\} \\ &\quad \le \min\Bigl\{ \sum_{\mathbf{u}' \in U \setminus \{\mathbf{u}\}} \mathsf{P}_{\mathcal{E}} \Bigl\{ \mathsf{Prob}(\mathbf{y}|\mathcal{E}(\mathbf{u}')) \ge \mathsf{Prob}(\mathbf{y}|\mathbf{c}) \Bigm| \mathcal{E}(\mathbf{u}){=}\mathbf{c}, \mathbf{y} \Bigr\}, 1 \Bigr\} \\ &\quad = \min\Bigl\{ \sum_{\mathbf{u}' \in U \setminus \{\mathbf{u}\}} \mathsf{P}_{\mathcal{E}} \Bigl\{ \mathsf{Prob}(\mathbf{y}|\mathcal{E}(\mathbf{u}')) \ge \mathsf{Prob}(\mathbf{y}|\mathbf{c}) \Bigr\}, 1 \Bigr\} , \end{aligned}$$

with the (last) equality following from the fact that for $\mathbf{u}' \ne U \setminus \{\mathbf{u}\}$, the value $\mathcal{E}(\mathbf{u}')$ is statistically independent of $\mathcal{E}(\mathbf{u})$ and, obviously, it is also independent of $\mathbf{y}$. Hence, for every $\mathbf{c} \in F^n$, $\mathbf{y} \in \Phi^n$, and $\varrho \in [0, 1]$,

$$\overline{\mathrm{P_{err}}(\mathbf{u} \,|\, \mathcal{E}(\mathbf{u}){=}\mathbf{c}, \mathbf{y})} \le \Bigl(\sum_{\mathbf{u}' \in U \setminus \{\mathbf{u}\}} \mathsf{P}_{\mathcal{E}} \Bigl\{ \mathsf{Prob}(\mathbf{y}|\mathcal{E}(\mathbf{u}')) \ge \mathsf{Prob}(\mathbf{y}|\mathbf{c}) \Bigr\} \Bigr)^{\varrho} . \qquad (4.27)$$

Next, we bound each term in the sum in (4.27): for every $\mathbf{u}' \in U \setminus \{\mathbf{u}\}$, $\mathbf{c} \in F^n$, $\mathbf{y} \in \Phi^n$, and $\sigma > 0$ we have

$$\begin{aligned} \mathsf{P}_{\mathcal{E}}\Big\{\mathsf{Prob}(\mathbf{y}|\mathcal{E}(\mathbf{u}')) \ge \mathsf{Prob}(\mathbf{y}|\mathbf{c})\Big\} &= \sum_{\substack{\mathbf{x}\in F^n: \\ \mathsf{Prob}(\mathbf{y}|\mathbf{x}) \ge \mathsf{Prob}(\mathbf{y}|\mathbf{c})}} P_n(\mathbf{x}) \\ &\le \sum_{\mathbf{x}\in F^n} P_n(\mathbf{x}) \cdot \left(\frac{\mathsf{Prob}(\mathbf{y}|\mathbf{x})}{\mathsf{Prob}(\mathbf{y}|\mathbf{c})}\right)^{\sigma} , \end{aligned}$$

and combining with (4.27) yields

$$\overline{\mathrm{P}_{\mathrm{err}}(\mathbf{u}\,|\,\mathcal{E}(\mathbf{u})=\mathbf{c}, \mathbf{y})} \le \left((M-1) \sum_{\mathbf{x}\in F^n} P_n(\mathbf{x}) \cdot \left(\frac{\mathsf{Prob}(\mathbf{y}|\mathbf{x})}{\mathsf{Prob}(\mathbf{y}|\mathbf{c})}\right)^{\sigma}\right)^{\varrho} .$$

Finally, we substitute the latter inequality into (4.26) to obtain

$$\begin{aligned} \overline{\mathrm{P}_{\mathrm{err}}(\mathbf{u})} &\le \sum_{\mathbf{y}\in\Phi^n} \sum_{\mathbf{c}\in F^n} P_n(\mathbf{c}) \cdot \mathsf{Prob}(\mathbf{y}|\mathbf{c}) \cdot \left((M-1) \sum_{\mathbf{x}\in F^n} P_n(\mathbf{x}) \cdot \left(\frac{\mathsf{Prob}(\mathbf{y}|\mathbf{x})}{\mathsf{Prob}(\mathbf{y}|\mathbf{c})}\right)^{\sigma}\right)^{\varrho} \\ &= (M-1)^{\varrho} \sum_{\mathbf{y}\in\Phi^n} \left(\sum_{\mathbf{c}\in F^n} P_n(\mathbf{c}) \cdot (\mathsf{Prob}(\mathbf{y}|\mathbf{c}))^{1-\varrho\sigma}\right) \left(\sum_{\mathbf{x}\in F^n} P_n(\mathbf{x}) \cdot (\mathsf{Prob}(\mathbf{y}|\mathbf{x}))^{\sigma}\right)^{\varrho} . \end{aligned}$$

The result follows by selecting σ so that $1 - \varrho\sigma = \sigma$, namely, $\sigma = 1/(\varrho + 1)$. □

Hereafter, we specialize to the case where S is a discrete memoryless channel (in short, DMC), namely, it satisfies

$$\mathsf{Prob}(y_1 y_2 \ldots y_n | x_1 x_2 \ldots x_n) = \prod_{j=1}^{n} \mathsf{Prob}(y_j | x_j)$$

for every $x_1 x_2 \ldots x_n \in F^n$ and $y_1 y_2 \ldots y_n \in \Phi^n$ (see Problem 1.8). We further assume that the probability distribution P_n takes the form

$$P_n(x_1 x_2 \ldots x_n) = \prod_{j=1}^{n} P(x_j)$$

for a probability distribution $P : F \to [0, 1]$. Denoting $|F|$ by q, the bound of Theorem 4.21 then becomes

$$\begin{aligned} \overline{\mathrm{P}_{\mathrm{err}}(\mathbf{u})} &\le (M-1)^{\varrho} \left(\sum_{y\in\Phi} \left(\sum_{x\in F} P(x) \cdot (\mathsf{Prob}(y|x))^{1/(\varrho+1)}\right)^{\varrho+1}\right)^{n} \\ &\le q^{-n(\mathbb{E}_S(\varrho, P) - \varrho R)} , \end{aligned} \tag{4.28}$$

where R can be any positive real such that $q^{nR} \ge M-1$ (e.g., R is the rate $(\log_q M)/n$) and

$$\mathbb{E}_S(\varrho, P) = -\log_q \left(\sum_{y\in\Phi} \left(\sum_{x\in F} P(x) \cdot (\mathsf{Prob}(y|x))^{1/(\varrho+1)}\right)^{\varrho+1}\right) .$$

Obviously, the upper bound (4.28) will be tightest if we maximize the expression $\mathbb{E}_S(\varrho, P) - \varrho R$ over ϱ and P. Thus,

$$\overline{\mathrm{P}_{\mathrm{err}}(\mathbf{u})} \le q^{-n\mathrm{E}_S(R)} ,$$

where

$$\mathrm{E}_S(R) = \max_{0 \le \varrho \le 1} \left\{ \max_P \{\mathbb{E}_S(\varrho, P)\} - \varrho R \right\} \tag{4.29}$$

and the inner maximization is taken over all distribution functions P over F.

The value $\mathrm{E}_S(R)$ is called *Gallager's random coding error exponent.* It is easy to see that this exponent is strictly positive for every rate R in the range

$$0 \le R < \max_{0 \le \varrho \le 1} \max_P \frac{\mathbb{E}_S(\varrho, P)}{\varrho} .$$

In particular, it is positive whenever

$$0 \le R < \max_P \lim_{\varrho \to 0} \frac{\mathbb{E}_S(\varrho, P)}{\varrho} .$$

Noting that $\mathbb{E}_S(0, P) = 0$, we can apply L'Hôpital rule to yield

$$\begin{aligned} \lim_{\varrho \to 0} \frac{\mathbb{E}_S(\varrho, P)}{\varrho} &= \frac{\partial}{\partial \varrho} \mathbb{E}_S(\varrho, P) \Big|_{\varrho=0} \\ &= \sum_{y \in \Phi} \sum_{x \in F} P(x) \cdot \mathsf{Prob}(y|x) \cdot \log_q \left(\frac{\mathsf{Prob}(y|x)}{\sum_{z \in F} P(z) \cdot \mathsf{Prob}(y|z)} \right) . \end{aligned}$$

Given a DMC S and a probability distribution $P : F \to [0, 1]$, let $Q : F \times \Phi \to [0, 1]$ be the probability distribution $Q(x, y) = P(x) \cdot \mathsf{Prob}(y|x)$; also, for $y \in \Phi$, denote by $\Psi(y)$ the marginal distribution $\sum_{x \in F} Q(x, y)$. Using this notation, we can write

$$\lim_{\varrho \to 0} \frac{\mathbb{E}_S(\varrho, P)}{\varrho} = \sum_{y \in \Phi} \sum_{x \in F} Q(x, y) \cdot \log_q \left(\frac{Q(x, y)}{P(x)\Psi(y)} \right) ,$$

with the right-hand side being identified as the *mutual information* $I(Q)$, which was defined in the notes on Section 1.4. We also recall from there that the capacity of the DMC S is given by

$$\mathsf{cap}(S) = \max_P I(Q) ,$$

where P ranges over all probability distributions on the elements of F and $Q(x, y) = P(x) \cdot \mathsf{Prob}(y|x)$. We conclude that $\mathrm{E}_S(R) > 0$ whenever

$$0 \le R < \max_P \lim_{\varrho \to 0} \frac{\mathbb{E}_S(\varrho, P)}{\varrho} = \max_P I(Q) = \mathsf{cap}(S) .$$

Hence, for this range of R, the average decoding error probability per message, $\overline{\mathrm{P}_{\mathrm{err}}(\mathbf{u})}$, in Theorem 4.21 decays to zero exponentially with n.

Observe, however, that we cannot infer from our analysis that such an exponential decay applies also to the average of the (overall) decoding error probability $\max_{\mathbf{u} \in U} \mathrm{P}_{\mathrm{err}}(\mathbf{u})$. Still, we can make such an inference if—instead of the average performance—we are interested only in an existence result. We demonstrate this in the next theorem.

Theorem 4.22 *Let $S = (F, \Phi, \mathsf{Prob})$ be a discrete memoryless channel where $|F| = q$ and let the real R be in the range $0 \le R < \mathsf{cap}(S)$. For every positive integer n there exists an $(n, M{=}\lceil q^{nR} \rceil)$ code $\mathcal{C}$ over F whose MLD (with respect to S) has decoding error probability $\mathrm{P}_{\mathrm{err}}(\mathcal{C})$ that satisfies*

$$\mathrm{P}_{\mathrm{err}}(\mathcal{C}) < 4q^{-n\mathrm{E}_S(R)} ,$$

where $\mathrm{E}_S(R)$ is given by (4.29).

Proof. Let (ϱ, P) be the pair of arguments of $\mathbb{E}_S(\cdot, \cdot)$ for which (4.29) is maximized and let U be a set of size $2M{-}2$ (hereafter we exclude the trivial case $M = 1$; we can also assume that $\varrho > 0$, or else $\mathrm{E}_S(R) = \mathbb{E}_S(0, P) = 0$). We apply Theorem 4.21 to the DMC S, where $\mathcal{E}$ is a random $(n, 2M{-}2)$ encoder whose distribution is induced by $P_n(x_1 x_2 \ldots x_n) = \prod_j P(x_j)$; this yields

$$\overline{\mathrm{P}_{\mathrm{err}}(\mathbf{u})} \le (2M{-}3)^{\varrho} \cdot q^{-n\mathbb{E}_S(\varrho,P)} \qquad \text{for every } \mathbf{u} \in U . \tag{4.30}$$

Denote by α the average of $\overline{\mathrm{P}_{\mathrm{err}}(\mathbf{u})}$ over all $\mathbf{u} \in U$, i.e.,

$$\alpha = \frac{1}{2M{-}2} \sum_{\mathbf{u} \in U} \overline{\mathrm{P}_{\mathrm{err}}(\mathbf{u})} = \frac{1}{2M{-}2} \sum_{\varepsilon} \mathsf{P}_{\mathcal{E}}\{\mathcal{E}{=}\varepsilon\} \sum_{\mathbf{u} \in U} \mathrm{P}_{\mathrm{err}}(\mathbf{u}|\varepsilon) .$$

From (4.30), we can bound α from above by

$$\alpha \le (2M{-}3)^{\varrho} \cdot q^{-n\mathbb{E}_S(\varrho,P)} .$$

We next claim that there must be a subset $U_0 \subseteq U$ of size M and an instance ε_0 of the random $(n, 2M{-}2)$ encoder $\mathcal{E}$ such that

$$\mathrm{P}_{\mathrm{err}}(\mathbf{u}|\varepsilon_0) \le 2\alpha \qquad \text{for every } \mathbf{u} \in U_0 ;$$

otherwise, every instance ε of $\mathcal{E}$ would satisfy $\mathrm{P}_{\mathrm{err}}(\mathbf{u}|\varepsilon) > 2\alpha$ for at least half the elements $\mathbf{u}$ of U, which, in turn, would yield the contradiction

$$\begin{aligned} \alpha &= \frac{1}{2M{-}2} \sum_{\varepsilon} \mathsf{P}_{\mathcal{E}}\{\mathcal{E}{=}\varepsilon\} \sum_{\mathbf{u} \in U} \mathrm{P}_{\mathrm{err}}(\mathbf{u}|\varepsilon) \\ &\ge \frac{1}{2M{-}2} \sum_{\varepsilon} \mathsf{P}_{\mathcal{E}}\{\mathcal{E}{=}\varepsilon\} \underbrace{\sum_{\mathbf{u}\,:\,\mathrm{P}_{\mathrm{err}}(\mathbf{u}|\varepsilon) > 2\alpha} \mathrm{P}_{\mathrm{err}}(\mathbf{u}|\varepsilon)}_{>(M-1)\cdot 2\alpha} > \alpha . \end{aligned}$$

Having ε_0 and U_0 at hand, we let $\mathcal{C}$ be the (n, M) code over F that is given by the set $\{\varepsilon_0(\mathbf{u}) : \mathbf{u} \in U_0\}$ (when $\alpha < 1/4$, the restriction of ε_0 to the domain U_0 is necessarily one-to-one; if this were not so, there would be a message $\mathbf{u}' \in U_0$ for which $(2\alpha \ge)\ \mathrm{P}_{\mathrm{err}}(\mathbf{u}'|\varepsilon_0) \ge 1/2$). We have,

$$\begin{aligned} \mathrm{P}_{\mathrm{err}}(\mathcal{C}) \le 2\alpha &\le 2 \cdot (2M{-}3)^{\varrho} \cdot q^{-n\mathbb{E}_S(\varrho,P)} \\ &< 4 \cdot \underbrace{(M{-}1)}_{<q^{nR}}{}^{\varrho} \cdot q^{-n\mathbb{E}_S(\varrho,P)} < 4 \cdot q^{-n(\mathbb{E}_S(\varrho,P) - \varrho R)} = 4 \cdot q^{-n\mathrm{E}_S(R)} . \end{aligned}$$

□

Example 4.7 Let S be the memoryless q-ary symmetric channel with crossover probability p. We compute $\mathrm{E}_S(R)$ using (4.29), by first finding the probability distribution $P : F \to [0,1]$ that maximizes $\mathbb{E}_S(\varrho, P)$, for any fixed ϱ. This is equivalent to minimizing the expression

$$\begin{aligned} q^{-\mathbb{E}_S(\varrho,P)} &= \sum_{y\in\Phi}\Bigl(\sum_{x\in F} P(x)\cdot(\mathsf{Prob}(y|x))^{1/(\varrho+1)}\Bigr)^{\varrho+1} \\ &= \sum_{y\in\Phi}\Bigl((1-p)^{1/(\varrho+1)}P(y) + (p/(q-1))^{1/(\varrho+1)} \sum_{x\in F\setminus\{y\}} P(x)\Bigr)^{\varrho+1} \end{aligned}$$

over all the nonnegative real vectors $(P(x))_{x\in F}$ that satisfy the constraint

$$\sum_{x\in F} P(x) = 1 \, .$$

The minimum is attained (rather expectedly) when $P(x) = 1/q$ for every $x \in F$, and for this distribution we get

$$q^{-\mathbb{E}_S(\varrho,P)} = q^{-\varrho}\left((1-p)^{1/(\varrho+1)} + ((q-1)^{\varrho}p)^{1/(\varrho+1)}\right)^{\varrho+1}$$

or

$$\mathbb{E}_S(\varrho, P) = \varrho - (\varrho+1)\log_q\left(\tau(\varrho) + \omega(\varrho)\right) \, ,$$

where

$$\tau(\varrho) = \tau(\varrho, p) = (1-p)^{1/(\varrho+1)} \quad \text{and} \quad \omega(\varrho) = \omega(\varrho, p, q) = ((q-1)^{\varrho}p)^{1/(\varrho+1)} \, .$$

Turning again to (4.29), the value of $\mathrm{E}_S(R)$ equals the maximum of the expression

$$\mathbb{E}_{S,R}(\varrho) = \varrho(1-R) - (\varrho+1)\log_q\left(\tau(\varrho) + \omega(\varrho)\right)$$

over all $\varrho \in [0,1]$. So, we write

$$\frac{d\mathbb{E}_{S,R}(\varrho)}{d\varrho} = 0$$

and, consequently, end up with the equality

$$1 - R = \mathsf{H}_q\left(\frac{\omega(\varrho)}{\tau(\varrho) + \omega(\varrho)}\right) \, . \tag{4.31}$$

Let $\theta = \theta(R)$ denote the value $\mathsf{H}_q^{-1}(1-R)$. We can solve (4.31) for ϱ to obtain

$$\varrho = \frac{\log((1-p)/p) - \log((1-\theta)/\theta)}{\log(q-1) + \log((1-\theta)/\theta)} \, .$$

One can see that this solution increases with θ and, therefore, decreases with R. Yet, this solution for ϱ is valid as long as it lies within the interval $[0,1]$: the upper boundary of this interval, $\varrho = 1$, is obtained as a solution when

$$\theta = \frac{1}{1 + \sqrt{(1-p)/(p(q-1))}} \, ,$$

and the respective rate value, $1 - \mathsf{H}_q(\theta)$, is called the *critical rate* and is denoted by R_{cr}. The lower boundary, $\varrho = 0$, is obtained as a solution when $\theta = p$, and thus corresponds to the rate value $R = 1 - \mathsf{H}_q(p) = \mathsf{cap}(S)$.

For rates R in the range $[R_{\mathrm{cr}}, \mathsf{cap}(S))$, we can plug the solution for ϱ into $\mathbb{E}_{S,R}(\varrho)$ to yield

$$\mathrm{E}_S(R) = \mathbb{E}_{S,R}(\varrho) = \mathsf{D}_q\left(\theta(R) \| p\right) .$$

On the other hand, for $R \in [0, R_{\mathrm{cr}})$, we need to substitute $\varrho = 1$ in $\mathbb{E}_{S,R}(\varrho)$, thereby yielding the following expression for $\mathrm{E}_S(R)$ (which is a linear function in R):

$$\mathrm{E}_S(R) = \mathbb{E}_{S,R}(1) = 1 - 2\log_q\left(\sqrt{1-p} + \sqrt{p(q-1)}\right) - R .$$

Summarizing, we have

$$\mathrm{E}_S(R) = \begin{cases} 1 - 2\log_q\left(\sqrt{1-p} + \sqrt{p(q-1)}\right) - R & \text{if } 0 \le R < R_{\mathrm{cr}} \\ \mathsf{D}_q\left(\mathsf{H}_q^{-1}(1-R) \| p\right) & \text{if } R_{\mathrm{cr}} \le R < \mathsf{cap}(S) \end{cases} ,$$

where

$$R_{\mathrm{cr}} = 1 - \mathsf{H}_q\left(\frac{1}{1 + \sqrt{(1-p)/(p(q-1))}}\right) .$$

Figure 4.2 depicts the curve $R \mapsto \mathrm{E}_S(R)$ for $q = 2$ and $p = 0.1$; for these parameters,

$$\mathsf{cap}(S) = 1 - \mathsf{H}_2(0.1) \approx 0.5310 , \quad R_{\mathrm{cr}} = 1 - \mathsf{H}_2(0.25) \approx 0.1887 ,$$

and

$$\mathrm{E}_S(0) = \mathbb{E}_{S,0}(1) = 1 - 2\log_2(\sqrt{0.9} + \sqrt{0.1}) \approx 0.3219 .$$

We remark that—except for trivial cases—Gallager's exponent $\mathrm{E}_S(R)$ is larger than the exponent $\mathrm{E}_q(p, R)$ of random *linear* codes that was obtained in Theorem 4.17 and Corollary 4.18. Yet, $\mathrm{E}_q(p, R)$ can be improved to be equal to $\mathrm{E}_S(R)$ and even to surpass it for low rates; see Barg and Forney [27], Gallager [140, Section 6.2], Peterson and Weldon [278, Section 4.2], and Viterbi and Omura [374, Section 3.10]. □

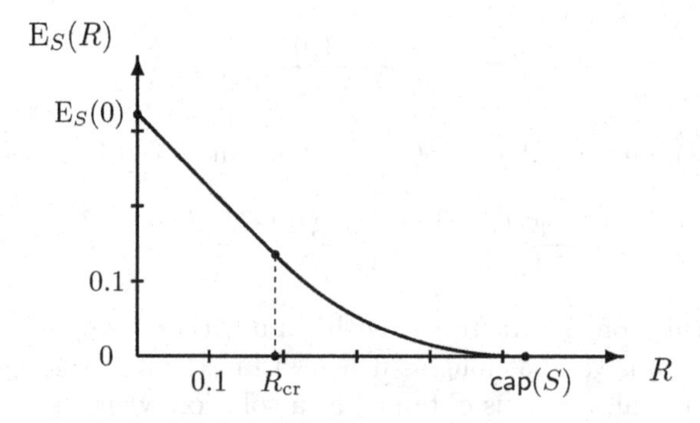

Figure 4.2. Gallager's random coding error exponent for the binary symmetric channel with crossover probability $p = 0.1$.

Chapter 5

Reed–Solomon and Related Codes

Generalized Reed–Solomon (in short, GRS) codes and their derivative codes are probably the most extensively-used codes in practice. This may be attributed to several advantages that these codes have. First, GRS codes are maximum distance separable, namely, they attain the Singleton bound. Secondly, being linear codes, they can be encoded efficiently; furthermore, as we see in this chapter, encoders for the sub-class of conventional Reed–Solomon (in short, RS) codes can be implemented by particularly simple hardware circuits. Thirdly, we will show in Chapters 6 and 9 that GRS codes can also be decoded efficiently.

As their names suggest, RS codes pre-dated their GRS counterparts. Nevertheless, we find it more convenient herein to define GRS codes first and prove several properties thereof; we then present RS codes as a special class of GRS codes.

One seeming limitation of GRS codes is the fact that their length is bounded from above by the size of the field over which they are defined. This could imply that these codes might be useful only when the application calls for a field size that is relatively large, e.g., when the field is $\mathrm{GF}(2^8)$ and the symbols are bytes. Still, we show that GRS codes can serve as building blocks to derive new codes over small alphabets as well. We present two methods for doing so. The first technique is called concatenation and is based on two stages of encoding, the first of which is a GRS encoder. In the second method, we first construct a GRS code over a sufficiently large extension field of the target field F; we then extract from this code only the codewords that lie wholly in F. The codes over F thus obtained are called alternant codes, and when the underlying GRS code is an RS code then the resulting codes are called Bose–Chaudhuri–Hocquenghem (in short, BCH) codes.

5.1 Generalized Reed–Solomon codes

5.1.1 Definition

Let $F = \mathrm{GF}(q)$, let $\alpha_1, \alpha_2, \ldots, \alpha_n$ be distinct nonzero elements of F, and let $v_1, v_2, \ldots, v_n$ be nonzero elements of F (which do not have to be distinct). A *generalized Reed–Solomon* (in short, GRS) *code* over F is a linear $[n, k, d]$ code $\mathcal{C}_{\mathrm{GRS}}$ over F with a parity-check matrix

$$H_{\mathrm{GRS}} = \begin{pmatrix} 1 & 1 & \ldots & 1 \\ \alpha_1 & \alpha_2 & \ldots & \alpha_n \\ \alpha_1^2 & \alpha_2^2 & \ldots & \alpha_n^2 \\ \vdots & \vdots & \vdots & \vdots \\ \alpha_1^{n-k-1} & \alpha_2^{n-k-1} & \ldots & \alpha_n^{n-k-1} \end{pmatrix} \begin{pmatrix} v_1 & & & 0 \\ & v_2 & & \\ 0 & & \ddots & \\ & & & v_n \end{pmatrix} . \qquad (5.1)$$

The elements α_j are called *code locators* and the values v_j are called *column multipliers*. The definition of GRS codes requires that the length n be at most $q-1$. (The requirement that the code locators be nonzero is not necessary for many of the properties of GRS codes; yet, the decoding algorithm that we will present in Chapter 6 assumes that each code locator has a multiplicative inverse in F.) The matrix H_{GRS} is called a *canonical* parity-check matrix of $\mathcal{C}_{\mathrm{GRS}}$.

We remark that, typically, the same GRS code can be defined through more than one list of code locators (see Problem 5.4); therefore, a canonical parity-check matrix is not unique—not even up to scaling of the column multipliers.

We have already established in Proposition 4.2 the following result.

Proposition 5.1 *Every $[n, k, d]$ GRS code over F is MDS, namely, $d = n-k+1$.*

(Strictly speaking, we assumed in Proposition 4.2 that the column multipliers are all 1. However, multiplying each column of a parity-check matrix by a nonzero element of F does not affect the minimum distance of the code.)

Next, we turn to the dual codes of GRS codes. We know from Problem 4.1 that a code is MDS if and only if its dual code is. Therefore, the dual code of a GRS code is necessarily MDS. In fact, we can make an even stronger statement.

Proposition 5.2 *The dual code of an $[n, k{<}n]$ GRS code is an $[n, n-k]$ GRS code; furthermore, both codes can be defined through the same list of code locators.*

Proof. Let $\mathcal{C}_{\mathrm{GRS}}$ be an $[n, k]$ GRS code over F and let a canonical parity-check matrix H_{GRS} of $\mathcal{C}_{\mathrm{GRS}}$ be given by (5.1). Consider the $k \times n$ matrix

$$G_{\mathrm{GRS}} = \begin{pmatrix} 1 & 1 & \ldots & 1 \\ \alpha_1 & \alpha_2 & \ldots & \alpha_n \\ \alpha_1^2 & \alpha_2^2 & \ldots & \alpha_n^2 \\ \vdots & \vdots & \vdots & \vdots \\ \alpha_1^{k-1} & \alpha_2^{k-1} & \ldots & \alpha_n^{k-1} \end{pmatrix} \begin{pmatrix} v'_1 & & & 0 \\ & v'_2 & & \\ 0 & & \ddots & \\ & & & v'_n \end{pmatrix},$$

where each v'_j is an element of F. We prove that there is a choice of *nonzero* values v'_j such that $G_{\mathrm{GRS}} H_{\mathrm{GRS}}^T = 0$. This will show that the dual code $\mathcal{C}_{\mathrm{GRS}}^{\perp}$ is a GRS code, with the same code locators as $\mathcal{C}_{\mathrm{GRS}}$ and with column multipliers v'_j.

For each $i = 0, 1, \ldots, k-1$ and $\ell = 0, 1 \ldots, n-k-1$, we require that the scalar product of row i in G_{GRS} and row ℓ in H_{GRS} be zero, namely,

$$\sum_{j=1}^{n} v_j v'_j \alpha_j^{i+\ell} = 0 . \tag{5.2}$$

Now, $i + \ell$ ranges between 0 and $n-2$. Hence, $G_{\mathrm{GRS}} H_{\mathrm{GRS}}^T = 0$ if and only if

$$\sum_{j=1}^{n} v_j v'_j \alpha_j^{r} = 0 , \quad 0 \le r \le n-2 ,$$

or, in matrix notation,

$$\begin{pmatrix} 1 & 1 & \ldots & 1 \\ \alpha_1 & \alpha_2 & \ldots & \alpha_n \\ \alpha_1^2 & \alpha_2^2 & \ldots & \alpha_n^2 \\ \vdots & \vdots & \vdots & \vdots \\ \alpha_1^{n-2} & \alpha_2^{n-2} & \ldots & \alpha_n^{n-2} \end{pmatrix} \begin{pmatrix} v_1 & & & 0 \\ & v_2 & & \\ 0 & & \ddots & \\ & & & v_n \end{pmatrix} \begin{pmatrix} v'_1 \\ v'_2 \\ \vdots \\ v'_n \end{pmatrix} = \mathbf{0} .$$

The possible solutions for $(v'_1 \; v'_2 \; \ldots \; v'_n)$ are therefore the nonzero codewords of an $[n, 1, n]$ GRS code over F (that has the same code locators and column multipliers as $\mathcal{C}_{\mathrm{GRS}}$). These codewords all have Hamming weight n, i.e., each v'_j is nonzero. $\square$

We will identify certain GRS codes by special names.

When $n = q-1$, we say that the GRS code is *primitive*. In this case, the code locators range over all the nonzero elements of F.

A GRS code is called *normalized* if its column multipliers are all equal to 1.

In a *narrow-sense GRS code*, each column multiplier is equal to the respective code locator, i.e., $v_j = \alpha_j$ for all $1 \le j \le n$.

When one of the code locators is allowed to be zero, the resulting code is called a *(singly-)extended GRS code*. The longest such code has length q, in which case it is referred to as a singly-extended primitive GRS code. Note that we cannot allow the zero element to be among the code locators of a narrow-sense GRS code (or else the parity-check matrix would have an all-zero column).

Example 5.1 Let $v_1, v_2, \ldots, v_n$ be the column multipliers of a primitive GRS code over $F = \mathrm{GF}(q)$. We verify that the dual GRS code has column multipliers α_j/v_j. Let α be a primitive element in F. Substituting $v'_j = \alpha_j/v_j$ in the left-hand side of (5.2), for every $0 \le r \le n-2$ we get

$$\sum_{j=1}^{n} v_j v'_j \alpha_j^r = \sum_{j=1}^{n} \alpha_j^{r+1} = \sum_{j=1}^{n} \alpha^{(j-1)(r+1)} = \frac{\alpha^{n(r+1)} - 1}{\alpha^{r+1} - 1} = 0 ,$$

where the last equality follows from $\alpha^n = \alpha^{q-1} = 1$ (see Problem 3.22). In particular, the dual code of a normalized primitive GRS code is a narrow-sense primitive GRS code. □

5.1.2 Polynomial interpretation of GRS codes

Let C_{GRS} be an $[n, k, d]$ GRS code with a generator matrix

$$G_{\mathrm{GRS}} = \begin{pmatrix} 1 & 1 & \ldots & 1 \\ \alpha_1 & \alpha_2 & \ldots & \alpha_n \\ \alpha_1^2 & \alpha_2^2 & \ldots & \alpha_n^2 \\ \vdots & \vdots & \vdots & \vdots \\ \alpha_1^{k-1} & \alpha_2^{k-1} & \ldots & \alpha_n^{k-1} \end{pmatrix} \begin{pmatrix} v'_1 & & & 0 \\ & v'_2 & & \\ & & \ddots & \\ 0 & & & v'_n \end{pmatrix} ,$$

and suppose we encode an information word $\mathbf{u} = (u_0 \; u_1 \; \ldots \; u_{k-1})$ using the mapping

$$\mathbf{u} \mapsto \mathbf{u} G_{\mathrm{GRS}} .$$

We associate with $\mathbf{u}$ the polynomial $u(x) = u_0 + u_1 x + \ldots + u_{k-1} x^{k-1} \in F_k[x]$ and write

$$\mathbf{u} G_{\mathrm{GRS}} = \Big(v'_1 u(\alpha_1) \;\; v'_2 u(\alpha_2) \;\; \ldots \;\; v'_n u(\alpha_n) \Big) . \tag{5.3}$$

That is, the entries of the codeword associated with $\mathbf{u}$ are obtained by evaluating the polynomial $u(x)$ at the code locators α_j, where each value $u(\alpha_j)$ is scaled by v'_j (the scaling does not depend on $\mathbf{u}$).

This representation of GRS codes provides yet another argument why these codes are MDS. If $u(x)$ is a nonzero polynomial in $F_k[x]$, then it has at most $k-1$ distinct zeros in F. Therefore, there are at most $k-1$ zero entries in the codeword (5.3) and, so, the minimum distance of the code is at least $n - (k-1)$.

Suppose that $d = 2t + 1$ for some integer t, in which case $n = k + 2t$. The decoding problem of up to $(d-1)/2$ errors when using GRS codes can be formulated as follows: given the word in the right-hand side of (5.3), possibly with up to t altered entries, reconstruct the word $\mathbf{u}$. For the case $t = 0$ (and $n = k$), this is known as the polynomial interpolation problem: every polynomial in $F_k[x]$ can be reconstructed from its values at k distinct points (see Problem 3.14). For general t we have a *noisy* interpolation problem: every polynomial in $F_k[x]$ can be reconstructed from its values at $k + 2t$ distinct points, even when at most t of those values are wrong.

5.2 Conventional Reed–Solomon codes

Conventional Reed–Solomon (in short, RS) *codes* over $F = \mathrm{GF}(q)$ are special cases of GRS codes obtained as follows. Let n be a positive integer dividing $q-1$ and let α be an element of multiplicative order n in F. Also, let b be an integer. An $[n, k]$ RS code over F is a GRS code $\mathcal{C}_{\mathrm{RS}}$ with code locators

$$\alpha_j = \alpha^{j-1} , \quad 1 \le j \le n ,$$

and column multipliers

$$v_j = \alpha^{b(j-1)} , \quad 1 \le j \le n .$$

A canonical parity-check matrix of an RS code is given by

$$H_{\mathrm{RS}} = \begin{pmatrix} 1 & \alpha^b & \ldots & \alpha^{(n-1)b} \\ 1 & \alpha^{b+1} & \ldots & \alpha^{(n-1)(b+1)} \\ \vdots & \vdots & \vdots & \vdots \\ 1 & \alpha^{b+d-2} & \ldots & \alpha^{(n-1)(b+d-2)} \end{pmatrix}$$

(the number of rows in H_{RS} is $d-1 = n-k$).

Associate each word

$$\mathbf{c} = (c_0 \; c_1 \; \ldots \; c_{n-1})$$

in F^n with the polynomial

$$c(x) = c_0 + c_1 x + \ldots + c_{n-1} x^{n-1}$$

in $F_n[x]$. Clearly, for every $\mathbf{c} \in F^n$,

$$\mathbf{c} \in \mathcal{C}_{\mathrm{RS}} \quad \Longleftrightarrow \quad H_{\mathrm{RS}} \mathbf{c}^T = \mathbf{0} ,$$

and from the special form of H_{RS} we thus obtain the following characterization of $\mathcal{C}_{\mathrm{RS}}$:

$$\mathbf{c} \in \mathcal{C}_{\mathrm{RS}} \quad \Longleftrightarrow \quad c(\alpha^\ell) = 0 \quad \text{for } \ell = b, b{+}1, \ldots, b{+}d{-}2 \,. \tag{5.4}$$

The elements $\alpha^b, \alpha^{b+1}, \ldots, \alpha^{b+d-2}$ are called the *roots* of $\mathcal{C}_{\mathrm{RS}}$, and (5.4) can be rephrased as follows: $\mathbf{c} \in \mathcal{C}_{\mathrm{RS}}$ if and only if the roots of $\mathcal{C}_{\mathrm{RS}}$ are all roots of $c(x)$.

Define the *generator polynomial* $g(x)$ of $\mathcal{C}_{\mathrm{RS}}$ by the product

$$g(x) = (x - \alpha^b)(x - \alpha^{b+1}) \cdots (x - \alpha^{b+d-2}) \,.$$

By (5.4) it follows that for every $\mathbf{c} \in F^n$,

$$\mathbf{c} \in \mathcal{C}_{\mathrm{RS}} \quad \Longleftrightarrow \quad g(x) \,|\, c(x) \,.$$

This leads to yet another characterization of (the polynomial representations of the codewords of) the code $\mathcal{C}_{\mathrm{RS}}$:

$$\mathcal{C}_{\mathrm{RS}} = \{\, u(x)g(x) \;:\; u(x) \in F_k[x] \,\} \,. \tag{5.5}$$

Note that $\deg g(x) = d{-}1 = n{-}k$ and, so, $u(x)g(x) \in F_n[x]$ for every $u(x) \in F_k[x]$.

We also mention that every root of $g(x)$ is also a root of $x^n - 1$. Therefore, each one of the (distinct) linear factors of $g(x)$ divides $x^n - 1$ and, hence, $g(x) \,|\, x^n - 1$.

The terms primitive, normalized, and narrow-sense RS codes inherit their meanings from their GRS counterparts. In particular, normalized and narrow-sense RS codes correspond to $b = 0$ and $b = 1$, respectively. A *(singly-)extended RS code* is an extended GRS code whose parity-check matrix is obtained from H_{RS} by adding the column $(1\,0\,0\,\ldots\,0)^T$.

5.3 Encoding of RS codes

We can encode GRS codes—and hence RS codes—as any other linear $[n, k, d]$ code over F by mapping a vector $\mathbf{u} \in F^k$ to $\mathbf{u}G$, where G is a generator matrix of the code.

In the special case of RS codes, we can also use their characterization (5.5) for encoding: we can define an encoding mapping $F_k[x] \to \mathcal{C}_{\mathrm{RS}}$ by $u(x) \mapsto u(x)g(x)$, where $g(x)$ is the generator polynomial of $\mathcal{C}_{\mathrm{RS}}$. Writing $g(x) = g_0 + g_1 x + \ldots + g_{n-k}x^{n-k}$ (where $g_{n-k} = 1$), such a mapping can be represented as $\mathbf{u} \mapsto \mathbf{u}G$, where G is the (non-systematic) $k \times n$ generator matrix

$$G = \begin{pmatrix} g_0 & g_1 & \dots & g_{n-k} & & & & 0 \\ & g_0 & g_1 & \dots & g_{n-k} & & & \\ 0 & & \ddots & \ddots & \dots & \ddots & & \\ & & & g_0 & g_1 & \dots & g_{n-k} \end{pmatrix} .$$

An implementation of the mapping $u(x) \mapsto u(x)g(x)$ is given by the multiplication circuit in Figure 5.1. Each of the $n-k$ boxes represents a delay unit, which can store an element of F and is initially reset to zero. The delay units are synchronous through the control of a clock. A circle labeled g_i represents a multiplication by the constant g_i, and the circled "+" represents addition in F. The encoding process lasts n clock ticks and produces for every polynomial $u(x) = u_0 + u_1x + \ldots + u_{k-1}x^{k-1}$ a codeword $c(x) = c_0 + c_1x + \ldots + c_{n-1}x^{n-1} = u(x)g(x)$ as follows. At the ℓth clock tick, the encoder is fed with the coefficient $u_{k-\ell}$ (or with zero if $\ell > k$). Given that the contents of the delay units (from left to right) at that time are $u_{k-\ell+1}, u_{k-\ell+2}, \ldots, u_{n-\ell}$ (where $u_i = 0$ for $i < 0$), the encoder then generates the output $c_{n-\ell} = \sum_{i=k}^{n} u_{i-\ell}g_{n-i}$. And $c_{n-\ell}$ is indeed the coefficient of $x^{n-\ell}$ in the product $u(x)g(x)$.

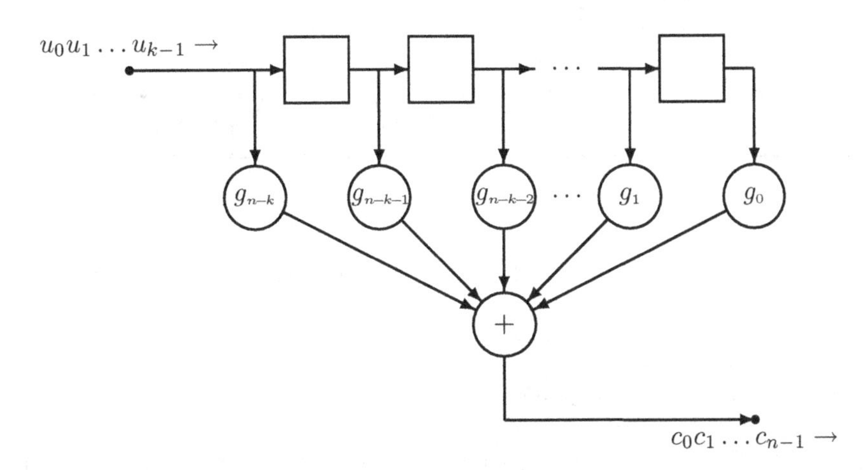

Figure 5.1. Multiplication circuit.

A second encoding scheme for RS codes can be obtained by *remaindering*. For every $u(x) \in F_k[x]$, let $r_u(x)$ be the unique polynomial in $F_{n-k}[x]$ such that

$$x^{n-k}u(x) \equiv r_u(x) \pmod{g(x)} .$$

Clearly, for every $u(x) \in F_k[x]$, the polynomial $x^{n-k}u(x) - r_u(x)$ is a codeword in $\mathcal{C}_{\mathrm{RS}}$. It is easy to see that the encoding mapping defined by $u(x) \mapsto x^{n-k}u(x) - r_u(x)$ is a linear systematic mapping from $F_k[x]$ to $\mathcal{C}_{\mathrm{RS}}$;

that is, it can be represented as $\mathbf{u} \mapsto \mathbf{u}G$, where G is a systematic generator matrix of $\mathcal{C}_{\mathrm{RS}}$.

An encoding circuit that implements the mapping $u(x) \mapsto x^{n-k}u(x) - r_u(x)$ is shown in Figure 5.2. In addition to the components that we have seen in Figure 5.1, this circuit also contains two switches that can be in one of two positions, marked A and B. The $n-k$ delay units are initially reset to zero. The encoding process again lasts n clock ticks. During the first k clock ticks, both switches are in position A; this means that the output equals the input and the feedback line is closed. One can show that for $\ell = 1, 2, \ldots, k$, the contents of the $n-k$ delay units right after the ℓth clock tick equal the remainder obtained when dividing the polynomial $x^{n-k}\sum_{i=1}^{\ell} u_{k-i}x^{\ell-i}$ by $g(x)$ (Problem 5.16). In the remaining $n-k$ clock ticks, both switches are in position B and the input is assumed to be zero (as the multipliers by g_i are now disconnected, their output is assumed to be zero also); during that period, the contents of the delay units are flushed to the output.

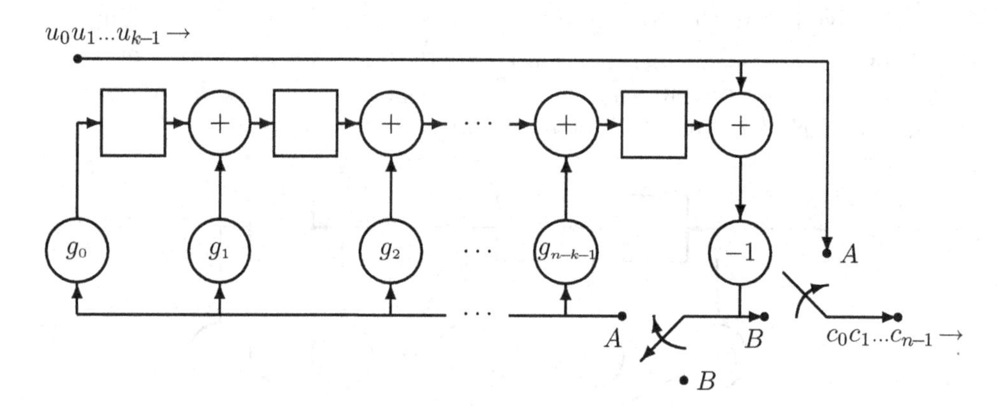

Figure 5.2. Remaindering circuit.

5.4 Concatenated codes

Recall that the length of GRS codes must be smaller than the field size. Still, we can construct codes over small fields using GRS codes as building blocks.

Let $\mathcal{C}_{\mathrm{in}}$ be a linear $[n, k, d]$ code over $F = \mathrm{GF}(q)$ and let $\mathcal{C}_{\mathrm{out}}$ be a linear $[N, K, D]$ code over $\Phi = \mathrm{GF}(q^k)$; namely, the extension degree $[\Phi : F]$ equals the dimension of $\mathcal{C}_{\mathrm{in}}$. A *(linearly-)concatenated code* with an *inner code* $\mathcal{C}_{\mathrm{in}}$ and an *outer code* $\mathcal{C}_{\mathrm{out}}$ is a code $\mathcal{C}_{\mathrm{cont}}$ over F that is defined as follows. Fix some one-to-one (and onto) mapping

$$\mathcal{E}_{\mathrm{in}} : \Phi \rightarrow \mathcal{C}_{\mathrm{in}}$$

that is linear over F; such a mapping can be specified through a basis $\Omega =$

$(\omega_1 \, \omega_2 \, \ldots \, \omega_k)$ of Φ over F and a generator matrix G_{in} of $\mathcal{C}_{\text{in}}$ by

$$\mathcal{E}_{\text{in}}(\Omega \mathbf{u}^T) = \mathbf{u}G_{\text{in}} \, , \quad \mathbf{u} \in F^k \, . \tag{5.6}$$

The code $\mathcal{C}_{\text{cont}}$ consists of all words in F^{nN} of the form

$$(\mathcal{E}_{\text{in}}(z_1) \,|\, \mathcal{E}_{\text{in}}(z_2) \,|\, \ldots \,|\, \mathcal{E}_{\text{in}}(z_N)) \, ,$$

where $(z_1 \, z_2 \, \ldots \, z_N)$ ranges over all the codewords in $\mathcal{C}_{\text{out}}$.

This definition implies a two-step encoder for $\mathcal{C}_{\text{cont}}$, by which the information word is first mapped to a codeword $(z_j)_{j=1}^N$ of $\mathcal{C}_{\text{out}}$ using some encoder of $\mathcal{C}_{\text{out}}$; a second encoding step then maps each entry z_j to the codeword $\mathcal{E}_{\text{in}}(z_j)$ of $\mathcal{C}_{\text{in}}$. The generated sequence of codewords of $\mathcal{C}_{\text{in}}$ is then transmitted—as one codeword of $\mathcal{C}_{\text{cont}}$—through the channel. Thus, the adjectives "inner" and "outer" describe the positions of the encoders of $\mathcal{C}_{\text{in}}$ and $\mathcal{C}_{\text{out}}$, as seen by the channel.

Example 5.2 Let $\mathcal{C}_{\text{in}}$ be the $[7, 3, 4]$ simplex code over $F = \text{GF}(2)$: this code, which was presented in Problem 2.18, is the dual code of the binary $[7, 4, 3]$ Hamming code and is therefore generated by the matrix

$$G_{\text{in}} = \begin{pmatrix} 0 & 0 & 0 & 1 & 1 & 1 & 1 \\ 0 & 1 & 1 & 0 & 0 & 1 & 1 \\ 1 & 0 & 1 & 0 & 1 & 0 & 1 \end{pmatrix} . \tag{5.7}$$

Let Φ be the field $F[\xi]/(\xi^3 + \xi + 1)$ and $\mathcal{C}_{\text{out}}$ be the $[6, 2, 5]$ GRS code over Φ with a generator matrix

$$G_{\text{GRS}} = \begin{pmatrix} 1 & 1 & 1 & 1 & 1 & 1 \\ \xi & \xi^2 & \xi^3 & \xi^4 & \xi^5 & \xi^6 \end{pmatrix} .$$

We now define $\mathcal{C}_{\text{cont}}$ to be the concatenated code $\mathcal{C}_{\text{cont}}$ over F, with $\mathcal{C}_{\text{in}}$ as the inner code and $\mathcal{C}_{\text{out}}$ as the outer code, where $\mathcal{E}_{\text{in}}$ is specified by (5.6) using the basis $\Omega = (1 \; \xi \; \xi^2)$ of Φ over F and the generator matrix G_{in} in (5.7).

To encode to a codeword of $\mathcal{C}_{\text{cont}}$, we first map the information word to a codeword of $\mathcal{C}_{\text{out}}$, say to the codeword

$$(z_1 \, z_2 \, \ldots \, z_6) = (\xi \;\; 1)G_{\text{GRS}} = (0 \;\; \xi{+}\xi^2 \;\; \xi{+}\xi^3 \;\; \xi{+}\xi^4 \;\; \xi{+}\xi^5 \;\; \xi{+}\xi^6)$$

(see Figure 5.3). Next, for $j = 1, 2, \ldots, 6$, we compute $\mathcal{E}_{\text{in}}(z_j)$ by first solving

$$z_j = \Omega \mathbf{u}_j^T$$

for the unique representation $\mathbf{u}_j \in F^3$. From Table 3.2 we obtain

$$\mathbf{u}_1 = (000) \, , \quad \mathbf{u}_2 = (011) \, , \quad \mathbf{u}_3 = (100) \, , \quad \mathbf{u}_4 = (001) \, ,$$
$$\mathbf{u}_5 = (101) \, , \quad \text{and} \quad \mathbf{u}_6 = (111) \, .$$

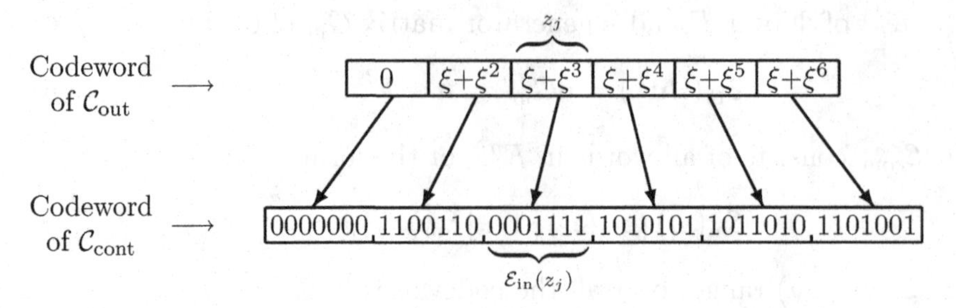

Figure 5.3. Codeword of a concatenated code.

We then compute the respective sub-blocks $\mathcal{E}_{\mathrm{in}}(z_j) = \mathbf{u}_j G_{\mathrm{in}}$ (which are all codewords of $\mathcal{C}_{\mathrm{in}}$); this results in

$$\underbrace{0000000}_{\mathcal{E}_{\mathrm{in}}(0)}, \quad \underbrace{1100110}_{\mathcal{E}_{\mathrm{in}}(\xi+\xi^2)}, \quad \underbrace{0001111}_{\mathcal{E}_{\mathrm{in}}(\xi+\xi^3)}, \quad \underbrace{1010101}_{\mathcal{E}_{\mathrm{in}}(\xi+\xi^4)}, \quad \underbrace{1011010}_{\mathcal{E}_{\mathrm{in}}(\xi+\xi^5)}, \quad \text{and} \quad \underbrace{1101001}_{\mathcal{E}_{\mathrm{in}}(\xi+\xi^6)} .$$

Finally, the concatenation of these sub-blocks yields a codeword of $\mathcal{C}_{\mathrm{cont}}$ of length $7 \cdot 6 = 42$ over F. □

We now state several properties of $\mathcal{C}_{\mathrm{cont}}$. Clearly, $\mathcal{C}_{\mathrm{cont}}$ is a code of length nN over F. In addition, the definition of $\mathcal{C}_{\mathrm{cont}}$ induces in effect a one-to-one correspondence between codewords of $\mathcal{C}_{\mathrm{out}}$ and codewords of $\mathcal{C}_{\mathrm{cont}}$; therefore, both of these codes have the same size. Furthermore, one can verify that $\mathcal{C}_{\mathrm{cont}}$ is a linear code over F (Problem 5.17). As such, its dimension is given by

$$\log_q |\mathcal{C}_{\mathrm{cont}}| = \log_q |\mathcal{C}_{\mathrm{out}}| = kK .$$

Turning to the minimum distance of $\mathcal{C}_{\mathrm{out}}$, a nonzero codeword

$$(\mathcal{E}_{\mathrm{in}}(z_1) \,|\, \mathcal{E}_{\mathrm{in}}(z_2) \,|\, \ldots \,|\, \mathcal{E}_{\mathrm{in}}(z_N))$$

of $\mathcal{C}_{\mathrm{cont}}$ contains at least D nonzero sub-blocks $\mathcal{E}_{\mathrm{in}}(z_j)$, and each such sub-block has Hamming weight at least d. Hence, the minimum distance of $\mathcal{C}_{\mathrm{cont}}$ is at least dD. We therefore summarize that $\mathcal{C}_{\mathrm{cont}}$ is a linear $[nN, kK, {\geq}dD]$ code over F.

To maximize D, the code $\mathcal{C}_{\mathrm{out}}$ is typically taken as a GRS code, which is possible whenever $N \leq q^k$ (equality is attained by singly-extended primitive GRS codes). So, even when q is fixed, we can obtain arbitrarily long codes. A special case of the concatenated code construction is where $\mathcal{C}_{\mathrm{in}}$ is taken as the $[n, n, 1]$ code F^n: here we can assume without loss of generality that G_{in} is the $n \times n$ identity matrix and, so, $\mathcal{E}_{\mathrm{in}}$ maps the entries $z_1, z_2, \ldots, z_N$ of a given codeword of $\mathcal{C}_{\mathrm{out}}$ to their respective vector representations $\mathbf{u}_1, \mathbf{u}_2, \ldots, \mathbf{u}_N$ in F^n, according to some basis of $\Phi = \mathrm{GF}(q^n)$ over $F = \mathrm{GF}(q)$. The resulting codeword of $\mathcal{C}_{\mathrm{cont}}$ is then the concatenation of these representations, namely,

$(\mathbf{u}_1 \,|\, \mathbf{u}_2 \,|\, \ldots \,|\, \mathbf{u}_N)$. The code $\mathcal{C}_{\mathrm{cont}}$ in this case is a linear $[nN, nK, {\geq}D]$ code over F.

Example 5.3 Let $\mathcal{C}_{\mathrm{cont}}$ be the concatenated code over GF(2) as in Example 5.2. From the parameters of $\mathcal{C}_{\mathrm{in}}$ and $\mathcal{C}_{\mathrm{out}}$, which are $[n, k, d] = [7, 3, 4]$ and $[N, K, D] = [6, 2, 5]$, we get that $\mathcal{C}_{\mathrm{cont}}$ is a linear $[42, 6, {\geq}20]$ code over GF(2). Here the minimum distance is exactly 20, since the codeword shown in Example 5.2 has Hamming weight 20. In fact, no linear $[42, 6]$ code over GF(2) can have minimum distance greater than 20: by the Griesmer bound (see part 2 of Problem 4.4), the shortest linear code of dimension 6 and minimum distance 21 over GF(2) must have length at least $\sum_{i=0}^{5} \lceil 21/2^i \rceil = 44$. □

A more comprehensive treatment of concatenated codes will be given in Chapter 12; in particular, we present there an efficient decoding algorithm for correcting any error pattern whose Hamming weight is less than $dD/2$. We also demonstrate the following theoretical significance of concatenated codes: a sequence of such codes can be effectively constructed for increasing lengths, where both the rate $(kK)/(nN)$ and (the lower bound on) the relative minimum distance $(dD)/(nN)$ remain bounded away from zero. Moreover, we show that by using two levels of concatenation, one can effectively realize codes with the properties guaranteed by the Shannon Coding Theorem for the q-ary symmetric channel (Theorem 4.17).

5.5 Alternant codes

We present in this section a second construction that is derived from GRS codes.

Let $F = \mathrm{GF}(q)$ and let $\mathcal{C}_{\mathrm{GRS}}$ be an $[N, K, D]$ GRS code over $\Phi = \mathrm{GF}(q^m)$. The intersection $\mathcal{C}_{\mathrm{GRS}} \cap F^N$ is called an *alternant code* and is denoted by $\mathcal{C}_{\mathrm{alt}}$.

The code $\mathcal{C}_{\mathrm{alt}}$ is a linear $[n, k]$ code over F with length $n = N$; furthermore, if $k > 0$ then the minimum distance of $\mathcal{C}_{\mathrm{alt}}$ is at least D. The parameter D is called the *designed minimum distance* of $\mathcal{C}_{\mathrm{alt}}$.

The attributes primitive, normalized, or narrow-sense are carried over to alternant codes from their underlying GRS codes. Similarly, we define a (singly-)extended alternant code over F as the intersection of an $[N, K]$ singly-extended GRS code with F^N.

We next show how a parity-check matrix of $\mathcal{C}_{\mathrm{alt}}$ can be obtained from a parity-check matrix of the underlying GRS code $\mathcal{C}_{\mathrm{GRS}}$. Let

$$H = (H_{i,j})_{i=1}^{D-1}\,{}_{j=1}^{N}$$

be a $(D-1) \times N$ parity-check matrix over Φ of $\mathcal{C}_{\mathrm{GRS}}$ (say, H is a canonical parity-check matrix H_{GRS}). Then, by the definition of $\mathcal{C}_{\mathrm{alt}}$, a word $\mathbf{c} = (c_1\, c_2\, \ldots\, c_n)$ over F is a codeword in $\mathcal{C}_{\mathrm{alt}}$ if and only if

$$H\mathbf{c}^T = \mathbf{0} ,$$

or, in scalar notation,

$$\sum_{j=1}^{n} H_{i,j} c_j = 0 , \quad 1 \le i \le D-1 . \tag{5.8}$$

Fix $\Omega = (\omega_1\, \omega_2\, \ldots\, \omega_m)$ to be a basis of Φ over F and for each entry $H_{i,j}$ of H, let the column vector $\mathbf{h}_{i,j} \in F^m$ be the representation of that entry according to the basis Ω, namely, $H_{i,j} = \Omega \mathbf{h}_{i,j}$. Then, condition (5.8) can be re-written as

$$\sum_{j=1}^{n} (\Omega \mathbf{h}_{i,j}) c_j = 0 , \quad 1 \le i \le D-1 ,$$

or

$$\Omega \Bigl(\sum_{j=1}^{n} \mathbf{h}_{i,j} c_j \Bigr) = 0 , \quad 1 \le i \le D-1 .$$

Yet, the m entries of Ω are linearly independent over F; therefore, (5.8) is equivalent to

$$\sum_{j=1}^{n} \mathbf{h}_{i,j} c_j = \mathbf{0} , \quad 1 \le i \le D-1 . \tag{5.9}$$

Shifting back to matrix notation, the left-hand side of (5.9) can be re-written as a product of the $m \times n$ matrix $(\, \mathbf{h}_{i,1}\, \mathbf{h}_{i,2}\, \ldots\, \mathbf{h}_{i,n}\,)$ by the vector $\mathbf{c}^T$ (both the matrix and the vector are over F); thus, (5.9) is equivalent to

$$(\, \mathbf{h}_{i,1}\, \mathbf{h}_{i,2}\, \ldots\, \mathbf{h}_{i,n}\,) \mathbf{c}^T = \mathbf{0} , \quad 1 \le i \le D-1 ,$$

or to

$$H_{\mathrm{alt}} \mathbf{c}^T = \mathbf{0} ,$$

where H_{alt} is the following $((D-1)m) \times n$ matrix over F:

$$H_{\mathrm{alt}} = \begin{pmatrix} \mathbf{h}_{1,1} & \mathbf{h}_{1,2} & \ldots & \mathbf{h}_{1,n} \\ \mathbf{h}_{2,1} & \mathbf{h}_{2,2} & \ldots & \mathbf{h}_{2,n} \\ \vdots & \vdots & \vdots & \vdots \\ \mathbf{h}_{D-1,1} & \mathbf{h}_{D-1,2} & \ldots & \mathbf{h}_{D-1,n} \end{pmatrix} .$$

In other words, a parity-check matrix H_{alt} of $\mathcal{C}_{\mathrm{alt}}$ can be obtained from H by replacing every entry in H with a representation of that entry as a column vector in F^m, according to some fixed basis of Φ over F.

Since the redundancy of $\mathcal{C}_{\mathrm{alt}}$ is bounded from above by the number of rows in H_{alt}, we obtain the inequality

$$n - k \leq (D{-}1)m \;, \tag{5.10}$$

which readily implies the following lower bound on the dimension of $\mathcal{C}_{\mathrm{alt}}$:

$$k \geq n - (D{-}1)m \;.$$

Example 5.4 Let F be the field $\mathrm{GF}(2)$ and Φ be the extension field $F[\xi]/(\xi^4 + \xi + 1)$; recall from Example 3.8 that ξ is a primitive element in Φ. Let $\mathcal{C}_{\mathrm{GRS}}$ be an $[N{=}15, K{=}13, D{=}3]$ primitive GRS code over Φ whose code locators and column multipliers are given by $\alpha_j = \xi^{j-1}$ and $v_j = \xi^{3(j-1)}$, respectively ($\mathcal{C}_{\mathrm{GRS}}$ is, in fact, a conventional RS code). A canonical parity-check matrix of $\mathcal{C}_{\mathrm{GRS}}$ is given by

$$H_{\mathrm{GRS}} = \begin{pmatrix} 1 & \xi^3 & \xi^6 & \xi^9 & \xi^{12} & 1 & \xi^3 & \xi^6 & \xi^9 & \xi^{12} & 1 & \xi^3 & \xi^6 & \xi^9 & \xi^{12} \\ 1 & \xi^4 & \xi^8 & \xi^{12} & \xi & \xi^5 & \xi^9 & \xi^{13} & \xi^2 & \xi^6 & \xi^{10} & \xi^{14} & \xi^3 & \xi^7 & \xi^{11} \end{pmatrix} .$$

We now construct a primitive alternant code $\mathcal{C}_{\mathrm{alt}}$ over F by taking the intersection $\mathcal{C}_{\mathrm{GRS}} \cap F^{15}$. A parity-check matrix of $\mathcal{C}_{\mathrm{alt}}$ can be obtained by replacing the entries in H_{GRS} with their column-vector representations in F^4 according to the basis (say) $\Omega = (1 \;\xi\; \xi^2 \;\xi^3)$. On doing that, we get from Table 3.3 the following parity-check matrix of $\mathcal{C}_{\mathrm{alt}}$:

$$H_{\mathrm{alt}} = \left(\begin{array}{ccccccccccccccc} 1 & 0 & 0 & 0 & 1 & 1 & 0 & 0 & 0 & 1 & 1 & 0 & 0 & 0 & 1 \\ 0 & 0 & 0 & 1 & 1 & 0 & 0 & 0 & 1 & 1 & 0 & 0 & 0 & 1 & 1 \\ 0 & 0 & 1 & 0 & 1 & 0 & 0 & 1 & 0 & 1 & 0 & 0 & 1 & 0 & 1 \\ 0 & 1 & 1 & 1 & 1 & 0 & 1 & 1 & 1 & 1 & 0 & 1 & 1 & 1 & 1 \\ \hline 1 & 1 & 1 & 1 & 0 & 0 & 0 & 1 & 0 & 0 & 1 & 1 & 0 & 1 & 0 \\ 0 & 1 & 0 & 1 & 1 & 1 & 1 & 0 & 0 & 0 & 1 & 0 & 0 & 1 & 1 \\ 0 & 0 & 1 & 1 & 0 & 1 & 0 & 1 & 1 & 1 & 1 & 0 & 0 & 0 & 1 \\ 0 & 0 & 0 & 1 & 0 & 0 & 1 & 1 & 0 & 1 & 0 & 1 & 1 & 1 & 1 \end{array} \right) .$$

The matrix H_{alt} has $(D{-}1)m = 2 \cdot 4 = 8$ rows and, so, the dimension of $\mathcal{C}_{\mathrm{alt}}$ is at least $15 - 8 = 7$. The minimum distance of $\mathcal{C}_{\mathrm{alt}}$ is at least $D = 3$ (the true minimum distance turns out to be 5 in this case; in fact, it follows from Problem 5.25 that $\mathcal{C}_{\mathrm{alt}}$ is the same code as in Example 3.11). $\square$

When designing an alternant code, the specifications provided are the field size q, the code length n, and the desired minimum distance. We use the latter as the value of the designed minimum distance D. We now need to select an extension degree m and an $[n, n{-}D{+}1, D]$ GRS code over $\mathrm{GF}(q^m)$. Based on the upper bound (5.10) on the redundancy, we will tend to select values of m as small as possible. Since we must also have $n \le q^m$, the smallest possible m is $\lceil \log_q n \rceil$.

The upper bound (5.10) can be improved in certain (interesting) cases, as we show in the next examples.

Example 5.5 Suppose that $F = \mathrm{GF}(2)$ and that the designed minimum distance D is odd. We construct an $[n, k, d{\ge}D]$ narrow-sense alternant code using an $[n, n{-}D{+}1, D]$ narrow-sense GRS code $\mathcal{C}_{\mathrm{GRS}}$ over $\Phi = \mathrm{GF}(2^m)$. Letting H_{GRS} be a canonical parity-check of $\mathcal{C}_{\mathrm{GRS}}$, for every $\mathbf{c} \in F^n$,

$$\mathbf{c} \in \mathcal{C}_{\mathrm{alt}} \qquad \Longleftrightarrow \qquad H_{\mathrm{GRS}}\mathbf{c}^T = \mathbf{0}$$

or, equivalently,

$$\mathbf{c} \in \mathcal{C}_{\mathrm{alt}} \qquad \Longleftrightarrow \qquad \sum_{j=1}^{n} c_j \alpha_j^i = 0 \quad \text{for } i = 1, 2, 3, \ldots, D{-}1 \,. \tag{5.11}$$

Recall, however, that

$$\sum_{j=1}^{n} c_j \alpha_j^i = 0 \qquad \Longleftrightarrow \qquad \sum_{j=1}^{n} c_j^2 \alpha_j^{2i} = 0 \qquad \Longleftrightarrow \qquad \sum_{j=1}^{n} c_j \alpha_j^{2i} = 0 \,.$$

This means that the equalities in (5.11) for even values of i are redundant, since they are implied by the equalities that correspond to odd values of i. So, (5.11) is equivalent to

$$\mathbf{c} \in \mathcal{C}_{\mathrm{alt}} \qquad \Longleftrightarrow \qquad \sum_{j=1}^{n} c_j \alpha_j^i = 0 \quad \text{for } i = 1, 3, 5, \ldots, D{-}2 \,.$$

It follows that a parity-check of $\mathcal{C}_{\mathrm{alt}}$ can be obtained by replacing every entry in the matrix

$$\begin{pmatrix} \alpha_1 & \alpha_2 & \ldots & \alpha_n \\ \alpha_1^3 & \alpha_2^3 & \ldots & \alpha_n^3 \\ \alpha_1^5 & \alpha_2^5 & \ldots & \alpha_n^5 \\ \vdots & \vdots & \vdots & \vdots \\ \alpha_1^{D-2} & \alpha_2^{D-2} & \ldots & \alpha_n^{D-2} \end{pmatrix}$$

with its representation as a column vector in F^m, according to some fixed basis of Φ over F. This, in turn, implies the inequality

$$n - k \le \frac{(D{-}1)m}{2} \,,$$

which applies to every binary narrow-sense alternant code with an odd designed minimum distance D. $\square$

The double-error-correcting code that was studied in Section 3.8 coincides with the code in Example 5.5 for $D = 5$.

Example 5.6 Suppose now that $F = \mathrm{GF}(2)$ and that the designed minimum distance D is even. We construct an $[n, k, d{\geq}D]$ normalized alternant code using an $[n, n{-}D{+}1, D]$ normalized GRS code $\mathcal{C}_{\mathrm{GRS}}$ over $\Phi = \mathrm{GF}(2^m)$. A parity-check matrix H_{alt} (over F) of the respective code $\mathcal{C}_{\mathrm{alt}}$ can be obtained by representing the entries in the matrix

$$\begin{pmatrix} 1 & 1 & \ldots & 1 \\ \alpha_1 & \alpha_2 & \ldots & \alpha_n \\ \alpha_1^3 & \alpha_2^3 & \ldots & \alpha_n^3 \\ \alpha_1^5 & \alpha_2^5 & \ldots & \alpha_n^5 \\ \vdots & \vdots & \vdots & \vdots \\ \alpha_1^{D-3} & \alpha_2^{D-3} & \ldots & \alpha_n^{D-3} \end{pmatrix}$$

as column vectors in F^m. Notice that the elements of the first row will all be represented in H_{alt} by the same (nonzero) vector in F^m. Thus, the first m rows in H_{alt} form an $m \times n$ matrix over F whose rank is 1 and, so, the rank of H_{alt} is bounded from above by $(D/2 - 1)m + 1$. This leads to the inequality

$$n - k \leq 1 + \left(\frac{D}{2} - 1\right)m ,$$

which holds for every binary (possibly singly-extended) normalized alternant code with an even designed minimum distance. □

Example 5.7 We construct an extended primitive alternant code $\mathcal{C}_{\mathrm{alt}}$ of length $n = 64$ and designed minimum distance 12 over $F = \mathrm{GF}(2)$. We take $m = 6$ (so that $n = 2^m$) and select the underlying $[64, 53, 12]$ extended primitive GRS code to be normalized. By the improved bound of Example 5.6 it follows that the redundancy of $\mathcal{C}_{\mathrm{alt}}$ is bounded from above by $1 + (D/2 - 1)m = 31$.

For the code at hand we can tighten the bound on the redundancy even further, as we show next. Letting $\{\alpha_1, \alpha_2, \ldots, \alpha_{64}\}$ be the elements of $\Phi = \mathrm{GF}(2^6)$, a parity-check matrix H_{alt} of $\mathcal{C}_{\mathrm{alt}}$ is obtained by representing the entries in the matrix

$$H = \begin{pmatrix} 1 & 1 & \ldots & 1 \\ \alpha_1 & \alpha_2 & \ldots & \alpha_{64} \\ \alpha_1^3 & \alpha_2^3 & \ldots & \alpha_{64}^3 \\ \alpha_1^5 & \alpha_2^5 & \ldots & \alpha_{64}^5 \\ \alpha_1^7 & \alpha_2^7 & \ldots & \alpha_{64}^7 \\ \alpha_1^9 & \alpha_2^9 & \ldots & \alpha_{64}^9 \end{pmatrix}$$

as column vectors in F^6.

As pointed out in Example 5.6, the first row in H contributes only 1 to the rank of H_{alt}. Consider now the last row in H. Each nonzero entry in that row has a multiplicative order which is either 1 or 7. Therefore, those entries and the zero element are roots (in Φ) of the polynomial $Q(x) = x^8 - x$. Since the mapping $x \mapsto Q(x)$ is linear over F and the size of its kernel is at most $\deg Q(x) = 8$, it follows that the roots of $Q(x)$ form a linear space over F with dimension at most 3; in fact, those roots form the field $\mathrm{GF}(2^3)$ as a subfield of Φ. Hence, when representing the last row in H as column vectors in F^6, those vectors will span a linear space whose dimension is (at most) 3. This means that the last six rows in H_{alt} contribute at most 3 to its rank and, so, the redundancy of $\mathcal{C}_{\mathrm{alt}}$ is at most 28. We thus conclude that $\mathcal{C}_{\mathrm{alt}}$ is a linear $[64, \geq 36, \geq 12]$ code over $\mathrm{GF}(2)$. $\square$

The bound (5.10) is useful only when D is relatively small: when D grows with n faster than $n/\log_q n$ $(\geq n/m)$, then this bound becomes trivial. And the same applies also to the improved bounds in Examples 5.5 and 5.6. On the other hand, while D serves as a lower bound on the minimum distance of an alternant code, the true minimum distance may sometimes be much larger. In fact, there exist alternant codes that attain the Gilbert–Varshamov bound (see Problem 5.29).

5.6 BCH codes

Bose–Chaudhuri–Hocquenghem (in short, BCH) *codes* are alternant codes whose underlying GRS codes are conventional RS codes. That is, if $F = \mathrm{GF}(q)$ and $\mathcal{C}_{\mathrm{RS}}$ is an $[N, K, D]$ RS code over $\Phi = \mathrm{GF}(q^m)$, then $\mathcal{C}_{\mathrm{RS}} \cap F^N$ is a BCH code, which we denote by $\mathcal{C}_{\mathrm{BCH}}$.

Recall from Section 5.2 that N must divide $q^m - 1$, which readily implies that $\gcd(N, q) = 1$. Conversely, if $\gcd(N, q) = 1$, then q (or rather its remainder upon division by N) belongs to the multiplicative group modulo N (see Problem A.17). So, given a code length $n = N$, the smallest possible value of m is the order of q in that group.

We can summarize the definition of BCH codes as follows. Let $F = \mathrm{GF}(q)$ and let n be a positive integer such that $\gcd(n, q) = 1$. Let m be [the smallest] positive integer such that $n \mid q^m - 1$ and α be an element of multiplicative order n in $\Phi = \mathrm{GF}(q^m)$. Also, let b and D be integers where $0 < D \leq n$. Given those parameters, a BCH code $\mathcal{C}_{\mathrm{BCH}}$ consists of all polynomials $c(x) = c_0 + c_1 x + \ldots + c_{n-1} x^{n-1} \in F_n[x]$ such that

$$c(\alpha^\ell) = 0 \quad \text{for } \ell = b, b+1, \ldots, b+D-2 .$$

The list

$$\alpha^b, \alpha^{b+1}, \ldots, \alpha^{b+D-2} ,$$

which consists of the roots of the underlying RS code, is called the *consecutive root sequence* of $\mathcal{C}_{\mathrm{BCH}}$. Like for every alternant code, the parameter D is the designed minimum distance of $\mathcal{C}_{\mathrm{BCH}}$.

Example 5.8 We design a BCH code of length $n = 85$ over $F = \mathrm{GF}(2)$ that can correct three errors. We take the value of m to be the smallest positive integer such that 85 divides 2^m-1, resulting in $m = 8$. We select $b = 1$ so that the code is of the narrow-sense type, thereby allowing us to use the improved bound of Example 5.5. The designed minimum distance is $D = 7$ and the code redundancy is at most $(D-1)m/2 = 24$. Therefore, the resulting BCH code is a linear $[85, \geq 61, \geq 7]$ code over F. Letting α be an element of multiplicative order 85 in $\mathrm{GF}(2^8)$, a 24×85 parity-check matrix of the code can be obtained by representing the entries in the matrix

$$\begin{pmatrix} 1 & \alpha & \alpha^2 & \ldots & \alpha^j & \ldots & \alpha^{83} & \alpha^{84} \\ 1 & \alpha^3 & \alpha^6 & \ldots & \alpha^{3j} & \ldots & \alpha^{79} & \alpha^{82} \\ 1 & \alpha^5 & \alpha^{10} & \ldots & \alpha^{5j} & \ldots & \alpha^{75} & \alpha^{80} \end{pmatrix}$$

as column vectors in F^8. □

Example 5.9 Let α be a primitive element in $\mathrm{GF}(2^6)$. The code in Example 5.7 becomes an extended primitive BCH code over $\mathrm{GF}(2)$ if we order the code locators so that $\alpha_j = \alpha^{j-1}$ for $1 \leq j \leq 63$ and $\alpha_{64} = 0$. □

We will learn more about the properties of BCH codes in Chapter 8.

Problems

[Section 5.1]

Problem 5.1 Show that the dual code of a $[q, k{<}q]$ singly-extended normalized primitive GRS code over $\mathrm{GF}(q)$ is a $[q, q-k]$ singly-extended normalized primitive GRS code.

Problem 5.2 (Doubly-extended GRS codes) Let $\mathcal{C}$ be a linear $[n, k{=}n-r, d]$ code over F defined by a parity-check matrix

$$H = \begin{pmatrix} 1 & 1 & \ldots & 1 & 0 \\ \alpha_1 & \alpha_2 & \ldots & \alpha_{n-1} & 0 \\ \alpha_1^2 & \alpha_2^2 & \ldots & \alpha_{n-1}^2 & 0 \\ \vdots & \vdots & \vdots & \vdots & \vdots \\ \alpha_1^{r-2} & \alpha_2^{r-2} & \ldots & \alpha_{n-1}^{r-2} & 0 \\ \alpha_1^{r-1} & \alpha_2^{r-1} & \ldots & \alpha_{n-1}^{r-1} & 1 \end{pmatrix} \begin{pmatrix} v_1 & & & 0 \\ & v_2 & & \\ 0 & & \ddots & \\ & & & v_n \end{pmatrix},$$

where $\alpha_1, \alpha_2, \ldots, \alpha_{n-1}$ are distinct elements of F and $v_1, v_2, \ldots, v_n$ are nonzero elements of F. (The code $\mathcal{C}$ is called a *doubly-extended GRS code*, and the last column in H is said to correspond to the code locator ∞, i.e., the "infinity" element.)

1. Show that $\mathcal{C}$ is MDS.
2. Assuming that $k < n$, show that the dual code $\mathcal{C}^\perp$ is an $[n, n-k]$ doubly-extended GRS code that can be defined through the same code locators as $\mathcal{C}$.

Problem 5.3 Let $\mathcal{C}$ be a linear $[n, k=n-r, d]$ code over F defined by a parity-check matrix

$$H = \begin{pmatrix} 1 & 1 & \ldots & 1 & 0 \\ \alpha_1 & \alpha_2 & \ldots & \alpha_{n-1} & 0 \\ \alpha_1^2 & \alpha_2^2 & \ldots & \alpha_{n-1}^2 & 0 \\ \vdots & \vdots & \vdots & \vdots & \vdots \\ \alpha_1^{r-3} & \alpha_2^{r-3} & \ldots & \alpha_{n-1}^{r-3} & 0 \\ \alpha_1^{r-2} & \alpha_2^{r-2} & \ldots & \alpha_{n-1}^{r-2} & 1 \\ \alpha_1^{r-1} & \alpha_2^{r-1} & \ldots & \alpha_{n-1}^{r-1} & \delta \end{pmatrix},$$

where $\alpha_1, \alpha_2, \ldots, \alpha_{n-1}$ are distinct elements of F and $\delta \in F$. The purpose of this problem is to find conditions on δ so that $\mathcal{C}$ is MDS.

1. Consider the $r \times r$ matrix

$$B = \begin{pmatrix} 1 & 1 & \ldots & 1 & 0 \\ \beta_1 & \beta_2 & \ldots & \beta_{r-1} & 0 \\ \beta_1^2 & \beta_2^2 & \ldots & \beta_{r-1}^2 & 0 \\ \vdots & \vdots & \vdots & \vdots & \vdots \\ \beta_1^{r-3} & \beta_2^{r-3} & \ldots & \beta_{r-1}^{r-3} & 0 \\ \beta_1^{r-2} & \beta_2^{r-2} & \ldots & \beta_{r-1}^{r-2} & 1 \\ \beta_1^{r-1} & \beta_2^{r-1} & \ldots & \beta_{r-1}^{r-1} & 0 \end{pmatrix},$$

where $\beta_1, \beta_2, \ldots, \beta_{r-1}$ are elements of F. Show that

$$\det(B) = -\Bigl(\sum_{i=1}^{r-1} \beta_i\Bigr) \prod_{\substack{(i,j):\\ 1 \le i < j < r}} (\beta_j - \beta_i)$$

(a product over an empty set is defined as 1).

Hint: Show that $\det(B)$ is the coefficient of x^{r-2} in the polynomial $b(x) \in F_r[x]$ defined by

$$b(x) = \det \begin{pmatrix} 1 & 1 & \ldots & 1 & 1 \\ \beta_1 & \beta_2 & \ldots & \beta_{r-1} & x \\ \beta_1^2 & \beta_2^2 & \ldots & \beta_{r-1}^2 & x^2 \\ \vdots & \vdots & \vdots & \vdots & \vdots \\ \beta_1^{r-2} & \beta_2^{r-2} & \ldots & \beta_{r-1}^{r-2} & x^{r-2} \\ \beta_1^{r-1} & \beta_2^{r-1} & \ldots & \beta_{r-1}^{r-1} & x^{r-1} \end{pmatrix} = \Bigl(\prod_{i=1}^{r-1} (x - \beta_i)\Bigr) \prod_{\substack{(i,j):\\ 1 \le i < j < r}} (\beta_j - \beta_i) .$$

(See also Problem 3.13.)

2. Show that $\mathcal{C}$ is MDS if and only if there are no $r-1$ distinct elements in $\{\alpha_1, \alpha_2, \ldots, \alpha_{n-1}\}$ whose sum is δ.
3. Show that the condition in part 2 still applies when the "infinity" column $(0\ 0\ \ldots\ 0\ 1)^T$ is appended to H.
4. What is the minimum distance d of $\mathcal{C}$ when it is not MDS?
5. (Triply-extended GRS codes) Show that when $F = \mathrm{GF}(q)$ and q is even, there is a linear $[q+2, q-1]$ MDS code over F.

Problem 5.4 Let $\mathcal{C}_{\mathrm{GRS}}$ be an $[n, k]$ GRS code over F with code locators $\alpha_1, \alpha_2, \ldots, \alpha_n$ and column multipliers $v_1, v_2, \ldots, v_n$.

1. Fix μ, ν, and η to be elements of F where $\mu, \eta \neq 0$. Show that $\mathcal{C}_{\mathrm{GRS}}$ is identical to the $[n, k]$ GRS code $\mathcal{C}'_{\mathrm{GRS}}$ over F defined by the code locators

$$\alpha'_j = \mu\alpha_j + \nu\ , \quad j = 1, 2, \ldots, n\ ,$$

and column multipliers

$$v'_j = \eta v_j\ , \quad j = 1, 2, \ldots, n\ .$$

(Given μ, there are certain choices of ν for which $\mathcal{C}'_{\mathrm{GRS}}$ will in fact be singly-extended, i.e., one of the code locators α'_j will be zero; still, if $n < |F|$, one can select ν so that each α'_j is nonzero, even when $\mathcal{C}_{\mathrm{GRS}}$ is singly-extended.)

Hint: Show that each row in a canonical parity-check matrix of $\mathcal{C}'_{\mathrm{GRS}}$ can be written as a linear combination of some rows in a canonical parity-check matrix of $\mathcal{C}_{\mathrm{GRS}}$.

2. Show that $\mathcal{C}_{\mathrm{GRS}}$ is identical to an $[n, k]$ GRS code over F with code locators $\alpha_1^{-1}, \alpha_2^{-1}, \ldots, \alpha_n^{-1}$ (and with properly selected column multipliers). Verify that this holds also when $\mathcal{C}_{\mathrm{GRS}}$ is singly-extended or doubly-extended (see Problem 5.2), i.e., when the code locators of $\mathcal{C}_{\mathrm{GRS}}$ include the zero element or the "infinity" element, regarding one to be the multiplicative inverse of the other.
3. Let μ, ν, σ, and τ be elements of F such that $\mu\tau \neq \sigma\nu$. Based on parts 1 and 2, show that $\mathcal{C}_{\mathrm{GRS}}$ is identical to an $[n, k]$ GRS code over F with code locators

$$\alpha'_j = \frac{\mu\alpha_j + \nu}{\sigma\alpha_j + \tau}\ , \quad j = 1, 2, \ldots, n\ ,$$

and with properly selected column multipliers. Verify that this applies to singly-extended and doubly-extended codes as well (with $\alpha_j = \infty$ being mapped to $\alpha'_j = \mu/\sigma$ and $\alpha_j = -\tau/\sigma$ to $\alpha'_j = \infty$).

Problem 5.5 Let $\mathcal{C}_{\mathrm{GRS}}$ be an $[n, k, d]$ GRS code over F with $1 < k < n$.

1. Show that $\mathcal{C}_{\mathrm{GRS}}$ is a proper subset (sub-code) of an $[n, k+1, d-1]$ GRS code over F.
2. The Hamming distance of a word $\mathbf{y} \in F^n$ from $\mathcal{C}_{\mathrm{GRS}}$ is defined as the Hamming distance between $\mathbf{y}$ and a nearest codeword in $\mathcal{C}_{\mathrm{GRS}}$ to $\mathbf{y}$. Show that there is a word in F^n whose Hamming distance from $\mathcal{C}_{\mathrm{GRS}}$ is at least $d-1$.

3. Show that the *covering radius* of $\mathcal{C}_{\mathrm{GRS}}$, which is the largest distance from $\mathcal{C}_{\mathrm{GRS}}$ of any word in F^n, equals $d-1$ (see Problem 4.13).

   Hint: Show that the covering radius of every MDS code must be smaller than its minimum distance.

Problem 5.6 Let $\alpha_1, \alpha_2, \ldots, \alpha_n$ be distinct elements in a field F and define the $n \times n$ Vandermonde matrix X by

$$X = \begin{pmatrix} 1 & 1 & \ldots & 1 \\ \alpha_1 & \alpha_2 & \ldots & \alpha_n \\ \alpha_1^2 & \alpha_2^2 & \ldots & \alpha_n^2 \\ \vdots & \vdots & \vdots & \vdots \\ \alpha_1^{n-1} & \alpha_2^{n-1} & \ldots & \alpha_n^{n-1} \end{pmatrix} .$$

For $\ell = 1, 2, \ldots, n$, let $b_\ell(x) = \sum_{i=0}^{n-1} b_{\ell,i} x^i$ be the polynomial in $F_n[x]$ that is defined by

$$b_\ell(x) = b_{\ell,0} + b_{\ell,1}x + \ldots + b_{\ell,n-1}x^{n-1} = \prod_{\substack{1 \le j \le n: \\ j \ne \ell}} \frac{x - \alpha_j}{\alpha_\ell - \alpha_j} ;$$

notice that $b_\ell(x)$ is the unique polynomial in $F_n[x]$ that interpolates through the n points

$$\{(\alpha_\ell, 1)\} \cup \{(\alpha_j, 0) \; : \; 1 \le j \le n \, , \; j \ne \ell\}$$

(see Problem 3.14), namely,

$$b_\ell(\alpha_j) = \begin{cases} 1 & \text{if } j = \ell \\ 0 & \text{otherwise} \end{cases} .$$

Let B be the $n \times n$ matrix obtained from the coefficients of the polynomials $b_\ell(x)$ by

$$B = \begin{pmatrix} b_{1,0} & b_{1,1} & \ldots & b_{1,n-1} \\ b_{2,0} & b_{2,1} & \ldots & b_{2,n-1} \\ \vdots & \vdots & \vdots & \vdots \\ b_{n,0} & b_{n,1} & \ldots & b_{n,n-1} \end{pmatrix} .$$

1. Show that

$$B = X^{-1} .$$

2. Let $\Psi(x)$ denote the polynomial $\prod_{j=1}^{n}(x - \alpha_j)$ over F and let $\Psi'(x)$ be the formal derivative of $\Psi(x)$ (see Section 3.7 and Problem 3.38). Show that

$$b_\ell(x) = \frac{\Psi(x)}{\Psi'(\alpha_\ell)(x - \alpha_\ell)} , \quad 1 \le \ell \le n ,$$

   and deduce that the entries along the last column of B are given by

$$b_{\ell,n-1} = \frac{1}{\Psi'(\alpha_\ell)} , \quad 1 \le \ell \le n .$$

Problem 5.7 Let $\mathcal{C}_{\mathrm{GRS}}$ be an $[n,k,d]$ normalized GRS code with code locators $\alpha_1, \alpha_2, \ldots, \alpha_n$ over F and let

$$H_{\mathrm{GRS}} = (\, \alpha_j^i \,)_{i=0}^{d-2}\,{}_{j=1}^{n} \quad \text{and} \quad G_{\mathrm{GRS}} = (\, v'_j \alpha_j^i \,)_{i=0}^{k-1}\,{}_{j=1}^{n}$$

be a parity-check matrix and a generator matrix of $\mathcal{C}_{\mathrm{GRS}}$, respectively. Define the $n \times n$ matrices X and B for the elements $\alpha_1, \alpha_2, \ldots, \alpha_n$ as in Problem 5.6.

1. Based on Problem 5.6, show that the values v'_j can be taken as

$$v'_j = \frac{1}{\Psi'(\alpha_j)} = \Bigl(\prod_{\substack{1 \le m \le n: \\ m \ne j}} (\alpha_j - \alpha_m) \Bigr)^{-1}, \quad 1 \le j \le n\,.$$

   Is this choice unique?

   Hint: Let $\hat{X}$ denote the $(n-1) \times n$ matrix that consists of the first $n-1$ rows of X. Recall from the proof of Proposition 5.2 that the elements v'_j satisfy

$$\hat{X} \begin{pmatrix} v'_1 \\ v'_2 \\ \vdots \\ v'_n \end{pmatrix} = \mathbf{0}\,.$$

   Then compute the product of $\hat{X}$ with the last column of B.

2. Find the value of each of the last k columns in the $(d-1) \times n$ matrix $H_{\mathrm{GRS}}B$.

3. Show that the last k rows in B^T form a generator matrix of $\mathcal{C}_{\mathrm{GRS}}$.

4. Suppose now that $\mathcal{C}_{\mathrm{GRS}}$ is not necessarily normalized, and let $v_1, v_2, \ldots, v_n$ be the column multipliers of $\mathcal{C}_{\mathrm{GRS}}$. Show that in this case, the values v'_j can be taken as

$$v'_j = \frac{1}{v_j \Psi'(\alpha_j)} = \Bigl(v_j \prod_{\substack{1 \le m \le n: \\ m \ne j}} (\alpha_j - \alpha_m) \Bigr)^{-1}, \quad 1 \le j \le n\,.$$

Problem 5.8 (Systematic generator matrices of GRS codes) Let $\mathcal{C}_{\mathrm{GRS}}$ be an $[n,k]$ GRS code over F (possibly singly-extended) with a (canonical) generator matrix

$$G_{\mathrm{GRS}} = \bigl(v'_j \alpha_j^i \bigr)_{i=0}^{k-1}\,{}_{j=1}^{n}\,.$$

Let $(\, I \,|\, A \,)$ be a systematic generator matrix of $\mathcal{C}_{\mathrm{GRS}}$, where A is a $k \times (n-k)$ matrix.

1. Based on Problem 5.6, show that the (i,j)th entry in A is given by

$$\frac{v'_{j+k}}{v'_i} \cdot \prod_{\substack{1 \le s \le k: \\ s \ne i}} \frac{\alpha_{j+k} - \alpha_s}{\alpha_i - \alpha_s}, \quad 1 \le i \le k\,, \quad 1 \le j \le n-k\,.$$

2. Define
$$\eta_i = v'_i \cdot \prod_{\substack{1 \le s \le k: \\ s \ne i}} (\alpha_i - \alpha_s)\,, \quad 1 \le i \le k\,,$$
and
$$\eta_j = v'_j \cdot \prod_{1 \le s \le k} (\alpha_j - \alpha_s)\,, \quad k < j \le n\,.$$
Show that the (i,j)th entry in A is given by
$$\frac{\eta_{j+k}/\eta_i}{\alpha_{j+k} - \alpha_i}\,, \quad 1 \le i \le k\,, \quad 1 \le j \le n{-}k\,.$$

3. Let $\mathcal{C}'_{\mathrm{GRS}}$ be an $[n{+}1, k]$ doubly-extended GRS code whose generator matrix is obtained by appending the "infinity" column $\eta_{n+1} \cdot (0\ 0\ \ldots\ 0\ 1)^T$ to G_{GRS} for some nonzero element η_{n+1} in F (see Problem 5.2). Show that a systematic generator matrix of $\mathcal{C}'_{\mathrm{GRS}}$ is given by $(\, I \,|\, B\,)$, where B is obtained from A by appending the column
$$\eta_{n+1} \cdot \left(\eta_1^{-1}\ \eta_2^{-1}\ \ldots\ \eta_k^{-1}\right)^T .$$

Problem 5.9 A *(generalized) Cauchy matrix* over a field F is a $k \times r$ matrix whose (i,j)th entry is given by
$$\frac{\eta_{j+k}/\eta_i}{\alpha_{j+k} - \alpha_i}\,, \quad 1 \le i \le k\,, \quad 1 \le j \le r\,,$$
where $\alpha_1, \alpha_2, \ldots, \alpha_{r+k}$ are distinct elements of F and $\eta_1, \eta_2, \ldots, \eta_{r+k}$ are nonzero elements of F.

A $k \times (r{+}1)$ *extended* Cauchy matrix is obtained from a Cauchy matrix by appending the column
$$\eta_{r+k+1} \cdot \left(\eta_1^{-1}\ \eta_2^{-1}\ \ldots\ \eta_k^{-1}\right)^T$$
for some nonzero element η_{r+k+1} in F.

1. Based on Problem 5.8, show that the following two conditions are equivalent:

   (i) A is a Cauchy matrix or an extended Cauchy matrix.

   (ii) $(\, I \,|\, A\,)$ is a generator matrix of a GRS code, possibly singly-extended or doubly-extended.

2. Show that every square sub-matrix of a Cauchy matrix is nonsingular.

   Hint: This can be shown by combining the MDS property of GRS codes with Problem 4.2. Alternatively, one can show that when $k = r$ and $\eta_i = 1$ for all $1 \le i \le 2k$, the determinant of a $k \times k$ Cauchy matrix is given by
$$\prod_{i=1}^{k} \frac{\prod_{j=i+1}^{k}\left((\alpha_i - \alpha_j)(\alpha_{j+k} - \alpha_{i+k})\right)}{\prod_{j=1}^{k}(\alpha_{j+k} - \alpha_i)}\,.$$

Similarly, the determinant of a $k \times k$ extended Cauchy matrix (with $\eta_i = 1$ for all i) is equal to

$$\prod_{i=1}^{k} \frac{\left(\prod_{j=i+1}^{k}(\alpha_i - \alpha_j)\right)\left(\prod_{j=i+1}^{k-1}(\alpha_{j+k} - \alpha_{i+k})\right)}{\prod_{j=1}^{k-1}(\alpha_{j+k} - \alpha_i)} .$$

3. Let A be a $k \times r$ matrix over F such that all the entries in A are nonzero. Let A^{c} be the $k \times r$ matrix whose (i,j)th entry is the multiplicative inverse of the (i,j)th entry of A. Show that the following two conditions are equivalent:

   (i) A is a Cauchy matrix or an extended Cauchy matrix.

   (ii) Every 2×2 sub-matrix of A^{c} is nonsingular and every 3×3 sub-matrix of A^{c} is singular (the latter condition on A^{c} is equivalent to saying that the rank of A^{c} is at most 2).

Problem 5.10 Let γ be a primitive element in $F = \mathrm{GF}(q)$ and consider the following triangular array over F:

$$\begin{array}{cccccccc}
1 & 1 & 1 & 1 & \cdots & 1 & 1 & 1 \\
1 & a_1 & a_2 & a_3 & \cdots & a_{q-3} & a_{q-2} & \\
1 & a_2 & a_3 & \cdots & a_{q-3} & a_{q-2} & & \\
1 & a_3 & \cdots & a_{q-3} & a_{q-2} & & & \\
\vdots & \cdots & a_{q-3} & a_{q-2} & & & & \\
1 & a_{q-3} & a_{q-2} & & & & & \\
1 & a_{q-2} & & & & & & \\
1 & & & & & & &
\end{array} ,$$

where

$$a_i = \frac{1}{1-\gamma^i} , \quad 1 \le i \le q-2 .$$

Show that every square sub-matrix in the array is nonsingular.

Hint: Identify the square sub-matrices as (possibly extended) Cauchy matrices (Problem 5.9).

Problem 5.11 Let $\mathcal{C}_{\mathrm{GRS}}$ be an $[n,k]$ GRS code over F with (nonzero) code locators $\alpha_1, \alpha_2, \ldots, \alpha_n$ and column multipliers $v_1, v_2, \ldots, v_n$. Also, let

$$f(x) = f_0 + f_1 x + \ldots + f_{n-k} x^{n-k}$$

be a polynomial of degree $n-k$ over F.

1. Show that for every selection of the coefficients $f_1, f_2, \ldots, f_{n-k}$ of $f(x)$ there exists at least one value for f_0 in F such that $f(\alpha_j) \neq 0$ for all $1 \le j \le n$.

Hereafter in this problem assume that $f(x)$ is such that $f(\alpha_j) \neq 0$ for all $1 \le j \le n$, and define $\vartheta_j(x)$ by

$$\vartheta_j(x) = -\frac{f(x) - f(\alpha_j)}{f(\alpha_j)(x - \alpha_j)} , \quad 1 \le j \le n .$$

2. Show that for $1 \le j \le n$,

$$\vartheta_j(x) = -\frac{1}{f(\alpha_j)} \sum_{\ell=0}^{n-k-1} x^\ell \sum_{i=\ell+1}^{n-k} f_i \alpha_j^{i-\ell-1} .$$

3. Show that for $1 \le j \le n$,

$$(x - \alpha_j)\vartheta_j(x) \equiv 1 \pmod{f(x)} .$$

4. Show that a word $(c_1\ c_2\ \ldots\ c_n) \in F^n$ is a codeword of $\mathcal{C}_{\mathrm{GRS}}$ if and only if

$$\sum_{j=1}^{n} c_j v_j f(\alpha_j) \cdot \vartheta_j(x) = 0 .$$

Hint: Consider the parity-check matrix

$$H = -\begin{pmatrix} f_{n-k} & 0 & 0 & \ldots & 0 \\ f_{n-k-1} & f_{n-k} & 0 & \ldots & 0 \\ \vdots & \vdots & \ddots & \ddots & 0 \\ f_2 & f_3 & \ldots & f_{n-k} & 0 \\ f_1 & f_2 & \ldots & f_{n-k-1} & f_{n-k} \end{pmatrix} H_{\mathrm{GRS}} ,$$

where H_{GRS} is a canonical parity-check matrix of $\mathcal{C}_{\mathrm{GRS}}$, and relate the contents of the jth column of H with the coefficients of $\vartheta_j(x)$.

5. Show that a word $(c_1\ c_2\ \ldots\ c_n) \in F^n$ is a codeword of $\mathcal{C}_{\mathrm{GRS}}$ if and only if

$$\sum_{j=1}^{n} \frac{c_j v_j f(\alpha_j)}{x - \alpha_j} \equiv 0 \pmod{f(x)} ,$$

where $1/(x - \alpha_j)$ denotes the multiplicative inverse of $x - \alpha_j$ in the ring $F[x]/f(x)$ (i.e., the ring of residues of polynomials in $F[x]$ modulo $f(x)$).

Hint: See parts 3 and 4.

6. Suppose that the column multipliers are given by

$$v_j = \frac{1}{f(\alpha_j)} , \quad 1 \le j \le n .$$

Show that a word $(c_1\ c_2\ \ldots\ c_n) \in F^n$ is a codeword of $\mathcal{C}_{\mathrm{GRS}}$ if and only if $f(x)$ divides the polynomial

$$\sum_{j=1}^{n} c_j \prod_{\substack{1 \le m \le n: \\ m \ne j}} (x - \alpha_m) .$$

7. Suppose that the column multipliers are now given by

$$v_j = \Bigl(f(\alpha_j) \prod_{\substack{1 \le m \le n: \\ m \ne j}} (\alpha_j - \alpha_m) \Bigr)^{-1} , \quad 1 \le j \le n .$$

With every word $\mathbf{c} = (c_1\ c_2\ \ldots\ c_n)$ in F^n, associate the interpolation polynomial

$$\lambda_{\mathbf{c}}(x) = \sum_{j=1}^{n} c_j \prod_{\substack{1 \le m \le n: \\ m \ne j}} \frac{x - \alpha_m}{\alpha_j - \alpha_m}$$

(that is, $\lambda_{\mathbf{c}}(x)$ is the unique polynomial in $F_n[x]$ that satisfies $\lambda_{\mathbf{c}}(\alpha_j) = c_j$ for $1 \le j \le n$; see Problem 3.14). Show that

$$\mathbf{c} \in \mathcal{C}_{\mathrm{GRS}} \iff f(x) \mid \lambda_{\mathbf{c}}(x) .$$

Hint: This can be proved based on part 5. Alternatively, use (5.3) and part 4 of Problem 5.7 to show that the codewords of $\mathcal{C}_{\mathrm{GRS}}$ take the form

$$\Big(f(\alpha_1)u(\alpha_1)\ \ f(\alpha_2)u(\alpha_2)\ \ \ldots\ \ f(\alpha_n)u(\alpha_n) \Big) ,$$

where $u(x)$ ranges over the elements of $F_k[x]$. Find the relation between the polynomials $u(x)$ and $\lambda_{\mathbf{c}}(x)$ that are associated with the same codeword $\mathbf{c}$.

Problem 5.12 Let $\mathcal{C}_{\mathrm{GRS}}$ be an $[n, k, 3]$ normalized GRS code with code locators $\alpha_1, \alpha_2, \ldots, \alpha_n$ over F.

1. A codeword of $\mathcal{C}_{\mathrm{GRS}}$ is transmitted through an additive channel (F, F, Prob) and a word $\mathbf{y} \in F^n$ is received with one error at location j. Let $(S_0\ S_1)^T$ be the syndrome of $\mathbf{y}$ with respect to a canonical parity-check matrix of $\mathcal{C}_{\mathrm{GRS}}$. Show that
$$\alpha_j = S_1/S_0$$
and that the error value is equal to S_0.

2. A codeword $\mathbf{c} = (c_1\ c_2\ \ldots\ c_n)$ of $\mathcal{C}_{\mathrm{GRS}}$ is transmitted through an erasure channel $(F, F \cup \{?\}, \mathsf{Prob})$ and a word $\mathbf{y} = (y_1\ y_2\ \ldots\ y_n) \in (F \cup \{?\})^n$ is received, where "?" stands for an erasure. The word $\mathbf{y}$ contains two erasures whose locations are denoted by i and j.

   The syndrome of $\mathbf{y}$ is computed as in part 1 where, for the purpose of this computation, the value 0 is substituted for y_i and y_j. Show that the entries of $\mathbf{c}$ at the erased locations are given by
$$c_i = \frac{S_1 - \alpha_j S_0}{\alpha_j - \alpha_i} \quad \text{and} \quad c_j = \frac{S_1 - \alpha_i S_0}{\alpha_i - \alpha_j} .$$

Problem 5.13 Let $\mathcal{C}_{\mathrm{GRS}}$ be an $[n, k, 4]$ normalized GRS code with code locators $\alpha_1, \alpha_2, \ldots, \alpha_n$ over F. A codeword $(c_1\ c_2\ \ldots\ c_n)$ of $\mathcal{C}_{\mathrm{GRS}}$ is transmitted through an erasure channel $(F, F \cup \{?\}, \mathsf{Prob})$ and a word $\mathbf{y} = (y_1\ y_2\ \ldots\ y_n) \in (F \cup \{?\})^n$ is received. The word $\mathbf{y}$ contains one erasure at location i and one error at location $j \ne i$.

Let $(S_0\ S_1\ S_2)^T$ be the syndrome of $\mathbf{y}$ with respect to a canonical parity-check matrix of $\mathcal{C}_{\mathrm{GRS}}$ where, for the purpose of computing the syndrome, the value 0 is substituted for y_i.

1. Show that α_j is related to S_0, S_1, S_2, and α_i by
$$\alpha_j = \frac{S_2 - \alpha_i S_1}{S_1 - \alpha_i S_0} .$$

2. Show that the error value, $e_j = y_j - c_j$, at location j is given by
$$e_j = \frac{S_1 - \alpha_i S_0}{\alpha_j - \alpha_i} = \frac{(S_1 - \alpha_i S_0)^2}{S_2 - 2\alpha_i S_1 + \alpha_i^2 S_0} .$$

[Section 5.2]

Problem 5.14 Let $\mathcal{C}_{\mathrm{RS}}$ be a $[9, 6]$ normalized RS code over $F = \mathrm{GF}(2^6)$ defined by an element α in F of multiplicative order 9.

1. Find all the roots of $\mathcal{C}_{\mathrm{RS}}$; express them as powers of α.
2. Find the values of the code locators and the column multipliers in a canonical parity-check matrix of $\mathcal{C}_{\mathrm{RS}}$.
3. Find the values of the code locators and the column multipliers in a canonical generator matrix of $\mathcal{C}_{\mathrm{RS}}$ (i.e., in a canonical parity-check matrix of the dual code of $\mathcal{C}_{\mathrm{RS}}$).
4. Show that $(1\ \alpha^3\ \alpha^{-3}\ 1\ \alpha^3\ \alpha^{-3}\ 1\ \alpha^3\ \alpha^{-3})$ is a codeword of $\mathcal{C}_{\mathrm{RS}}$.

Problem 5.15 Let $\mathcal{C}_{\mathrm{RS}}$ be a $[17, 15]$ normalized RS code over $F = \mathrm{GF}(2^m)$ defined by an element α in F of multiplicative order 17.

1. Find the smallest possible value of m for which this construction is possible.
2. Write the generator polynomial of $\mathcal{C}_{\mathrm{RS}}$ as a function of α.

[Section 5.3]

Problem 5.16 The circuit in Figure 5.2 is used for encoding an $[n, k]$ RS code with a generator polynomial $g(x) = \sum_{i=0}^{n-k} g_i x^i$. Show by induction on $\ell = 1, 2, \ldots, k$ that right after the ℓth clock tick, the contents of the $n-k$ delay units in the figure equal the remainder obtained when dividing the polynomial $x^{n-k} \sum_{i=1}^{\ell} u_{k-i} x^{\ell-i}$ by $g(x)$.

Hint: See Problem 3.9.

[Section 5.4]

Problem 5.17 Let $\mathcal{C}_{\mathrm{in}}$ be a linear $[n, k, d]$ code over $F = \mathrm{GF}(q)$ and let Φ be the field $\mathrm{GF}(q^k)$. A concatenated code $\mathcal{C}_{\mathrm{cont}}$ is constructed by a linear one-to-one mapping $\mathcal{E}_{\mathrm{in}} : \Phi \to \mathcal{C}_{\mathrm{in}}$ over F and a linear outer code $\mathcal{C}_{\mathrm{out}}$ over Φ. Show that $\mathcal{C}_{\mathrm{cont}}$ is a linear code over F.

Problem 5.18 Let $\mathcal{C}_{\text{cont}}$ be a concatenated code over $F = \text{GF}(q)$ obtained by taking an $[N, K, D]$ narrow-sense RS code over $\text{GF}(q^m)$ as an outer code and the $[m, m, 1]$ code F^m as an inner code. Show that $\mathcal{C}_{\text{cont}}$ contains codewords of Hamming weights $N, 2N, 3N, \ldots, mN$.

Hint: Show that the outer code contains the codeword $(1\ 1\ 1\ \ldots\ 1)$.

Problem 5.19 Let $\mathcal{C}_{\text{cont}}$ be a concatenated code over $K = \text{GF}(2^2)$ consisting of the following codes: the outer code is a $[9, 7]$ normalized RS code $\mathcal{C}_{\text{out}}$ over the smallest extension field Φ of K for which such a code exists, and the inner code is a linear code $\mathcal{C}_{\text{in}}$ of length 5 and minimum distance 3 over K.

1. Identify the field Φ.
2. Find the dimension of $\mathcal{C}_{\text{in}}$.
3. Suggest a code that can serve as the code $\mathcal{C}_{\text{in}}$.
4. Find the length and dimension of $\mathcal{C}_{\text{cont}}$.
5. Show that the minimum distance of $\mathcal{C}_{\text{cont}}$ allows one to recover correctly any pattern of up to four errors that occur in a codeword of $\mathcal{C}_{\text{cont}}$.

A codeword of $\mathcal{C}_{\text{cont}}$ has been transmitted through an additive channel $S = (K, K, \mathsf{Prob})$ and a word

$$\mathbf{y} = (\mathbf{y}_1 \,|\, \mathbf{y}_2 \,|\, \ldots \,|\, \mathbf{y}_9) \in K^{45}$$

has been received with at most four errors, where each sub-block $\mathbf{y}_j$ is in K^5. A nearest-codeword decoder for $\mathcal{C}_{\text{in}}$ is applied to each sub-block $\mathbf{y}_j$, thereby producing the word

$$\mathbf{x} = (\mathbf{x}_1 \,|\, \mathbf{x}_2 \,|\, \ldots \,|\, \mathbf{x}_9) \in K^{45} ,$$

where each sub-block $\mathbf{x}_j$ is a codeword of $\mathcal{C}_{\text{in}}$.

6. Assume that each sub-block $\mathbf{y}_j$ contains at most one error. Show that $\mathbf{x}$ is the correct transmitted codeword.
7. Assume now that at most one sub-block $\mathbf{y}_j$ contains two or more errors. Explain how the errors in $\mathbf{y}$ can be recovered from $\mathbf{x}$ while using the decoder for $\mathcal{C}_{\text{out}}$ in part 1 of Problem 5.12.
8. Next, assume that there are two sub-blocks $\mathbf{y}_j$ containing two errors each. Show how the errors in $\mathbf{y}$ can be corrected while using the decoder in part 2 of Problem 5.12.
9. How can the receiving end determine which of the three assumptions—in part 6, 7, or 8—is the one that actually took place?

Problem 5.20 Consider a concatenated code $\mathcal{C}_{\text{cont}}$ over $F = \text{GF}(2)$ obtained by taking a $[15, 11]$ GRS code over $\text{GF}(2^4)$ as an outer code and the $[4, 4, 1]$ code F^4 as an inner code.

1. Find the length and dimension of $\mathcal{C}_{\text{cont}}$.
2. Find a lower bound on the minimum distance of $\mathcal{C}_{\text{cont}}$.

3. Recall from Problem 2.21 that a *burst of length* ℓ is the event of having errors in a codeword such that the locations i and j of the first (leftmost) and last (rightmost) errors, respectively, satisfy $j-i = \ell-1$. Show that there is a decoder for $\mathcal{C}_{\mathrm{cont}}$ that can correct every burst of length 5 (note that the burst length is measured in elements of F, since the code $\mathcal{C}_{\mathrm{cont}}$ and the errors are over F).

Problem 5.21 Repeat Problem 5.20 where now the inner code is a $[7, 4, 3]$ Hamming code over $F = \mathrm{GF}(2)$. Show that with this inner code, every burst of length 11 or less can be corrected.

Problem 5.22 Repeat Problem 5.20 where now the inner code is a $[8, 4, 4]$ extended Hamming code over $F = \mathrm{GF}(2)$. Show that with this inner code, every burst of length 13 or less can be corrected.

[Section 5.5]

Problem 5.23 Let $\mathcal{C}$ be a linear $[n, k, d]$ code over $F = \mathrm{GF}(q)$ where $d \geq 3$. Show that there exists an $[n, n-1, 2]$ GRS code $\mathcal{C}_{\mathrm{GRS}}$ over $\mathrm{GF}(q^{n-k})$ such that $\mathcal{C} = \mathcal{C}_{\mathrm{GRS}} \cap F^n$; that is, $\mathcal{C}$ can be seen as an alternant code with designed minimum distance 2.

Problem 5.24 Let $\mathcal{C}_{\mathrm{GRS}}$ be an $[N, N-1, 2]$ narrow-sense GRS code over $\Phi = \mathrm{GF}(2^m)$ and let $\mathcal{C}_{\mathrm{alt}}$ be the alternant code over $F = \mathrm{GF}(2)$ defined by $\mathcal{C}_{\mathrm{GRS}} \cap F^N$. Show that there exists an $[N, N-2, 3]$ narrow-sense GRS code $\mathcal{C}'_{\mathrm{GRS}}$ over Φ such that

$$\mathcal{C}_{\mathrm{alt}} \subseteq \mathcal{C}'_{\mathrm{GRS}} \subset \mathcal{C}_{\mathrm{GRS}} .$$

Problem 5.25 Let $\mathcal{C}_{\mathrm{alt}}$ be the alternant code over $F = \mathrm{GF}(2)$ as in Example 5.4. Show that $\mathcal{C}_{\mathrm{alt}}$ can be written as the intersection $\mathcal{C}'_{\mathrm{GRS}} \cap F^{15}$, where $\mathcal{C}'_{\mathrm{GRS}}$ is an $[N{=}15, K'{=}11, D'{=}5]$ narrow-sense primitive GRS code over $\Phi = \mathrm{GF}(2^4)$ whose code locators are $\alpha_j = \xi^{j-1}$.

Hint: For every word $(c_0\, c_1\, \ldots\, c_{14})$ in F^{15},

$$\sum_{\ell=0}^{14} c_\ell \xi^{4\ell} = 0 \iff \sum_{\ell=0}^{14} c_\ell \xi^{8\ell} = 0 \iff \sum_{\ell=0}^{14} c_\ell \xi^{\ell} = 0 \iff \sum_{\ell=0}^{14} c_\ell \xi^{2\ell} = 0 .$$

Problem 5.26 (Generalizing Example 5.5 to arbitrary finite fields) Let $\mathcal{C}_{\mathrm{GRS}}$ be an $[N, N-D+1, D]$ narrow-sense GRS code over $\Phi = \mathrm{GF}(q^m)$ and let $\mathcal{C}_{\mathrm{alt}}$ be the $[n{=}N, k, {\geq}D]$ alternant code over $F = \mathrm{GF}(q)$ that is given by $\mathcal{C}_{\mathrm{GRS}} \cap F^N$. Show that

$$k \geq n - \left\lceil \tfrac{q-1}{q}(D-1) \right\rceil m .$$

Problem 5.27 (Subfield sub-codes) Let $F = \mathrm{GF}(q)$ and $\Phi = \mathrm{GF}(q^m)$, and let $\mathcal{C}$ be a linear $[N, K, D]$ code over Φ. The *subfield sub-code* of $\mathcal{C}$ over F is the intersection $\mathcal{C} \cap F^N$.

1. Show that $\mathcal{C} \cap F^N$ is a linear $[n{=}N, k]$ code over F with

$$k \geq N - m(N{-}K) = mK - (m{-}1)N$$

and that the minimum distance of $\mathcal{C} \cap F^N$ is at least D whenever $k > 0$.

Denote by $\mathrm{T}(x) = \mathrm{T}_{\Phi:F}(x)$ the trace polynomial over Φ with respect to F, as defined in Problem 3.31. For a vector $\mathbf{z} = (z_1 \; z_2 \; \ldots \; z_N)$ over Φ, define

$$\mathrm{T}(\mathbf{z}) = (\mathrm{T}(z_1) \; \mathrm{T}(z_2) \; \ldots \; \mathrm{T}(z_N)) \; .$$

Extend the definition to codes $\mathcal{C} \subseteq \Phi^N$ by

$$\mathrm{T}(\mathcal{C}) = \left\{ \mathbf{x} \in F^N \; : \; \mathbf{x} = \mathrm{T}(\mathbf{z}) \text{ for some } \mathbf{z} \in \mathcal{C} \right\}$$

(note that distinct codewords of $\mathcal{C}$ may be mapped by T to the same $\mathbf{x} \in F^n$).
Let $\mathcal{C}$ be a linear $[N, K, D]$ code over Φ.

2. Show that $\mathrm{T}(\mathcal{C})$ is a linear code over F.
3. (Dual codes of subfield sub-codes) Show that

$$\mathcal{C} \cap F^N = \left(\mathrm{T}(\mathcal{C}^\perp)\right)^\perp .$$

   Hint: To show the containment $\mathcal{C} \cap F^N \subseteq \left(\mathrm{T}(\mathcal{C}^\perp)\right)^\perp$, verify that for every $\mathbf{c} \in \mathcal{C} \cap F^N$ and $\mathbf{z} \in \mathcal{C}^\perp$,

$$\mathbf{c} \cdot (\mathrm{T}(\mathbf{z}))^T = \mathrm{T}(\mathbf{c} \cdot \mathbf{z}^T) = 0 .$$

   To show the converse containment $\left(\mathrm{T}(\mathcal{C}^\perp)\right)^\perp \subseteq \mathcal{C} \cap F^N$, let $(\beta_1 \; \beta_2 \; \ldots \; \beta_m)$ be a basis of Φ over F and let $(\lambda_1 \; \lambda_2 \; \ldots \; \lambda_m)$ be the respective dual basis, as defined in Problem 3.35. Assume that $\mathbf{c} \in \left(\mathrm{T}(\mathcal{C}^\perp)\right)^\perp$ and justify the following claims:

   (i) $\mathbf{c} \cdot (\mathrm{T}(\lambda_i \mathbf{z}))^T = 0$ for every $\mathbf{z} \in \mathcal{C}^\perp$ and $i = 1, 2, \ldots, m$,
   (ii) $\mathbf{c} \cdot \sum_{i=1}^m \beta_i (\mathrm{T}(\lambda_i \mathbf{z}))^T = 0$ for every $\mathbf{z} \in \mathcal{C}^\perp$, and
   (iii) $\mathbf{c} \cdot \mathbf{z}^T = 0$ for every $\mathbf{z} \in \mathcal{C}^\perp$.

4. (Dual codes of alternant codes) Let $\mathcal{C}_{\mathrm{GRS}}$ be an $[N, N{-}D{+}1, D]$ GRS code over Φ with code locators $\alpha_1, \alpha_2, \ldots, \alpha_N$ and column multipliers $v_1, v_2, \ldots, v_N$. Show that the dual code of the alternant code $\mathcal{C}_{\mathrm{alt}} = \mathcal{C}_{\mathrm{GRS}} \cap F^N$ is given by

$$\mathcal{C}_{\mathrm{alt}}^\perp = \left\{ \left(\mathrm{T}(v_1 u(\alpha_1)) \; \mathrm{T}(v_2 u(\alpha_2)) \; \ldots \; \mathrm{T}(v_N u(\alpha_N)) \right) \; : \; u(x) \in \Phi_{D-1}[x] \right\} .$$

Problem 5.28 Let $F = \mathrm{GF}(q)$ and $\Phi = \mathrm{GF}(q^m)$, and let $\alpha_1, \alpha_2, \ldots, \alpha_n$ be n distinct nonzero elements in Φ. Show that for every D in the range $1 \leq D \leq n$, there exists an $[n, n{-}D{+}1, D]$ GRS code over Φ with code locators $\alpha_1, \alpha_2, \ldots, \alpha_n$ (and a certain selection of column multipliers) such that the alternant code $\mathcal{C}_{\mathrm{GRS}} \cap F^n$ has minimum distance (exactly) D.

Problem 5.29 Let $F = \mathrm{GF}(q)$ and $\Phi = \mathrm{GF}(q^m)$, and let n and D be positive integers such that $(D-1)m < n < q^m$. Also, let $\alpha_1, \alpha_2, \ldots, \alpha_n$ be fixed nonzero elements in Φ. For every vector $\mathbf{v} = (v_1\ v_2\ \ldots v_n) \in (\Phi^*)^n$, denote by $\mathcal{C}_{\mathrm{GRS}}(\mathbf{v})$ the $[n, n-D+1, D]$ GRS code over Φ with code locators $\alpha_1, \alpha_2, \ldots, \alpha_n$ and column multipliers $v_1, v_2, \ldots, v_n$. The respective alternant code $\mathcal{C}_{\mathrm{alt}}(\mathbf{v})$ is defined by $\mathcal{C}_{\mathrm{GRS}}(\mathbf{v}) \cap F^n$.

The purpose of this problem is to show that while the codes $\mathcal{C}_{\mathrm{alt}}(\mathbf{v})$ all have designed minimum distance D, the true minimum distance may be much larger; in particular, at least one of these codes approaches the Gilbert–Varshamov bound (for $D = 2$, this result already follows from Problem 5.23).

1. Let $\mathbf{c}$ be a nonzero word in F^n. Show that there exist at most
$$(q^m - 1)^{n-D+1}$$
vectors $\mathbf{v} \in (\Phi^*)^n$ for which $\mathbf{c} \in \mathcal{C}_{\mathrm{alt}}(\mathbf{v})$.

   Hint: First argue that it suffices to consider words $\mathbf{c}$ whose Hamming weight is at least D. Then assume without loss of generality that the last $D-1$ entries in $\mathbf{c}$ are nonzero. Show that for every assignment of values to the first $n-D+1$ entries in $\mathbf{v}$, there is at most one way to set its remaining entries so that $\mathbf{c} \in \mathcal{C}_{\mathrm{alt}}(\mathbf{v})$.

2. Show that if d is a positive integer satisfying
$$V_q(n, d-1) \leq (q^m - 1)^{D-1}$$
(where $V_q(n, d-1) = \sum_{i=0}^{d-1} \binom{n}{i} (q-1)^i$) then there exists a vector $\mathbf{v} \in (\Phi^*)^n$ such that the minimum distance of $\mathcal{C}_{\mathrm{alt}}(\mathbf{v})$ is at least d.

   Hint: Using part 1, show that such a vector $\mathbf{v}$ exists if
$$(q^m - 1)^{n-D+1} (V_q(n, d-1) - 1) < |\Phi^*|^n .$$

3. Write
$$R = 1 - \frac{(D-1)m}{n}$$
and let δ be a real in $(0, 1-(1/q)]$ that satisfies
$$\mathsf{H}_q(\delta) \leq (1 - R)(1 - \epsilon(m, q)) ,$$
where $\mathsf{H}_q(\cdot)$ is the q-ary entropy function (defined in Section 4.5) and
$$\epsilon(m, q) = -(1/m) \log_q (1 - q^{-m})$$
(notice that $\lim_{m \to \infty} \epsilon(m, q) = 0$). Show that there exists an $[n, k{\geq}nR, d{\geq}\delta n]$ alternant code over F.

[Section 5.6]

Problem 5.30 Let $\mathcal{C}_{\mathrm{RS}}$ be a $[21, 17]$ normalized RS code over an extension field Φ of $F = \mathrm{GF}(2)$ and let $\mathcal{C}_{\mathrm{BCH}}$ be the BCH code $\mathcal{C}_{\mathrm{RS}} \cap F^{21}$ over F.

1. Find the smallest possible size of Φ.
2. Show that the dimension of $\mathcal{C}_{\mathrm{BCH}}$ is at least 8.
3. Show that the minimum distance of $\mathcal{C}_{\mathrm{BCH}}$ is at least 6.

Notes

[Sections 5.1 and 5.2]

Generalized Reed–Solomon codes were studied as combinatorial objects already in the early 1950s (see Bush [72]). As codes, they were first introduced by Reed and Solomon in [289], where the characterization of the codewords was given in the form (5.3). (More precisely, the definition in [289] yields a singly-extended normalized primitive GRS code over fields of even characteristic. The general definition was suggested by Delsarte in [96].)

One can encode a given $[n, k]$ GRS code $\mathcal{C}_{\mathrm{GRS}}$ over F by the mapping $\mathbf{u} \mapsto \mathbf{u}G_{\mathrm{GRS}}$, where G_{GRS} is a canonical generator matrix of $\mathcal{C}_{\mathrm{GRS}}$ (namely, G_{GRS} is a canonical parity-check matrix of the dual code of $\mathcal{C}_{\mathrm{GRS}}$). By (5.3), the encoding process is equivalent in this case to evaluating the information polynomial $u(x) \in F_k[x]$ at the code locators $\alpha_1, \alpha_2, \ldots, \alpha_n$ of $\mathcal{C}_{\mathrm{GRS}}$, and then multiplying each value $u(\alpha_j)$ by the respective column multiplier. The simultaneous evaluation of $u(x) \in F_k[x]$ at n elements of F, in turn, can be carried out using $O(n \log^2 n \log\log n)$ arithmetic operations in F; see Aho *et al.* [6, Section 8.5] and von zur Gathen and Gerhard [144, Section 10.1]. While the encoder $\mathbf{u} \mapsto \mathbf{u}G_{\mathrm{GRS}}$ is non-systematic, one can attain essentially the same time complexity also with a systematic encoder (see the notes on Section 6.6).

Problem 5.3 is taken from Roth and Lempel [300]. The codes considered in the problem were used by Khachiyan [212] and Vardy [370] as an ingredient when showing that finding the minimum distance of a code is an NP-complete problem.

The systematic form of the generator matrices of GRS codes in Problems 5.8 and 5.9 is taken from Roth and Lempel [301] and Roth and Seroussi [303]. See also Dür [111].

The interpretation of GRS decoding as a noisy interpolation problem, which was discussed in Section 5.1.2, can be applied also to Reed–Muller codes (see Problem 2.19). We demonstrate this next.

Let $\mathbf{x} = (x_0\ x_1\ \ldots\ x_{m-1})$ be a vector of m indeterminates over a field F, and denote by $F_t[\mathbf{x}] = F_t[x_0, x_1, \ldots, x_{m-1}]$ the set of all multivariate polynomials over F in the indeterminates $x_0, x_1, \ldots, x_{m-1}$, such that the power of each instance of every indeterminate is less than t; namely,

$$F_t[\mathbf{x}] = \left\{ u(\mathbf{x}) = \textstyle\sum_{\mathbf{e}} u_{\mathbf{e}} \mathbf{x}^{\mathbf{e}} \ :\ \mathbf{e} \in \{0, 1, 2, \ldots, t-1\}^m,\ u_{\mathbf{e}} \in F \right\},$$

where $\mathbf{x}^{\mathbf{e}}$ stands for the monomial $\prod_{i=0}^{m-1} x_i^{e_i}$. We define the *total degree* of such a monomial by

$$\deg \mathbf{x}^{\mathbf{e}} = \sum_i e_i\ ,$$

and the total degree of the multivariate polynomial $u(\mathbf{x}) = \sum_{\mathbf{e}} u_{\mathbf{e}} \mathbf{x}^{\mathbf{e}}$ is given by

$$\deg u(\mathbf{x}) = \max_{\mathbf{e}\,:\,u_{\mathbf{e}} \neq 0} \deg \mathbf{x}^{\mathbf{e}}\ .$$

We let $F_{t,h}[\mathbf{x}]$ denote the set $\{u(\mathbf{x}) \in F_t[\mathbf{x}] \ :\ \deg u(\mathbf{x}) < h\}$.

Now, recall from Problem 2.19 that the rth order Reed–Muller code of length 2^m over $F = \mathrm{GF}(2)$ is defined by

$$\mathcal{C}_{\mathrm{RM}}(m, r) = \left\{ (u(\mathbf{a}))_{\mathbf{a} \in F^m} \ :\ u(\mathbf{x}) \in F_{2,r+1}[\mathbf{x}] \right\}\ .$$

Thus, the problem of decoding Reed–Muller codes is equivalent to the problem of interpolating a multivariate polynomial in $F_{2,r+1}[\mathbf{x}]$ from its values at the elements of F^m, where some of these values may be erroneous. This generalizes in a natural manner to any finite field GF(q): the rth order (generalized) Reed–Muller code of length q^m over $F = \mathrm{GF}(q)$ is defined by

$$\mathcal{C}_{\mathrm{RM}}(m,r) = \{ (u(\mathbf{a}))_{\mathbf{a}\in F^m} \,:\, u(\mathbf{x}) \in F_{q,r+1}[\mathbf{x}] \}$$

(note that $\mathcal{C}_{\mathrm{RM}}(1,r)$ is a $[q,r{+}1]$ singly-extended normalized primitive GRS code over F). By arguments that are similar to those used in Example 5.1, it can be shown that $(\mathcal{C}_{\mathrm{RM}}(m,r))^{\perp} = \mathcal{C}_{\mathrm{RM}}(m,(q{-}1)m{-}r{-}1)$. For further properties of Reed–Muller codes over non-binary alphabets—including the computation of their dimension and minimum distance—see Assmus and Key [18, Section 5.5] and Berlekamp [36, Section 15.3].

[Section 5.3]

Among GRS codes, conventional RS codes (and codes that are obtained by shortening RS codes at a set of consecutive coordinates) are the most commonly used in practice, especially because they can be encoded efficiently: the multiplication circuit in Figure 5.1 and the remaindering circuit in Figure 5.2 lend themselves to fast implementations—either in hardware or in software. The fastest algorithms currently known for polynomial multiplication and polynomial remaindering require $O(n \log n \log\log n)$ arithmetic field operations; see Aho *et al.* [6, Section 8.3] and von zur Gathen and Gerhard [144, Sections 8.3 and 9.1].

For the application of shortened RS codes in optical storage, see the books by Immink [192, Chapter 2] and Pohlmann [283, Chapter 3].

[Section 5.4]

Concatenated codes were introduced by Forney in [129]. The decoding algorithm in Problem 5.19 exhibits an instance of a method for decoding concatenated codes, which is known as *generalized minimum distance* (in short, GMD) *decoding.* This method will be presented in detail in Section 12.2.

Let $\mathcal{C}_1$ be a linear $[n_1,k_1,d_1]$ code over $F = \mathrm{GF}(q)$ and let $\mathcal{C}_2$ be a linear $[n_2,k_2,d_2]$ code over the same field F. The *product code* $\mathcal{C}_1 * \mathcal{C}_2$ is defined as the set of all words

$$\mathbf{c} = (c_{1,1}\; c_{1,2}\; \ldots\; c_{1,n_1} \,|\, c_{2,1}\; c_{2,2}\; \ldots\; c_{2,n_1} \,|\, c_{n_2,1}\; c_{n_2,2}\; \ldots\; c_{n_2,n_1}) \tag{5.12}$$

in $F^{n_1 n_2}$ such that:

- $(c_{i,1}\; c_{i,2}\; \ldots\; c_{i,n_1}) \in \mathcal{C}_1$ for $i = 1, 2, \ldots, n_2$ and
- $(c_{1,j}\; c_{2,j}\; \ldots\; c_{n_2,j}) \in \mathcal{C}_2$ for $j = 1, 2, \ldots, n_1$.

Equivalently, associate with each word $\mathbf{c}$ as in (5.12) an $n_2 \times n_1$ matrix $\Gamma(\mathbf{c})$ over F whose ith row is given by the ith sub-block $(c_{i,1}\; c_{i,2}\; \ldots\; c_{i,n_1})$; then $\mathbf{c} \in \mathcal{C}_1 * \mathcal{C}_2$ if and only if each row in $\Gamma(\mathbf{c})$ is a codeword of $\mathcal{C}_1$ and each column is a codeword of $\mathcal{C}_2$. These codes were presented in Problem 2.21, where it was assumed without real loss of generality that $\mathcal{C}_1$ and $\mathcal{C}_2$ have systematic generator matrices. It was also demonstrated therein that $\mathcal{C}_1 * \mathcal{C}_2$ is a linear $[n_1 n_2, k_1 k_2, d_1 d_2]$ code over F.

We next verify that $\mathcal{C}_1 * \mathcal{C}_2$ is, in fact, a concatenated code. Let G_2 be a $k_2 \times n_2$ generator matrix of $\mathcal{C}_2$ and let $\mathcal{C}_{\text{out}}$ be the linear $[n_2, k_2]$ over $\Phi = \text{GF}(q^{k_1})$ that is generated by G_2. Fix a basis $\Omega = (\omega_1\ \omega_2\ \ldots\ \omega_{k_1})$ of Φ over F; note that the codewords of $\mathcal{C}_{\text{out}}$ are given by

$$\sum_{j=1}^{k_1} \omega_j (a_{1,j}\ a_{2,j}\ \ldots\ a_{n_2,j}) , \qquad (a_{1,j}\ a_{2,j}\ \ldots\ a_{n_2,j}) \in \mathcal{C}_2 .$$

In matrix notation, these codewords take the form ΩA^T, where $A = (a_{i,j})_{i=1}^{n_2}{}_{j=1}^{k_1}$ ranges over all $n_2 \times k_1$ arrays over F whose columns are codewords of $\mathcal{C}_2$. We denote this set of arrays by $\mathcal{C}_2^{k_1}$.

We now select an arbitrary generator matrix G_{in} of $\mathcal{C}_1$ and define the one-to-one linear mapping $\mathcal{E}_{\text{in}} : \Phi \to \mathcal{C}_1$ by

$$\mathcal{E}_{\text{in}}(\Omega \mathbf{u}^T) = \mathbf{u} G_{\text{in}} , \quad \mathbf{u} \in F^{k_1} .$$

When this mapping is applied to each of the n_2 coordinates of ΩA^T, we obtain the n_2 rows of an $n_2 \times n_1$ array AG_{in} over F, and the concatenation of these rows forms a word $\mathbf{c} \in F^{n_1 n_2}$ which, by definition, is a codeword of the concatenated code $\mathcal{C}_{\text{cont}}$ over F that is defined by the mapping $\mathcal{E}_{\text{in}} : \Phi \to \mathcal{C}_1$ and the outer code $\mathcal{C}_{\text{out}}$. At the same time, each row (respectively, column) in AG_{in} $(= \Gamma(\mathbf{c}))$ is a codeword of $\mathcal{C}_1$ (respectively, $\mathcal{C}_2$) and, so, $\mathbf{c} \in \mathcal{C}_1 * \mathcal{C}_2$. Finally, since the linear spaces $\mathcal{C}_2^{k_1}$, $\mathcal{C}_{\text{cont}}$, and $\mathcal{C}_1 * \mathcal{C}_2$ all have the same dimension—namely, $k_1 k_2$—over F, we conclude that $\mathcal{C}_{\text{cont}} = \mathcal{C}_1 * \mathcal{C}_2$.

For more on product codes and concatenated codes see Blokh and Zyablov [57], [58], Elias [114], Farrell [121], Hirasawa *et al.* [178]–[180], Kasahara *et al.* [205], Lin and Costello [230, Section 9.6], MacWilliams and Sloane [249, Sections 10.11, 18.2, and 18.5], Roth and Seroussi [305], Sugiyama *et al.* [350], Zinov'ev [397], [398], and Zinov'ev and Zyablov [400]–[402].

[Section 5.5]

Alternant codes were introduced and studied by Helgert [174].

The characterization of the dual codes of alternant codes in Problem 5.27 is due to Delsarte [96]. We next present properties of the weight distribution of the dual codes of certain primitive alternant codes, for the special case where the field size is a prime p. Letting ω be a root of order p of unity in the complex field $\mathbb{C}$, we recall from Problem 3.36 that the additive characters of $\Phi = \text{GF}(p^m)$ are the mappings $x \mapsto \omega^{\text{T}(\mu x)}$, where $\text{T}(x)$ is the trace polynomial over Φ with respect to $F = \text{GF}(p)$ and μ ranges over the elements of Φ (with $\mu = 0$ corresponding to the trivial character).

We quote without proof the following theorem by Carlitz and Uchiyama [73], which refines a well-known theorem of Weil [379] on character sums (see Lidl and Niederreiter [229, Section 5.4]).

Theorem 5.3 (The Carlitz–Uchiyama bound) *Let Φ be the field* $\text{GF}(p^m)$ *where p is a prime, and let $u(x)$ be a polynomial in $\Phi_r[x]$ that is not of the form $(f(x))^p -$*

$f(x)+b$ for any $f(x) \in \Phi[x]$ and $b \in \Phi$. Then for every nontrivial additive character $\chi : \Phi \to \mathbb{C}$,

$$\Big|\sum_{\beta \in \Phi} \chi(u(\beta))\Big| \le r \cdot p^{m/2} .$$

The Carlitz–Uchiyama bound leads to the following result.

Theorem 5.4 *Let F be the field $\mathrm{GF}(p)$ where p is a prime, and let $\mathcal{C}_{\mathrm{alt}}$ be a singly-extended normalized primitive alternant code over F with an underlying $[p^m, p^m - D+1, D]$ singly-extended GRS code over $\Phi = \mathrm{GF}(p^m)$. Then every codeword $\mathbf{c} \in \mathcal{C}_{\mathrm{alt}}^{\perp}$ is either a multiple of the all-one vector, or*

$$\big|(p-1)p^{m-1} - \mathsf{w}(\mathbf{c})\big| \le (D-1) \cdot (p-1) \cdot p^{(m/2)-1} .$$

In particular, the minimum distance of $\mathcal{C}_{\mathrm{alt}}^{\perp}$ is at least $\frac{p-1}{p}\big(p^m - (D-1) \cdot p^{m/2}\big)$.

Proof. Denote by $\alpha_1, \alpha_2, \ldots, \alpha_{p^m}$ the code locators of $\mathcal{C}_{\mathrm{alt}}$ and let $\mathbf{c} = (c_j)_{j=1}^{p^m}$ be a codeword of $\mathcal{C}_{\mathrm{alt}}^{\perp}$. By part 4 of Problem 5.27, there exists a polynomial $u(x) \in \Phi_{D-1}[x]$ such that

$$c_j = \mathrm{T}(u(\alpha_j)) , \quad 1 \le j \le p^m .$$

We distinguish between two cases.

Case 1: $u(x) = (f(x))^p - f(x) + b$ for some $f(x) \in \Phi[x]$ and $b \in \Phi$. In this case,

$$c_j = \mathrm{T}((f(\alpha_j))^p) - \mathrm{T}(f(\alpha_j)) + \mathrm{T}(b) = (\mathrm{T}(f(\alpha_j)))^p - \mathrm{T}(f(\alpha_j)) + \mathrm{T}(b) = \mathrm{T}(b) ,$$

i.e., $\mathbf{c}$ is a multiple of the all-one vector.

Case 2: $u(x) \ne (f(x))^p - f(x) + b$ for every $f(x) \in \Phi[x]$ and $b \in \Phi$. Letting ω be a root of order p of unity in $\mathbb{C}$, we know that

$$\sum_{a \in F^*} \omega^{a \cdot c_j} = \begin{cases} p-1 & \text{if } c_j = 0 \\ -1 & \text{otherwise} \end{cases} .$$

Therefore,

$$\begin{aligned} \sum_{a \in F^*} \sum_{j=1}^{p^m} \omega^{a \cdot c_j} = \sum_{j=1}^{p^m} \Big(\sum_{a \in F^*} \omega^{a \cdot c_j} \Big) &= (p-1)(p^m - \mathsf{w}(\mathbf{c})) - \mathsf{w}(\mathbf{c}) \\ &= (p-1)p^m - p \cdot \mathsf{w}(\mathbf{c}) . \end{aligned}$$

On the other hand, by Theorem 5.3 we have

$$\begin{aligned} \Big| \sum_{a \in F^*} \sum_{j=1}^{p^m} \omega^{a \cdot c_j} \Big| &= \Big| \sum_{a \in F^*} \sum_{\beta \in \Phi} \omega^{a \cdot \mathrm{T}(u(\beta))} \Big| \\ &\le \sum_{a \in F^*} \Big| \sum_{\beta \in \Phi} \omega^{a \cdot \mathrm{T}(u(\beta))} \Big| \\ &\le (p-1) \cdot (D-1) \cdot p^{m/2} . \end{aligned}$$

We conclude that

$$\left|(p-1)p^m - p \cdot \mathsf{w}(\mathbf{c})\right| \leq (D-1) \cdot (p-1) \cdot p^{m/2} ,$$

thereby yielding the desired result. □

It is fairly easy to see that the dual code of a (non-extended) normalized primitive alternant code is obtained from the code $\mathcal{C}_{\text{alt}}^{\perp}$ in Theorem 5.4 by puncturing at the coordinate that corresponds to the code locator 0 (Problem 2.3). Similarly, the dual code of a narrow-sense primitive alternant code is obtained from $\mathcal{C}_{\text{alt}}^{\perp}$ by shortening at that coordinate (Problem 2.14). So, for example, in the narrow-sense primitive case, we can state the following corollary.

Corollary 5.5 *Let $F = \text{GF}(p)$ where p is a prime, and let $\mathcal{C}_{\text{alt}}$ be a narrow-sense primitive alternant code over F with an underlying $[p^m-1, p^m-D, D]$ GRS code over $\text{GF}(p^m)$. Then every nonzero codeword $\mathbf{c} \in \mathcal{C}_{\text{alt}}^{\perp}$ satisfies*

$$\left|(p-1)p^{m-1} - \mathsf{w}(\mathbf{c})\right| \leq D \cdot (p-1) \cdot p^{(m/2)-1} .$$

In particular, the minimum distance of $\mathcal{C}_{\text{alt}}^{\perp}$ is at least $\frac{p-1}{p}\left(p^m - D \cdot p^{m/2}\right)$.

Clearly, the bounds in Theorem 5.4 and Corollary 5.5 are meaningful only when the designed minimum distance D is small, namely, when $D < p^{m/2} + 1$. In Problem 8.12, we demonstrate that in this range of D, we can write an exact expression for the dimension of $\mathcal{C}_{\text{alt}}^{\perp}$: in the singly-extended normalized case of Theorem 5.4, that dimension equals

$$1 + \left\lceil \tfrac{p-1}{p}(D-2) \right\rceil m ,$$

while in the narrow-sense case (Corollary 5.5) we get

$$\left\lceil \tfrac{p-1}{p}(D-1) \right\rceil m .$$

Finally, we point out that the choice of column multipliers may have a significant effect on the true minimum distance of $\mathcal{C}_{\text{alt}}$ and $\mathcal{C}_{\text{alt}}^{\perp}$. Specifically, it follows from Problem 5.28 that for certain column multipliers, the minimum distance of an alternant code equals its designed minimum distance. On the other hand, Problem 5.29 demonstrates that for properly selected column multipliers, alternant codes attain the Gilbert–Varshamov bound (see also Problem 12.11 in Chapter 12). As for $\mathcal{C}_{\text{alt}}^{\perp}$, we see that in the cases covered by Theorem 5.4 and Corollary 5.5, the relative minimum distance of $\mathcal{C}_{\text{alt}}^{\perp}$ is close to $(p-1)/p$, provided that the designed minimum distance of $\mathcal{C}_{\text{alt}}$ is small. On the other hand, if all but one of the column multipliers are selected to have trace 0, then we get from part 4 of Problem 5.27 that $\mathcal{C}_{\text{alt}}^{\perp}$ contains a codeword whose Hamming weight is 1.

[Section 5.6]

BCH codes over GF(2) were introduced by Bose and Ray-Chaudhuri in [61] and [62], and independently by Hocquenghem in [187]. The non-binary case was treated by Gorenstein and Zierler in [160]. BCH codes are useful when the number of errors is small compared to the code length. Indeed, it was shown by Berlekamp [35] and Lin and Weldon [231] that the (true) relative minimum distance of primitive BCH codes approaches zero as the code length increases.

Other than BCH codes, another widely-studied subset of alternant codes is the family of (classical) *Goppa codes*. In these codes, the underlying $[N, K, D]$ GRS code over Φ has column multipliers that are related to the code locators by

$$v_j = \frac{1}{f(\alpha_j)}, \quad 1 \le j \le N,$$

where $f(x)$ is a polynomial of degree $D-1$ over Φ such that $f(\alpha_j) \neq 0$ for $1 \le j \le N$. See Goppa [156]–[158], MacWilliams and Sloane [249, Chapter 12], part 6 of Problem 5.11, and Problem 12.11.

Chapter 6

Decoding of Reed–Solomon Codes

In this chapter, we present an efficient decoding algorithm for GRS codes that recovers all error patterns whose Hamming weight is less than half the minimum distance of the code. We develop a set of equations, referred to collectively as the key equation of GRS decoding; solving the key equation is the core step of the decoding algorithm described herein. This step is preceded by syndrome computation and is followed by evaluating certain polynomials at the multiplicative inverses of the code locators—thereby producing the error locations and error values.

We present two efficient techniques for solving the key equation. The first technique is based on Euclid's algorithm for polynomials, except that we apply here a different stopping rule in that algorithm, compared to its standard use. The second technique, known as the Berlekamp–Massey algorithm, is in effect an efficient method for computing the shortest linear recurrence that is satisfied by a given sequence; in our case, that sequence is formed by the syndrome values.

6.1 Introduction

Let $\mathcal{C}_{\mathrm{GRS}}$ be a given $[n, k, d]$ GRS code over F with a canonical parity-check matrix

$$H_{\mathrm{GRS}} = \begin{pmatrix} 1 & 1 & \ldots & 1 \\ \alpha_1 & \alpha_2 & \ldots & \alpha_n \\ \alpha_1^2 & \alpha_2^2 & \ldots & \alpha_n^2 \\ \vdots & \vdots & \vdots & \vdots \\ \alpha_1^{d-2} & \alpha_2^{d-2} & \ldots & \alpha_n^{d-2} \end{pmatrix} \begin{pmatrix} v_1 & & & \\ & v_2 & & \text{\huge 0} \\ \text{\huge 0} & & \ddots & \\ & & & v_n \end{pmatrix} ,$$

where the code locators—$\alpha_1, \alpha_2, \ldots, \alpha_n$—are distinct nonzero elements of F, and the column multipliers—$v_1, v_2, \ldots, v_n$—are nonzero elements of F. We describe an algorithm for decoding up to $\lfloor \frac{1}{2}(d-1) \rfloor$ errors, given that a codeword of $\mathcal{C}_{\mathrm{GRS}}$ is transmitted through an additive channel (F, F, Prob).

Let $\mathbf{c}$ be the transmitted codeword and $\mathbf{y} = (y_1\ y_2\ \ldots\ y_n)$ be the received word. The error word is given by

$$\mathbf{e} = (e_1\ e_2\ \ldots\ e_n) = \mathbf{y} - \mathbf{c}\ ,$$

and we let J be the set of error locations; that is,

$$e_\kappa \neq 0 \qquad \Longleftrightarrow \qquad \kappa \in J\ .$$

The number of errors is $|J|$, and we assume that this number does not exceed $\frac{1}{2}(d-1)$.

The steps of the decoding algorithm are described and analyzed in Sections 6.2–6.6 below.

6.2 Syndrome computation

The first decoding step is the computation of the syndrome of $\mathbf{y}$ with respect to H_{GRS}:

$$\begin{pmatrix} S_0 \\ S_1 \\ \vdots \\ S_{d-2} \end{pmatrix} = H_{\mathrm{GRS}}\mathbf{y}^T\ .$$

That is, the individual syndrome entries are given by

$$S_\ell = \sum_{j=1}^{n} y_j v_j \alpha_j^\ell\ , \qquad \ell = 0, 1, \ldots, d-2\ .$$

Example 6.1 In the special case of RS codes we have $\alpha_j = \alpha^{j-1}$ and $v_j = \alpha^{b(j-1)}$, where α has multiplicative order n in F and b is an integer. In this case,

$$S_\ell = \sum_{j=1}^{n} y_j \alpha^{(j-1)(b+\ell)}\ , \qquad \ell = 0, 1, \ldots, d-2\ ,$$

which means that S_ℓ is obtained by evaluating the polynomial $y(x) = y_1 + y_2 x + \ldots + y_n x^{n-1}$ at $x = \alpha^{b+\ell}$. □

We will introduce certain polynomials over F, which are computed throughout the course of the decoding algorithm. The first is the *syndrome polynomial*, denoted by $S(x)$, whose coefficients are given by the syndrome entries, namely,

$$S(x) = \sum_{\ell=0}^{d-2} S_\ell x^\ell .$$

We next recall that the syndrome of the error word $\mathbf{e}$ equals that of the received word $\mathbf{y}$; so,

$$S_\ell = \sum_{j=1}^{n} e_j v_j \alpha_j^\ell , \quad \ell = 0, 1, \ldots, d-2 ,$$

or,

$$S_\ell = \sum_{j \in J} e_j v_j \alpha_j^\ell , \quad \ell = 0, 1, \ldots, d-2 \tag{6.1}$$

(a sum over an empty set is read as 0). Hence, the syndrome polynomial can be expressed in terms of the error word $\mathbf{e}$ as follows:

$$S(x) = \sum_{\ell=0}^{d-2} x^\ell \sum_{j \in J} e_j v_j \alpha_j^\ell = \sum_{j \in J} e_j v_j \sum_{\ell=0}^{d-2} (\alpha_j x)^\ell . \tag{6.2}$$

Consider the ring $F[x]/x^{d-1}$, i.e., the ring of residues of the polynomials in $F[x]$ modulo x^{d-1}. The elements in that ring that are not divisible by x form a group under the multiplication of the ring. The polynomial $1 - \alpha_j x$ belongs to that group and its multiplicative inverse is the polynomial $\sum_{\ell=0}^{d-2} (\alpha_j x)^\ell$; indeed,

$$(1 - \alpha_j x) \sum_{\ell=0}^{d-2} (\alpha_j x)^\ell = 1 - (\alpha_j x)^{d-1} \equiv 1 \pmod{x^{d-1}} .$$

Based on (6.2), we can thus write the following relationship between the syndrome polynomial $S(x)$ and the error word $\mathbf{e}$:

$$S(x) \equiv \sum_{j \in J} \frac{e_j v_j}{1 - \alpha_j x} \pmod{x^{d-1}} . \tag{6.3}$$

6.3 Key equation of GRS decoding

We next associate two additional polynomials with the error word $\mathbf{e}$. Define the *error-locator polynomial* (in short, ELP) by

$$\Lambda(x) = \prod_{j \in J} (1 - \alpha_j x)$$

and the *error-evaluator polynomial* (in short, EEP) by

$$\Gamma(x) = \sum_{j \in J} e_j v_j \prod_{m \in J \setminus \{j\}} (1 - \alpha_m x)$$

(a product over an empty set is read as 1).

First, observe that

$$\Lambda(\alpha_\kappa^{-1}) = 0 \qquad \Longleftrightarrow \qquad \kappa \in J$$

(recall that we assume that all the code locators α_j are nonzero). Thus, the roots of the ELP tell us where the errors are. On the other hand, for every $\kappa \in J$,

$$\Gamma(\alpha_\kappa^{-1}) = e_\kappa v_\kappa \prod_{m \in J \setminus \{\kappa\}} (1 - \alpha_m \alpha_\kappa^{-1}) \neq 0 \, .$$

Hence,

$$\boxed{\gcd(\Lambda(x), \Gamma(x)) = 1} \, . \tag{6.4}$$

In particular, $\Gamma(x) = 0 \Longrightarrow \Lambda(x) = 1$, which corresponds to the case where no errors have occurred, i.e., $S(x) = 0$.

Secondly, the degrees of $\Lambda(x)$ and $\Gamma(x)$ satisfy

$$\deg \Lambda = |J| \quad \text{and} \quad \deg \Gamma < |J| \, .$$

Since $|J| \leq \frac{1}{2}(d-1)$ we therefore have

$$\boxed{\deg \Gamma < \deg \Lambda \leq \tfrac{1}{2}(d-1)} \, . \tag{6.5}$$

Thirdly, we can relate the ELP and the EEP by

$$\Gamma(x) = \Lambda(x) \sum_{j \in J} \frac{e_j v_j}{1 - \alpha_j x} \, .$$

Hence, from (6.3) we obtain

$$\boxed{\Lambda(x) S(x) \equiv \Gamma(x) \pmod{x^{d-1}}} \, . \tag{6.6}$$

Equations (6.4)–(6.6) together form the *key equation* of GRS decoding.

Our next decoding step will be solving the key equation for $\Lambda(x)$ and $\Gamma(x)$. Once we know the ELP, we can exhaustively check which among the elements $\alpha_1^{-1}, \alpha_2^{-1}, \ldots, \alpha_n^{-1}$ is a root of $\Lambda(x)$, thereby determining the set J (this method for finding the roots of $\Lambda(x)$ is called a *Chien search*). At this point, the set of equations (6.1) becomes *linear* in the error values e_j.

6.3.1 Solving the key equation

We next demonstrate that solving the key equation is, in principle, equivalent to solving a set of linear equations, and that the solution to the key equation is essentially unique. (We remark that the presentation in this section is meant to convince that the solution of the key equation is a conceptually simple task. To this end, we include here an algorithm for solving (6.4)–(6.6), yet this algorithm is not the fastest known. More efficient algorithms will be presented in Sections 6.4 and 6.7.)

Let τ stand for $\lfloor\frac{1}{2}(d-1)\rfloor$; by the degree constraints (6.5) we can write

$$\Lambda(x) = \sum_{m=0}^{\tau} \Lambda_m x^m \qquad \text{and} \qquad \Gamma(x) = \sum_{m=0}^{\tau-1} \Gamma_m x^m ,$$

and from the polynomial congruence (6.6) we get that the coefficients $(\Lambda_m)_{m=0}^{\tau}$ and $(\Gamma_m)_{m=0}^{\tau-1}$ solve the following set of $d-1$ linear equations in the variables $(\lambda_m)_{m=0}^{\tau}$ and $(\gamma_m)_{m=0}^{\tau-1}$:

$$\left(\begin{array}{ccccc} S_0 & 0 & 0 & \ldots & 0 \\ S_1 & S_0 & 0 & \ldots & 0 \\ \vdots & \vdots & \ddots & \ddots & \vdots \\ S_{\tau-1} & S_{\tau-2} & \ldots & S_0 & 0 \\ \hline S_\tau & S_{\tau-1} & \ldots & S_1 & S_0 \\ S_{\tau+1} & S_\tau & \ldots & S_2 & S_1 \\ \vdots & \vdots & \ddots & \vdots & \vdots \\ S_{d-2} & S_{d-3} & \ldots & S_{d-\tau-1} & S_{d-\tau-2} \end{array}\right) \begin{pmatrix} \lambda_0 \\ \lambda_1 \\ \lambda_2 \\ \vdots \\ \lambda_\tau \end{pmatrix} = \left(\begin{array}{c} \gamma_0 \\ \gamma_1 \\ \vdots \\ \gamma_{\tau-1} \\ \hline 0 \\ 0 \\ \vdots \\ 0 \end{array}\right) . \tag{6.7}$$

Furthermore, the subset that consists of the last $d-1-\tau$ $(\geq \tau)$ equations in (6.7) involves only the variables $(\lambda_m)_{m=0}^{\tau}$. Thus, one can solve first for $(\lambda_m)_{m=0}^{\tau}$ using only this subset of equations, and then any such solution determines—by the first τ equations—a unique respective solution for $(\gamma_m)_{m=0}^{\tau-1}$.

Conversely, if $(\lambda_m)_{m=0}^{\tau}$ and $(\gamma_m)_{m=0}^{\tau-1}$ solve (6.7), then the respective polynomials,

$$\lambda(x) = \sum_{m=0}^{\tau} \lambda_m x^m \qquad \text{and} \qquad \gamma(x) = \sum_{m=0}^{\tau-1} \gamma_m x^m ,$$

satisfy the degree constraints

$$\deg \gamma < \deg \lambda \leq \tfrac{1}{2}(d-1) \tag{6.8}$$

and the polynomial congruence

$$\lambda(x)S(x) \equiv \gamma(x) \pmod{x^{d-1}} . \tag{6.9}$$

The next result characterizes the set of solutions to (6.8) and (6.9) (or, equivalently, to (6.7)), in terms of the ELP and EEP. This result also implies that the solution to the key equation is essentially unique.

Proposition 6.1 *Let $\lambda(x)$ and $\gamma(x)$ be polynomials over F that satisfy (6.8). Then the following conditions hold:*

(i) The polynomials $\lambda(x)$ and $\gamma(x)$ satisfy (6.9) if and only if there exists a polynomial $c(x) \in F[x]$ such that

$$\lambda(x) = c(x) \cdot \Lambda(x) \quad \textit{and} \quad \gamma(x) = c(x) \cdot \Gamma(x) .$$

(ii) The solution to (6.8) and (6.9) with a nonzero polynomial $\lambda(x)$ of smallest possible degree is unique, up to scaling by some nonzero constant $c \in F$, and is given by

$$\lambda(x) = c \cdot \Lambda(x) \quad \textit{and} \quad \gamma(x) = c \cdot \Gamma(x) .$$

(iii) The solution in part (ii) is also the unique solution to (6.8) and (6.9) for which

$$\gcd(\lambda(x), \gamma(x)) = 1 .$$

Proof. We prove part (i); parts (ii) and (iii) then immediately follow. Since $\Lambda(0) = 1$, the ELP has a multiplicative inverse in the ring $F[x]/x^{d-1}$ and, so, from (6.6) we obtain

$$S(x) \equiv \Gamma(x)(\Lambda(x))^{-1} \pmod{x^{d-1}} .$$

Hence, $\lambda(x)$ and $\gamma(x)$ satisfy (6.9) if and only if

$$\lambda(x) \cdot \Gamma(x)(\Lambda(x))^{-1} \equiv \gamma(x) \pmod{x^{d-1}}$$

or

$$\lambda(x)\Gamma(x) \equiv \Lambda(x)\gamma(x) \pmod{x^{d-1}} .$$

The latter congruence, in turn, can be replaced by an equality, as (6.5) and (6.8) yield that the degrees of $\lambda(x)\Gamma(x)$ and $\Lambda(x)\gamma(x)$ are both smaller than $d-1$. We therefore conclude that $\lambda(x)$ and $\gamma(x)$ satisfy (6.9) if and only if

$$\lambda(x)\Gamma(x) = \Lambda(x)\gamma(x) . \tag{6.10}$$

By (6.4) and Lemma 3.2, the equality (6.10) implies that $\Lambda(x) \mid \lambda(x)$. Hence, (6.10) is equivalent to having $\lambda(x) = c(x) \cdot \Lambda(x)$ and $\gamma(x) = c(x) \cdot \Gamma(x)$ for some polynomial $c(x) \in F[x]$. □

In summary, we have shown that the solution of the key equation boils down to solving (the last $d-1-\tau$ equations in) the set (6.7) for a nonzero polynomial $\lambda(x) = \sum_{m=0}^{\tau} \lambda_m x^m$ of smallest possible degree; this polynomial, up to scaling, is then equal to the ELP $\Lambda(x)$. As pointed out earlier, once the ELP is known, we can search for its roots among $\alpha_1^{-1}, \alpha_2^{-1}, \ldots, \alpha_n^{-1}$ to find the set J, and then solve a second set of linear equations—namely, (6.1)—for the error values e_j.

The two mentioned sets of equations—(6.1) and (6.7)—can be solved by applying standard Gaussian elimination, whose time complexity is *cubic* in d. The resulting decoding algorithm is known as the *Peterson–Gorenstein–Zierler algorithm*. It turns out, however, that these two particular sets of equations can be solved by algorithms whose time complexity is only *quadratic* in d. We present such an algorithm for solving the key equation in Section 6.4, and in Section 6.5 we show how the EEP $\Gamma(x)$ can be applied to compute the error values—also in quadratic time. A second efficient algorithm for solving the key equation will be presented in Section 6.7.

6.3.2 GRS decoding through infinite power series

In this section, we re-derive the congruence (6.6), yet in a somewhat different manner. This alternate derivation is not essential for understanding the upcoming sections of this chapter; still, it provides an interesting interpretation to the problem of decoding GRS codes. In addition, the concepts that we introduce next will be useful later on in Chapters 10 and 14.

For an infinite sequence $(a_i)_{i=0}^{\infty}$ over a (possibly infinite) field F and an indeterminate x, define the respective *(infinite) formal power series* over F by the expression

$$a(x) = \sum_{i=0}^{\infty} a_i x^i .$$

The set of all formal power series over F will be denoted by $F[[x]]$. Clearly, every polynomial over F can be regarded as an element of $F[[x]]$ where all but a finite number of its coefficients are zero.

For formal power series $a(x) = \sum_{i=0}^{\infty} a_i x^i$ and $b(x) = \sum_{i=0}^{\infty} b_i x^i$ over F, we define the sum $a(x) + b(x)$ and the product $a(x)b(x)$, respectively, as the following formal power series $f(x) = \sum_{i=0}^{\infty} f_i x^i$ and $g(x) = \sum_{i=0}^{\infty} g_i x^i$:

$$f_i = a_i + b_i \qquad \text{and} \qquad g_i = \sum_{j=0}^{i} a_j b_{i-j} , \qquad i \geq 0 .$$

Under those operations, $F[[x]]$ is an integral domain, with the elements 0 and 1 of F being the respective additive and multiplicative unity elements (Problem 6.5).

Given $a(x) = \sum_{i=0}^{\infty} a_i x^i$ in $F[[x]]$ and a positive integer t, we say that x^t divides $a(x)$ (and write $x^t \,|\, a(x)$), if $a_i = 0$ for $0 \leq i \leq t$. The notation $a(x) \equiv b(x) \pmod{x^t}$ is the same as saying that x^t divides $a(x) - b(x)$.

An element $a(x) = \sum_{i=0}^{\infty} a_i x^i$ in $F[[x]]$ is *invertible* (or is a *unit*) in $F[[x]]$ if it has a multiplicative inverse in $F[[x]]$; namely, there is an element $b(x) = \sum_{i=0}^{\infty} b_i x^i$ such that $a(x)b(x) = 1$. An element $a(x)$ is invertible if and only if x does not divide $a(x)$; when the latter condition holds, the coefficients of the inverse $b(x)$ can be computed iteratively by

$$b_0 = \frac{1}{a_0} \qquad \text{and} \qquad b_i = -\frac{1}{a_0} \sum_{j=1}^{i} a_j b_{i-j} \,, \quad i \geq 1$$

(Problem 6.6).

The inverse of an invertible $a(x) \in F[[x]]$ will be denoted by $1/a(x)$, and $c(x)/a(x)$ for $c(x) \in F[[x]]$ will be a shorthand notation for $c(x) \cdot (1/a(x))$.

Example 6.2 For an element $\beta \in F$, let $(a_i)_{i=0}^{\infty}$ be the sequence that is given by $a_i = \beta^i$. The respective formal power series is $a(x) = \sum_{i=0}^{\infty} (\beta x)^i$, and the inverse $b(x)$ of $a(x)$ in $F[[x]]$ is the polynomial $b(x) = 1 - \beta x$ (Problem 6.7). □

Turning back to GRS decoding, we associate with the error word $\mathbf{e}$ the formal power series $E(x) = \sum_{\ell=0}^{\infty} E_\ell x^\ell$ in $F[[x]]$ that is defined by

$$E_\ell = \sum_{j \in J} e_j v_j \alpha_j^\ell \,, \quad \ell = 0, 1, 2, \cdots \,.$$

The series $E(x)$ can be thought of as an "infinite syndrome" of $\mathbf{e}$, and it is easy to see from (6.1) that

$$S_\ell = E_\ell \,, \quad \ell = 0, 1, \ldots, d-2 \,,$$

or, equivalently,

$$S(x) \equiv E(x) \pmod{x^{d-1}} \,. \tag{6.11}$$

Thus, by computing the (ordinary) syndrome, the decoder recovers, in effect, the first $d-1$ coefficients of $E(x)$.

Now, similarly to (6.2), we have,

$$E(x) = \sum_{\ell=0}^{\infty} x^\ell \sum_{j \in J} e_j v_j \alpha_j^\ell = \sum_{j \in J} e_j v_j \sum_{\ell=0}^{\infty} (\alpha_j x)^\ell = \sum_{j \in J} \frac{e_j v_j}{1 - \alpha_j x} \,, \tag{6.12}$$

where the last equality follows from Example 6.2. Noting that the ELP $\Lambda(x)$ is invertible as a formal power series, we identify the rightmost sum in (6.12) as the ratio $\Gamma(x)/\Lambda(x)$; hence, in $F[[x]]$,

$$E(x) = \frac{\Gamma(x)}{\Lambda(x)} \,. \tag{6.13}$$

Combining the latter equality with (6.11) yields the congruence (6.6) in $F[[x]]$ and—since $S(x)$, $\Lambda(x)$, and $\Gamma(x)$ are elements of $F[x]$—that congruence should hold also in $F[x]$.

We conclude that the decoding of GRS codes amounts to extending $S(x)$ into an (infinite) formal power series that can be written in $F[[x]]$ as a reduced ratio $\Gamma(x)/\Lambda(x)$, where $\Lambda(x)$ and $\Gamma(x)$ are polynomials whose degrees are bounded by (6.5) (a ratio is called reduced if its numerator and denominator have no common divisors; this is precisely what we require in (6.4)).

6.4 Solving the key equation by Euclid's algorithm

The efficient algorithm that we present here for solving the key equation for the ELP and EEP, makes use of (the extended version of) *Euclid's algorithm*. Given polynomials $a(x)$ and $b(x)$ over a field F such that $a(x) \neq 0$ and $\deg a > \deg b$, the algorithm computes *remainders* $r_i(x)$, *quotients* $q_i(x)$, and *auxiliary polynomials* $s_i(x)$ and $t_i(x)$, as shown in Figure 6.1. (See Problem 3.3; the notation "$r_{i-2}(x)$ div $r_{i-1}(x)$" stands for the quotient obtained when $r_{i-2}(x)$ is divided by $r_{i-1}(x)$.)

$r_{-1}(x) \leftarrow a(x)$; $r_0(x) \leftarrow b(x)$;
$s_{-1}(x) \leftarrow 1$; $s_0(x) \leftarrow 0$;
$t_{-1}(x) \leftarrow 0$; $t_0(x) \leftarrow 1$;
for ($i \leftarrow 1$; $r_{i-1}(x) \neq 0$; i++) {
  $q_i(x) \leftarrow r_{i-2}(x)$ div $r_{i-1}(x)$;
  $r_i(x) \leftarrow r_{i-2}(x) - q_i(x) r_{i-1}(x)$;
  $s_i(x) \leftarrow s_{i-2}(x) - q_i(x) s_{i-1}(x)$;
  $t_i(x) \leftarrow t_{i-2}(x) - q_i(x) t_{i-1}(x)$;
}

Figure 6.1. Euclid's algorithm.

(We remark that while the polynomials $s_i(x)$ in Figure 6.1 will be used in the forthcoming analysis, they will not be required for the actual decoding. Also, the polynomials in Figure 6.1 have been tagged by subscripts which identify the loop iteration in which each polynomial is computed; in practice, however, there is no need to allocate separate space in every loop iteration. Instead, it suffices to use a queue for storing the computed polynomials of the previous two iterations only.)

Let ν denote the largest index i for which $r_i(x) \neq 0$. It is known (Problem 3.3) that

$$r_\nu(x) = \gcd(a(x), b(x)) .$$

Lemma 6.2 *Using the notation of Euclid's algorithm, the following conditions hold:*

(i) For $i = -1, 0, \ldots, \nu{+}1$,

$$s_i(x)a(x) + t_i(x)b(x) = r_i(x) .$$

(ii) For $i = 0, 1, \ldots, \nu{+}1$,

$$\deg t_i + \deg r_{i-1} = \deg a .$$

Proof. By induction on i (see parts 2 and 3 of Problem 3.3). □

The property that we present in the next proposition will serve as the basis for applying Euclid's algorithm in solving the key equation.

Proposition 6.3 *Using the notation of Euclid's algorithm, suppose that* $t(x)$ *and* $r(x)$ *are nonzero polynomials over* F *satisfying the following conditions:*

(C1) $\gcd(t(x), r(x)) = 1$.

(C2) $\deg t + \deg r < \deg a$.

(C3) $t(x)b(x) \equiv r(x) \pmod{a(x)}$.

Then there is an index $h \in \{0, 1, \ldots, \nu{+}1\}$ *and a constant* $c \in F$ *such that*

$$t(x) = c \cdot t_h(x) \quad \textit{and} \quad r(x) = c \cdot r_h(x) .$$

Proof. First observe that $\deg r_i$ strictly decreases with i. By condition (C2) we have $\deg r < \deg a = \deg r_{-1}$; so, there is a unique value $h \geq 0$ of the index i for which

$$\deg r_h \leq \deg r < \deg r_{h-1} . \tag{6.14}$$

From Lemma 6.2(i) we get that

$$s_h(x)a(x) + t_h(x)b(x) = r_h(x) . \tag{6.15}$$

By condition (C3) there exists a polynomial $s(x)$ such that

$$s(x)a(x) + t(x)b(x) = r(x) . \tag{6.16}$$

Multiplying (6.15) by $t(x)$ and (6.16) by $t_h(x)$ and subtracting the resulting equations, we obtain

$$(t(x)s_h(x) - t_h(x)s(x))a(x) = t(x)r_h(x) - t_h(x)r(x) . \tag{6.17}$$

Now, by (6.14) and condition (C2),

$$\deg t + \deg r_h \le \deg t + \deg r < \deg a \ ,$$

and by (6.14) and Lemma 6.2(ii),

$$\deg t_h + \deg r = \deg a - \deg r_{h-1} + \deg r < \deg a \ .$$

Hence, the degree of the right-hand side of (6.17) is less than $\deg a$. However, the left-hand side of (6.17) is a multiple of $a(x)$; therefore,

$$t(x)r_h(x) = t_h(x)r(x) \ . \tag{6.18}$$

Note that Lemma 6.2(ii) implies that $\deg t_h \ge 0$ and, so, both sides of (6.18) are nonzero. It follows from (6.18) and condition (C1) that $r(x)$ divides the nonzero polynomial $r_h(x)$. Combining this with (6.14), there is a constant c such that $r(x) = c \cdot r_h(x)$, and by (6.18) we also have $t(x) = c \cdot t_h(x)$. □

One can verify that Proposition 6.3 holds also when $r(x) = 0$: in this case, condition (C1) implies that $t(x)$ is a nonzero scalar of F, and from condition (C3) we have $b(x) = 0$. The corresponding index h will therefore be $\nu+1 = 0$.

Based on Proposition 6.3, we can solve the key equation for $\Lambda(x)$ and $\Gamma(x)$ as follows. We apply Euclid's algorithm with $a(x) \leftarrow x^{d-1}$ and $b(x) \leftarrow S(x)$ to produce $\Lambda(x) \leftarrow c \cdot t_h(x)$ and $\Gamma(x) \leftarrow c \cdot r_h(x)$, where the constant c will then be set so that $\Lambda(0) = 1$; note that the key equation implies conditions (C1)–(C3). However, we still need to determine the index h that is guaranteed by the proposition: the way we set its value in the proof—specifically, in (6.14)—is not too useful, since we assume there that the polynomials $t(x)$ and $r(x)$ are already given. To compute h while $t(x)$ and $r(x)$ are not known, we will make use of the next result, which assumes more information about $\deg t$ and $\deg r$ than just condition (C2); that additional information is indeed provided by the degree constraints (6.5) in the key equation.

Proposition 6.4 *Let $t(x)$ and $r(x)$ be as in Proposition 6.3, and assume in addition that*

$$\deg t \le \tfrac{1}{2} \deg a \qquad \textit{and} \qquad \deg r < \tfrac{1}{2} \deg a \ .$$

Then the value h in that proposition is the unique index for which the remainders in Euclid's algorithm satisfy

$$\deg r_h < \tfrac{1}{2} \deg a \le \deg r_{h-1} \ .$$

Proof. A smaller index i would result in a polynomial $c \cdot r_i(x)$ whose degree is too large. On the other hand, by Lemma 6.2(ii) we have for every $i > h$,

$$\deg t_i \geq \deg t_{h+1} = \deg a - \deg r_h > \tfrac{1}{2} \deg a .$$

So, for every $i > h$ we would end up with a polynomial $c \cdot t_i(x)$ whose degree is too large. □

Corollary 6.5 *The solution to the key equation is given by $\Lambda(x) = c \cdot t_h(x)$ and $\Gamma(x) = c \cdot r_h(x)$ for some nonzero constant $c \in F$, where $\{t_i(x)\}_i$ and $\{r_i(x)\}_i$ are obtained by an application of Euclid's algorithm to $a(x) \leftarrow x^{d-1}$ and $b(x) \leftarrow S(x)$, and h is the unique value of i for which*

$$\deg r_h < \tfrac{1}{2}(d-1) \leq \deg r_{h-1} .$$

Corollary 6.5 serves as an alternative to Proposition 6.1 for establishing the uniqueness of the solution to the key equation.

6.5 Finding the error values

Having found $\Lambda(x)$ and $\Gamma(x)$, we show an efficient way for computing the error values.

Recall from Section 3.7 that for a polynomial $a(x) = \sum_{m=0}^{s} a_m x^m$ over F we define the formal derivative by

$$a'(x) = \sum_{m=1}^{s} m a_m x^{m-1} .$$

As we saw in Problem 3.38, the formal derivative of a product of two polynomials obeys the rule

$$(a(x)b(x))' = a'(x)b(x) + a(x)b'(x) ,$$

and by repeated applications of this rule to $\Lambda(x)$, we get

$$\Lambda'(x) = \sum_{j \in J} (-\alpha_j) \prod_{m \in J \setminus \{j\}} (1 - \alpha_m x) .$$

Therefore, for every $\kappa \in J$,

$$\Lambda'(\alpha_\kappa^{-1}) = -\alpha_\kappa \prod_{m \in J \setminus \{\kappa\}} (1 - \alpha_m \alpha_\kappa^{-1}) .$$

On the other hand, for every $\kappa \in J$,

$$\Gamma(\alpha_\kappa^{-1}) = e_\kappa v_\kappa \prod_{m \in J \setminus \{\kappa\}} (1 - \alpha_m \alpha_\kappa^{-1}) .$$

Hence, we obtain the following expression for the error values for every $\kappa \in J$:

$$\boxed{e_\kappa = -\frac{\alpha_\kappa}{v_\kappa} \cdot \frac{\Gamma(\alpha_\kappa^{-1})}{\Lambda'(\alpha_\kappa^{-1})}} .$$

This formula is known as *Forney's algorithm* for computing the error values.

6.6 Summary of the GRS decoding algorithm

Figure 6.2 summarizes the decoding algorithm for an $[n, k, d]$ GRS code over F with code locators $\alpha_1, \alpha_2, \ldots, \alpha_n$ and column multipliers $v_1, v_2, \ldots, v_n$, assuming that the number of errors does not exceed $\tau = \lfloor \frac{1}{2}(d-1) \rfloor$.

A respective decoding circuit is presented by the schematic diagram in Figure 6.3. The received word $\mathbf{y}$ is read serially—entry by entry—into a

---

Input: received word $(y_1\ y_2\ \ldots\ y_n) \in F^n$.
Output: error word $(e_1\ e_2\ \ldots\ e_n) \in F^n$.

1. *Syndrome computation:* compute the polynomial $S(x) = \sum_{\ell=0}^{d-2} S_\ell x^\ell$ by

$$S_\ell = \sum_{j=1}^{n} y_j v_j \alpha_j^\ell , \quad \ell = 0, 1, \ldots, d-2 .$$

2. *Solving the key equation:* apply Euclid's algorithm to

$$a(x) \leftarrow x^{d-1} \quad \text{and} \quad b(x) \leftarrow S(x) ,$$

to produce

$$\Lambda(x) \leftarrow t_h(x) \quad \text{and} \quad \Gamma(x) \leftarrow r_h(x) ,$$

where h is the smallest index i for which $\deg r_i < \frac{1}{2}(d-1)$.

3. *Forney's algorithm:* compute the error locations and values by

$$e_j = \begin{cases} -\dfrac{\alpha_j}{v_j} \cdot \dfrac{\Gamma(\alpha_j^{-1})}{\Lambda'(\alpha_j^{-1})} & \text{if } \Lambda(\alpha_j^{-1}) = 0 \\ 0 & \text{otherwise} \end{cases} , \quad j = 1, 2, \ldots, n .$$

Figure 6.2. Decoding algorithm for GRS codes.

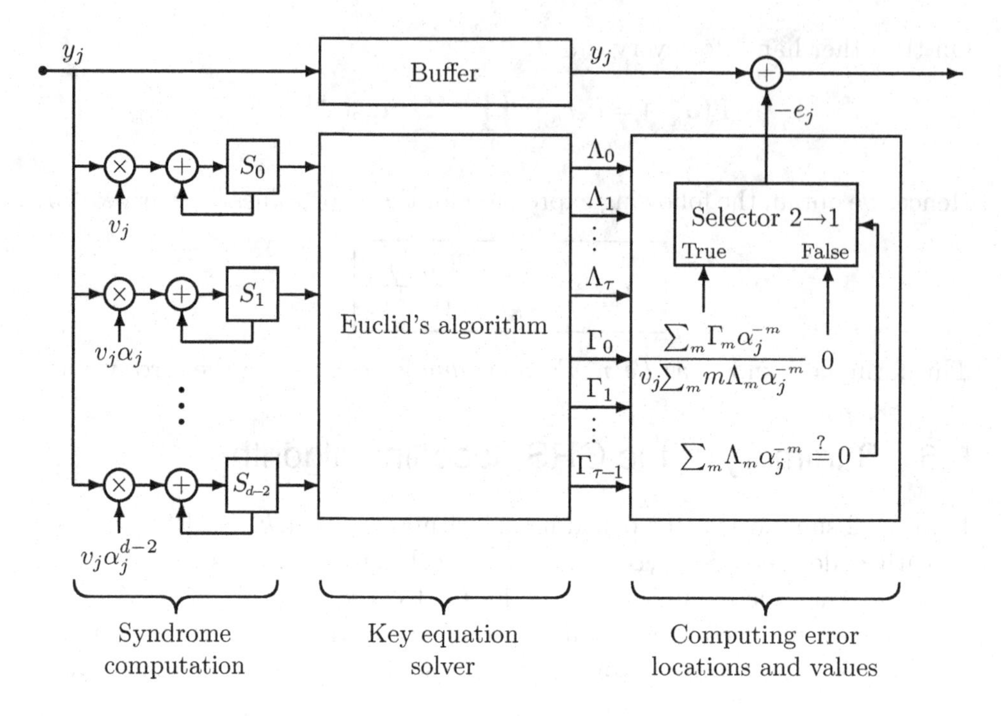

Figure 6.3. Decoding circuit for GRS codes.

buffer. While $\mathbf{y}$ is being read, the syndrome entries $S_0, S_1, \ldots, S_{d-2}$ are computed and stored in $d-1$ delay units (Step 1 in Figure 6.2). Those units are initially reset to zero, and when the jth entry of $\mathbf{y}$ is at the input, that entry is multiplied by $v_j\alpha_j^\ell$ for $\ell = 0, 1, \ldots, d-2$ and each result is added to the contents of the respective delay unit. The center part of the circuit solves the key equation and produces the polynomials $\Lambda(x)$ and $\Gamma(x)$ (Step 2 in Figure 6.2). Finally, the received word $\mathbf{y}$ is flushed out serially, during which the error locations and values are found (Step 3): when the jth entry of $\mathbf{y}$ is at the output, the values $\Lambda(\alpha_j^{-1})$ and $\Gamma(\alpha_j^{-1})/(v_j\alpha_j^{-1}\Lambda'(\alpha_j^{-1}))$ are computed; if the former is zero, then the latter is added to the value at the jth location.

Note that Step 2 in Figure 6.2 can begin only after the whole word $\mathbf{y}$ has been read and must be completed before $\mathbf{y}$ starts to be flushed out from the buffer. Therefore, the time complexity of Step 2 affects the delay, or latency, of the decoding process. Hence, it is desirable to have a fast implementation of this step, and in our decoding algorithm this step is realized with less than d iterations of Euclid's algorithm. One can verify that Steps 1–3 in Figure 6.2 can be carried out by a number of field operations that is proportional, respectively, to dn, $|J|d$, and $|J|n$, where $|J|$ is the number of actual errors (not to exceed $\frac{1}{2}(d-1)$); obviously, Steps 2 and 3 can be skipped when $|J| = 0$ (i.e., when the syndrome is zero).

An *alternant code* $\mathcal{C}_{\mathrm{alt}}$ can be decoded by applying a decoding algorithm for its underlying GRS codes (in which case d is the *designed* minimum distance of $\mathcal{C}_{\mathrm{alt}}$). Note that for alternant codes over GF(2), the error values will always be 1 and, hence, need not be computed.

6.7 The Berlekamp–Massey algorithm

In this section, we describe a second algorithm for solving the key equation.

6.7.1 Linear recurrence

Let $b(x) = \sum_{i=0}^{\deg b} b_i x^i$ be a polynomial over F and N be a nonnegative integer. (In what follows, the coefficients b_i for $i \geq N$ will be immaterial; hence, one might as well assume that $b(x)$ is a formal power series in $F[[x]]$, possibly with an infinite degree.) An *N-recurrence* of $b(x)$ over F is an (ordered) pair of polynomials $(\sigma(x), \omega(x))$ over F such that $\sigma(0) = 1$ and

$$\sigma(x)b(x) \equiv \omega(x) \pmod{x^N} .$$

The *recurrence order* of $(\sigma(x), \omega(x))$, denoted by $\mathrm{ord}(\sigma, \omega)$, is defined as

$$\mathrm{ord}(\sigma, \omega) = \max\{\deg \sigma, 1+\deg \omega\} .$$

The following result is given as an exercise (Problem 6.15).

Proposition 6.6 *Let $b(x)$ be a polynomial over F and let $(\sigma(x), \omega(x))$ be an N-recurrence of $b(x)$ over F whose recurrence order, $\mathrm{ord}(\sigma, \omega)$, is the smallest possible. Then the following conditions hold:*

(i) $\gcd(\sigma(x), \omega(x)) = 1$.

(ii) If $\mathrm{ord}(\sigma, \omega) \leq N/2$ then $(\sigma(x), \omega(x))$ is unique; namely, the recurrence order of every other N-recurrence of $b(x)$ is greater than $\mathrm{ord}(\sigma, \omega)$.

Let $(\sigma(x), \omega(x))$ be an N-recurrence of $b(x)$ and write $L = \mathrm{ord}(\sigma, \omega)$,

$$\sigma(x) = 1 + \sum_{m=1}^{L} \lambda_m x^m , \qquad \text{and} \qquad \omega(x) = \sum_{m=0}^{L-1} \gamma_m x^m .$$

Then,

$$b_i = \begin{cases} \gamma_i - \sum_{m=1}^{i} \lambda_m b_{i-m} & \text{for } 0 \leq i < L \\ -\sum_{m=1}^{L} \lambda_m b_{i-m} & \text{for } L \leq i < N \end{cases} \tag{6.19}$$

(where $b_i = 0$ for $i > \deg b$). That is, the sequence $(b_i)_{i=0}^{N-1}$ satisfies a linear recurrence of order L, and the coefficients of the recurrence are given by $\lambda_1, \lambda_2, \ldots, \lambda_L$; hence the terms N-recurrence and recurrence order.

Figure 6.4 shows a circuit that synthesizes a sequence $(b_i)_{i=0}^{N-1}$ that satisfies (6.19). The circuit, commonly referred to as an L-tap *linear-feedback shift register* (in short, LFSR), contains L cascaded delay units, which are controlled by a clock and are initially reset to zero. During clock ticks $0, 1, \ldots, L-1$, the switch is in position A and the coefficients of $\omega(x)$ are fed into the circuit, starting with γ_0. At clock tick L, the circuit becomes *autonomous*: the external input is disconnected by changing the switch to position B, and the input to the circuit from this point onward is fixed to be all-zero. At every clock tick $i = 0, 1, \ldots, N-1$, the contents of the L delay units are multiplied, respectively, by $\lambda_1, \lambda_2, \ldots, \lambda_L$, then summed up, and the result is fed back and subtracted from the input (γ_i or 0) to produce the output b_i. The contents of the delay units shifts to the right and the newly computed b_i is fed into the leftmost unit.

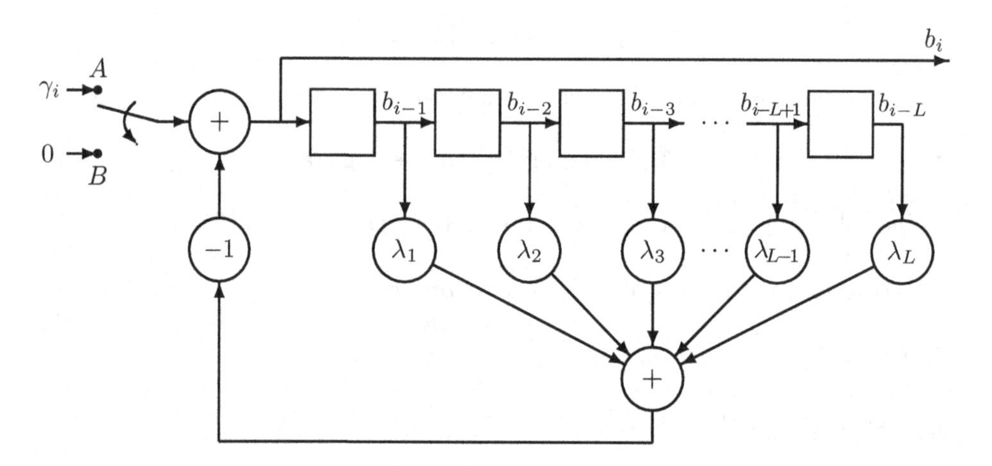

Figure 6.4. Linear-feedback shift register.

Turning back to GRS decoding, let $S(x) = \sum_{\ell=0}^{d-2} S_\ell x^\ell$ be the syndrome polynomial and $\Lambda(x)$ and $\Gamma(x)$ be the ELP and EEP, respectively. Assuming that the number of errors does not exceed $\frac{1}{2}(d-1)$, it follows from the key equation that $(\Lambda(x), \Gamma(x))$ is a $(d-1)$-recurrence of $S(x)$ and $\mathrm{ord}(\Lambda, \Gamma) \leq \frac{1}{2}(d-1)$ (and from (6.13) we get that $(\Lambda(x), \Gamma(x))$ is an N-recurrence of $E(x)$ for every $N \geq 0$).

Given a polynomial $b(x)$ and a nonnegative integer N, we present in Section 6.7.3 an algorithm that computes iteratively i-recurrences $(\sigma_i(x), \omega_i(x))$ of $b(x)$ for $i = 0, 1, 2, \ldots, N$, and for each i, the recurrence order $\mathrm{ord}(\sigma_i, \omega_i)$ is the smallest possible. In particular, if we apply the algorithm to $N \leftarrow d-1$ and $b(x) \leftarrow S(x)$, we will end up with a $(d-1)$-recurrence $(\sigma_{d-1}(x), \omega_{d-1}(x))$

of $S(x)$ such that

$$\text{ord}(\sigma_{d-1}, \omega_{d-1}) \le \text{ord}(\Lambda, \Gamma) \le \tfrac{1}{2}(d-1)$$

and $\gcd(\sigma_{d-1}(x), \omega_{d-1}(x)) = 1$ (see Proposition 6.6). That is, the polynomials $\sigma_{d-1}(x)$ and $\omega_{d-1}(x)$ solve the key equation; so, by Proposition 6.1 (or Corollary 6.5), they must be the ELP and EEP, respectively.

6.7.2 Lower bound on the recurrence order

We precede the description and analysis of the algorithm by the following lemma, which provides for every $i \ge 0$ a recursive lower bound on the recurrence order of any i-recurrence of a given polynomial $b(x)$.

Lemma 6.7 *Let $b(x)$ be a polynomial over F and for every $i \ge 0$, let L_i be the smallest recurrence order of any i-recurrence of $b(x)$. Then $L_0 = 0$, and for every $i \ge 0$,*

$$L_{i+1} \ge L_i \; ;$$

furthermore, if some i-recurrence of $b(x)$ with recurrence order L_i is not an $(i+1)$-recurrence, then

$$L_{i+1} \ge \max\{L_i, i+1-L_i\} \; .$$

Proof. Obviously, $(1, 0)$ is a 0-recurrence of $b(x)$. Also, L_i is non-decreasing with i and, so, $L_{i+1} \ge L_i$.

Next, suppose that $L_{i+1} < i+1-L_i$ and let $(\sigma(x), \omega(x))$ be an i-recurrence of $b(x)$ with $\text{ord}(\sigma, \omega) = L_i$. We show that $(\sigma(x), \omega(x))$ is necessarily an $(i+1)$-recurrence.

Let $(\hat{\sigma}(x), \hat{\omega}(x))$ be an $(i+1)$-recurrence with $\text{ord}(\hat{\sigma}, \hat{\omega}) = L_{i+1}$; obviously, $(\hat{\sigma}(x), \hat{\omega}(x))$ is also an i-recurrence. Both $\sigma(x)$ and $\hat{\sigma}(x)$ have multiplicative inverses in $F[x]/x^i$ and, so, we can write

$$b(x) \equiv (\sigma(x))^{-1}\omega(x) \equiv (\hat{\sigma}(x))^{-1}\hat{\omega}(x) \pmod{x^i}$$

or

$$\hat{\sigma}(x)\omega(x) \equiv \sigma(x)\hat{\omega}(x) \pmod{x^i} \; . \tag{6.20}$$

Since we assume that

$$\max\{\deg \sigma, 1 + \deg \omega\} + \max\{\deg \hat{\sigma}, 1 + \deg \hat{\omega}\} = L_i + L_{i+1} \le i \; ,$$

it follows that

$$\deg \sigma + \deg \hat{\omega} < i \quad \text{and} \quad \deg \hat{\sigma} + \deg \omega < i \; .$$

Hence, the congruence (6.20) holds with equality, namely,

$$\hat{\sigma}(x)\omega(x) = \sigma(x)\hat{\omega}(x) .$$

But from Proposition 6.6(i) we also have

$$\gcd(\sigma(x), \omega(x)) = \gcd(\hat{\sigma}(x), \hat{\omega}(x)) = 1 .$$

Therefore, $(\sigma(x), \omega(x)) = (\hat{\sigma}(x), \hat{\omega}(x))$. □

6.7.3 The algorithm

Figure 6.5 presents an algorithm, known as the *Berlekamp–Massey* algorithm, for computing an N-recurrence of a polynomial $b(x)$ over F. Given $b(x)$ and N, the algorithm computes pairs of polynomials $(\sigma_i(x), \omega_i(x))$ for $i = 0, 1, 2, \ldots, N$, the properties of which are summarized in the lemmas below.

Input: polynomial $b(x) \in F[x]$, nonnegative integer N.
Output: pair of polynomials $(\sigma_N(x), \omega_N(x))$ over F.

$\sigma_{-1}(x) \leftarrow 0$; $\sigma_0(x) \leftarrow 1$;
$\omega_{-1}(x) \leftarrow -x^{-1}$; $\omega_0(x) \leftarrow 0$;
$\mu \leftarrow -1$; $\delta_{-1} \leftarrow 1$;
for ($i \leftarrow 0$; $i < N$; i++) {
  $\delta_i \leftarrow$ coefficient of x^i in $\sigma_i(x)b(x)$;
  $\sigma_{i+1}(x) \leftarrow \sigma_i(x) - (\delta_i/\delta_\mu) \cdot x^{i-\mu} \cdot \sigma_\mu(x)$;
  $\omega_{i+1}(x) \leftarrow \omega_i(x) - (\delta_i/\delta_\mu) \cdot x^{i-\mu} \cdot \omega_\mu(x)$;
  if (($\delta_i \neq 0$) and ($2\,\mathrm{ord}(\sigma_i, \omega_i) \leq i$))
    $\mu \leftarrow i$;
}

Figure 6.5. Berlekamp–Massey algorithm.

For the analysis of the algorithm, it will be convenient to define the *reverse degree* of $b(x)$, denoted by $\mathrm{rdeg}(b)$, as the largest integer t, if any, such that $x^t \mid b(x)$; when no such integer exists—i.e., when $b(x) = 0$—define $\mathrm{rdeg}(b) = \infty$.

Lemma 6.8 *Using the notation of the Berlekamp–Massey algorithm (Figure 6.5), the pair $(\sigma_i(x), \omega_i(x))$ is an i-recurrence of $b(x)$ for every $i = 0, 1, 2, \ldots, N$.*

Proof. We distinguish between three types of iterations of the main loop in Figure 6.5, according to the value of the loop variable i.

Case 1: $0 \le i < \mathrm{rdeg}(b)$. By a simple induction on i it follows that in each such iteration, the computed value of δ_i is zero; so, $(\sigma_{i+1}(x), \omega_{i+1}(x)) = (1, 0)$, which is readily an $(i+1)$-recurrence of $b(x)$ for the assumed range of i (and also for $i = -1$).

Case 2: $i = \mathrm{rdeg}(b)$. Here we get $\delta_i \ne 0$ (for the first time), and the particular selection for the values of $\sigma_{-1}(x)$ and $\omega_{-1}(x)$ yields

$$(\sigma_{i+1}(x), \omega_{i+1}(x)) = (1, b_i x^i) ,$$

where b_i is the (nonzero) coefficient of x^i in $b(x)$. It follows that the computed pair $(\sigma_{i+1}(x), \omega_{i+1}(x))$ is an $(i+1)$-recurrence for $b(x)$. Observe that the variable μ is updated in this iteration.

Case 3: $\mathrm{rdeg}(b) < i < N$. Using Cases 1 and 2 as an induction base, we prove by induction on $i \ge 1 + \mathrm{rdeg}(b)$ that $(\sigma_i(x), \omega_i(x))$ is an i-recurrence of $b(x)$ and that $\deg \omega_i(x) < i$. By the induction hypothesis we have

$$\sigma_i(x) b(x) \equiv \omega_i(x) \pmod{x^i} \tag{6.21}$$

and

$$\sigma_\mu(x) b(x) \equiv \omega_\mu(x) \pmod{x^\mu} ,$$

where we refer here to the value of the variable μ taken at the beginning of iteration i (recall that $\mu \ge 0$, as it has already been updated at least once in previous loop iterations). Multiply the last congruence by $(\delta_i/\delta_\mu) \cdot x^{i-\mu}$ to obtain

$$(\delta_i/\delta_\mu) x^{i-\mu} \sigma_\mu(x) b(x) \equiv (\delta_i/\delta_\mu) x^{i-\mu} \omega_\mu(x) \pmod{x^i} .$$

Next, subtract the resulting congruence from (6.21); this yields

$$\sigma_{i+1}(x) b(x) \equiv \omega_{i+1}(x) \pmod{x^i} .$$

Hence, the proof will be complete once we verify that the coefficients of x^i in $\sigma_{i+1}(x) b(x)$ and $\omega_{i+1}(x)$ are equal; in fact, we show that they are both zero. Indeed, the coefficients of x^i in $\sigma_i(x) b(x)$ and $(\delta_i/\delta_\mu) x^{i-\mu} \sigma_\mu(x) b(x)$ both equal δ_i, and by the induction hypothesis on $\deg \omega_i$ we obtain

$$\deg \omega_{i+1} \le \max\{\deg \omega_i, i - \mu + \deg \omega_\mu\} < i \quad (< i+1) ,$$

where the first inequality follows from the expression for $\omega_{i+1}(x)$ in the algorithm. □

6.7.4 Minimality of the recurrence order

The next lemma provides a recursive upper bound on the recurrence order of the pairs $(\sigma_i(x), \omega_i(x))$ that are computed throughout the course of the algorithm in Figure 6.5.

Lemma 6.9 *For* $i = 0, 1, \ldots, N$, *let* U_i *be the value* $\mathrm{ord}(\sigma_i, \omega_i)$ *in the Berlekamp–Massey algorithm. Then* $U_0 = 0$, *and for every* $0 \le i < N$,

$$U_{i+1} \le \max\{U_i, i+1-U_i\} \; ;$$

furthermore, if $\delta_i = 0$*—i.e., if* $(\sigma_i(x), \omega_i(x))$ *is an* $(i+1)$ *recurrence of* $b(x)$*—then,*

$$U_{i+1} = U_i \; .$$

Proof. We first verify that the claim holds for $i \le \mathrm{rdeg}(b)$: $U_{i+1} = U_i = 0$ for $0 \le i < \mathrm{rdeg}(b)$, while for $i = \mathrm{rdeg}(b)$ we have $\delta_i \ne 0$ and $U_{i+1} = i+1 = i+1-U_i$.

Next, we prove the claim for a given iteration $i > \mathrm{rdeg}(b)$ under an induction hypothesis on all previous iterations. By the way $(\sigma_{i+1}(x), \omega_{i+1}(x))$ is computed in Figure 6.5, we have

$$U_{i+1} = U_i \quad \text{when } \delta_i = 0$$

and

$$U_{i+1} \le \max\{U_i, i-\mu+U_\mu\} \quad \text{when } \delta_i \ne 0 \; .$$

Hence, to complete the proof it suffices to show that $i-\mu+U_\mu \le i+1-U_i$, or that

$$U_i \le \mu+1-U_\mu \; . \tag{6.22}$$

Consider iteration μ (≥ 0) of the main loop in Figure 6.5. From the “if” statement in the loop we have $2U_\mu \le \mu$ and, so, an application of the induction hypothesis to iteration μ yields

$$U_{\mu+1} \le \max\{U_\mu, \mu+1-U_\mu\} = \mu+1-U_\mu \; .$$

Next, consider iteration j of the main loop where $\mu < j < i$. If $\delta_j \ne 0$ then by the “if” statement we must have $2U_j > j$, i.e., $\max\{U_j, j+1-U_j\} = U_j$. The induction hypothesis on iteration j thus implies

$$U_{j+1} \le U_j \quad \text{for } \mu < j < i$$

(regardless of whether δ_j is zero or not). Writing the chain of inequalities

$$U_i \le U_{i-1} \le \ldots \le U_{\mu+2} \le U_{\mu+1} \le \mu+1-U_\mu \; ,$$

we obtain (6.22). □

Lemmas 6.7 through 6.9 lead to the following result.

Proposition 6.10 *Using the notations of Lemmas 6.7 and 6.9,*

$$L_i = U_i \ , \quad 0 \le i \le N \ ,$$

and

$$U_{i+1} = \begin{cases} \max\{U_i, i+1-U_i\} & \text{if } \delta_i \ne 0 \\ U_i & \text{if } \delta_i = 0 \end{cases} \ , \quad 0 \le i < N \ .$$

Proof. We prove the claim by induction on i. The case $i = 0$ is immediate, so we proceed with the induction step and assume that $L_i = U_i$ for a given $i < N$. We distinguish between two cases.

Case 1: some i-recurrence of $b(x)$ with recurrence order L_i is not an $(i+1)$-recurrence; in particular, this case includes the event $\delta_i \ne 0$. Here,

$$L_{i+1} \ge \max\{L_i, i+1-L_i\} = \max\{U_i, i+1-U_i\} \ge U_{i+1} \ge L_{i+1} \ ,$$

where the first inequality follows from Lemma 6.7, the last two inequalities follow, respectively, from Lemmas 6.9 and 6.8, and the equality in between follows from the induction hypothesis. We conclude that all the inequalities are, in fact, equalities.

Case 2: every i-recurrence of $b(x)$ with recurrence order L_i is also an $(i+1)$-recurrence. This means that δ_i is necessarily zero, and by Lemmas 6.7–6.9 we obtain

$$L_{i+1} \ge L_i = U_i = U_{i+1} \ge L_{i+1} \ .$$

The inequalities thus must hold with equality. □

It follows from Proposition 6.10 that

$$U_{i+1} > U_i \qquad \Longleftrightarrow \qquad (\delta_i \ne 0 \ \text{ and } \ 2U_i \le i) \ ;$$

that is, the recurrence order increases in the Berlekamp–Massey algorithm precisely when the condition in the "if" statement is satisfied.

6.7.5 Summary of the properties of the algorithm

The next corollary is obtained by substituting $i = N$ in Lemma 6.8 and Proposition 6.10.

Corollary 6.11 *Given a polynomial $b(x) \in F[x]$ and a nonnegative integer N as input to the Berlekamp–Massey algorithm, the output produced is an N-recurrence of $b(x)$ over F with the smallest possible recurrence order among all N-recurrences of $b(x)$.*

As discussed earlier, the Berlekamp–Massey algorithm can be used for solving the key equation during the decoding of GRS codes: apply the algorithm to $b(x) \leftarrow S(x)$ and $N \leftarrow d-1$ to produce $(\sigma_{d-1}(x), \omega_{d-1}(x)) = (\Lambda(x), \Gamma(x))$. Hence, it can replace Euclid's algorithm in Step 2 of Figure 6.2 and in the center part of the diagram in Figure 6.3. When decoding binary alternant codes, we do not need the EEP and, so, it is unnecessary to compute the polynomials $\omega_i(x)$. Still, the "if" statement in the Berlekamp–Massey algorithm does require keeping track of the value $U_i = \text{ord}(\sigma_i, \omega_i)$, and Proposition 6.10 shows how this can be done without computing $\omega_i(x)$ explicitly.

One can verify that when the Berlekamp–Massey algorithm is used in the decoding of GRS codes, it takes a number of field operations that is proportional to $|J|d$, where $|J|$ is the number of actual errors; as such, this algorithm ties with Euclid's (see also Problem 6.17). We also point out that the subscripts i in $(\sigma_i(x), \omega_i(x))$ and δ_i have been inserted in Figure 6.5 only to allow easier reference in the analysis; an actual implementation of the algorithm does not require adding new space in each loop iteration. Nevertheless, when the variable μ is updated, we also need to store $(\sigma_\mu(x), \omega_\mu(x))$ and δ_μ, since these quantities may be required in the computation of subsequent iterations (until the next update of μ).

Problems

[Section 6.2]

Problem 6.1 Let H_{GRS} be a canonical parity-check matrix of an $[n, k, d]$ GRS code over $F = \text{GF}(q)$ where $0 < k \leq n-2$. Given a word $\mathbf{e} \in F^n$, denote by $(S_0\ S_1\ \ldots\ S_{d-2})^T$ its syndrome with respect to H_{GRS}.

1. Let $\mathbf{e}$ be a nonzero word in F^n of Hamming weight t. Show that the longest run of 0's in the sequence

$$S_0, S_1, \ldots, S_{d-2}$$

has length less than t; that is, for every i in the range $0 \leq i < d-t$ there is a j in the range $i \leq j < i+t$ such that $S_j \neq 0$.

2. Show that the bound in part 1 is tight in the following sense. For every t in the range $0 < t \leq d$ and every i in the range $0 \leq i \leq d-t$, there is a word $\mathbf{e} \in F^n$ of Hamming weight t whose syndrome satisfies

$$S_j = 0 \quad \text{for } i \leq j < i+t-1$$

(in fact, for every subset $J \subseteq \{1, 2, \ldots, n\}$ of size t there is such a word whose support is J).

3. Express the syndrome entries of a word of Hamming weight 1 in terms of S_0 and S_1 only.

[Section 6.3]

Problem 6.2 Let F be the field GF(2) and let $\mathcal{C}_{\rm RS}$ be a $[15, 11]$ normalized primitive RS code over $\Phi = F[\xi]/(\xi^4 + \xi + 1)$ with a canonical parity-check matrix

$$H_{\rm RS} = \left(\xi^{ij} \right)_{i=0}^{3} {}_{j=0}^{14} .$$

The word

$$\mathbf{y} = (\xi \;\; \xi \;\; \xi^2 \; \xi^2 \; \xi^4 \; \xi^5 \; \xi^6 \; \xi^7 \; \xi^8 \; \xi^9 \; \xi^{10} \; \xi^{11} \; \xi^{12} \; \xi^{13} \; \xi^{14})$$

has been received as a result of transmitting a codeword of $\mathcal{C}_{\rm RS}$ through an additive channel $(\Phi, \Phi, \mathsf{Prob})$.

1. Compute the syndrome of $\mathbf{y}$ with respect to the parity-check matrix $H_{\rm RS}$.
2. Compute the ELP and EEP of the error word under the assumption that at most two errors have occurred.

Problem 6.3 Let α be a primitive element in $\Phi = {\rm GF}(2^m)$ and let $\mathcal{C}_{\rm BCH}$ be a narrow-sense primitive BCH code (of length $n = 2^m - 1$) over $F = {\rm GF}(2)$ whose parity-check matrix is obtained by representing each entry in

$$\begin{pmatrix} 1 & \alpha & \alpha^2 & \ldots & \alpha^{n-1} \\ 1 & \alpha^3 & \alpha^6 & \ldots & \alpha^{3(n-1)} \end{pmatrix}$$

as a column vector in F^m.

1. Show that
$$\mathcal{C}_{\rm BCH} = \mathcal{C}_{\rm RS} \cap F^n ,$$
where $\mathcal{C}_{\rm RS}$ is an $[n, n-4, 5]$ narrow-sense RS code over Φ.

   Hint: See Example 5.5.

A codeword $\mathbf{c} \in \mathcal{C}_{\rm BCH}$ has been transmitted through an additive channel (F, F, Prob) and a word $\mathbf{y} \in F^n$ with at most two errors has been received.

2. Let $(S_0 \; S_1 \; S_2 \; S_3)^T$ be the syndrome $H_{\rm RS}\mathbf{y}^T$, where $H_{\rm RS}$ is a canonical parity-check matrix of the code $\mathcal{C}_{\rm RS}$ in part 1. Show that $S_1 = S_0^2$ and $S_3 = S_0^4$.

   Hint: Recall that $\mathbf{y}$ is over F.

3. Assuming that exactly two errors have occurred, show that the EEP is the constant polynomial $\Gamma(x) = S_0$.
4. Assuming that exactly two errors have occurred, derive from the key equation expressions for the coefficients of the ELP $\Lambda(x) = 1 + \Lambda_1 x + \Lambda_2 x^2$. Write those expressions in terms of S_0 and S_2 only. Compare the result with the polynomial in Equation (3.6).

Problem 6.4 Let $\mathcal{C}_{\rm GRS}$ be a narrow-sense GRS code of length n over an extension field Φ of $F = {\rm GF}(2)$. A codeword of the alternant code $\mathcal{C}_{\rm GRS} \cap F^n$ is transmitted through an additive channel (F, F, Prob). Show that the EEP and the ELP are related by $\Gamma(x) = \Lambda'(x)$, where $(\cdot)'$ stands for a formal derivative (as defined in Section 3.7). Deduce that $\Gamma(x)$ is a square of a polynomial in $\Phi[x]$.

Problem 6.5 Let F be a field. Show that $F[[x]]$ is an integral domain under the definition of sum and product of elements in $F[[x]]$ (see the Appendix).

Problem 6.6 Let F be a field.

1. Show that an element $a(x) \in F[[x]]$ is invertible in $F[[x]]$ if and only if x does not divide $a(x)$.

2. Show that if $a(x) = \sum_{i=0}^{\infty} a_i x^i$ is invertible in $F[[x]]$ then its multiplicative inverse, $b(x) = \sum_{i=0}^{\infty} b_i x^i$, is unique and can be iteratively computed by

$$b_0 = \frac{1}{a_0} \qquad \text{and} \qquad b_i = -\frac{1}{a_0} \sum_{j=1}^{i} a_j b_{i-j} \,, \quad i \geq 1 \,.$$

Problem 6.7 Let β be an element in a field F. Show that the multiplicative inverse of $1 - \beta x$ in $F[[x]]$ equals $\sum_{i=0}^{\infty} (\beta x)^i$.

Problem 6.8 (Periodic sequences) A sequence $(a_i)_{i=0}^{\infty}$ over a field F is called *periodic* if there is an integer $e > 0$ such that $a_{i+e} = a_i$ for all $i \geq 0$. The smallest such e is called the *period* of the sequence. The period of a non-periodic sequence is defined as infinity. These definitions extend also to the respective formal power series $a(x) = \sum_{i=0}^{\infty} a_i x^i$.

1. Let $(a_i)_{i=0}^{\infty}$ be a periodic sequence with period e over a field F and suppose that for some positive integer ℓ,

$$a_{i+\ell} = a_i \quad \text{for all } i \geq 0 \,.$$

Show that $e \,|\, \ell$.

Hint: Show that if there were a nonzero remainder r when dividing ℓ by e, then r would be a period.

2. Show that an element $a(x) \in F[[x]]$ is periodic if and only if there is an integer $e > 0$ and a polynomial $c(x) \in F_e[x]$ such that

$$a(x) = \frac{c(x)}{x^e - 1} \,.$$

Furthermore, show that if $a(x)$ is periodic, then its period is the smallest $e > 0$ for which the last equality holds for some $c(x) \in F_e[x]$.

Problem 6.9 Let $\sigma(x)$ be a polynomial over a field F such that $\gcd(\sigma(x), x) = 1$. The *exponent* of $\sigma(x)$, denoted by $\exp \sigma(x)$, is the smallest integer $e > 0$, if any, such that

$$\sigma(x) \mid x^e - 1 \,.$$

If no such integer exists then define $\exp \sigma(x) = \infty$.

1. Show that if F is a finite field then $\exp \sigma(x) < \infty$.

Hint: Consider the multiplicative order of x in the (finite) ring $F[x]/\sigma(x)$.

2. Show that when F is a finite field and $\deg \sigma(x) > 0$, then the element $1/\sigma(x)$ in $F[[x]]$ is periodic and that the period equals $\exp \sigma(x)$ (see Problem 6.8).

[Section 6.4]

Problem 6.10 (Generalization of Proposition 6.3) Let $a(x)$ and $b(x)$ be polynomials over F such that $a(x) \neq 0$ and $\deg a > \deg b$. Suppose that $t(x)$ and $r(x)$ are polynomials over F that satisfy the following two conditions (which are the same as conditions (C2) and (C3) in Proposition 6.3):

- $\deg t + \deg r < \deg a$.
- $t(x)b(x) \equiv r(x) \pmod{a(x)}$.

Using the notation of Euclid's algorithm in Figure 6.1, show that there is an index $h \in \{0, 1, \ldots, \nu{+}1\}$ and a polynomial $c(x) \in F[x]$ such that

$$t(x) = c(x) \cdot t_h(x) \quad \text{and} \quad r(x) = c(x) \cdot r_h(x) .$$

Hint: First verify that the proof of Proposition 6.3 still holds until (and including) the equality (6.18). Conclude from (6.17) and (6.18) that

$$t(x) s_h(x) = t_h(x) s(x) .$$

Then, based on part 1 of Problem 3.3, argue that $\gcd(s_h(x), t_h(x)) = 1$ and deduce that $t_h(x)$ divides $t(x)$.

Problem 6.11 (Decoding errors and erasures) Let $\mathcal{C}_{\mathrm{GRS}}$ be an $[n, k, d]$ GRS code over $F = \mathrm{GF}(q)$ with nonzero code locators $\alpha_1, \alpha_2, \ldots, \alpha_n$ and column multipliers $v_1, v_2, \ldots, v_n$. A codeword $\mathbf{c} \in \mathcal{C}_{\mathrm{GRS}}$ has been transmitted through an erasure channel $(F, F \cup \{?\}, \mathsf{Prob})$ and a word $\mathbf{y} = (y_1\ y_2\ \ldots y_n) \in (F \cup \{?\})^n$ has been received (where the symbol "?" stands for an erasure). Denote by K the set of erasure locations in $\mathbf{y}$ and let $\rho = |K|$. The set of error locations in $\mathbf{y}$ is denoted by J where $J \cap K = \emptyset$. Assume hereafter that $2|J| + \rho \leq d{-}1$.

For the purpose of syndrome computation, the values at the erased locations are assumed to be zero: letting $\mathbf{z} = (z_1\ z_2\ \ldots z_n) \in F^n$ stand for

$$z_j = \begin{cases} y_j & \text{if } y_j \neq ? \\ 0 & \text{otherwise} \end{cases} ,$$

the syndrome is obtained by $(S_0\ S_1\ \ldots\ S_{d-2})^T = H_{\mathrm{GRS}} \mathbf{z}^T$, where H_{GRS} is a canonical parity-check matrix of $\mathcal{C}_{\mathrm{GRS}}$. The vector $\mathbf{e} = (e_1\ e_2\ \ldots\ e_n)$ will denote the difference $\mathbf{z} - \mathbf{c}$.

Define the syndrome polynomial, error-locator polynomial, *erasure-locator polynomial*, and *error–erasure-evaluator polynomial* by

$$S(x) = \sum_{\ell=0}^{d-2} S_\ell x^\ell , \qquad \Lambda(x) = \prod_{j \in J} (1 - \alpha_j x) , \qquad M(x) = \prod_{j \in K} (1 - \alpha_j x) ,$$

and

$$\Gamma(x) = \sum_{j \in K \cup J} e_j v_j \prod_{m \in (K \cup J) \setminus \{j\}} (1 - \alpha_m x) ,$$

respectively. Note that both $S(x)$ and $M(x)$ are known to the decoder. The *modified syndrome polynomial*, denoted by $\tilde{S}(x)$, is the unique polynomial in $F_{d-1}[x]$ that satisfies

$$\tilde{S}(x) \equiv M(x) S(x) \pmod{x^{d-1}} .$$

1. Show that $\tilde{S}(x) = 0$ if and only if $\mathbf{e} = \mathbf{0}$.

2. Show that
$$\gcd(\Lambda(x), \Gamma(x)) = 1 .$$

3. Show that $\Gamma(x) = 0$ if and only if $\mathbf{e} = \mathbf{0}$.

4. Show that
$$\deg \Gamma < \rho + \deg \Lambda \leq \tfrac{1}{2}(d+\rho-1) .$$

5. Show that
$$S(x) \equiv \sum_{j \in K \cup J} \frac{e_j v_j}{1 - \alpha_j x} \pmod{x^{d-1}} .$$

6. Show that
$$\Lambda(x)\tilde{S}(x) \equiv \Gamma(x) \pmod{x^{d-1}} .$$

7. Show that by applying Euclid's algorithm in Figure 6.1 to $a(x) \leftarrow x^{d-1}$ and $b(x) \leftarrow \tilde{S}(x)$ one obtains $\Lambda(x) = c \cdot t_h(x)$ and $\Gamma(x) = c \cdot r_h(x)$, where h is the unique index for which
$$\deg r_h < \tfrac{1}{2}(d+\rho-1) \leq \deg r_{h-1} .$$

[Section 6.5]

Problem 6.12 Let $\mathcal{C}_{\mathrm{GRS}}$ be an $[n, k, d]$ GRS code over F with code locators $\alpha_1, \alpha_2, \ldots, \alpha_n$ and column multipliers $v_1, v_2, \ldots, v_n$. Also, let
$$f(x) = f_0 + f_1 x + \ldots + f_{d-1} x^{d-1}$$
be a polynomial of degree $d-1$ $(= n-k)$ over F such that $f(\alpha_j) \neq 0$ for all $1 \leq j \leq n$. Define the polynomials $\vartheta_j(x) \in F_{d-1}[x]$ by
$$\vartheta_j(x) = -\frac{1}{f(\alpha_j)} \sum_{\ell=0}^{d-2} x^{\ell} \sum_{i=\ell+1}^{d-1} f_i \alpha_j^{i-\ell-1} .$$

The proof of the following properties was given as an exercise in Problem 5.11:

(i) $\vartheta_j(x)$ is the multiplicative inverse of $x - \alpha_j$ in the ring $F[x]/f(x)$.

(ii) $\mathcal{C}_{\mathrm{GRS}}$ consists of all words $(c_1 \; c_2 \; \ldots \; c_n) \in F^n$ such that
$$\sum_{j=1}^{n} c_j v_j f(\alpha_j) \cdot \vartheta_j(x) = 0 .$$

(iii) $\mathcal{C}_{\mathrm{GRS}}$ consists of all words $(c_1 \; c_2 \; \ldots \; c_n) \in F^n$ such that
$$\sum_{j=1}^{n} \frac{c_j v_j f(\alpha_j)}{x - \alpha_j} \equiv 0 \pmod{f(x)} .$$

A codeword $\mathbf{c} \in \mathcal{C}_{\mathrm{GRS}}$ is transmitted through an additive channel (F, F, Prob) and a word $\mathbf{y} = (y_1\ y_2\ \ldots\ y_n)$ over F is received. Associate with $\mathbf{y}$ the following polynomial $Z_f(x) \in F_{d-1}[x]$:

$$Z_f(x) = \sum_{j=1}^{n} y_j v_j f(\alpha_j) \cdot \vartheta_j(x) .$$

Denote by $\mathbf{e} = (e_1\ e_2\ \ldots\ e_n)$ the error word $\mathbf{y}-\mathbf{c}$ and by J the set of error locations (i.e., the support of $\mathbf{e}$), and assume that $|J| \le \frac{1}{2}(d-1)$.

1. Show that
$$Z_f(x) \equiv \sum_{j \in J} \frac{e_j v_j f(\alpha_j)}{x - \alpha_j} \pmod{f(x)} .$$

2. Define the polynomials $\mathrm{V}(x)$ and $\mathrm{L}(x)$ by
$$\mathrm{V}(x) = \prod_{j \in J} (x - \alpha_j)$$
and
$$\mathrm{L}(x) = \sum_{j \in J} e_j v_j f(\alpha_j) \prod_{m \in J \setminus \{j\}} (x - \alpha_m)$$
(note that $\mathrm{V}(x)$ is obtained by reversing the order of coefficients of the ELP $\Lambda(x)$, namely, $\mathrm{V}(x) = x^{|J|}\Lambda(x^{-1})$). Show that the following three conditions hold:

   (a) $\gcd(\mathrm{V}(x), \mathrm{L}(x)) = 1$.
   (b) $\deg \mathrm{L}(x) < \deg \mathrm{V}(x) \le \frac{1}{2}(d-1)$.
   (c) $\mathrm{V}(x) Z_f(x) \equiv \mathrm{L}(x) \pmod{f(x)}$.

3. Based on part 2, show how $\mathrm{V}(x)$ and $\mathrm{L}(x)$ can be computed by applying Euclid's algorithm in Figure 6.1 to $a(x) \leftarrow f(x)$ and $b(x) \leftarrow Z_f(x)$.

4. Show that the error values are given by
$$e_\kappa = \frac{1}{v_\kappa f(\alpha_\kappa)} \cdot \frac{\mathrm{L}(\alpha_\kappa)}{\mathrm{V}'(\alpha_\kappa)} , \qquad \kappa \in J .$$
Hint: Notice the resemblance to Forney's algorithm.

5. Let $S(x)$ be the syndrome polynomial, whose coefficients are the syndrome entries of the received word $\mathbf{y}$ with respect to the canonical parity-check matrix $\left(v_j \alpha_j^i \right)_{i=0}^{d-2}{}_{j=1}^{n}$; that is,
$$S_\ell = \sum_{j=1}^{n} y_j v_j \alpha_j^\ell , \quad \ell = 0, 1, \ldots, d-2 .$$
Show that when $f(x) = x^{d-1}$,
$$Z_f(x) = -x^{d-2} S(x^{-1}) .$$

(It is interesting to observe that the decoding algorithm in parts 3 and 4 applies to *any* polynomial $f(x)$ of degree $d-1$ that does not vanish at any of the code locators; thus, the required relationship between $f(x)$ and the code $\mathcal{C}_{\mathrm{GRS}}$ is rather weak. The particular selection of the polynomial $f(x)$ may now be dictated by complexity criteria; for example, one can choose a "simple" polynomial, such as the polynomial x^{d-1} in part 5. Alternatively, one can first select the polynomial $f(x)$ and then set the column multipliers v_j to

$$v_j = \frac{1}{f(\alpha_j)} , \quad 1 \le j \le n ,$$

in which case the expression for the error values in part 4 is simplified (see also part 6 of Problem 5.11). Finally, notice that the decoding algorithm in parts 3 and 4 can be applied also to singly-extended GRS codes, where one of the code locators is 0: when $d \ge 3$ and $f(x)$ is taken as an irreducible polynomial of degree $d-1$ over F, one is guaranteed that, regardless of the choice of the code locators, none of them is a root of $f(x)$.)

Problem 6.13 (The Welch–Berlekamp equations) Let $\mathcal{C}_{\mathrm{GRS}}$ be an $[n, k, d]$ GRS code over $F = \mathrm{GF}(q)$ with a generator matrix

$$G_{\mathrm{GRS}} = \begin{pmatrix} 1 & 1 & \ldots & 1 \\ \alpha_1 & \alpha_2 & \ldots & \alpha_n \\ \alpha_1^2 & \alpha_2^2 & \ldots & \alpha_n^2 \\ \vdots & \vdots & \vdots & \vdots \\ \alpha_1^{k-1} & \alpha_2^{k-1} & \ldots & \alpha_n^{k-1} \end{pmatrix} ,$$

namely, the dual code $\mathcal{C}_{\mathrm{GRS}}^{\perp}$ is normalized; thus, from (5.3), the code can be equivalently described by

$$\mathcal{C}_{\mathrm{GRS}} = \Big\{ (u(\alpha_1) \ u(\alpha_2) \ \ldots \ u(\alpha_n)) \ : \ u(x) \in F_k[x] \Big\} .$$

For $u(x) \in F_k[x]$, let

$$\mathbf{c} = (u(\alpha_1) \ u(\alpha_2) \ \ldots \ u(\alpha_n))$$

be the codeword of $\mathcal{C}_{\mathrm{GRS}}$ that is transmitted through a channel (F, F, Prob) and denote by $\mathbf{y} = (y_1 \ y_2 \ \ldots \ y_n)$ the received word. Assume that the support, J, of the error word, $(e_1 \ e_2 \ \ldots \ e_n) = \mathbf{y} - \mathbf{c}$, satisfies $|J| \le \frac{1}{2}(d-1)$.

Let $\tilde{u}(x)$ be the (unique) polynomial in $F_k[x]$ that interpolates through the points $\{(\alpha_j, y_j)\}_{j=d}^{n}$, i.e.,

$$\tilde{u}(\alpha_j) = y_j , \quad d \le j \le n .$$

The *re-encoded codeword* is the codeword $\tilde{\mathbf{c}} \in \mathcal{C}_{\mathrm{GRS}}$ that is given by

$$\tilde{\mathbf{c}} = (\tilde{u}(\alpha_1) \ \tilde{u}(\alpha_2) \ \ldots \ \tilde{u}(\alpha_n)) .$$

Note that $\tilde{\mathbf{c}}$ may differ from $\mathbf{c}$, yet its last k $(= n-d+1)$ entries are identical to the respective entries in $\mathbf{y}$. Denote the difference $\mathbf{y} - \tilde{\mathbf{c}}$ by $\tilde{\mathbf{y}} = (\tilde{y}_1 \ \tilde{y}_2 \ \ldots \ \tilde{y}_n)$, where $\tilde{y}_j = 0$ for $d \le j \le n$.

1. Show that a canonical parity-check matrix of $\mathcal{C}_{\mathrm{GRS}}$ is given by

$$H_{\mathrm{GRS}} = \left(v_j \alpha_j^i \right)_{i=0}^{d-2} {}_{j=1}^{n} ,$$

where the column multipliers are

$$v_j = -\Big(\prod_{\substack{1 \le m \le n: \\ m \ne j}} (\alpha_j - \alpha_m) \Big)^{-1} , \quad 1 \le j \le n .$$

Hint: See Problem 5.7.

2. Verify that $\mathbf{y}$ and $\tilde{\mathbf{y}}$ belong to the same coset of $\mathcal{C}_{\mathrm{GRS}}$ in F^n.

3. Let $\hat{\mathbf{c}}$ be the closest codeword in $\mathcal{C}_{\mathrm{GRS}}$ to $\tilde{\mathbf{y}}$. How is $\hat{\mathbf{c}}$ related to the transmitted codeword $\mathbf{c}$?

4. Show that for $j \in \{1, 2, \ldots, n\}$,

$$\tilde{y}_j = u(\alpha_j) - \tilde{u}(\alpha_j) \quad \Longleftrightarrow \quad j \ne J .$$

5. Let $A(x)$ be the polynomial over F that is given by

$$A(x) = \prod_{j=1}^{d-1} (x - \alpha_j) ,$$

and denote by $A'(x)$ the formal derivative of $A(x)$. Show that there exists a nonzero polynomial pair $(\mathrm{V}(x), \mathrm{N}(x))$ over F that satisfies the *degree constraints*

$$\boxed{\deg \mathrm{V} < \tfrac{1}{2}(d+1) \quad \text{and} \quad \deg \mathrm{N} < \tfrac{1}{2}(d-1)} ,$$

along with the following set of $d-1$ linear homogeneous equations (whose variables are the coefficients of $\mathrm{V}(x)$ and $\mathrm{N}(x)$):

$$\boxed{\mathrm{N}(\alpha_j) = \tilde{y}_j v_j A'(\alpha_j) \cdot \mathrm{V}(\alpha_j)} , \quad 1 \le j < d .$$

(The polynomial pair $(\mathrm{V}(x), \mathrm{N}(x))$ is said to be nonzero if at least one of the constituent polynomials is nonzero; yet notice that the degree constraints and the linear equations together force $\mathrm{N}(x)$ to be zero whenever $\mathrm{V}(x)$ is. Therefore, a pair is nonzero if and only if $\mathrm{V}(x) \ne 0$.)

Hint: Verify that the degree constraints still allow the pair $(\mathrm{V}(x), \mathrm{N}(x))$ to have more than $d-1$ significant coefficients.

(The linear equations—with the respective degree constraints—in part 5 are called the *Welch–Berlekamp equations* of GRS decoding. These equations can be viewed as a relaxed (or weak) version of the following rational interpolation problem: find a nonzero pair $(\mathrm{V}(x), \mathrm{N}(x))$ that satisfies the interpolation constraint

$$\left. \frac{\mathrm{N}(x)}{\mathrm{V}(x)} \right|_{x=\alpha_j} = \tilde{y}_j v_j A'(\alpha_j) , \quad 1 \le j < d ,$$

subject to

$$\max\{\deg \mathrm{V}, 1+\deg \mathrm{N}\} < \tfrac{1}{2}(d+1) .$$

Now, in part 5, the polynomial $\mathrm{V}(x)$ is allowed to vanish at some of the interpolation abscissas α_j, and the interpolation constraint is then relaxed to only requiring that $\mathrm{N}(x)$ vanish at each such abscissa: while a reduced ratio can be obtained by clearing the common factor, $x - \alpha_j$, from the numerator $\mathrm{N}(x)$ and the denominator $\mathrm{V}(x)$, no constraint is imposed on the value taken by that ratio at $x = \alpha_j$. The set of linear equations in part 5 will be referred to as the *weak interpolation constraint.*)

A nonzero polynomial pair $(\mathrm{V}(x), \mathrm{N}(x))$ is called *feasible* if it satisfies the degree constraints and the weak interpolation constraint in part 5. A feasible pair is called *minimal* if $\mathrm{V}(x)$ is (nonzero and) monic and the value

$$\max\{\deg \mathrm{V}, 1+\deg \mathrm{N}\}$$

is the smallest possible among all the nonzero feasible pairs.

6. Let $\Omega(x)$ denote the polynomial $\prod_{j=d}^{n}(x - \alpha_j)$. Show that if $(\mathrm{V}(x), \mathrm{N}(x))$ is a feasible pair, then

$$\mathrm{N}(\alpha_j)\Omega(\alpha_j) + \mathrm{V}(\alpha_j)\tilde{y}_j = 0 , \quad 1 \le j \le n .$$

Hint: $\Omega(\alpha_j) = -1/(v_j A'(\alpha_j))$ for every $1 \le j < d$.

7. Show that for every feasible pair $(\mathrm{V}(x), \mathrm{N}(x))$,

$$\mathrm{N}(x)\Omega(x) = \mathrm{V}(x)(\tilde{u}(x) - u(x)) .$$

Hint: Consider the polynomial

$$Q(x) = \mathrm{N}(x)\Omega(x) + \mathrm{V}(x)(u(x) - \tilde{u}(x)) .$$

By combining parts 4 and 6, show that $Q(x)$ has at least $n-|J|$ roots among the set of code locators. On the other hand, show that $\deg Q < n-|J|$ and, so, $Q(x)$ must be identically zero.

8. Show that if $(\mathrm{V}(x), \mathrm{N}(x))$ is a feasible pair, then $\mathrm{V}(x)$ is divisible by the polynomial

$$\prod_{j \in J}(x - \alpha_j) .$$

Hint: Deduce from parts 6 and 7 that

$$\mathrm{V}(\alpha_j)(\tilde{y}_j + \tilde{u}(\alpha_j) - u(\alpha_j)) = 0 , \quad 1 \le j \le n ,$$

and then use part 4.

9. Show that a *minimal* feasible pair is unique and is given by

$$\mathrm{V}(x) = \prod_{j \in J}(x - \alpha_j) \qquad \text{and} \qquad \mathrm{N}(x) = \frac{\mathrm{V}(x)(\tilde{u}(x) - u(x))}{\Omega(x)} .$$

In particular,

$$\deg \mathrm{N} < \deg \mathrm{V} = |J| .$$

10. Let $B(x)$ be the (unique) polynomial in $F_{d-1}[x]$ that interpolates through the points $\{(\alpha_j, \tilde{y}_j v_j A'(\alpha_j))\}_{j=1}^{d-1}$, namely,

$$B(\alpha_j) = \tilde{y}_j v_j A'(\alpha_j) , \quad 1 \le j < d .$$

Show that every feasible pair $(\mathrm{V}(x), \mathrm{N}(x))$ satisfies

$$\mathrm{V}(x)B(x) \equiv \mathrm{N}(x) \pmod{A(x)} .$$

11. Based on part 10, show how the minimal feasible pair can be computed by applying Euclid's algorithm in Figure 6.1 to $a(x) \leftarrow A(x)$ and $b(x) \leftarrow B(x)$.

Hint: Similarly to what was done in Section 6.4, use Problem 6.10 to show that for every feasible pair $(\mathrm{V}(x), \mathrm{N}(x))$ there is an index $h \ge 0$ and a polynomial $C(x) \in F[x]$ such that

$$\mathrm{V}(x) = C(x) \cdot t_h(x) \quad \text{and} \quad \mathrm{N}(x) = C(x) \cdot r_h(x) .$$

Furthermore, show that h must be the smallest index for which $\deg r_h < \frac{1}{2}(d-1)$. Finally, verify that $(t_h(x), r_h(x))$ is a feasible pair and conclude that the minimal feasible pair is obtained when $C(x)$ is taken as a (particular) nonzero scalar of F.

The remaining parts of the problem demonstrate how the error values can be computed from the minimal feasible pair. To this end, partition the set J into the following two subsets:

$$J' = J \cap \{1, 2, \ldots, d-1\} \quad \text{and} \quad J'' = J \cap \{d, d+1, \ldots, n\} .$$

(The subset J'' can be viewed as the set of error locations within the information word, in the case where the encoding is carried out by a systematic encoder that places the information word in the part of the codeword that is indexed by $\{d, d+1, \ldots, n\}$. The errors within the remaining $d-1$ coordinates are then indexed by J'.)

Hereafter in this problem, $(\mathrm{V}(x), \mathrm{N}(x))$ stands for the minimal feasible pair.

12. Show that

$$\gcd(\mathrm{V}(x), \mathrm{N}(x)) = \prod_{j \in J'} (x - \alpha_j) .$$

Hint: Show that for $\kappa \in J''$, the multiplicity of $x - \alpha_\kappa$ in the right-hand side of the equality in part 7 is (exactly) 1.

13. Show that when $\kappa \in J''$, the error value e_κ is given by

$$e_\kappa = -\frac{\mathrm{N}(\alpha_\kappa)}{v_\kappa A(\alpha_\kappa) \cdot \mathrm{V}'(\alpha_\kappa)} .$$

Hint: Take the formal derivative of both sides of the equality in part 7, and substitute $x = \alpha_\kappa$. Note that $\Omega'(\alpha_j) = -1/(v_j A(\alpha_j))$ for every $d \le j \le n$.

14. Show that when $\kappa \in J'$, the error value e_κ is given by

$$e_\kappa = \tilde{y}_\kappa - \frac{\mathrm{N}'(\alpha_\kappa)}{v_\kappa A'(\alpha_\kappa) \cdot \mathrm{V}'(\alpha_\kappa)} .$$

Hint: Follow the steps of part 13, yet now recall from part 12 that $\mathrm{N}(\alpha_\kappa) = 0$ when $\kappa \in J'$.

[Section 6.7]

Problem 6.14 Let L be a positive integer and $\sigma(x) = 1 + \sum_{m=1}^{L} \lambda_m x^m$ be a polynomial in $F_{L+1}[x]$.

1. Let $(a_i)_{i=0}^{\infty}$ be an infinite sequence over F that satisfies the linear recurrence

$$a_i = -\sum_{m=1}^{L} \lambda_m a_{i-m} = 0 , \quad \text{for every } i \geq L ,$$

and denote by $a(x)$ the formal power series $\sum_{i=0}^{\infty} a_i x^i$. Show that there exists a unique polynomial $\omega(x) \in F_L[x]$ such that

$$a(x) = \frac{\omega(x)}{\sigma(x)}$$

in $F[[x]]$ (thus, $(\sigma(x), \omega(x))$ is an N-recurrence of $a(x)$ for every $N \geq 0$).

2. Conversely, show that for every $\omega(x) \in F_L[x]$, the infinite sequence of coefficients of the formal power series $a(x) = \omega(x)/\sigma(x)$ satisfies the linear recurrence in part 1.

3. Show that when F is a finite field, there are $|F|^L$ distinct sequences $(a_i)_{i=0}^{\infty}$ that satisfy the linear recurrence in part 1.

Problem 6.15 (Smallest recurrence order of finite sequences) Let $b(x)$ be a given polynomial over a field F and let $(\sigma(x), \omega(x))$ be an N-recurrence of $b(x)$ over F whose recurrence order is the smallest possible.

1. Show that $\gcd(\sigma(x), \omega(x)) = 1$.

2. Show that if $\mathrm{ord}(\sigma, \omega) \leq N/2$ then $(\sigma(x), \omega(x))$ is unique.

   Hint: See the proof of Proposition 6.1.

Problem 6.16 (Minimal recurrence of infinite sequences) Let $\sigma(x)$ and $\omega(x)$ be polynomials over a field F such that $\gcd(\sigma(x), x) = 1$ and consider the formal power series $a(x) = \omega(x)/\sigma(x)$ in $F[[x]]$.

1. Show that there exists a unique pair of polynomials $(t(x), r(x))$ over F that satisfies the following two conditions:

   (i) $\gcd(t(x), r(x)) = 1$.

   (ii) $(t(x), r(x))$ is an N-recurrence of $a(x)$ over F for every $N \geq 0$.

Hereafter in this problem, the pair $(t(x), r(x))$ is as in part 1.

2. Show that for every integer $N \geq 2\,\mathrm{ord}(t, r)$, the pair $(t(x), r(x))$ is an N-recurrence of $a(x)$ with the smallest possible recurrence order.

   Hint: See Problem 6.15.

3. Show that $a(x)$ is periodic if and only if the exponent of $t(x)$ is finite and $\deg r < \deg t$ (see Problems 6.8 and 6.9). What is then the period of $a(x)$?

4. Show that if F is a finite field and $\deg r < \deg t$, then $a(x)$ is necessarily periodic.

Problem 6.17 (Early stopping of the Berlekamp–Massey algorithm) Suppose that the Berlekamp–Massey algorithm in Figure 6.5 is used in the decoding of a GRS code with minimum distance d and that no more than $\frac{1}{2}(d-1)$ errors have occurred. Let i be the first loop iteration of the algorithm for which $i \geq \lfloor \frac{1}{2}(d-1) \rfloor + \mathrm{ord}(\sigma_i, \omega_i)$. Show that $(\Lambda(x), \Gamma(x)) = (\sigma_i(x), \omega_i(x))$.

Notes

[Sections 6.1–6.3]

The first polynomial-time algorithm for decoding GRS codes was introduced by Peterson [277] and Gorenstein and Zierler [160]. The discussion in Section 6.3.1 is based on their algorithm.

Given an $[n, k, d]$ GRS code, a direct computation of the syndrome requires $O(dn)$ arithmetic field operations. The roots of the ELP can be found by a Chien search (see [79]) with time complexity $O(|J|n)$, where $|J|$ is the number of actual errors.

[Section 6.4]

The decoding algorithm for GRS codes that uses Euclid's algorithm is due to Sugiyama *et al.* [348]. A direct application of Euclid's algorithm to solving the key equation requires $O(|J|d)$ arithmetic field operations, but there are ways to accelerate this algorithm so that the asymptotic time complexity becomes $O(d \log^2 d \log\log d)$ arithmetic operations; see Aho *et al.* [6, Section 8.9] and von zur Gathen and Gerhard [144, Section 11.1].

[Section 6.5]

The formula for computing the error values out of $\Lambda'(x)$ and $\Gamma(x)$ is due to Forney [128].

The Welch–Berlekamp equations, which are presented in Problem 6.13, form the basis of another decoding algorithm for GRS codes, not covered in this chapter. That algorithm is known as the *Welch–Berlekamp algorithm* and is described in [37] and [380].

[Section 6.6]

We next consider the asymptotic time complexity of GRS decoding when the ratio d/n is bounded away from zero.

As mentioned earlier, the key equation can be solved in time complexity $O(d \log^2 d \log\log d) = O(n \log^2 n \log\log n)$. In fact, this is also the time complexity of Forney's algorithm, since there exists an algorithm with time complexity $O(n \log^2 n \log\log n)$ for evaluating a polynomial in $F_n[x]$ simultaneously at n elements of F [6, Section 8.5], [144, Section 10.1]. As for the initial syndrome computation step, by a result of Kaminski *et al.* [203] we get that its complexity is essentially the same as that of the evaluation of a polynomial at n points. One can therefore conclude that the overall complexity of GRS decoding is $O(n \log^2 n \log\log n)$ arithmetic field operations; see Justesen [200] and Sarwate [317]. By Problem 6.11 it follows that one gets the same time complexity also when erasures are present. And this is also the time complexity of systematic encoding of GRS codes, since the encoding can be seen as a special case of erasure decoding.

While the algorithm in Figure 6.2 assumes that the code locators are all nonzero, we can use this algorithm also for the decoding of singly-extended GRS codes, as described next (see also the remark at the end of Problem 6.12).

Let $\mathcal{C}_{\mathrm{GRS}}$ be an $[n, k, d]$ singly-extended GRS code over $F = \mathrm{GF}(q)$ with a canonical parity-check matrix

$$H_{\mathrm{GRS}} = (\, v_j \alpha_j^i \,)_{i=0}^{d-2}\,{}_{j=1}^{n} \,,$$

where $\alpha_n = 0$. Given a received word $\mathbf{y} = (y_1\ y_2\ \ldots\ y_n)$, we first apply Step 1 in Figure 6.2 to compute the syndrome

$$(S_0\ S_1\ S_2\ \ldots\ S_{d-2})^T = H_{\mathrm{GRS}} \mathbf{y}^T \,.$$

Next, we execute Steps 2 and 3 in two rounds. In the first round, we assume that the entry y_n is error-free and apply these two steps to the computed syndrome, while replacing n in Step 3 by $n-1$. This means that we effectively use Figure 6.2 as a decoder for the $[n-1, k-1, d]$ GRS code $\mathcal{C}'_{\mathrm{GRS}}$ that is obtained by shortening $\mathcal{C}_{\mathrm{GRS}}$ at the nth coordinate (see Problem 2.14): a parity-check matrix of $\mathcal{C}'_{\mathrm{GRS}}$ is obtained from H_{GRS} by deleting the nth column.

In the second round, we assume that the entry y_n is in error: we replace d, n, and the column multipliers in Steps 2 and 3 by $d-1$, $n-1$, and $v_j \alpha_j$, respectively, and apply these steps to the truncated syndrome

$$(S_1\ S_2\ \ldots\ S_{d-2})^T \,.$$

In other words, we now use Figure 6.2 as a decoder for the $[n-1, k, d-1]$ GRS code $\mathcal{C}''_{\mathrm{GRS}}$ obtained by puncturing $\mathcal{C}_{\mathrm{GRS}}$ at the nth coordinate (see Problem 2.3): a parity-check matrix of $\mathcal{C}''_{\mathrm{GRS}}$ is obtained from H_{GRS} by deleting the first row and the nth column. The second round ends by recovering the error value at the nth coordinate from the syndrome entry S_0.

Obviously, if the number of errors in $\mathbf{y}$ does not exceed $\lfloor (d-1)/2 \rfloor$, one (and only one) of these two rounds will end up with a codeword whose Hamming distance from $\mathbf{y}$ is at most $\lfloor (d-1)/2 \rfloor$.

The decoder that we have just described can be generalized to handle erasures as well, using the algorithm presented in Problem 6.11.

[Section 6.7]

The Berlekamp–Massey algorithm is due to Berlekamp [36, Section 7.4] and Massey [254]. Blahut [46, Sections 11.6 and 11.7] describes a method for accelerating the Berlekamp–Massey algorithm through recursion, resulting in an algorithm that requires $O(d \log^2 d \log\log d)$ arithmetic field operations.

On properties of the sequences that are generated by linear-feedback shift registers see the book by Golomb [152].

Extensions of the Berlekamp–Massey algorithm to multi-dimensional recurrences are described by Sakata [314], [315]; see also Fitzpatrick and Norton [126]. Feng and Tzeng [124], [125] study the problem of synthesizing a linear recurrence that is satisfied simultaneously by several given sequences. Roth and Ruckenstein [302] describe an algorithm for generating linear sliding-block transformations on several given sequences such that when each transformation is applied to the respective sequence, the resulting images sum to zero. Reeds and Sloane [290] extend the Berlekamp–Massey algorithm to sequences over the ring $\mathbb{Z}_m$, and Fitzpatrick and Norton [127] consider the extension of the algorithm to unique factorization domains. Deutsch [103] studies the application of the Berlekamp–Massey algorithm to the decoding of certain RS-type codes over the ring $F[x]/(1+x+\ldots+x^{p-1})$, where $F = \mathrm{GF}(q)$ and p is a prime (these codes will be presented in Problem 11.7).

The correspondence between the steps of the Berlekamp–Massey algorithm and those of Euclid's algorithm has been studied by Cheng [77], Dornstetter [108], Heydtmann and Jensen [175], Mills [262], and Welch and Scholtz [381].

There are other known efficient algorithms for decoding GRS codes: the Welch–Berlekamp algorithm mentioned earlier [37], [380] and Blahut's time-domain decoding algorithm [46, Section 9.5], [47]. The latter algorithm is derived by applying the inverse of the Fourier transform to the steps of the Berlekamp–Massey algorithm (Problem 3.27). See also Hasan *et al.* [173].

Chapter 7

Structure of Finite Fields

In this chapter, we make a second pause in the treatment of codes and continue with our study of finite fields, which started in Chapter 3. Given a finite field $F = \mathrm{GF}(q)$ and an extension field $\Phi = \mathrm{GF}(q^n)$, we show that Φ can be partitioned into subsets, which we call conjugacy classes, each forming the set of roots in Φ of some irreducible polynomial over F whose degree divides the extension degree n; conversely, the set of roots in Φ of every such polynomial forms a conjugacy class. This result leads to a closed expression for the number of monic irreducible polynomials of a given degree over any finite field. Another key result to be shown is that all finite fields of the same size are isomorphic.

7.1 Minimal polynomials

Throughout this section, we fix F to be the finite field $\mathrm{GF}(q)$ and let Φ be an extension field of F with extension degree $[\Phi : F] = n$.

Two elements $\alpha, \beta \in \Phi$ are called *conjugates* (with respect to F) if there is a nonnegative integer r such that

$$\beta = \alpha^{q^r} .$$

Without loss of generality we can assume that $r < n$; indeed, writing $r = cn + d$ where $0 \le d < n$, we have

$$\alpha^{q^{cn}} = \alpha^{q^n \cdot q^{(c-1)n}} = (\alpha^{q^n})^{q^{(c-1)n}} = \alpha^{q^{(c-1)n}} = \ldots = \alpha^{q^{0 \cdot n}} = \alpha ,$$

and, so,

$$\alpha^{q^r} = \alpha^{q^{cn+d}} = (\alpha^{q^{cn}})^{q^d} = \alpha^{q^d} .$$

The next proposition presents a basic property of the conjugacy relation.

Proposition 7.1 *Conjugacy is an equivalence relation.*

Proof. This follows from reflexivity ($\alpha = \alpha^{q^0}$), symmetry (if $\beta = \alpha^{q^r}$ for some $0 \le r < n$, then $\alpha = \beta^{q^{n-r}}$), and transitivity (if $\beta = \alpha^{q^r}$ and $\gamma = \beta^{q^s}$, then $\gamma = \alpha^{q^{r+s}}$). □

Being an equivalence relation, the conjugacy relation partitions Φ into equivalence classes, which will be referred to as *conjugacy classes*. A conjugacy class that contains an element $\alpha \in \Phi$ will be denoted by C_α. The next proposition characterizes a typical class C_α.

Proposition 7.2 *The conjugacy class (with respect to F) of an element $\alpha \in \Phi$ is given by*

$$C_\alpha = \{ \alpha, \alpha^q, \alpha^{q^2}, \ldots, \alpha^{q^{m-1}} \} ,$$

where m is the smallest positive integer such that $\alpha^{q^m} = \alpha$.

Proof. Clearly, the elements in C_α are the conjugates of α. We verify that they are all distinct. Assume to the contrary that $\alpha^{q^i} = \alpha^{q^j}$ for some $0 < i < j < m$. Raising to the power q^{m-j} yields $\alpha^{q^{m-j+i}} = \alpha^{q^m} = \alpha$, contradicting the minimality of m. □

Let α be an element of Φ and denote the size of the conjugacy class C_α by $m = m_\alpha$. The *minimal polynomial* (with respect to F) of α is defined by

$$M_\alpha(x) = \prod_{\gamma \in C_\alpha} (x - \gamma) = \prod_{i=0}^{m-1} (x - \alpha^{q^i}) .$$

Clearly, $\deg M_\alpha(x) = m = m_\alpha$, and $M_\beta(x) = M_\alpha(x)$ for every $\beta \in C_\alpha$.

Example 7.1 Let $F = \mathrm{GF}(2)$ and consider the extension field $\Phi = F[\xi]/(\xi^3 + \xi + 1)$, which we constructed in Example 3.6. The conjugacy classes (with respect to F) of the elements of Φ are

$$C_0 = \{0\} , \quad C_1 = \{1\} , \quad C_\xi = \{\xi, \xi^2, \xi^4\} , \quad \text{and} \quad C_{\xi^3} = \{\xi^3, \xi^6, \xi^5\} .$$

Using Table 3.2, we can compute the minimal polynomials of the elements of Φ, as follows:

$$\begin{aligned}
M_0(x) &= x \\
M_1(x) &= x - 1 \\
M_\xi(x) &= (x - \xi)(x - \xi^2)(x - \xi^4) = x^3 + x + 1 = M_{\xi^2}(x) = M_{\xi^4}(x) \\
M_{\xi^3}(x) &= (x - \xi^3)(x - \xi^6)(x - \xi^5) = x^3 + x^2 + 1 = M_{\xi^6}(x) = M_{\xi^5}(x) .
\end{aligned}$$

Notice that the coefficients of the minimal polynomials all lie in the field $F = \mathrm{GF}(2)$. □

We exhibit several properties of minimal polynomials through a sequence of propositions. In the first proposition, we show that the phenomenon observed in Example 7.1 is not a coincidence: while minimal polynomials are defined over the extension field $\Phi = \mathrm{GF}(q^n)$, they are in fact polynomials over the ground field $F = \mathrm{GF}(q)$.

Proposition 7.3 *For every* $\alpha \in \Phi$,

$$M_\alpha(x) \in F[x] .$$

Proof. Since q is a power of the characteristic of Φ, we have,

$$(M_\alpha(x))^q = \prod_{\gamma \in C_\alpha} (x - \gamma)^q = \prod_{\gamma \in C_\alpha} (x^q - \gamma^q) .$$

Now, when γ ranges over all the conjugates of α, so does γ^q. Therefore,

$$(M_\alpha(x))^q = \prod_{\gamma \in C_\alpha} (x^q - \gamma^q) = \prod_{\gamma \in C_\alpha} (x^q - \gamma) = M_\alpha(x^q) . \tag{7.1}$$

Write $M_\alpha(x) = \sum_{i=0}^{m} a_i x^i$ where $m = m_\alpha$. We next compute the leftmost and rightmost expressions in (7.1): the former equals

$$(M_\alpha(x))^q = \Big(\sum_{i=0}^{m} a_i x^i\Big)^q = \sum_{i=0}^{m} a_i^q x^{iq} ,$$

while the latter is

$$M_\alpha(x^q) = \sum_{i=0}^{m} a_i x^{iq} .$$

Thus,

$$\sum_{i=0}^{m} a_i^q x^{iq} = \sum_{i=0}^{m} a_i x^{iq} ,$$

which readily implies that $a_i^q - a_i = 0$ for every $0 \le i \le m$. Hence, by Problem 3.11, each coefficient a_i is an element of F. □

Proposition 7.4 *Let* $\alpha \in \Phi$ *and* $b(x) \in F[x]$ *be such that* $b(\alpha) = 0$. *Then,*

$$M_\alpha(x) \mid b(x) .$$

Proof. Since α is a root of $b(x)$, so is every conjugate α^{q^r} (Problem 3.30). Therefore, $b(x)$ must be divisible by $\prod_{\gamma \in C_\alpha}(x - \gamma)$. □

The previous proposition provides the reason for the term "minimal polynomial": the proposition implies that $M_\alpha(x)$ is a nonzero polynomial of smallest degree in $F[x]$ that vanishes at $x = \alpha$ (see also Problem 3.12).

Proposition 7.5 *The polynomial $M_\alpha(x)$ is irreducible over F for every $\alpha \in \Phi$.*

Proof. Since α is a root of $M_\alpha(x)$, it must be a root of at least one of the irreducible factors, say, $a(x)$, of $M_\alpha(x)$ over F. But then, by Proposition 7.4, $M_\alpha(x) \,|\, a(x)$. Hence, $M_\alpha(x)$ is a scalar multiple of the irreducible factor $a(x)$. □

Example 7.2 Shifting momentarily from finite fields to the infinite case, consider the complex field $\mathbb{C}$, which is an extension field of the real field $\mathbb{R}$ with extension degree 2 (see Example 3.5). Denote by $\imath$ the square root of -1 in $\mathbb{C}$, and recall that for every $a, b \in \mathbb{R}$, the elements $\alpha = a + b\imath$ and $\alpha^* = a - b\imath$ are conjugates in $\mathbb{C}$ (with respect to $\mathbb{R}$). Thus, a conjugacy class of an element $\alpha \in \mathbb{C}$ has size 1 (if $\alpha \in \mathbb{R}$) or 2 (otherwise). Extending the definition of minimal polynomials to elements of $\mathbb{C}$, we get that for every $a \in \mathbb{R}$ and $b \in \mathbb{R} \setminus \{0\}$, the minimal polynomial of $\alpha = a + b\imath$ (with respect to $\mathbb{R}$) is

$$\begin{aligned} M_\alpha(x) &= (x - \alpha)(x - \alpha^*) \\ &= (x - a - b\imath)(x - a + b\imath) \\ &= x^2 - 2ax + (a^2 + b^2) \,. \end{aligned}$$

This polynomial is irreducible over $\mathbb{R}$. □

Lemma 7.6 *For every $\alpha \in \Phi$ and every positive integer s,*

$$\alpha^{q^s} = \alpha \qquad \Longleftrightarrow \qquad m_\alpha \mid s \,.$$

Proof. Write $m = m_\alpha$ and $s = cm + d$, where $0 \le d < m$. Now,

$$\alpha^{q^{cm}} = \alpha^{q^m \cdot q^{(c-1)m}} = (\alpha^{q^m})^{q^{(c-1)m}} = \alpha^{q^{(c-1)m}} = \ldots = \alpha \,;$$

so, $\alpha^{q^s} = \alpha^{q^{cm+d}} = (\alpha^{q^{cm}})^{q^d} = \alpha^{q^d}$. Hence, by the definition of m_α it follows that $\alpha^{q^s} = \alpha$ if and only if $d = 0$. □

Proposition 7.7 *For every $\alpha \in \Phi$,*

$$m_\alpha \mid n \,.$$

Proof. Every $\alpha \in \Phi$ satisfies $\alpha^{q^n} = \alpha$. The result now follows from Lemma 7.6. □

Denote by $Q(x)$ the polynomial $x^{q^n} - x$ over F. Since $Q(\alpha) = 0$ for every $\alpha \in \Phi$, the polynomial $Q(x)$ factors over Φ as follows:

$$Q(x) = \prod_{\alpha \in \Phi} (x - \alpha) \,. \tag{7.2}$$

Proposition 7.8

$$Q(x) = \prod_{M_\alpha(x)} M_\alpha(x) ,$$

where $M_\alpha(x)$ ranges over all distinct minimal polynomials of the elements $\alpha \in \Phi$.

Proof. By (7.2),

$$Q(x) = \prod_{C_\alpha} \prod_{\gamma \in C_\alpha} (x - \gamma) ,$$

where C_α ranges over all the (distinct and disjoint) equivalence classes C_α. □

Example 7.3 For $F = \mathrm{GF}(2)$ and $\Phi = F[\xi]/(\xi^3 + \xi + 1)$:

$$\begin{aligned} x^8 - x &= x \cdot (x-1) \cdot (x-\xi)(x-\xi^2)(x-\xi^4) \cdot (x-\xi^3)(x-\xi^6)(x-\xi^5) \\ &= M_0(x)M_1(x)M_\xi(x)M_{\xi^3}(x) \\ &= x(x-1)(x^3+x+1)(x^3+x^2+1) . \end{aligned}$$

□

Proposition 7.9 *The minimal polynomials of the elements in Φ are all the monic irreducible polynomials over F with degrees dividing n.*

Proof. By Propositions 7.5 and 7.7, every minimal polynomial of an element in Φ is an irreducible polynomial with degree dividing n.

As for the other direction, suppose that $a(x)$ is a monic irreducible polynomial over F whose degree, m, divides n. Consider the extension field $K = F[\xi]/a(\xi)$ of F. Every $\beta \in K$ satisfies $\beta^{q^m} = \beta$, and, since m divides n, we also have $\beta^{q^n} - \beta = 0$. Therefore, by Proposition 7.4, every minimal polynomial of an element in K divides $Q(x)$. In particular, this applies to $a(x)$, which is the minimal polynomial of the element $\beta = \xi$ in K (Problem 7.2). By Proposition 7.8, $a(x)$ divides—and is therefore equal to—a minimal polynomial of an element in Φ. □

By combining Propositions 7.8 and 7.9, we end up with the following theorem, which characterizes the irreducible factorization of $Q(x)$ over F.

Theorem 7.10

$$Q(x) = \prod_{a(x)} a(x) ,$$

where $a(x)$ ranges over all monic irreducible polynomials over F with degrees dividing n.

It is interesting to note that the statement in Theorem 7.10 involves only polynomials over the ground field F, even though the proof of the theorem does rely on the existence of the extension field Φ and on properties of the minimal polynomials of the elements of Φ. The field Φ, in turn, is guaranteed to exist by Proposition 3.16 for every positive integer n; as a matter of fact, that proposition uses the very same polynomial $Q(x)$ to construct such a field.

Example 7.4 For $F = \mathrm{GF}(2)$ and $n = 1, 2, 3, 4$, we use Table 3.1 to obtain the following irreducible factorization of $x^{2^n} - x$ over F:

$$\begin{aligned}
x^2 - x &= x(x-1) \\
x^4 - x &= x(x-1)(x^2+x+1) \\
x^8 - x &= x(x-1)(x^3+x+1)(x^3+x^2+1) \\
x^{16} - x &= \underbrace{x(x-1)(x^2+x+1)}_{x^4-x}(x^4+x+1)(x^4+x^3+1)(x^4+x^3+x^2+x+1) .
\end{aligned}$$

Notice that each irreducible factor of $x^2 - x$ is also an irreducible factor of $x^4 - x$; similarly, the irreducible factors of $x^4 - x$ all divide $x^{16} - x$. □

Example 7.5 Let $F = \mathrm{GF}(2)$ and $P_1(x) = x^4 + x + 1$, and consider the field $\Phi = F[\xi]/P_1(\xi)$. The conjugacy classes of the elements of Φ are

$$C_0 = \{0\} , \quad C_1 = \{1\} , \quad C_\xi = \{\xi, \xi^2, \xi^4, \xi^8\} , \quad C_{\xi^3} = \{\xi^3, \xi^6, \xi^{12}, \xi^9\} ,$$

$$C_{\xi^5} = \{\xi^5, \xi^{10}\} , \quad \text{and} \quad C_{\xi^7} = \{\xi^7, \xi^{14}, \xi^{13}, \xi^{11}\} .$$

The conjugacy classes of size 1 correspond to the elements of the subfield F of Φ, and the respective minimal polynomials are x and $x - 1$.

There is one conjugacy class of size 2 in Φ, consisting of the elements ξ^5 and ξ^{10}. The minimal polynomial of those elements is $x^2 + x + 1$, as we have seen in Example 3.4 that this is the only irreducible polynomial of degree 2 over F. Now,

$$x(x-1)(x-\xi^5)(x-\xi^{10}) = x(x-1)(x^2+x+1) = x^4 - x .$$

That is, $F \cup C_{\xi^5}$ is the set of roots of $x^4 - x$ in Φ, and this set forms a subfield of Φ of size 4 (refer to the proof of Proposition 3.16); this subfield, which we denote by K, is in fact $\mathrm{GF}(2^2)$. The containment relationships between F, K, and Φ are shown in Figure 7.1. Note that the only possible sizes of proper subfields of Φ are 2 and 4 (for any other size the extension degree of Φ would not be an integer).

There are 12 elements in Φ that do not belong to the proper subfields of Φ; i.e., they are not roots of $x^4 - x$. Those elements form three conjugacy

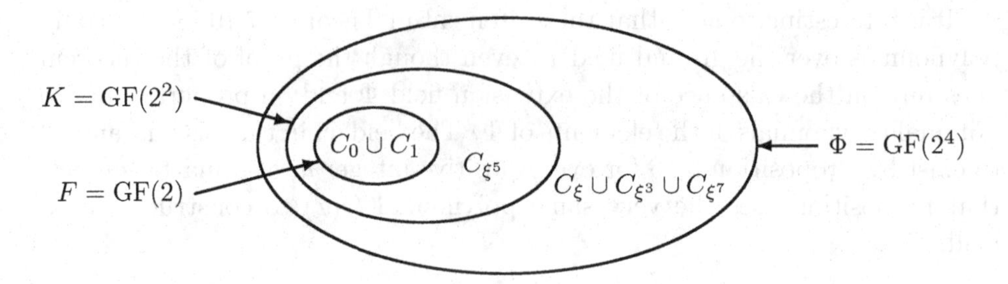

Figure 7.1. Subfields of $\Phi = F[\xi]/(\xi^4 + \xi + 1)$.

classes, C_ξ, C_{ξ^3}, and C_{ξ^7}, each of size 4. The element $\xi \in \Phi$ is a root of the irreducible polynomial $P_1(x) = x^4 + x + 1$; therefore, $P_1(x)$ is the minimal polynomial of the elements in C_ξ. Next, we show that the irreducible polynomial $P_2(x) = x^4 + x^3 + 1$ is the minimal polynomial of the elements in C_{ξ^7}. To this end, it suffices to check that one of the elements in C_{ξ^7}, say $\xi^{14} = \xi^{-1}$, is a root of $P_2(x)$. Indeed,

$$P_2(\xi^{-1}) = \xi^{-4}(1 + \xi + \xi^4) = \xi^{-4} P_1(\xi) = 0 \,.$$

There is one remaining irreducible polynomial of degree 4 over F, namely, $P_3(x) = x^4 + x^3 + x^2 + x + 1$. We conclude that $P_3(x)$ is the minimal polynomial of the elements in C_{ξ^3}. $\square$

7.2 Enumeration of irreducible polynomials

In this section, we use Theorem 7.10 to obtain a formula for the number of monic irreducible polynomials of any given degree over any given finite field.

Let $\mu : \mathbb{Z}^+ \to \{-1, 0, 1\}$ denote the *Möbius function*, whose value at any given positive integer n is determined as follows (see Problem A.2). Let $n = \prod_{j=1}^{s} p_j^{e_j}$ be the factorization of n into distinct primes $p_1, p_2, \ldots, p_s$. Then,

$$\mu(n) = \begin{cases} 1 & \text{if } n = 1 \\ (-1)^s & \text{if } e_j = 1 \text{ for } 1 \le j \le s \\ 0 & \text{otherwise} \end{cases} .$$

The Möbius function is known for the following property.

Proposition 7.11 (The Möbius inversion formula) *Let $h : \mathbb{Z}^+ \to \mathbb{R}$ and $H : \mathbb{Z}^+ \to \mathbb{R}$ be two real-valued functions defined over the domain of positive integers. The following two conditions are equivalent:*

(i) $H(n) = \sum_{m \mid n} h(m)$ *for every* $n \in \mathbb{Z}^+$.

(ii) $h(n) = \sum_{m \mid n} \mu(m) H(n/m)$ *for every* $n \in \mathbb{Z}^+$.

(The summations are taken over all positive integers m that divide n.)

The proof of Proposition 7.11 is given as an exercise in Problem A.2. We next use this proposition in our main result of this section.

Theorem 7.12 *Let* $\mathcal{I}(n,q)$ *denote the number of monic irreducible polynomials of degree n over* $F = \mathrm{GF}(q)$. *Then,*

$$\mathcal{I}(n,q) = \frac{1}{n} \sum_{m \mid n} \mu(m) \cdot q^{n/m} .$$

Proof. By Theorem 7.10,

$$q^n = \deg Q(x) = \sum_{a(x)} \deg a(x) ,$$

where $a(x)$ ranges over all monic irreducible polynomials over F with degrees dividing n. The sum $\sum_{a(x)} \deg a(x)$, in turn, equals $\sum_{m \mid n} m \cdot \mathcal{I}(m,q)$. Hence, for every $n \in \mathbb{Z}^+$,

$$q^n = \sum_{m \mid n} m \cdot \mathcal{I}(m,q) .$$

The result now follows by applying Proposition 7.11 to $h(n) = n \cdot \mathcal{I}(n,q)$ and $H(n) = q^n$. □

The values of $\mathcal{I}(n,2)$ for $n = 1,2,3,4$ are computed in Table 7.1 (the rightmost column in the table is taken from Table 3.1).

Table 7.1. Values of $\mathcal{I}(n,2)$ for $n = 1,2,3,4$.

n	$\mathcal{I}(n,2)$		Irreducible polynomials of degree n
1	$\frac{1}{1} \cdot \mu(1) \cdot 2$	$= 2$	$x, \ x+1$
2	$\frac{1}{2} (\mu(1) \cdot 4 + \mu(2) \cdot 2)$	$= 1$	x^2+x+1
3	$\frac{1}{3} (\mu(1) \cdot 8 + \mu(3) \cdot 2)$	$= 2$	$x^3+x+1, \ x^3+x^2+1$
4	$\frac{1}{4} (\mu(1) \cdot 16 + \mu(2) \cdot 4 + \mu(4) \cdot 2)$	$= 3$	$x^4+x+1, x^4+x^3+1, x^4+x^3+x^2+x+1$

Example 7.6 Let Φ be an extension field of size 2^6 of $F = \text{GF}(2)$. The minimal polynomials (with respect to F) of the elements of Φ have degrees 1, 2, 3, or 6. As was the case in Example 7.5, the polynomial $x^4 - x$ has four roots in Φ, which form a subfield, K, of Φ. The minimal polynomials of the elements of K are x, $x - 1$, and $x^2 + x + 1$, among which the first two polynomials are the minimal polynomials of the elements of the subfield F of Φ.

Since there are two irreducible polynomials of degree 3 over F, there must be two respective conjugacy classes of size 3 in Φ. The elements in those classes, along with the elements of F, are roots of the polynomial

$$x^8 - x = x(x-1)(x^3 + x + 1)(x^3 + x^2 + 1) .$$

The eight roots of $x^8 - x$, in turn, form a subfield of Φ, which we denote by J. The intersection of J with K is the set of roots of

$$\gcd(x^8 - x, x^4 - x) = x \cdot \gcd(x^7 - 1, x^3 - 1) = x(x-1) = x^2 - x ,$$

that is, $J \cap K = F$.

The containment relationships between F, K, J, and Φ are shown in Figure 7.2. One can verify—as in Example 7.5—that F, K, and J are the only proper subfields of Φ.

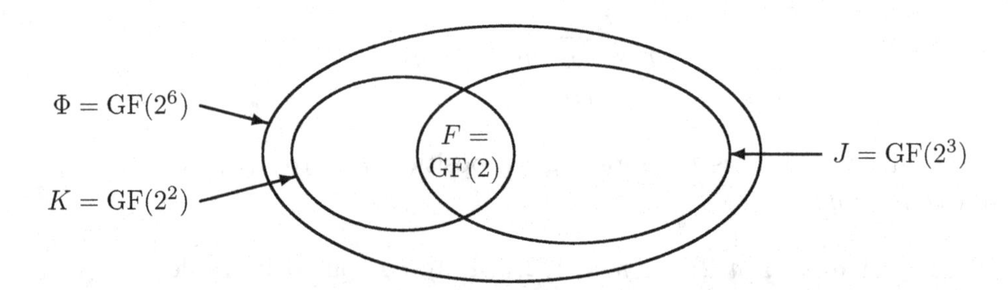

Figure 7.2. Subfields of a field Φ of size 2^6.

The elements in Φ whose minimal polynomials have degree 6 are those that do not belong to any of the proper subfields of Φ. The number of those elements is given by

$$|\Phi| - |J| - |K| + |\underbrace{J \cap K}_{F}| = 2^6 - 2^3 - 2^2 + 2 = 54 .$$

Since each minimal polynomial of degree 6 is shared by six such elements, we obtain from Proposition 7.9 that the number of monic irreducible polynomials of degree 6 over F equals 9. This is precisely what we get when we compute that number using Theorem 7.12:

$$\mathcal{I}(6,2) = \frac{1}{6} \sum_{m \mid 6} \mu(m) \cdot 2^{6/m} = \frac{1}{6} \cdot (2^6 - 2^3 - 2^2 + 2) = 9 .$$

□

7.3 Isomorphism of finite fields

The next theorem was already mentioned in Section 3.3, when we explained why the standard notation for finite fields specifies only the field size. We are now in a position to prove this theorem.

Theorem 7.13 *All finite fields of the same size are isomorphic.*

Proof. Let Φ and K be finite fields of the same size. Both fields have the same characteristic p; therefore, their respective fields of integers are both isomorphic to $\mathrm{GF}(p)$. This allows us to assume hereafter in the proof that Φ and K are extension fields of the same field F and have the same extension degree $[\Phi : F] = [K : F] = n$.

Fix an irreducible polynomial $a(x)$ of degree n over F. By Proposition 7.9, this polynomial is a minimal polynomial of an element $\alpha \in \Phi$ and also of an element $\beta \in K$. From Problem 7.6 we get that

$$\Phi = \{\, u(\alpha) \;:\; u(x) \in F_n[x] \,\} \quad \text{and} \quad K = \{\, u(\beta) \;:\; u(x) \in F_n[x] \,\} .$$

Define the mapping $\psi : K \to \Phi$ by

$$\psi(u(\beta)) = u(\alpha) \,, \quad u(x) \in F_n[x] .$$

We show that ψ is an isomorphism. It is easy to see that ψ is additive, one-to-one, and onto; it remains to prove that it is also multiplicative. Let $u(x)$ and $v(x)$ be polynomials in $F_n[x]$ and let $r(x)$ denote the remainder of $u(x)v(x)$ when divided by $a(x)$. Since $a(\alpha) = 0$ in Φ and $a(\beta) = 0$ in K, we have

$$\psi(u(\beta) \cdot v(\beta)) = \psi(r(\beta)) = r(\alpha) = u(\alpha) \cdot v(\alpha) = \psi(u(\beta)) \cdot \psi(v(\beta)) .$$

□

7.4 Primitive polynomials

Let $a(x)$ be a polynomial over $F = \mathrm{GF}(q)$ such that $\gcd(a(x), x) = 1$. Recall from Problem 6.9 that the exponent of $a(x)$, denoted by $\exp a(x)$, is the smallest positive integer e such that

$$a(x) \mid x^e - 1 .$$

In other words, the exponent e is the multiplicative order of x in the ring $F[x]/a(x)$. It turns out that when $a(x)$ is irreducible, e is also the multiplicative order of the roots of $a(x)$ in any extension field of F.

Proposition 7.14 *Let $a(x)$ be an irreducible polynomial over $F = \mathrm{GF}(q)$ other than a scalar multiple of x. The exponent of $a(x)$ equals the multiplicative order of every root of $a(x)$ in any extension field of F (thus, all of these roots have the same multiplicative order).*

Proof. Assume without loss of generality that $a(x)$ is monic and let α be a root of $a(x)$ in an extension field of F. Then $a(x)$ is the minimal polynomial of α (with respect to F); so, by Proposition 7.4, $\alpha^e = 1$ if and only if $a(x) \mid x^e - 1$. □

The next proposition relates the multiplicative order of an element to the degree of its minimal polynomial.

Proposition 7.15 *Let α be a nonzero element of multiplicative order e in an extension field of $F = \mathrm{GF}(q)$. The degree of the minimal polynomial of α (with respect to F) is the smallest positive integer m such that $e \mid q^m - 1$.*

Proof. The degree of the minimal polynomial of α is the smallest integer $m > 0$ such that $\alpha^{q^m} = \alpha$, or equivalently, $\alpha^{q^m - 1} = 1$. The latter equality holds if and only if $e \mid q^m - 1$. □

By combining Propositions 7.14 and 7.15, we get the following relation between the degree of an irreducible polynomial and its exponent.

Proposition 7.16 *Let $a(x) \neq x$ be a monic irreducible polynomial of degree m over $F = \mathrm{GF}(q)$ and let e be the exponent of $a(x)$. Then m is the smallest positive integer such that $e \mid q^m - 1$.*

Example 7.7 Let $F = \mathrm{GF}(2)$ and $\Phi = F[\xi]/(\xi^4 + \xi + 1)$. Table 7.2 lists the multiplicative orders of the elements of Φ. Note that elements in the same conjugacy class share the same minimal polynomial; so, they also have the same multiplicative order. □

An irreducible polynomial of degree n over $\mathrm{GF}(q)$ is called a *primitive polynomial* if its exponent equals $q^n - 1$.

Proposition 7.17 *The monic primitive polynomials of degree n over $\mathrm{GF}(q)$ are the minimal polynomials of the primitive elements in $\mathrm{GF}(q^n)$.*

Table 7.2. Multiplicative orders of the elements in $\mathrm{GF}(2^4)$.

Conjugacy class	Minimal polynomial	Order/Exponent
$\{0\}$	x	–
$\{1\}$	$x+1$	1
$\{\xi, \xi^2, \xi^4, \xi^8\}$	x^4+x+1	15
$\{\xi^3, \xi^6, \xi^{12}, \xi^9\}$	$x^4+x^3+x^2+x+1$	5
$\{\xi^5, \xi^{10}\}$	x^2+x+1	3
$\{\xi^7, \xi^{14}, \xi^{13}, \xi^{11}\}$	x^4+x^3+1	15

Proof. Let α be a nonzero element in $\Phi = \mathrm{GF}(q^n)$ with $\deg M_\alpha(x) = m$. By Proposition 7.14 it follows that α is primitive in Φ if and only if $\exp M_\alpha(x) = q^n-1$. And the latter equality implies by Proposition 7.16 that $m = n$. □

Recall that the number of primitive elements in $\mathrm{GF}(q^n)$ is $\phi(q^n-1)$, where $\phi(\cdot)$ is the Euler function. The next enumeration result follows from Proposition 7.17.

Theorem 7.18 *Let $\mathcal{P}(n,q)$ denote the number of monic primitive polynomials of degree n over $F = \mathrm{GF}(q)$. Then,*

$$\mathcal{P}(n,q) = \frac{1}{n}\,\phi(q^n-1)\ .$$

Table 7.3 summarizes the values of $\mathcal{P}(n,2)$ for $n = 1,2,3,4$.

Table 7.3. Values of $\mathcal{P}(n,2)$ for $n = 1,2,3,4$.

n	$\mathcal{P}(n,2)$	Primitive polynomials of degree n
1	$\frac{1}{1}\cdot\phi(1) = 1$	$x+1$
2	$\frac{1}{2}\cdot\phi(3) = 1$	x^2+x+1
3	$\frac{1}{3}\cdot\phi(7) = 2$	$x^3+x+1,\ \ x^3+x^2+1$
4	$\frac{1}{4}\cdot\phi(15) = 2$	$x^4+x+1,\ \ x^4+x^3+1$

7.5 Cyclotomic cosets

In this section, we consider the properties of polynomials of the form $x^e - 1$ over $\mathrm{GF}(q)$, where $\gcd(e,q) = 1$. These properties are summarized in the next two propositions.

Proposition 7.19 *Let $F = \mathrm{GF}(q)$ and let e be a positive integer relatively prime to q. Then the following conditions hold:*

(i) The roots of the polynomial $x^e - 1$ in any extension field of F are simple.

(ii) The splitting field of $x^e - 1$ over F is $\mathrm{GF}(q^m)$, where m is the smallest positive integer such that $e \mid q^m - 1$.

Proof. The formal derivative of $x^e - 1$ equals ex^{e-1}, which is a nonzero polynomial due to our assumption on e. Therefore, $\gcd(x^e - 1, ex^{e-1}) = 1$ and, so, by Lemma 3.15, the roots of $x^e - 1$ are simple, thereby proving part (i).

As for part (ii), the polynomial $x^e - 1$ has e (simple) roots in $\mathrm{GF}(q^r)$ if and only if $x^e - 1 \mid x^{q^r} - x$, or, equivalently, $x^e - 1 \mid x^{q^r - 1} - 1$. By Problem 3.4, the latter condition holds if and only if $e \mid q^r - 1$. □

Proposition 7.20 *Given $F = \mathrm{GF}(q)$, let e be a positive integer relatively prime to q and let m be the smallest positive integer such that $e \mid q^m - 1$. Then*

$$x^e - 1 = \prod_{M_\alpha(x)} M_\alpha(x) ,$$

where $M_\alpha(x)$ ranges over all distinct minimal polynomials (with respect to F) of the elements $\alpha \in \mathrm{GF}(q^m)$ with multiplicative orders dividing e.

Proof. The roots of $x^e - 1$ in $\Phi = \mathrm{GF}(q^m)$ are the e elements in Φ whose multiplicative order divides e. These roots are all simple and form a union of conjugacy classes in Φ (with respect to F). The proof continues as in Proposition 7.8. □

Let q, e, and m be as in Proposition 7.20 and let α be an element of multiplicative order e in $\mathrm{GF}(q^m)$. Every conjugacy class of roots of $x^e - 1$ has the form

$$\{\alpha^s, \alpha^{sq}, \alpha^{sq^2}, \ldots, \alpha^{sq^{t-1}}\}$$

for some integer s, where t is the smallest positive integer such that $\alpha^{sq^t} = \alpha^s$. Equivalently, t is the smallest positive integer such that $sq^t \equiv s \pmod{e}$. The respective set of exponents,

$$\{s, sq, sq^2, \ldots, sq^{t-1}\} ,$$

with each value taken modulo e, is called a *cyclotomic coset* modulo e over $\mathrm{GF}(q)$.

Example 7.8 Consider the polynomial $x^{15} - 1$ over $F = \mathrm{GF}(2)$. By Proposition 7.19, the splitting field of this polynomial over F is $\mathrm{GF}(2^4)$. Letting α be an element of multiplicative order 15 in $\mathrm{GF}(2^4)$ (i.e., a primitive element), the conjugacy classes of the roots of $x^{15}-1$ in $\mathrm{GF}(2^4)$ are as shown in Table 7.4 (compare with Table 7.2). □

Table 7.4. Conjugacy classes of the roots of $x^{15} - 1$ in $\mathrm{GF}(2^4)$ over $\mathrm{GF}(2)$.

Conjugacy class	Cyclotomic coset	Order of elements
$\{1\}$	$\{0\}$	1
$\{\alpha, \alpha^2, \alpha^4, \alpha^8\}$	$\{1, 2, 4, 8\}$	15
$\{\alpha^3, \alpha^6, \alpha^{12}, \alpha^9\}$	$\{3, 6, 12, 9\}$	5
$\{\alpha^5, \alpha^{10}\}$	$\{5, 10\}$	3
$\{\alpha^7, \alpha^{14}, \alpha^{13}, \alpha^{11}\}$	$\{7, 14, 13, 11\}$	15

Example 7.9 The splitting field of the polynomial $x^{21} - 1$ over $F = \mathrm{GF}(2)$ is $\mathrm{GF}(2^6)$. Selecting α to be an element of multiplicative order 21 in $\mathrm{GF}(2^6)$, the conjugacy classes of the roots of $x^{21} - 1$ in $\mathrm{GF}(2^6)$ are as shown in Table 7.5.

Table 7.5. Conjugacy classes of the roots of $x^{21} - 1$ in $\mathrm{GF}(2^6)$ over $\mathrm{GF}(2)$.

Conjugacy class	Cyclotomic coset	Order of elements
$\{1\}$	$\{0\}$	1
$\{\alpha, \alpha^2, \alpha^4, \alpha^8, \alpha^{16}, \alpha^{11}\}$	$\{1, 2, 4, 8, 16, 11\}$	21
$\{\alpha^3, \alpha^6, \alpha^{12}\}$	$\{3, 6, 12\}$	7
$\{\alpha^5, \alpha^{10}, \alpha^{20}, \alpha^{19}, \alpha^{17}, \alpha^{13}\}$	$\{5, 10, 20, 19, 17, 13\}$	21
$\{\alpha^7, \alpha^{14}\}$	$\{7, 14\}$	3
$\{\alpha^9, \alpha^{18}, \alpha^{15}\}$	$\{9, 18, 15\}$	7

Based on Table 7.5, we next apply Proposition 7.20 to find the irreducible factorization of $x^{21} - 1$ over F. Clearly, $M_1(x) = x - 1$ and, since $x^2 + x + 1$ is the unique irreducible polynomial of degree 2 over F, we must have

$$M_{\alpha^7}(x) = x^2 + x + 1 .$$

Similarly, $x^3 + x + 1$ and $x^3 + x^2 + 1$ are the only irreducible polynomials of degree 3 over F; so, one of them must be $M_{\alpha^3}(x)$ and the other is $M_{\alpha^9}(x)$.

Therefore,

$$M_\alpha(x)M_{\alpha^5}(x) = \frac{x^{21}-1}{M_1(x)M_{\alpha^7}(x)M_{\alpha^3}(x)M_{\alpha^9}(x)}$$
$$= x^{12}+x^{11}+x^9+x^8+x^6+x^4+x^3+x+1 .$$

Write $M_\alpha(x) = a(x) = \sum_{i=0}^{6} a_i x^i$. Since $\alpha^{-1} \in C_{\alpha^5}$, it follows from Problem 7.3 that $M_{\alpha^5}(x)$ is obtained by reversing the order of coefficients of $M_\alpha(x)$; namely, $M_{\alpha^5}(x) = \sum_{i=0}^{6} a_{6-i} x^i$. From

$$\Big(\sum_{i=0}^{6} a_i x^i\Big)\Big(\sum_{i=0}^{6} a_{6-i} x^i\Big) = x^{12}+x^{11}+x^9+x^8+x^6+x^4+x^3+x+1$$

we obtain equations in the unknown values $a_0, a_1, \ldots, a_6$, resulting in two solutions for $a(x)$: $x^6+x^4+x^2+x+1$ and its reverse $x^6+x^5+x^4+x^2+1$ (see Problem 7.13). This leads to the factorization

$$x^{21}-1 = (x-1)(x^2+x+1)(x^3+x+1)(x^3+x^2+1)$$
$$\cdot (x^6+x^4+x^2+x+1)(x^6+x^5+x^4+x^2+1) .$$

□

Problems

[Sections 7.1 and 7.2]

Problem 7.1 Let α be a primitive element in $\mathrm{GF}(q^m)$ and let s be a positive integer.

1. Show that when $1 \le s \le q^{\lceil m/2 \rceil}$, the degree of the minimal polynomial of α^s with respect to $\mathrm{GF}(q)$ equals m.

   Hint: Let $\mathbf{s} = (s_0\ s_1\ \ldots\ s_{m-1})$ be the coefficients of the q-ary representation of the integer s, i.e., $s = \sum_{j=0}^{m-1} s_j q^j$, where $0 \le s_j < q$. Verify that the remainder of sq^i when divided by $q^m - 1$ is an integer whose q-ary representation is obtained by shifting the contents of $\mathbf{s}$ cyclically i times to the right. Show that when $1 \le s \le q^{\lceil m/2 \rceil}$, these cyclic shifts are distinct for all $0 \le i < m$.

2. What is the degree of the minimal polynomial of α^s when m is even and $s = q^{m/2}+1$?

Problem 7.2 Let $a(x)$ be a monic irreducible polynomial over a finite field F. Show that $a(x)$ is the minimal polynomial of the element ξ in the extension field $F[\xi]/a(\xi)$.

Problem 7.3 Let α be a nonzero element in an extension field Φ of $F = \mathrm{GF}(q)$.

1. Show that the conjugacy classes of α and α^{-1} (with respect to F) are of the same size.

2. Let $M_\alpha(x) = \sum_{i=0}^{m} a_i x^i$ be the minimal polynomial of α (with respect to F). Show that the minimal polynomial of α^{-1} is given by

$$M_{\alpha^{-1}}(x) = \frac{1}{a_0} \sum_{i=0}^{m} a_{m-i} x^i .$$

   That is, up to scaling by some nonzero element of F, the minimal polynomial of α^{-1} is obtained by reversing the order of coefficients of the minimal polynomial of α.

Problem 7.4 Let $F = \mathrm{GF}(q)$ and consider the polynomial $L(x) = x^q - ax$, where a is a nonzero element in F with multiplicative order n. Denote by Φ the splitting field of $L(x)$ over F.

1. Suppose that $L(x)$ has a nonzero root in (the ground field) F. What can be said about the value of a? What is the splitting field over F in this case?

2. Let α be a root of $L(x)$ in Φ. Show that $\alpha^{q^r} = a^r \alpha$ for every positive integer r.

3. Find the degrees of the irreducible factors of $L(x)$ over F. How many irreducible factors are there of any given degree?

   Hint: Use part 2 to find the size of the conjugacy class (with respect to F) of each nonzero root of $L(x)$ in Φ.

4. Show that the equation

$$b^{(q-1)/n} = a$$

   has $(q-1)/n$ distinct solutions for b in F.

   Hint: Write $a = \gamma^{s(q-1)/n}$, where γ is a primitive element in F and s is an integer such that $\gcd(s, n) = 1$. Express the solutions for b in terms of γ, s, and n.

5. Let b be an element of F that satisfies $b^{(q-1)/n} = a$. Show that the polynomial $L(x) = x^q - ax$ is divisible by the polynomial $x^n - b$.

6. What are the irreducible factors of $L(x) = x^q - ax$ over F?

   Hint: Show that each polynomial $x^n - b$ in part 5 is a minimal polynomial (with respect to F) of some nonzero root of $L(x)$ in Φ.

Problem 7.5 Let K be the field $\mathrm{GF}(2^3)$ and let β be an element of multiplicative order 9 in the extension field Φ of K with extension degree $[\Phi : K] = 2$.

1. Partition the set of powers

$$\{1, \beta, \beta^2, \ldots, \beta^8\}$$

   into conjugacy classes (with respect to K).

2. For each element β^i, find the degree of the minimal polynomial $M_{\beta^i}(x)$ (with respect to K).
3. For each minimal polynomial $M_{\beta^i}(x)$, find its constant coefficient (i.e., the value of $M_{\beta^i}(x)$ at $x = 0$).
4. Show that the elements $\beta^i + \beta^{-i}$ are in K.
5. Find the value of $\beta^3 + \beta^{-3}$.

   Hint: Show that this value belongs to a proper subfield of K.
6. Identify the minimal polynomials $M_{\beta^i}(x)$ (with respect to K) whose coefficients are in GF(2).

[Section 7.3]

Problem 7.6 Let $F = \mathrm{GF}(q)$ and $\Phi = \mathrm{GF}(q^n)$, and let α be an element of Φ whose minimal polynomial (with respect to F) has degree (exactly) n. Define the mapping $\varphi : F_n[x] \to \Phi$ by

$$\varphi(u(x)) = u(\alpha) , \quad u(x) \in F_n[x] .$$

Show that the mapping φ is one-to-one and onto.

Problem 7.7 Let p be a prime and let Φ and K be extension fields of GF(p) with the same extension degree n.

1. Let $\psi : K \to \Phi$ be an isomorphism. Show that the minimal polynomial (with respect to GF(p)) of every element $\beta \in K$ is the same as the minimal polynomial of the element $\psi(\beta) \in \Phi$.
2. Show that every isomorphism $\psi : K \to \Phi$ is completely defined by the value of ψ at an element $\beta \in K$ that does not belong to any proper subfield of K.
3. Show that there are exactly n distinct isomorphisms $\psi : K \to \Phi$.

   Hint: Given an element $\beta \in K$ that does not belong to any proper subfield of K, what are the possible values that $\psi(\beta)$ may take?
4. Show that the automorphisms $\psi : \Phi \to \Phi$ are the Frobenius mappings $f_m : x \mapsto x^{p^m}$.

Problem 7.8 The purpose of this problem is to show that the polynomial

$$B_n(x) = \sum_{j=0}^{n-1} x^j$$

is irreducible over the rational field $\mathbb{Q}$ if and only if n is a prime.

1. Show that $B_n(x)$ is reducible over $\mathbb{Q}$ whenever n is a composite integer.

   Hint: Using Problem 3.4, show that $B_r(x)$ divides $B_n(x)$ for every positive divisor r of n.

Assume hereafter in this problem that n is a prime and write $B_n(x) = a(x)b(x)$, where $a(x)$ and $b(x)$ are monic polynomials in $\mathbb{Q}[x]$.

2. Show that $a(x)$ and $b(x)$ have integer coefficients.

   Hint: By clearing denominators from $a(x)$ and $b(x)$, write

$$c \cdot B_n(x) = \hat{a}(x)\hat{b}(x) ,$$

   where c is an integer and $\hat{a}(x)$ and $\hat{b}(x)$ have integer coefficients. Argue that each coefficient in the product $\hat{a}(x)\hat{b}(x)$ must be divisible by every prime divisor p of c, and then recall from Problem 3.2 that the ring of polynomials over $\mathrm{GF}(p)$ is an integral domain.

3. Show that either $a(1) = 1$ or $b(1) = 1$.

Without loss of generality assume hereafter that $a(1) = 1$.

4. Show that there is an integer $m \in \{0, 1, \ldots, n-1\}$ such that

$$a(x) \equiv (x-1)^m \pmod{n}$$

   (where the congruence holds for every two coefficients in $a(x)$ and $(x-1)^m$, respectively, that multiply the same power of x).

   Hint: Show that $B_n(x) \equiv (x-1)^{n-1} \pmod{n}$.

5. Show that if $\deg a > 0$ then $n \mid a(1)$.

6. Deduce that $a(x)$ has degree 0 and, so, $B_n(x)$ is irreducible over $\mathbb{Q}$.

(It follows from this problem that when n is a prime, the ring of residues $\mathbb{Q}[\omega]/B_n(\omega)$ is an extension field of $\mathbb{Q}$. Several properties of this field are summarized in the notes on Section 7.3.)

[Section 7.4]

Problem 7.9 Let m be a positive integer. Show that every irreducible polynomial of degree m over $\mathrm{GF}(2)$ is primitive if and only if 2^m-1 is a prime.

Hint: Use part 1 of Problem 7.1 to show the "only if" part. See also Problem 3.19.

Problem 7.10 For a positive integer m, let α be an element of multiplicative order $3 \cdot (4^m + 1)$ in an extension field Φ of $F = \mathrm{GF}(2)$.

1. Show that $3 \cdot (4^m + 1)$ divides $2^{4m} - 1$.

2. Find the smallest positive integer ℓ such that $3 \cdot (4^m + 1)$ divides $2^\ell - 1$.

   Hint: The sought ℓ is the multiplicative order of 2 in the ring of integer residues modulo $3 \cdot (4^m + 1)$.

3. What is the smallest possible size of Φ?

4. What is the size of the conjugacy class of α (with respect to F)?

5. What is the multiplicative order of α^3 in Φ?

6. What is the size of the conjugacy class of α^3?
7. Are α and α^3 in the same conjugacy class?
8. Are α^3 and α^{-3} in the same conjugacy class?
9. Show that α and α^{-1} are not in the same conjugacy class (with respect to F).

   Hint: Verify that if $\alpha^{-1} = \alpha^{2^\ell}$ for some nonnegative integer $\ell < 4m$ then $3 \cdot (4^m + 1)$ divides $2^\ell + 1$. Deduce that $2m \,|\, \ell$ and, so, $\ell \geq 4m$, thereby reaching a contradiction.

Problem 7.11 The purpose of this problem is to show that the polynomial

$$U_n(x) = x^{2 \cdot 3^n} + x^{3^n} + 1$$

is irreducible over $F = \mathrm{GF}(2)$ for every $n \geq 0$.

1. Show that $U_n(x)$ divides $x^{3^{n+1}} - 1$.
2. Let z_n denote the integer 2^{3^n}. Show by induction on n that $z_n + 1$ is divisible by 3^{n+1}, yet it is not divisible by 3^{n+2}.

   Hint: Write
   $$z_{n+1} + 1 = (z_n + 1)(z_n^2 - z_n + 1)$$
   and show that
   $$z_n^2 - z_n + 1 \equiv 3 \pmod 9 .$$
3. Deduce from part 2 that 3^{n+1} divides $2^{2 \cdot 3^n} - 1 \; (= z_n^2 - 1)$, yet for $n > 0$ it divides neither $2^{3^n} - 1 \; (= z_n - 1)$ nor $2^{2 \cdot 3^{n-1}} - 1 \; (= z_{n-1}^2 - 1)$.
4. Show that the smallest positive integer ℓ such that $3^{n+1} \,|\, 2^\ell - 1$ equals $2 \cdot 3^n$.

   Hint: Use part 3 (and see the hint in part 2 of Problem 7.10).
5. Let α be a root of $U_n(x)$ in the splitting field of $U_n(x)$ over F. Show that the multiplicative order of α equals 3^{n+1}.

   Hint: If $\mathcal{O}(\alpha)$ divided 3^n, then $U_n(\alpha)$ would be 1.
6. Let α be as in part 5. Show that the conjugacy class of α (with respect to F) contains $2^{2 \cdot 3^n}$ elements of the splitting field. Deduce that $U_n(x)$ is the minimal polynomial of α (with respect to F).

Problem 7.12 (More on linear-recurring sequences) Let $\sigma(x) = 1 + \sum_{i=1}^{m} \sigma_i x^i$ be an irreducible polynomial of degree m over $F = \mathrm{GF}(q)$ and let α be a root of $\sigma(x)$ in $\Phi = \mathrm{GF}(q^m)$.

Recall from Problem 3.31 that the trace polynomial over Φ with respect to F is defined by

$$\mathrm{T}(x) = \mathrm{T}_{\Phi:F}(x) = x + x^q + x^{q^2} + \ldots + x^{q^{m-1}} .$$

Fix η to be an element of Φ, and consider the infinite *trace sequence* $\mathbf{a} = (a_i)_{i=0}^{\infty}$ whose elements are given by

$$a_i = \mathrm{T}(\eta \alpha^{-i}) \,, \quad i \geq 0 \,.$$

1. Show that the sequence $\mathbf{a}$ satisfies the linear recurrence

$$a_i = -\sum_{j=1}^{m} \sigma_j a_{i-j} , \quad \text{for every } i \geq m .$$

2. Show that distinct values of η yield distinct sequences $\mathbf{a}$; so, when η ranges over the elements of Φ, the infinite sequence $\left(\mathrm{T}(\eta\alpha^{-i})\right)_{i=0}^{\infty}$ ranges over all the distinct sequences that satisfy the linear recurrence in part 1.

   Hint: See part 3 of Problem 6.14.

3. A sequence $(b_i)_{i=0}^{\infty}$ is called a *phase* of $\mathbf{a}$ if there exists a nonnegative integer τ such that

$$b_i = a_{i+\tau} , \quad i \geq 0 .$$

   Show that for each phase $(b_i)_{i=0}^{\infty}$ of $\mathbf{a}$ there exists an element $\beta \in \Phi$ such that

$$b_i = \mathrm{T}(\beta\alpha^{-i}) , \quad i \geq 0 .$$

   Express β in terms of α, η, and τ.

4. The sequence $\mathbf{a}$ is said to be of *natural phase* if $a_{iq} = a_i$ for every $i \geq 0$. Show that $\mathbf{a}$ is of natural phase if and only if $\eta \in F$.

5. Recall from Problem 6.8 that the period of $\mathbf{a}$ is the least integer $e > 0$, if any, such that $a_{i+e} = a_i$ for all $i \geq 0$. Show that when $\eta \neq 0$, the period of $\mathbf{a}$ equals the multiplicative order of α in Φ (which, by Proposition 7.14, is also the exponent of $\sigma(x)$; see also part 3 of Problem 6.16).

   (Thus, for the special case where $\sigma(x)$ is a primitive polynomial, the period of $\mathbf{a}$ is q^m-1. The sequence $\mathbf{a}$ is then called a *maximal-length (linear-recurring) sequence* or, in short, an *M-sequence*. The respective primitive polynomial $\sigma(x)$ is said to represent a *maximal-length linear-feedback shift register*. See also Problem 3.37 and Section 6.7.1.)

6. How many distinct phases does $\mathbf{a}$ have?

Associate with $\mathbf{a}$ the formal power series $a(x) = \sum_{i=0}^{\infty} a_i x^i$ in $F[[x]]$ as defined in Section 6.3.2.

7. Show that the formal power series $a(x)$ can be expressed as

$$a(x) = \sum_{j=0}^{m-1} \frac{\eta^{q^j}}{1 - \alpha^{-q^j} x} .$$

8. Recall from Problem 6.14 that there exists a unique polynomial $\omega(x) \in F_m[x]$ such that

$$a(x) = \frac{\omega(x)}{\sigma(x)} .$$

   Show that η is related to $\sigma(x)$ and $\omega(x)$ by

$$\eta = -\frac{\omega(\alpha)}{\alpha \cdot \sigma'(\alpha)} ,$$

where $\sigma'(x)$ denotes the formal derivative of $\sigma(x)$. Conclude that the coefficients of $\omega(x)$ form the representation of the element $-\eta \cdot \alpha \cdot \sigma'(\alpha)$ according to the basis $\Omega = (1 \; \alpha \; \alpha^2 \; \ldots \; \alpha^{m-1})$ of Φ over F.

Hint: Based on part 7, write

$$\omega(x) = \sum_{j=0}^{m-1} \frac{\eta^{q^j} \sigma(x)}{1 - \alpha^{-q^j} x} = \sum_{j=0}^{m-1} \eta^{q^j} \cdot \prod_{\substack{0 \le k < m: \\ k \ne j}} (1 - \alpha^{-q^k} x) ,$$

and substitute $x = \alpha$.

9. Assume now that $\mathbf{a}$ is an M-sequence (that is, $\sigma(x)$ is a primitive polynomial over F and $\eta \ne 0$). Show that when i ranges over $\{0, 1, \ldots, q^m-2\}$, the m-tuple $(a_i \; a_{i+1} \; \ldots \; a_{i+m-1})$ ranges over all the nonzero elements of F^m. Conclude that each nonzero element of F appears in the finite sequence $(a_i)_{i=0}^{q^m-2}$ exactly q^{m-1} times, while the zero element appears $q^{m-1}-1$ times (this property follows also from part 5 of Problem 3.31).

[Section 7.5]

Problem 7.13 Let $a(x) = \sum_{i=0}^{6} a_i x^i$ be a polynomial over $F = \mathrm{GF}(2)$ that satisfies the equality

$$\Bigl(\sum_{i=0}^{6} a_i x^i\Bigr)\Bigl(\sum_{i=0}^{6} a_{6-i} x^i\Bigr) = x^{12} + x^{11} + x^9 + x^8 + x^6 + x^4 + x^3 + x + 1 .$$

By comparing respective coefficients of powers of x in both sides of this equality, show that $a(x)$ equals either $x^6 + x^4 + x^2 + x + 1$ or $x^6 + x^5 + x^4 + x^2 + 1$.

Problem 7.14 Consider the polynomial

$$P_n(x) = x^n - 1$$

over GF(2). Answer parts 1–6 for the following values of n: (i) $n = 7$, (ii) $n = 11$, and (iii) $n = 17$.

1. Find the splitting field Φ of $P_n(x)$ over GF(2).

2. Find the number of distinct roots of $P_n(x)$ in Φ.

3. Find the multiplicative orders of the roots of $P_n(x)$ in Φ and the number of roots of each order.

4. Let α be an element of multiplicative order n in Φ. Find the conjugates of α in Φ with respect to GF(2) (express the conjugates as powers α^i, where $0 \le i < n$).

5. For an element α as in part 4, find the conjugates of α^3 in Φ with respect to GF(2).

6. Determine whether the polynomial

$$B_n(x) = \sum_{j=0}^{n-1} x^j = \frac{P_n(x)}{x-1}$$

is irreducible over GF(2).

Problem 7.15 Consider the polynomial

$$B_n(x) = \sum_{j=0}^{n-1} x^j$$

over $F = \mathrm{GF}(q)$.

1. Show that $B_n(x)$ is reducible over F whenever n is a composite number.

   Hint: See part 1 of Problem 7.8.

2. Suppose now that n is a prime. Show that $B_n(x)$ is irreducible over F if and only if the integer q, when taken modulo n, is a primitive element in GF(n).

   Hint: See Problem 7.14.

Problem 7.16 (Legendre sequence as a linear-recurring sequence) Let p be a prime and suppose that the multiplicative order of 2 in GF(p) is $m = \frac{1}{2}(p-1)$. Also, let α be an element in $\Phi = \mathrm{GF}(2^m)$ of multiplicative order p (why does such an element exist?).

1. Show that 2 is a generator of the set of quadratic residues in GF(p), when this set is regarded as a cyclic subgroup of the multiplicative group of GF(p) (see Problems 3.23 and 3.26).

2. Show that Φ is the splitting field of the polynomial $x^p - 1$ over GF(2).

3. Show that the polynomial $x^p - 1$ factors over $F = \mathrm{GF}(2)$ into

   $$x^p - 1 = (x-1)\sigma(x)\tau(x) ,$$

   where both $\sigma(x)$ and $\tau(x)$ are irreducible polynomials of degree m over F, with $\sigma(x)$ being the minimal polynomial of α with respect to F. Furthermore, show that the set of roots of $\sigma(x)$ in Φ is

   $$\mathcal{Q} = \left\{\alpha^i \ : \ i \text{ is a quadratic residue modulo } p\right\} ,$$

   while that of $\tau(x)$ is

   $$\mathcal{N} = \left\{\alpha^i \ : \ i \text{ is a quadratic non-residue modulo } p\right\} .$$

4. Let $\mathrm{T}(x)$ be the trace polynomial $\mathrm{T}_{\Phi:F}(x) = x + x^2 + x^{2^2} + \ldots + x^{2^{m-1}}$ over Φ with respect to $F = \mathrm{GF}(2)$. Show that the mapping $x \mapsto \mathrm{T}(x)$ is constant when restricted to the domain $\mathcal{Q}$, and is also constant when restricted to $\mathcal{N}$. On the other hand, show that the values taken by $\mathrm{T}(x)$ on these two domains are distinct: one value is 0, while the other is 1.

Hint: Show that for any $\beta \in \mathcal{Q}$,

$$\mathrm{T}(\beta) = \mathrm{T}(\alpha) ,$$

and for any $\beta \in \mathcal{N}$,

$$1 + \mathrm{T}(\alpha) + \mathrm{T}(\beta) = \sum_{i=0}^{p-1} \alpha^i = 0 .$$

5. Suppose that α is such that $\mathrm{T}(\alpha^{-1}) = 0$ (argue why, indeed, one could select α in the first place so that this condition holds), and define the infinite sequence $\mathbf{a} = (a_i)_{i=0}^{\infty}$ by

$$a_i = \mathrm{T}(\alpha^{-i}) , \quad i \geq 0 .$$

Show that for every $i \geq 0$,

$$a_i = \begin{cases} 0 & \text{if } p \text{ divides } i \text{ and } m \text{ is even} \\ 1 & \text{if } p \text{ divides } i \text{ and } m \text{ is odd} \\ 0 & \text{if } i \text{ is a quadratic residue modulo } p \\ 1 & \text{otherwise} \end{cases} .$$

(Thus, when $p \equiv 3 \pmod 4$ and $\mathcal{O}(2) = \frac{1}{2}(p-1)$ in $\mathrm{GF}(p)$, the infinite integer sequence $(x_i)_{i=0}^{\infty}$ that is given by

$$x_i = (-1)^{a_i} , \quad i \geq 0 ,$$

coincides with the Legendre sequence, which was defined in Problem 3.26. As a linear-recurring sequence, $\mathbf{a}$ has a minimal linear recurrence order $\deg \sigma(x) = \frac{1}{2}(p-1)$, which is large—almost half its period p; see Problems 6.16 and 7.12.)

Notes

[Sections 7.1 and 7.2]

The book by Lidl and Niederreiter [229] contains an encyclopedic treatment of finite fields. The structure of finite fields is studied in the first three chapters of that book.

Let C_α be a conjugacy class of elements of $\Phi = \mathrm{GF}(q^m)$ (with respect to $F = \mathrm{GF}(q)$). Sometimes the elements of C_α form a basis of the linear space Φ over F. When this happens, we say that the elements of C_α form a *normal basis* of Φ over F (clearly, this occurs only if $|C_\alpha| = m$). It can be shown that a normal basis always exists; see Lidl and Niederreiter [229, Theorems 2.35 and 3.73].

[Section 7.3]

Let p be a prime and $B_p(x)$ be the polynomial $\sum_{j=0}^{p-1} x^i$ over the rational field $\mathbb{Q}$. It follows from Problem 7.8 that the ring of residues $K_p = \mathbb{Q}[\omega]/B_p(\omega)$ is a field, which is called the *pth cyclotomic extension* of $\mathbb{Q}$, and $[K_p : \mathbb{Q}] = p-1$. The element $\omega \in K_p$, being a root of $B_p(x)$, is a root of order p of unity in the complex field $\mathbb{C}$.

Fix a primitive element g in GF(p). It can be shown that the mapping $\varphi : K_p \to K_p$, which is defined by

$$\varphi\left(\sum_{i=0}^{p-2} a_i \omega^i\right) = \sum_{i=0}^{p-2} a_i \omega^{ig} ,$$

is an automorphism of K_p (note that raising ω to the power ig is well-defined, since the multiplicative order of ω in $\mathbb{C}$ is equal to the modulus p). In fact, φ is a generator of the automorphism group $\{\varphi^\ell\}_{\ell=0}^{p-2}$ of K_p, where

$$\varphi^\ell\left(\sum_{i=0}^{p-2} a_i \omega^i\right) = \sum_{i=0}^{p-2} a_i \omega^{ig^\ell} , \quad \ell \geq 0 .$$

The terms conjugacy class and minimal polynomial can be defined also for elements of K_p, with the automorphism $x \mapsto \varphi(x)$ playing the same role as the Frobenius mapping $x \mapsto x^q$ does for GF(q^n). Specifically, the conjugacy class (with respect to $\mathbb{Q}$) of an element $\alpha \in K_p$ is defined by

$$C_\alpha = \left\{\alpha, \varphi(\alpha), \varphi^2(\alpha), \ldots, \varphi^{m-1}(\alpha)\right\} ,$$

where $m = m_\alpha$ is the smallest positive integer such that $\varphi^m(\alpha) = \alpha$. The minimal polynomial (with respect to $\mathbb{Q}$) of α is defined by $M_\alpha(x) = \prod_{\gamma \in C_\alpha}(x - \gamma)$, and this polynomial is a monic irreducible polynomial over $\mathbb{Q}$ whose degree, m_α, divides $p-1$.

[Sections 7.4 and 7.5]

An extensive treatment of M-sequences (which are defined in Problem 7.12)—and sequences in general—can be found in the book by Golomb [152]. The polynomials in Problem 7.11 are mentioned by Golomb in [152, p. 96].

For more on the linear-recurring properties of Legendre sequences (Problem 7.16), see Ding *et al.* [105], Kim and Song [213], and No *et al.* [272]. It is still an open problem whether there exist infinitely many primes p such that either 2 is primitive in GF(p) (in which case the polynomial $B_p(x)$ in Problem 7.15 is irreducible over GF(2)) or 2 has multiplicative order $\frac{1}{2}(p-1)$ in GF(p) (which is the condition assumed in Problem 7.16).

Chapter 8

Cyclic Codes

Cyclic codes form a class of linear codes that have two major advantages: the codes in this class can be encoded by simple hardware circuits, and their structure lends itself to a more extensive analysis of their parameters, compared to general linear codes. Conventional Reed–Solomon codes and BCH codes are prominent examples of cyclic codes; we will revisit BCH codes in this chapter, now reviewing them through the lens of cyclic codes.

With each cyclic code we associate two polynomials, which are referred to as the generator polynomial and the check polynomial. For the analysis of the parameters of cyclic codes, we will examine the set of roots of the generator polynomial in its splitting field. In particular, we obtain a lower bound on the minimum distance by looking for the largest possible subset of roots that form a sequence of consecutive powers of some element whose multiplicative order equals the code length.

8.1 Definition

A linear $[n, k]$ code $\mathcal{C}$ over a field F is called *cyclic* if every cyclic shift of a codeword in $\mathcal{C}$ is also a codeword; namely,

$$(c_0 \, c_1 \, \ldots \, c_{n-1}) \in \mathcal{C} \quad \Longrightarrow \quad (c_{n-1} \, c_0 \, \ldots \, c_{n-2}) \in \mathcal{C}$$

(in the context of cyclic codes, it will be convenient to use an indexing convention whereby the first entry in a word is indexed by 0).

Let $a(x)$ and $b(x)$ be polynomials in $F[x]$ such that $\deg a(x) = m \geq 0$. We denote by $b(x) \text{ MOD } a(x)$ the remainder in $F_m[x]$ obtained when dividing $b(x)$ by $a(x)$ (as an operation on polynomials, MOD will have the same precedence as multiplication and division).

Hereafter in this chapter, we will commonly use polynomial notation for words, by associating a word $\mathbf{c} = (c_0 \, c_1 \, \ldots \, c_{n-1})$ in F^n with a polynomial

$$c(x) = c_0 + c_1 x + \ldots + c_{n-1} x^{n-1}$$

in $F_n[x]$. In this notation, the cyclic shift of $c(x)$ is given by

$$\begin{aligned} c_{n-1} + c_0 x + \ldots + c_{n-2} x^{n-1} &= x \cdot c(x) - c_{n-1} \cdot (x^n - 1) \\ &= x \cdot c(x) \text{ MOD } (x^n - 1) , \end{aligned}$$

and a linear code $\mathcal{C}$ is cyclic if and only if

$$c(x) \in \mathcal{C} \quad \Longrightarrow \quad x \cdot c(x) \text{ MOD } (x^n - 1) \in \mathcal{C} .$$

It follows that when $c(x)$ is a codeword in a cyclic code, so are the words $x^i \cdot c(x) \text{ MOD } (x^n - 1)$ for every $i \geq 0$. By linearity we thus conclude that in a cyclic code $\mathcal{C}$,

$$c(x) \in \mathcal{C} \quad \Longrightarrow \quad \sum_i u_i x^i c(x) \text{ MOD } (x^n - 1) \in \mathcal{C}$$

for every (finitely many) $u_0, u_1, u_2, \ldots$ in F; equivalently,

$$c(x) \in \mathcal{C} \quad \Longrightarrow \quad u(x) c(x) \text{ MOD } (x^n - 1) \in \mathcal{C} \tag{8.1}$$

for every $u(x) \in F[x]$. Hence, $\mathcal{C}$ is an *ideal* in the ring $F[x]/(x^n - 1)$ (see the Appendix).

Example 8.1 The parity code and the repetition code are cyclic. □

Example 8.2 We show that conventional RS codes over $F = \mathrm{GF}(q)$ are cyclic. Let n be a positive integer dividing $q-1$ and let α be an element of multiplicative order n in F. Also, let b and d be integers such that $0 < d \leq n$. These parameters define an $[n, k{=}n{-}d{+}1, d]$ RS code $\mathcal{C}_{\mathrm{RS}}$ over F, which consists of all polynomials $c(x) \in F_n[x]$ such that

$$c(\alpha^\ell) = 0 \quad \text{for } \ell = b, b{+}1, \ldots, b{+}d{-}2$$

(see Equation (5.4)).

Let $c(x) = c_0 + c_1 x + \ldots + c_{n-1} x^{n-1}$ be a codeword in $\mathcal{C}_{\mathrm{RS}}$. Its cyclic shift is given by

$$\tilde{c}(x) = x \cdot c(x) \text{ MOD } (x^n - 1) = x \cdot c(x) - c_{n-1} \cdot (x^n - 1) ,$$

and when we substitute α^ℓ in $\tilde{c}(x)$ we obtain

$$\tilde{c}(\alpha^\ell) = \alpha^\ell \cdot c(\alpha^\ell) - c_{n-1} \cdot ((\alpha^\ell)^n - 1) = \alpha^\ell \cdot c(\alpha^\ell) .$$

Hence,

$$\tilde{c}(\alpha^\ell) = 0 \quad \text{for } \ell = b, b{+}1, \ldots, b{+}d{-}2 ,$$

thus implying that $\tilde{c}(x) \in \mathcal{C}_{\mathrm{RS}}$. □

Example 8.3 Recall from Section 5.6 that a BCH code over a field F consists of the codewords over F of an RS code over an extension field of F. Since RS codes are cyclic, so are BCH codes. □

Example 8.4 We find sufficient conditions for having a cyclic Hamming code over $F = \mathrm{GF}(q)$. Given an integer $m > 1$, let $n = (q^m - 1)/(q-1)$ and let α be an element in $\mathrm{GF}(q^m)$ with multiplicative order n. Consider a BCH code $\mathcal{C}_{\mathrm{BCH}}$ of length n over F with an $m \times n$ parity-check matrix obtained by representing each entry in

$$(1 \; \alpha \; \alpha^2 \; \ldots \; \alpha^{n-1})$$

as a column vector in F^m, according to some basis of $\mathrm{GF}(q^m)$ over F. To have minimum distance at least 3, we require that every two columns in the resulting parity-check matrix be linearly independent over F. Equivalently, we require that for every $0 \le i < j \le n-1$ and every $u \in F$,

$$\alpha^j \neq u \cdot \alpha^i .$$

This, in turn, holds if and only if

$$\alpha^\ell \notin F \quad \text{for } \ell = 1, 2, \ldots, n-1 .$$

Now, recall that $\alpha^\ell \in F$ if and only if $(\alpha^\ell)^{q-1} = 1$, and the latter equality holds if and only if the multiplicative order n of α divides $\ell(q-1)$. We thus conclude that $\mathcal{C}_{\mathrm{BCH}}$ has minimum distance at least 3 if and only if

$$n \text{ does not divide } \ell(q-1) \text{ for } \ell = 1, 2, \ldots, n-1 ,$$

and this happens if and only if

$$\gcd(n, q-1) = 1 .$$

So, the latter condition guarantees that $\mathcal{C}_{\mathrm{BCH}}$ is a linear $[n, k{\ge}n-m, d{\ge}3]$ code over $\mathrm{GF}(q)$, and we can recognize this code as the $[n, k{=}n-m, d{=}3]$ Hamming code. □

8.2 Generator polynomial and check polynomial

In this section, we introduce two polynomials that can be associated with a given cyclic code. The definition of the first polynomial is based on the following property of cyclic codes.

Proposition 8.1 *Let $\mathcal{C}$ be a cyclic $[n, k]$ code over F with $k > 0$. Then there is a unique monic polynomial $g(x)$ such that for every $c(x) \in F_n[x]$,*

$$c(x) \in \mathcal{C} \qquad \Longleftrightarrow \qquad g(x) \mid c(x) .$$

Proof. First, from the requirements on $g(x)$ it follows that if $g(x)$ exists then it must be a codeword of $\mathcal{C}$ (since obviously $g(x) \mid g(x)$) and it is unique (since it divides all other monic codewords in $\mathcal{C}$).

Now, select $g(x)$ to be a monic nonzero codeword with a smallest degree in $\mathcal{C}$. For every $u(x) \in F[x]$ we have $u(x)g(x) \text{ MOD } (x^n - 1) \in \mathcal{C}$. In particular, for every $u(x) \in F_{n-\deg g}[x]$ we have $u(x)g(x) \in \mathcal{C}$. Therefore, all the polynomial multiples of $g(x)$ in $F_n[x]$ are codewords of $\mathcal{C}$.

Conversely, let $c(x) \in \mathcal{C}$ and write $c(x) = u(x)g(x) + r(x)$ where $\deg r < \deg g$. Since both $c(x)$ and $u(x)g(x)$ are in $\mathcal{C}$ then, by linearity, so is $r(x) = c(x) - u(x)g(x)$. From the minimality of $\deg g$ we get that $r(x) = 0$, i.e., $c(x)$ is divisible by $g(x)$. $\square$

The polynomial $g(x)$ in Proposition 8.1 is called the *generator polynomial* of $\mathcal{C}$. (We remark that the trivial case $k = 0$ was excluded from Proposition 8.1. The generator polynomial can be formally defined in this case as $x^n - 1$.)

It follows from Proposition 8.1 that a cyclic $[n, k]$ code $\mathcal{C}$ over F can be written as

$$\mathcal{C} = \{\, u(x)g(x) \;:\; u(x) \in F_{n-\deg g}[x] \,\} \;, \tag{8.2}$$

where $g(x)$ is the generator polynomial of $\mathcal{C}$. This also implies that

$$\deg g = n - k \;.$$

Writing $g(x) = g_0 + g_1 x + \ldots + g_{n-k} x^{n-k}$ (and assuming that $k > 0$), we obtain from (8.2) the following $k \times n$ generator matrix of the code $\mathcal{C}$:

$$G = \begin{pmatrix} g_0 & g_1 & \ldots & g_{n-k} & & & & \\ & g_0 & g_1 & \ldots & g_{n-k} & & \text{\Large 0} & \\ & \text{\Large 0} & \ddots & \ddots & \ldots & \ddots & & \\ & & & g_0 & g_1 & \ldots & g_{n-k} \end{pmatrix} . \tag{8.3}$$

We have seen Equations (8.2) and (8.3) already in Sections 5.2 and 5.3, for the special case of RS codes: using the notation of Example 8.2, the generator polynomial of $\mathcal{C}_{\mathrm{RS}}$ is given by

$$g(x) = (x - \alpha^b)(x - \alpha^{b+1}) \cdots (x - \alpha^{b+d-2}) \;.$$

In fact, cyclic codes can be encoded using the same multiplication and remaindering circuits as shown in Figures 5.1 and 5.2. Such encoding circuits lend themselves to simple hardware implementations.

The next proposition provides the basis for the definition of a second polynomial that we associate with a given cyclic code.

Proposition 8.2 *Let $g(x)$ be the generator polynomial of a cyclic $[n,k]$ code over F. Then*
$$g(x) \mid x^n - 1 .$$

Proof. Write $x^n - 1 = h(x)g(x) + r(x)$ where $\deg r < \deg g$. We have
$$r(x) = -h(x)g(x) \text{ MOD } (x^n - 1)$$
and, so, from the property (8.1) of cyclic codes it follows that $r(x) \in \mathcal{C}$. This means that $r(x)$ is zero, since no other codeword in $\mathcal{C}$ can have degree smaller than $\deg g$. □

We can also state a converse to Proposition 8.2: if $g(x)$ is a polynomial over F that divides $x^n - 1$, then the set (8.2) is a cyclic code (Problem 8.1).

Let $\mathcal{C}$ be a cyclic $[n,k]$ code with a generator polynomial $g(x)$. The *check polynomial* of $\mathcal{C}$, denoted as $h(x)$, is the monic polynomial of degree k obtained by
$$h(x) = \frac{x^n - 1}{g(x)} .$$

Proposition 8.3 *Let $\mathcal{C}$ be a cyclic $[n,k]$ code over F with a check polynomial $h(x) = h_0 + h_1 x + \ldots + h_k x^k$. Then the following $(n-k) \times n$ matrix*
$$H = \begin{pmatrix} h_k & h_{k-1} & \ldots & h_0 & & & & \\ & h_k & h_{k-1} & \ldots & h_0 & & & 0 \\ 0 & & \ddots & \ddots & \ldots & \ddots & & \\ & & & h_k & h_{k-1} & \ldots & h_0 \end{pmatrix}$$
is a parity-check matrix of $\mathcal{C}$.

Proof. First, observe that $\mathrm{rank}(H) = n-k$. Next, let G be obtained by (8.3) from the generator polynomial $g(x)$ of $\mathcal{C}$; we verify that $GH^T = 0$. For every $i = 0, 1, \ldots, k-1$ and $\ell = 0, 1 \ldots, n-k-1$, the scalar product of row i in G and row ℓ in H is given by
$$\sum_{j=0}^{n-1} g_{j-i} h_{k+\ell-j} , \tag{8.4}$$
where $g_j = 0$ for $j \notin \{0, 1, \ldots, n-k\}$ and $h_j = 0$ for $j \notin \{0, 1, \ldots, k\}$. The expression in (8.4) is the coefficient of $x^{k+\ell-i}$ in the product $g(x)h(x) = x^n - 1$. Now, for the range of values of i and ℓ we have
$$1 \le k+\ell-i \le n-1$$
and, so, the respective coefficients in $g(x)h(x)$ are zero. □

The following property immediately follows from (8.3) and Proposition 8.3.

Corollary 8.4 *Let $\mathcal{C}$ be a cyclic $[n, k]$ code over F and let $h(x) = \sum_{j=0}^{k} h_j x^j$ be the check polynomial of $\mathcal{C}$. Then the dual code of $\mathcal{C}$ is a cyclic $[n, n-k]$ code over F whose generator polynomial is*

$$g^{\perp}(x) = \frac{1}{h_0} \sum_{j=0}^{k} h_{k-j} x^j = \frac{x^k h(x^{-1})}{h(0)} .$$

Example 8.5 Let $\mathcal{C}$ be a cyclic code of length 7 over $F = \mathrm{GF}(2)$ with a generator polynomial $g(x) = x^3 + x + 1$. The check polynomial is given by

$$h(x) = \frac{x^7 - 1}{x^3 + x + 1} = x^4 + x^2 + x + 1 ,$$

and the respective generator and parity-check matrices are given by

$$G = \begin{pmatrix} 1 & 1 & 0 & 1 & 0 & 0 & 0 \\ 0 & 1 & 1 & 0 & 1 & 0 & 0 \\ 0 & 0 & 1 & 1 & 0 & 1 & 0 \\ 0 & 0 & 0 & 1 & 1 & 0 & 1 \end{pmatrix} \quad \text{and} \quad H = \begin{pmatrix} 1 & 0 & 1 & 1 & 1 & 0 & 0 \\ 0 & 1 & 0 & 1 & 1 & 1 & 0 \\ 0 & 0 & 1 & 0 & 1 & 1 & 1 \end{pmatrix} .$$

The parity-check matrix H indicates that $\mathcal{C}$ is a $[7, 4, 3]$ Hamming code. □

8.3 Roots of a cyclic code

Let $a(x)$ be a polynomial over $F = \mathrm{GF}(q)$ such that $\gcd(a(x), x) = 1$. Recall from Problem 6.9 and Section 7.4 that the exponent of $a(x)$, denoted by $\exp a(x)$, is the smallest positive integer e such that

$$a(x) \mid x^e - 1 .$$

The exponent of $a(x)$ equals the multiplicative order of x in the ring $F[x]/a(x)$; as such, it satisfies the following property.

Proposition 8.5 *Let $a(x)$ be a polynomial over $F = \mathrm{GF}(q)$ such that $\gcd(a(x), x) = 1$. Then $a(x) \mid x^{\ell} - 1$ if and only if $\exp a(x) \mid \ell$.*

Let $\mathcal{C}$ be a cyclic $[n, k]$ code over $F = \mathrm{GF}(q)$ and let $g(x)$ be the generator polynomial of $\mathcal{C}$. It follows from Proposition 8.2 that $g(x) \mid x^n - 1$ and, so, $\gcd(g(x), x) = 1$. By Proposition 8.5 we have

$$\exp g(x) \mid n .$$

Suppose first that $\exp g(x) < n$. In this case, the word $x^{\exp g(x)} - 1$ is a codeword in $\mathcal{C}$, which means that the minimum distance of $\mathcal{C}$ is at most 2.

Hence, we will be mainly interested in the case where $\exp g(x) = n$. In addition, we will assume in this section that

$$\gcd(n, q) = 1 .$$

This, in turn, implies that $\gcd(\exp g(x), q) = 1$.

The next proposition, which generalizes Proposition 7.19, will be used in the sequel to obtain some properties of the roots of generator polynomials.

Proposition 8.6 *Let $a(x)$ be a polynomial over $F = \mathrm{GF}(q)$ such that $\gcd(a(x), x) = 1$ and let $e = \exp a(x)$ be such that $\gcd(e, q) = 1$. Then the following conditions hold:*

(i) The roots of $a(x)$ in any extension field of F are simple.

(ii) The splitting field of $a(x)$ over F is $\mathrm{GF}(q^m)$, where m is the smallest positive integer such that $e \mid q^m - 1$.

Proof. From Proposition 7.19 we get that all the roots of $x^e - 1$ in any extension field of F are simple, and the same must therefore hold for the roots of the divisor $a(x)$ of $x^e - 1$. This proves part (i).

Turning to part (ii), $a(x)$ has $\deg a(x)$ (simple) roots in $\mathrm{GF}(q^m)$ if and only if $a(x) \mid x^{q^m} - x$, or, equivalently, $a(x) \mid x^{q^m - 1} - 1$. By Proposition 8.5, the latter condition holds if and only if $e \mid q^m - 1$. □

Let $\mathcal{C}$ be a cyclic $[n, k]$ code over $F = \mathrm{GF}(q)$. The *roots* of $\mathcal{C}$ are the roots of its generator polynomial $g(x)$ in the splitting field $\Phi = \mathrm{GF}(q^m)$ of $g(x)$ over F.

If $\gcd(n, q) = 1$ then, by Proposition 8.6, $g(x)$ has $n-k$ distinct roots $\beta_1, \beta_2 \ldots, \beta_{n-k}$ in Φ and

$$g(x) = \prod_{\ell=1}^{n-k} (x - \beta_\ell) .$$

Now, an element in Φ is a root of $g(x)$ if and only if so are all its conjugates (with respect to F). Therefore, we can partition the set of roots of $\mathcal{C}$ into distinct conjugacy classes

$$\{\beta_1, \beta_2, \ldots, \beta_{n-k}\} = C_{\gamma_1} \cup C_{\gamma_2} \cup \ldots \cup C_{\gamma_t} ,$$

where C_{γ_i} is a conjugacy class that contains (and is represented by) the root γ_i of $\mathcal{C}$. Hence,

$$g(x) = \prod_{i=1}^{t} M_{\gamma_i}(x) ,$$

where $M_{\gamma_i}(x)$ is the minimal polynomial (with respect to F) of each element in C_{γ_i}. The extension degree m of Φ and the number of conjugacy classes t can be related to n and k by

$$n - k = \deg g(x) = \sum_{i=1}^{t} \deg M_{\gamma_i}(x) \leq tm \ .$$

Observe that for every $c(x) \in F_n[x]$,

$$c(x) \in \mathcal{C} \qquad \Longleftrightarrow \qquad M_{\gamma_i}(x) \mid c(x) \quad \text{for } i = 1, 2, \ldots, t \ ;$$

equivalently,

$$c(x) \in \mathcal{C} \qquad \Longleftrightarrow \qquad c(\gamma_i) = 0 \quad \text{for } i = 1, 2, \ldots, t \ .$$

Thus, we can obtain a $tm \times n$ parity-check matrix of $\mathcal{C}$ by representing each entry in the matrix

$$\begin{pmatrix} 1 & \gamma_1 & \gamma_1^2 & \ldots & \gamma_1^{n-1} \\ 1 & \gamma_2 & \gamma_2^2 & \ldots & \gamma_2^{n-1} \\ \vdots & \vdots & \vdots & \vdots & \vdots \\ 1 & \gamma_t & \gamma_t^2 & \ldots & \gamma_t^{n-1} \end{pmatrix}$$

as a column vector in F^m, according to some basis of Φ over F.

Example 8.6 Let $\mathcal{C}$ be a cyclic code of length 7 over $F = \mathrm{GF}(2)$ with a generator polynomial $g(x) = x^4 + x^3 + x^2 + 1$. The polynomial $g(x)$ factors over F into $g(x) = (x - 1)(x^3 + x + 1)$. From the factorization of $x^8 - x$ $(= x(x^7 - 1))$ over F, as seen in Example 7.4, we get that

$$x^7 - 1 = (x - 1)(x^3 + x + 1)(x^3 + x^2 + 1) \ ,$$

namely, $g(x)$ divides $x^7 - 1$. Therefore, $\exp g(x) \mid 7$, which means that $\exp g(x)$ actually equals 7. The splitting field of $g(x)$ over F is $\Phi = \mathrm{GF}(2^3)$, since 3 is the smallest positive integer m such that $7 \mid 2^m - 1$.

Denote by α a root of $x^3 + x + 1$ in $\mathrm{GF}(2^3)$ (in which case $M_\alpha(x) = x^3 + x + 1$). The roots of $\mathcal{C}$, partitioned into conjugacy classes, are given by

$$\{1\} \cup \{\alpha, \alpha^2, \alpha^4\} \ .$$

A parity-check matrix (over F) of $\mathcal{C}$ can be obtained by representing each entry in the matrix

$$\begin{pmatrix} 1 & 1 & 1 & 1 & 1 & 1 & 1 \\ 1 & \alpha & \alpha^2 & \alpha^3 & \alpha^4 & \alpha^5 & \alpha^6 \end{pmatrix}$$

as a column vector in F^3. The elements of the first row (which are all represented by the same nonzero vector in F^3) will contribute only 1 to the rank of the parity-check matrix of $\mathcal{C}$. So, that matrix will have rank at most 4; in fact, since $n-k = \deg g(x) = 4$, the rank will be exactly 4. The parity-check matrix obtained when representing GF(2^3) by $F[\xi]/(\xi^3 + \xi + 1)$ and taking $\alpha = \xi$ is

$$\begin{pmatrix} 1 & 1 & 1 & 1 & 1 & 1 & 1 \\ 0 & 0 & 0 & 0 & 0 & 0 & 0 \\ 0 & 0 & 0 & 0 & 0 & 0 & 0 \\ \hline 1 & 0 & 0 & 1 & 0 & 1 & 1 \\ 0 & 1 & 0 & 1 & 1 & 1 & 0 \\ 0 & 0 & 1 & 0 & 1 & 1 & 1 \end{pmatrix},$$

and the two all-zero rows can obviously be deleted.

The generator polynomial $g(x)$ can be obtained by reversing the order of coefficients of the check polynomial in Example 8.5. Hence, by Corollary 8.4, the code in that example is the dual code of $\mathcal{C}$. □

8.4 BCH codes as cyclic codes

We have established in Example 8.3 that BCH codes are cyclic, so now we can analyze them as such. Let $F = \mathrm{GF}(q)$ and let n be a positive integer such that $\gcd(n, q) = 1$. Let $\Phi = \mathrm{GF}(q^m)$ be the splitting field of $x^n - 1$ over F; by Proposition 8.6 (or Proposition 7.19), m is the smallest positive integer such that $n \,|\, q^m - 1$. Also, let α be an element of multiplicative order n in Φ and let b and D be integers where $0 < D \leq n$. These parameters define a BCH code $\mathcal{C}_{\mathrm{BCH}}$ over F, which consists of all polynomials $c(x) \in F_n[x]$ such that

$$c(\alpha^\ell) = 0 \quad \text{for } \ell = b, b+1, \ldots, b+D-2 .$$

Equivalently, $\mathcal{C}_{\mathrm{BCH}}$ is a cyclic $[n, k]$ code over F whose set of roots consists of the $D-1$ elements of the *consecutive root sequence*

$$\alpha^b, \alpha^{b+1}, \ldots, \alpha^{b+D-2} \tag{8.5}$$

and their conjugates (see Section 5.6).

The elements in the consecutive root sequence (8.5) are the roots of the underlying $[n, n-D+1, D]$ RS code $\mathcal{C}_{\mathrm{RS}}$ over Φ, from which the BCH code $\mathcal{C}_{\mathrm{BCH}}$ was originally derived in Section 5.6 as the intersection $\mathcal{C}_{\mathrm{RS}} \cap F^n$. The latter characterization of $\mathcal{C}_{\mathrm{BCH}}$ also implies that when $k > 0$, the (true) minimum distance of $\mathcal{C}_{\mathrm{BCH}}$ is bounded from below by the designed minimum distance D (which is also the true minimum distance of $\mathcal{C}_{\mathrm{RS}}$). On the other hand, due to the closure under conjugacy, there may be roots of $\mathcal{C}_{\mathrm{BCH}}$ that do not belong to the sequence (8.5); such roots will be referred to as *excess roots* of the BCH code $\mathcal{C}_{\mathrm{BCH}}$.

The generator polynomial of $\mathcal{C}_{\mathrm{BCH}}$ is given by

$$g(x) = \prod_{C_{\alpha^\ell}} M_{\alpha^\ell}(x) ,$$

where the product is taken over the *distinct* conjugacy classes that contain the consecutive root sequence of $\mathcal{C}_{\mathrm{BCH}}$. Therefore,

$$n - k = \deg g(x) = \sum_{C_{\alpha^\ell}} \deg M_{\alpha^\ell}(x) .$$

Now, while the designed minimum distance D of $\mathcal{C}_{\mathrm{BCH}}$ is determined only by the number of elements in the consecutive root sequence (8.5), the dimension k depends also on the number of excess roots: the fewer we have of the latter, the larger k becomes. Specifically, the number of elements in (8.5) is $D-1$, and the number of excess roots equals

$$\deg g(x) - (D-1) = (n-D+1) - k .$$

Since (the set of elements in) the consecutive root sequence of $\mathcal{C}_{\mathrm{BCH}}$ is contained in no more than $D-1$ distinct conjugacy classes, we have

$$n - k \leq (D-1)m .$$

For the special case $q = 2$, $b = 1$, and $D = 2t + 1$ (i.e., binary narrow-sense BCH codes with an odd designed minimum distance), the consecutive root sequence consists of elements that belong to the t conjugacy classes

$$C_\alpha \cup C_{\alpha^3} \cup \ldots \cup C_{\alpha^{D-2}}$$

and, so,

$$n - k \leq tm = \frac{(D-1)m}{2} . \tag{8.6}$$

Similarly, for $q = 2$, $b = 0$, and $D = 2t$ (i.e., binary normalized BCH codes with an even designed minimum distance), the consecutive root sequence comprises elements which are all in

$$C_1 \cup C_\alpha \cup C_{\alpha^3} \cup \ldots \cup C_{\alpha^{D-3}} ,$$

where $\deg M_1(x) = |C_1| = 1$. Hence here

$$n - k \leq 1 + (t-1)m = 1 + \Bigl(\frac{D}{2} - 1\Bigr)m . \tag{8.7}$$

The bounds (8.6) and (8.7) are familiar from Examples 5.5 and 5.6, where we obtained them in the more general context of binary alternant codes.

Example 8.7 We construct a BCH code $\mathcal{C}_{\mathrm{BCH}}$ of length $n = 15$ over $F = \mathrm{GF}(2)$ with $b = 1$ and $D = 7$. Let α be an element of $\mathrm{GF}(2^4)$ whose minimal polynomial is

$$M_\alpha(x) = x^4 + x + 1 .$$

This polynomial is primitive over F; so, α is primitive in $\mathrm{GF}(2^4)$, namely, it has multiplicative order $n = 15$. The (designed) consecutive root sequence of $\mathcal{C}_{\mathrm{BCH}}$ is

$$\alpha, \alpha^2, \alpha^3, \alpha^4, \alpha^5, \alpha^6 ,$$

which means that the set of roots of $\mathcal{C}_{\mathrm{BCH}}$ consists of the union of the following conjugacy classes:

$$C_\alpha = \{\underline{\alpha}, \underline{\alpha}^2, \underline{\alpha}^4, \alpha^8\} , \quad C_{\alpha^3} = \{\underline{\alpha}^3, \underline{\alpha}^6, \alpha^{12}, \alpha^9\} , \quad C_{\alpha^5} = \{\underline{\alpha}^5, \alpha^{10}\}$$

(the underlined elements are part of the consecutive root sequence, and there are four excess roots, namely, α^8, α^9, α^{10}, and α^{12}). The four elements in C_{α^3} are non-primitive in $\mathrm{GF}(2^4)$ and, therefore, $M_{\alpha^3}(x)$ is given by the (only) non-primitive irreducible polynomial of degree 4 over F:

$$M_{\alpha^3}(x) = x^4 + x^3 + x^2 + x + 1 .$$

Similarly,

$$M_{\alpha^5}(x) = x^2 + x + 1 ,$$

which is the only irreducible polynomial of degree $|C_{\alpha^5}| = 2$ over F.

Thus, the generator polynomial of $\mathcal{C}_{\mathrm{BCH}}$ is given by

$$\begin{aligned} g(x) &= M_\alpha(x) M_{\alpha^3}(x) M_{\alpha^5}(x) \\ &= (x^4 + x + 1)(x^4 + x^3 + x^2 + x + 1)(x^2 + x + 1) \\ &= x^{10} + x^8 + x^5 + x^4 + x^2 + x + 1 ; \end{aligned}$$

that is, the Hamming weight of $g(x)$ is 7 and $n - k = \deg g(x) = 10$ (notice that in this case, the bound (8.6) is not tight). We conclude that the BCH code $\mathcal{C}_{\mathrm{BCH}}$ is a cyclic $[15, 5, 7]$ code.

In comparison, by the sphere-packing bound, every linear $[n{=}15, k, 7]$ code over F must satisfy

$$2^{n-k} \ge \binom{15}{0} + \binom{15}{1} + \binom{15}{2} + \binom{15}{3} = 576 ,$$

yielding the lower bound $n - k \ge 10$; this bound is attained by $\mathcal{C}_{\mathrm{BCH}}$.

A parity-check matrix of $\mathcal{C}_{\mathrm{BCH}}$ can be obtained by representing each entry in the 3×15 matrix

$$H = \begin{pmatrix} 1 & \alpha & \alpha^2 & \alpha^3 & \alpha^4 & \alpha^5 & \alpha^6 & \alpha^7 & \alpha^8 & \alpha^9 & \alpha^{10} & \alpha^{11} & \alpha^{12} & \alpha^{13} & \alpha^{14} \\ 1 & \alpha^3 & \alpha^6 & \alpha^9 & \alpha^{12} & 1 & \alpha^3 & \alpha^6 & \alpha^9 & \alpha^{12} & 1 & \alpha^3 & \alpha^6 & \alpha^9 & \alpha^{12} \\ 1 & \alpha^5 & \alpha^{10} & 1 & \alpha^5 & \alpha^{10} & 1 & \alpha^5 & \alpha^{10} & 1 & \alpha^5 & \alpha^{10} & 1 & \alpha^5 & \alpha^{10} \end{pmatrix}$$

as a column vector in F^4, resulting in a 12×15 matrix over F whose rank is $n-k=10$. Each of the first two rows of H contributes (at most) 4 to that rank, while the third row contributes only 2: the elements in this row all belong to the subfield $\mathrm{GF}(2^2)$ and so, their representations as column vectors in F^4 span a linear space of dimension (at most) 2. □

The generator polynomial in Example 8.7 is also a minimum-weight nonzero codeword of the code. While this happens in many cyclic codes, there are codes—including BCH codes—where it does not (but see Problem 8.5).

8.5 The BCH bound

The following result provides a lower bound on the minimum distance of cyclic codes.

Proposition 8.7 (The BCH bound) *Let $\mathcal{C}$ be a cyclic $[n,k,d]$ code over $F=\mathrm{GF}(q)$ where $\gcd(n,q)=1$. Let α be an element of multiplicative order n in the splitting field of x^n-1 over F and suppose that $\alpha^b, \alpha^{b+1}, \ldots, \alpha^{b+D-2}$ belong to the set of roots of $\mathcal{C}$ for some integers b and $D \ge 2$. Then $d \ge D$.*

Proof. The result follows by observing that $\mathcal{C}$ is a subset of the BCH code $\mathcal{C}_{\mathrm{BCH}}$ over F defined by the consecutive root sequence $\alpha^b, \alpha^{b+1}, \ldots, \alpha^{b+D-2}$; the minimum distance of $\mathcal{C}_{\mathrm{BCH}}$, in turn, is at least D. □

Example 8.8 Let $\mathcal{C}$ be a cyclic $[2^m-1, 2^m-1-m, d]$ code over $F=\mathrm{GF}(2)$ whose generator polynomial is a primitive polynomial $P(x)$ of degree m over F (a special case of this construction, for $m=3$, was presented in Example 8.5). The polynomial $P(x)$ is a minimal polynomial of an element α of multiplicative order 2^m-1 in the splitting field, $\mathrm{GF}(2^m)$, of $x^{2^m-1}-1$ over F. The roots of $\mathcal{C}$ form the conjugacy class

$$\{\underline{\alpha}, \underline{\alpha^2}, \alpha^4, \ldots, \alpha^{2^{m-1}}\} ,$$

and the underlined elements allow us to apply the BCH bound with $b=1$ and $D=3$, thereby yielding the lower bound $d \ge 3$. In fact, $\mathcal{C}$ is a Hamming code and d is exactly 3. □

Example 8.9 Let $F=\mathrm{GF}(q)$ and $n=(q^m-1)/(q-1)$, where $m>1$. Clearly, $\gcd(n,q)=1$ and the splitting field of x^n-1 over F is $\Phi=\mathrm{GF}(q^m)$. We further assume here that $\gcd(n, q-1)=1$.

Consider a cyclic code $\mathcal{C}$ of length n over F whose generator polynomial, $g(x)$, is a minimal polynomial of an element $\beta \in \mathrm{GF}(q^m)$ whose multiplicative order is n. The set of roots of $g(x)$ is given by

$$C_\beta = \{\beta, \beta^q, \beta^{q^2}, \ldots, \beta^{q^{m-1}}\} .$$

A direct application of the BCH bound to this set (with $b = 1$ and $D = 2$) would yield a lower bound of 2 on the minimum distance d of $\mathcal{C}$. Yet, this lower bound can be improved if we express the roots of $g(x)$ as powers of the element $\alpha = \beta^{q-1}$; note that by Problem A.9, the multiplicative order of α is also n:

$$\mathcal{O}(\alpha) = \frac{\mathcal{O}(\beta)}{\gcd(\mathcal{O}(\beta), q{-}1)} = \frac{n}{\gcd(n, q{-}1)} = n .$$

Specifically, let s be the multiplicative inverse of $q{-}1$ in the ring of integer residues modulo n; such an inverse indeed exists since $\gcd(n, q{-}1) = 1$. Then,

$$\beta = (\beta^{q-1})^s = \alpha^s$$

and

$$\beta^q = \beta^{q-1} \cdot \beta = \alpha \cdot \alpha^s = \alpha^{s+1} .$$

We now apply the BCH bound with $b = s$ and $D = 3$ to yield $d \geq 3$. The code $\mathcal{C}$ coincides with the cyclic Hamming code in Example 8.4 (see also Problem 8.19). □

Example 8.10 Let $P(x)$ be a primitive polynomial of degree $m \geq 3$ over $F = \mathrm{GF}(2)$ and let $\widehat{P}(x)$ be a polynomial of degree m over F obtained from $P(x)$ by reversing the order of coefficients of $P(x)$; namely, $\widehat{P}(x) = x^m P(x^{-1})$. It is easy to see that $\widehat{P}(x)$ is also primitive: if $P(x)$ is a minimal polynomial of a (primitive) element $\alpha \in \mathrm{GF}(2^m)$, then $\widehat{P}(x)$ is a minimal polynomial of the (primitive) element α^{-1} (see Problem 7.3). Note that when $m \geq 3$, the element α^{-1} is not in the conjugacy class $C_\alpha = \{\alpha, \alpha^2, \ldots, \alpha^{2^{m-1}}\}$; this means that $P(x)$ and $\widehat{P}(x)$ are distinct polynomials and, in particular, the product $P(x)(x+1)\widehat{P}(x)$ divides $x^{2^m-1} - 1$.

Let $\mathcal{C}$ be a cyclic $[2^m{-}1, k, d]$ code over $F = \mathrm{GF}(2)$ whose generator polynomial is $g(x) = P(x)(x+1)\widehat{P}(x)$. Clearly, $\deg g(x) = 2m+1$ and, so, $k = 2^m - 2m - 2$ (which is positive when $m > 3$). The set of roots of $\mathcal{C}$ is given by

$$C_\alpha \cup \{1\} \cup C_{\alpha^{-1}} = \{\underline{\alpha}, \underline{\alpha}^2, \ldots, \alpha^{2^{m-1}}\} \cup \{\underline{1}\} \cup \{\underline{\alpha}^{-1}, \underline{\alpha}^{-2}, \ldots, \alpha^{-2^{m-1}}\} .$$

Applying the BCH bound with $b = -2$ and $D = 6$ yields the lower bound $d \geq 6$. Hence, $\mathcal{C}$ is a cyclic $[2^m{-}1, 2^m{-}2m{-}2, {\geq}6]$ code over F for every $m > 3$. □

Example 8.11 Let $\mathcal{C}$ be a cyclic $[9, 2, d]$ code over $F = \mathrm{GF}(2)$ with the generator polynomial

$$g(x) = x^7 + x^6 + x^4 + x^3 + x + 1 = (x + 1)(x^6 + x^3 + 1)$$

(it is easy to verify that $g(x)$ divides $x^9 - 1$ and, therefore, such a code $\mathcal{C}$ indeed exists). Next, we find the value of d.

By Proposition 8.6, the splitting field of $x^9 - 1$ over F is $\mathrm{GF}(2^6)$. The roots of $x^9 - 1$ in this field are the powers of an element α of multiplicative order 9 in $\mathrm{GF}(2^6)$. Those powers belong to three conjugacy classes as follows:

$$C_1 = \{1\}, \quad C_\alpha = \{\alpha, \alpha^2, \alpha^4, \alpha^8, \alpha^7, \alpha^5\}, \quad C_{\alpha^3} = \{\alpha^3, \alpha^6\} .$$

The elements in C_{α^3} have multiplicative order 3, which means that these elements belong to $\mathrm{GF}(2^2)$ and, as such, their minimal polynomial is $x^2 + x + 1$. Obviously, the minimal polynomial of the element in C_1 is $x - 1$. It follows that the minimal polynomial of the elements in C_α is given by

$$\frac{x^9 - 1}{(x - 1)(x^2 + x + 1)} = \frac{x^9 - 1}{x^3 - 1} = x^6 + x^3 + 1 .$$

We conclude that the set of roots of $\mathcal{C}$ is $C_1 \cup C_\alpha$. In particular, the elements $\alpha^7, \alpha^8, 1, \alpha, \alpha^2$ are roots of $\mathcal{C}$; thus, by the BCH bound we obtain $d \geq 6$. In fact, we have equality since the Hamming weight of $g(x)$ is 6. □

Example 8.12 The binary Golay code, which was defined in the notes on Section 4.2, is a cyclic $[23, 12, 7]$ code over $F = \mathrm{GF}(2)$. We next check how the BCH bound performs on this code. By Proposition 8.6, the splitting field of $x^{23} - 1$ over F is $\mathrm{GF}(2^{11})$. Let α be an element of multiplicative order 23 in this field. The conjugacy class (with respect to F) of α is given by

$$C_\alpha = \{\underline{\alpha}, \underline{\alpha^2}, \underline{\alpha^4}, \alpha^8, \alpha^{16}, \alpha^9, \alpha^{18}, \alpha^{13}, \underline{\alpha^3}, \alpha^6, \alpha^{12}\} \tag{8.8}$$

and the remaining powers of α belong to $C_1 = \{1\}$ and

$$C_{\alpha^{-1}} = \{\alpha^5, \alpha^{10}, \underline{\alpha^{20}}, \alpha^{17}, \alpha^{11}, \underline{\alpha^{22}}, \underline{\alpha^{21}}, \underline{\alpha^{19}}, \alpha^{15}, \alpha^7, \alpha^{14}\} \tag{8.9}$$

(here $\alpha^{-1} = \alpha^{22}$). Hence, the irreducible factorization of $x^{23} - 1$ over F is given by

$$x^{23} - 1 = M_\alpha(x)(x + 1)M_{\alpha^{-1}}(x) ,$$

where $\deg M_\alpha(x) = \deg M_{\alpha^{-1}}(x) = 11$; in fact, $M_{\alpha^{-1}}(x)$ is obtained by reversing the order of coefficients of $M_\alpha(x)$ (Problem 7.3). It can be verified that the factors of $x^{23} - 1$ of degree 11 are

$$x^{11} + x^{10} + x^6 + x^5 + x^4 + x^2 + 1 \quad \text{and} \quad x^{11} + x^9 + x^7 + x^6 + x^5 + x + 1 .$$

The generator polynomial $g(x)$ of the binary Golay code, being of degree $23 - 12 = 11$ and dividing $x^{23} - 1$, must therefore equal either $M_\alpha(x)$ or $M_{\alpha^{-1}}(x)$. In either case, the BCH bound yields a lower bound of 5 on the minimum distance (the underlined elements in (8.8) or in (8.9) form consecutive root sequences of length 4). However, the true minimum distance of the binary Golay code is 7, which makes it a perfect code. □

Example 8.13 We analyze the ternary Golay code, which is a cyclic $[11, 6, 5]$ code over $F = \mathrm{GF}(3)$ (also mentioned in the notes on Section 4.2). Let α be an element of multiplicative order 11 in the field $\mathrm{GF}(3^5)$, which is the splitting field of $x^{11} - 1$ over F. The powers of α belong to three conjugacy classes (with respect to F), namely,

$$C_\alpha = \{\alpha, \underline{\alpha}^3, \alpha^9, \underline{\alpha}^5, \underline{\alpha}^4\}, \quad C_1 = \{1\}, \quad \text{and} \quad C_{\alpha^{-1}} = \{\alpha^2, \underline{\alpha}^6, \underline{\alpha}^7, \alpha^{10}, \underline{\alpha}^8\}.$$

The irreducible factorization of $x^{11} - 1$ over F is given by

$$\begin{aligned} x^{11} - 1 &= M_\alpha(x)(x-1)M_{\alpha^{-1}}(x) \\ &= (x^5 + x^4 - x^3 + x^2 - 1)(x-1)(x^5 - x^3 + x^2 - x - 1), \end{aligned}$$

and the generator polynomial of the code is either $M_\alpha(x)$ or $M_{\alpha^{-1}}(x)$. While the BCH bound yields a lower bound of 4 on the minimum distance in this case, the true minimum distance turns out to be 5. Thus, the ternary Golay code is a perfect code. □

Problems

[Section 8.2]

Problem 8.1 Show that if $g(x)$ is a polynomial over F that divides $x^n - 1$, then the set

$$\mathcal{C} = \{ u(x)g(x) \ : \ u(x) \in F_{n-\deg g}[x] \}$$

is a cyclic code over F (what is the generator polynomial of $\mathcal{C}$?).

Problem 8.2 Let $\mathcal{C}$ be a cyclic $[n, k{>}0]$ code over F. Show that every set of k consecutive columns in every generator matrix of $\mathcal{C}$ is linearly independent. Deduce that $\mathcal{C}$ has a systematic generator matrix.

Problem 8.3 Let $\mathcal{C}_1$ and $\mathcal{C}_2$ be cyclic codes of length n over F and let $g_1(x)$ and $g_2(x)$ be their generator polynomials, respectively. Show that each of the following sets is a cyclic code of length n over F and find its generator polynomial.

1. $\mathcal{C}_1 \cap \mathcal{C}_2$.
2. $\mathcal{C}_1 + \mathcal{C}_2 = \{ \mathbf{c}_1 + \mathbf{c}_2 \ : \ \mathbf{c}_1 \in \mathcal{C}_1 \text{ and } \mathbf{c}_2 \in \mathcal{C}_2 \}$.
3. $\{ c(x) \in F_n[x] \ : \ c(x) \equiv g_2(x)c_1(x) \pmod{x^n - 1} \text{ for some } c_1(x) \in \mathcal{C}_1 \}$.

Hint: In part 3, assume first that $\gcd(g_1(x), g_2(x)) = 1$.

Problem 8.4 Let $\mathcal{C}$ be a cyclic $[n, k]$ code over F.

1. Find the largest integer t such that every burst of length up to t that occurs in a codeword of $\mathcal{C}$ can be detected.

   (Recall from Problem 2.21 that a burst of length ℓ is the event of having errors in a codeword such that the locations i and j of the first and last errors, respectively, satisfy $j-i = \ell-1$.)

2. What is the respective value of t so that every burst *erasure* of length up to t can be recovered?

   (A burst erasure of length ℓ is the event where erasures occur in ℓ consecutive locations within a codeword.)

Problem 8.5 Let $\mathcal{C}$ be an $[n, k, d=n-k+1]$ RS code over $F = \mathrm{GF}(q)$ with a generator polynomial

$$g(x) = (x - \alpha^b)(x - \alpha^{b+1}) \cdots (x - \alpha^{b+d-2}) ,$$

where α is an element of multiplicative order n in F and b is an integer.

1. Show that the coefficients of $1, x, x^2, \ldots, x^{n-k}$ in $g(x)$ are all nonzero.

   Hint: Do *not* expand the expression for $g(x)$.

2. Assuming that $d > 1$, show that the dual code of $\mathcal{C}$ is an $[n, d-1, k+1]$ RS code over F with a generator polynomial

$$g^{\perp}(x) = (x - \alpha^{1-b})(x - \alpha^{2-b}) \cdots (x - \alpha^{k-b}) .$$

Problem 8.6 (Cyclic codes of length q over $\mathrm{GF}(q)$) Let $F = \mathrm{GF}(q)$ and let $\mathcal{C}$ be a cyclic $[q, k]$ code over F. Denote by p the characteristic of F.

1. Show that the generator polynomial of $\mathcal{C}$ is $(x - 1)^{q-k}$ (therefore, $\mathcal{C}$ is the unique cyclic $[q, k]$ code over F).

2. Show that for $i = 0, 1, \ldots, p-1$, the polynomial

$$c_i(x) = \sum_{j=0}^{q-1} j^i x^j$$

   (where $0^0 = 1$) is divisible by $(x - 1)^{q-i-1}$.

   Hint: Verify that for $i > 0$, the polynomial $c_i(x)$ is related to the formal derivative of $c_{i-1}(x)$ by

$$c_i(x) = x \cdot c'_{i-1}(x) .$$

   Then apply induction.

3. Show that when $q = p$, a generator matrix of $\mathcal{C}$ is given by

$$G = \begin{pmatrix} 1 & 1 & 1 & \ldots & 1 \\ 0 & 1 & 2 & \ldots & p-1 \\ 0 & 1^2 & 2^2 & \ldots & (p-1)^2 \\ \vdots & \vdots & \vdots & \vdots & \vdots \\ 0 & 1^{k-1} & 2^{k-1} & \ldots & (p-1)^{k-1} \end{pmatrix}$$

   (namely, $\mathcal{C}$ is a singly-extended GRS code).

4. Show that when $q \neq p$, the code $\mathcal{C}$ is MDS if and only if $k \in \{1, q-1, q\}$.

   Hint: Distinguish between two ranges of k. Assume first that $q/p \leq k \leq q-(q/p)$ and show that in this case, the code $\mathcal{C}$ contains the codeword $(x^{q/p}-1)^{p-1}$ (what is the Hamming weight of this codeword?). Then assume that $k < q/p$ or $k > q-(q/p)$ and show that either $\mathcal{C}$ or $\mathcal{C}^{\perp}$ contains the codeword $x^{q/p}-1$.

[Section 8.3]

Problem 8.7 Let $\mathcal{C}_1$ be a cyclic $[15, 15-r]$ code over $F = \mathrm{GF}(2)$ and let $\mathcal{C}_2$ be a cyclic $[17, 17-r]$ code over F (with the same redundancy as $\mathcal{C}_1$). The following parts will lead to the values of r for which such code pairs $(\mathcal{C}_1, \mathcal{C}_2)$ indeed exist.

1. Find the splitting field of the polynomial $x^{17}-1$ over F.
2. Let α be a root of $x^{17}-1$ other than 1 in the splitting field of $x^{17}-1$ over F. Find the conjugates of α and determine the degree of the minimal polynomial of α with respect to F.
3. Find the degrees of the irreducible factors of $x^{17}-1$ over F and how many irreducible factors there are of each degree.
4. Show that there exist pairs of cyclic codes $(\mathcal{C}_1, \mathcal{C}_2)$ that satisfy the given specifications when $r = 9$. What is the number of such pairs?

   Hint: The degrees of the irreducible factors of $x^{15}-1$ over F can be deduced from Example 7.4.
5. Find all other values of r for which pairs $(\mathcal{C}_1, \mathcal{C}_2)$ exist.

Problem 8.8 Let F be the field $\mathrm{GF}(2^2)$ and $\mathcal{C}$ be a cyclic $[85, 81, d]$ code over F. Denote by $g(x)$ the generator polynomial of $\mathcal{C}$ and by Φ the splitting field of $g(x)$ over F.

1. Show that $g(x)$ is either irreducible over F or is a product of two irreducible polynomials of degree 2 over F.
2. For each one of the cases in part 1, determine the possible values of the multiplicative orders of the roots of $g(x)$ in Φ.
3. In which of the cases in part 1 is x^5+1 a codeword of $\mathcal{C}$?
4. Identify the field Φ in the case where the minimum distance d is 3.
5. Determine the possible values of the multiplicative orders of the roots of $g(x)$ in Φ in the case where $d = 3$.

[Section 8.4]

Problem 8.9 Let α be a primitive element in $\Phi = \mathrm{GF}(2^m)$ and let $\mathcal{C}_{\mathrm{BCH}}$ be a narrow-sense primitive BCH code (of length $n = 2^m-1$) over $F = \mathrm{GF}(2)$ whose parity-check matrix is obtained by representing each entry in

$$\begin{pmatrix} 1 & \alpha & \alpha^2 & \ldots & \alpha^{n-1} \\ 1 & \alpha^3 & \alpha^6 & \ldots & \alpha^{3(n-1)} \end{pmatrix}$$

as a column vector in F^m. For the given code $\mathcal{C}_{\mathrm{BCH}}$, let $\mathcal{C}_{\mathrm{RS}}$ be an $[N{=}n, K, D]$ code over Φ that satisfies

$$\mathcal{C}_{\mathrm{BCH}} = \mathcal{C}_{\mathrm{RS}} \cap F^N .$$

1. Show that α^2, α^4, and α^6 are roots of $\mathcal{C}_{\mathrm{BCH}}$.

2. Write a parity-check matrix of $\mathcal{C}_{\mathrm{RS}}$, assuming that $D = 5$ and that α^3 is a root of $\mathcal{C}_{\mathrm{RS}}$.

3. Repeat part 2, except that now $D = 4$ (and α^3 is still a root of $\mathcal{C}_{\mathrm{RS}}$). Is the solution unique?

4. Find the largest possible dimension of any RS code $\mathcal{C}_{\mathrm{RS}}$ over Φ that satisfies $\mathcal{C}_{\mathrm{BCH}} = \mathcal{C}_{\mathrm{RS}} \cap F^N$.

Problem 8.10 Let α be a primitive element in $\Phi = \mathrm{GF}(2^m)$ and let $\mathcal{C}_{\mathrm{BCH}}$ be a normalized primitive BCH code (of length $n = 2^m{-}1$) over $F = \mathrm{GF}(2)$ whose parity-check matrix is obtained by representing each entry in

$$\begin{pmatrix} 1 & 1 & 1 & \ldots & 1 \\ 1 & \alpha & \alpha^2 & \ldots & \alpha^{n-1} \\ 1 & \alpha^3 & \alpha^6 & \ldots & \alpha^{3(n-1)} \end{pmatrix}$$

as a column vector in F^m. Define the codes $\widehat{\mathcal{C}}_{\mathrm{BCH}}$ and $\mathcal{C}$ by

$$\widehat{\mathcal{C}}_{\mathrm{BCH}} = \{(c_0\ c_1\ \ldots\ c_{n-2}\ c_{n-1}) \ :\ (c_{n-1}\ c_{n-2}\ \ldots\ c_1\ c_0) \in \mathcal{C}_{\mathrm{BCH}}\}$$

and

$$\mathcal{C} = \mathcal{C}_{\mathrm{BCH}} \cap \widehat{\mathcal{C}}_{\mathrm{BCH}} .$$

1. Verify that $\widehat{\mathcal{C}}_{\mathrm{BCH}}$ is a BCH code and express its generator polynomial as a function of the generator polynomial of $\mathcal{C}_{\mathrm{BCH}}$.

2. Show that the dimension of $\mathcal{C}$ is at least $2^m - 4m - 2$.

3. Find the exact dimension of $\mathcal{C}$ when $m = 4$, and write the check polynomial in this case.

4. Find the exact dimension of $\mathcal{C}$ when $m = 5$.

Problem 8.11 Let F and Φ be the finite fields $\mathrm{GF}(2^2)$ and $\mathrm{GF}(2^8)$, respectively.

1. Let α be an element in Φ of multiplicative order 17. Find the size of the conjugacy class of α with respect to the field F.

2. Find the splitting field of the polynomial $x^{17} - 1$ over F.

3. Find the degrees of the irreducible factors of $x^{17} - 1$ over F and how many monic irreducible factors there are of each degree.

4. Show that if $g(x) = \sum_{i=0}^{17-k} g_i x^i$ is a generator polynomial of a cyclic code of length 17 over F then $g_i = g_{17-k-i}$ for $i = 0, 1, \ldots, 17{-}k$.

Denote by $\mathcal{C}_{\mathrm{BCH}}$ a BCH code with the largest possible dimension among all BCH codes of length 17 over F with designed minimum distance 7.

5. Express the roots of the code $\mathcal{C}_{\mathrm{BCH}}$ as powers of an element α of multiplicative order 17 in Φ and find the dimension of $\mathcal{C}_{\mathrm{BCH}}$.

6. Let $\mathcal{C}_{\mathrm{RS}}$ be an RS code of length 17 over Φ with the largest possible dimension such that $\mathcal{C}_{\mathrm{BCH}} = \mathcal{C}_{\mathrm{RS}} \cap F^{17}$. Find the dimension of $\mathcal{C}_{\mathrm{RS}}$ and compute the possible sets of roots of $\mathcal{C}_{\mathrm{RS}}$ (as powers of α).

Problem 8.12 Let $\mathcal{C}_{\mathrm{BCH}}$ be an $[n{=}q^m{-}1, k, d{\ge}D]$ primitive BCH code over $F = \mathrm{GF}(q)$ with designed minimum distance D.

1. Show that when $\mathcal{C}_{\mathrm{BCH}}$ is a narrow-sense BCH code and $D \le q^{\lceil m/2 \rceil} + 1$ then

$$n - k = \left\lceil \frac{q-1}{q}(D-1) \right\rceil m \,.$$

Hint: See Problem 7.1.

2. Show that when $\mathcal{C}_{\mathrm{BCH}}$ is normalized and $D \le q^{\lceil m/2 \rceil} + 2$ then

$$n - k = 1 + \left\lceil \frac{q-1}{q}(D-2) \right\rceil m \,.$$

Problem 8.13 The purpose of this problem is to show that shortened binary Reed–Muller codes are cyclic codes.

Let $F = \mathrm{GF}(2)$ and for $0 \le r \le m$, denote by $\mathcal{S}(m,r)$ the set of all words in F^m whose Hamming weight is at most r. Recall from Problem 2.19 that the rth order Reed–Muller code of length 2^m over F, denoted by $\mathcal{C}_{\mathrm{RM}}(m,r)$, is generated by the following $|\mathcal{S}(m,r)| \times 2^m$ matrix $G_{\mathrm{RM}}(m,r)$ over F: for every $\mathbf{e} = (e_0\ e_1\ \ldots\ e_{m-1})$ in $\mathcal{S}(m,r)$ and $\mathbf{a} = (a_0\ a_1\ \ldots\ a_{m-1})$ in F^m, the entry of $G_{\mathrm{RM}}(m,r)$ that is indexed by $(\mathbf{e}, \mathbf{a})$ is given by

$$\mathbf{a}^{\mathbf{e}} = a_0^{e_0} a_1^{e_1} \cdots a_{m-1}^{e_{m-1}}$$

(where $0^0 = 1^0 = 1^1 = 1$ and $0^1 = 0$). It is also known that $(\mathcal{C}_{\mathrm{RM}}(m,r))^{\perp} = \mathcal{C}_{\mathrm{RM}}(m, m{-}r{-}1)$.

Let ξ be an indeterminate over the extension field $\Phi = \mathrm{GF}(2^m)$, and for every vector $\mathbf{a} = (a_0\ a_1\ \ldots\ a_{m-1})$ in F^m associate the polynomial

$$a(\xi) = \sum_{h=0}^{m-1} a_h \xi^h$$

and the integer

$$\imath(\mathbf{a}) = \sum_{h=0}^{m-1} a_h 2^h \,.$$

1. Show that for every $\mathbf{s} \in \mathcal{S}(m,r)$ and $\mathbf{a} \in F^m$,

$$(a(\xi))^{\imath(\mathbf{s})} = \sum_{j=0}^{(m-1)\imath(\mathbf{s})} \Bigl(\sum_{\mathbf{e} \in \mathcal{S}(m,r)} u_{\mathbf{s},\mathbf{e},j} \mathbf{a}^{\mathbf{e}} \Bigr) \xi^j \,,$$

where each $u_{\mathbf{s},\mathbf{e},j}$ is an element of F that depends on $\mathbf{s}$, $\mathbf{e}$, and j, but not on $\mathbf{a}$.

Hint: Writing $\mathbf{s} = (s_0\ s_1\ \ldots\ s_{m-1})$, show that

$$(a(\xi))^{\imath(\mathbf{s})} = \prod_{h\,:\,s_h=1} a(\xi^{2^h}) ,$$

and then check how the coefficients of the powers of ξ in the right-hand side depend on $\mathbf{a}$.

2. Let β be a root in Φ of an irreducible polynomial of degree m over F. Show that for every $\mathbf{s} \in \mathcal{S}(m, r)$ and $\mathbf{a} \in F^m$,

$$(a(\beta))^{\imath(\mathbf{s})} = \sum_{j=0}^{m-1} \Bigl(\sum_{\mathbf{e} \in \mathcal{S}(m,r)} u'_{\mathbf{s},\mathbf{e},j} \mathbf{a}^{\mathbf{e}} \Bigr) \beta^j ,$$

where each $u'_{\mathbf{s},\mathbf{e},j}$ is an element of F that depends on $\mathbf{s}$, $\mathbf{e}$, and j, but not on $\mathbf{a}$.

3. Let H be the $|\mathcal{S}(m, r)| \times 2^m$ matrix over Φ whose rows and columns are indexed by the elements of $\mathcal{S}(m, r)$ and Φ, respectively, and

$$H = \left(\gamma^{\imath(\mathbf{s})}\right)_{\mathbf{s} \in \mathcal{S}(m,r),\ \gamma \in \Phi} .$$

Let β be as in part 2 and denote by H_F the $(m|\mathcal{S}(m, r)|) \times 2^m$ matrix over F that is obtained by representing each entry in H as a column vector in F^m, according to the basis $(1\ \beta\ \beta^2\ \ldots\ \beta^{m-1})$ of Φ over F. Show that—up to a permutation of columns of H_F (or H)—the rows of H_F are spanned by the rows of $G_{\mathrm{RM}}(m, r)$.

Let H and H_F be as in part 3, and let H^* and H_F^* be obtained from H and H_F, respectively, by deleting the column that is indexed by the zero element of Φ. Thus, without loss of generality,

$$H^* = \left(\alpha^{\imath(\mathbf{s}) \cdot j}\right)_{\mathbf{s} \in \mathcal{S}(m,r),\ j \in \{0,1,\ldots,2^m-2\}} ,$$

where α is a primitive element in Φ. It follows from part 3 that up to a permutation of columns, the rows of H_F^* are spanned by the rows of the matrix $G^*_{\mathrm{RM}}(m, r)$, which is obtained from $G_{\mathrm{RM}}(m, r)$ by deleting the column that is indexed by $\mathbf{a} = \mathbf{0}$.

4. Show that H_F^* is a parity-check matrix of a cyclic code of length $2^m - 1$ over F whose set of roots is

$$\left\{\alpha^{\imath(\mathbf{s})}\ :\ \mathbf{s} \in \mathcal{S}(m, r)\right\} .$$

Hint: Show that this set is a union of (whole) conjugacy classes with respect to F.

5. Let $\mathcal{C}^*_{\mathrm{RM}}(m, m-r-1)$ be the cyclic code in part 4. Show that a parity-check matrix of $\mathcal{C}^*_{\mathrm{RM}}(m, m-r-1)$ can be obtained by permuting the columns of $G^*_{\mathrm{RM}}(m, r)$; i.e., $\mathcal{C}^*_{\mathrm{RM}}(m, m-r-1)$ is obtained from $\mathcal{C}_{\mathrm{RM}}(m, m-r-1)$ by a permutation of coordinates and shortening.

Hint: Deduce from part 4 that $\mathrm{rank}(H_F^*) = |\mathcal{S}(m, r)| = \mathrm{rank}(G^*_{\mathrm{RM}}(m, r))$.

6. Show that $\mathcal{C}^*_{\mathrm{RM}}(m, m-r-1)$ is a sub-code of a normalized primitive binary BCH code whose designed minimum distance is 2^r. Deduce that $\mathcal{C}_{\mathrm{RM}}(m, m-r-1)$ is a sub-code of the respective extended primitive BCH code.

[Section 8.5]

Problem 8.14 Let α be an element of multiplicative order 9 in $\Phi = \mathrm{GF}(2^6)$ and let $\mathcal{C}_{\mathrm{RS}}$ be an RS code of length 9 over $\Phi = \mathrm{GF}(2^6)$ whose set of roots is $\{\alpha^2, \alpha^3, \alpha^4\}$. Denote by $\mathcal{C}_{\mathrm{BCH}}$ the BCH code of length 9 over $F = \mathrm{GF}(2^3)$ that is obtained by the intersection $\mathcal{C}_{\mathrm{RS}} \cap F^9$.

1. What is the dimension and minimum distance of $\mathcal{C}_{\mathrm{RS}}$?
2. Find all the roots of $\mathcal{C}_{\mathrm{BCH}}$; express those roots as powers of α.
3. Find the dimension of $\mathcal{C}_{\mathrm{BCH}}$.
4. Show that $\mathcal{C}_{\mathrm{BCH}}$ is MDS.

Problem 8.15 (Cyclic codes of length $q+1$ over $\mathrm{GF}(q)$) Let $F = \mathrm{GF}(q)$ and let α be an element of multiplicative order $q+1$ in $\mathrm{GF}(q^2)$. For $i = 0, 1, \ldots, q$, denote by $M_{\alpha^i}(x)$ the minimal polynomial of α^i over F.

1. Show that $\mathrm{GF}(q^2)$ is the splitting field of the polynomial $x^{q+1} - 1$ over F.
2. Show that $M_{\alpha^i}(x) = (x - \alpha^i)(x - \alpha^{-i})$ for every $i \notin \{0, (q+1)/2\}$.
3. Show that when q is odd, there exist cyclic $[q+1, k]$ MDS codes over F for every odd value of k in the range $1 \le k \le q$.

   Hint: Consider the cyclic $[q+1, k]$ code whose generator polynomial is

$$\prod_{i=0}^{(q-k)/2} M_{\alpha^i}(x) .$$

4. Show that when q is even, there exist cyclic $[q+1, k]$ MDS codes over F for every $k \in \{1, 2, \ldots, q+1\}$. (An example of such a code is presented in Problem 8.14.)

Problem 8.16 Let Φ be the splitting field of the polynomial $x^{13} - 1$ over the field $\mathrm{GF}(3)$.

1. Identify the field Φ.
2. Find all the possible values for the dimension of a cyclic code of length 13 over $\mathrm{GF}(3)$ and the number of such codes for every given dimension.

Let α be an element of multiplicative order 13 in Φ and let $\mathcal{C}$ be a cyclic code with the largest possible dimension among all cyclic codes of length 13 over $\mathrm{GF}(3)$ whose set of roots contains the elements 1, α, α^2, and α^4.

3. Find all the roots of $\mathcal{C}$.
4. Find the dimension of $\mathcal{C}$.

5. Using the BCH bound, compute a lower bound on the minimum distance of $\mathcal{C}$.

6. Using the Griesmer bound from part 2 of Problem 4.4, show that the bound in part 5 is tight.

Problem 8.17 Let $\mathcal{C}$ be the cyclic code of length 15 over $F = \mathrm{GF}(2)$ with a generator polynomial $g(x) = (x^5+1)(x^4+x+1)$ (why does such a cyclic code exist?). Also, let $\widehat{\mathcal{C}}$ be defined by

$$\widehat{\mathcal{C}} = \{(c_0\ c_1\ \ldots\ c_{14}\ c_{15})\ :\ (c_{15}\ c_{14}\ \ldots\ c_1\ c_0) \in \mathcal{C}\}\,.$$

1. Find the dimension of $\mathcal{C}$.

2. Show that the Hamming weight of $g(x)$ attains the BCH bound on the minimum distance of $\mathcal{C}$.

3. Show that the word $(1\,1\,0\,1\,1\,0\,1\,1\,0\,1\,1\,0\,1\,1\,0)$ is contained in the intersection $\mathcal{C} \cap \widehat{\mathcal{C}}$.

4. Find the dimension and minimum distance of $\mathcal{C} \cap \widehat{\mathcal{C}}$.

Problem 8.18 Let α be a primitive element in $\Phi = \mathrm{GF}(2^4)$ and let $\mathcal{C}_{\mathrm{RS}}^{(1)}$ be a primitive RS code over Φ whose set of roots is $\{\alpha^2, \alpha^3, \alpha^4, \alpha^5\}$. Let $\mathcal{C}_{\mathrm{parity}}$ be the parity code of length 15 over $F = \mathrm{GF}(2^2)$ and define the code $\mathcal{C}$ over F by

$$\mathcal{C} = \mathcal{C}_{\mathrm{RS}}^{(1)} \cap \mathcal{C}_{\mathrm{parity}}\,.$$

1. Show that $\mathcal{C}$ is a BCH code over F.

2. Find the set of roots of $\mathcal{C}$ in Φ.

3. Using the BCH bound, show that the minimum distance of $\mathcal{C}$ is at least 7.

4. Find a primitive RS code $\mathcal{C}_{\mathrm{RS}}^{(2)}$ with minimum distance 7 over Φ such that $\mathcal{C} = \mathcal{C}_{\mathrm{RS}}^{(2)} \cap F^{15}$.

5. Let $H_{\mathrm{RS}}^{(1)}$ and $H_{\mathrm{RS}}^{(2)}$ denote canonical parity-check matrices of $\mathcal{C}_{\mathrm{RS}}^{(1)}$ and $\mathcal{C}_{\mathrm{RS}}^{(2)}$, respectively. A codeword of $\mathcal{C}$ is transmitted through a channel (F, F, Prob) and a word $\mathbf{y} = (y_0\ y_1\ \ldots,\ y_{14})$ that contains at most three errors is received. The word $\mathbf{y}$ is corrected using the following decoding steps:

   (i) An element $u \in F$ is computed as the sum $u = \sum_{j=0}^{14} y_j$.

   (ii) A vector $\mathbf{v} \in \Phi^4$ is computed from $\mathbf{y}$ by $\mathbf{v} = H_{\mathrm{RS}}^{(1)} \mathbf{y}^T$.

   (iii) A syndrome polynomial $S(x) = \sum_{\ell=0}^{5} S_\ell x^\ell$ of $\mathbf{y}$, with respect to $H_{\mathrm{RS}}^{(2)}$, is computed from u and $\mathbf{v}$.

   (iv) A decoder for $\mathcal{C}_{\mathrm{RS}}^{(2)}$ for correcting up to three errors is applied to the syndrome polynomial $S(x)$.

   Explain how $S(x)$ can be computed in Step (iii) from u and $\mathbf{v}$ (rather than computing it directly from $\mathbf{y}$).

Problem 8.19 Let $F = \mathrm{GF}(q)$ and $n = (q^m - 1)/(q - 1)$, where $m > 1$, and let $\mathcal{C}$ be a cyclic $[n, n{-}m, d]$ code over F with a generator polynomial $g(x)$. The purpose of this problem is to show that $d \geq 3$ only if the following three conditions hold:

(i) $\exp g(x) = n$.

(ii) $g(x)$ is irreducible over F.

(iii) $\gcd(n, q{-}1) = 1$.

(Note that the respective "if" direction was shown in Examples 8.4 and 8.9.) Assume hereafter in this problem that $d \geq 3$. The condition $\exp g(x) = n$ then follows already from the discussion in Section 8.3.

1. Let $m_1, m_2, \ldots, m_s$ be positive integers and let e be the least common multiplier of the values
$$q^{m_1}{-}1, q^{m_2}{-}1, \ldots, q^{m_s}{-}1$$
(namely, e is the smallest positive integer that is divisible by all of these values). Show that if $s > 1$ then e necessarily divides
$$\frac{1}{q{-}1} \prod_{i=1}^{s} (q^{m_i} - 1) .$$

2. Let $m_1, m_2, \ldots, m_s$ be positive integers. Show that
$$\prod_{i=1}^{s} (q^{m_i} - 1) \leq q^{m_1+m_2+\ldots+m_s} - 1 ,$$
with equality holding if and only if $s = 1$.

3. Let
$$g(x) = \prod_{i=1}^{s} a_i(x)$$
be the factorization of $g(x)$ into irreducible polynomials over F. Show that the polynomials $a_i(x)$ are all distinct.

4. Let the polynomials $a_i(x)$ be as in part 3 and let e be defined as in part 1 for $m_i = \deg a_i$. Show that $a_i(x) \,|\, x^e - 1$ for all $1 \leq i \leq s$.

5. Deduce from parts 3 and 4 that $n \,|\, e$.

6. Conclude from parts 1, 2, and 5 that $s = 1$, namely, that $g(x)$ is irreducible.

7. Show that the splitting field of $g(x)$ over F is $\mathrm{GF}(q^m)$.

8. Let α be a root of $g(x)$ in $\mathrm{GF}(q^m)$. Show that
$$\alpha^\ell \notin F \quad \text{for } \ell = 1, 2, \ldots, n{-}1 .$$
Hint: The set of roots of $g(x)$ forms one conjugacy class (with respect to F); consequently, an $m \times n$ parity-check matrix of $\mathcal{C}$ is obtained by representing each entry in
$$(\,1\; \alpha\; \alpha^2\; \ldots\; \alpha^{n-1}\,)$$
as a column vector in F^m. Argue that every two distinct columns are linearly independent over F.

9. Based on part 8, show that $\gcd(n, q{-}1) = 1$.

Notes

[Section 8.2]

Cyclic codes of length q over GF(q) (Problem 8.6) were studied by Berman [40], Falkner *et al.* [119], Massey *et al.* [255], Roth and Seroussi [304], and Zehendner [393].

[Section 8.4]

The discussion in this section concentrated on cyclic codes whose roots are all simple. However, as Problem 8.6 suggests, one may get interesting cyclic codes by allowing the roots of the generator polynomial to have multiplicity greater than 1, in which case the code is called a *repeated-root* cyclic code; see Castagnoli *et al.* [76] and van Lint [234].

The property of Reed–Muller codes that is presented in Problem 8.13 is due to Kasami *et al.* [207] and Kolesnik and Mironchikov [218]. For a generalization to non-binary Reed–Muller codes, see also Assmus and Key [18, Section 5.4] and Berlekamp [36, Section 15.3].

[Section 8.5]

There are some known generalizations of the BCH bound that improve on the latter: the Hartmann–Tzeng bound [172] and the Roos bound [296].

It is still unknown whether there exists an infinite family of cyclic codes over a given field GF(q) such that both their rate and relative minimum distance are bounded away from zero. On the other hand, for the case of primitive BCH codes with rate bounded away from zero, it is known that the relative minimum distance must approach zero as the code length increases: see Berlekamp [35] and Lin and Weldon [231].

Cyclic codes of length $q+1$ over GF(q) (Problem 8.15) were studied by Dür [112], Falkner *et al.* [119], and Georgiades [145]. See also MacWilliams and Sloane [249, Section 11.5].

Chapter 9

List Decoding of Reed–Solomon Codes

In Chapter 6, we introduced an efficient decoder for GRS codes, yet we assumed that the number of errors does not exceed $\lfloor (d-1)/2 \rfloor$, where d is the minimum distance of the code. In this chapter, we present a decoding algorithm for GRS codes, due to Guruswami and Sudan, where this upper limit is relaxed.

When a decoder attempts to correct more than $\lfloor (d-1)/2 \rfloor$ errors, the decoding may sometimes not be unique; therefore, we consider here a more general model of decoding, allowing the decoder to return a list of codewords, rather than just one codeword. In this more general setting, a decoding is considered successful if the computed list of codewords contains the transmitted codeword. The (maximum) number of errors that a list decoder can successfully handle is called the decoding radius of the decoder.

The approach that leads to the Guruswami–Sudan list decoder is quite different from the GRS decoder which was introduced in Chapter 6. Specifically, the first decoding step now computes from the received word a certain bivariate polynomial $Q(x, z)$ over the ground field, F, of the code. Regarding $Q(x, z)$ as a univariate polynomial in the indeterminate z over the ring $F[x]$, a second decoding step computes the roots of $Q(x, z)$ in $F[x]$; these roots are then mapped to codewords which, in turn, form the returned list. Both steps can be implemented in a time complexity that is polynomially large in the code length and the list size.

We also present a generalization, due to Koetter and Vardy, of the Guruswami–Sudan algorithm. The Koetter–Vardy algorithm provides an improvement especially when used as a decoder for alternant codes.

We end this chapter by presenting a lower bound on the largest decoding radius of any list decoder for any given code, as a function of the code parameters and the size of the returned list. It turns out that for a given length, minimum distance, and list size, the lower bound on the decoding radius of the Koetter–Vardy decoder is, in fact, a lower bound on the largest decoding radius attainable for every code. Still, for the special case of GRS codes, we have at hand a list decoder that has a polynomial-time implementation.

9.1 List decoding

Let $\mathcal{C}$ be an (n, M) code over an alphabet F and let $S = (F, \Phi, \mathsf{Prob})$ be a channel. Recall that a decoder for $\mathcal{C}$ with respect to S is defined as a mapping $\mathcal{D} : \Phi^n \to \mathcal{C} \cup \{\text{"e"}\}$; i.e., the decoder either returns a codeword or an indicator of error detection. If we rename the latter indicator by the empty set, we can say that a decoder returns a subset of $\mathcal{C}$ of size at most 1.

We next consider a more general family of decoders where the return value is a set (or "list") of codewords, and the size of the set can be greater than 1. Denote by $2^{\mathcal{C}}$ the set of all the subsets of $\mathcal{C}$. Given an (n, M) code $\mathcal{C}$ over F, a channel $S = (F, \Phi, \mathsf{Prob})$, and a positive integer ℓ, define a *list-ℓ decoder* (of $\mathcal{C}$ with respect to S) to be a mapping

$$\mathcal{D} : \Phi^n \to 2^{\mathcal{C}} ,$$

where $|\mathcal{D}(\mathbf{y})| \le \ell$ for every $\mathbf{y} \in \Phi^n$.

In the framework of list-ℓ decoders, a decoding success will occur when the returned list contains the transmitted codeword. Therefore, the decoding error probability $\mathrm{P_{err}}$ of a list-ℓ decoder $\mathcal{D}$ is defined by

$$\mathrm{P_{err}} = \max_{\mathbf{c} \in \mathcal{C}} \sum_{\mathbf{y} \in \Phi^n \,:\, \mathbf{c} \notin \mathcal{D}(\mathbf{y})} \mathsf{Prob}\{\, \mathbf{y} \text{ received} \mid \mathbf{c} \text{ transmitted} \,\} .$$

Even when the correct codeword is included in the returned list, the receiving end still needs to identify that codeword within the list. This can be done by, say, selecting the codeword that maximizes the conditional probability

$$\mathsf{Prob}\{\, \mathbf{y} \text{ received} \mid \mathbf{c} \text{ transmitted} \,\}$$

among all codewords $\mathbf{c}$ in the list $\mathcal{D}(\mathbf{y})$; such a selection criterion guarantees that a list-ℓ decoder will do no worse than a maximum-likelihood decoder, *under the assumption that the correct codeword is indeed in the list.* Some side information about codewords—such as their *a priori* probabilities—may also be incorporated into the selection procedure.

We assume hereafter in this chapter that $S = (F, F, \mathsf{Prob})$ is an additive channel; namely, both the input and output alphabets equal F. We say that a positive integer τ is a *decoding radius* of a list-ℓ decoder $\mathcal{D} : F^n \to 2^{\mathcal{C}}$ if for every word $\mathbf{y} \in F^n$ and every codeword $\mathbf{c} \in \mathcal{C}$,

$$\mathsf{d}(\mathbf{y}, \mathbf{c}) \le \tau \quad \Longrightarrow \quad \mathbf{c} \in \mathcal{D}(\mathbf{y}) .$$

That is, the list returned by the decoder contains all codewords whose Hamming distance from the received word is at most τ. Observe that if the number of errors that actually occurred is τ or less, then the returned list is guaranteed to contain the transmitted codeword. Obviously, a nearest-codeword decoder for an (n, M, d) code $\mathcal{C}$ is a list-1 decoder with decoding radius $\lfloor (d-1)/2 \rfloor$. On the other hand, it follows from Problem 1.10 that the decoding radius of a list-ℓ decoder for $\mathcal{C}$ may exceed $\lfloor (d-1)/2 \rfloor$ only if $\ell > 1$.

9.2 Bivariate polynomials

The decoders to be discussed in the sequel involve operations on bivariate polynomials. Therefore, we precede our further discussion on list decoders by reviewing several concepts that relate to such polynomials.

Given a field F, denote by $F[x, z]$ the set of all bivariate polynomials over F in the indeterminates x and z; that is,

$$F[x, z] = \left\{ a(x, z) = \textstyle\sum_{i,j=0}^{m} a_{i,j} x^i z^j \;:\; 0 \le m < \infty,\; a_{i,j} \in F \right\} .$$

We will hereafter regard the elements of $F[x, z]$ as elements of the ring $F[x][z]$—namely, as univariate polynomials over $F[x]$ in the indeterminate z; such a characterization of the elements of $F[x, z]$ implies in a natural way the definition of addition and multiplication in $F[x, z]$.

Let μ and ν be nonnegative integers and let $a(x, z) = \sum_{i,j} a_{i,j} x^i z^j$ be a nonzero bivariate polynomial in $F[x, z]$. The (μ, ν)-*degree* of $a(x, z)$, denoted by $\deg_{\mu,\nu} a(x, z)$, is defined as

$$\deg_{\mu,\nu} a(x, z) = \max_{i,j \,:\, a_{i,j} \ne 0} \{ i\mu + j\nu \} .$$

In particular, $\deg_{0,1} a(x, z)$ is the ordinary degree of $a(x, z)$ when regarded as an element of $F[x][z]$. The (μ, ν)-degree of the zero polynomial is defined as $-\infty$.

Let $a(x, z)$ and $b(x, z)$ be elements of $F[x, z]$ where $a(x, z) \ne 0$. We say that $a(x, z)$ *divides* $b(x, z)$ (in $F[x, z]$) or that $a(x, z)$ is a *factor* of $b(x, z)$ if there exists $c(x, z) \in F[x, z]$ such that $a(x, z) = c(x, z)b(x, z)$. A *linear* factor of $Q(x, z)$ is a factor of $Q(x, z)$ of the form $z - f(x)$ where $f(x) \in F[x]$.

We say that $f(x) \in F[x]$ is a z-*root* of $Q(x, z) \in F[x, z]$ if the univariate polynomial $Q(x, f(x))$ is identically zero.

Lemma 9.1 *A polynomial $f(x) \in F[x]$ is a z-root of $Q(x, z) \in F[x, z]$ if and only if $z - f(x)$ is a (linear) factor of $Q(x, z)$.*

Proof. Let $F(x)$ denote the field of rational functions over F: the elements of $F(x)$ are all the expressions of the form $a(x)/b(x)$, where

$a(x), b(x) \in F[x]$, the polynomial $b(x)$ is nonzero and normalized to some prescribed form (e.g., it is monic), and $\gcd(a(x), b(x)) = 1$. Addition and multiplication in $F(x)$ are defined in the conventional way, with an extra step of clearing common factors of the numerator and denominator from the result.

Regarding the bivariate polynomial $Q(x, z)$ as a univariate polynomial in $F(x)[z]$, we have (by Proposition 3.5) that $f(x)$ is a z-root of $Q(x, z)$ if and only if $Q(x, z)$ is divisible by $z - f(x)$ in the ring $F(x)[z]$. Furthermore, since $z - f(x)$ is monic in $F[x][z]$, we get that

$$z - f(x) \text{ divides } Q(x, z) \text{ in } F(x)[z]$$

if and only if

$$z - f(x) \text{ divides } Q(x, z) \text{ in } F[x, z]$$

(Problem 9.2). □

9.3 GRS decoding through bivariate polynomials

Let $\mathcal{C}_{\mathrm{GRS}}$ be an $[n, k, d]$ GRS code over $F = \mathrm{GF}(q)$. In Chapter 6, we have seen efficient implementations of list-1 decoders for $\mathcal{C}_{\mathrm{GRS}}$ with decoding radius $\lfloor \frac{1}{2}(d-1) \rfloor$. To obtain list-$\ell$ decoders with larger decoding radii (and necessarily larger list sizes ℓ), we follow a different approach—one that is based on interpolation of bivariate polynomials. We demonstrate this approach by first applying it to obtain yet another implementation of a GRS list-1 decoder with decoding radius $\lfloor \frac{1}{2}(d-1) \rfloor$.

Let $\alpha_1, \alpha_2, \ldots, \alpha_n$ be the code locators of $\mathcal{C}_{\mathrm{GRS}}$. For simplicity, we will further assume that the column multipliers (of the parity-check matrix) of $\mathcal{C}_{\mathrm{GRS}}$ are such that the generator matrix of the code is given by

$$G_{\mathrm{GRS}} = \begin{pmatrix} 1 & 1 & \ldots & 1 \\ \alpha_1 & \alpha_2 & \ldots & \alpha_n \\ \alpha_1^2 & \alpha_2^2 & \ldots & \alpha_n^2 \\ \vdots & \vdots & \vdots & \vdots \\ \alpha_1^{k-1} & \alpha_2^{k-1} & \ldots & \alpha_n^{k-1} \end{pmatrix} . \tag{9.1}$$

Associating each vector $\mathbf{u} = (u_0 \ u_1 \ \ldots \ u_{k-1})$ in F^k with a polynomial $u(x) = u_0 + u_1 x + \ldots + u_{k-1} x^{k-1} \in F_k[x]$, the codewords of $\mathcal{C}_{\mathrm{GRS}}$ are then given by

$$\mathcal{C}_{\mathrm{GRS}} = \left\{ \mathbf{u} G_{\mathrm{GRS}} = (\, u(\alpha_1) \ u(\alpha_2) \ \ldots \ u(\alpha_n) \,) \ : \ u(x) \in F_k[x] \right\} .$$

Let $\mathbf{c} = (c_1 \ c_2 \ \ldots \ c_n)$ be the transmitted codeword and $\mathbf{y} = (y_1 \ y_2 \ \ldots \ y_n)$ be the received word, where $\mathsf{d}(\mathbf{y}, \mathbf{c}) \leq \frac{1}{2}(d-1)$. Also, let $u(x)$

be the (unique) polynomial in $F_k[x]$ such that $c_j = u(\alpha_j)$ for $1 \le j \le n$; clearly, reconstructing the codeword $\mathbf{c}$ is equivalent to finding the polynomial $u(x)$.

Given the received word $\mathbf{y}$, we first compute a nonzero bivariate polynomial $Q(x,z) \in F[x,z]$ that satisfies the *degree constraints*

$$\deg_{0,1} Q(x,z) \le 1 \tag{9.2}$$

and

$$\deg_{1,k-1} Q(x,z) < n - \tfrac{1}{2}(d-1) , \tag{9.3}$$

as well as the *interpolation constraint*

$$Q(\alpha_j, y_j) = 0 , \quad j = 1,2,\ldots,n . \tag{9.4}$$

Note that conditions (9.2) and (9.3) simply mean that $Q(x,z)$ has the form

$$Q(x,z) = Q_0(x) + zQ_1(x) ,$$

where

$$\deg Q_0(x) < n - \tfrac{1}{2}(d-1) \quad \text{and} \quad \deg Q_1(x) < \tfrac{1}{2}(d+1) .$$

These degree constraints on $Q_0(x)$ and $Q_1(x)$ still allow $Q(x,z)$ to have

$$\lceil n - \tfrac{1}{2}(d-1) \rceil + \lceil \tfrac{1}{2}(d+1) \rceil \ge n+1$$

significant coefficients; on the other hand, (9.4) is a set of n linear homogeneous equations in these (unknown) coefficients. Hence, there is at least one nonzero solution $Q(x,z) \in F[x,z]$ that satisfies (9.2)–(9.4).

Let $Q(x,z)$ be any such nonzero solution and consider the univariate polynomial

$$\varphi(x) = Q(x, u(x)) = Q_0(x) + u(x)Q_1(x) . \tag{9.5}$$

Denote by J the set of error locations; that is,

$$y_j \ne c_j \quad \Longleftrightarrow \quad j \in J .$$

On the one hand, for every location $j \notin J$ we have

$$\varphi(\alpha_j) = Q(\alpha_j, u(\alpha_j)) = Q(\alpha_j, c_j) = Q(\alpha_j, y_j) = 0 , \tag{9.6}$$

namely, $\varphi(x)$ has at least $n - |J|$ distinct roots in F. On the other hand,

$$\deg \varphi(x) \le \max\{\deg Q_0(x), \deg u(x) + \deg Q_1(x)\} < n - \tfrac{1}{2}(d-1) \le n - |J| .$$

It follows that $\varphi(x)$ has more distinct roots in F than its degree, which means that it must be identically zero. Thus, we can solve (9.5) for $u(x)$ to obtain

$$u(x) = -\frac{Q_0(x)}{Q_1(x)} .$$

From $\varphi(x)$ being identically zero we also conclude that (9.6) holds for $j \in J$. Therefore,

$$Q_1(\alpha_j)(y_j - c_j) = Q(\alpha_j, y_j) - Q(\alpha_j, c_j) = 0 \,, \quad j \in J \,,$$

or

$$Q_1(\alpha_j) = 0 \,, \quad j \in J \,.$$

This means that $Q_1(x)$ is divisible by the polynomial

$$\mathrm{V}(x) = \prod_{j \in J} (x - \alpha_j) \,,$$

which is obtained by reversing the order of coefficients of the error-locator polynomial $\Lambda(x) = \prod_{j \in J}(1 - \alpha_j x)$ (see Section 6.3); that is, $\mathrm{V}(x) = x^{|J|}\Lambda(x^{-1})$.

Conversely, it is easy to check that the bivariate polynomial

$$Q(x, z) = \mathrm{V}(x)(z - u(x))$$

is a nonzero solution to (9.2)–(9.4). Here $Q_1(x)$ is actually equal to $\mathrm{V}(x)$ and, so, this solution has the smallest possible $(1, k{-}1)$-degree; furthermore, such an extremal solution is unique up to a scalar multiple. (Refer to Problem 9.4 to see the connection between the list-1 decoder presented herein and the Welch–Berlekamp equations that were introduced in Problem 6.13.)

9.4 Sudan's algorithm

We now turn to generalizing the decoding method of Section 9.3 to larger list sizes ℓ. Given the parameters $[n, k, d]$ of the GRS code $\mathcal{C}_{\mathrm{GRS}}$, we will find it convenient to introduce the notation

$$R' = \frac{k{-}1}{n} \,.$$

The value R', which is typically very close to the rate of $\mathcal{C}_{\mathrm{GRS}}$, is related to the relative minimum distance $\delta = d/n$ of $\mathcal{C}_{\mathrm{GRS}}$ by $R' = 1 - \delta$.

Assume a prescribed list size ℓ. The list-ℓ decoder to be presented in this section has decoding radius $\lceil n\Theta_{\ell,1}(R')\rceil - 1$, where

$$\boxed{\Theta_{\ell,1}(R') = \tfrac{\ell}{\ell+1} - \tfrac{\ell}{2}R'}$$

(the reason for the additional subscript 1 will become apparent in Section 9.5). If we regard R' momentarily as a real variable, then the function $R' \mapsto \Theta_{\ell,1}(R')$ represents a line in the real plane.

Example 9.1 For $\ell = 1$ we have $\Theta_{1,1}(R') = (1 - R')/2 = \delta/2$ and, so, the decoding radius in this case equals the familiar value

$$\lceil d/2 \rceil - 1 = \lfloor (d{-}1)/2 \rfloor .$$

For $\ell = 2$ we have $\Theta_{2,1}(R') = \frac{2}{3} - R'$; therefore, the decoding radius is

$$\lceil \tfrac{2}{3} n \rceil - k = \lfloor \tfrac{2}{3}(n{+}1) \rfloor - k .$$

Note, however, that when $R' > \frac{1}{3}$,

$$\Theta_{1,1}(R') = \tfrac{1}{2}(1 - R') > \tfrac{2}{3} - R' = \Theta_{2,1}(R') .$$

Hence, when $R' > \frac{1}{3}$, there is no point in selecting $\ell = 2$ over $\ell = 1$. □

As the last example indicates, there is a range of values of R' for which the decoding radius can be made *larger* by selecting a *smaller* list size ℓ; in such circumstances we will prefer the smaller ℓ. It follows that the value of ℓ that should be selected for a given R' (or, alternatively, the value of R' that should be selected for a given ℓ) is such that

$$\Theta_{\ell,1}(R') \ge \Theta_{\ell-1,1}(R') .$$

The latter inequality holds if and only if

$$R' \le \frac{2}{\ell(\ell{+}1)}$$

(see Problem 9.5).

The list-ℓ decoder to be presented is based on the following two lemmas.

Lemma 9.2 (Bivariate interpolation lemma) *Given the* $[n, k{=}nR'{+}1]$ *GRS code over* F *that is generated by (9.1), let* ℓ *and* τ *be positive integers such that* $\tau < n\Theta_{\ell,1}(R')$. *For every vector* $(y_1 \; y_2 \; \ldots \; y_n)$ *in* F^n *there exists a nonzero bivariate polynomial* $Q(x, z) \in F[x, z]$ *that satisfies the degree constraints*

$$\deg_{0,1} Q(x, z) \le \ell \tag{9.7}$$

and

$$\deg_{1,k-1} Q(x, z) < n - \tau , \tag{9.8}$$

and the interpolation constraint

$$Q(\alpha_j, y_j) = 0 , \quad j = 1, 2, \ldots, n . \tag{9.9}$$

Proof. Condition (9.9) defines a set of n linear homogeneous equations in the unknown coefficients of $Q(x, z)$. By (9.7)–(9.8), the number of these coefficients is at least

$$\begin{aligned}\sum_{t=0}^{\ell}\Big((n-\tau) - t(k-1)\Big) &= (\ell+1)(n-\tau) - \tbinom{\ell+1}{2}(k-1) \\ &= (\ell+1)(n-\tau) - \tbinom{\ell+1}{2}nR' \\ &= (\ell+1)\Big(n\Theta_{\ell,1}(R') - \tau\Big) + n > n \,,\end{aligned}$$

where the last inequality follows from $\tau < n\Theta_{\ell,1}(R')$. Hence, (9.9) has a nontrivial solution. □

Lemma 9.3 (Factorization lemma) *Given the $[n, k]$ GRS code over F that is generated by (9.1), let a nonzero $Q(x, z) \in F[x, z]$ satisfy (9.8) and (9.9) for some positive integer τ and a vector $\mathbf{y} = (y_1 \; y_2 \; \ldots \; y_n)$ in F^n. Suppose that there exists $u(x) \in F_k[x]$ such that the respective codeword, $\mathbf{c} = (\, u(\alpha_1) \;\; u(\alpha_2) \;\; \ldots \;\; u(\alpha_n)\,)$, satisfies $\mathsf{d}(\mathbf{y}, \mathbf{c}) \le \tau$. Then $z - u(x)$ divides $Q(x, z)$.*

Proof. Let J be the set of location indexes j where $y_j \ne u(\alpha_j)$, and consider the univariate polynomial $\varphi(x) = Q(x, u(x))$. On the one hand,

$$\deg \varphi(x) = \deg Q(x, u(x)) \le \deg_{1,k-1} Q(x, z) < n - \tau \le n - |J| \,,$$

where the penultimate inequality follows from (9.8) and the last inequality follows from the assumption that $\mathsf{d}(\mathbf{y}, \mathbf{c}) \le \tau$. On the other hand, for every location index $j \notin J$ we have

$$\varphi(\alpha_j) = Q(\alpha_j, u(\alpha_j)) = Q(\alpha_j, y_j) = 0 \,,$$

with the last equality implied by (9.9). We conclude that $\varphi(x)$, having more distinct roots in F than its degree, is identically zero. Thus, $u(x)$ is a z-root of $Q(x, z)$ and the result now follows from Lemma 9.1. □

Our analysis in Section 9.3 can be seen as a restricted version of the last two proofs, for the special case $\ell = 1$.

We are now in a position where we can describe our algorithm for implementing a list-ℓ decoder for $\mathcal{C}_{\mathrm{GRS}}$ and verify its correctness, based on Lemmas 9.2 and 9.3. Specifically, we apply these lemmas with ℓ being the prescribed list size; the parameter τ, which can be any positive integer smaller than $n\Theta_{\ell,1}(R')$, will serve as the decoding radius. In particular, we can take $\tau = \lceil n\Theta_{\ell,1}(R') \rceil - 1$.

Let $\mathbf{y} = (y_1\ y_2\ \ldots\ y_n)$ be the received word. We first compute a nonzero bivariate polynomial $Q(x,z) \in F[x,z]$ that satisfies (9.7)–(9.9). This involves solving a set of linear homogeneous equations, and Lemma 9.2 guarantees that a nontrivial solution indeed exists.

Having computed $Q(x,z)$, we next compute all the factors of $Q(x,z)$ in $F[x,z]$ of the form $z - f(x)$, where $f(x) \in F_k[x]$. Since $\deg_{0,1} Q(x,z) \le \ell$, the number of such factors cannot exceed ℓ. And by Lemma 9.3 we are guaranteed to find in this way all the factors $z - u(x)$ that correspond to codewords $(\, u(\alpha_1)\ u(\alpha_2)\ \ldots\ u(\alpha_n) \,)$ within Hamming distance τ from $\mathbf{y}$.

The list decoder that we have just described is known as *Sudan's algorithm*. Figure 9.1 summarizes the algorithm for a given $[n, nR'{+}1]$ GRS code $\mathcal{C}_{\mathrm{GRS}}$ over F that is generated by (9.1), where τ is assumed to be $\lceil n\Theta_{\ell,1}(R') \rceil - 1$.

Input: received word $\mathbf{y} = (y_1\ y_2\ \ldots\ y_n) \in F^n$, list size ℓ.
Output: list of up to ℓ codewords $\mathbf{c} \in \mathcal{C}_{\mathrm{GRS}}$.

1. *Interpolation step:* find a nonzero bivariate polynomial $Q(x,z) \in F[x,z]$ that satisfies

$$\deg_{0,1} Q(x,z) \le \ell\,, \qquad \deg_{1,nR'} Q(x,z) \le n\,(1 - \Theta_{\ell,1}(R'))\,,$$

   and

$$Q(\alpha_j, y_j) = 0\,, \quad j = 1, 2, \ldots, n\,.$$

2. *Factorization step:* compute the set U of all the polynomials $f(x) \in F_{nR'+1}[x]$ such that $z - f(x)$ is a factor of $Q(x,z)$ in $F[x,z]$.

3. Output all the codewords $\mathbf{c} = (\, u(\alpha_1)\ u(\alpha_2)\ \ldots\ u(\alpha_n) \,)$ that correspond to $u(x) \in U$ such that $\mathsf{d}(\mathbf{y},\mathbf{c}) < n\Theta_{\ell,1}(R')$.

Figure 9.1. Sudan's list-decoding algorithm for GRS codes.

The algorithm in Figure 9.1 can be realized in time complexity that is polynomially large in the code length and list size. Specifically, Step 1 in Figure 9.1 can be implemented by using Gaussian elimination to find a nontrivial solution to (9.7)–(9.9), thereby requiring $O(n^3)$ operations in F. An efficient procedure for implementing Step 2 will be described in Section 9.7, and a straightforward implementation of Step 3 takes $O(|U|kn) = O(\ell kn)$ operations in F.

Example 9.2 We consider a list-4 decoder for the $[18,2]$ GRS code over $F = \mathrm{GF}(19)$ with code locators $\alpha_j = j$ for $1 \le j \le 18$ (while such a field size and dimension are hardly ever found in practice, these parameters were selected to make this example easy to follow). For $\ell = 4$ we have $18 \cdot \Theta_{4,1}(\frac{1}{18}) = 12.4$; so, the decoding radius is $\tau = 12$.

Let the transmitted codeword correspond to the polynomial $u(x) = 18 + 14x$; that is,

$$\begin{aligned}\mathbf{c} &= (\, u(1)\ u(2)\ \ldots\ u(18)\,) \\ &= (\ 13\ 8\ 3\ 17\ 12\ 7\ 2\ 16\ 11\ 6\ 1\ 15\ 10\ 5\ 0\ 14\ 9\ 4\ )\,,\end{aligned}$$

and let the error word and the received word be given by

$$\mathbf{e} = (\ 11\ 16\ 17\ 12\ 17\ 0\ 0\ 2\ 14\ 0\ 0\ 0\ 3\ 0\ 14\ 8\ 11\ 15\ )$$

and

$$\mathbf{y} = (\ 5\ 5\ 1\ 10\ 10\ 7\ 2\ 18\ 6\ 6\ 1\ 15\ 13\ 5\ 14\ 3\ 1\ 0\ )\,,$$

respectively.

A nonzero solution to (9.7)–(9.9) is given by

$$\begin{aligned}Q(x,z) &= 4 + 12x + 5x^2 + 11x^3 + 8x^4 + 13x^5 \\ &\quad + (14 + 14x + 9x^2 + 16x^3 + 8x^4)z \\ &\quad + (14 + 13x + x^2)z^2 \\ &\quad + (2 + 11x + x^2)z^3 \\ &\quad + 17z^4\,,\end{aligned}$$

and one can verify that

$$\begin{aligned}Q(x,z) &= 17(z - 18 - 14x)(z - 8 - 8x)(z - 14 - 16x) \\ &\qquad \cdot (z - 18 - 15x - 10x^2)\,.\end{aligned}$$

Therefore, $Q(x,z)$ has three z-roots in $F_k[x]$:

$$u_1(x) = 18 + 14x\,, \quad u_2(x) = 8 + 8x\,, \quad \text{and} \quad u_3(x) = 14 + 16x\,.$$

Both $u_1(x)$ $(= u(x))$ and $u_2(x)$ correspond to codewords at Hamming distance 12 from $\mathbf{y}$, while $u_3(x)$ corresponds to a codeword at Hamming distance 15 from $\mathbf{y}$. So, in this case, Sudan's algorithm will produce two codewords.

An alternate nonzero solution to (9.7)–(9.9) is given by

$$\begin{aligned}\hat{Q}(x,z) &= 8 + 12x^2 + 9x^3 + 8x^4 \\ &\quad + (5 + 14x + 7x^2 + 15x^3 + 4x^4)z \\ &\quad + (12 + 12x + 15x^2 + 4x^3)z^2 \\ &\quad + (9 + 10x + 14x^2)z^3 \\ &\quad + (13 + x)z^4\,,\end{aligned}$$

and one can verify that the complete factorization of $Q(x,z)$ in $F[x,z]$ is

$$\hat{Q}(x,z) = (z-18-14x)(z-8-8x)((13+x)z^2 + (5+18x+17x^2)z+(18+6x+15x^2))$$

(the third factor, of $(0,1)$-degree 2, has no linear factors in $F[x,z]$). In this case, $\hat{Q}(x,z)$ has only two z-roots in $F_k[x]$: $u_1(x) = 18+14x$ and $u_2(x) = 8+8x$. □

9.5 The Guruswami–Sudan algorithm

The decoding radius in Sudan's algorithm can be increased by considering also the derivatives of the bivariate polynomial $Q(x,z)$, as we show next.

Let $a(x,z)$ be a polynomial in $F[x,z]$. The (s,t)*th Hasse derivative* of $a(x,z)$, denoted by $a^{[s,t]}(x,z)$, is defined as

$$a^{[s,t]}(x,z) = \sum_{i,j} \binom{i}{s}\binom{j}{t} a_{i,j} x^{i-s} z^{j-t} ,$$

where a binomial coefficient $\binom{h}{m}$ is defined to be zero when $h<m$. This definition of Hasse derivatives is a natural extension of its univariate counterpart, which was introduced in Problem 3.40.

Let $\mathrm{T}(r)$ denote the set

$$\mathrm{T}(r) = \{(s,t) \ : \ s,t \in \mathbb{N} \, , \ \ s+t<r\} ,$$

where $\mathbb{N}$ stands for the set of nonnegative integers.

Lemma 9.4 *Given $u(x) \in F[x]$ and $a(x,z) \in F[x,z]$, let β and γ be elements of F such that $u(\beta)=\gamma$ and*

$$a^{[s,t]}(x,z)|_{(x,z)=(\beta,\gamma)} = 0 \quad \textit{for all } (s,t) \in \mathrm{T}(r) .$$

Then $(x-\beta)^r \,|\, a(x,u(x))$.

Proof. Define the polynomial $b(v,w) = \sum_{s,t} b_{s,t} v^s w^t \in F[v,w]$ by

$$b(v,w) = a(v+\beta, w+\gamma) . \tag{9.10}$$

By comparing the coefficients of $v^s w^t$ on both sides of (9.10) we get that

$$b_{s,t} = \sum_{i,j} \binom{i}{s}\binom{j}{t} a_{i,j} \beta^{i-s} \gamma^{j-t} = a^{[s,t]}(x,z)|_{x=\beta,z=\gamma}$$

and, so, $b_{s,t} = 0$ for every $(s,t) \in \mathrm{T}(r)$. Hence,

$$a(x, u(x)) = b(x - \beta, u(x) - \gamma) = \sum_{s,t\,:\,s+t \geq r} b_{s,t}(x-\beta)^s (u(x) - \gamma)^t .$$

The result now follows by observing that $(x - \beta) \mid (u(x) - \gamma)$. □

Let ℓ be the prescribed list size and r be a positive integer not greater than ℓ. Define

$$\boxed{\Theta_{\ell,r}(R') = \frac{1}{(\ell+1)r}\left(\binom{\ell+1}{2}(1-R') - \binom{\ell+1-r}{2}\right)} .$$

The expression $\Theta_{\ell,r}(R')$—which will be used in our subsequent analysis—can also be written as

$$\Theta_{\ell,r}(R') = 1 - \frac{r+1}{2(\ell+1)} - \frac{\ell}{2r} \cdot R' .$$

In particular, the function $R' \mapsto \Theta_{\ell,r}(R')$, when viewed over the real field, represents a line in the real plane. It is easy to see that when $r = 1$, the expression $\Theta_{\ell,r}(R')$ coincides with the definition of $\Theta_{\ell,1}(R')$ in Section 9.4.

The next two lemmas generalize Lemmas 9.2 and 9.3 to the case where not only does the bivariate polynomial $Q(x, z)$ vanish at the points $\{(\alpha_j, y_j)\}_{j=1}^n$, but so do also some of its Hasse derivatives.

Lemma 9.5 *Given the $[n, k{=}nR'{+}1]$ GRS code over F that is generated by (9.1), let ℓ, r, and τ be positive integers such that $r \leq \ell$ and $\tau < n\Theta_{\ell,r}(R')$. For every vector $(y_1\ y_2\ \ldots\ y_n)$ in F^n there exists a nonzero bivariate polynomial $Q(x, z) \in F[x, z]$ that satisfies*

$$\deg_{0,1} Q(x, z) \leq \ell , \tag{9.11}$$

$$\deg_{1,k-1} Q(x, z) < r(n - \tau) , \tag{9.12}$$

and

$$Q^{[s,t]}(x, z)|_{(x,z)=(\alpha_j, y_j)} = 0 , \quad j = 1, 2, \ldots, n , \quad (s,t) \in \mathrm{T}(r) . \tag{9.13}$$

Proof. The proof is similar to that of Lemma 9.2, except that now (9.13) defines a set of $\binom{r+1}{2}n$ linear homogeneous equations in the unknown coefficients of $Q(x, z)$. By (9.11) and (9.12) we obtain that the number of these coefficients is at least

$$\sum_{t=0}^{\ell}\Big(r(n-\tau) - t(k-1)\Big) \quad = \quad (\ell+1)r(n-\tau) - \binom{\ell+1}{2}(k-1)$$

$$
\begin{aligned}
&= \binom{\ell+1}{2} n(1-R') + \left((\ell+1)r - \binom{\ell+1}{2} \right) n - (\ell+1)r\tau \\
&= \binom{\ell+1}{2} n(1-R') - \left(\binom{\ell+1-r}{2} - \binom{r+1}{2} \right) n - (\ell+1)r\tau \\
&= (\ell+1)r \Big(n\,\Theta_{\ell,r}(R') - \tau \Big) + \binom{r+1}{2} n \\
&> \binom{r+1}{2} n ,
\end{aligned}
$$

where the last inequality follows from $\tau < n\,\Theta_{\ell,r}(R')$. Hence, (9.13) has a nontrivial solution. □

Lemma 9.6 *Given the $[n,k]$ GRS code over F that is generated by (9.1), let a nonzero $Q(x,z) \in F[x,z]$ satisfy (9.12) and (9.13) for positive integers r and τ and a vector $\mathbf{y} = (y_1\ y_2\ \ldots\ y_n)$ in F^n. Suppose that there exists $u(x) \in F_k[x]$ such that the respective codeword, $\mathbf{c} = (\,u(\alpha_1)\ \ u(\alpha_2)\ \ \ldots\ \ u(\alpha_n)\,)$, satisfies $\mathsf{d}(\mathbf{y},\mathbf{c}) \le \tau$. Then $z - u(x)$ divides $Q(x,z)$.*

Proof. Let $\overline{J}$ be the set of indexes j for which $u(\alpha_j) = y_j$. By (9.13) and Lemma 9.4 we obtain

$$
(x-\alpha_j)^r \,|\, Q(x,u(x)) , \quad j \in \overline{J} ,
$$

and, so,

$$
\Big(\prod_{j \in \overline{J}} (x-\alpha_j)^r \Big) \;\Big|\; Q(x,u(x)) . \tag{9.14}
$$

On the other hand, by (9.12) we have

$$
\deg Q(x,u(x)) \le \deg_{1,k-1} Q(x,z) < r(n-\tau) \le r|\overline{J}| .
$$

Combining this with (9.14) we conclude that $Q(x,u(x))$ is identically zero. The result now follows from Lemma 9.1. □

Based on the last two lemmas, we can now modify Sudan's algorithm by using (9.11)–(9.13) instead of (9.7)–(9.9). The resulting list decoding algorithm, which is shown in Figure 9.2, is known as the *Guruswami–Sudan algorithm*: its decoding radius is $\lceil n\,\Theta_{\ell,r}(R') \rceil - 1$, and it reduces to Sudan's algorithm when $r = 1$.

The additional parameter r in the Guruswami–Sudan algorithm allows us to increase the decoding radius (compared to Sudan's algorithm) by maximizing over r. Specifically, we can now attain a decoding radius $\lceil n\,\Theta_{\ell}(R') \rceil - 1$, where

$$
\Theta_\ell(R') = \max_{1 \le r \le \ell} \Theta_{\ell,r}(R') . \tag{9.15}
$$

Input: received word $\mathbf{y} = (y_1\ y_2\ \ldots\ y_n) \in F^n$, list size ℓ.
Output: list of up to ℓ codewords $\mathbf{c} \in \mathcal{C}_{\mathrm{GRS}}$.

1. *Interpolation step:* find a nonzero bivariate polynomial $Q(x, z) \in F[x, z]$ that satisfies

$$\deg_{0,1} Q(x, z) \le \ell , \qquad \deg_{1,nR'} Q(x, z) \le rn\,(1 - \Theta_{\ell,r}(R')) ,$$

and

$$Q^{[s,t]}(x, z)|_{(x,z)=(\alpha_j,y_j)} = 0 , \quad j = 1, 2, \ldots, n , \quad (s, t) \in \mathrm{T}(r) .$$

2. *Factorization step:* compute the set U of all the polynomials $f(x) \in F_{nR'+1}[x]$ such that $z - f(x)$ is a factor of $Q(x, z)$ in $F[x, z]$.

3. Output all the codewords $\mathbf{c} = (\,u(\alpha_1)\ \ u(\alpha_2)\ \ \ldots\ \ u(\alpha_n)\,)$ that correspond to $u(x) \in U$ such that $\mathsf{d}(\mathbf{y}, \mathbf{c}) < n\Theta_{\ell,r}(R')$.

Figure 9.2. The Guruswami–Sudan list-decoding algorithm for GRS codes.

(It should be noted, however, that the parameter r also affects the time complexity of the algorithm: this complexity increases (polynomially) with r.)

We next characterize the value of r that achieves the maximum in (9.15). For $1 \le r \le \ell{+}1$ define

$$\Upsilon_{\ell,r} = \frac{r(r-1)}{\ell(\ell+1)} .$$

It can be shown (Problem 9.6) that

$$\Theta_{\ell,r}(R') \ge \Theta_{\ell,r-1}(R') \quad \Longleftrightarrow \quad R' \ge \Upsilon_{\ell,r}$$

(i.e., the lines $R' \mapsto \Theta_{\ell,r-1}(R')$ and $R' \mapsto \Theta_{\ell,r}(R')$ intersect in the real plane when $R' = \Upsilon_{\ell,r}$). So, given ℓ and R', we reach a maximum in (9.15) when r equals the (unique) integer $r_0 = r_0(\ell, R')$ that satisfies

$$\Upsilon_{\ell,r_0} \le R' < \Upsilon_{\ell,r_0+1} .$$

It follows that the function $R' \mapsto \Theta_\ell(R')$, when viewed over the real interval $[0, 1)$, is continuous and piecewise linear in R' (for fixed ℓ) and is given by

$$\Theta_\ell(R') = \begin{cases} \Theta_{\ell,1}(R') & \text{for } \Upsilon_{\ell,1} \le R' < \Upsilon_{\ell,2} \\ \Theta_{\ell,2}(R') & \text{for } \Upsilon_{\ell,2} \le R' < \Upsilon_{\ell,3} \\ \quad\vdots & \qquad\qquad \vdots \\ \Theta_{\ell,\ell}(R') & \text{for } \Upsilon_{\ell,\ell} \le R' < \Upsilon_{\ell,\ell+1} \end{cases}$$

(where $\Upsilon_{\ell,1} = 0$ and $\Upsilon_{\ell,\ell+1} = 1$).

We have seen in Example 9.1 that the value $\Theta_{\ell,1}(R')$ may sometimes decrease as ℓ increases; the value $\Theta_\ell(R')$, on the other hand, is better behaved in the sense that it is always non-decreasing with ℓ, and it can be shown to converge to the limit

$$\Theta_\infty(R') = \lim_{\ell \to \infty} \Theta_\ell(R') = 1 - \sqrt{R'}$$

(see Problem 9.7).

The functions $R' \mapsto \Theta_\ell(R')$ for $\ell = 1, 4, \infty$ are plotted in Figure 9.3.

9.6 List decoding of alternant codes

The Guruswami–Sudan algorithm is applicable also to the list decoding of alternant codes over $F = \mathrm{GF}(q)$: we simply use the list decoder for the underlying GRS code. Recall that this strategy was already suggested in Section 6.6 for the special case of list-1 decoding. Namely, if d is the *designed*

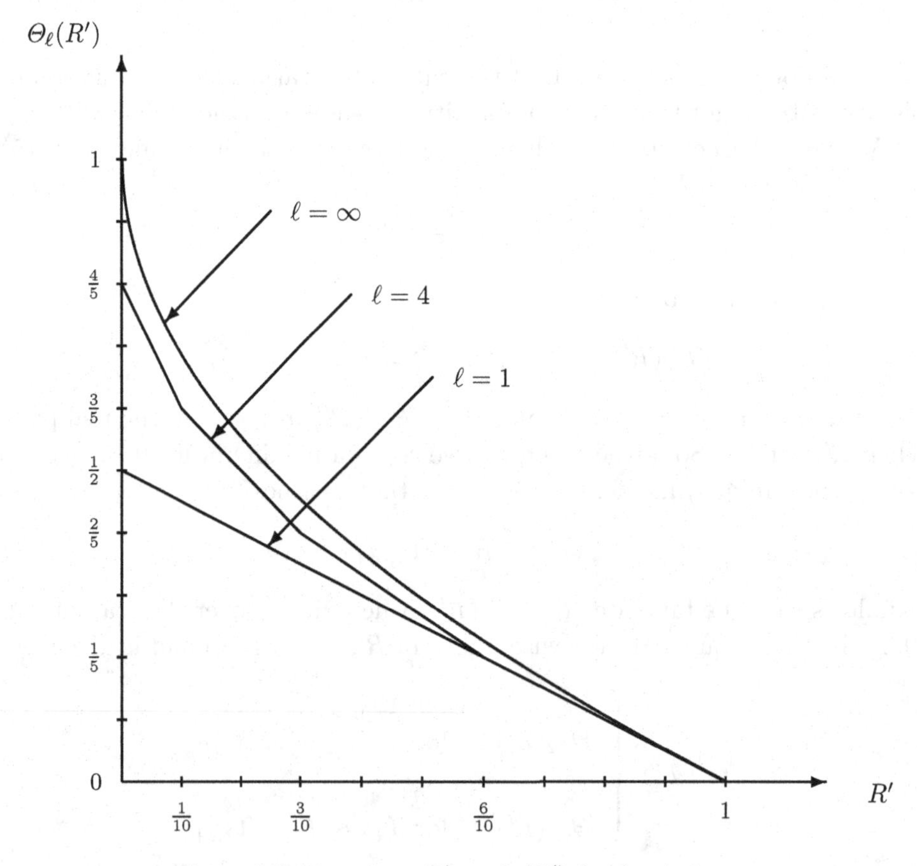

Figure 9.3. Functions $R' \mapsto \Theta_\ell(R')$ for $\ell = 1, 4, \infty$.

minimum distance of a given alternant code $\mathcal{C}_{\text{alt}}$, then the list-1 decoders of Chapter 6 will correct up to $\lfloor (d-1)/2 \rfloor$ errors, since d is the minimum distance of the underlying GRS code. And if the designed minimum distance d equals the true minimum distance of $\mathcal{C}_{\text{alt}}$, then we cannot expect to be able to correct more errors by a list-1 decoder.

It turns out that when the list size goes beyond 1, we may gain in the decoding radius of $\mathcal{C}_{\text{alt}}$ (compared to its underlying GRS counterpart) by taking into account the fact that the entries of the transmitted codeword and the received word are restricted to the field F of $\mathcal{C}_{\text{alt}}$. Incorporating this restriction into the Guruswami–Sudan algorithm requires some modification of the algorithm, as we now describe.

Throughout this section, we fix $\mathcal{C}_{\text{GRS}}$ to be an $[n, n-d+1, d]$ GRS code over an extension field Φ of $F = \text{GF}(q)$, and the generator matrix of $\mathcal{C}_{\text{GRS}}$ is assumed to take the form

$$G_{\text{GRS}} = \left(\alpha_j^i \right)_{i=0}^{n-d} {}_{j=1}^{n} .$$

We let $\mathcal{C}_{\text{alt}}$ be the respective alternant code $\mathcal{C}_{\text{GRS}} \cap F^n$ that is used for the transmission and assume a prescribed list size ℓ. (As we will see, the decoding will depend on the field F and on the parameters n and d of the underlying code $\mathcal{C}_{\text{GRS}}$, yet not on the dimension and true minimum distance of $\mathcal{C}_{\text{alt}}$. Therefore, hereafter we use the notation $n-d$ instead of $k-1$ to avoid confusion with the dimension of $\mathcal{C}_{\text{alt}}$. Accordingly, δ and R' should be understood as d/n and $1-(d/n)$, respectively.)

Recall that the change from Sudan's algorithm to the Guruswami–Sudan algorithm involved the introduction of the parameter r and requiring that the bivariate polynomial $Q(x, z)$ satisfies (9.11)–(9.13), instead of (9.7)–(9.9). We now add a second nonnegative integer parameter $\overline{r} < r$ and replace (9.11)–(9.13) by the degree constraints

$$\deg_{0,1} Q(x, z) \leq \ell , \tag{9.16}$$

$$\deg_{1,n-d} Q(x, z) < r(n-\tau) + \overline{r}\tau , \tag{9.17}$$

and the interpolation constraint

$$Q^{[s,t]}(x, z)|_{(x,z)=(\alpha_j,\gamma)} = 0 , \quad \text{for all} \quad j = 1, 2, \ldots, n ,$$
$$\gamma \in F , \quad \text{and} \quad (s, t) \in \begin{cases} \text{T}(r) & \text{if } \gamma = y_j \\ \text{T}(\overline{r}) & \text{if } \gamma \neq y_j \end{cases} . \tag{9.18}$$

That is, the bivariate polynomial $Q(x, z)$ (which is now assumed to be over the field Φ) and its (s, t)th Hasse derivatives are required to vanish also at points (α_j, γ) where $\gamma \neq y_j$, yet (s, t) ranges for these points over $\text{T}(\overline{r})$ rather than $\text{T}(r)$. Notice that (9.18) is where we take into account the ground field F of $\mathcal{C}_{\text{alt}}$: the elements γ should range over F (and not Φ).

Define

$$\boxed{\Theta_{\ell,r,\overline{r}}(R',q) = \frac{1}{(\ell+1)(r-\overline{r})}\Bigl(\tbinom{\ell+1}{2}(1-R') - \tbinom{\ell+1-r}{2} - \tbinom{\overline{r}+1}{2}(q-1)\Bigr)} \ .$$

Lemmas 9.7 and 9.8 below replace Lemmas 9.5 and 9.6 of Section 9.5. As the proofs are similar, we include here only the proof of Lemma 9.8.

Lemma 9.7 *Let Φ be an extension field of $F = \mathrm{GF}(q)$. Given an alternant code of length n and designed minimum distance $d = n(1-R')$ over F, with the underlying GRS code being over Φ and generated by (9.1), let ℓ, r, $\overline{r}$, and τ be integers such that $0 \le \overline{r} < r \le \ell$ and $0 < \tau < n\Theta_{\ell,r,\overline{r}}(R',q)$. For every vector $(y_1\ y_2\ \ldots\ y_n)$ in Φ^n there exists a nonzero bivariate polynomial $Q(x,z) \in \Phi[x,z]$ that satisfies (9.16)–(9.18).*

Lemma 9.8 *Let Φ be an extension field of F. Given an alternant code over F with the underlying $[n, n-d+1, d]$ GRS code being over Φ and generated by (9.1), let a nonzero $Q(x,z) \in \Phi[x,z]$ satisfy (9.17)–(9.18) for positive integers r and τ, a nonnegative integer $\overline{r} \le r$, and a vector $\mathbf{y} = (y_1\ y_2\ \ldots\ y_n)$ in F^n. Suppose that there exists $u(x) \in \Phi_{n-d+1}[x]$ such that the respective codeword, $\mathbf{c} = (\,u(\alpha_1)\ \ u(\alpha_2)\ \ \ldots\ \ u(\alpha_n)\,)$, is in F^n and $\mathsf{d}(\mathbf{y},\mathbf{c}) \le \tau$. Then $z - u(x)$ divides $Q(x,z)$.*

Proof. We use the same notation as in the proof of Lemma 9.6. By (9.18) and Lemma 9.4 it follows that

$$(x-\alpha_j)^{\overline{r}} \mid Q(x,u(x)) \ , \quad j = 1,2,\ldots,n \ .$$

Combining this with (9.14) we obtain

$$\Bigl(\prod_{j\in\overline{J}}(x-\alpha_j)^r \prod_{j\in J}(x-\alpha_j)^{\overline{r}}\Bigr) \mid Q(x,u(x)) \ ,$$

where J stands for the set of indexes j for which $u(\alpha_j) \ne y_j$. On the other hand, by (9.17) we have

$$\deg Q(x,u(x)) \le \deg_{1,n-d} Q(x,z) < r(n-\tau) + \overline{r}\tau \le r|\overline{J}| + \overline{r}|J| \ .$$

Hence, $Q(x,u(x))$ is identically zero and the result follows by Lemma 9.1. □

The previous two lemmas imply a list-decoding algorithm, which is shown in Figure 9.4. This algorithm is known as the *Koetter–Vardy algorithm* for decoding alternant codes. The Koetter–Vardy algorithm reduces to the Guruswami–Sudan algorithm when $\overline{r} = 0$.

Input: received word $\mathbf{y} = (y_1 \; y_2 \; \ldots \; y_n) \in F^n$, list size ℓ.
Output: list of up to ℓ codewords $\mathbf{c} \in \mathcal{C}_{\mathrm{alt}}$.

1. *Interpolation step:* find a nonzero bivariate polynomial $Q(x,z) \in \Phi[x,z]$ that satisfies

$$\deg_{0,1} Q(x,z) \le \ell\,, \qquad \deg_{1,n-d} Q(x,z) \le rn - (r-\overline{r})n\,\Theta_{\ell,r,\overline{r}}(R')\,,$$

   and

$$Q^{[s,t]}(x,z)|_{(x,z)=(\alpha_j,\gamma)} = 0\,, \quad j = 1,2,\ldots,n\,, \quad \gamma \in F\,,$$
$$(s,t) \in \begin{cases} \mathrm{T}(r) & \text{if } \gamma = y_j \\ \mathrm{T}(\overline{r}) & \text{if } \gamma \ne y_j \end{cases}$$

   (where $R' = 1 - (d/n)$).

2. *Factorization step:* compute the set U of all the polynomials $f(x) \in \Phi_{n-d+1}[x]$ such that $z - f(x)$ is a factor of $Q(x,z)$ in $\Phi[x,z]$.

3. Output all the codewords $\mathbf{c} = (\, u(\alpha_1) \; u(\alpha_2) \; \ldots \; u(\alpha_n)\,)$ that correspond to $u(x) \in U$ such that $\mathbf{c} \in F^n$ and $\mathsf{d}(\mathbf{y},\mathbf{c}) < n\Theta_{\ell,r,\overline{r}}(R')$.

Figure 9.4. The Koetter–Vardy list-decoding algorithm for alternant codes.

By maximizing over r and $\overline{r}$, the Koetter–Vardy algorithm can reach a decoding radius $\lceil n\Theta_\ell(R',q)\rceil - 1$, where

$$\Theta_\ell(R',q) = \max_{0 \le \overline{r} < r \le \ell} \Theta_{\ell,r,\overline{r}}(R',q)\,. \tag{9.19}$$

Clearly, this maximum is never smaller than $\Theta_\ell(R')$, which is the respective maximum for the Guruswami–Sudan algorithm, as defined by (9.15). In fact, there are instances where $(q-1)R'$ is an integer and

$$\lceil (q-1)\,\Theta_\ell(R')\rceil < \lceil (q-1)\,\Theta_\ell(R',q)\rceil$$

(e.g., take $q = 64$, $R' = 1/7$, and $\ell = 102$). When this happens, the decoding radius of the Koetter–Vardy algorithm will be larger than its Guruswami–Sudan counterpart for codes $\mathcal{C}_{\mathrm{alt}}$ that are as short as $n = q-1$. Such a code length, in turn, is attainable by GRS codes over F, i.e., we can select $\Phi = F$ and $\mathcal{C}_{\mathrm{alt}} = \mathcal{C}_{\mathrm{GRS}}$. Hence, the improvement of the Koetter–Vardy algorithm can sometimes be seen not only for alternant codes, but for GRS codes as well.

A further analysis of the function $R' \mapsto \Theta_\ell(R',q)$ is included in Problems 9.9 and 9.10.

9.7 Finding linear bivariate factors

In this section, we present an efficient algorithm for finding the z-roots in $F_k[x]$ of a nonzero bivariate polynomial $Q(x,z) \in F[x,z]$; this algorithm, in turn, can be used to implement the factorization step (Step 2) in Figures 9.1, 9.2, or 9.4 (in the latter figure, the roles of F and k are played by Φ and $n-d+1$, respectively).

Our algorithm makes use of the recursive procedure BiRoot in Figure 9.5 (the numbers that are assigned to the lines in the figure will be used in the sequel for reference). The procedure is initially called with the parameters $(Q, k, 0)$, where $Q = Q(x,z)$ is a nonzero bivariate polynomial in $F[x,z]$ and k is a positive integer that defines the space, $F_k[x]$, where the z-roots are to be found. A third parameter to BiRoot—denoted by λ—is used for keeping track of the recursion level, with the initial call corresponding to level 0. In addition, two global values are assumed, one of which is a set U that will ultimately contain the z-roots of $Q(x,z)$ in $F_k[x]$. This set is initialized in lines 1 and 2 in BiRoot upon the first call to the procedure.

In line 5 of BiRoot, we assume access to a "black box" that computes the set of all distinct roots in F of a (univariate) polynomial $T(0,z) \in F[z]$. An exhaustive search over all the elements of F—sometimes referred to as

---

```
BiRoot(Q(x,z) ∈ F[x,y], k ∈ ℕ, λ ∈ ℕ)
/*      Global variables:
            set U ⊆ F_k[x],
            polynomial g(x) = Σ_{s=0}^{k-1} g_s x^s ∈ F_k[x].
        Call procedure initially with Q(x,z) ≠ 0, k > 0, and λ = 0.
*/
        if (λ == 0)                                              /* 1 */
            U ← ∅;                                               /* 2 */
        m ← largest integer such that x^m divides Q(x,z);        /* 3 */
        T(x,z) ← x^{-m} Q(x,z);                                  /* 4 */
        Z ← set of all distinct (z-)roots of T(0,z) in F;        /* 5 */
        for each γ ∈ Z do {                                      /* 6 */
            g_λ ← γ;                                             /* 7 */
            if (λ < k−1)                                         /* 8 */
                BiRoot(T(x, xz + γ), k, λ+1);                    /* 9 */
            else                                                 /* 10 */
                if (Q(x, g_{k−1}) == 0)                          /* 11 */
                    U ← U ∪ {g(x)};                              /* 12 */
        }                                                        /* 13 */
```

Figure 9.5. Recursive procedure for finding the set of z-roots of $Q(x,z)$ in $F_k[x]$.

a *Chien search*—is one way to implement such a black box. The time complexity of such a search is $O(q \deg T(0,z))$ operations in F; this complexity can be considered efficient when q is of the order of $\deg_{1,k-1} Q(x,z)$ (e.g., when decoding primitive GRS codes, in which case q exceeds the code length only by 1). Alternatively, there are known probabilistic algorithms for root extraction with expected time complexity which grows polynomially with $\log q$ (see the notes on Section 3.2).

We now start analyzing the procedure BiRoot. We first notice that the check made in line 8 guarantees that the level of recursion cannot go beyond $k-1$. Now, every sequence of executions of line 7 along a (finite) recursion descent is associated with a unique polynomial

$$f(x) = f_0 + f_1 x + f_2 x^2 + \dots ,$$

which stands for the contents of the global polynomial $g(x)$ that is computed by that sequence. For $i \geq 0$, we let $Q_i(x,z)$ and $T_i(x,z)$ denote the values of the bivariate polynomials $Q(x,z)$ and $T(x,z)$, respectively, after the execution of line 4 during recursion level $\lambda = i$, along the particular descent that computes some given $f(x)$. Clearly, $Q_0(x,z) = Q(x,z)$ $(\neq 0)$, and for every $i \geq 0$,

$$T_i(x,z) = x^{-m_i} Q_i(x,z) \quad \text{and} \quad Q_{i+1}(x,z) = T_i(x, xz + f_i) ,$$

where m_i is the largest integer m such that x^m divides $Q_i(x,z)$. Observe that the $(0,1)$-degrees of the polynomials $Q_i(x,z)$ are the same for all i and, so, $Q_i(x,z) \neq 0$ and m_i is well-defined. Also, since x does not divide $T_i(x,z)$ then $T_i(0,z)$ is not identically zero. This means that the set Z, which is computed in line 5 in BiRoot, is always finite—even in applications where the field F is taken to be infinite. We readily conclude that the number of recursive calls to BiRoot is finite and, so, this procedure always halts.

In the next proposition, we show that upon termination of the recursive calls to BiRoot, the set U consists (only) of z-roots in $F_k[x]$ of the bivariate polynomial $Q(x,z)$ which BiRoot is initially called with.

Proposition 9.9 *Let $Q(x,z)$ be a nonzero bivariate polynomial in $F[x,z]$ and let U be the set that is computed by the call* BiRoot$(Q,k,0)$. *Every element of U is a z-root of $Q(x,z)$.*

Proof. Let $f(x) = f_0 + f_1 x + \ldots + f_{k-1} x^{k-1}$ be an element of U and let $Q_i(x,z)$ and $T_i(x,z)$ be the values of $Q(x,z)$ and $T(x,z)$ during recursion level $\lambda = i$ along the descent that computes $f(x)$. For $0 \leq i < k$, define the polynomial $\psi_i(x)$ by

$$\psi_i(x) = f_i + f_{i+1} x + f_{i+2} x^2 + \ldots + f_{k-1} x^{k-1-i} .$$

We show by a backward induction on $i = k-1, k-2, \ldots, 0$ that $\psi_i(x)$ is a z-root of $Q_i(x, z)$.

The induction base $i = k-1$ follows from the check made in line 11 in BiRoot. As for the induction step, suppose that $\psi_{i+1}(x)$ is a z-root of $Q_{i+1}(x, z)$. Then,

$$\begin{aligned} x^{-m_i} Q_i(x, \psi_i(x)) &= T_i(x, \psi_i(x)) \\ &= T_i(x, x \cdot \psi_{i+1}(x) + f_i) \\ &= Q_{i+1}(x, \psi_{i+1}(x)) = 0 . \end{aligned}$$

That is, $\psi_i(x)$ is a z-root of $Q_i(x, z)$.

In particular, for $i = 0$ we obtain that $\psi_0(x) = f(x)$ is a z-root of $Q_0(x, z) = Q(x, z)$. □

The next lemma will help us prove that upon termination of BiRoot, the set U in fact contains *all* the z-roots of $Q(x, z)$ in $F_k[x]$.

Lemma 9.10 *Let $Q(x, z)$ be a nonzero bivariate polynomial in $F[x, z]$ and let $f(x) = f_0 + f_1 x + \ldots + f_{k-1} x^{k-1}$ be a z-root of $Q(x, z)$ in $F[x]$. For $0 \le i < k$, define $Q_i(x, z)$ and $T_i(x, z)$ inductively by $Q_0(x, z) = Q(x, z)$ and*

$$T_i(x, z) = x^{-m_i} Q_i(x, z) \quad \text{and} \quad Q_{i+1}(x, z) = T_i(x, xz + f_i) ,$$

where m_i is the largest integer m such that x^m divides $Q_i(x, z)$. Then the following conditions hold for every $0 \le i < k$:

(i) The polynomial

$$\psi_i(x) = f_i + f_{i+1} x + f_{i+2} x^2 + \ldots + f_{k-1} x^{k-1-i}$$

is a z-root of $Q_i(x, z)$.

(ii) $T_i(0, f_i) = 0$.

Proof. We prove part (i) by induction on i. The induction base $i = 0$ is obvious. As for the induction step, if $\psi_i(x)$ is a z-root of $Q_i(x, z)$ then $\psi_{i+1}(x) = (\psi_i(x) - f_i)/x$ is a z-root of $Q_i(x, xz + f_i)$ and, hence, of $Q_{i+1}(x, z) = T_i(x, xz + f_i) = x^{-m_i} Q_i(x, xz + f_i)$. This completes the proof of part (i).

Now, substituting $z = \psi_i(x)$ in $T_i(x, z)$, we get from part (i) that

$$T_i(x, \psi_i(x)) = x^{-m_i} \underbrace{Q_i(x, \psi_i(x))}_{0} = 0 .$$

In particular,

$$T_i(0, f_i) = T_i(x, \psi_i(x))|_{x=0} = 0 ,$$

thereby proving part (ii). □

We are now in a position to prove the following converse of Proposition 9.9.

Proposition 9.11 *Let $Q(x,z)$ be a nonzero bivariate polynomial in $F[x,z]$ and let U be the set that is computed by the call* BIROOT$(Q,k,0)$. *Every z-root of $Q(x,z)$ in $F_k[x]$ is contained in U.*

Proof. Let $f(x) = f_0 + f_1 x + \ldots + f_{k-1}x^{k-1}$ be a z-root of $Q(x,z)$ in $F_k[x]$ and define $Q_i(x,z)$ and $T_i(x,z)$ as in Lemma 9.10. We prove by induction on $i = 0,1,\ldots,k-1$ that there is a recursion descent in BIROOT along which recursion level i is called with the parameters (Q_i,k,i).

The induction base $i = 0$ is again obvious. Turning to the induction step, consider an execution of BIROOT at recursion level $\lambda = i < k$ with the parameters (Q_i,k,i). The result of the computation in line 4 in BIROOT is then $T_i(x,z)$, and by Lemma 9.10(ii) we get that f_i is inserted into the set Z in line 5. This means that γ equals f_i in one of the iterations of the loop in lines 6–13. If $i < k-1$, the recursive call in line 9 is made during that iteration with the parameters $(T_i(x,xz+f_i),k,\lambda{+}1) = (Q_{i+1}(x,z),k,i{+}1)$, thereby completing the proof of the induction step. This proof also shows that the contents of the global variable $g(x)$ equals $f(x)$ upon one of the executions of line 7 at level $i = k-1$ of the recursion descent; by Lemma 9.10(i) we then have $Q_{k-1}(x,g_{k-1}) = Q_{k-1}(x,f_{k-1}) = 0$, which means that line 12 inserts $f(x)$ into U. □

We have already seen that the number of recursive calls to BIROOT is always finite; we next compute an upper bound on this number. Consider a particular recursion descent of BIROOT and let $T_i(x,z)$ denote the result of the computation in line 4 during the execution of recursion level $i < k-1$ along that descent; the number of recursive calls in line 9 then equals the number of (distinct) roots of $T_i(0,z)$. It might seem at first that the total number of recursive calls made in a given level i, across all recursion descents, could grow exponentially with i. However, we show in the next lemma that along each recursion descent, the degree of $T_i(0,z)$ (for $i > 0$) is bounded from above by the largest multiplicity of any root of $T_{i-1}(0,z)$. Thus, if $T_i(0,z)$ happens to have many roots, then they will be compensated for by $T_{i-1}(0,z)$, as the latter polynomial will necessarily have only a few distinct roots.

Lemma 9.12 *Let $T_{i-1}(x,z)$ be a bivariate polynomial in $F[x,z]$ such that $T_{i-1}(0,z)$ is not identically zero, and let γ be a (z-)root of multiplicity h of $T_{i-1}(0,z)$ in F. Define $T_i(x,z) = x^{-m}T_{i-1}(x,xz+\gamma)$, where m is the largest integer such that x^m divides $T_{i-1}(x,xz+\gamma)$. Then* $\deg T_i(0,z) \le h$.

Proof. Denote $A(x,z) = \sum_t z^t A^{(t)}(x) = T_{i-1}(x, z+\gamma)$. Since γ is a root of multiplicity h of $T_{i-1}(0,z)$ then $z = 0$ is a root of multiplicity h of $A(0,z)$. Therefore, $A^{(t)}(0) = 0$ for $0 \le t < h$ and $A^{(h)}(0) \ne 0$; equivalently, x divides $A^{(t)}(x)$ for $0 \le t < h$ but it does not divide $A^{(h)}(x)$. Noting that $A(x,xz) = \sum_t z^t x^t A^{(t)}(x)$, it follows that x divides $A(x,xz)$ but x^{h+1} does not. The largest integer m such that x^m divides $A(x,xz)$ thus satisfies $1 \le m \le h$. Now, for this m,

$$T_i(x,z) = x^{-m} T_{i-1}(x, xz+\gamma) = x^{-m} A(x, xz)$$

and, so,

$$\begin{aligned} T_i(0,z) = (x^{-m} A(x,xz))|_{x=0} &= \sum_t z^t (x^{t-m} A^{(t)}(x))|_{x=0} \\ &= \sum_{t \le m} z^t (x^{t-m} A^{(t)}(x))|_{x=0} \, . \end{aligned}$$

Hence, $\deg T_i(0,z) \le m \le h$. □

The following proposition is our final result of this section, providing an upper bound on the number of recursive calls throughout the execution of BiRoot. This result also implies that BiRoot has polynomial-time complexity, if so does the root extractor called in line 5.

Proposition 9.13 *Suppose that* BiRoot *is initially called with the parameters* $(Q, k, 0)$, *where* $Q = Q(x,z)$ *is a nonzero bivariate polynomial in* $F[x,z]$ *with* $(0,1)$*-degree* ℓ. *Then the overall number of recursive calls made to* BiRoot *is at most* $\ell(k-1)$.

Proof. Fix some i in the range $1 \le i \le k-1$ and let $T_{i-1}(x,z)$ and Z_{i-1} denote, respectively, the values of the polynomial $T(x,z)$ and the set Z during the execution of lines 4 and 5 at level $i-1$ of some recursion descent. In that level, we make a recursive call to BiRoot in line 9 for each element γ in Z_{i-1}. Each such recursive call, in turn, computes a new value for the polynomial $T(x,z)$. Denoting that value by $T_{i,\gamma}(x,z)$, we get from Lemma 9.12 that

$$\sum_{\gamma \in Z_{i-1}} \deg T_{i,\gamma}(0,z) \le \deg T_{i-1}(0,z) \, . \tag{9.20}$$

Based on this observation, we next bound from above the sum of the degrees of the polynomials $T(0,z)$ to which line 5 is applied during level i, *across all recursion descents*; note that this sum bounds from above the total number of recursive calls made during level i (across all recursion descents). Letting ζ_i stand for that sum, for $i = 0$ we clearly have

$$\zeta_0 \le \deg T_0(0,z) \le \deg_{0,1} Q(x,z) \le \ell \, ,$$

where $T_0(x, z)$ is the polynomial computed in the first execution of line 4 in BiRoot. For $i > 0$ we get from (9.20) that

$$\zeta_{k-2} \leq \zeta_{k-3} \leq \ldots \leq \zeta_1 \leq \zeta_0 \leq \ell \, .$$

Therefore, the total number of recursive calls to BiRoot is at most $\sum_{i=0}^{k-2} \zeta_i \leq \ell(k-1)$. □

9.8 Bounds on the decoding radius

For an (n, M, d) code $\mathcal{C}$ over an alphabet F, denote by $\Delta_\ell(\mathcal{C})$ the largest decoding radius of any list-ℓ decoder for $\mathcal{C}$. Equivalently, $\Delta_\ell(\mathcal{C})$ is the largest integer τ such that all Hamming spheres with radius τ in F^n contain at most ℓ codewords of $\mathcal{C}$.

When $\ell \geq M$, we have a trivial list-ℓ decoder $\mathbf{y} \mapsto \mathcal{D}(\mathbf{y})$ that returns the list $\mathcal{D}(\mathbf{y}) = \mathcal{C}$ for every $\mathbf{y} \in F^n$. Thus, when $\ell \geq M$, the value $\Delta_\ell(\mathcal{C})$ can be defined to be any number between n and ∞. In the remaining part of this section, we focus on the case $\ell < M$.

It follows from the Koetter–Vardy algorithm that for every $[n, k, \delta n]$ GRS code $\mathcal{C}$ over $F = \mathrm{GF}(q)$,

$$\Delta_\ell(\mathcal{C}) \geq n\Theta_\ell(1-\delta, q) - 1 \, ,$$

where $\Theta_\ell(R', q)$ is given by (9.19). We will demonstrate that this inequality is, in fact, a special case of a more general result, which applies to every $(n, M, \delta n)$ code (linear or nonlinear) over an alphabet of size q.

For positive integers M and q and a real $\theta \in [0, 1]$, define

$$\mathcal{J}(M, \theta, q) = \frac{(M-\rho-\sigma+1)M\theta + \binom{\rho}{2} + \binom{\sigma}{2}(q-1)}{\binom{M}{2}} \, , \tag{9.21}$$

where $\rho = \lceil M\theta \rceil$ and $\sigma = \lceil \rho/(q-1) \rceil$. This expression was introduced in Problem 4.27 while presenting a stronger version of the Johnson bound (see Proposition 4.11). Specifically, the following result was given as an exercise in Problem 4.27.

Proposition 9.14 (Improvement on the Johnson bound) *Let $\mathcal{C}$ be an $(n, M, \delta n)$ code over an Abelian group of size q and let θn be the largest Hamming weight of any codeword in $\mathcal{C}$. If $\theta \leq 1 - (1/q)$ then*

$$\delta \leq \mathcal{J}(M, \theta, q) \, .$$

We also recall from Problem 4.27 the following properties of the mapping

$$\theta \mapsto \mathcal{J}(M, \theta, q) ,$$

when viewed as a function over the real interval $[0, 1]$: it is continuous and piecewise linear, and it is strictly increasing for $0 < \theta < 1 - (\lceil M/q \rceil / M)$. Restricting this function now to the interval $\theta \in [0, 1 - (\lceil M/q \rceil / M)]$, the inverse function,

$$\delta \mapsto \mathcal{J}^{-1}(M, \delta, q) ,$$

is well-defined. In fact, it turns out that this inverse is very closely related to the function $\Theta_\ell(R', q)$ in (9.19): the former can be expressed in terms of the latter as

$$\mathcal{J}^{-1}(M, \delta, q) = \Theta_{M-1}(1-\delta, q) .$$

The proof of this equality is given as a guided exercise in Problem 9.9. Thus, we infer from Proposition 9.14 the following result.

Proposition 9.15 *Let $\mathcal{C}$ be an $(n, M, \delta n)$ code over an Abelian group of size q and suppose that there is an integer τ such that each codeword in $\mathcal{C}$ has Hamming weight at most τ. Then*

$$\tau \geq \lceil n \Theta_{M-1}(1-\delta, q) \rceil .$$

We next use Proposition 9.15 to obtain a lower bound on $\Delta_\ell(\mathcal{C})$ for the range $\ell < M$.

Theorem 9.16 *For every $(n, M, \delta n)$ code $\mathcal{C}$ over an alphabet of size q and every positive integer $\ell < M$,*

$$\Delta_\ell(\mathcal{C}) \geq \lceil n \Theta_\ell(1-\delta, q) \rceil - 1 .$$

Proof. Let τ be $\Delta_\ell(\mathcal{C}) + 1$ and F be the alphabet of $\mathcal{C}$; without loss of generality we can assume that F is an Abelian group. By the definition of $\Delta_\ell(\mathcal{C})$ it follows that there is a word $\mathbf{y} \in F^n$ and $\ell+1$ codewords in $\mathcal{C}$ within Hamming distance τ from $\mathbf{y}$. By translation, we can assume that $\mathbf{y} = \mathbf{0}$, thereby implying that $\mathcal{C}$ contains a subset $\mathcal{C}'$ of $\ell+1$ codewords, all of which have Hamming weight at most τ. The result is now obtained by applying Proposition 9.15 to the $(n, \ell+1, \geq \delta n)$ code $\mathcal{C}'$. □

As said earlier, for the special case of GRS codes (and also for alternant codes whose designed minimum distance equals their true minimum distance), the lower bound of Theorem 9.16 is already implied by the results of Section 9.6. Furthermore, Section 9.6 provides not just a bound, but also a polynomial-time decoding algorithm.

Problems

[Section 9.1]

Problem 9.1 Let $F = \mathrm{GF}(q)$ and let $\mathcal{C}$ be a Hamming code of length $n = (q^m - 1)/(q-1)$ over F. Consider the mapping $\mathcal{D} : F^n \to 2^{\mathcal{C}}$ that is defined for every $\mathbf{y} \in F^n$ by

$$\mathcal{D}(\mathbf{y}) = \{\mathbf{c} \in \mathcal{C} \,:\, \mathsf{d}(\mathbf{y}, \mathbf{c}) \le 2\} .$$

Show that $\mathcal{D}$ is a list-$\left(\frac{1}{2}(q^m - q) + 1\right)$ decoder (with decoding radius 2).

Hint: First argue that it suffices to consider only words $\mathbf{y} \in F^n$ whose Hamming weight is 0 or 1. Then show that when $\mathsf{w}(\mathbf{y}) = 1$, there are exactly $\frac{1}{2}(q^m - q) = \frac{1}{2}(n-1)(q-1)$ codewords in $\mathcal{C}$ of Hamming weight 3 that are at distance 2 from $\mathbf{y}$.

[Section 9.2]

Problem 9.2 Let $a(x, z)$ be a monic nonzero polynomial in $F[x][z]$. Show that for every $Q(x, z) \in F[x, z]$,

$$a(x, z) \text{ divides } Q(x, z) \text{ in } F(x)[z] \quad \Longleftrightarrow \quad a(x, z) \text{ divides } Q(x, z) \text{ in } F[x, z] .$$

Problem 9.3 Let $Q(x, z) = \sum_{t=0}^{\ell} z^t Q_t(x)$ be a nonzero polynomial in $F[x, z]$ with $\deg_{0,1} Q(x, z) = \ell$, and suppose that $f(x) \in F[x]$ is a z-root of $Q(x, z)$.

1. Show that
$$\deg f \le \max_{0 \le t < \ell} \frac{\deg Q_t - \deg Q_\ell}{\ell - t} .$$
2. Show that if $f(x)$ is identically zero then so is $Q_0(x)$, and if $f(x)$ is nonzero then $f(x) \,|\, Q_0(x)$.

[Section 9.3]

Problem 9.4 Let $\mathcal{C}_{\mathrm{GRS}}$ be an $[n, k, d]$ GRS code over $F = \mathrm{GF}(q)$, and assume that the generator matrix of $\mathcal{C}_{\mathrm{GRS}}$ takes the form (9.1). The purpose of this problem is to show how the Welch–Berlekamp equations in Problem 6.13 can be obtained from the degree constraints (9.2) and (9.3), and the interpolation constraint (9.4), by reducing the number of equations from n to $d-1$.

Recalling the notation from Problem 6.13, let

$$\mathbf{c} = (\, u(\alpha_1) \;\; u(\alpha_2) \;\; \ldots \;\; u(\alpha_n) \,)$$

be the transmitted codeword where $u(x) \in F_k[x]$, and denote by $\mathbf{y} = (y_1 \; y_2 \; \ldots \; y_n)$ the received word. Let $\tilde{u}(x)$ be the unique polynomial in $F_k[x]$ that satisfies

$$\tilde{u}(\alpha_j) = y_j \, , \quad d \le j \le n \, ,$$

and define the re-encoded codeword to be

$$\tilde{\mathbf{c}} = (\, \tilde{u}(\alpha_1) \;\; \tilde{u}(\alpha_2) \;\; \ldots \;\; \tilde{u}(\alpha_n) \,) .$$

Denote the difference $\mathbf{y} - \tilde{\mathbf{c}}$ by $\tilde{\mathbf{y}} = (\tilde{y}_1 \; \tilde{y}_2 \; \ldots \; \tilde{y}_n)$, where $\tilde{y}_j = 0$ for $d \le j \le n$.

Let $Q(x, z) = Q_0(x) + zQ_1(x)$ be a nonzero bivariate polynomial in $F[x, z]$ that satisfies the degree constraints (9.2) and (9.3), as well as the interpolation constraint (9.4) when applied to $\tilde{\mathbf{y}}$, namely,

$$Q(\alpha_j, \tilde{y}_j) = 0 , \quad j = 1, 2, \ldots, n .$$

1. Show that $Q_0(x)$ is divisible by the polynomial

$$\Omega(x) = \prod_{j=d}^{n} (x - \alpha_j) .$$

2. Write $Q_1(x) = \mathrm{V}(x)$ and $Q_0(x) = \mathrm{N}(x)\Omega(x)$, and let $A(x)$ be given by

$$A(x) = \prod_{j=1}^{d-1} (x - \alpha_j) .$$

   Also, denote by $A'(x)$ the formal derivative of $A(x)$. Show that the polynomial pair $(\mathrm{V}(x), \mathrm{N}(x))$ satisfies the Welch–Berlekamp equations in part 5 of Problem 6.13, namely,

$$\deg \mathrm{V} < \tfrac{1}{2}(d+1) \quad \text{and} \quad \deg \mathrm{N} < \tfrac{1}{2}(d-1)$$

   and

$$\mathrm{N}(\alpha_j) = \tilde{y}_j v_j A'(\alpha_j) \cdot \mathrm{V}(\alpha_j) , \quad 1 \leq j < d ,$$

   where $v_1, v_2, \ldots, v_n$ denote the column multipliers of the canonical parity-check matrix $(v_j \alpha_j^i)_{i=0}^{d-2}{}_{j=1}^{n}$ of $\mathcal{C}_{\mathrm{GRS}}$, as in part 1 of Problem 6.13.

   (It follows from part 6 of Problem 6.13 that the following converse also holds: if a nonzero pair $(\mathrm{V}(x), \mathrm{N}(x))$ satisfies the Welch–Berlekamp equations, then $Q(x, z) = \mathrm{V}(x) + z\,\mathrm{N}(x)\Omega(x)$ satisfies (9.2)–(9.4), with (9.4) applied to $\tilde{\mathbf{y}}$.)

[Section 9.4]

Problem 9.5 Show that the expression $\Theta_{\ell,1}(R')$, which appears in Sudan's algorithm, satisfies

$$\Theta_{\ell,1}(R') \geq \Theta_{\ell-1,1}(R') \quad \Longleftrightarrow \quad R' \leq \frac{2}{\ell(\ell+1)} .$$

[Section 9.5]

Problem 9.6 Verify the following properties of the expression $\Theta_{\ell,r}(R')$, which appears in the Guruswami–Sudan algorithm:

1. For $2 \leq r \leq \ell$,

$$\Theta_{\ell,r}(R') \geq \Theta_{\ell,r-1}(R') \quad \Longleftrightarrow \quad R' \geq \Upsilon_{\ell,r} .$$

2. For $2 \leq r \leq \ell$,

$$\Theta_{\ell,r}(R') \geq \Theta_{\ell-1,r-1}(R') \quad \Longleftrightarrow \quad R' \geq \Upsilon_{\ell,r} .$$

3. For $1 \le r \le \ell{-}1$,

$$\Theta_{\ell,r}(R') \ge \Theta_{\ell-1,r}(R') \quad \Longleftrightarrow \quad R' \le \Upsilon_{\ell,r+1} .$$

Problem 9.7 Verify the following properties of the expression $\Theta_\ell(R')$ defined in (9.15):

1. For $\ell > 1$,

$$\Theta_\ell(R') \ge \Theta_{\ell-1}(R') .$$

2. (Limit when $\ell \to \infty$)

$$\lim_{\ell\to\infty} \Theta_\ell(R') = 1 - \sqrt{R'} .$$

[Section 9.7]

Problem 9.8 Suppose that line 11 is deleted from the procedure BiRoot in Figure 9.5, namely, the polynomial $g(x)$ is inserted in line 12 to U when λ reaches the value $k{-}1$, regardless of whether $Q(x, g_{k-1})$ is zero.

1. Show that the size of U is (still) bounded from above by $\deg_{0,1} Q(x, z)$.

   Hint: See the proof of Proposition 9.13.

2. Verify that when the modified BiRoot is applied to the polynomial $Q(x, z)$ in Example 9.2 with $k = 2$, the procedure will return in U the following four polynomials:

   $u_1(x) = 18{+}14x\,,\ u_2(x) = 8{+}8x\,,\ u_3(x) = 14{+}16x\,,$ and $u_4(x) = 18{+}15x\,.$

   (So, the change in the algorithm may produce polynomials that are not z-roots of $Q(x, z)$.)

3. Verify that when the modified BiRoot is applied to the polynomial $\hat{Q}(x, z)$ in Example 9.2, the procedure will return the polynomials

   $\hat{u}_1(x) = 18 + 14x\,,\ \hat{u}_2(x) = 8 + 8x\,,\ \hat{u}_3(x) = 13 + 9x\,,$ and $\hat{u}_4(x) = 10 + x\,.$

[Section 9.8]

Problem 9.9 For a real $\theta \in [0, 1]$, positive integers M and q, and nonnegative integers μ and ν, define

$$\mathcal{J}_{\mu,\nu}(M, \theta, q) = \frac{(M{-}\mu{-}\nu{+}1)M\theta + \binom{\mu}{2} + \binom{\nu}{2}(q{-}1)}{\binom{M}{2}} .$$

In particular, if $\rho = \lceil M\theta \rceil$ and $\sigma = \lceil \rho/(q{-}1) \rceil$ for some $\theta \in [0, 1]$ then

$$\mathcal{J}_{\rho,\sigma}(M, \theta, q) = \mathcal{J}(M, \theta, q) ,$$

where the expression $\mathcal{J}(M, \theta, q)$ is given by (9.21).

1. Show that for every real $\theta \in [0,1]$,

$$\mathcal{J}_{\mu,\nu}(M,\theta,q) \le \mathcal{J}_{\mu+1,\nu}(M,\theta,q) \iff \mu \ge M\theta$$

   and

$$\mathcal{J}_{\mu,\nu}(M,\theta,q) \le \mathcal{J}_{\mu,\nu+1}(M,\theta,q) \iff \nu \ge M\theta/(q-1) .$$

2. Show that for every $\theta \in [0,1]$,

$$\mathcal{J}(M,\theta,q) = \min_{\mu,\nu \in \mathbb{N}} \mathcal{J}_{\mu,\nu}(M,\theta,q) .$$

3. For nonnegative integers μ and ν such that $\mu+\nu \le M$, let $\delta \mapsto \mathcal{J}_{\mu,\nu}^{-1}(M,\delta,q)$ be the inverse of the function that is defined for $\theta \in [0,1]$ by

$$\theta \mapsto \mathcal{J}_{\mu,\nu}(M,\theta,q) .$$

   Also, let $\delta \mapsto \mathcal{J}^{-1}(M,\delta,q)$ be the inverse of $\theta \mapsto \mathcal{J}(M,\theta,q)$, where the latter function is restricted to the interval $\theta \in [0, 1-(\lceil M/q \rceil/M)]$. Denote by $\delta_{M,q}$ the value of $\mathcal{J}(M,\theta,q)$ at $\theta = 1-(\lceil M/q \rceil/M)$ (recall from part 4 of Problem 4.27 that $\delta_{M,q}$ is the maximum value of $\mathcal{J}(M,\theta,q)$ over $[0,1]$). Show that for every $\delta \in [0,\delta_{M,q}]$,

$$\mathcal{J}^{-1}(M,\delta,q) = \max_{\mu,\nu \in \mathbb{N}\,:\,\mu+\nu \le M} \mathcal{J}_{\mu,\nu}^{-1}(M,\delta,q) .$$

4. Show that for every two integers r and $\overline{r}$ such that $0 \le \overline{r} < r < M$,

$$\mathcal{J}_{M-r,\overline{r}+1}^{-1}(M,\delta,q) = \Theta_{M-1,r,\overline{r}}(1-\delta,q) ,$$

   where $\Theta_{\ell,r,\overline{r}}(R',q)$ is the expression that appears in the Koetter–Vardy algorithm.

5. Show that for every $\delta \in (0,\delta_{M,q}]$,

$$\mathcal{J}^{-1}(M,\delta,q) = \Theta_{M-1}(1-\delta,q) ,$$

   where $\Theta_\ell(R',q)$ is given by (9.19).

Problem 9.10 Using the notation of Problem 9.9, verify the following properties of the expression $\Theta_\ell(R',q)$ in (9.19):

1. For every real R' in $[1-\delta_{\ell+1,q}, 1)$,

$$\Theta_\ell(R',q) = \mathcal{J}^{-1}(\ell+1, 1-R', q) .$$

   (The expression $\Theta_\ell(R',q)$ is formally defined also for $R' \in [0, 1-\delta_{\ell+1,q})$, where it may even take values that are greater than 1; yet this range of R' is not too interesting, as argued next. By the improved version of the Plotkin bound (as presented in Problem 4.29), it follows that over an alphabet of size q, there can be no codes of size $\ell+1$ and relative minimum distance greater than $\delta_{\ell+1,q}$. In particular, an alternant code $\mathcal{C}_{\mathrm{alt}}$ over GF(q) can have a designed relative minimum distance $\delta > \delta_{\ell+1,q}$ only if $|\mathcal{C}_{\mathrm{alt}}| \le \ell$. Therefore, when R' $(=1-\delta)$ is smaller than $1-\delta_{\ell+1,q}$, one can realize a trivial list-ℓ decoder for $\mathcal{C}_{\mathrm{alt}}$ simply by returning the whole code $\mathcal{C}_{\mathrm{alt}}$ as the list.)

2. For every real R' in $[1-\delta_{\ell+1,q}, 1)$,

$$\Theta_\ell(R', q) \geq \frac{q-1}{q} \left(1 - \sqrt{\frac{\ell q R' + q - \ell - 1}{(\ell+1)(q-1)}} \right) .$$

Hint: Comparing the Johnson bound in Proposition 4.11 with its improved version in Problem 4.27, verify that for $\theta \in [0, \delta_{M,q}]$,

$$\mathcal{J}(M, \theta, q) \leq \frac{M}{M-1} \cdot (2\theta - \tfrac{q}{q-1}\theta^2) .$$

3. (Limit when $\ell \to \infty$) For every real $R' \in (1/q, 1)$,

$$\lim_{\ell \to \infty} \Theta_\ell(R', q) = \frac{q-1}{q} \left(1 - \sqrt{\frac{qR' - 1}{q - 1}} \right) .$$

Hint: Verify that $\lim_{\ell \to \infty} \delta_{\ell+1,q} = 1 - (1/q)$ and that

$$\lim_{M \to \infty} \mathcal{J}(M, \theta, q) = 2\theta - \tfrac{q}{q-1}\theta^2 .$$

4. (Limit when $q \to \infty$)

$$\lim_{q \to \infty} \Theta_\ell(R', q) = \Theta_\ell(R') ,$$

where $\Theta_\ell(R')$ is given by (9.15).

Notes

[Section 9.1]

The notion of list decoding was first studied by Elias and Wozencraft in the late 1950s (see [116]).

[Section 9.2]

Properties of multivariate polynomials and related algorithms can be found in the book by Zippel [403].

[Section 9.3]

As demonstrated in Problem 9.4, the GRS list-1 decoder through bivariate polynomials is, in fact, equivalent to solving the Welch–Berlekamp equations; see Berlekamp [37], Blackburn [45], Dabiri and Blake [89], Ma and Wang [244], and Welch and Berlekamp [380].

[Sections 9.4 and 9.5]

Sudan's decoder is taken from [346] and the Guruswami–Sudan algorithm is taken from [168].

While Step 1 in these algorithms can be implemented using Gaussian elimination, there are faster methods for computing the polynomial $Q(x, z)$, by taking advantage of the particular form of the equations involved. See Alekhnovich [7], Feng [123], Nielsen and Høholdt [269], O'Keeffe and Fitzpatrick [274], Olshevsky and Shokrollahi [275], Roth and Ruckenstein [302], and Sakata *et al.* [316]. The fastest implementation currently known is that of Alekhnovich [7], with a time complexity of $(\ell/R')^{O(1)} n \log^2 n \log\log n$ operations in F.

[Section 9.6]

The presentation of the Koetter–Vardy algorithm in this section is an adaptation of their results in [216] to the hard-decision list decoding of GRS codes and alternant codes (for fixed list sizes ℓ). See also Tal and Roth [353].

[Section 9.7]

The algorithm BIROOT is taken from Roth and Ruckenstein [302], where a full complexity analysis of the algorithm can be found. Denoting $N = \deg_{1,k-1} Q(x, y)$, Alekhnovich shows in [7] how by a divide-and-conquer implementation of BIROOT, one can achieve a time complexity of $\ell^{O(1)} N \log N$ operations in F, plus $O(\ell N)$ calls to the univariate root extractor. See also Augot and Pecquet [25], Feng [123], and Gao and Shokrollahi [142].

[Section 9.8]

Somewhat weaker versions of Theorem 9.16 can be found in Goldreich *et al.* [151, Section 4.1] (where the theorem is stated with $\Theta_\ell(1-\delta, q)$ replaced by the bound in part 2 of Problem 9.10), Ruckenstein [309], and Ruckenstein and Roth [310] (where $\Theta_\ell(1-\delta, q)$ is replaced by $\Theta_\ell(1-\delta)$); see also Tal and Roth [353].

Recall from Problem 4.28 that for any given Abelian group F of size q, positive integer ℓ, and rational $\theta \in (0, 1-(1/q)]$, there always exists an $(n, \ell+1, \delta n)$ code $\mathcal{C}$ over F whose codewords all have Hamming weight θn and

$$\delta = \mathcal{J}(\ell+1, \theta, q) .$$

The decoding radius of every list-ℓ decoder $\mathcal{D}$ for such a code $\mathcal{C}$ is necessarily smaller than θn, or else we would have $|\mathcal{D}(\mathbf{0})| = |\mathcal{C}| > \ell$; therefore,

$$\Delta_\ell(\mathcal{C}) \le \lceil \theta n \rceil - 1 = \lceil n\Theta_\ell(1-\delta, q) \rceil - 1 .$$

Hence, the bound in Theorem 9.16 is tight in the sense that for every q and ℓ and every rational δ in $(0, 1-(1/q)]$, there is an $(n, M{>}\ell, \delta n)$ code over an alphabet of size q for which the bound holds with equality; see also Goldreich *et al.* [151, Section 4.3] and Justesen and Høholdt [201]. On the other hand, the references [309] and [310] identify a range of values of ℓ, n, R', and q for which

$$\Delta_\ell(\mathcal{C}_{\mathrm{GRS}}) > \lceil n\Theta_\ell(R', q) \rceil - 1 ,$$

for every $[n, nR'+1]$ GRS code $\mathcal{C}_{\mathrm{GRS}}$ over $\mathrm{GF}(q)$; namely, for the mentioned range of parameters, the Koetter–Vardy algorithm does not attain the largest possible decoding radius (and, hence, neither does the Guruswami–Sudan algorithm).

Chapter 10

Codes in the Lee Metric

The study of error-correcting codes concentrates primarily on codes in the Hamming metric. Such codes are designed to correct a prescribed number of errors, where by an error we mean a change of an entry in the transmitted codeword, irrespective of the (nonzero) error value. The assignment of the same weight to each nonzero error value is reflected also in the model of the q-ary symmetric channel, where all nonzero error values occur with the same probability.

In this chapter, we consider codes in the Lee metric. This metric is defined over the ring of integer residues modulo q and it corresponds to an error model where a change of an entry in a codeword by ± 1 is counted as one error. This type of errors is found in noisy channels that use phase-shift keying (PSK) modulation, or in channels that are susceptible to synchronization errors.

Our focus herein will be on GRS codes and alternant codes: we first study their distance properties in the Lee metric, and then present an efficient decoding algorithm for these codes, which corrects any error pattern whose Lee weight is less than half the designed minimum Lee distance of the code.

We also describe another family of codes in the Lee metric, due to Berlekamp. For certain parameters, these codes are shown to be perfect in that metric; namely, they attain the Lee-metric analog of the sphere-packing bound. The latter bound and a Gilbert–Varshamov-type bound conclude our treatment of this metric. Several more bounds are included in the problems and the notes at the end of this chapter.

10.1 Lee weight and Lee distance

Let $\mathbb{Z}_q$ denote the ring of integer residues modulo the positive integer q. For an element $\alpha \in \mathbb{Z}_q$, denote by $\langle \alpha \rangle$ the smallest nonnegative integer m such that $\alpha = m \cdot 1$, where 1 stands for the multiplicative unity in $\mathbb{Z}_q$.

The *Lee weight* of an element $\alpha \in \mathbb{Z}_q$, denoted by $\mathsf{w}_{\mathcal{L}}(\alpha)$ or $|\alpha|$, takes nonnegative integer values and is defined by

$$\mathsf{w}_{\mathcal{L}}(\alpha) = |\alpha| = \begin{cases} \langle \alpha \rangle & \text{if } 0 \leq \langle \alpha \rangle \leq q/2 \\ q - \langle \alpha \rangle & \text{otherwise} \end{cases} .$$

We refer to the elements $1, 2, \ldots, \lfloor q/2 \rfloor$ as the "positive" elements of $\mathbb{Z}_q$, for which $\langle \alpha \rangle = |\alpha|$; the remaining elements in $\mathbb{Z}_q \setminus \{0\}$ are the "negative" elements of the ring. The set of positive (respectively, negative) elements of $\mathbb{Z}_q$ will be denoted by $\mathbb{Z}_q^+$ (respectively, $\mathbb{Z}_q^-$).

Example 10.1 In $\mathbb{Z}_8$ we have

$$|0| = 0 , \quad |1| = |7| = 1 , \quad |2| = |6| = 2 , \quad |3| = |5| = 3 , \quad \text{and} \quad |4| = 4 .$$

The sets $\mathbb{Z}_8^+$ and $\mathbb{Z}_8^-$ are given by $\{1, 2, 3, 4\}$ and $\{5, 6, 7\}$, respectively. $\square$

Even though $\mathbb{Z}_q$ is not necessarily a field, we will use the vector notation $(x_1 \; x_2 \; \ldots \; x_n)$ for words in $\mathbb{Z}_q^n$. Clearly, $\mathbb{Z}_q^n$ is an Abelian group, with the addition of two words being their sum, component by component, over $\mathbb{Z}_q$. The multiplication of a vector over $\mathbb{Z}_q$ by a scalar in $\mathbb{Z}_q$ is defined similarly to fields.

For a word $\mathbf{c} = (c_1 \; c_2 \; \ldots \; c_n)$ in $\mathbb{Z}_q^n$, define the Lee weight by

$$\mathsf{w}_{\mathcal{L}}(\mathbf{c}) = \sum_{j=1}^{n} |c_j|$$

(with the summation being taken over the integers). The *Lee distance* between two words $\mathbf{x}, \mathbf{y} \in \mathbb{Z}_q^n$ is defined as $\mathsf{w}_{\mathcal{L}}(\mathbf{x} - \mathbf{y})$; we denote that distance by $\mathsf{d}_{\mathcal{L}}(\mathbf{x}, \mathbf{y})$. One can verify (Problem 10.1) that the Lee distance satisfies the following properties of a metric for every $\mathbf{x}, \mathbf{y}, \mathbf{z} \in \mathbb{Z}_q^n$: (a) $\mathsf{d}_{\mathcal{L}}(\mathbf{x}, \mathbf{y}) \geq 0$, with equality holding if and only if $\mathbf{x} = \mathbf{y}$, (b) $\mathsf{d}_{\mathcal{L}}(\mathbf{x}, \mathbf{y}) = \mathsf{d}_{\mathcal{L}}(\mathbf{y}, \mathbf{x})$, and (c) $\mathsf{d}_{\mathcal{L}}(\mathbf{x}, \mathbf{y}) \leq \mathsf{d}_{\mathcal{L}}(\mathbf{x}, \mathbf{z}) + \mathsf{d}_{\mathcal{L}}(\mathbf{z}, \mathbf{y})$ (the triangle inequality).

The *minimum Lee distance* of an (n, M) code $\mathcal{C}$ over $\mathbb{Z}_q$ with $M > 1$ is defined by

$$\mathsf{d}_{\mathcal{L}}(\mathcal{C}) = \min_{\mathbf{c}_1, \mathbf{c}_2 \in \mathcal{C} \,:\, \mathbf{c}_1 \neq \mathbf{c}_2} \mathsf{d}_{\mathcal{L}}(\mathbf{c}_1, \mathbf{c}_2) .$$

A code $\mathcal{C}$ of length n over $\mathbb{Z}_q$ is called a *group code* if it is an (Abelian) subgroup of $\mathbb{Z}_q^n$ under the addition operation in $\mathbb{Z}_q^n$; a code $\mathcal{C}$ of length n is *linear* over $\mathbb{Z}_q$ if $\mathcal{C}$ is a group code over $\mathbb{Z}_q$ and $\mathbf{c} \in \mathcal{C} \Longrightarrow a \cdot \mathbf{c} \in \mathcal{C}$ for every $a \in \mathbb{Z}_q$ (see Problems 2.9 and 2.20). The minimum Lee distance of a group code $\mathcal{C} \neq \{\mathbf{0}\}$ over $\mathbb{Z}_q$ is the minimum Lee weight of any nonzero codeword in $\mathcal{C}$ (Problem 10.2).

Example 10.2 Consider the $(2, 15)$ code $\mathcal{C}$ over $\mathbb{Z}_{15}$ that is defined by

$$\mathcal{C} = \{(a \;\; 6a) \; : \; a \in \mathbb{Z}_{15}\} \; .$$

This code is linear over $\mathbb{Z}_{15}$. Examining the 14 nonzero codewords in $\mathcal{C}$ yields that those with the smallest Lee weight are

$$\pm(2 \;\; -3) \quad \text{and} \quad \pm(5 \;\; 0) \; .$$

Therefore, $\mathsf{d}_{\mathcal{L}}(\mathcal{C}) = 5$. □

10.2 Newton's identities

In upcoming sections, we will see how GRS codes and alternant codes perform as Lee-metric codes over $\mathbb{Z}_q$, where q is a prime. Our analysis will make use of properties of formal power series over fields, as we summarize next. Some of these properties have already been described in Section 6.3.2 and are repeated here for completeness.

Recall from Section 6.3.2 that the set of formal power series over a field Φ is defined by

$$\Phi[[x]] = \left\{ a(x) = \textstyle\sum_{i=0}^{\infty} a_i x^i \; : \; a_i \in F \right\} .$$

This set forms an integral domain, and an element $\sum_{i=0}^{\infty} a_i x^i$ is invertible in $\Phi[[x]]$ if and only if $a_0 \neq 0$. Given $a(x) = \sum_{i=0}^{\infty} a_i x^i$ and $b(x) = \sum_{i=0}^{\infty} b_i x^i$ in $\Phi[[x]]$, we write $a(x) \equiv b(x) \pmod{x^t}$ if $a_i = b_i$ for $0 \leq i < t$.

The formal derivative of $a(x) = \sum_{i=0}^{\infty} a_i x^i \in \Phi[[x]]$ is defined by $a'(x) = \sum_{i=1}^{\infty} i a_i x^{i-1}$. Given $a(x), b(x) \in \Phi[[x]]$, the following rules of differentiation,

$$(a(x) + b(x))' = a'(x) + b'(x)$$

and

$$(a(x)b(x))' = a'(x)b(x) + a(x)b'(x) \; ,$$

extend easily from their polynomial counterparts. The next lemma provides the rule for differentiating the ratio of two elements of $\Phi[[x]]$.

Lemma 10.1 *Let $a(x)$ and $b(x)$ be elements in $\Phi[[x]]$ where $b(x)$ is invertible. Then,*

$$\left(\frac{a(x)}{b(x)}\right)' = \frac{a'(x)b(x) - a(x)b'(x)}{b^2(x)} \; .$$

The proof is left as an exercise (Problem 10.4).

The following lemma presents a useful relation between a polynomial and the power sums of its roots. This relation is commonly referred to as *Newton's identities*.

Lemma 10.2 (Newton's identities) *Let $\beta_1, \beta_2, \ldots, \beta_h$ be (not necessarily distinct) nonzero elements in a field Φ and define the polynomial $\sigma(x) = \sum_{i=0}^{h} \sigma_i x^i \in \Phi[x]$ by*

$$\sigma(x) = \prod_{j=1}^{h} (1 - \beta_j x) .$$

Let $S(x) = \sum_{\ell=1}^{\infty} S_\ell x^\ell \in \Phi[[x]]$ be defined by

$$S_\ell = \sum_{j=1}^{h} \beta_j^\ell , \quad \ell \geq 1 .$$

Then $\sigma(x)$ and $S(x)$ are related by

$$\sigma(x)S(x) = -x\sigma'(x) ,$$

or, equivalently,

$$\sum_{\ell=0}^{i-1} \sigma_\ell S_{i-\ell} = -i\sigma_i , \quad i \geq 1 .$$

Proof. By the definition of $S(x)$ we have,

$$S(x) = \sum_{\ell=1}^{\infty} \Big(\sum_{j=1}^{h} \beta_j^\ell \Big) x^\ell = \sum_{j=1}^{h} \sum_{\ell=1}^{\infty} (\beta_j x)^\ell = \sum_{j=1}^{h} (\beta_j x) \sum_{i=0}^{\infty} (\beta_j x)^i = \sum_{j=1}^{h} \frac{\beta_j x}{1 - \beta_j x}$$

(see Example 6.2). Hence,

$$\sigma(x)S(x) = \sum_{j=1}^{h} (\beta_j x) \prod_{\substack{1 \leq m \leq h: \\ m \neq j}} (1 - \beta_m x) = -x\sigma'(x) ,$$

thereby completing the proof. □

The main result of this section is given by the next lemma, which generalizes Lemma 10.2.

Lemma 10.3 *Let $\beta_1, \beta_2, \ldots, \beta_h, \gamma_1, \gamma_2, \ldots, \gamma_k$ be (not necessarily distinct) nonzero elements in a field Φ and let $\psi(x)$ be the (unique) formal power series in $\Phi[[x]]$ that satisfies*

$$\psi(x) = \frac{\prod_{j=1}^{h} (1 - \beta_j x)}{\prod_{j=1}^{k} (1 - \gamma_j x)} .$$

Define the formal power series $S(x) = \sum_{\ell=1}^{\infty} S_\ell x^\ell \in \Phi[[x]]$ *by*

$$S_\ell = \Bigl(\sum_{j=1}^{h} \beta_j^\ell\Bigr) - \Bigl(\sum_{j=1}^{k} \gamma_j^\ell\Bigr) , \qquad \ell \ge 1 .$$

Then $\psi(x)$ *and* $S(x)$ *are related by*

$$\psi(x) S(x) = -x\psi'(x) .$$

Proof. Let the formal power series $S^+(x) = \sum_{\ell=1}^{\infty} S_\ell^+ x^\ell$ and $S^-(x) = \sum_{\ell=1}^{\infty} S_\ell^- x^\ell$ be defined by

$$S_\ell^+ = \sum_{j=1}^{h} \beta_j^\ell \qquad \text{and} \qquad S_\ell^- = \sum_{j=1}^{k} \gamma_j^\ell , \qquad \ell \ge 1 ,$$

and let the polynomials $\sigma^+(x)$ and $\sigma^-(x)$ be given by

$$\sigma^+(x) = \prod_{j=1}^{h} (1 - \beta_j x) \qquad \text{and} \qquad \sigma^-(x) = \prod_{j=1}^{k} (1 - \gamma_j x) .$$

By Lemma 10.2 we have

$$\sigma^+(x) S^+(x) = -x(\sigma^+(x))' \qquad \text{and} \qquad \sigma^-(x) S^-(x) = -x(\sigma^-(x))' . \tag{10.1}$$

Now multiply the first equality in (10.1) by $1/\sigma^-(x)$ and the second equality by $\sigma^+/(\sigma^-(x))^2$, then subtract one resulting equation from the other; this yields

$$\frac{\sigma^+(x)}{\sigma^-(x)} (S^+(x) - S^-(x)) = -x \, \frac{(\sigma^+(x))'\sigma^-(x) - \sigma^+(x)(\sigma^-(x))'}{(\sigma^-(x))^2} .$$

Recalling that $S(x) = S^+(x) - S^-(x)$ and that $\psi(x) = \sigma^+(x)/\sigma^-(x)$, the result is obtained from Lemma 10.1. □

10.3 Lee-metric alternant codes and GRS codes

Let F be the field GF(p) where p is a prime. Throughout this section, we fix an $[n, n-\varrho]$ *normalized* GRS code $\mathcal{C}_{\mathrm{GRS}}$ over a finite extension field Φ of F with (nonzero) code locators $\alpha_1, \alpha_2, \ldots, \alpha_n$ and redundancy ϱ. Denote by $\mathcal{C}_{\mathrm{alt}}$ the respective alternant code over F; namely, $\mathcal{C}_{\mathrm{alt}} = \mathcal{C}_{\mathrm{GRS}} \cap F^n$.

The next theorem is the first step in our study of the performance of alternant codes in the Lee metric: this theorem provides a lower bound on the minimum Lee distance of $\mathcal{C}_{\mathrm{alt}}$.

Theorem 10.4 *If* $\mathcal{C}_{\mathrm{alt}} \neq \{\mathbf{0}\}$ *and* $1 \le \varrho \le p/2$ *then*

$$\mathsf{d}_{\mathcal{L}}(\mathcal{C}_{\mathrm{alt}}) \ge 2\varrho\,.$$

Proof. Assume that $\mathbf{c}$ is a codeword of $\mathcal{C}_{\mathrm{alt}}$ with $\mathsf{w}_{\mathcal{L}}(\mathbf{c}) < 2\varrho$. We show that $\mathbf{c} = \mathbf{0}$. Letting F^+ (respectively, F^-) denote the set of positive (respectively, negative) elements of F, define the index sets J^+ and J^- by

$$J^{\pm} = \left\{ j \in \{1, 2, \ldots, n\} \;:\; c_j \in F^{\pm} \right\},$$

and let the polynomials $\sigma^+(x), \sigma^-(x) \in \Phi[x]$ be given by

$$\sigma^{\pm}(x) = \prod_{j \in J^{\pm}} (1 - \alpha_j x)^{|c_j|}\,.$$

Observe that

$$\mathsf{w}_{\mathcal{L}}(\mathbf{c}) = \sum_{j=1}^{n} |c_j| = \deg \sigma^+ + \deg \sigma^-\,. \tag{10.2}$$

Define the formal power series $S(x) = \sum_{\ell=1}^{\infty} S_\ell x^\ell \in \Phi[[x]]$ by

$$S_\ell = \sum_{j=1}^{n} c_j \alpha_j^\ell = \Bigl(\sum_{j \in J^+} |c_j| \alpha_j^\ell \Bigr) - \Bigl(\sum_{j \in J^-} |c_j| \alpha_j^\ell \Bigr), \quad \ell \ge 1\,.$$

By Lemma 10.3 we obtain

$$\frac{\sigma^+(x)}{\sigma^-(x)} S(x) = -x \left(\frac{\sigma^+(x)}{\sigma^-(x)} \right)'\,.$$

Let $\varphi(x) = 1 + \sum_{i=1}^{2\varrho-1} \varphi_i x^i$ be the unique polynomial in $\Phi_{2\varrho}[x]$ such that

$$\varphi(x) \equiv \frac{\sigma^+(x)}{\sigma^-(x)} \pmod{x^{2\varrho}}\,. \tag{10.3}$$

Taking derivatives of both sides of the congruence (10.3), we obtain

$$\varphi'(x) \equiv \left(\frac{\sigma^+(x)}{\sigma^-(x)} \right)' \pmod{x^{2\varrho-1}}$$

(note the degree of the modulus). From the last three equations we get

$$\varphi(x) S(x) \equiv -x \varphi'(x) \pmod{x^{2\varrho}}\,. \tag{10.4}$$

Now, $\mathbf{c} \in \mathcal{C}_{\mathrm{alt}}$ implies that $S(x) \equiv 0 \pmod{x^\varrho}$; therefore, by (10.4) we obtain that $x^{\varrho-1} \,|\, \varphi'(x)$. Hence, $i\varphi_i = 0$ for $1 \le i < \varrho\ (\le p)$, or

$$\varphi(x) \equiv 1 \pmod{x^\varrho}\,. \tag{10.5}$$

Next, consider the (integer) difference

$$\deg \sigma^+ - \deg \sigma^- = \Big(\sum_{j \in J^+} |c_j| \Big) - \Big(\sum_{j \in J^-} |c_j| \Big) .$$

Since $\mathbf{c} \in \mathcal{C}_{\mathrm{alt}}$ implies that $\sum_{j=1}^{n} c_j = 0$, this difference satisfies

$$\deg \sigma^+ - \deg \sigma^- \equiv 0 \pmod{p} ,$$

from which we deduce that

$$\text{either} \quad \deg \sigma^+ = \deg \sigma^- \quad \text{or} \quad |\deg \sigma^+ - \deg \sigma^-| \geq p .$$

We next examine each of these two cases.

Case 1: $\deg \sigma^+ = \deg \sigma^-$. By (10.2) and our assumption on the Lee weight of $\mathbf{c}$, we have

$$\deg \sigma^+ = \deg \sigma^- = \tfrac{1}{2} \mathsf{w}_{\mathcal{L}}(\mathbf{c}) < \varrho$$

and, so, from (10.3) and (10.5) we obtain $\sigma^+(x) = \sigma^-(x)$. However, $J^+ \cap J^- = \emptyset$ implies that $\gcd(\sigma^+(x), \sigma^-(x)) = 1$; so, $\sigma^+(x) = \sigma^-(x) = 1$, i.e., $\mathbf{c} = \mathbf{0}$.

Case 2: $|\deg \sigma^+ - \deg \sigma^-| \geq p$. We again use (10.2) and our assumption on $\mathsf{w}_{\mathcal{L}}(\mathbf{c})$: here we get

$$p \leq \deg \sigma^+ + \deg \sigma^- = \mathsf{w}_{\mathcal{L}}(\mathbf{c}) < 2\varrho ,$$

yet this is impossible given the condition that $\varrho \leq p/2$. □

We mention that Theorem 10.4 would not hold in general if we removed the condition $\varrho \leq p/2$: it turns out that for certain code locators, the minimum Lee distance of $\mathcal{C}_{\mathrm{alt}}$ is bounded from *above* by p. Indeed, suppose that $[\Phi : F] \geq 2$ and $n \geq p$ and let the first p code locators be given by $\alpha_j = \beta + \gamma_j$, where $\beta \in \Phi \setminus F$ and γ_j ranges over all the elements of F. Now, $\sum_{j=1}^{p} \gamma_j^r = 0$ for every $0 \leq r < p-1$ (Problem 3.22). Therefore, for $0 \leq \ell < p-1$,

$$\sum_{j=1}^{p} \alpha_j^\ell = \sum_{j=1}^{p} (\beta + \gamma_j)^\ell = \sum_{j=1}^{p} \sum_{i=0}^{\ell} \binom{\ell}{i} \beta^i \gamma_j^{\ell-i} = \sum_{i=0}^{\ell} \binom{\ell}{i} \beta^i \sum_{j=1}^{p} \gamma_j^{\ell-i} = 0 .$$

It follows that for every $\varrho < p$, the code $\mathcal{C}_{\mathrm{alt}}$ contains the codeword

$$(\underbrace{1\,1\,\ldots\,1}_{p}\underbrace{0\,0\,\ldots\,0}_{n-p}) ,$$

thus implying the upper bound $\mathsf{d}_{\mathcal{L}}(\mathcal{C}_{\mathrm{alt}}) \leq p$.

Still, the condition $\varrho \leq p/2$ can be removed from Theorem 10.4 in the special case where ($\Phi = F$ and) $\mathcal{C}_{\mathrm{alt}} = \mathcal{C}_{\mathrm{GRS}}$. The next theorem handles this case.

Theorem 10.5 *If $\Phi = F$ and $1 \le \varrho < n\ (< p)$ then*

$$\mathsf{d}_{\mathcal{L}}(\mathcal{C}_{\mathrm{GRS}}) \ge 2\varrho\ .$$

Proof. The range $\varrho \le p/2$ has already been covered by Theorem 10.4, so we can assume here that $\varrho > p/2$. Using the same notation as in the proof of Theorem 10.4, we will amend the last paragraph of that proof for the case where $|\deg \sigma^+ - \deg \sigma^-| = p$. Without loss of generality we assume that $\deg \sigma^+ - \deg \sigma^- = p$ (or else apply the proof to $-\mathbf{c}$).

Multiplying both sides of (10.5) by $S(x)$ yields

$$\varphi(x)S(x) \equiv S(x) \pmod{x^{2\varrho}}\ ;$$

hence, by (10.4), we obtain

$$S(x) \equiv -x\varphi'(x) \pmod{x^{2\varrho}}\ . \tag{10.6}$$

Since each code locator α_j is in F, we have $\alpha_j^{p-1} = 1$ and, therefore,

$$S_{\ell+p-1} = \sum_{j=1}^{n} c_j \alpha_j^{\ell+p-1} = \sum_{j=1}^{n} c_j \alpha_j^{\ell} = S_\ell\ , \quad \ell \ge 0\ ,$$

where $S_0 = \sum_{j=1}^n c_j = 0$; namely, $(S_\ell)_{\ell=0}^\infty$ is a *periodic* sequence whose period divides $p-1$ (see Problem 6.8). This implies that $S_\ell = 0$ for $p-1 \le \ell < p+\varrho-1$, which, with (10.6), leads to

$$i\varphi_i = 0 \quad \text{for} \quad p-1 \le i < 2\varrho$$

or

$$\varphi_i = 0 \quad \text{for} \ \ i = p-1 \ \text{ or } \ p < i < 2\varrho\ .$$

In particular, $\deg \varphi(x) \le p$. It follows that

$$\deg(\sigma^-(x)\varphi(x)) \le p + \deg \sigma^- = \deg \sigma^+ \le \mathsf{w}_{\mathcal{L}}(\mathbf{c}) < 2\varrho\ ;$$

so, the congruence (10.3) can be replaced by the equality

$$\sigma^-(x)\varphi(x) = \sigma^+(x)\ .$$

Recalling that $\gcd(\sigma^+(x), \sigma^-(x)) = 1$, we get that $\sigma^-(x) = 1$ and $\sigma^+(x) = \varphi(x)$ with

$$\deg \varphi = \deg \sigma^+ = p + \deg \sigma^- = p\ .$$

Observe that since $\deg \sigma^+ > 0$, the codeword $\mathbf{c}$ is nonzero and, therefore, so must be the formal power series $S(x)$. We let t be the smallest positive integer such that $S_t \ne 0$. By periodicity we have $S_{p-1} = S_p = \ldots =$

$S_{p+t-2} = 0$; consequently, we can increase the power of x in the modulus in (10.6) to obtain

$$S(x) \equiv -x\varphi'(x) \pmod{x^{p+t-1}},$$

and then apply Lemma 10.2 to yield

$$\varphi(x)S(x) \equiv S(x) \pmod{x^{p+t-1}}.$$

Now, from the definition of t we also have $S(x) \not\equiv 0 \pmod{x^{t+1}}$; hence, by the last equation we get

$$\varphi(x) \equiv 1 \pmod{x^{p-1}}.$$

Furthermore, we have shown that $\varphi_{p-1} = 0$, so we are left with

$$\sigma^+(x) = \varphi(x) = 1 + \varphi_p x^p = (1 + \varphi_p x)^p,$$

where $\varphi_p \neq 0$. This means that $\sigma^+(x)$ has a root of multiplicity p, thereby contradicting the fact that the multiplicity of each root of $\sigma^+(x)$ must be a valid Lee weight of some element in F. We thus reach the conclusion that it is impossible to have $|\deg \sigma^+ - \deg \sigma^-| = p$. $\square$

10.4 Decoding alternant codes in the Lee metric

Let $\mathcal{C}_{\mathrm{alt}}$ be an alternant code over $F = \mathrm{GF}(p)$, p prime, with an underlying $[n, n-\varrho]$ normalized GRS code $\mathcal{C}_{\mathrm{GRS}}$ with redundancy $\varrho > 0$ over a finite extension field Φ of F. In this section, we present a decoding procedure for $\mathcal{C}_{\mathrm{alt}}$, based upon Euclid's algorithm, that will correct all error words with Lee weight less than ϱ whenever $\varrho \leq \frac{1}{2}(p+1)$, and detect all error words of Lee weight ϱ whenever $\varrho \leq p/2$.

We first establish some notation. Denote by $\mathbf{c} = (c_1\ c_2\ \ldots\ c_n)$ the codeword in $\mathcal{C}_{\mathrm{alt}}$ that is transmitted and by $\mathbf{y} = (y_1\ y_2\ \ldots\ y_n)$ the word in F^n that is received, with the error word given by $\mathbf{e} = (e_1\ e_2\ \ldots\ e_n) = \mathbf{y} - \mathbf{c}$. Define the index sets J^+ and J^- by

$$J^{\pm} = \left\{ j \in \{1, 2, \ldots, n\} \ : \ e_j \in F^{\pm} \right\}.$$

Namely, J^+ (respectively, J^-) is the set of locations of the positive (respectively, negative) entries in $\mathbf{e}$. We assume that $\mathsf{w}_{\mathcal{L}}(\mathbf{e}) < \varrho$.

Let the infinite sequence $(S_\ell)_{\ell=0}^{\infty}$ be defined by

$$S_\ell = \sum_{j=1}^{n} e_j \alpha_j^\ell = \sum_{j \in J^+} |e_j| \alpha_j^\ell - \sum_{j \in J^-} |e_j| \alpha_j^\ell, \quad \ell \geq 0.$$

The first ϱ elements of this sequence form the syndrome, $(S_0\ S_1\ \ldots\ S_{\varrho-1})^T$, of $\mathbf{y}$ (and $\mathbf{e}$) with respect to a canonical parity-check matrix,

$$H_{\mathrm{GRS}} = (\,\alpha_j^\ell\,)_{\ell=0}^{\varrho-1}\,{}_{j=1}^{n}\,,$$

of $\mathcal{C}_{\mathrm{GRS}}$. The formal power series $S(x)$ is then defined as

$$S(x) = \sum_{\ell=1}^{\infty} S_\ell x^\ell$$

(note that the first element of the syndrome, S_0, is excluded from $S(x)$).

We also associate with $\mathbf{e}$ positive and negative *error-locator polynomials* $\Lambda(x)$ and $\mathrm{V}(x)$, which are defined by

$$\Lambda(x) = \prod_{j\in J^+} (1-\alpha_j x)^{|e_j|} \qquad \text{and} \qquad \mathrm{V}(x) = \prod_{j\in J^-} (1-\alpha_j x)^{|e_j|}\,.$$

We will write these two polynomials formally as an expression $\Lambda(x) : \mathrm{V}(x)$ and refer to the latter as the *error-locator ratio*. In fact, we will find it convenient to extend this notation to any two polynomials $a(x), b(x) \in \Phi[x]$ and write $a(x) : b(x)$ instead of $(a(x), b(x))$. (The notation $a(x) : b(x)$ should not be confused with $a(x)/b(x)$; the latter stands for an element of $\Phi[[x]]$, which is defined whenever $\gcd(b(x), x) = 1$.) If an irreducible polynomial $P(x)$ over Φ has multiplicity s and t, respectively, in the irreducible factorization of $a(x)$ and $b(x)$ over Φ, then the *multiplicity* of $P(x)$ in $a(x) : b(x)$ is defined as the integer $s - t$. So, if m_j is the multiplicity of $1 - \alpha_j x$ in the error-locator ratio $\Lambda(x) : \mathrm{V}(x)$, then $|m_j| \le p/2$ and $e_j = m_j \cdot 1$.

Since $J^+ \cap J^- = \emptyset$, we have

$$\boxed{\gcd(\Lambda(x), \mathrm{V}(x)) = 1}\,, \tag{10.7}$$

and from $\left(\sum_{j\in J^+} |e_j|\right) + \left(\sum_{j\in J^-} |e_j|\right) = \sum_{j=1}^n |e_j| = \mathsf{w}_{\mathcal{L}}(\mathbf{e}) < \varrho$ we obtain

$$\boxed{\deg \Lambda + \deg \mathrm{V} < \varrho}\,. \tag{10.8}$$

Noting that $\mathrm{V}(x)$ is invertible in the ring $\Phi[x]/x^\varrho$, there is a unique polynomial $\Psi(x) = \sum_{i=0}^{\varrho-1} \Psi_i x^i$ in $\Phi_\varrho[x]$ such that

$$\boxed{\Psi(x) \equiv \frac{\Lambda(x)}{\mathrm{V}(x)} \pmod{x^\varrho}}\,. \tag{10.9}$$

Furthermore, from $\Lambda(0) = \mathrm{V}(0) = 1$ we have $\Psi(0) = \Psi_0 = 1$. Finally, the equality

$$\Big(\sum_{j\in J^+} |e_j| - \sum_{j\in J^-} |e_j|\Big) \cdot 1 = \sum_{j=1}^n e_j = S_0$$

yields

$$\boxed{\deg \Lambda - \deg \mathrm{V} \equiv \langle S_0 \rangle \pmod{p}} \tag{10.10}$$

(recall the notation $\langle \cdot \rangle$ from the beginning of Section 10.1).

Equations (10.7)–(10.10) form the *key equation* of the Lee-metric decoding of GRS codes and alternant codes. One can readily see the resemblance to the key equation that we obtained for the Hamming metric in Section 6.3, and this similarity will be reflected also in the decoding algorithm that we present. We do point out one notable difference, however: only one coefficient of the syndrome—namely, S_0—seems to appear explicitly in the new key equation. The dependence of the equation on the remaining syndrome coefficients has now become implicit through the polynomial $\Psi(x)$. Still, this polynomial can be uniquely computed from the syndrome coefficients $S_1, S_2, \ldots, S_{\varrho-1}$; we demonstrate this next.

By Lemma 10.3 we obtain

$$\frac{\Lambda(x)}{\mathrm{V}(x)} S(x) = -x \left(\frac{\Lambda(x)}{\mathrm{V}(x)} \right)' ,$$

and from (10.9) we have

$$\Psi'(x) \equiv \left(\frac{\Lambda(x)}{\mathrm{V}(x)} \right)' \pmod{x^{\varrho-1}} .$$

The last two equations, along with (10.9), yield

$$\boxed{\Psi(x) S(x) \equiv -x \Psi'(x) \pmod{x^{\varrho}}} , \tag{10.11}$$

and (10.11), in turn, can be re-written as

$$\sum_{\ell=1}^{i} \Psi_{i-\ell} S_\ell = -i \Psi_i , \quad 1 \le i < \varrho . \tag{10.12}$$

When $\varrho \le p$, the index i in (10.12) ranges over invertible integers modulo p. Starting with $\Psi_0 = 1$, we can therefore apply (10.12) iteratively to solve (uniquely) for the values Ψ_i for $i = 1, 2, \ldots, \varrho-1$. Hence, (10.12) induces a mapping

$$(S_1 \; S_2 \; \ldots \; S_{\varrho-1}) \mapsto \Psi(x) ;$$

furthermore, since $\Psi(x)$ is invertible in the ring $\Phi[x]/x^{\varrho}$, this mapping is one-to-one: from (10.11) we get that the coefficients $S_1, S_2, \ldots, S_{\varrho-1}$ are uniquely determined by

$$\sum_{\ell=1}^{\varrho-1} S_\ell x^\ell \equiv -\frac{x \Psi'(x)}{\Psi(x)} \pmod{x^{\varrho}} .$$

It follows that whenever $\mathsf{d}_{\mathcal{L}}(\mathcal{C}_{\mathrm{alt}}) \ge 2\varrho$, distinct error words $\mathbf{e}$ with $\mathsf{w}_{\mathcal{L}}(\mathbf{e}) < \varrho$ correspond to distinct syndromes $(S_0\ S_1\ S_2\ \ldots\ S_{\varrho-1})^T$ and, therefore, to distinct pairs $(S_0, \Psi(x))$.

We next proceed as in Section 6.4 and solve the key equation for $\Lambda(x)$: $\mathrm{V}(x)$ by making use of (the extended version of) Euclid's algorithm. For the sake of completeness, we recall here the algorithm and several properties thereof. Given polynomials $a(x)$ and $b(x)$ over a field Φ such that $a(x) \ne 0$ and $\deg a > \deg b$, the algorithm computes remainders $r_i(x)$, quotients $q_i(x)$, and auxiliary polynomials $t_i(x)$, as shown in Figure 10.1 (the algorithm here is the same as in Figure 6.1, except that we have omitted the second set of auxiliary polynomials, $s_i(x)$, which are not needed for the decoding).

$r_{-1}(x) \leftarrow a(x)$; $r_0(x) \leftarrow b(x)$;
$t_{-1}(x) \leftarrow 0$; $t_0(x) \leftarrow 1$;
for $(i \leftarrow 1;\ r_{i-1}(x) \ne 0;\ i{+}{+})$ {
  $q_i(x) \leftarrow r_{i-2}(x)$ div $r_{i-1}(x)$;
  $r_i(x) \leftarrow r_{i-2}(x) - q_i(x) r_{i-1}(x)$;
  $t_i(x) \leftarrow t_{i-2}(x) - q_i(x) t_{i-1}(x)$;
}

Figure 10.1. Euclid's algorithm.

Let ν denote the largest index i for which $r_i(x) \ne 0$.

Lemma 10.6 *Using the notation of Euclid's algorithm,* $\deg r_i - \deg t_i$ *strictly decreases for* $i = 0, 1, \ldots, \nu{+}1$.

Proof. On the one hand, the degrees of r_i strictly decrease. On the other hand, by part 3 of Problem 3.3,

$$\deg t_i + \deg r_{i-1} = \deg a\ , \quad i = 0, 1, \ldots, \nu{+}1\ ;$$

i.e., the degrees of t_i strictly increase. □

We have also shown the following result (see Proposition 6.3).

Proposition 10.7 *Using the notation of Euclid's algorithm, suppose that* $t(x)$ *and* $r(x)$ *are nonzero polynomials over* Φ *satisfying the following conditions:*

(C1) $\gcd(t(x), r(x)) = 1$.

(C2) $\deg t + \deg r < \deg a$.

(C3) $t(x)b(x) \equiv r(x) \pmod{a(x)}$.

Then there is an index $h \in \{0, 1, \ldots, \nu{+}1\}$ *and a constant* $c \in \Phi$ *such that*

$$t(x) \leftarrow c \cdot t_h(x) \quad \textit{and} \quad r(x) \leftarrow c \cdot r_h(x) .$$

Comparing Equations (10.7)–(10.9) with conditions (C1)–(C3) in Proposition 10.7, we can solve for $\Lambda(x) : \mathrm{V}(x)$ by applying Euclid's algorithm to

$$a(x) \leftarrow x^{\varrho} \quad \text{and} \quad b(x) \leftarrow \Psi(x) ,$$

to produce

$$\Lambda(x) \leftarrow c \cdot r_h(x) \quad \text{and} \quad \mathrm{V}(x) \leftarrow c \cdot t_h(x) .$$

The value of h is determined by the degree constraint (10.10); specifically, if $\varrho \le \frac{1}{2}(p{+}1)$ then

$$|\deg r_h - \deg t_h| = |\deg \Lambda(x) - \deg \mathrm{V}(x)| \le \mathrm{w}_{\mathcal{L}}(\mathbf{e}) < \varrho \le \tfrac{1}{2}(p{+}1)$$

and, so, (10.10) implies that h is an index such that

$$\deg r_h - \deg t_h = \begin{cases} \langle S_0 \rangle & \text{if } 0 \le \langle S_0 \rangle < \varrho \\ \langle S_0 \rangle - p & \text{if } p{-}\varrho < \langle S_0 \rangle < p \end{cases} .$$

And Lemma 10.6 then guarantees that such an index h is unique.

Note that when $\varrho \le p/2$, there is a nonempty range of values of $\langle S_0 \rangle$, namely, $\varrho \le \langle S_0 \rangle \le p{-}\varrho$, which corresponds to detectable but uncorrectable error patterns. Uncorrectable errors are detected also when the computed ratio $\Lambda(x) : \mathrm{V}(x)$ violates (10.8), or when $\Lambda(x)$ or $\mathrm{V}(x)$ does not factor into linear terms $1 - \alpha_j x$ for code locators α_j. When $\varrho \le p/2$, an (uncorrectable) error word that has Lee weight exactly ϱ will always be detected.

Having determined the error-locator ratio $\Lambda(x) : \mathrm{V}(x)$, we can now solve for the error word $\mathbf{e} = (e_1 \; e_2 \; \ldots \; e_n)$ by finding for $j = 1, 2, \ldots, n$ the multiplicity, m_j, of $1 - \alpha_j x$ in $\Lambda(x) : \mathrm{V}(x)$; the value e_j then equals $m_j \cdot 1$. If α_j^{-1} is a root of $\Lambda(x)$, then m_j equals the multiplicity of α_j^{-1} as such a root; we compute m_j by finding the smallest integer $i \ge 0$ for which the ith *Hasse derivative* of $\Lambda(x)$,

$$\Lambda^{[i]}(x) = \sum_{\ell \ge i} \binom{\ell}{i} \Lambda_\ell x^{\ell - i} ,$$

does not vanish at $x = \alpha_j^{-1}$ (see Problem 3.40; here Λ_ℓ stands for the coefficient of x^ℓ in $\Lambda(x)$). Otherwise, if α_j^{-1} is a root of $\mathrm{V}(x)$, then m_j is the negative integer whose absolute value equals the multiplicity of α_j^{-1} as a root of $\mathrm{V}(x)$; this multiplicity, in turn, is computed by evaluating the Hasse derivatives of $\mathrm{V}(x)$ at $x = \alpha_j^{-1}$. If α_j^{-1} is neither a root of $\Lambda(x)$ nor of $\mathrm{V}(x)$ then $m_j = 0$.

Input: received word $(y_1\ y_2\ \ldots\ y_n) \in F^n$.
Output: error word $(e_1\ e_2\ \ldots\ e_n) \in F^n$, or an error-detection indicator "e".

1. Compute the syndrome values $S_\ell \leftarrow \sum_{j=1}^{n} y_j \alpha_j^\ell$ for $0 \le \ell < \varrho$.
2. Compute the polynomial $\Psi(x) = \sum_{i=0}^{\varrho-1} \Psi_i x^i$ iteratively by

$$\Psi_0 \leftarrow 1 \qquad \text{and} \qquad \Psi_i \leftarrow -\frac{1}{i} \sum_{\ell=1}^{i} \Psi_{i-\ell} S_\ell\,, \quad 1 \le i < \varrho\,.$$

3. Apply Euclid's algorithm to the polynomials $a(x) \leftarrow x^\varrho$ and $b(x) \leftarrow \Psi(x)$ to obtain ratios $r_i(x) : t_i(x)$, $i = 0, 1, 2, \ldots$, until $\deg r_i - \deg t_i \le \langle S_0 \rangle - p$.
4. For an integer h such that $\deg r_h - \deg t_h \in \{\langle S_0 \rangle, \langle S_0 \rangle - p\}$ and $\deg r_h + \deg t_h < \varrho$ do:
   (a) let $\Lambda(x) : \mathrm{V}(x) \leftarrow r_h(x) : t_h(x)$;
   (b) using Hasse derivatives find, for $j = 1, 2, \ldots, n$, the multiplicity m_j of $1 - \alpha_j x$ in $\Lambda(x) : \mathrm{V}(x)$;
   (c) if $\sum_{j=1}^{n} |m_j| = \deg \Lambda + \deg \mathrm{V}$, set $e_j \leftarrow m_j \cdot 1$ for $j = 1, 2, \ldots, n$.
5. If no integer h satisfies the condition in Step 4, or if the values e_j were not set in Step 4c, return "e".

Figure 10.2. Lee-metric decoding algorithm for alternant codes.

Figure 10.2 presents a decoding algorithm for an alternant code $\mathcal{C}_{\mathrm{alt}}$ over $F = \mathrm{GF}(p)$ with an underlying $[n, n{-}\varrho]$ normalized GRS code whose code locators are $\alpha_1, \alpha_2, \ldots, \alpha_n$. When $\varrho \le \frac{1}{2}(p{+}1)$, there can be at most one integer h that satisfies the condition

$$\deg r_h - \deg t_h \in \Big\{\langle S_0 \rangle, \langle S_0 \rangle - p\Big\} \quad \text{and} \quad \deg r_h + \deg t_h < \varrho\,, \qquad (10.13)$$

which is tested in Step 4 in Figure 10.2. Therefore, when $\varrho \le \frac{1}{2}(p{+}1)$, the algorithm will recover correctly any error word $\mathbf{e}$ with $\mathsf{w}_{\mathcal{L}}(\mathbf{e}) < \varrho$. Furthermore, when $\varrho \le p/2$, the algorithm will detect all error words with Lee weight ϱ.

An attempt to apply the algorithm in Figure 10.2 to alternant codes with $\frac{1}{2}(p{+}1) < \varrho \le p$ may result in an incorrect decoding. Still, for this range of ϱ, there can be no more than *two* integers h that satisfy the condition (10.13). Hence, when $\varrho \le p$, the decoding algorithm is a Lee-metric *list-2 decoder* for alternant codes (see Chapter 9).

10.5 Decoding GRS codes in the Lee metric

We next consider the decoding of normalized GRS codes; recall that by Theorem 10.5, the lower bound 2ϱ on the minimum Lee distance applies in this case also when $\varrho > p/2$. Yet, as we have pointed out, when $\varrho > \frac{1}{2}(p+1)$, the stopping rule (10.13) might become ambiguous. We illustrate this in the following example.

Example 10.3 Consider the $[p-1, p-1-\varrho]$ normalized GRS code over $F = \mathrm{GF}(p)$ with $p = 7$, $\varrho = 5$, and $\alpha_1 = 1$, and assume the error word

$$\mathbf{e} = (e_1 \; e_2 \; \ldots \; e_6) = (4\,0\,0\,0\,0\,0) \,.$$

The syndrome of $\mathbf{e}$ is given by

$$(S_0 \; S_1 \; \ldots \; S_4)^T = (4\;4\;\ldots\;4)^T \,,$$

and the respective polynomial $\Psi(x)$ equals $1 + 3x + 6x^2 + 3x^3 + x^4$. Now, the stopping rule (10.13) is satisfied for $h = 0$, yielding

$$\Lambda(x) : \mathrm{V}(x) = \; r_0(x) : t_0(x) \; = (1-x)^4 : 1 \,,$$

and also for $h = 4$, yielding

$$\hat{\Lambda}(x) : \hat{\mathrm{V}}(x) = 4t_4(x) : 4r_4(x) = 1 : (1-x)^3 \,.$$

Both ratios, $\Lambda : \mathrm{V}$ and $\hat{\Lambda} : \hat{\mathrm{V}}$, satisfy all four equations (10.7)–(10.10). However, the multiplicity, 4, of $1-x$ in the irreducible factorization of $\Lambda(x)$ is not a valid Lee weight (even though it does equal $\langle e_1 \rangle$ in this case). Disregarding this inconsistency, both error-locator ratios correspond to the same true error word. □

This (seeming) ambiguity in the stopping rule is resolved by the next proposition.

Proposition 10.8 *Given an $[n, n-\varrho]$ normalized GRS code C_{GRS} over $F = \mathrm{GF}(p)$, p prime, let $\mathbf{e} = (e_1 \; e_2 \; \ldots \; e_n)$ be an error word in F^n such that $\mathsf{w}_{\mathcal{L}}(\mathbf{e}) < \varrho$ and let $\Psi(x)$ be obtained from the syndrome $(S_0 \; S_1 \; \ldots \; S_{\varrho-1})^T$ of $\mathbf{e}$ by (10.12).*

Suppose that $\Lambda(x)$ and $\mathrm{V}(x)$ are polynomials that factor into linear terms over F and satisfy (10.7)–(10.10). For any code locator α_j of C_{GRS}, let μ_j be the multiplicity of $1 - \alpha_j x$ in the ratio $\Lambda(x) : \mathrm{V}(x)$. Then,

$$e_j = \mu_j \cdot 1 \,, \quad 1 \le j \le n \,.$$

Proof. First observe that if $p-1-n$ zeros are appended to each codeword of $\mathcal{C}_{\mathrm{GRS}}$, the resulting words form a subset of a primitive GRS code over F (that is, $\mathcal{C}_{\mathrm{GRS}}$ is obtained from a primitive GRS code by shortening—see Problem 2.14). Hence, it suffices to prove the result for primitive GRS codes, and we assume hereafter in the proof that $\mathcal{C}_{\mathrm{GRS}}$ is primitive, with its code locators ranging over all the elements of F^*.

Let the ratio $\Lambda(x) : \mathrm{V}(x)$ satisfy the conditions in the proposition. By (10.7) and (10.9) (and since $\Psi(0) = 1$), we can rule out the possibility of having the term x as one of the linear factors of $\Lambda(x)$ or $\mathrm{V}(x)$; thus, we can write

$$\Lambda(x) = \prod_{j\,:\,\mu_j>0} (1-\alpha_j x)^{\mu_j} \qquad \text{and} \qquad \mathrm{V}(x) = \prod_{j\,:\,\mu_j<0} (1-\alpha_j x)^{-\mu_j} .$$

For every code locator $\alpha_j \in F^*$ define

$$m_j = \begin{cases} \mu_j & \text{if } |\mu_j| \le p/2 \\ \mu_j - p & \text{if } \mu_j > p/2 \\ \mu_j + p & \text{if } \mu_j < -p/2 \end{cases} , \tag{10.14}$$

and let the ratio $\hat{\Lambda}(x) : \hat{\mathrm{V}}(x)$ be given by

$$\hat{\Lambda}(x) = \prod_{j\,:\,m_j>0} (1-\alpha_j x)^{m_j} \qquad \text{and} \qquad \hat{\mathrm{V}}(x) = \prod_{j\,:\,m_j<0} (1-\alpha_j x)^{-m_j} .$$

It is easy to see that $\hat{\Lambda}(x) : \hat{\mathrm{V}}(x)$ is the error-locator ratio of the error word $\hat{\mathbf{e}} = (m_1\ m_2\ \ldots\ m_n) \cdot 1$, and

$$\mathsf{w}_{\mathcal{L}}(\hat{\mathbf{e}}) = \deg \hat{\Lambda} + \deg \hat{\mathrm{V}} = \sum_{j=1}^{n} |m_j| \le \sum_{j=1}^{n} |\mu_j| = \deg \Lambda + \deg \mathrm{V} < \varrho .$$

Let $(\hat{S}_\ell)_{\ell=0}^{\varrho-1}$ be the syndrome of $\hat{\mathbf{e}}$ and let $\hat{\Psi}(x)$ be the polynomial in $\Phi_\varrho[x]$ that satisfies (10.9) with respect to $\hat{\Lambda} : \hat{\mathrm{V}}$. Now, by construction we have

$$\frac{\Lambda(x)}{\hat{\Lambda}(x)} \cdot \frac{\hat{\mathrm{V}}(x)}{\mathrm{V}(x)} = \frac{\prod_{j\,:\,m_j=\mu_j-p}(1-\alpha_j x)^p}{\prod_{j\,:\,m_j=\mu_j+p}(1-\alpha_j x)^p} = \frac{\prod_{j\,:\,m_j=\mu_j-p}(1-\alpha_j x^p)}{\prod_{j\,:\,m_j=\mu_j+p}(1-\alpha_j x^p)} \tag{10.15}$$

and, so,

$$\frac{\Lambda(x)}{\mathrm{V}(x)} \equiv \frac{\hat{\Lambda}(x)}{\hat{\mathrm{V}}(x)} \pmod{x^p} .$$

It follows from the latter equality and (10.9) that $\hat{\Psi}(x) = \Psi(x)$. Also, from (10.10) and (10.15) we have

$$\langle S_0 \rangle \equiv \deg \Lambda - \deg \mathrm{V} \equiv \deg \hat{\Lambda} - \deg \hat{\mathrm{V}} \equiv \langle \hat{S}_0 \rangle \pmod{p} ,$$

i.e., $\hat{S}_0 = S_0$. Therefore, $(\hat{S}_0, \hat{\Psi}(x)) = (S_0, \Psi(x))$, which implies that the syndromes of $\mathbf{e}$ and $\hat{\mathbf{e}}$ must be the same. Hence, $\hat{\mathbf{e}}$ is equal to $\mathbf{e}$ and the result now follows from (10.14). □

We can conclude from Proposition 10.8 that when we apply the algorithm in Figure 10.2 to $[n, n-\varrho]$ normalized GRS codes over prime fields GF(p), the decoding of less than ϱ errors will be correct also when $\varrho > \frac{1}{2}(p+1)$, provided that we insert the following change into the algorithm: if Step 4c fails to produce an error word, then Steps 4a–4c are applied also to the second integer h (if any) that satisfies (10.13). (There are examples—such as the one presented in Problem 10.5—where two integers h satisfy (10.13), yet only one of them produces an error word.)

10.6 Berlekamp codes

So far we have considered codes whose minimum Lee distance is guaranteed to be at least some prescribed even number. In this section, we introduce codes whose designed minimum Lee distance is odd.

Let $F = \mathrm{GF}(p)$ where p is a prime and let $\beta_1, \beta_2, \ldots, \beta_n$ be distinct nonzero elements in a finite extension field Φ of F such that

$$\beta_i + \beta_j \neq 0 \qquad \text{for all } 1 \leq i < j \leq n \, . \tag{10.16}$$

Consider the linear $[n, n-\tau]$ code $\mathcal{C}$ over Φ with a parity-check matrix

$$H_{\mathrm{Ber}} = \begin{pmatrix} \beta_1 & \beta_2 & \ldots & \beta_n \\ \beta_1^3 & \beta_2^3 & \ldots & \beta_n^3 \\ \beta_1^5 & \beta_2^5 & \ldots & \beta_n^5 \\ \vdots & \vdots & \vdots & \vdots \\ \beta_1^{2\tau-1} & \beta_2^{2\tau-1} & \ldots & \beta_n^{2\tau-1} \end{pmatrix} . \tag{10.17}$$

The intersection $\mathcal{C} \cap F^n$ is called a *Berlekamp code* and will be denoted by $\mathcal{C}_{\mathrm{Ber}}$. When $p = 2$, Berlekamp codes coincide with narrow-sense alternant codes. Note that when $p > 2$, the requirement (10.16) bounds the code length n from above by $\frac{1}{2}(|\Phi|-1)$ (Problem 10.6).

Next, we show that when the transmitted codewords are taken from $\mathcal{C}_{\mathrm{Ber}}$ and $\tau < p/2$, one can recover correctly every error word with Lee weight $\leq \tau$; this, in turn, will imply that $\mathsf{d}_{\mathcal{L}}(\mathcal{C}_{\mathrm{Ber}}) \geq 2\tau + 1$. In fact, we will transform the decoding problem of $\mathcal{C}_{\mathrm{Ber}}$ into that of some alternant code $\mathcal{C}_{\mathrm{alt}}$. We start by specifying this alternant code.

Let $N = 2n$ and $\varrho = 2\tau+1$ $(\leq p)$ and let the elements $\alpha_1, \alpha_2, \ldots, \alpha_N \in \Phi$ be defined by

$$\alpha_j = \begin{cases} \beta_j & \text{for } 1 \leq j \leq n \\ -\beta_{j-n} & \text{for } n < j \leq N \end{cases} . \tag{10.18}$$

By (10.16), all of these elements are nonzero and distinct. We let $\mathcal{C}_{\mathrm{alt}}$ be the alternant code $\mathcal{C}_{\mathrm{GRS}} \cap F^n$, where $\mathcal{C}_{\mathrm{GRS}}$ is the $[N, N-\varrho]$ normalized GRS code over Φ with code locators $\alpha_1, \alpha_2, \ldots, \alpha_N$.

Suppose that a codeword of $\mathcal{C}_{\mathrm{Ber}}$ is transmitted, and denote by $\mathbf{y}$ and $\mathbf{e} = (e_1 \; e_2 \; \ldots \; e_n)$ the received word and the error word, respectively, both in F^n. We assume that $\mathsf{w}_{\mathcal{L}}(\mathbf{e}) \leq \tau < p/2$.

Let $(S_1 \; S_3 \; \ldots \; S_{2\tau-1})^T$ be the syndrome of $\mathbf{y}$ (and $\mathbf{e}$) with respect to the parity-check matrix H_{Ber}, namely,

$$S_\ell = \sum_{j=1}^{n} e_j \beta_j^\ell \; , \quad \ell = 1, 3, \ldots, 2\tau{-}1 \; .$$

Define the word $\hat{\mathbf{e}} = (\hat{e}_1 \; \hat{e}_2 \; \ldots \; \hat{e}_N) \in F^N$ by

$$\hat{e}_j = \begin{cases} e_j & \text{for } 1 \leq j \leq n \\ -e_{j-n} & \text{for } n < j \leq N \end{cases} ; \tag{10.19}$$

clearly, $\mathsf{w}_{\mathcal{L}}(\hat{\mathbf{e}}) = 2\,\mathsf{w}_{\mathcal{L}}(\mathbf{e}) \leq 2\tau < \varrho$. We regard $\hat{\mathbf{e}}$ as if it were an error word in a transmission of a codeword of $\mathcal{C}_{\mathrm{alt}}$, and we let $(\hat{S}_\ell)_{\ell=0}^{\varrho-1}$ be the syndrome of $\hat{\mathbf{e}}$ with respect to a canonical parity-check matrix of $\mathcal{C}_{\mathrm{GRS}}$; that is,

$$\hat{S}_\ell = \sum_{j=1}^{N} \hat{e}_j \alpha_j^\ell \; , \quad 0 \leq \ell < \varrho \; .$$

By (10.18) and (10.19) we have the following relation between the syndromes $(\hat{S}_\ell)_{\ell=0}^{\varrho-1}$ and $(S_{2u-1})_{u=1}^{\tau}$:

$$\begin{aligned} \hat{S}_\ell &= \sum_{j=1}^{n} \hat{e}_j \alpha_j^\ell + \sum_{j=n+1}^{N} \hat{e}_j \alpha_j^\ell \\ &= \sum_{j=1}^{n} e_j \left(\beta_j^\ell - (-\beta_j)^\ell \right) = \begin{cases} 2S_\ell & \text{for } \ell = 1, 3, \ldots, \varrho{-}2 \\ 0 & \text{for } \ell = 0, 2, \ldots, \varrho{-}1 \end{cases} . \end{aligned}$$

Noting that $\mathsf{w}_{\mathcal{L}}(\hat{\mathbf{e}}) < \varrho$, we can now apply the decoding algorithm in Figure 10.2 to the code $\mathcal{C}_{\mathrm{alt}}$ with the syndrome $(\hat{S}_\ell)_{\ell=0}^{\varrho-1}$. Indeed, it turns out that the decoding will be successful also when $\varrho > (p{+}1)/2$: since the syndrome entry $\hat{S}_0$ equals zero, the condition (10.13) reduces to the unambiguous stopping rule

$$\deg r_h - \deg t_h = 0 \; , \tag{10.20}$$

whenever $(2\tau{+}1 =) \; \varrho \leq p$. We are therefore able to decode $\hat{\mathbf{e}}$ and, hence, $\mathbf{e}$.

Since every error word of Lee weight τ or less is correctable, it follows (from a Lee-metric counterpart of Problem 1.10) that the minimum Lee distance of $\mathcal{C}_{\mathrm{Ber}}$ is at least $2\tau{+}1$. We summarize this in the next theorem.

Theorem 10.9 *Let $\mathcal{C}_{\mathrm{Ber}} \neq \{\mathbf{0}\}$ be defined over $F = \mathrm{GF}(p)$, p prime, by the parity-check matrix (10.17). If $\tau < p/2$ then*

$$\mathsf{d}_{\mathcal{L}}(\mathcal{C}_{\mathrm{Ber}}) \geq 2\tau+1 .$$

The particular choice of the code locators α_j in (10.18) and the form of the word $\hat{\mathbf{e}}$ in (10.19) allow a significant shortcut in the algorithm of Figure 10.2. Specifically, let the index sets J^+ and J^- be defined by

$$J^{\pm} = \left\{ j \in \{1, 2, \ldots, n\} \; : \; e_j \in F^{\pm} \right\} ;$$

then the error-locator ratio $\Lambda : \mathrm{V}$ that is associated with $\hat{\mathbf{e}}$ is given by

$$\Lambda(x) = \prod_{j \in J^+} (1 - \beta_j x)^{|e_j|} \cdot \prod_{j \in J^-} (1 + \beta_j x)^{|e_j|}$$

and

$$\mathrm{V}(x) = \prod_{j \in J^+} (1 + \beta_j x)^{|e_j|} \cdot \prod_{j \in J^-} (1 - \beta_j x)^{|e_j|} = \Lambda(-x) .$$

Hence, it suffices to compute only one polynomial, say $\Lambda(x)$, and its roots—each with its multiplicity—determine the error word $\mathbf{e}$ completely, as follows. Let m_j be the multiplicity of either β_j^{-1} or $-\beta_j^{-1}$ whenever one of these elements is a root of $\Lambda(x)$ (note that $\Lambda(x)$ cannot have both as roots); then

$$e_j = \begin{cases} \pm m_j \cdot 1 & \text{if } \Lambda(\pm\beta_j^{-1}) = 0 \\ 0 & \text{otherwise} \end{cases} .$$

Sufficing to compute $\Lambda(x)$ only, a simpler version of Euclid's algorithm can be used where $t_i(x)$ need not be computed, and by Lemma 10.6 we can express the stopping rule (10.20) only in terms of the remainders $r_i(x)$ as

$$\deg r_h + \deg r_{h-1} = 2\tau+1 .$$

Equivalently, h is the unique value of i for which

$$\deg r_h \leq \tau < \deg r_{h-1} .$$

Figure 10.3 summarizes the decoding algorithm for a Berlekamp code $\mathcal{C}_{\mathrm{Ber}}$ over $F = \mathrm{GF}(p)$ with a parity-check matrix (10.17).

10.7 Bounds for codes in the Lee metric

In this section, we develop two bounds on the parameters of codes in the Lee metric. Specifically, we present the Lee-metric counterparts of the sphere-packing bound and the Gilbert–Varshamov bound. More bounds can be

---

Input: received word $(y_1\ y_2\ \ldots\ y_n) \in F^n$.
Output: error word $(e_1\ e_2\ \ldots\ e_n) \in F^n$, or an error-detection indicator "e".

1. Compute the syndrome values $S_\ell \leftarrow \sum_{j=1}^{n} y_j \beta_j^\ell$ for $\ell = 1, 3, 5, \ldots, 2\tau{-}1$.

2. Compute the polynomial $\Psi(x) = \sum_{i=0}^{\varrho-1} \Psi_i x^i$ iteratively by

$$\Psi_0 \leftarrow 1 \qquad \text{and} \qquad \Psi_i \leftarrow -\frac{2}{i} \sum_{u=1}^{\lceil i/2 \rceil} \Psi_{i+1-2u} S_{2u-1} \ , \qquad 1 \le i \le 2\tau \ .$$

3. Apply Euclid's algorithm to $a(x) \leftarrow x^{2\tau+1}$ and $b(x) \leftarrow \Psi(x)$ and compute remainders $r_i(x)$, $i = 0, 1, 2, \ldots$, until $\deg r_i \le \tau$; let h be the last value of i.

4. If $\deg r_h + \deg r_{h-1} = 2\tau{+}1$ then do:

   (a) let $\Lambda(x) \leftarrow r_h(x)$;

   (b) using Hasse derivatives find, for $j = 1, 2, \ldots, n$, the multiplicity m_j of either β_j^{-1} or $-\beta_j^{-1}$ as a root of $\Lambda(x)$; if both are roots let $m_j \leftarrow \infty$;

   (c) if $\sum_{j=1}^{n} |m_j| = \deg \Lambda$, set

$$e_j \leftarrow \begin{cases} \pm m_j \cdot 1 & \text{if } \Lambda(\pm\beta_j^{-1}) = 0 \\ 0 & \text{otherwise} \end{cases} \ , \qquad j = 1, 2, \ldots, n \ .$$

5. If $\deg r_h + \deg r_{h-1} \ne 2\tau{+}1$, or if the values e_j were not set in Step 4c, return "e".

Figure 10.3. Decoding algorithm for Berlekamp codes.

found in Problems 10.14 and 10.15, and in the notes on this section at the end of this chapter.

For a word $\mathbf{c} \in \mathbb{Z}_q^n$, denote by $\mathcal{S}_\mathcal{L}(\mathbf{c}, t)$ the *Lee sphere* with radius t in $\mathbb{Z}_q^n$ that is centered at $\mathbf{c}$; that is,

$$\mathcal{S}_\mathcal{L}(\mathbf{c}, t) = \left\{ \mathbf{y} \in \mathbb{Z}_q^n \ : \ \mathsf{d}_\mathcal{L}(\mathbf{y}, \mathbf{c}) \le t \right\} \ .$$

Equivalently,

$$\mathcal{S}_\mathcal{L}(\mathbf{c}, t) = \left\{ \mathbf{c} + \mathbf{e} \ : \ \mathbf{e} \in \mathbb{Z}_q^n \ , \ \mathsf{w}_\mathcal{L}(\mathbf{e}) \le t \right\} \ .$$

We see from the latter characterization of $\mathcal{S}_\mathcal{L}(\mathbf{c}, t)$ that the size of $\mathcal{S}_\mathcal{L}(\mathbf{c}, t)$ does not depend on the center $\mathbf{c}$; we denote this size by $V_{\mathcal{L}|q}(n, t)$ (we will sometimes omit the subscript q if it can be understood from the context).

As was the case in the Hamming metric, sizes of Lee spheres will appear in our bounds. The next proposition is therefore useful.

Proposition 10.10 *The size of a Lee sphere in $\mathbb{Z}_q^n$ with radius $t < q/2$ is given by*

$$V_{\mathcal{L}}(n,t) = \sum_{i=0}^{n} 2^i \binom{n}{i}\binom{t}{i}$$

(where $\binom{t}{i} = 0$ if $i > t$).

Proof. Fix a subset $J \subseteq \{1,2,\ldots,n\}$ of size $|J| = i$ and let the set $\mathcal{S}_J$ consist of all the words in $\mathcal{S}_{\mathcal{L}}(\mathbf{0},t)$ whose support is J; namely, an entry of a word in $\mathcal{S}_J$ is nonzero if and only if that entry is indexed by some $j \in J$. Also, let $\mathcal{S}_J^+$ be the set of all words in $\mathcal{S}_J$ whose nonzero entries are all in $\mathbb{Z}_q^+$. From $t < q/2$ it follows that

$$|\mathcal{S}_J| = 2^i \cdot |\mathcal{S}_J^+| \, .$$

Now, $|\mathcal{S}_J^+|$ equals the number of words of length i over the (strictly) positive integers such that the sum of entries in each word is at most t; namely, $|\mathcal{S}_J^+| = |\mathcal{Q}(t,i)|$, where

$$\mathcal{Q}(t,i) = \left\{ (m_1 \; m_2 \; \ldots \; m_i) \mid m_1, m_2, \ldots, m_i \in \mathbb{Z}^+ \, , \quad \textstyle\sum_{s=1}^{i} m_s \le t \right\} \, .$$

The size of $\mathcal{Q}(t,i)$, in turn, equals $\binom{t}{i}$: as $(m_1 \; m_2 \; \ldots \; m_i)$ ranges over the elements of $\mathcal{Q}(t,i)$, the set

$$\{m_1, \; m_1{+}m_2, \; \ldots, \; m_1{+}m_2{+}\cdots{+}m_i\}$$

ranges over all the subsets of $\{1,2,\ldots,t\}$ of size i. We thus have,

$$\begin{aligned} V_{\mathcal{L}}(n,t) = |\mathcal{S}_{\mathcal{L}}(\mathbf{0},t)| &= \sum_{J \subseteq \{1,2,\ldots,n\}} |\mathcal{S}_J| = \sum_J 2^{|J|} \cdot |\mathcal{S}_J^+| \\ &= \sum_{i=0}^{n} \sum_{J\,:\,|J|=i} 2^i \tbinom{t}{i} = \sum_{i=0}^{n} 2^i \tbinom{n}{i}\tbinom{t}{i} \, , \end{aligned}$$

as claimed. $\square$

The triangle inequality was the basis for proving the sphere-packing bound in the Hamming metric (Theorem 4.3); we now use essentially the same proof also for the Lee metric.

Theorem 10.11 (Sphere-packing bound in the Lee metric) *Let C be an (n,M) code of size $M > 1$ over $\mathbb{Z}_q$ and let $t = \lfloor \frac{1}{2}(\mathsf{d}_{\mathcal{L}}(C) - 1) \rfloor$. Then,*

$$M \cdot V_{\mathcal{L}}(n,t) \le q^n \, .$$

Proof. By the triangle inequality we have $\mathcal{S}_{\mathcal{L}}(\mathbf{c}_1, t) \cap \mathcal{S}_{\mathcal{L}}(\mathbf{c}_2, t) = \emptyset$ for every two distinct codewords $\mathbf{c}_1, \mathbf{c}_2 \in \mathcal{C}$. Hence,

$$M \cdot V_{\mathcal{L}}(n,t) = \sum_{\mathbf{c} \in \mathcal{C}} |\mathcal{S}_{\mathcal{L}}(\mathbf{c},t)| = \Big| \bigcup_{\mathbf{c} \in \mathcal{C}} \mathcal{S}_{\mathcal{L}}(\mathbf{c},t) \Big| \le q^n ,$$

as claimed. □

A code that attains the bound of Theorem 10.11 is called a *perfect code* in the Lee metric.

Example 10.4 Let $F = \mathrm{GF}(p)$ and $\Phi = \mathrm{GF}(p^m)$ for an odd prime p and a positive integer m, and construct the $[n,k]$ Berlekamp code $\mathcal{C}_{\mathrm{Ber}}$ over F by the parity-check matrix (10.17) with $\tau = 1$ and $n = \frac{1}{2}(p^m - 1)$. For this code we have $n-k \le m$ and, so,

$$|\mathcal{C}_{\mathrm{Ber}}| \cdot V_{\mathcal{L}}(n,1) = |\mathcal{C}_{\mathrm{Ber}}| \cdot (1+2n) \ge p^{n-m} \cdot (1+2n) = p^n ,$$

where the first equality follows from Proposition 10.10. Since $\mathsf{d}_{\mathcal{L}}(\mathcal{C}_{\mathrm{Ber}}) \ge 3$, we conclude that a Berlekamp code with $\tau = 1$ and $n = \frac{1}{2}(|\Phi| - 1)$ is a perfect code in the Lee metric. □

While the construction in Example 10.4 is seemingly restricted to alphabets of prime size, it can in fact be generalized to every alphabet $\mathbb{Z}_q$ of odd size q (see Problem 10.13).

Example 10.5 Let the ring $\mathbb{Z}_q$ be such that $q = 2t^2 + 2t + 1$ for a positive integer t. Consider the following linear code

$$\mathcal{C} = \Big\{ c \cdot (1 \;\; 2t+1) \mid c \in \mathbb{Z}_q \Big\}$$

(of length 2 and size q) over $\mathbb{Z}_q$. Notice that $-(2t+1)$ is the multiplicative inverse of $2t+1$ in $\mathbb{Z}_q$; therefore, $\mathbf{c} = (c_1 \; c_2)$ belongs to $\mathcal{C}$ if and only if $-(2t+1)\mathbf{c} = (-c_2 \; c_1)$ does.

We claim that $\mathsf{d}_{\mathcal{L}}(\mathcal{C}) \ge 2t+1$. Suppose to the contrary that there is a nonzero codeword $\mathbf{c} = (c_1 \; c_2)$ in $\mathcal{C}$ such that $\mathsf{w}_{\mathcal{L}}(\mathbf{c}) = |c_1| + |c_2| < 2t+1$. Then either $|c_1| \le t$ or $|c_2| \le t$. Without loss of generality we can assume that $|c_1| \le t$: otherwise, select $(-c_2 \; c_1)$ for the codeword $\mathbf{c}$. By possibly taking $-\mathbf{c}$ instead of $\mathbf{c}$, we can further assume that $c_1 \in \mathbb{Z}_q^+$.

Write $s = \langle c_1 \rangle$; since $1 \le s \le t$ we have

$$\langle c_2 \rangle = \langle (2t+1)c_1 \rangle = (2t+1)s$$

and, so,

$$|c_2| = \begin{cases} (2t+1)s & \text{if } 1 \le s \le t/2 \\ q - (2t+1)s & \text{if } t/2 < s \le t \end{cases} .$$

Hence,

$$\mathsf{w}_{\mathcal{L}}(\mathbf{c}) = |c_1| + |c_2| = \begin{cases} (2t+2)s & \text{if } 1 \le s \le t/2 \\ (2t^2+2t+1) - 2ts & \text{if } t/2 < s \le t \end{cases} .$$

In either case we reach the contradiction $\mathsf{w}_{\mathcal{L}}(\mathbf{c}) \ge 2t+1$. We thus conclude that $\mathsf{d}_{\mathcal{L}}(\mathcal{C}) \ge 2t+1$.

The code $\mathcal{C}$ is perfect in the Lee metric: by Proposition 10.10 we have

$$V_{\mathcal{L}}(2,t) = 1 + 2t + 2t^2 = q$$

and, so, $|\mathcal{C}| \cdot V_{\mathcal{L}}(2,t) = q^2$. □

Our next theorem is a Gilbert–Varshamov-type bound in the Lee metric over prime fields of odd size.

Theorem 10.12 (Gilbert–Varshamov bound in the Lee metric) *Let p be an odd prime and let n, k, and d be positive integers such that*

$$\frac{p^{n-k+1}-1}{p-1} > \frac{V_{\mathcal{L}|p}(n,d-1)-1}{2} .$$

Then there exists a linear $[n,k]$ code $\mathcal{C}$ over $F = \mathrm{GF}(p)$ with $\mathsf{d}_{\mathcal{L}}(\mathcal{C}) \ge d$.

Proof. Starting with $\mathcal{C}_0 = \{\mathbf{0}\}$, we construct iteratively a sequence of codes $\mathcal{C}_1, \mathcal{C}_2, \ldots, \mathcal{C}_k$ such that each code $\mathcal{C}_i$ is a linear $[n,i]$ code over F with $\mathsf{d}_{\mathcal{L}}(\mathcal{C}_i) \ge d$.

Suppose that we have constructed the codes $\mathcal{C}_1, \mathcal{C}_2, \ldots, \mathcal{C}_{i-1}$ for some $i \le k$. For a word $\mathbf{e} \in F^n \setminus \mathcal{C}_{i-1}$, let the set $\mathcal{C}_{i-1}(\mathbf{e})$ be defined by

$$\mathcal{C}_{i-1}(\mathbf{e}) = \left\{ \mathbf{c} + a \cdot \mathbf{e} \;:\; \mathbf{c} \in \mathcal{C}_{i-1} ,\; a \in F^* \right\} .$$

Note that $\mathcal{C}_{i-1}(\mathbf{e})$ is a union of $p-1$ distinct cosets of $\mathcal{C}_{i-1}$ in F^n and that distinct sets $\mathcal{C}_{i-1}(\mathbf{e})$ must be disjoint. Ranging over all $\mathbf{e} \in F^n \setminus \mathcal{C}_{i-1}$, it follows that the number of distinct sets $\mathcal{C}_{i-1}(\mathbf{e})$ is given by

$$\frac{1}{p-1}\left(\frac{p^n}{|\mathcal{C}_{i-1}|} - 1\right) = \frac{p^{n-i+1}-1}{p-1} > \frac{V_{\mathcal{L}|p}(n,d-1)-1}{2} .$$

The right-hand side of the last inequality is the size of the set, $\mathcal{S}^*_{\mathcal{L}}(\mathbf{0},d-1)$, of all nonzero words in $\mathcal{S}_{\mathcal{L}}(\mathbf{0},d-1)$ whose leading nonzero entry is in F^+. Therefore, there is at least one set $\mathcal{C}_{i-1}(\mathbf{e}_0)$ for which $\mathcal{C}_{i-1}(\mathbf{e}_0) \cap \mathcal{S}^*_{\mathcal{L}}(\mathbf{0},d-1) = \emptyset$. Moreover, $\mathbf{c} \in \mathcal{C}_{i-1}(\mathbf{e}_0) \iff -\mathbf{c} \in \mathcal{C}_{i-1}(\mathbf{e}_0)$, and $\mathbf{0} \notin \mathcal{C}_{i-1}(\mathbf{e}_0)$; so, $\mathcal{C}_{i-1}(\mathbf{e}_0) \cap \mathcal{S}_{\mathcal{L}}(\mathbf{0},d-1) = \emptyset$. The union $\mathcal{C}_{i-1} \cup \mathcal{C}_{i-1}(\mathbf{e}_0)$ thus forms a linear $[n,i]$ code, $\mathcal{C}_i$, with minimum Lee distance at least d. □

Problems

[Section 10.1]

Problem 10.1 Consider the Lee distance mapping $(\mathbf{x}, \mathbf{y}) \mapsto \mathsf{d}_{\mathcal{L}}(\mathbf{x}, \mathbf{y})$ over $\mathbb{Z}_q \times \mathbb{Z}_q$.

1. Show that the Lee distance satisfies the properties of a metric.

2. Verify that for $q = 2, 3$, the Lee distance is identical to the Hamming distance.

Problem 10.2 Show that the minimum Lee distance of a group code $\mathcal{C} \neq \{\mathbf{0}\}$ over $\mathbb{Z}_q$ equals the minimum Lee weight of any nonzero codeword in $\mathcal{C}$.

Problem 10.3 Let m and h be positive integers and write $q = m^h$. A *Gray mapping* (commonly known as a "Gray code") is a one-to-one mapping $\Gamma : \mathbb{Z}_q \rightarrow \mathbb{Z}_m^h$ that satisfies the following distance-preserving property for every $x, y \in \mathbb{Z}_q$:

$$|x - y| = 1 \quad \Longrightarrow \quad \mathsf{d}(\Gamma(x), \Gamma(y)) = 1$$

(here $|\cdot|$ stands for the Lee weight in $\mathbb{Z}_q$ and $\mathsf{d}(\cdot, \cdot)$ stands for the *Hamming* distance in $\mathbb{Z}_m^h$).

1. Show that for every two elements $x, y \in \mathbb{Z}_q$ and a Gray mapping $\Gamma : \mathbb{Z}_q \rightarrow \mathbb{Z}_m^h$,

$$\mathsf{d}(\Gamma(x), \Gamma(y)) \leq |x - y| .$$

2. For an element $x \in \mathbb{Z}_q$, denote by $\vec{x} = (x_1\ x_2\ \ldots\ x_h)$ the word in $\mathbb{Z}_m^h$ whose entries form the h-digit representation of the integer $\langle x \rangle$ to the base m, with the first entry in $\vec{x}$ being the least significant digit of the representation; i.e., the entries x_j are determined by

$$\langle x \rangle = \sum_{j=1}^{h} \langle x_j \rangle \cdot m^{j-1} .$$

Consider the mapping $\Gamma : \mathbb{Z}_q \rightarrow \mathbb{Z}_m^h$ where for every $x \in \mathbb{Z}_q$, the value $\Gamma(x) = (y_1\ y_2\ \ldots\ y_h)$ is obtained from $\vec{x} = (x_1\ x_2\ \ldots\ x_h)$ by

$$y_j = \begin{cases} x_j - x_{j+1} & \text{if } 1 \leq j < h \\ x_h & \text{if } j = h \end{cases}$$

(where the subtraction is in $\mathbb{Z}_m$). Show that $x \mapsto \Gamma(x)$ is a Gray mapping.

3. Let $\mathcal{C}$ be an (nh, M, d) code over $\mathbb{Z}_m$. Define the (n, M) code $\hat{\mathcal{C}}$ over $\mathbb{Z}_q$ by

$$\hat{\mathcal{C}} = \Big\{ (c_1\ c_2\ \ldots\ c_n) \in \mathbb{Z}_q^n \ : \ (\Gamma(c_1) \,|\, \Gamma(c_2) \,|\, \ldots \,|\, \Gamma(c_n)) \in \mathcal{C} \Big\} ,$$

where $(\cdot|\cdot)$ denotes concatenation of words. Show that $\mathsf{d}_{\mathcal{L}}(\hat{\mathcal{C}}) \geq d$.

[Section 10.2]

Problem 10.4 Let Φ be a field. Prove the following properties of the formal derivative, for every $a(x), b(x) \in \Phi[[x]]$ and $c \in \Phi$:

1. $(a(x) + b(x))' = a'(x) + b'(x)$.
2. $(c \cdot a(x))' = c \cdot a'(x)$.
3. $(a(x)b(x))' = a'(x)b(x) + a(x)b'(x)$.
4. If $b(x)$ is invertible then

$$\left(\frac{a(x)}{b(x)}\right)' = \frac{a'(x)b(x) - a(x)b'(x)}{b^2(x)} .$$

[Section 10.5]

Problem 10.5 Let $\mathcal{C}_{\mathrm{GRS}}$ be an $[n, n{-}\varrho]$ normalized GRS code over $F = \mathrm{GF}(p)$, where p is a prime such that $p \equiv 3 \pmod 4$ and

$$\varrho = \tfrac{1}{2}(p{+}3) \le n < p .$$

Denote the code locators of $\mathcal{C}_{\mathrm{GRS}}$ by $\alpha_1, \alpha_2, \ldots, \alpha_n$.

1. Show that there must be at least two indexes r and s such that $\alpha_s = \alpha_r + 1$.

   Hint: Assume without loss of generality that the code locators α_j are ordered so that $\langle \alpha_j \rangle < \langle \alpha_{j+1} \rangle$ for $1 \le j < n$. Then show that there is an index r such that $\langle \alpha_{r+1} \rangle = \langle \alpha_r \rangle + 1$.

2. Let m be the smallest positive integer such that $\varrho \,|\, (p^m{-}1)$ (why does such an integer exist?) and let $\beta \in \mathrm{GF}(p^m)$ be such that $\mathcal{O}(\beta) = \varrho$. Show that $\beta^\ell \in F$ if and only if $\varrho \,|\, \ell$.

   Hint: $\gcd(\varrho, p{-}1) = 1$.

3. Show that the field element $\frac{3}{2}$ (three times the multiplicative inverse of 2) is in $\mathbb{Z}_p^-$ and that $|\frac{3}{2}| = \frac{1}{2}(p{-}3)$.

4. Let r and s be as in part 1 and let the error word $\mathbf{e} = (e_1\ e_2\ \ldots\ e_n)$ be given by

$$e_j = \begin{cases} \frac{3}{2} & \text{if } j = r \\ -1 & \text{if } j = s \\ 0 & \text{otherwise} \end{cases} .$$

   Show that when applying the algorithm of Figure 10.2 to the decoding of $\mathbf{e}$, there are two integers h that satisfy (10.13), yet only one of them will produce an error word in Step 4c.

   Hint: Let m and β be as in part 2 and denote by $\hat{\mathcal{C}}_{\mathrm{GRS}}$ the $[n{+}\varrho{-}1, n{-}1]$ normalized GRS code over $\mathrm{GF}(p^m)$ whose code locators are

$$\hat{\alpha}_j = \begin{cases} \alpha_j & \text{if } 1 \le j \le n \\ \beta^{j-n} + \alpha_r & \text{if } n < j < n{+}\varrho \end{cases} .$$

Consider the error word $\hat{\mathbf{e}} = (\hat{e}_1\ \hat{e}_2\ \ldots\ \hat{e}_{n+\varrho-1})$ whose entries are given by

$$\hat{e}_j = \begin{cases} 1 & \text{if } n < j < n+\varrho \\ 0 & \text{otherwise} \end{cases} .$$

Show that with respect to a canonical parity-check matrix of $\hat{\mathcal{C}}_{\mathrm{GRS}}$, the word $\hat{\mathbf{e}}$ has the same syndrome as the error word that is obtained by appending $\varrho-1$ zeros to $\mathbf{e}$. Deduce that when applying the algorithm of Figure 10.2 to that syndrome, the value h for which $\deg r_h - \deg t_h = \langle S_0 \rangle = \frac{1}{2}(p+1)$ will generate in Step 4a polynomials that do not factor into linear terms over F.

[Section 10.6]

Problem 10.6 Let Φ be a finite field with odd characteristic and let B be a set of distinct nonzero elements of Φ such that every two elements $\beta, \gamma \in B$ satisfy $\beta+\gamma \neq 0$. Show that $|B| \leq \frac{1}{2}(|\Phi|-1)$, and construct sets B for which this inequality holds with equality.

Problem 10.7 (Negacyclic codes) A linear $[n, k]$ code $\mathcal{C}$ over a field F is called *negacyclic* if

$$(c_0\ c_1\ \ldots\ c_{n-1}) \in \mathcal{C} \quad \Longrightarrow \quad (-c_{n-1}\ c_0\ c_1\ \ldots\ c_{n-2}) \in \mathcal{C} .$$

Let $\mathcal{C}$ be a negacyclic $[n, k>0]$ code over a field F, and associate each codeword $(c_0\ c_1\ \ldots\ c_{n-1})$ in $\mathcal{C}$ with the polynomial

$$c(x) = c_0 + c_1 x + \ldots + c_{n-1} x^{n-1}$$

in $F_n[x]$.

1. Show that $\mathcal{C}$ is an ideal in the ring $F[x]/(x^n + 1)$.
2. Show that there is a unique monic polynomial $g(x) \in F[x]$ such that for every $c(x) \in F_n[x]$,
$$c(x) \in \mathcal{C} \quad \Longleftrightarrow \quad g(x) \mid c(x)$$
(compare with Proposition 8.1). The polynomial $g(x)$ is called the *generator polynomial* of the negacyclic code $\mathcal{C}$.
3. Show that $\deg g = n - k$.
4. Show that $g(x) \mid x^n + 1$.
5. Show that every monic polynomial $g(x) \in F_n[x]$ that divides $x^n + 1$ is a generator polynomial of a negacyclic $[n, n - \deg g]$ code over F.
6. Suppose that $F = \mathrm{GF}(p)$ for an odd prime p. Let n be a positive integer not divisible by p and let β be an element of multiplicative order $2n$ in a finite extension field Φ of F (verify that such a field Φ always exists). Define the $\tau \times n$ matrix H over Φ by
$$H = \begin{pmatrix} 1 & \beta & \ldots & \beta^{n-1} \\ 1 & \beta^3 & \ldots & \beta^{3(n-1)} \\ 1 & \beta^5 & \ldots & \beta^{5(n-1)} \\ \vdots & \vdots & \vdots & \vdots \\ 1 & \beta^{2\tau-1} & \ldots & \beta^{(2\tau-1)(n-1)} \end{pmatrix} ,$$

and let $\mathcal{C}$ be the code

$$\mathcal{C} = \{ \mathbf{c} \in F^n \ : \ H\mathbf{c}^T = \mathbf{0} \} .$$

Show that $\mathcal{C}$ is both a negacyclic code and a Berlekamp code.

Problem 10.8 (Chiang–Wolf codes) Let $F = \mathrm{GF}(p)$, p prime, and let n be an odd integer not divisible by p. Select an element β of multiplicative order n in a finite extension field of F and let $g(x)$ be the product of the distinct minimal polynomials (with respect to F) of the elements $\beta, \beta^3, \ldots, \beta^{2t-1}$. Show that $g(x)$ generates a cyclic $[n, n - \deg g]$ code over F and that this code is a Berlekamp code.

[Section 10.7]

Problem 10.9 Show that $V_{\mathcal{L}}(n,t)$ is strictly smaller than $\sum_{i=0}^{n} 2^i \binom{n}{i}\binom{t}{i}$ whenever $t \geq q/2$ (and $n > 0$).

Problem 10.10 Let δ be a nonnegative real number. Show that for every two positive integers n and m,

$$V\left(n{+}m, \lfloor \delta(n{+}m) \rfloor\right) \geq V\left(n, \lfloor \delta n \rfloor\right) \cdot V\left(m, \lfloor \delta m \rfloor\right) ,$$

where $V(\cdot,\cdot)$ stands for a sphere volume either in the Hamming metric or in the Lee metric.

Problem 10.11 Let $\mathbb{Z}_q$ be such that $q = (2t{+}1)M$ for positive integers t and $M > 1$. Construct a $(1, M)$ code $\mathcal{C}$ over $\mathbb{Z}_q$ with $\mathsf{d}_{\mathcal{L}}(\mathcal{C}) \geq 2t{+}1$ and show that $\mathcal{C}$ is perfect in the Lee metric.

Problem 10.12 Let q be divisible by $m = 2t^2 + 2t + 1$ for some positive integer t. Construct a $(2, q^2/m)$ code $\mathcal{C}$ over $\mathbb{Z}_q$ with $\mathsf{d}_{\mathcal{L}}(\mathcal{C}) \geq 2t{+}1$ and show that $\mathcal{C}$ is perfect in the Lee metric.

Hint: Let $\hat{\mathcal{C}}$ be obtained by applying the construction in Example 10.5 to the ring $\mathbb{Z}_m$, and consider the code

$$\mathcal{C} = \left\{ (a_1 m{+}\langle c_1 \rangle \ \ a_2 m{+}\langle c_2 \rangle) \cdot 1 \ : \ (c_1 \ c_2) \in \hat{\mathcal{C}}, \quad 0 \leq a_1, a_2 < q/m \right\} .$$

Problem 10.13 Given an odd integer $q > 2$ and a positive integer m, write $n = \frac{1}{2}(q^m{-}1)$ and let $\mathbf{h}_1, \mathbf{h}_2, \ldots, \mathbf{h}_n$ be the nonzero column vectors in $\mathbb{Z}_q^m$ whose leading nonzero entries are in $\mathbb{Z}_q^+$. Define the code $\mathcal{C}$ over $\mathbb{Z}_q$ by

$$\mathcal{C} = \left\{ (c_1 \ c_2 \ \ldots c_n) \in \mathbb{Z}_q^n \ : \ \textstyle\sum_{j=1}^{n} c_j \mathbf{h}_j = \mathbf{0} \right\} .$$

1. Verify that $\mathcal{C}$ is a linear code over $\mathbb{Z}_q$.
2. Show that $\mathsf{d}_{\mathcal{L}}(\mathcal{C}) \geq 3$.
3. Show that $|\mathcal{C}| = q^{n-m}$.
4. Show that $\mathcal{C}$ is a perfect code in the Lee metric.

Problem 10.14 (Eigenvalues of the Lee adjacency matrix) A $q \times q$ matrix $A = (a_{i,j})_{i,j=0}^{q-1}$ over a field K is called *circulant* if $a_{i,j} = a_{0,j-i}$ for every $0 \le i, j < q$, with indexes taken modulo q. Associate with a circulant matrix A the polynomial $\vartheta_A(x) = \sum_{j=0}^{q-1} a_{0,j} x^j$ (i.e., the coefficients of $\vartheta_A(x)$ form the first row of A), and suppose that K contains an element ω with multiplicative order q.

1. Show that the eigenvalues of A are given by $\vartheta_A(\omega^m)$, $0 \le m < q$, with the respective associated right eigenvectors $(1\ \omega^m\ \omega^{2m}\ \ldots\ \omega^{(q-1)m})^T$.
2. Show that $\mathrm{rank}(A) = n - \deg \gcd(\vartheta_A(x), x^q - 1)$.
3. For an integer $q > 1$, let D_q be the following $q \times q$ matrix over the real field $\mathbb{R}$: the rows and columns of D_q are indexed by the elements of $\mathbb{Z}_q$, and entry (a, b) in D_q equals $|a-b|$ for every $a, b \in \mathbb{Z}_q$; e.g., for $q = 5$,

$$D_5 = \begin{pmatrix} 0 & 1 & 2 & 2 & 1 \\ 1 & 0 & 1 & 2 & 2 \\ 2 & 1 & 0 & 1 & 2 \\ 2 & 2 & 1 & 0 & 1 \\ 1 & 2 & 2 & 1 & 0 \end{pmatrix} .$$

The matrix D_q is called the *Lee adjacency matrix* for $\mathbb{Z}_q$.
Let ω be a root of order q of unity in the complex field $\mathbb{C}$; that is, $\omega = \mathrm{e}^{2\pi\imath/q}$, where $\mathrm{e} = 2.71828\cdots$ is the base of natural logarithms, $\pi = 3.14159\cdots$, and $\imath = \sqrt{-1}$. Show that the eigenvalues, $(\lambda_a)_{a\in\mathbb{Z}_q}$, of D_q are given by

$$\lambda_0 = \vartheta_{D_q}(1) = \begin{cases} \frac{1}{4}(q^2-1) & \text{if } q \text{ is odd} \\ \frac{1}{4}q^2 & \text{if } q \text{ is even} \end{cases}$$

and

$$\lambda_{m\cdot 1} = \vartheta_{D_q}(\omega^m) = \begin{cases} \dfrac{(-1)^m \cos(\pi m/q) - 1}{2\sin^2(\pi m/q)} & \text{if } q \text{ is odd} \\ \dfrac{(-1)^m - 1}{2\sin^2(\pi m/q)} & \text{if } q \text{ is even} \end{cases}, \quad 1 \le m < q .$$

Hint: Consider the polynomial $f(x) \in \mathbb{R}_q[x]$ that satisfies

$$f(x) \equiv \left(\tfrac{1}{2}(x + x^{-1}) - 1\right) \cdot \vartheta_{D_q}(x) \pmod{(x^q - 1)} .$$

Denoting by $\mathbf{d}_a$ the row of D_q that corresponds to the element $a \in \mathbb{Z}_q$, show that the vector of coefficients of $f(x)$ is given by $\frac{1}{2}(\mathbf{d}_1 + \mathbf{d}_{-1}) - \mathbf{d}_0$ and, so,

$$f(x) = \begin{cases} 1 - \frac{1}{2}\left(x^{(q-1)/2} + x^{(q+1)/2}\right) & \text{if } q \text{ is odd} \\ 1 - x^{q/2} & \text{if } q \text{ is even} \end{cases} .$$

Recalling that $\omega^{m(q-1)/2}$ and $\omega^{m(q+1)/2}$ are conjugate elements in $\mathbb{C}$, deduce that

$$f(\omega^m) = \begin{cases} 1 - \mathrm{Re}\left\{\omega^{m(q-1)/2}\right\} & \text{if } q \text{ is odd} \\ 1 - (-1)^m & \text{if } q \text{ is even} \end{cases}, \quad 0 \le m < q ,$$

where $\text{Re}\{y\}$ stands for the real value of $y \in \mathbb{C}$. Finally, recall that $\text{Re}\{e^{\imath\theta}\} = \cos\theta$ and apply known trigonometric identities to

$$\vartheta_{D_q}(\omega^m) = \frac{f(\omega^m)}{\frac{1}{2}(\omega^m + \omega^{-m}) - 1} = \frac{f(\omega^m)}{\text{Re}\{\omega^m\} - 1}, \quad 1 \le m < q.$$

4. Let $P = (P_{a,b})$ denote the $q \times q$ complex matrix whose rows and columns are indexed by the elements of $\mathbb{Z}_q$, and its entries are given by

$$P_{a,b} = \sqrt{1/q} \cdot \omega^{\langle ab \rangle}, \quad a, b \in \mathbb{Z}_q.$$

Let P^* denote the $q \times q$ conjugate transpose of P; that is,

$$P^*_{a,b} = \sqrt{1/q} \cdot \omega^{-\langle ab \rangle}, \quad a, b \in \mathbb{Z}_q.$$

Show that $P^*P = I$ and that D_q can be written as $P\Lambda P^*$, where Λ is a $q \times q$ diagonal matrix with its main diagonal equaling to $(\lambda_a)_{a \in \mathbb{Z}_q}$.

Problem 10.15 (Plotkin bound in the Lee metric) Let $\mathcal{C}$ be an $(n, M{>}1)$ code over $\mathbb{Z}_q$.

1. For $a \in \mathbb{Z}_q$ and $j \in \{1, 2, \ldots, n\}$, denote by $x_{a,j}$ the number of codewords in $\mathcal{C}$ whose jth entry equals a. Show that when ranging over all pairs of codewords in $\mathcal{C}$, the total sum of the Lee distances between the codewords satisfies

$$\sum_{\mathbf{c}_1, \mathbf{c}_2 \in \mathcal{C}} \mathsf{d}_{\mathcal{L}}(\mathbf{c}_1, \mathbf{c}_2) = \sum_{j=1}^{n} \sum_{a,b \in \mathbb{Z}_q} x_{a,j} x_{b,j} |a{-}b|.$$

2. Let D_q be the Lee adjacency matrix for $\mathbb{Z}_q$ as defined in Problem 10.14. Denote by $\mathbf{x}_j$ the row vector $(x_{a,j})_{a \in \mathbb{Z}_q}$ in $\mathbb{R}^q$; namely, the entries of $\mathbf{x}_j$ are indexed by the elements of $\mathbb{Z}_q$, and the entry that is indexed by a equals $x_{a,j}$. Show that

$$\sum_{\mathbf{c}_1, \mathbf{c}_2 \in \mathcal{C}} \mathsf{d}_{\mathcal{L}}(\mathbf{c}_1, \mathbf{c}_2) = \sum_{j=1}^{n} \mathbf{x}_j D_q \mathbf{x}_j^T.$$

3. Show that for every $j = 1, 2, \ldots, n$,

$$\mathbf{x}_j D_q \mathbf{x}_j^T \le \chi_{\mathcal{L}}(q) \cdot M^2,$$

where $\chi_{\mathcal{L}}(q)$ is the average Lee weight of the elements of $\mathbb{Z}_q$, namely,

$$\chi_{\mathcal{L}}(q) = \begin{cases} \dfrac{q^2-1}{4q} & \text{if } q \text{ is odd} \\ \dfrac{q}{4} & \text{if } q \text{ is even} \end{cases}.$$

Hint: Let $D_q = P\Lambda P^*$ be the decomposition in part 4 of Problem 10.14. Define the row vector $\mathbf{y}_j = (y_{a,j})_{a \in \mathbb{Z}_q} = \mathbf{x}_j P$ and let $\mathbf{y}_j^*$ be the column vector $(y^*_{a,j})_{a \in \mathbb{Z}_q}$ where $y^*_{a,j}$ denotes the complex conjugate of $y_{a,j}$. Verify that $y_{0,j} = M/\sqrt{q}$ and that all the eigenvalues of D_q other than λ_0 are (real and) nonpositive. Conclude by justifying the equalities and the inequality in the following chain:

$$\mathbf{x}_j D_q \mathbf{x}_j^T = \mathbf{y}_j \Lambda \mathbf{y}_j^* = \sum_{a \in \mathbb{Z}_q} \lambda_a |y_{a,j}|^2 \leq \lambda_0 y_{0,j}^2 = \chi_{\mathcal{L}}(q) \cdot M^2 .$$

4. Show that

$$\frac{\mathsf{d}_{\mathcal{L}}(\mathcal{C})}{n} \leq \frac{\chi_{\mathcal{L}}(q)}{1 - (1/M)} .$$

Hint: Bound $\mathsf{d}_{\mathcal{L}}(\mathcal{C})$ from above by

$$\frac{1}{M(M-1)} \sum_{\mathbf{c}_1, \mathbf{c}_2 \in \mathcal{C}} \mathsf{d}_{\mathcal{L}}(\mathbf{c}_1, \mathbf{c}_2) .$$

5. Verify that when q is an odd prime, the following codes attain the bound in part 4:

(a) The shortened first-order Reed–Muller code over $\mathbb{Z}_q$, which is the linear $[q^m-1, m]$ code over $\mathbb{Z}_q$ with an $m \times (q^m-1)$ generator matrix whose columns range over all the nonzero vectors in $\mathbb{Z}_q^m$.

(b) A linear code over $\mathbb{Z}_q$ of length $n = \frac{1}{2}(q^m-1)$ and dimension m, where the columns of the generator matrix range over the nonzero vectors in $\mathbb{Z}_q^m$ whose leading nonzero entries are in $\mathbb{Z}_q^+$.

Notes

[Section 10.1]

The Lee metric was developed as an alternative to the Hamming metric for certain noisy channels—primarily channels that use phase-shift keying (PSK) modulation (see Nakamura [266]). Codes in the Lee metric have also been proposed in channels that are susceptible to *synchronization errors*. In this error model, symbols (which are typically over the binary alphabet) may be deleted from—or inserted into—the transmitted sequence. Looking at the sequence of run-lengths of identical symbols, a synchronization error within a run translates into adding ± 1 to the respective run-length; hence, if the run-lengths are set according to the entries of a codeword of a Lee-metric code (over the proper alphabet), then a prescribed number of synchronization errors can be corrected. A similar coding method can be applied to channels that are prone to *bit-shift* (or *peak-shift*) errors, where a deletion (respectively, insertion) within a run is followed by an insertion (respectively, deletion) in the next run. Coding methods for channels with synchronization or bit-shift errors were studied by Hilden *et al.* [176], Iizuka *et al.* [191], Kuznetsov and Vinck [220], Levenshtein [225]–[227], Levenshtein and Vinck [228], Tanaka and Kasai [355], Tenengolts [358], [359], and Ullman [366], [367]. For the application of Lee-metric codes in such channels see Bours [63], Roth and Siegel [306], Saitoh [312], and Saitoh *et al.* [313]. Orlitsky presents in [276] an application of Lee-metric codes in the area of interactive communication.

Codes in the Lee metric were first described in the late 1950s by Lee [222] and Ulrich [368]. Constructions of Lee-metric codes have been obtained since by quite a few authors: Astola [23], Berlekamp [36, Chapter 9], Chiang and Wolf [78] (see Problem 10.8), Golomb and Welch [154], [155], Nakamura [266], Roth and Siegel [306], and Satyanarayana [319]. The construction of Lee-metric codes from Hamming-metric codes through a Gray mapping (Problem 10.3) was suggested by Orlitsky [276] (the Gray mapping in Problem 10.3 is due to Sharma and Khanna [332]); see also Davydov [92]. Hammons *et al.* showed in [170] that certain known families of nonlinear codes over GF(2) become linear over $\mathbb{Z}_4$ under the Gray mapping; these families include (a modified version of) the $(n{=}2^{2m}, M{=}2^{n-4m}, d{=}6)$ Preparata codes [285], and the family of the Delsarte–Goethals codes [98], with their special case of the $(n{=}2^{2m}, M{=}2^{4m}, d{=}2^{m-1}(2^m{-}1))$ Kerdock codes [209].

The Lee metric can be defined also over the integer ring, with the Lee weight of $a \in \mathbb{Z}$ being the ordinary absolute value of a. The Lee weight of a word in $\mathbb{Z}^n$ then coincides with the L_1-norm of that word when the latter is regarded as an element of $\mathbb{R}^n$.

[Section 10.2]

Newton's identities (Lemma 10.2) are also referred to as the *Newton–Girard formulas*. See Lidl and Niederreiter [229, pp. 29–30] or Problem 5.18 in van der Waerden [377, Section 5.7].

[Sections 10.3–10.5]

The Lee-metric properties of alternant codes and GRS codes are taken from Roth and Siegel [306], where one can also find the decoding algorithm in Figure 10.2. There are other known lower bounds on $\mathsf{d}_{\mathcal{L}}(\mathcal{C}_{\mathrm{GRS}})$ that improve on Theorem 10.5 for values of ϱ that are close to p. For example, it is shown in [306] that

$$\mathsf{d}_{\mathcal{L}}(\mathcal{C}_{\mathrm{GRS}}) \geq \frac{\varrho{+}1}{2} + \frac{(\varrho{+}1)^2}{4(p{-}1{-}\varrho)} .$$

This bound is better than Theorem 10.5 for $\varrho \geq \frac{6}{7}p$, and it becomes quadratic in ϱ for $\varrho = p - O(1)$. Mazur [257] has proved that

$$\mathsf{d}_{\mathcal{L}}(\mathcal{C}_{\mathrm{GRS}}) \geq \tfrac{1}{4}\left(p^2{-}1 - (p{-}\varrho{-}2) \cdot p^{3/2}\right) ,$$

and this bound becomes quadratic in ϱ when $\varrho = p - O(\sqrt{p})$.

List decoding of alternant codes in the Lee metric was studied by Tal [352], [353].

One can obtain a counterpart of an alternant code for the case where the underlying ring is the integer ring $\mathbb{Z}$, by starting with an $[n, n{-}\varrho]$ normalized GRS code $\mathcal{C}_{\mathrm{GRS}}$ over the rational field $\mathbb{Q}$ (i.e., by letting the code locators be distinct nonzero elements of $\mathbb{Q}$) and then considering the intersection $\mathcal{C} = \mathcal{C}_{\mathrm{GRS}} \cap \mathbb{Z}^n$. We have $\mathsf{d}_{\mathcal{L}}(\mathcal{C}) \geq \mathsf{d}_{\mathcal{L}}(\mathcal{C}_{\mathrm{GRS}}) \geq 2\varrho$, and the decoding algorithm in Figure 10.2 also fits the code $\mathcal{C}$, except that in Step 4, one only needs to consider the case where the difference $\deg r_h - \deg t_h$ equals S_0.

Suppose now that the code $\mathcal{C}_{\mathrm{GRS}}$ (over $\mathbb{Q}$) has code locators $\alpha_j = j$ for $1 \le j \le n$. By performing elementary linear operations on the rows of H_{GRS}, we can get an alternative parity-check matrix of $\mathcal{C}_{\mathrm{GRS}}$ which takes the form

$$H = \begin{pmatrix} 1 & 1 & 1 & \dots & 1 \\ \binom{0}{1} & \binom{1}{1} & \binom{2}{1} & \dots & \binom{n-1}{1} \\ \binom{0}{2} & \binom{1}{2} & \binom{2}{2} & \dots & \binom{n-1}{2} \\ \vdots & \vdots & \vdots & \vdots & \vdots \\ \binom{0}{\varrho-1} & \binom{1}{\varrho-1} & \binom{2}{\varrho-1} & \dots & \binom{n-1}{\varrho-1} \end{pmatrix}$$

(where $\binom{j}{i} = 0$ if $i > j$). Associating every word $(c_0 \; c_1 \; \dots \; c_{n-1})$ in $\mathbb{Z}^n$ with the polynomial

$$c(x) = c_0 + c_1 x + \dots + c_{n-1} x^{n-1}$$

in $\mathbb{Z}_n[x]$, it is easy to verify that the entries of $H\mathbf{c}^T$ equal the values of the first ϱ Hasse derivatives, $c(x), c^{[1]}(x), \dots, c^{[\varrho-1]}(x)$, at $x = 1$. We thus get the following equivalent characterization for the respective alternant code $\mathcal{C} = \mathcal{C}_{\mathrm{GRS}} \cap \mathbb{Z}^n$:

$$\mathcal{C} = \left\{ c(x) \in \mathbb{Z}_n[x] \; : \; (x-1)^\varrho \,|\, c(x) \right\} .$$

A word in $\mathcal{C}$ with coefficients in $\{+1, -1\}$ is said to have an *ϱth order spectral null at zero frequency.* The requirement of having words with a prescribed spectral null at zero frequency appears in several recording applications; see Immink [192], [193] and Karabed and Siegel [204]. An *ϱth order spectral-null code* is a subset of $\mathcal{X}(n, \varrho) = \mathcal{C} \cap \{\pm 1\}^n$. It can be easily verified that

$$\mathsf{d}(\mathcal{X}(n, \varrho)) = \tfrac{1}{2} \mathsf{d}_{\mathcal{L}}(\mathcal{X}(n, \varrho)) \ge \mathsf{d}_{\mathcal{L}}(\mathcal{C}) \ge 2\varrho$$

whenever $\mathcal{X}(n, \varrho) \ne \emptyset$.

Determining the exact value of $\mathsf{d}(\varrho) = \min_{n \,:\, \mathcal{X}(n,\varrho) \ne \emptyset} \mathsf{d}(\mathcal{X}(n, \varrho))$ is equivalent to a known problem in number theory referred to as the *Prouhet–Tarry problem*: given ϱ, find the smallest positive integer $\mathsf{d}(\varrho)$ for which there exist two disjoint subsets $\mathcal{A}, \mathcal{B} \in \mathbb{Z}^+$ of total size $\mathsf{d}(\varrho)$ such that

$$\sum_{\alpha \in \mathcal{A}} \alpha^\ell = \sum_{\beta \in \mathcal{B}} \beta^\ell , \quad 0 \le \ell < \varrho$$

(for $\ell = 0$ this equality becomes $|\mathcal{A}| = |\mathcal{B}|$). It is known that $\mathsf{d}(\varrho)$ equals 2ϱ for $\varrho \le 10$ and is bounded from above by $\frac{1}{2}\varrho(\varrho-1) + 1$ for larger ϱ; see Hardy and Wright [171, p. 329] and Hua [189, p. 507]. Roth *et al.* [307] showed that $\mathcal{X}(n, \varrho)$ is nonempty only if n is divisible by $2^{\lfloor \log_2 \varrho \rfloor + 1}$. Freiman and Litsyn [134] then showed that this condition is also sufficient for every fixed ϱ and large enough n; furthermore, they proved that for every fixed ϱ,

$$\liminf_{n \to \infty} \frac{n - \log_2 |\mathcal{X}(n, \varrho)|}{\log_2 n} = \frac{\varrho^2}{2} .$$

Determining the smallest value of n for which $\mathcal{X}(n, \varrho)$ is nonempty remains an open problem. This value equals 2^{ϱ} for $\varrho \le 5$, yet for $\varrho = 6$ the set $\mathcal{X}(n, 6)$ is already nonempty for $n = 48$; see Boyd [64], [65]. For more on spectral-null codes, see Eleftheriou and Cideciyan [113], Immink [192], [193], Immink and Beenker [194], Karabed and Siegel [204], Monti and Pierobon [264], Roth [298], Roth *et al.* [307], Skachek *et al.* [341], and Tallini and Bose [354].

[Section 10.6]

Berlekamp codes were obtained in [36, Chapter 9], where their description is given in the form of negacyclic codes (see Problem 10.7). While the analysis of Berlekamp codes is based here on the decoding algorithm of alternant codes, Berlekamp codes and their decoding algorithm were obtained much earlier than the respective work on alternant codes.

[Section 10.7]

The sphere-packing bound in the Lee metric (stated in Theorem 10.11) is due to Berlekamp [36, Chapter 13] and Golomb and Welch [154], [155]; these references also contain the constructions in Example 10.5 and Problems 10.11–10.13. From Problem 10.1 it follows that all perfect codes in the Hamming metric over GF(2) and GF(3) are also perfect codes in the Lee metric; this includes the Hamming codes and the Golay codes over these fields. It is conjectured that for $q > 3$, there are no perfect codes in the Lee metric over $\mathbb{Z}_q$ with code length greater than 2 and minimum Lee distance greater than 3. See Astola [19]–[22], Bassalygo [31], Gravier *et al.* [162], Lepistö [223], [224], Post [284], and Riihonen [295].

The Gilbert–Varshamov bound in the Lee metric (Theorem 10.12) is from Berlekamp [36, Chapter 13], and the Plotkin bound (part 4 of Problem 10.15) is due to Wyner and Graham [390]. The properties of circulant matrices , which are used in the proof of the Plotkin bound, can be found in the book by Davis [91].

The Lee-metric counterpart of the Johnson bound, due to Berlekamp [36, Chapter 13], takes the following form.

Proposition 10.13 (Johnson bound in the Lee metric) *Let $\mathcal{C}$ be an $(n, M{>}1)$ code over $\mathbb{Z}_q$ and denote by $\chi_{\mathcal{L}}(q)$ the average Lee weight of the elements of $\mathbb{Z}_q$, namely,*

$$\chi_{\mathcal{L}}(q) = \begin{cases} (q^2-1)/(4q) & \text{if } q \text{ is odd} \\ q/4 & \text{if } q \text{ is even} \end{cases} .$$

Suppose that there is a real $\theta \in (0, \chi_{\mathcal{L}}(q)]$ such that each codeword in $\mathcal{C}$ has Lee weight at most θn. Then,

$$\frac{\mathsf{d}_{\mathcal{L}}(\mathcal{C})}{n} \le \frac{M}{M-1} \cdot \left(2\theta - \frac{\theta^2}{\chi_{\mathcal{L}}(q)} \right) .$$

Proof. Denote by $\boldsymbol{\varepsilon}$ and $\mathbf{1}$ the column vectors $(1\ 0\ 0\ \ldots\ 0)^T$ and $(1\ 1\ \ldots\ 1)^T$, respectively, in $\mathbb{R}^q$. Using the notation in part 3 of Problem 10.15, for every $j \in \{1, 2, \ldots, n\}$ we have,

$$\sum_{a \in \mathbb{Z}_q} x_{a,j} |a| = \mathbf{x}_j D_q \boldsymbol{\varepsilon} = \mathbf{x}_j P \Lambda P^* \boldsymbol{\varepsilon} = \sqrt{1/q} \cdot \mathbf{y}_j \Lambda \mathbf{1} = \sqrt{1/q} \cdot \sum_{a \in \mathbb{Z}_q} \lambda_a y_{a,j} \,. \quad (10.21)$$

Now, it is assumed that $\mathsf{w}_{\mathcal{L}}(\mathbf{c}) \le \theta n$ for every $\mathbf{c} \in \mathcal{C}$; hence,

$$\sum_{j=1}^{n} \sum_{a \in \mathbb{Z}_q} x_{a,j} |a| = \sum_{\mathbf{c} \in \mathcal{C}} \mathsf{w}_{\mathcal{L}}(\mathbf{c}) \le M \theta n ,$$

which, by (10.21), translates into

$$\sum_{j=1}^{n} \sum_{a \in \mathbb{Z}_q} \lambda_a \mathrm{Re}\,\{y_{a,j}\} \le M \theta n \sqrt{q} . \tag{10.22}$$

On the other hand,

$$M(M-1) \cdot \mathsf{d}_{\mathcal{L}}(\mathcal{C}) \le \sum_{\mathbf{c}_1, \mathbf{c}_2 \in \mathcal{C}} \mathsf{d}_{\mathcal{L}}(\mathbf{c}_1, \mathbf{c}_2) = \sum_{j=1}^{n} \mathbf{x}_j D_q \mathbf{x}_j^T \tag{10.23}$$

(see part 2 of Problem 10.15). Now, for every real v,

$$\mathbf{x}_j D_q \mathbf{x}_j^T = \sum_{a \in \mathbb{Z}_q} \lambda_a |y_{a,j}|^2 = \sum_{a \in \mathbb{Z}_q} \lambda_a |y_{a,j} - v|^2 + 2v \sum_{a \in \mathbb{Z}_q} \lambda_a \mathrm{Re}\,\{y_{a,j}\} - v^2 \sum_{a \in \mathbb{Z}_q} \lambda_a .$$

Recall that the sum of the eigenvalues of a matrix equals its trace (namely, the sum of elements on its main diagonal) and, so $\sum_{a \in \mathbb{Z}_q} \lambda_a = 0$. Furthermore, all the eigenvalues λ_a other than λ_0 are (real and) nonpositive, and $y_{0,j} = M/\sqrt{q}$; therefore,

$$\begin{aligned}
\mathbf{x}_j D_q \mathbf{x}_j^T &\le \lambda_0 |y_{0,j} - v|^2 + 2v \sum_{a \in \mathbb{Z}_q} \lambda_a \mathrm{Re}\,\{y_{a,j}\} \\
&= \lambda_0 \left(\frac{M}{\sqrt{q}} - v \right)^2 + 2v \sum_{a \in \mathbb{Z}_q} \lambda_a \mathrm{Re}\,\{y_{a,j}\} .
\end{aligned}$$

Summing over $j \in \{1, 2, \ldots, n\}$ we obtain from (10.22) and (10.23) that for every $v \ge 0$,

$$\begin{aligned}
M(M-1) \cdot \mathsf{d}_{\mathcal{L}}(\mathcal{C}) \le \sum_{j=1}^{n} \mathbf{x}_j D_q \mathbf{x}_j^T &\le \lambda_0 n \left(\frac{M}{\sqrt{q}} - v \right)^2 + 2v \sum_{j=1}^{n} \sum_{a \in \mathbb{Z}_q} \lambda_a \mathrm{Re}\,\{y_{a,j}\} \\
&\le \lambda_0 n \left(\frac{M}{\sqrt{q}} - v \right)^2 + 2v M \theta n \sqrt{q} .
\end{aligned}$$

Finally, substituting

$$v \leftarrow \frac{M}{\sqrt{q}} \left(1 - \frac{q\theta}{\lambda_0} \right) = \frac{M}{\sqrt{q}} \left(1 - \frac{\theta}{\chi_{\mathcal{L}}(q)} \right)$$

yields the desired result. □

We next turn to the Elias bound in the Lee metric, also due to Berlekamp [36, Chapter 13]. Let $(a_n)_{n=0}^{\infty}$ be an infinite all-positive real sequence. We say that this sequence is *super-multiplicative* if $a_{n+m} \ge a_n a_m$ for every $n, m \ge 0$. The following lemma is by Kingman [214] (see also Seneta [328, p. 249]).

Lemma 10.14 *If $(a_n)_{n=0}^{\infty}$ is a super-multiplicative sequence then the limit*

$$a = \lim_{n \to \infty} a_n^{1/n}$$

exists, and $a_n \le a^n$ for every $n \ge 0$.

It follows from the last lemma and Problem 10.10 that for every nonnegative real δ, the limit

$$\mathsf{H}_{\mathcal{L}|q}(\delta) = \lim_{n \to \infty} \frac{1}{n} \cdot \log_q V_{\mathcal{L}|q}(n, \lfloor \delta n \rfloor)$$

exists. Combining this with Proposition 10.13, we obtain the next theorem.

Theorem 10.15 (Elias bound in the Lee metric) *Let C be an (n, q^{nR}) code over $\mathbb{Z}_q$ with $\mathsf{d}_{\mathcal{L}}(C) = \delta n$ where $\delta \le \chi_{\mathcal{L}}(q)$. Then,*

$$R \le 1 - \mathsf{H}_{\mathcal{L}|q}\left(\chi_{\mathcal{L}}(q) \cdot \left(1 - \sqrt{1 - (\delta / \chi_{\mathcal{L}}(q))} \right) \right) + o(1) ,$$

where $o(1)$ stands for an expression that goes to zero as $n \to \infty$ (this expression may depend on q or δ).

We omit the proof, as it is very similar to that of the Hamming-metric Elias bound (Theorem 4.12). The Elias bound in the Lee metric was also studied by Astola [22], [24] and Lepistö [223].

Chiang and Wolf [78] obtained a bound that can be viewed as the Lee-metric version of the Singleton bound.

Chapter 11

MDS Codes

In Section 4.1, we defined MDS codes as codes that attain the Singleton bound. This chapter further explores their properties. The main topic to be covered here is the problem of determining for a given positive integer k and a finite field $F = \mathrm{GF}(q)$, the largest length of any linear MDS code of dimension k over F. This problem is still one of the most notable unresolved questions in coding theory, as well as in other disciplines, such as combinatorics and projective geometry over finite fields. The problem has been settled so far only for a limited range of dimensions k. Based on the partial proved evidence, it is believed that within the range $2 \leq k \leq q-1$ (and with two exceptions for even values of q), linear $[n, k]$ MDS codes exist over F if and only if $n \leq q+1$. One method for proving this conjecture for certain values of k is based on identifying a range of parameters for which MDS codes are necessarily extended GRS codes. To this end, we will devote a part of this chapter to reviewing some of the properties of GRS codes and their extensions.

11.1 Definition revisited

We start by recalling the Singleton bound from Section 4.1. We will prove it again here, using a certain characterization of the minimum distance of a code, as provided by the following lemma.

Lemma 11.1 *Let F be an alphabet of size q and $\mathcal{C}$ be an (n, M, d) code over F. Denote by T the $M \times n$ array whose rows form the codewords of $\mathcal{C}$, and let ℓ be the smallest integer such that the rows in every $M \times \ell$ sub-array of T are all distinct. Then,*

$$d = n - \ell + 1 .$$

Proof. Since no two distinct codewords in $\mathcal{C}$ agree on any $n-d+1$ coordinates, we have $\ell \le n-d+1$. On the other hand, there exist two distinct codewords that agree on $n-d$ coordinates; hence, $\ell > n-d$. □

Corollary 11.2 (The Singleton bound) *For any* (n, M, d) *code over an alphabet of size* q,

$$d \le n - (\log_q M) + 1 .$$

Proof. The definition of ℓ in Lemma 11.1 requires that $q^\ell \ge M$. □

Codes that attain the bound in Corollary 11.2 are called maximum distance separable (MDS). The following are examples of MDS codes of length n over any finite Abelian group F of size q.

- The $(n, q^n, 1)$ code F^n.
- The $(n, q^{n-1}, 2)$ parity code $\{c_1 c_2 \ldots c_n \in F^n : \sum_{i=1}^n c_i = 0\}$, with the summation taken in the group F.
- The (n, q, n) repetition code $\{aa \ldots a : a \in F\}$.

We will refer to these codes as the *trivial* MDS constructions.

Let T be an $M \times n$ array over an alphabet F of size q and let k be a positive integer. We say that T is an *orthogonal array* with parameters (M, n, q, k) (in short, $\mathrm{OA}(M, n, q, k)$) if each element of F^k appears as a row in every $M \times k$ sub-array of T exactly M/q^k times (in particular, q^k must divide M). The value M/q^k is called the *index* of the orthogonal array. Our interest will be focused on orthogonal arrays with index 1, in which case $M = q^k$ and the rows of every $q^k \times k$ sub-array of T range over the elements of F^k.

We can use the notion of orthogonal arrays to obtain an equivalent definition of MDS codes. We demonstrate this through the next proposition.

Proposition 11.3 *Let* F *be an alphabet of size* q *and* $\mathcal{C}$ *be an* (n, M, d) *code over* F. *Denote by* T *the* $M \times n$ *array whose rows form the codewords of* $\mathcal{C}$. *Then* $\mathcal{C}$ *is MDS if and only if* $M = q^k$ *for some integer* k *and* T *is an* $\mathrm{OA}(q^k, n, q, k)$ *(with index* 1*).*

Proof. Clearly, $\mathcal{C}$ is MDS only if $\log_q M$ is an integer. Now, suppose that $M = q^k$ for an integer k and let ℓ be as in Lemma 11.1; note that $\ell \ge \log_q M = k$. By Lemma 11.1 we obtain that $\mathcal{C}$ is MDS if and only if $\ell = k$. The latter equality, in turn, is equivalent to saying that the rows in every $q^k \times k$ sub-array of T are all distinct, and that can happen if and only if T is an $\mathrm{OA}(q^k, n, q, k)$. □

We next turn to the linear case and state necessary and sufficient conditions that a given linear code is MDS. One of the conditions makes use of the following definition: a $k \times r$ matrix A over a field F is called *super-regular* if every square sub-matrix in A is nonsingular.

Proposition 11.4 *Let $\mathcal{C}$ be a linear $[n, k{<}n, d]$ code over a field F. Each of the following conditions is necessary and sufficient for the code $\mathcal{C}$ to be MDS:*

(i) Every set of $n-k$ columns in any parity-check matrix of $\mathcal{C}$ is linearly independent.

(ii) Every set of k columns in any generator matrix of $\mathcal{C}$ is linearly independent.

(iii) The dual code $\mathcal{C}^\perp$ is MDS.

(iv) The code $\mathcal{C}$ has a systematic generator matrix $(\, I \,|\, A \,)$ where A is super-regular.

Proof. Part (i) follows from the characterization of the minimum distance of $\mathcal{C}$ as the largest integer d such that every set of $d-1$ columns in a parity-check matrix of $\mathcal{C}$ is linearly independent (Theorem 2.2). Parts (ii) and (iii) were given as an exercise in Problem 4.1 (alternatively, part (ii) can be obtained by recasting Proposition 11.3 to the linear case, and part (iii) follows from (ii) by applying part (i) to the dual code $\mathcal{C}^\perp$). Finally, part (iv) was given as an exercise in Problem 4.2. □

11.2 GRS codes and their extensions

Our primary interest in this chapter will be determining, for a given positive integer k and field size q, the largest n for which there exist linear $[n, k]$ MDS codes over $F = \mathrm{GF}(q)$. The GRS code construction readily yields linear $[n, k]$ MDS codes over F for every n and k such that $1 \le k \le n \le q-1$. A code length $n = q$ can be attained with a *singly-extended* GRS code, in which one of the code locators is zero.

As shown in Problem 5.2, GRS codes can be further extended by one additional symbol while preserving the MDS property. Specifically, let $\alpha_1, \alpha_2, \ldots, \alpha_{n-1}$ be distinct elements of F and $v_1, v_2, \ldots, v_n$ be elements of F^*. An $[n, k]$ *doubly-extended* GRS code is the linear code over F with a parity-check matrix

$$\begin{pmatrix} 1 & 1 & \ldots & 1 & 0 \\ \alpha_1 & \alpha_2 & \ldots & \alpha_{n-1} & 0 \\ \alpha_1^2 & \alpha_2^2 & \ldots & \alpha_{n-1}^2 & 0 \\ \vdots & \vdots & \vdots & \vdots & \vdots \\ \alpha_1^{n-k-2} & \alpha_2^{n-k-2} & \ldots & \alpha_{n-1}^{n-k-2} & 0 \\ \alpha_1^{n-k-1} & \alpha_2^{n-k-1} & \ldots & \alpha_{n-1}^{n-k-1} & 1 \end{pmatrix} \begin{pmatrix} v_1 & & & \\ & v_2 & & 0 \\ 0 & & \ddots & \\ & & & v_n \end{pmatrix} .$$

This code is MDS, and its dual code is also a doubly-extended GRS code.

Hereafter, by an *extended* GRS code we will mean either a singly-extended or doubly-extended GRS code. In fact, every (ordinary) GRS code $\mathcal{C}_{\mathrm{GRS}}$ is also an extended GRS code, since the zero element can always be assumed to be one of the code locators of $\mathcal{C}_{\mathrm{GRS}}$ (see Problem 5.4).

Recall from Problem 5.9 that a *(generalized) Cauchy matrix* over a field F is a $k \times r$ matrix whose (i, j)th entry is given by

$$\frac{\eta_{j+k}/\eta_i}{\alpha_{j+k} - \alpha_i} , \quad 1 \le i \le k , \quad 1 \le j \le r ,$$

where $\alpha_1, \alpha_2, \ldots, \alpha_{r+k}$ are distinct elements of F and $\eta_1, \eta_2, \ldots, \eta_{r+k}$ are elements of F^* (strictly speaking, Cauchy matrices are commonly defined with all the values η_i being 1; however, herein we will use the more general definition where each η_i can take any nonzero element of F). A $k \times (r+1)$ *extended* Cauchy matrix is obtained from a Cauchy matrix by appending the column

$$\eta_{r+k+1} \cdot \left(\eta_1^{-1} \; \eta_2^{-1} \; \ldots \; \eta_k^{-1}\right)^T$$

for some element η_{r+k+1} in F^*; the $(r+k+1)$st column is said to correspond to $\alpha_{r+k+1} = \infty$ (i.e., the "infinity" code locator). Every Cauchy matrix is an extended Cauchy matrix (Problem 11.8).

Let A be a $k \times r$ matrix over F such that all the entries in A are nonzero and let A^{c} be the $k \times r$ matrix whose (i, j)th entry is the multiplicative inverse of the (i, j)th entry of A. We have the following result, which describes the form of the systematic matrices of extended GRS codes.

Proposition 11.5 *Let A be a $k \times r$ matrix whose entries are nonzero elements of a field F. The following three conditions are equivalent:*

(i) A is an extended Cauchy matrix.

(ii) Every 2×2 sub-matrix of A^{c} is nonsingular and every 3×3 sub-matrix of A^{c} is singular.

(iii) The matrix $G = (I \,|\, A)$ generates an $[r+k, k]$ extended GRS code over F.

The proof of the proposition is given as an exercise (Problem 5.9).

Let A be an extended Cauchy matrix over a field F. Combining Proposition 11.4(iv) with Proposition 11.5, we deduce that A is super-regular (see also Problem 5.9). This, in turn, forces every 1×1 and 2×2 sub-matrix in A^{c} to be nonsingular. Yet, other than the nonsingularity of these small sub-matrices in A^{c}, we get from Proposition 11.5 that A^{c} is "highly singular": every 3×3 sub-matrix in A^{c} is singular, which means that $\mathrm{rank}(A^{\mathrm{c}}) = \min\{k, r, 2\}$; so, every $t \times t$ sub-matrix of A^{c} is singular also for all $t \geq 3$.

Proposition 11.5 provides a useful criterion for testing whether a given matrix generates an extended GRS code.

Example 11.1 Let α_1, α_2, and α_3 be distinct nonzero elements of a field F such that $\alpha_i + \alpha_j \neq 0$ for $1 \leq i < j \leq 3$. Consider the 3×3 Vandermonde matrix

$$A = \begin{pmatrix} 1 & 1 & 1 \\ \alpha_1 & \alpha_2 & \alpha_3 \\ \alpha_1^2 & \alpha_2^2 & \alpha_3^2 \end{pmatrix}$$

over F. It is easy to see that A is super-regular; so, by Proposition 11.4, the matrix $(\,I \,|\, A\,)$ generates a linear $[6, 3, 4]$ MDS code $\mathcal{C}$ over F. On the other hand, the matrix A^{c} (which is also of a Vandermonde form) has rank 3, thereby implying by Proposition 11.5 that A is not an extended Cauchy matrix. Hence, $\mathcal{C}$ is not an extended GRS code, even though it is MDS. $\square$

When F is a finite field of even size, doubly-extended GRS codes can be further extended in certain cases, while still maintaining the MDS property. Specifically, let F be the finite field $\mathrm{GF}(2^m)$ and for $n > 3$, let $\alpha_1, \alpha_2, \ldots, \alpha_{n-2}$ be distinct elements of F and $v_1, v_2, \ldots, v_n$ be elements of F^*. An $[n, n-3]$ *triply-extended* GRS code is the linear code over F with a parity-check matrix

$$\begin{pmatrix} 1 & 1 & \ldots & 1 & 0 & 0 \\ \alpha_1 & \alpha_2 & \ldots & \alpha_{n-2} & 0 & 1 \\ \alpha_1^2 & \alpha_2^2 & \ldots & \alpha_{n-2}^2 & 1 & 0 \end{pmatrix} \begin{pmatrix} v_1 & & & 0 \\ & v_2 & & \\ & & \ddots & \\ 0 & & & v_n \end{pmatrix} .$$

This code is MDS (see Problem 5.3) and, by Proposition 11.4(iii), so is its $[n, 3]$ dual code.

Table 11.1 summarizes the range of values of n, k, and q for which there exist linear $[n, k]$ MDS codes over $\mathrm{GF}(q)$ that are obtained either by extensions of GRS codes or by the trivial constructions of Section 11.1.

Table 11.1. Constructions of linear $[n, k]$ MDS codes over GF(q).

Range of n	Range of k	Construction	Remarks
$n \geq 1$	$k = n$	Whole space	—
$n > 1$	$k = n-1$	Parity code	—
$n > 1$	$k = 1$	Repetition code	—
$3 < n \leq q+1$	$2 \leq k \leq n-2$	Doubly-extended GRS code	—
$4 < n \leq q+2$	$k = n-3$	Triply-extended GRS code	q even
$4 < n \leq q+2$	$k = 3$	Dual of triply-extended code	q even

11.3 Bounds on the length of linear MDS codes

Given a finite field $F = \mathrm{GF}(q)$ and a positive integer k, denote by $L_q(k)$ the largest length of any linear MDS code of dimension k over F; if such codes exist for arbitrarily large lengths, define $L_q(k) = \infty$. By part (ii) of Proposition 11.4, we can define $L_q(k)$ equivalently as follows:

> $L_q(k)$ is the size of the largest subset $S \subseteq F^k$ such that every k elements in S form a basis of F^k.

(The subset S corresponds to the set of columns of the generator matrix in that proposition.) An alternate definition of $L_q(k)$ can be inferred from part (iv) of Proposition 11.4:

> $L_q(k)-k$ is the largest number of columns, r, in any $k \times r$ super-regular matrix A over F.

In this section, we present several bounds on $L_q(k)$.

Based on the constructions in Table 11.1, we can bound $L_q(k)$ from below as follows.

Proposition 11.6

$$L_q(k) \geq \begin{cases} \infty & \textit{when } k = 1 \\ q+1 & \textit{when } 2 \leq k \leq q-1 \\ q+2 & \textit{when } k \in \{3, q-1\} \textit{ and } q \textit{ is even} \\ k+1 & \textit{when } k \geq q \end{cases} .$$

Proof. The values ∞, $q+1$, $q+2$, and $k+1$ correspond, respectively, to the longest possible repetition code, doubly-extended GRS code, triply-extended GRS code (or its dual code), and parity code. □

We next show that for certain values of k, the lower bound in Proposition 11.6 is tight. We do this through a sequence of lemmas.

Lemma 11.7

$$L_q(2) \leq q+1 .$$

Proof. Let G be a $2 \times n$ generator matrix of a linear $[n, 2]$ MDS code over $F = \mathrm{GF}(q)$. Partition the set $F^2 \setminus \{\mathbf{0}\}$ into equivalence classes, where two nonzero vectors are considered to be equivalent if one is a scalar multiple of the other. By Proposition 11.4(ii), the columns of G must belong to distinct classes. Hence, n is bounded from above by the number of these classes, which is $(q^2-1)/(q-1) = q+1$. □

Lemma 11.8

$$L_q(k+1) \leq L_q(k) + 1 .$$

Proof. Let H be an $(n-k) \times (n+1)$ parity-check matrix of a linear $[n+1, k+1]$ MDS code of length $n+1 = L_q(k+1)$ over $\mathrm{GF}(q)$. By Proposition 11.4(i), every set of $n-k$ columns in H is linearly independent. Therefore, the first n columns in H form a parity-check matrix of a linear $[n, k]$ MDS code over F and, so, $L_q(k+1) - 1 = n \leq L_q(k)$. □

Lemma 11.9 *Given $F = \mathrm{GF}(q)$, let each of the sequences*

$$\beta_1, \beta_2, \ldots, \beta_{q-1} \qquad \text{and} \qquad \gamma_1, \gamma_2, \ldots, \gamma_{q-1}$$

consist of all the elements of F^*. Assume in addition that the $q-2$ ratios

$$\beta_1/\gamma_1,\ \beta_2/\gamma_2,\ \ldots,\ \beta_{q-2}/\gamma_{q-2}$$

are all distinct. Then these ratios range over all the elements of $F^* \setminus \{-\beta_{q-1}/\gamma_{q-1}\}$.

Proof. Recall from Problem 3.21 that the product of all the elements of F^* is -1. Therefore,

$$\prod_{j=1}^{q-1} \frac{\beta_j}{\gamma_j} = \frac{\prod_{j=1}^{q-1} \beta_j}{\prod_{j=1}^{q-1} \gamma_j} = \frac{-1}{-1} = 1$$

and, so,

$$\left(\prod_{j=1}^{q-2} \frac{\beta_j}{\gamma_j} \right) \cdot \left(-\frac{\beta_{q-1}}{\gamma_{q-1}} \right) = -1 .$$

Denote by δ the (only) element of F^* that is missing from the ratios β_j/γ_j, $1 \le j \le q-2$. Using Problem 3.21 again, we have,

$$\left(\prod_{j=1}^{q-2} \frac{\beta_j}{\gamma_j} \right) \cdot \delta = -1 .$$

The last two equations imply that $\delta = -\beta_{q-1}/\gamma_{q-1}$. □

Lemma 11.10

$$L_q(3) \le \begin{cases} q+1 & \textit{when } q \textit{ is odd} \\ q+2 & \textit{when } q \textit{ is even} \end{cases} .$$

Proof. By combining Lemmas 11.7 and 11.8 we obtain that $L_q(3) \le q+2$. It remains to be shown that $L_q(3) \le q+1$ when q is odd.

Suppose to the contrary that there exists a linear $[q+2, q-1]$ MDS code over F and let $(I \mid A)$ be a $3 \times (q+2)$ systematic generator matrix of $\mathcal{C}$. By Proposition 11.4(iv), every 1×1 and 2×2 sub-matrix in A must be nonsingular. Clearly, this property is maintained if each column of A is multiplied by some element of F^*; therefore, we can assume without loss of generality that

$$A = \begin{pmatrix} 1 & 1 & \ldots & 1 \\ \beta_1 & \beta_2 & \ldots & \beta_{q-1} \\ \gamma_1 & \gamma_2 & \ldots & \gamma_{q-1} \end{pmatrix} ,$$

where the $q-1$ elements β_j are nonzero and distinct for $1 \le j \le q-1$, and so are the $q-1$ elements γ_j and the $q-1$ ratios β_j/γ_j. Yet, by Lemma 11.9 we get that β_{q-1}/γ_{q-1} (which differs from $-\beta_{q-1}/\gamma_{q-1}$) equals β_j/γ_j for some $j < q-1$, thereby reaching a contradiction. □

Lemma 11.11

$$L_q(k) \le \begin{cases} q+k-2 & \textit{when } k \ge 3 \textit{ and } q \textit{ is odd} \\ q+k-1 & \textit{when } k \ge 3 \textit{ and } q \textit{ is even} \end{cases} .$$

Proof. Combine Lemmas 11.8 and 11.10. □

(We point out that the inequality $L_q(k) \le q+k-1$ for $k > 1$ can be obtained also from the Griesmer bound (see part 2 of Problem 4.4): this bound states that for every linear $[n, k, d]$ code $\mathcal{C}$ over GF(q),

$$n \ge \sum_{i=0}^{k-1} \left\lceil \frac{d}{q^i} \right\rceil ,$$

or, equivalently,

$$n-d \geq \sum_{i=1}^{k-1} \left\lceil \frac{d}{q^i} \right\rceil .$$

Now, if $\mathcal{C}$ is MDS then $n-d = k-1$ and, so, the bound becomes

$$k-1 \geq \sum_{i=1}^{k-1} \left\lceil \frac{d}{q^i} \right\rceil .$$

But the latter inequality can hold only if $\lceil d/q^i \rceil = 1$ for every $1 \leq i < k$; in particular, we must have $(n-k+1 =)\ d \leq q$, i.e., $n \leq q+k-1$.)

Lemma 11.12 *For $k \geq 2$ and $n \geq L_q(k)$,*

$$L_q(n-k+1) \leq n .$$

Proof. Suppose to the contrary that $n+1 \leq L_q(n-k+1)$. Then there exists a linear $[n+1, n-k+1]$ MDS code $\mathcal{C}$ over F. Its dual code, $\mathcal{C}^{\perp}$, is therefore a linear $[n+1, k]$ MDS code whose length, $n+1$, absurdly exceeds $L_q(k)$. □

Lemma 11.13 *For $k \geq q$,*

$$L_q(k) \leq k+1 .$$

Proof. Substituting $k = 2$ and $n = q+1$ $(\geq L_q(2))$ in Lemma 11.12 yields $L_q(q) \leq q+1$. The result for dimensions larger than q is obtained from Lemma 11.8. □

Lemma 11.14 *For odd q,*

$$L_q(q-1) \leq q+1 .$$

Proof. Substitute $k = 3$ and $n = q+1$ $(\geq L_q(3))$ in Lemma 11.12. □

Lemmas 11.7, 11.10, 11.13, and 11.14 identify values of k for which the lower bound in Proposition 11.6 is tight. We summarize these values in the next proposition.

Proposition 11.15

$$L_q(k) = \begin{cases} \infty & \text{when } k = 1 \\ q+1 & \text{when } k = 2 \\ q+1 & \text{when } k \in \{3, q-1\} \text{ and } q \text{ is odd} \\ q+2 & \text{when } k = 3 \text{ and } q \text{ is even} \\ k+1 & \text{when } k \geq q \end{cases} .$$

Determining $L_q(k)$ for general values of k and q is still an open problem. It is believed that $L_q(k)$ is always attained by one of the constructions in Table 11.1. This can be posed as follows.

Conjecture 11.16 (The MDS conjecture)

$$L_q(k) = \begin{cases} \infty & \textit{when } k = 1 \\ q{+}1 & \textit{when } k \in \{2\} \cup \{4, 5, \ldots, q{-}2\} \\ q{+}1 & \textit{when } k \in \{3, q{-}1\} \textit{ and } q \textit{ is odd} \\ q{+}2 & \textit{when } k \in \{3, q{-}1\} \textit{ and } q \textit{ is even} \\ k{+}1 & \textit{when } k \geq q \end{cases} .$$

In the next two sections, we present an approach that adds to Proposition 11.15 more cases for which the conjecture is proved (refer also to the notes on this section at the end of the chapter for additional information about the current state of this conjecture).

11.4 GRS codes and the MDS conjecture

One method for proving the MDS conjecture in certain cases is based on showing that for a range of values of k and q, every linear MDS code is necessarily an extended GRS code. We describe this method in this section and then demonstrate how it is used to prove that for q odd,

$$L_q(4) = L_q(q{-}2) = q{+}1 .$$

Let $F = \mathrm{GF}(q)$ and for $2 \leq k \leq q{-}1$, let $\Gamma_q(k)$ be the smallest integer, if any, such that every linear $[n, k]$ MDS code over F with $n \geq \Gamma_q(k)$ is an extended GRS code. If no such integer exists, define $\Gamma_q(k) = q{+}2$.

The next lemma readily follows from the definition of $\Gamma_q(k)$.

Lemma 11.17 *Suppose that $\Gamma_q(k) \leq q{+}1$ for some k in the range $2 \leq k \leq q{-}1$. Then,*

$$L_q(k) = q{+}1 .$$

In the sequence of lemmas that follows, we present several properties of the value $\Gamma_q(k)$.

Lemma 11.18 *For $q \geq 3$,*

$$\Gamma_q(2) = 2 .$$

Proof. Every linear $[n, 2]$ MDS code over GF(q) is necessarily an extended GRS code (refer to the proof of Lemma 11.7 for the characterization of the generator matrices of linear $[n, 2]$ MDS codes). $\square$

The proof of the next two lemmas is given as an exercise (Problems 11.13–11.15).

Lemma 11.19 *Let $\mathcal{C}$ be a linear $[n, k]$ MDS code over F where $n > k+3$. For $i = 1, 2, \ldots, n$, denote by $\mathcal{C}_i$ the code over F obtained by puncturing $\mathcal{C}$ at the ith coordinate, namely,*

$$\mathcal{C}_i = \{(c_1 \; c_2 \; \ldots \; c_{i-1} \; c_{i+1} \; \ldots \; c_n) \; : \; (c_1 \; c_2 \; \ldots \; c_n) \in \mathcal{C}\} \; .$$

Suppose that there are two distinct indexes i and j for which $\mathcal{C}_i$ and $\mathcal{C}_j$ are extended GRS codes over F. Then $\mathcal{C}$ is an extended GRS code as well.

Lemma 11.20 *Let $\mathcal{C}$ be a linear $[n, k]$ MDS code over F where $k > 3$. For $i = 1, 2, \ldots, n$, denote by $\mathcal{C}^{(i)}$ the code obtained by shortening $\mathcal{C}$ at the ith coordinate, namely,*

$$\mathcal{C}^{(i)} = \{(c_1 \; c_2 \; \ldots \; c_{i-1} \; c_{i+1} \; \ldots \; c_n) \; : \; (c_1 \; c_2 \; \ldots \; c_{i-1} \; 0 \; c_{i+1} \; \ldots \; c_n) \in \mathcal{C}\} \; .$$

Suppose that there are two distinct indexes i and j for which $\mathcal{C}^{(i)}$ and $\mathcal{C}^{(j)}$ are extended GRS codes over F. Then $\mathcal{C}$ is an extended GRS code.

Lemma 11.21 *Let k and N be such that $2 \leq k \leq q-2$ and $k+3 \leq N \leq q+1$, and suppose that every linear $[N, k]$ MDS code over* GF(q) *is an extended GRS code. Then,*

$$\Gamma_q(k) \leq N \leq q+1 \; .$$

Proof. We show by induction on $n = N, N+1, N+2, \cdots$ that every linear $[n, k]$ MDS code over $F = \text{GF}(q)$ is an extended GRS code.

The induction base $n = N$ follows from the assumption of the lemma. Turning to the induction step, let n be such that every linear $[n, k]$ MDS code over F is an extended GRS code. Given a linear $[n+1, k]$ MDS code $\mathcal{C}$ over F, let $\mathcal{C}_i$ be the result of puncturing $\mathcal{C}$ at the ith coordinate, where $i \in \{1, 2\}$. It follows from Problem 2.3 that a generator matrix of $\mathcal{C}_i$ is obtained by deleting the ith column from a generator matrix of $\mathcal{C}$; hence, $\mathcal{C}_i$ is a linear $[n, k]$ MDS code. By the induction hypothesis on n we deduce that each $\mathcal{C}_i$ is, in fact, an extended GRS code, and by Lemma 11.19 we conclude that so is $\mathcal{C}$. $\square$

Lemma 11.22 *For $3 \leq k \leq q-2$,*

$$\Gamma_q(k+1) \leq \Gamma_q(k) + 1 \; .$$

Proof. We assume that $\Gamma_q(k) \le q$, since otherwise the result is obvious. Let $\mathcal{C}$ be a linear $[n, k+1]$ MDS code $\mathcal{C}$ over $\mathrm{GF}(q)$ where $n \ge \Gamma_q(k) + 1$, and for $i = 1, 2$, let $\mathcal{C}^{(i)}$ be the result of shortening $\mathcal{C}$ at the ith coordinate. A parity-check matrix of each code $\mathcal{C}^{(i)}$ is obtained by deleting a column from a parity-check matrix of $\mathcal{C}$ (see Problem 2.14); so, $\mathcal{C}^{(i)}$ is a linear $[n-1, k]$ MDS code. Furthermore, the length, $n-1$, of $\mathcal{C}^{(i)}$, is at least $\Gamma_q(k)$; hence, each code $\mathcal{C}^{(i)}$ is an extended GRS code. The result now follows from Lemma 11.20. □

Lemma 11.23 *Let k and n be such that $3 \le k \le n-2$ and $\Gamma_q(k) \le n \le q+2$. Then*

$$\Gamma_q(n-k) \le n \, .$$

Proof. The result is obvious when $n = q+2$, so we assume that n is in the range $\Gamma_q(k) \le n \le q+1$. Let $\mathcal{C}$ be a linear $[n, n-k]$ MDS code over $F = \mathrm{GF}(q)$. Its dual code, $\mathcal{C}^\perp$, being a linear $[n, k]$ MDS code of length $n \ge \Gamma_q(k)$, is necessarily an extended GRS code; hence, so is $\mathcal{C}$. By Lemma 11.21 we thus obtain $\Gamma_q(n-k) \le n$. □

(At this point, it is interesting to observe the seeming parallels between the quantities $L_q(k)$ and $\Gamma_q(k)$: compare Lemma 11.8 with Lemma 11.22, and Lemma 11.12 with Lemma 11.23.)

The previous sequence of lemmas leads to the following result.

Proposition 11.24 *Let K be such that $3 \le K \le q-1$ and $\Gamma_q(K) \le q+1$, and define the value J by*

$$J = \max\{\Gamma_q(K) - K, 2\} \, .$$

Then the following conditions hold:

(i) $\Gamma_q(k) \le q+1$ whenever k belongs to any of the following two integer intervals:

$$K \le k \le q+1-J \qquad \text{or} \qquad J \le k \le q+1-K \, .$$

(ii) $L_q(k) = q+1$ whenever k belongs to any of the following two integer intervals:

$$K \le k \le q+2-J \qquad \text{or} \qquad J \le k \le q+2-K \, .$$

Proof. By repeatedly applying Lemma 11.22 we obtain

$$\Gamma_q(k) \le \Gamma_q(K) + k - K \le J + k \le q+1 \, ,$$

whenever $K \le k \le q{+}1{-}J$. This proves part (i) for the first integer interval therein. The proof for the second interval is now obtained by substituting $n = q{+}1$ in Lemma 11.23.

Turning to part (ii), by combining part (i) with Lemma 11.17 we get the desired result except possibly for $k = q{+}2{-}J$ or $k = q{+}2{-}K$. These two exceptions are handled by applying Lemma 11.12 with $n = q{+}1$ to $k = J$ and $k = K$, respectively. □

Our results in this section can be used to prove the MDS conjecture for certain instances of k and q, in the following manner. As the first (and most crucial) step, we identify fields $F = \mathrm{GF}(q)$ and positive integers K and N in the range $K{+}3 \le N \le q{+}1$, such that every linear $[N, K]$ MDS code over F is an extended GRS code. Having found such q, K, and N, we then get by Lemma 11.21 that

$$\Gamma_q(K) \le N .$$

Finally, we combine this inequality with Proposition 11.24(ii) and infer that $L_q(k) = q{+}1$ whenever

$$K \le k \le q{+}2{-}N{+}K \qquad \text{or} \qquad N{-}K \le k \le q{+}2{-}K .$$

Example 11.2 In Proposition 11.25 below, we show that for odd field size $q \ge 5$, every linear $[q{+}1, 3]$ MDS code over $\mathrm{GF}(q)$ is a doubly-extended GRS code. Substituting $k = 3$ and $N = q{+}1$ in Lemma 11.21, we obtain that for odd $q \ge 5$,

$$\Gamma_q(3) \le q{+}1$$

and, so, by Proposition 11.24(ii),

$$L_q(3) = L_q(4) = L_q(q{-}2) = L_q(q{-}1) = q{+}1 .$$

□

The technique that we have now described for proving instances of the MDS conjecture, is illustrated in Figure 11.1 for the case of $K = 3$ and q odd. (Note that when q is even, Proposition 11.24 is vacuous for $K = 3$: we already know that for such q, the value $L_q(3)$ equals $q{+}2$ and—by Lemma 11.17—so does $\Gamma_q(3)$.)

The horizontal and vertical axes in Figure 11.1 correspond, respectively, to the dimension k and the redundancy r of linear MDS codes over $F = \mathrm{GF}(q)$. Equivalently, k and r stand, respectively, for the number of rows and columns of super-regular matrices over F. The line $r : k \mapsto q{+}1{-}k$ marks the upper boundary of the existence range of $[r{+}k, k]$ extended GRS codes (or $k \times r$ extended Cauchy matrices) over F, for $2 \le k \le q{-}1$. The two shaded right-angled triangles in the figure form a set of pairs (k, r) for which

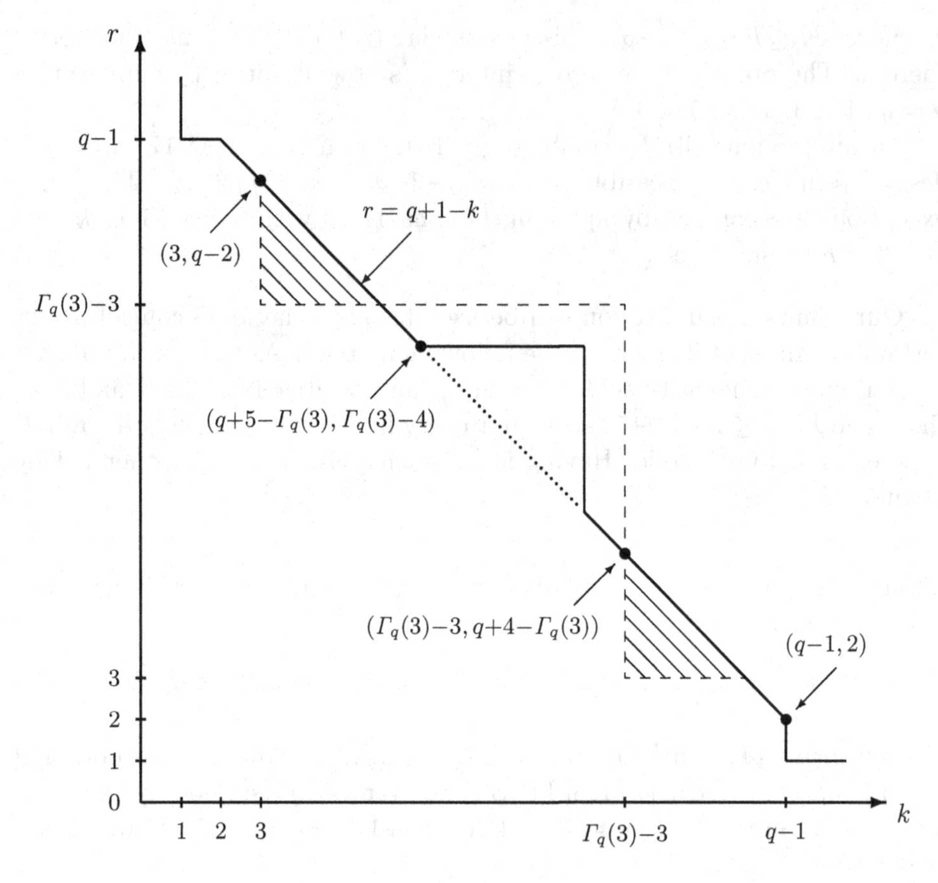

Figure 11.1. Existence range of linear $[r+k, k]$ MDS codes over GF(q), for odd q.

every linear $[r+k, k]$ MDS code over F is necessarily an extended GRS code: the horizontal leg of the upper-left triangle is drawn based on Lemma 11.22, and the other triangle is obtained by duality.

The solid line intervals in Figure 11.1 that pass through the points

$$(3, q-2) \longrightarrow (q+5-\mathit{\Gamma}_q(3), \mathit{\Gamma}_q(3)-4)$$

and

$$(\mathit{\Gamma}_q(3)-3, q+4-\mathit{\Gamma}_q(3)) \longrightarrow (q-1, 2)$$

depict part (ii) of Proposition 11.24: for the respective values of k, these lines mark the upper bound, $q+1-k$, on the redundancy r of every linear $[r+k, k]$ MDS code over F. Incorporating now also Lemma 11.8 and Proposition 11.15, we obtain an upper bound on $L_q(k)-k$, which is represented by the solid piecewise linear line in the figure.

11.5 Uniqueness of certain MDS codes

In this section, we prove the following result.

Proposition 11.25 *Let* $F = \mathrm{GF}(q)$ *for odd* $q \geq 5$. *Then every linear* $[q{+}1, 3]$ *MDS code over* F *is a doubly-extended GRS code.*

Proof. Let $G = (\, A \,|\, I \,)$ be a generator matrix of a linear $[q{+}1, 3]$ MDS code $\mathcal{C}$ over F. Without loss of generality we can assume that

$$G = (\, A \,|\, I \,) = \left(\begin{array}{cccc|ccc} 1 & 1 & \ldots & 1 & 1 & 0 & 0 \\ a_1 & a_2 & \ldots & a_{q-2} & 0 & 1 & 0 \\ b_1 & b_2 & \ldots & b_{q-2} & 0 & 0 & 1 \end{array} \right) , \tag{11.1}$$

where the entries a_j are distinct elements of F^*, the entries b_j are distinct elements of F^*, and so are the ratios a_j/b_j, $1 \leq j \leq q{-}2$. Let a_0 and b_0 be the elements of F^* that are missing from the second and third rows of A, respectively. By Lemma 11.9, the ratios a_j/b_j range over all the elements of F^* except $-a_0/b_0$.

Consider the codewords of Hamming weight q in $\mathcal{C}$ (these are the codewords of $\mathcal{C}$ that contain exactly one zero entry). As shown in Problem 11.17, there are $(q{+}1)(q{-}1)$ such codewords in $\mathcal{C}$, and for every index $\ell \in \{1, 2, \ldots q{+}1\}$, there exists a unique (up to scaling) codeword $\mathbf{c}_\ell$ of Hamming weight q whose ℓth entry is zero. We next find the linear combination of the rows of G that yields the codeword $\mathbf{c}_\ell$ for every $\ell \in \{1, 2, \ldots, q{-}2\}$, in two different ways.

(a) For $\ell \in \{1, 2, \ldots, q{-}2\}$, let the 3×3 matrix P_ℓ be given by

$$P_\ell = \begin{pmatrix} 1 & 0 & 1 \\ 0 & 1 & a_\ell \\ 0 & 0 & b_\ell \end{pmatrix} .$$

The inverse of P_ℓ is

$$P_\ell^{-1} = \begin{pmatrix} 1 & 0 & -1/b_\ell \\ 0 & 1 & -a_\ell/b_\ell \\ 0 & 0 & 1/b_\ell \end{pmatrix} .$$

Consider the generator matrix $P_\ell^{-1} G$ of $\mathcal{C}$. The columns of $P_\ell^{-1} G$ that are indexed by $\{\ell, q{-}1, q\}$ form a 3×3 identity matrix (with column ℓ being $(0\ 0\ 1)^T$), and the remaining columns in $P_\ell^{-1} G$ form a $3 \times (q{-}2)$ matrix, which we denote by B. A column in B is given either by

$$P_\ell^{-1} \begin{pmatrix} 1 \\ a_j \\ b_j \end{pmatrix} = \frac{b_j}{b_\ell} \begin{pmatrix} (b_\ell/b_j){-}1 \\ (b_\ell a_j/b_j){-}a_\ell \\ 1 \end{pmatrix} \quad \text{when } j \notin \{\ell, q{-}1, q, q{+}1\} \tag{11.2}$$

or by

$$P_\ell^{-1} \begin{pmatrix} 0 \\ 0 \\ 1 \end{pmatrix} = \frac{1}{b_\ell} \begin{pmatrix} -1 \\ -a_\ell \\ 1 \end{pmatrix} \quad \text{when } j = q{+}1 . \tag{11.3}$$

In what follows, we find it convenient to associate with this last column the "infinity" element $b_{q+1} = \infty$ and define the expression $1/b_{q+1}$ to be zero.

Using this convention, we now let j range over the $q{-}2$ values in the set $\{1, 2, \ldots, q{+}1\} \setminus \{\ell, q{-}1, q\}$ (which is the set of column indexes of B). The elements $1/b_j$ then range over all the elements of F except $1/b_\ell$ and $1/b_0$; accordingly, the elements

$$\lambda_j = (b_\ell / b_j) - 1$$

range over all the elements of F^* except $\lambda_0 = (b_\ell / b_0) - 1$.

Similarly, when j ranges over the column indexes of B, the ratios a_j/b_j range over all the elements of F except a_ℓ/b_ℓ and $-a_0/b_0$; therefore, the elements

$$\mu_j = b_\ell (a_j / b_j) - a_\ell$$

range over all the elements of F^* except $\mu_0 = -b_\ell (a_0/b_0) - a_\ell$.

From (11.2) and (11.3) we get that the ratios between the elements in the first row of B and the respective elements in the second row range over the $q{-}2$ values

$$\frac{\lambda_j}{\mu_j} = \frac{b_\ell / b_j - 1}{b_\ell (a_j / b_j) - a_\ell} , \quad j \notin \{\ell, q{-}1, q\} ,$$

and these ratios must all be distinct (or else $P_\ell^{-1} G$ would not generate an MDS code). By Lemma 11.9, these ratios exhaust all the elements of F^* except

$$-\frac{\lambda_0}{\mu_0} = \frac{b_\ell / b_0 - 1}{b_\ell (a_0/b_0) + a_\ell} = \frac{b_\ell {-} b_0}{b_\ell a_0 {+} a_\ell b_0} .$$

Recalling that column ℓ in $P_\ell^{-1} G$ equals $(0\ 0\ 1)^T$, we conclude that $\mathbf{c}_\ell$ is a linear combination of the first two rows of $P_\ell^{-1} G$; specifically,

$$\mathbf{c}_\ell = \mathbf{y}_\ell P_\ell^{-1} G = \mathbf{u}_\ell G ,$$

where, up to scaling,

$$\mathbf{y}_\ell = (\mu_0 \ \lambda_0 \ 0) = -\frac{1}{b_0} \Big(b_\ell a_0 {+} a_\ell b_0 \quad b_0 {-} b_\ell \quad 0 \Big)$$

and

$$\mathbf{u}_\ell = \mathbf{y}_\ell P_\ell^{-1} = -\frac{1}{b_0} \Big(b_\ell a_0 {+} a_\ell b_0 \quad b_0 {-} b_\ell \quad a_\ell {-} a_0 {-} 2\frac{a_\ell}{b_\ell} b_0 \Big) . \tag{11.4}$$

Thus, we have identified the coefficient vector $\mathbf{u}_\ell$ in the linear combination, $\mathbf{u}_\ell G$, which yields $\mathbf{c}_\ell$.

(b) We next obtain a second expression for the linear combination of the rows of G that yields the codeword $\mathbf{c}_\ell$. For $\ell \in \{1, 2, \ldots, q-2\}$, define the 3×3 matrix

$$Q_\ell = \begin{pmatrix} 1 & 0 & 0 \\ a_\ell & 1 & 0 \\ b_\ell & 0 & 1 \end{pmatrix} .$$

The inverse Q_ℓ^{-1} is given by

$$Q_\ell^{-1} = \begin{pmatrix} 1 & 0 & 0 \\ -a_\ell & 1 & 0 \\ -b_\ell & 0 & 1 \end{pmatrix} .$$

The columns of $Q_\ell^{-1}G$ that are indexed by $\{\ell, q, q+1\}$ form a 3×3 identity matrix (where now column ℓ is $(1\,0\,0)^T$), while the remaining columns form a $3 \times (q-2)$ matrix, denoted as C; the columns of C are given by

$$Q_\ell^{-1} \begin{pmatrix} 1 \\ a_j \\ b_j \end{pmatrix} = \begin{pmatrix} 1 \\ a_j - a_\ell \\ b_j - b_\ell \end{pmatrix} , \quad j \notin \{\ell, q, q+1\} ,$$

where a_{q-1} and b_{q-1} are defined to be zero. The elements

$$\rho_j = a_j - a_\ell$$

in the second row of C range over all the elements of F^* except $\rho_0 = a_0 - a_\ell$, and the elements

$$\sigma_j = b_j - b_\ell$$

in the third row of C range over all the elements of F^* except $\sigma_0 = b_0 - b_\ell$.

It follows from Lemma 11.9 that the $q-2$ ratios

$$\frac{\rho_j}{\sigma_j} = \frac{a_j - a_\ell}{b_j - b_\ell} , \quad j \notin \{\ell, q, q+1\} ,$$

range over all the elements of F^* except

$$-\frac{\rho_0}{\sigma_0} = -\frac{a_0 - a_\ell}{b_0 - b_\ell} .$$

Combining this with the fact that column ℓ in $Q_\ell^{-1}G$ equals $(1\ 0\ 0)^T$, we conclude that $\mathbf{c}_\ell$ is a linear combination of the last two rows of $Q_\ell^{-1}G$; namely,

$$\mathbf{c}_\ell = \mathbf{z}_\ell Q_\ell^{-1} G = \mathbf{v}_\ell G ,$$

where, up to scaling,

$$\mathbf{z}_\ell = (0 \;\sigma_0 \;\rho_0) = \begin{pmatrix} 0 & b_0 - b_\ell & a_0 - a_\ell \end{pmatrix}$$

and

$$\mathbf{v}_\ell = \mathbf{z}_\ell Q_\ell^{-1} = \begin{pmatrix} (-a_\ell b_0 + 2a_\ell b_\ell - b_\ell a_0) & b_0 - b_\ell & a_0 - a_\ell \end{pmatrix} . \tag{11.5}$$

This completes our second method for expressing $\mathbf{c}_\ell$ as a linear combination of the rows of G.

The two coefficient vectors that we have found, $\mathbf{u}_\ell$ and $\mathbf{v}_\ell$, must be equal up to scaling. Comparing the second entry in the right-hand side of (11.4) with the respective entry in (11.5), we find that the scaling factor equals $-b_0$; that is, $\mathbf{v}_\ell = -b_0 \mathbf{u}_\ell$ and, so,

$$-a_\ell b_0 + 2a_\ell b_\ell - b_\ell a_0 = b_\ell a_0 + a_\ell b_0$$

and

$$a_0 - a_\ell = a_\ell - a_0 - 2\frac{a_\ell}{b_\ell} b_0 .$$

Either one of the last two equations yields the equality

$$2a_0 b_\ell + 2a_\ell b_0 = 2a_\ell b_\ell . \tag{11.6}$$

Since q is odd, we can divide both sides of (11.6) by $2a_\ell b_\ell$, thus obtaining

$$\frac{a_0}{a_\ell} + \frac{b_0}{b_\ell} = 1 , \quad 1 \le \ell \le q-2 .$$

Equivalently,

$$(-1 \;a_0 \;b_0) A^{\mathrm{c}} = \mathbf{0} ,$$

where A^{c} is the $3 \times (q-2)$ matrix whose entries are the multiplicative inverses of the entries of the matrix A, which was defined by (11.1). Therefore, $\mathrm{rank}(A^{\mathrm{c}}) \le 2$, which means that every 3×3 sub-matrix of A^{c} is singular. On the other hand, every 2×2 sub-matrix of A is nonsingular and, therefore, so is every 2×2 sub-matrix of A^{c}. Hence, by Proposition 11.5, G generates a doubly-extended GRS code. □

Having proved Proposition 11.25, we can now use the arguments made in Example 11.2 to reach the following conclusion.

Proposition 11.26 *For odd field size $q \ge 5$,*

$$L_q(4) = L_q(q-2) = q+1 .$$

We mention without proof that Proposition 11.25 can be further improved as follows.

Proposition 11.27 *Let* $F = \mathrm{GF}(q)$ *for odd* $q \geq 7$. *Then for*

$$N = q+2 - \left\lceil \tfrac{1}{4}\sqrt{q} - \tfrac{9}{16} \right\rceil ,$$

every linear $[N, 3]$ *MDS code over* F *is an extended GRS code.*

By Lemma 11.21 it follows that for odd $q \geq 7$,

$$\Gamma_q(3) \leq q - \tfrac{1}{4}\sqrt{q} + \tfrac{41}{16} ,$$

and from Proposition 11.24 we obtain the following result.

Corollary 11.28 *Let* $F = \mathrm{GF}(q)$ *for odd* $q \geq 7$. *Then the following conditions hold:*

(i) $\Gamma_q(k) \leq q+1$ *whenever*

$$2 \leq k < \tfrac{1}{4}\sqrt{q} + \tfrac{39}{16} \qquad \text{or} \qquad q - \tfrac{1}{4}\sqrt{q} - \tfrac{23}{16} < k \leq q-1 .$$

(ii) $L_q(k) = q+1$ *whenever*

$$2 \leq k < \tfrac{1}{4}\sqrt{q} + \tfrac{55}{16} \qquad \text{or} \qquad q - \tfrac{1}{4}\sqrt{q} - \tfrac{23}{16} < k \leq q-1 .$$

Problems

[Section 11.1]

Problem 11.1 A *Latin square* of order q is a $q \times q$ array Λ over an alphabet F of q elements such that the elements in each row of Λ are distinct, and so are the elements in each column. For example, the 4×4 array

$$\Lambda = \begin{pmatrix} a & b & c & d \\ d & c & b & a \\ b & a & d & c \\ c & d & a & b \end{pmatrix}$$

over $F = \{a, b, c, d\}$ is a Latin square of order 4. As a convention, the rows and columns of a Latin square will be indexed by the elements of F.

Two Latin squares $\Lambda = (\Lambda_{i,j})$ and $\Lambda' = (\Lambda'_{i,j})$ of order q are said to be *orthogonal* if the q^2 pairs $(\Lambda_{i,j}, \Lambda'_{i,j})$ are all distinct when i and j range over F, i.e.,

$$\left\{(\Lambda_{i,j}, \Lambda'_{i,j})\right\}_{i,j \in F} = F \times F .$$

For example, the two Latin squares

$$\Lambda = \begin{pmatrix} a & b & c & d \\ d & c & b & a \\ b & a & d & c \\ c & d & a & b \end{pmatrix} \quad \text{and} \quad \Lambda' = \begin{pmatrix} a & b & c & d \\ b & a & d & c \\ c & d & a & b \\ d & c & b & a \end{pmatrix}$$

are orthogonal.

1. Let $\Lambda^{(1)}, \Lambda^{(2)}, \ldots, \Lambda^{(m)}$ be m pairwise orthogonal Latin squares of order q over an alphabet F. For $i, j \in F$, define the word

$$\mathbf{c}(i,j) = i \; j \; \Lambda^{(1)}_{i,j} \; \Lambda^{(2)}_{i,j} \; \ldots \; \Lambda^{(m)}_{i,j}$$

in F^{m+2}. Show that the set

$$\{\mathbf{c}(i,j) \; : \; i, j \in F\}$$

forms an $(m+2, q^2, m+1)$ MDS code over F.

2. Let $\mathcal{C}$ be an $(n, q^2, n-1)$ MDS code over an alphabet F of size q, where $n > 2$. For $i, j \in F$, denote by $\mathbf{c}(i,j)$ the unique codeword $c_1 c_2 \ldots c_n$ in $\mathcal{C}$ for which $c_1 = i$ and $c_2 = j$ (verify that such a codeword indeed exists). For $\ell = 1, 2, \ldots, n-2$, let $\Lambda^{(\ell)}$ be the $q \times q$ array over F whose entry $\Lambda^{(\ell)}_{i,j}$, which is indexed by $(i,j) \in F \times F$, equals the $(\ell+2)$nd entry in $\mathbf{c}(i,j)$. Show that the arrays $\Lambda^{(1)}, \Lambda^{(2)}, \ldots, \Lambda^{(n-2)}$ are pairwise orthogonal Latin squares of order q.

Problem 11.2 Let $\mathcal{C}$ be an (n, M, d) MDS code over an Abelian group F of size q, and suppose that $\mathcal{C}$ contains the all-zero word as a codeword. Show that for every set J of d coordinates there are exactly $q-1$ codewords in $\mathcal{C}$ whose support is J. Conclude that the number of codewords of Hamming weight d in $\mathcal{C}$ is $\binom{n}{d}(q-1)$.

Hint: Let J' be a set of $n-d+1$ coordinates that intersects with J at one coordinate. Consider the codewords of $\mathcal{C}$ that are zero on each of the coordinates in J', except for the coordinate in $J \cap J'$.

Problem 11.3 (Constant-weight codes with $d = w$ revisited) Let F be an Abelian group of size q. Recall from Problem 4.10 that an (n, M, d) code $\mathcal{C}$ over F is called an $(n, M, d; w)$ *constant-weight code* if each codeword in $\mathcal{C}$ has Hamming weight w.

1. Show that for every $(n, M, d; w{=}d)$ constant-weight code over F,

$$M \leq \binom{n}{d}(q-1) \, .$$

Hint: Show that no q codewords in the code may have the same support.

2. Show that given n and d, the bound in part 1 is attained by some $(n, M, d; d)$ constant-weight code over F, whenever there is an (n, q^{n-d+1}, d) MDS code over F.

Hint: See Problem 11.2.

Problem 11.4 (Singleton bound in the rank metric) Let F be the field GF(q) and denote by $F^{m \times n}$ the set of all $m \times n$ matrices over F. The *rank distance* between two matrices A and B in $F^{m \times n}$ is defined as rank($A-B$) and denoted by $\mathsf{d}_{\mathrm{rank}}(A, B)$. Recall from Problem 1.3 that the rank distance is a metric.

An $(m \times n, M)$ *(array) code* over F is a nonempty subset of $F^{m \times n}$ whose size is M. Given an $(m \times n, M)$ code $\mathcal{C}$ over F with size $M > 1$, the *minimum rank distance* μ of $\mathcal{C}$ is defined by

$$\mu = \min_{A,B \in \mathcal{C} \,:\, A \neq B} \mathsf{d}_{\mathrm{rank}}(A, B) .$$

An $(m \times n, M)$ code $\mathcal{C}$ with minimum rank distance μ will be called a μ-$(m \times n, M)$ code. If, in addition, $\mathcal{C}$ is a linear subspace of $F^{m \times n}$ with dimension k $(= \log_q M)$, then $\mathcal{C}$ is said to be a linear μ-$[m \times n, k]$ code over F.

1. Show that for every linear μ-$[m \times n, k]$ code $\mathcal{C}$ over F, the minimum rank distance can be written as

$$\mu = \min_{A \in \mathcal{C} \setminus \{0\}} \mathrm{rank}(A) .$$

2. Let m and n be positive integers where $m \leq n$. Show that for every μ-$(m \times n, M)$ code over F,

$$\log_q M \leq n(m-\mu+1) .$$

Hint: Assume the contrary and show that this implies that the array code necessarily contains two matrices A and B whose first $m-\mu+1$ rows are identical.

3. A polynomial $f(x)$ over GF(q^n) is called a *linearized polynomial* with respect to F if it has the form $\sum_i f_i x^{q^i}$. The mapping GF(q^n) $\rightarrow$ GF(q^n) that is defined by $x \mapsto f(x)$ is a linear transformation over F (see Problem 3.32). Denote by $F_r^{[n]}[x]$ the set of all linearized polynomials over GF(q^n) (with respect to F) whose degree is less than q^r.

Fix a basis $\Omega = (\omega_1 \; \omega_2 \; \ldots \; \omega_n)$ of GF(q^n) over F, and for every polynomial $f(x) \in F_n^{[n]}[x]$, denote by A_f the $n \times n$ matrix representation over F of the mapping $x \mapsto f(x)$; that is, if $\mathbf{u}$ is a column vector in F^n that represents an element $u \in$ GF(q^n) by $u = \Omega\mathbf{u}$, then $A_f\mathbf{u}$ is a vector representation of $f(u)$, namely, $f(u) = \Omega A_f \mathbf{u}$.

Given a positive integer $\mu \leq n$, consider the array code

$$\mathcal{C} = \left\{ A_f \in F^{n \times n} \;:\; f(x) \in F_{n-\mu+1}^{[n]}[x] \right\} .$$

Show that $\mathcal{C}$ is a linear μ-$[n \times n, k]$ code where $k = n(n-\mu+1)$ (as such, $\mathcal{C}$ attains the bound in part 2).

Hint: $|\ker(A_f)| \leq \deg f(x)$ for every nonzero $f(x) \in F_n^{[n]}[x]$.

4. Show that the bound in part 2 can be attained for every $1 \leq \mu \leq m \leq n$.

Hint: Let $\mathcal{C}$ be the construction in part 3 and consider a sub-code of $\mathcal{C}$ that consists of the matrices in $\mathcal{C}$ whose first $n-m$ rows are all zero.

Problem 11.5 (Singleton bound in the cover metric) Let F be a finite Abelian group of size q and denote by $F^{m\times n}$ the set of all $m \times n$ arrays (matrices) over F. A *cover* of an array $A = (a_{i,j})_{i=1}^{m}{}_{j=1}^{n}$ in $F^{m\times n}$ is a pair (X, Y) that consists of subsets $X \subseteq \{1, 2, \ldots, m\}$ and $Y \subseteq \{1, 2, \ldots, n\}$ such that the rows of A that are indexed by X and the columns that are indexed by Y contain all the nonzero entries in A; that is, for every entry $a_{i,j}$ in A,

$$a_{i,j} \neq 0 \quad \Longrightarrow \quad (i \in X) \text{ or } (j \in Y) .$$

The size of a cover (X, Y) is defined by $|X| + |Y|$. The *cover weight* of A, denoted by $\mathsf{w}_{\mathsf{cov}}(A)$, is the size of a smallest cover of A.

For example, the 4×4 array

$$A = \begin{pmatrix} 1 & 0 & 0 & 0 \\ 0 & 1 & 0 & 0 \\ 1 & 0 & 1 & 1 \\ 0 & 1 & 0 & 0 \end{pmatrix}$$

over GF(2) has two covers of size 3, namely, $(\{1, 3\}, \{2\})$ and $(\{3\}, \{1, 2\})$. Furthermore, since the three nonzero elements on the main diagonal of A belong to distinct rows and columns, the cover weight of A must be at least 3. Therefore, $\mathsf{w}_{\mathsf{cov}}(A) = 3$.

The *cover distance* between two arrays A and B in $F^{m\times n}$ is defined as $\mathsf{w}_{\mathsf{cov}}(A-B)$ and denoted by $\mathsf{d}_{\mathsf{cov}}(A, B)$.

An $(m \times n, M)$ (array) code over an Abelian group F is a nonempty subset $\mathcal{C}$ of $F^{m\times n}$ of size M. When $M > 1$, define the *minimum cover distance* d of $\mathcal{C}$ by

$$d = \min_{A,B\in\mathcal{C}\,:\,A\neq B} \mathsf{d}_{\mathsf{cov}}(A, B) .$$

An $(m \times n, M, d)$ code is an $(m \times n, M)$ code with $M > 1$ and minimum cover distance d. A linear $[m \times n, k, d]$ code over the field $F = \mathrm{GF}(q)$ is an $(m \times n, q^k, d)$ code over F that forms a k-dimensional linear subspace of $F^{m\times n}$.

1. Show that the cover distance is a metric (see Section 1.3).

2. Show that if F is a field then for every two matrices A and B in $F^{m\times n}$,

$$\mathsf{w}_{\mathsf{cov}}(A) \geq \mathrm{rank}(A)$$

and

$$\mathsf{d}_{\mathsf{cov}}(A, B) \geq \mathsf{d}_{\mathsf{rank}}(A, B) ,$$

where $\mathsf{d}_{\mathsf{rank}}(A, B) = \mathrm{rank}(A-B)$.

Hint: Show that a matrix A can be written as a sum of $\mathsf{w}_{\mathsf{cov}}(A)$ matrices of rank 1.

3. Suppose that $1 \leq m \leq n$ and let $\mathcal{C}_1$ be an (ordinary) (m, M_1, d) code over F. Consider the array code $\mathcal{C}_2$ over F that is defined by

$$\mathcal{C}_2 = \Big\{ (a_{i,j})_{i=1}^{m}{}_{j=1}^{n} \in F^{m\times n} \; : \; (a_{1,t+1}\; a_{2,t+2}\; \ldots\; a_{m,t+m}) \in \mathcal{C}_1 \quad \text{for all } 0 \leq t < n \Big\}$$

(when an index $t+i$ exceeds n it should be read as $t+i-n$). Show that $\mathcal{C}_2$ is an $(m \times n, M_1^n, d)$ code over F.

4. Assuming that $1 \le m \le n$, show that for every $(m \times n, M, d)$ code over F,

$$\log_q M \le n(m-d+1) .$$

   Hint: See the hint in part 2 of Problem 11.4.

5. Show that when the code $\mathcal{C}_1$ in part 3 is taken to be MDS, then the respective array code $\mathcal{C}_2$ attains the bound in part 4.

6. Show that when $F = \mathrm{GF}(q)$, the bound in part 4 can be attained for every $1 \le d \le m \le n$ (this holds even when (m, q^{m-d+1}, d) MDS codes over F do not exist!).

   Hint: Consider the construction in parts 3 and 4 of Problem 11.4.

Problem 11.6 Let $F = \mathrm{GF}(q)$ and $\Phi = \mathrm{GF}(q^m)$, and let $H = (\mathbf{h}_1\ \mathbf{h}_2\ \ldots\ \mathbf{h}_n)$ be an $m \times n$ parity-check matrix of a linear $[n, k{=}n{-}m, d{\ge}2]$ code over F. Fix a basis $\Omega = (\omega_1\ \omega_2\ \ldots\ \omega_m)$ of Φ over F, and let the elements $\alpha_1, \alpha_2, \ldots, \alpha_n$ of Φ be defined by

$$\alpha_j = \Omega \mathbf{h}_j\ , \quad 1 \le j \le n .$$

Denote by $\mathcal{C}(H, \Omega)$ the linear $[N{=}n, K, D]$ code over Φ with a parity-check matrix

$$\begin{pmatrix} \alpha_1 & \alpha_2 & \ldots & \alpha_n \\ \alpha_1^q & \alpha_2^q & \ldots & \alpha_n^q \\ \alpha_1^{q^2} & \alpha_2^{q^2} & \ldots & \alpha_n^{q^2} \\ \vdots & \vdots & \vdots & \vdots \\ \alpha_1^{q^{d-2}} & \alpha_2^{q^{d-2}} & \ldots & \alpha_n^{q^{d-2}} \end{pmatrix} .$$

1. Show that $D = d$ and that $\mathcal{C}(H, \Omega)$ is MDS.

   Hint: Recall from Problem 3.33 that an $r \times r$ matrix over Φ of the form

$$\left(\beta_j^{q^i}\right)_{i=0\ j=1}^{r-1\ \ r}$$

   is nonsingular if and only if $\beta_1, \beta_2, \ldots, \beta_r$, when regarded as elements of the linear space Φ over F, are linearly independent over F.

2. Identify the code $\mathcal{C}(H, \Omega) \cap F^N$.

Problem 11.7 (MDS codes over polynomial rings) Let $F = \mathrm{GF}(q)$ and p be a prime not dividing q. Denote by $B_p(x)$ the polynomial $1 + x + x^2 + \ldots + x^{p-1}$ and by Φ the *ring* $F[\xi]/B_p(\xi)$ (the polynomial $B_p(x)$ is not necessarily irreducible over F—see Problem 7.15). For a positive integer $\varrho < p$, let H be the following $\varrho \times p$ matrix over Φ:

$$H = \begin{pmatrix} 1 & 1 & 1 & \ldots & 1 \\ 1 & \xi & \xi^2 & \ldots & \xi^{p-1} \\ 1 & \xi^2 & \xi^4 & \ldots & \xi^{2(p-1)} \\ \vdots & \vdots & \vdots & \vdots & \vdots \\ 1 & \xi^{\varrho-1} & \xi^{2(\varrho-1)} & \ldots & \xi^{(p-1)(\varrho-1)} \end{pmatrix} .$$

Define the code $\mathcal{C}$ over Φ by

$$\mathcal{C} = \left\{ \mathbf{c} \in \Phi^p \, : \, H\mathbf{c}^T = \mathbf{0} \right\} \, ;$$

this code is linear over Φ (see Problem 2.20).

1. Show that the determinant of every $\varrho \times \varrho$ sub-matrix of H has a multiplicative inverse in Φ; therefore, every set of ϱ columns in H is linearly independent over Φ.

   Hint: Recall from Problem 3.13 that each determinant is a product of terms of the form $\xi^j - \xi^i = \xi^i(\xi^{j-i} - 1)$ where $0 \leq i < j < p$. Show that $\gcd(x^j - x^i, x^p - 1) = x - 1$ and, so, $\gcd(x^j - x^i, B_p(x)) = 1$ (see Problem 3.5).

2. Show that $|\mathcal{C}| = q^{p-\varrho}$ and that $\mathsf{d}(\mathcal{C}) \geq \varrho+1$; deduce that $\mathcal{C}$ is MDS.

3. Let Ω denote the basis $(1 \; \xi \; \xi^2 \; \ldots \; \xi^{p-2})$ of Φ as a vector space over F, and consider the following subset $\hat{\mathcal{C}}$ of the set of all $(p-1) \times p$ matrices over F:

   $$\hat{\mathcal{C}} = \left\{ A \in F^{(p-1)\times p} \, : \, \Omega A \in \mathcal{C} \right\} .$$

   Show that $\hat{\mathcal{C}}$ is a linear subspace of $F^{(p-1)\times p}$ over F and that a matrix $A = (a_{i,j})_{i=1}^{p-1}{}_{j=0}^{p-1}$ over F belongs to $\hat{\mathcal{C}}$ if and only if

   $$\sum_{j=0}^{p-1} a_{t-j\ell,j} = 0 \, , \quad 1 \leq t \leq p \, , \quad 0 \leq \ell < \varrho \, ,$$

   where each index $t-j\ell$ is read as the remainder (in $\{0, 1, \ldots, p-1\}$) obtained when that index is divided by p, and $a_{0,j}$ is defined as zero.

   Hint: If $u_0, u_1, \ldots, u_{p-1}$ are elements in F such that $\sum_{t=0}^{p-1} u_t \xi^t = 0$ (in Φ), then $u_0 = u_1 = \ldots = u_{p-1}$.

[Section 11.2]

Problem 11.8 Show that every Cauchy matrix is an extended Cauchy matrix.

Hint: Use Problem 5.4 to claim that the last code locator in an extended GRS code can always be assumed to be infinity. Then apply Problems 5.8 and 5.9.

Problem 11.9 Let $F = \mathrm{GF}(q)$ and let A be a $k \times k$ circulant matrix of the form

$$A = \begin{pmatrix} a_0 & a_1 & \ldots & a_{k-2} & a_{k-1} \\ a_{k-1} & a_0 & a_1 & \ldots & a_{k-2} \\ a_{k-2} & a_{k-1} & a_0 & a_1 & \ldots \\ \ldots & \ddots & \ddots & \ddots & \ddots \\ a_1 & \ldots & a_{k-2} & a_{k-1} & a_0 \end{pmatrix} ,$$

where $a_0, a_1, \ldots, a_{k-1}$ are nonzero elements of F. Denote by a_k and a_{k+1} the elements a_0 and a_1, respectively.

1. Show that the matrix $(I \mid A)$ generates a $[2k, k]$ extended GRS code if and only if the following two conditions hold:

   (i) The ratios
$$a_{j+1}/a_j \,, \quad 0 \le j < k \,,$$
   are all distinct.

   (ii) There exist $\sigma, \tau \in F$ such that
$$a_{j+2}^{-1} + \sigma a_{j+1}^{-1} + \tau a_j^{-1} = 0 \,, \quad 0 \le j < k \,.$$

2. Let $F = \mathrm{GF}(9)$ and let δ be a root in F of the polynomial $x^2 + 2x + 2$. Define the 5×5 matrix A as in part 1, where
$$(a_0 \; a_1 \; a_2 \; a_3 \; a_4) = (1 \; \delta^7 \; \delta^5 \; \delta^5 \; \delta^7) \,.$$
   Show that the matrix $(I \mid A)$ generates a $[10, 5, 6]$ doubly-extended GRS code over F.

3. For F and δ as in part 2, let now A be the 5×5 circulant matrix that is defined by
$$(a_0 \; a_1 \; a_2 \; a_3 \; a_4) = (\delta^7 \; \delta^5 \; 1 \; 1 \; \delta^5) \,.$$
   Show that $G = (I \mid A)$ does not generate a $[10, 5, 6]$ doubly-extended GRS code (still, one can verify that G generates a linear $[10, 5, 6]$ MDS code).

Problem 11.10 Let $\mathcal{C}$ be a linear $[n, n-3]$ code over $F = \mathrm{GF}(q)$ with a parity-check matrix
$$\begin{pmatrix} 1 & 1 & \ldots & 1 & u \\ \alpha_1 & \alpha_2 & \ldots & \alpha_{n-1} & v \\ \alpha_1^2 & \alpha_2^2 & \ldots & \alpha_{n-1}^2 & w \end{pmatrix} ,$$
where $\alpha_1, \alpha_2, \ldots, \alpha_{n-1}$ are distinct elements of F. The purpose of this problem is to find conditions on the last column $(u \; v \; w)^T$ so that $\mathcal{C}$ is MDS. Assume hereafter that $(u \; v \; w)$ is nonzero.

1. Let β and γ be distinct elements of F and consider the 3×3 matrix
$$A = \begin{pmatrix} 1 & 1 & u \\ \beta & \gamma & v \\ \beta^2 & \gamma^2 & w \end{pmatrix}$$
   over F. Show that A is singular if and only if
$$u\beta\gamma - v(\beta + \gamma) + w = 0 \,.$$

2. Show that $\mathcal{C}$ is MDS if and only if for every distinct i and j in the range $1 \le i, j < n$,
$$u\alpha_i\alpha_j - v(\alpha_i + \alpha_j) + w \ne 0 \,.$$

Assume from now on in this problem that $v^2 \ne uw$ (note that $v^2 = uw$ if and only if $(u \; v \; w)$ is a scalar multiple of either $(0 \; 0 \; 1)$ or $(1 \; \delta \; \delta^2)$ for some $\delta \in F$; hence, when $v^2 = uw$, the code $\mathcal{C}$ is MDS if and only if either $u = v = 0$ or $u\alpha_j \ne v$ for every $1 \le j < n$). Furthermore, when q is even assume that either $u \ne 0$ or $w \ne 0$.

3. Show that there are at least $\lceil (q-3)/2 \rceil$ disjoint subsets $\{\beta, \gamma\} \subseteq F$, each consisting of two distinct elements β and γ that satisfy

$$u\beta\gamma - v(\beta+\gamma) + w = 0 .$$

Hint: Exclude the (up to three) elements $\beta \in F$ that satisfy either $u\beta^2 - 2v\beta + w = 0$ or $u\beta - v = 0$. For each of the remaining elements β consider the subset $\{\beta, \gamma\}$ where

$$\gamma = \frac{v\beta - w}{u\beta - v} .$$

Verify that the condition $v^2 \neq uw$ guarantees that any two distinct subsets thus obtained are disjoint.

4. Show that when $n \geq \lfloor (q+7)/2 \rfloor$ (and with the current assumptions on $(u\ v\ w)$), the code $\mathcal{C}$ is *not* MDS.

   Hint: Show by a counting argument that at least one of the subsets in part 3 is wholly contained in $\{\alpha_1, \alpha_2, \ldots, \alpha_{n-1}\}$.

Problem 11.11 Let F be the finite field GF(q) where $q = 2^m$ and $m > 1$, and denote by $\alpha_1, \alpha_2, \ldots, \alpha_{q-1}$ the nonzero elements of F. For $\ell \in \{1, 2, \ldots, m-1\}$, let $\mathcal{C}_\ell$ be the linear $[q+1, 3]$ code over F that is generated by the $3 \times (q+1)$ matrix

$$G_\ell = \begin{pmatrix} 1 & 1 & \ldots & 1 & 1 & 0 \\ \alpha_1 & \alpha_2 & \ldots & \alpha_{q-1} & 0 & 0 \\ \alpha_1^{2^\ell} & \alpha_2^{2^\ell} & \ldots & \alpha_{q-1}^{2^\ell} & 0 & 1 \end{pmatrix} .$$

1. Show that $\mathcal{C}_\ell$ is MDS if and only if $\gcd(\ell, m) = 1$.

   Hint: $\gcd(2^\ell - 1, 2^m - 1) = 1$ if and only if $\gcd(\ell, m) = 1$.

2. Show that $\mathcal{C}_\ell$ is an extended GRS code if and only if $\ell = 1$.

   Hint: First use Problem 5.4 to show that if $\mathcal{C}_\ell$ is an extended GRS code then, without loss of generality, one can assume that $\mathcal{C}_\ell$ is generated by

$$G_{\mathrm{GRS}} = \begin{pmatrix} 1 & 1 & \ldots & 1 & 1 & 0 \\ \beta_1 & \beta_2 & \ldots & \beta_{q-1} & 0 & 0 \\ \beta_1^2 & \beta_2^2 & \ldots & \beta_{q-1}^2 & 0 & 1 \end{pmatrix} \begin{pmatrix} v_1 & & & \\ & v_2 & & 0 \\ 0 & & \ddots & \\ & & & v_{q+1} \end{pmatrix} ,$$

   where $\beta_1, \beta_2, \ldots, \beta_{q-1}$ range over the nonzero elements of F. Then, by identifying the codewords in $\mathcal{C}_\ell$ that end with two zeros, argue that $v_1, v_2, \ldots, v_{q-1}$ can be assumed to take the values

$$v_j = \frac{\alpha_j}{\beta_j} , \quad 1 \leq j \leq q-1 .$$

   By expressing the first and third rows of G_{GRS} as linear combinations of the rows of G_ℓ, show that there exist two polynomials over F,

$$f(x) = f_0 + f_1 x \quad \text{and} \quad g(x) = g_1 x + g_2 x^{2^\ell} ,$$

such that

$$\frac{\alpha_j}{\beta_j} = f(\alpha_j) \quad \text{and} \quad \alpha_j \beta_j = g(\alpha_j) , \quad 1 \le j \le q{-}1 ,$$

and, so,

$$f(\alpha_j) g(\alpha_j) = \alpha_j^2 , \qquad 1 \le j \le q{-}1 .$$

Deduce that

$$f(x)g(x) = x^2$$

and, finally, conclude that $f(x)$ must be a constant.

3. Let $\hat{G}_\ell$ be the $3 \times (q{+}2)$ matrix obtained by appending the column $(0\ 1\ 0)^T$ to G_ℓ. Show that when $\gcd(\ell, m) = 1$, the code that is generated by $\hat{G}_\ell$ is a linear $[q{+}2, 3]$ MDS code over F.

4. For $\ell \in \{1, 2, \ldots, m{-}1\}$, let $\mathcal{C}'_\ell$ be the linear $[q{+}1, 4]$ code over F that is generated by the $4 \times (q{+}1)$ matrix

$$\begin{pmatrix} 1 & 1 & \ldots & 1 & 1 & 0 \\ \alpha_1 & \alpha_2 & \ldots & \alpha_{q-1} & 0 & 0 \\ \alpha_1^{2^\ell} & \alpha_2^{2^\ell} & \ldots & \alpha_{q-1}^{2^\ell} & 0 & 0 \\ \alpha_1^{2^\ell+1} & \alpha_2^{2^\ell+1} & \ldots & \alpha_{q-1}^{2^\ell+1} & 0 & 1 \end{pmatrix} .$$

Show that $\mathcal{C}'_\ell$ is MDS if and only if $\gcd(\ell, m) = 1$.

[Section 11.3]

Problem 11.12 ("Almost converse" to Lemma 11.12) Show that for $1 < k < n \le L_q(k)$,

$$n \le L_q(n{-}k) .$$

Hint: Since $n \le L_q(k)$, there exists a linear $[n, k]$ MDS code over $F = \mathrm{GF}(q)$. Consider its dual code.

[Section 11.4]

Problem 11.13 Let A be a $k \times r$ super-regular matrix over a field F where $r > 3$. Suppose that A contains at least two $k \times (r{-}1)$ sub-matrices which are both extended Cauchy matrices. Show that A is an extended Cauchy matrix. Does the result hold also for $r = 3$?

Hint: Denote by A_1 and A_2 the two $k \times (r{-}1)$ extended Cauchy sub-matrices. Show that $\mathrm{rank}(A^{\mathrm{c}}) \le 2$, taking into account that A_1 and A_2 share at least two columns.

Problem 11.14 Let $\mathcal{C}$ be a linear $[n, k]$ MDS code over $F = \mathrm{GF}(q)$ where $n > k{+}3$. For $i = 1, 2, \ldots, n$, denote by $\mathcal{C}_i$ the code over F obtained by puncturing $\mathcal{C}$ at the ith coordinate. Suppose that there are two distinct indexes i and j for which $\mathcal{C}_i$ and $\mathcal{C}_j$ are extended GRS codes. Show that $\mathcal{C}$ is an extended GRS code.

Hint: Assume without loss of generality that $i, j > k$ and consider a systematic generator matrix $(\, I \,|\, A \,)$ of $\mathcal{C}$. Then use Problem 11.13.

Problem 11.15 Let $\mathcal{C}$ be a linear $[n, k]$ MDS code over $F = \mathrm{GF}(q)$ where $k > 3$. For $i = 1, 2, \ldots, n$, denote by $\mathcal{C}^{(i)}$ the code over F obtained by shortening $\mathcal{C}$ at the ith coordinate. Suppose that there are two distinct indexes i and j for which $\mathcal{C}^{(i)}$ and $\mathcal{C}^{(j)}$ are extended GRS codes. Show that $\mathcal{C}$ is an extended GRS code.

Hint: Assume that $i, j > n-k$ and consider a parity-check matrix of $\mathcal{C}$ of the form $(\, I \,|\, A \,)$. Then use Problem 11.13.

Problem 11.16 ("Almost converse" to Lemma 11.23) Show that for $k \geq 2$ and $k+3 < n \leq \Gamma_q(k)$,

$$n \leq \Gamma_q(n-k-1) \, .$$

Hint: Suppose to the contrary that $(q+1 \geq)\ n-1 \geq \Gamma_q(n-k-1)$, namely, that every linear $[n-1, n-k-1]$ MDS code over F is an extended GRS code. Apply duality to deduce that the same holds for every linear $[n-1, k]$ MDS code. Then use Lemma 11.21.

[Section 11.5]

Problem 11.17 Let $\mathcal{C}$ be a linear $[q+1, 3]$ MDS code over $F = \mathrm{GF}(q)$.

1. Let $\mathbf{c}$ and $\mathbf{c}'$ be two codewords in $\mathcal{C}$ of Hamming weight q, both having their (unique) zero entry at the same location. Show that $\mathbf{c}$ and $\mathbf{c}'$ are linearly dependent.

   Hint: Suppose to the contrary that $\mathbf{c}$ and $\mathbf{c}'$ are linearly independent, and consider a $3 \times (q+1)$ generator matrix G of $\mathcal{C}$ whose first two rows are $\mathbf{c}$ and $\mathbf{c}'$. Show that G contains a set of three linearly dependent columns.

2. Show that the number of codewords of Hamming weight q in $\mathcal{C}$ equals $(q+1)(q-1)$.

   Hint: Compute W_q in Example 4.6.

3. Show that for every index $\ell \in \{1, 2, \ldots q+1\}$ there are $q-1$ codewords of Hamming weight q whose ℓth entry is zero (and these codewords are all scalar multiples of the same codeword).

Problem 11.18 Let $\mathcal{C}$ be a linear $[q+1, 3]$ MDS code over $F = \mathrm{GF}(q)$ where q is odd, and let $G = (\, A \,|\, I \,)$ be a generator matrix of $\mathcal{C}$ as defined by (11.1). The purpose of this problem is to obtain a necessary and sufficient algebraic condition on the coefficients $u_0, u_1, u_2 \in F$ that correspond to codewords $(u_0 \ u_1 \ u_2)G$ of Hamming weight q in $\mathcal{C}$.

1. Let $(u_0 \ u_1 \ u_2)G$ be a codeword of Hamming weight q in $\mathcal{C}$. Show that

   $$(u_0 + a_0 u_1 + b_0 u_2)^2 = 4 a_0 b_0 u_1 u_2 \, ,$$

   where a_0 and b_0 are the elements of F^* that are missing from the second and third rows of A, respectively.

   Hint: Let ℓ denote the location of the zero entry in the codeword, and assume first that $\ell \leq q-2$. Use (11.4) and (11.5) to show that for some $t \in F^*$,

   $$u_0 = (b_\ell a_0 + a_\ell b_0)t \, , \quad u_1 = (b_0 - b_\ell)t \, , \quad \text{and} \quad u_2 = (a_0 - a_\ell)t \, .$$

Deduce that

$$u_0 + a_0 u_1 + b_0 u_2 = 2a_0 b_0 t$$

and that

$$u_1 u_2 = (b_0 - b_\ell)(a_0 - a_\ell)t^2 = a_0 b_0 t^2 ,$$

and then eliminate t from the last two equations. Finally, consider the codewords whose zero entry is indexed by $q-1$, q, or $q+1$.

2. Show that the solutions of the equation in part 1 for triples $(u_0\ u_1\ u_2)$ over F with $u_1 u_2 \neq 0$ are all the $(q-1)^2$ triples over F such that

$$u_2 = a_0 b_0 s / u_1 \quad \text{and} \quad u_0 = -a_0 u_1 - b_0 u_2 \pm 2a_0 b_0 \sqrt{s} ,$$

where s ranges over all the $(q-1)/2$ quadratic residues (i.e., squares) in F^* and u_1 ranges over all the elements of F^*.

3. Show that the equation in part 1 has $(q+1)(q-1)$ nontrivial solutions for $(u_0\ u_1\ u_2)$ over F. Deduce that there is a one-to-one correspondence between the solutions of that equation and the codewords of Hamming weight q in $\mathcal{C}$ (see Problem 11.17).

Notes

[Section 11.1]

The treatment of MDS codes in this chapter is inspired to a great extent by the exposition of MacWilliams and Sloane [249, Chapter 11]. MDS codes are of interest not only in coding theory, but in other areas as well—primarily projective geometry over finite fields and combinatorics; see the books by Hirschfeld [181] and Hirschfeld and Thas [186], and the *Handbook of Combinatorial Designs* [84, Part II].

We next present a characterization of linear MDS codes through the language of projective geometry. Recalling from the notes on Section 2.3, let $\mathrm{PG}(k-1, q)$ denote the $(k-1)$-dimensional projective geometry over $F = \mathrm{GF}(q)$. An *n-arc* in $\mathrm{PG}(k-1, q)$ is a set of n points in $\mathrm{PG}(k-1, q)$ such that no k points in the set belong to the same $(k-2)$-dimensional hyper-plane in $\mathrm{PG}(k-1, q)$. Let S be the set of columns of a $k \times n$ matrix G over F, where each column is nonzero and is regarded as a projective point in $\mathrm{PG}(k-1, q)$. Then S is an n-arc if and only if every k columns in G are linearly independent. Equivalently, S is an n-arc if and only if G is a parity-check matrix (and, by Proposition 11.4(ii), also a generator matrix) of a linear MDS code over F. See Thas [361].

The connection between Latin squares and MDS codes (Problem 11.1) was made by Golomb and Posner in [153]. For more on Latin squares, see Brualdi and Ryser [67, Chapter 8], Colbourn and Dinitz (and other authors) [84, Chapters II.1–3], and Dénes and Keedwell [101], [102]. Orthogonal arrays are treated by Bierbrauer and Colbourn in [84, Chapters II.4–5].

The Singleton bound in the rank metric (Problem 11.4) was first obtained by Delsarte [97], along with the attaining construction. See also Gabidulin [135] and Roth [297], [299].

The cover metric (Problem 11.5) was introduced in connection with handling *crisscross errors*. Under this error model, entries in an array may become erroneous only within a prescribed number of rows or columns (or both). Roth [297] lists several applications where such error patterns may be found. Constructions of array codes that attain the Singleton bound in the cover metric have been obtained by Blaum and Bruck [51], Gabidulin [136], Gabidulin and Korzhik [137], and Roth [297], [299].

Given a finite Abelian group F, the *term rank* of an array $A = (a_{i,j})_{i=1}^{m}{}_{j=1}^{n}$ in $F^{m \times n}$ equals the largest number of nonzero entries in A, no two of which belong to the same row or column of A. By König's Theorem, the term rank of A equals its cover weight (see the book by Brualdi and Ryser [67, p. 6]). Associate with $A = (a_{i,j})$ the following undirected *bipartite graph* $\mathcal{G} = \mathcal{G}_A$ (see Section 13.1): the vertices of $\mathcal{G}$ are given by the rows and columns of A, and connect row i with column j by an edge in $\mathcal{G}$ if and only if $a_{i,j} > 0$. A *matching* in $\mathcal{G}$ is a largest set of edges no two of which are incident with the same vertex. It follows from König's Theorem that the cover weight of A equals the size of the largest matching in $\mathcal{G}$. There is an efficient algorithm for finding a largest matching in a bipartite graph (see Biggs [44, Section 17.5]); hence, one can efficiently compute the cover weight of a given array A.

The codes in Problem 11.7 are taken from Blaum and Roth [53]. See also the related work by Deutsch [103] and Keren and Litsyn [210], [211]. Let $\mathcal{C}$ be a code obtained by the construction in Problem 11.7 for given $F = \mathrm{GF}(q)$, p, and ϱ. Each $(p-1) \times p$ array A in the respective code $\hat{\mathcal{C}}$ (as defined in part 3 of that problem) can be transformed into a vector in $F^{(p-1)p}$ simply by concatenating the columns of A. When doing so, the set of resulting vectors forms a linear $[p(p-1), (p-\varrho)(p-1)]$ code $\mathcal{C}_F$ over F. The code $\mathcal{C}_F$ can be shown to have a $\varrho(p-1) \times p(p-1)$ parity-check matrix H_F over F with at most $2\varrho - 1$ nonzero entries in each column. Sparse parity-check matrices in turn, are desirable as they allow efficient syndrome computation. This motivates the problem of designing for a given field F and "byte length" b, an MDS code over F^b that is linear over F and has the sparsest possible parity-check matrix; one may require, in addition, that the sparsest parity-check matrix be also systematic—in which case the code will have a sparsest systematic *generator* matrix as well (for given code parameters, the aforementioned matrix H_F is apparently neither the sparsest possible, nor is it systematic). The design problem of such MDS codes has been dealt with by Blaum *et al.* [50], [52], Blaum and Roth [54], Louidor [237], Louidor and Roth [238], and Zaitsev *et al.* [392]. (Linear codes that have sparse parity-check matrices are referred to as *low-density parity-check*—in short, LDPC—*codes*.)

[Section 11.2]

In projective geometry, the particular $(q+1)$-arc in $\mathrm{PG}(k-1, q)$ that is formed by the columns of a generator matrix of a $[q+1, k]$ doubly-extended GRS code is called a *normal rational curve*. See, for example, Hirschfeld and Thas [186, Chapter 27] and Thas [361].

The characterization of circulant extended Cauchy matrices in Problem 11.9 is taken from Roth and Lempel [301].

[Section 11.3]

Table 11.2 presents a range of values for which the MDS conjecture is proved (see also the notes on Section 11.5). In addition, the conjecture has been confirmed to hold in every field GF(q) for $k \le 5$ (and, thus, also for $k \ge q-3$).

Table 11.2. Values of q and k for which the MDS conjecture is proved.

$q = p^m$, p prime	Range of k		
$p = 2$	$k < \frac{1}{2}\sqrt{q} + \frac{15}{4}$	or	$k > q - \frac{1}{2}\sqrt{q} - \frac{7}{4}$
$p = 3$, even $m \ge 8$	$k < \frac{1}{2}\sqrt{q} + 2$	or	$k > q - \frac{1}{2}\sqrt{q}$
$p \ge 5$	$k < \frac{1}{2}\sqrt{q}$	or	$k > q - \frac{1}{2}\sqrt{q} + 2$
$p \ge 3$, odd m	$k < \frac{1}{4}\sqrt{pq} - \frac{29}{16}p + 4$	or	$k > q - \frac{1}{4}\sqrt{pq} + \frac{29}{16}p - 2$
$m = 1$	$k < \frac{1}{45}q + \frac{37}{9}$	or	$k > q - \frac{1}{45}q - \frac{19}{9}$

Let $\mathcal{C}$ be a linear $[n, k, d]$ code over $F = \mathrm{GF}(q)$. We say that $\mathcal{C}$ is *almost MDS* (in short, AMDS) if $d = n-k$; that is, we allow a slack of 1 between d and the Singleton bound. The code $\mathcal{C}$ is *near-MDS* (in short, NMDS) if both $\mathcal{C}$ and $\mathcal{C}^\perp$ are AMDS. The $[7, 4, 3]$ Hamming code over GF(2) and the $[11, 6, 5]$ ternary Golay code over GF(3) are examples of NMDS codes. It can be shown that when $n > k + q$, every linear $[n, k]$ AMDS code over GF(q) is NMDS. Constructions and bounds for AMDS and NMDS codes were obtained by de Boer [59], Dodunekov and Landjev [106], [107], Faldum and Willems [118], and Marcugini *et al.* [250].

[Section 11.4]

The presentation in this section is taken primarily from Roth and Lempel [301].

In projective geometry, an n-arc in PG$(k-1, q)$ is called *complete* if it is not a subset of an $(n+1)$-arc in PG$(k-1, q)$. One of the problems studied in projective geometry is characterizing the set of complete n-arcs in PG$(k-1, q)$. When $\Gamma_q(k) \le q+1$, every normal rational curve in PG$(k-1, q)$ is complete, and $\Gamma_q(k) - 1$ then equals the size of the largest complete n-arc in PG$(k-1, q)$, if any, that is not a normal rational curve.

It was shown by Seroussi and Roth in [329] that when $2 \le k \le \lfloor q/2 \rfloor + 2$, then—with the exception of $k = 3$ when q is even—a normal rational curve in PG$(k-1, q)$ is a complete $(q+1)$-arc; furthermore, for $n \le q+1$ and $2 \le k \le n - \lfloor (q-1)/2 \rfloor$, any n-arc that consists of n points of a normal rational curve is contained in a unique complete $(q+1)$-arc, which is a normal rational curve. When $k = 3$ and q is even, such an n-arc is contained in the (complete) $(q+2)$-arc in PG$(2, q)$ that consists of the columns of the parity-check matrix of a $[q+2, q-1]$ triply-extended GRS code over GF(q). Problem 11.10 is based on the analysis of [329] for the special case $k = 3$. See also Storme and Thas [344].

Note, however, that there are values of k and q for which normal rational curves in PG$(k-1, q)$ are provably complete, yet $\Gamma_q(k) = q+2$ as a result of the existence of other $(q+1)$-arcs in PG$(k-1, q)$. One such example, due to Glynn [147], is given in part 3 of Problem 11.9: while it is known that $L_9(5) = 10$, there is a complete 10-arc in PG(4, 9) that is not a normal rational curve and, so, $\Gamma_9(5) = 11$. Another example, taken from Casse and Glynn [75] and Lüneburg [243, Section 44], is shown in part 4 of Problem 11.11; the code $\mathcal{C}'_\ell$ therein, which is MDS whenever $\gcd(\ell, m) = 1$, is an extended GRS code only when $\ell \in \{1, m-1\}$.

[Section 11.5]

Proposition 11.25 was originally proved by Segre [325] using geometric arguments. Proposition 11.27 and Corollary 11.28 are due to Thas [360] (see also Segre [326]). The results of Thas have since been improved for most values of odd q: Table 11.3 summarizes known upper bounds on the values of $\Gamma_q(3)$, as a function of q. By Proposition 11.24, we obtain that the MDS conjecture is proved for odd q whenever

$$k \le q + 5 - \Gamma_q(3) \qquad \text{or} \qquad k \ge \Gamma_q(3) - 3 ,$$

thereby accounting for the last four entries in Table 11.2. See also Hirschfeld [181], Hirschfeld and Storme [184], [185], Hirschfeld and Thas [186, Chapter 27], and Thas [361].

Table 11.3. Upper bounds on $\Gamma_q(3)$.

$q = p^m$, p an odd prime	$\Gamma_q(3) \le$	Reference
$p = 3$, even $m \ge 8$	$q - \frac{1}{2}\sqrt{q} + 4$	Hirschfeld and
$p \ge 5$	$q - \frac{1}{2}\sqrt{q} + 6$	Korchmáros [182], [183]
odd m	$q - \frac{1}{4}\sqrt{pq} + \frac{29}{16}p + 2$	Voloch [376]
$m = 1$	$\frac{44}{45}q + \frac{17}{9}$	Voloch [375]

Turning to even values of q, we obviously have $\Gamma_q(3) = q{+}2$ since $L_q(3) > q{+}1$. In addition, it follows from the known properties of the code in part 4 of Problem 11.11 that $\Gamma_q(4) = q{+}2$ for even $q \ge 128$ (and also for $q = 32$). Nevertheless, $L_q(4) = L_q(q{-}2) = q{+}1$, as shown by Casse [74] and Gulati and Kounias [164]; furthermore, Storme and Thas have shown in [345] that for even $q \ge 8$,

$$\Gamma_q(5) \le q - \tfrac{1}{2}\sqrt{q} + \tfrac{17}{4} .$$

The first entry in Table 11.2 is obtained by combining these results with Proposition 11.24.

Chapter 12

Concatenated Codes

In this chapter, we continue the discussion on concatenated codes, which was initiated in Section 5.4. The main message to be conveyed in this chapter is that by using concatenation, one can obtain codes with favorable asymptotic performance—in a sense to be quantified more precisely—while the complexity of constructing these codes and decoding them grows polynomially with the code length.

We first present a decoding algorithm for concatenated codes, due to Forney. This algorithm, referred to as a generalized minimum distance (in short, GMD) decoder, corrects any error pattern whose Hamming weight is less than half the product of the minimum distances of the inner and outer codes (we recall that this product is a lower bound on the minimum distance of the respective concatenated code). A GMD decoder consists of a nearest-codeword decoder for the inner code, and a combined error–erasure decoder for the outer code. It then enumerates over a threshold value, marking the output of the inner decoder as erasure if that decoder returns an inner codeword whose Hamming distance from the respective received sub-word equals or exceeds that threshold. We show that under our assumption on the overall Hamming weight of the error word, there is at least one threshold for which the outer decoder recovers the correct codeword. If the outer code is taken as a GRS code, then a GMD decoder has an implementation with time complexity that is at most quadratic in the length of the concatenated code.

We then turn to analyzing the asymptotic attainable performance of concatenated codes, as their length goes to infinity. We do this first by computing a lower bound on the attainable rate of these codes, as a function of the relative minimum distance and the field size. Such a bound, which we call the Zyablov bound, is obtained by assuming that the inner code achieves the Gilbert–Varshamov bound and the outer code is a GRS code. Since the length of the inner code is significantly smaller than that of the overall concatenated code, it will follow that for every fixed rate, a generator matrix of the resulting concatenated code can be computed in time complexity that grows polynomially with the code length. The search for an inner code that attains the Gilbert–Varshamov bound can be avoided by using varying inner codes taken from a relatively small ensemble which is known to achieve that bound. This approach leads to the construction of Justesen codes, which also attain the Zyablov bound for a certain range of values of the relative minimum distance.

Finally, we turn to a second analysis of the asymptotic performance of concatenated codes, now in the framework of transmission through the memoryless q-ary symmetric channel. We show that by using two levels of concatenation, one can obtain a sequence of codes and respective decoders with the following properties: the code rates approach the capacity of the channel, the codes can be constructed and decoded in time complexity that is at most quadratic in their lengths, and the decoding error probability decays exponentially with the code length.

12.1 Definition revisited

We start by presenting a definition of concatenated codes, which will include our earlier definition in Section 5.4 as a special case. The construction of a concatenated code over a finite alphabet F uses the following ingredients:

- a one-to-one (and onto) mapping
$$\mathcal{E}_{\mathrm{in}} : \Phi \to \mathcal{C}_{\mathrm{in}} ,$$
where Φ is a finite set and $\mathcal{C}_{\mathrm{in}}$ is an $(n, |\Phi|, d)$ *inner code* over F, and
- an (N, M, D) *outer code* $\mathcal{C}_{\mathrm{out}}$ over Φ.

The respective concatenated code $\mathcal{C}_{\mathrm{cont}} = (\mathcal{E}_{\mathrm{in}}, \mathcal{C}_{\mathrm{out}})$ consists of all words in F^{nN} of the form
$$(\mathcal{E}_{\mathrm{in}}(z_1) \,|\, \mathcal{E}_{\mathrm{in}}(z_2) \,|\, \ldots \,|\, \mathcal{E}_{\mathrm{in}}(z_N)) ,$$
where $(z_1\, z_2\, \ldots\, z_N)$ ranges over all the codewords in $\mathcal{C}_{\mathrm{out}}$. One can readily verify that $\mathcal{C}_{\mathrm{cont}}$ is an $(nN, M, {\geq}dD)$ code over F (Problem 12.1).

The concatenated codes that we presented in Section 5.4 are a special case of this construction, where

- F is a finite field;
- Φ is an extension field of F;
- $\mathcal{E}_{\mathrm{in}}$ is a linear mapping over F (thereby implying that $\mathcal{C}_{\mathrm{in}}$ is a linear $[n, k, d]$ code over F with $k = [\Phi : F]$); and

- $\mathcal{C}_{\mathrm{out}}$ is a linear $[N, K, D]$ code over Φ.

If these four conditions are met, we say that the resulting code $\mathcal{C}_{\mathrm{cont}}$ is a *linearly-concatenated code*. Such a code is then a linear $[nN, kK, \geq dD]$ code over F. Note, however, that $\mathcal{C}_{\mathrm{cont}}$ may turn out to be linear also under weaker conditions: for example, we can relax the requirement on $\mathcal{C}_{\mathrm{out}}$ so that it is a linear space over F (rather than over Φ).

To maximize D, the code $\mathcal{C}_{\mathrm{out}}$ is typically taken to be an MDS code; e.g., in the linearly-concatenated case, we may select $\mathcal{C}_{\mathrm{out}}$ to be an (extended) GRS code, which is possible whenever $N \leq |\Phi|$.

12.2 Decoding of concatenated codes

Let $\mathcal{C}_{\mathrm{cont}}$ be an $(nN, M, \geq dD)$ concatenated code over F that is constructed using an (N, M, D) outer code $\mathcal{C}_{\mathrm{out}}$ over Φ and a one-to-one mapping $\mathcal{E}_{\mathrm{in}} : \Phi \rightarrow \mathcal{C}_{\mathrm{in}}$ onto an $(n, |\Phi|, d)$ inner code $\mathcal{C}_{\mathrm{in}}$ over F. We present in this section a decoding algorithm for correcting any error pattern with less than $dD/2$ errors.

Suppose that a codeword

$$\mathbf{c} = (\mathbf{c}_1 \,|\, \mathbf{c}_2 \,|\, \ldots \,|\, \mathbf{c}_N)$$

of $\mathcal{C}_{\mathrm{cont}}$ has been transmitted through an additive channel $S = (F, F, \mathsf{Prob})$. A word

$$\mathbf{y} = (\mathbf{y}_1 \,|\, \mathbf{y}_2 \,|\, \ldots \,|\, \mathbf{y}_N) \in F^{nN}$$

has been received, where each sub-block $\mathbf{y}_j$ is in F^n. Assume hereafter that $\mathbf{y}$ and $\mathbf{c}$ differ—as words in F^{nN}—on less than $dD/2$ coordinates. We next show how $\mathbf{c}$ can be decoded correctly from $\mathbf{y}$ using a nearest-codeword decoder for $\mathcal{C}_{\mathrm{in}}$ and an error–erasure decoder for $\mathcal{C}_{\mathrm{out}}$.

For $j = 1, 2, \ldots, N$, denote by z_j the value $\mathcal{E}_{\mathrm{in}}^{-1}(\mathbf{c}_j)$. Since $\mathbf{c}$ is a codeword of $\mathcal{C}_{\mathrm{cont}}$, we have by construction that

$$\mathbf{z} = (z_1 \; z_2 \; \ldots \; z_N)$$

is a codeword of $\mathcal{C}_{\mathrm{out}}$.

For $j = 1, 2, \ldots, N$, let $\hat{\mathbf{c}}_j$ be a nearest codeword in $\mathcal{C}_{\mathrm{in}}$ to $\mathbf{y}_j$ and let $\hat{z}_j$ be the value $\mathcal{E}_{\mathrm{in}}^{-1}(\hat{\mathbf{c}}_j)$. Denote by $\Theta(d)$ the set $\{1, 2, \ldots, \lceil d/2 \rceil\}$. Given an integer $\vartheta \in \Theta(d)$, define a word

$$\mathbf{x} = \mathbf{x}(\vartheta) = (x_1 \; x_2 \; \ldots \; x_N) \in (\Phi \cup \{?\})^N ,$$

where

$$x_j = \begin{cases} \hat{z}_j & \text{if } \mathsf{d}(\mathbf{y}_j, \hat{\mathbf{c}}_j) < \vartheta \\ ? & \text{otherwise} \end{cases} . \tag{12.1}$$

The value ϑ will serve as a threshold for the number of errors that we attempt to correct in each sub-block $\mathbf{y}_j$: we will decode $\mathbf{y}_j$ to a nearest codeword $\hat{\mathbf{c}}_j$ only if $\mathsf{d}(\mathbf{y}_j, \hat{\mathbf{c}}_j) < \vartheta$; otherwise, we mark that sub-block as an erasure ("?"). Note that this decoding process is local in the sense that it does not take into account the dependence between different sub-blocks that is induced by the outer code $\mathcal{C}_{\mathrm{out}}$. Such a dependence will be exploited in subsequent steps of the decoding, by determining the threshold ϑ and by applying a decoder for $\mathcal{C}_{\mathrm{out}}$ to $\mathbf{x}(\vartheta)$.

Specifically, let ρ_ϑ denote the number of erasures in $\mathbf{x}(\vartheta)$ and let τ_ϑ be the number of non-erased coordinates (with entries taking values in Φ) on which $\mathbf{x}(\vartheta)$ differs from $\mathbf{z}$. We will show that there exists a threshold value $\vartheta \in \Theta(d)$ for which

$$2\tau_\vartheta + \rho_\vartheta < D . \tag{12.2}$$

Indeed, when this inequality holds then, by Theorem 1.7, the outer codeword $\mathbf{z}$ can be recovered correctly from $\mathbf{x}(\vartheta)$ using a combined error–erasure decoder for $\mathcal{C}_{\mathrm{out}}$. The existence of such a threshold ϑ will be established by proving that—with respect to a certain probability measure—the average of $2\tau_\vartheta + \rho_\vartheta$ when ϑ ranges over $\Theta(d)$, is less than D.

Our proof makes use of the following definitions. For $j = 1, 2, \ldots, N$, let

$$w_j = \mathsf{d}(\mathbf{y}_j, \hat{\mathbf{c}}_j) ,$$

and for $\vartheta \in \Theta(d)$ define

$$\chi_j(\vartheta) = \begin{cases} 0 & \text{if } \hat{\mathbf{c}}_j = \mathbf{c}_j \text{ and } w_j < \vartheta \\ 1 & \text{if } \hat{\mathbf{c}}_j \neq \mathbf{c}_j \text{ and } w_j < \vartheta \\ \frac{1}{2} & \text{if } w_j \geq \vartheta \end{cases} .$$

We may think of $\chi_j(\vartheta)$ as a decoding penalty at the jth sub-block of $\mathbf{y}$ (or the jth coordinate of $\mathbf{x}$), given the threshold ϑ: the penalty is 1 if our local decoding at that sub-block resulted in an incorrect codeword of $\mathcal{C}_{\mathrm{in}}$, and is $\frac{1}{2}$ if that sub-block was marked as an erasure. This observation leads to the following result.

Lemma 12.1 *For every $\vartheta \in \Theta(d)$,*

$$\tau_\vartheta + \frac{\rho_\vartheta}{2} = \sum_{j=1}^{N} \chi_j(\vartheta) .$$

Next, we regard ϑ as a random variable taking values in $\Theta(d)$ and introduce the following probability measure over $\Theta(d)$:

$$\mathsf{P}_\vartheta\left\{\vartheta = x\right\} = \begin{cases} 2/d & \text{if } x \in \{1, 2, \ldots, \lfloor d/2 \rfloor\} \\ 1/d & \text{if } d \text{ is odd and } x = \lceil d/2 \rceil \end{cases} .$$

Note that, indeed, $\sum_{x \in \Theta(d)} \mathsf{P}_\vartheta \{\vartheta = x\} = 1$. We use hereafter the notation $\mathsf{E}_\vartheta \{\cdot\}$ for the expected value with respect to the measure P_ϑ.

Lemma 12.2 *For every $j \in \{1, 2, \ldots, N\}$,*

$$\mathsf{E}_\vartheta \{\chi_j(\vartheta)\} \le \frac{\mathsf{d}(\mathbf{y}_j, \mathbf{c}_j)}{d} .$$

Proof. We distinguish between two cases.

Case 1: $\hat{\mathbf{c}}_j = \mathbf{c}_j$ or $w_j \ge d/2$. Here $\chi_j(\vartheta)$ takes the value 0 (when $\vartheta > w_j$) or $\frac{1}{2}$ (when $\vartheta \le w_j$), and it never takes the value 1 (in particular, when $w_j \ge d/2$, the value of $\chi_j(\vartheta)$ is identically $\frac{1}{2}$ for every $\vartheta \in \Theta(d)$, even when $\hat{\mathbf{c}}_j \ne \mathbf{c}_j$). Therefore,

$$\mathsf{E}_\vartheta \{\chi_j(\vartheta)\} = \tfrac{1}{2} \mathsf{P}_\vartheta \{\vartheta \le w_j\} \le \frac{w_j}{d} = \frac{\mathsf{d}(\mathbf{y}_j, \hat{\mathbf{c}}_j)}{d} \le \frac{\mathsf{d}(\mathbf{y}_j, \mathbf{c}_j)}{d} ,$$

where the last step follows from $\hat{\mathbf{c}}_j$ being a nearest codeword in $\mathcal{C}_{\mathrm{in}}$ to $\mathbf{y}_j$.

Case 2: $\hat{\mathbf{c}}_j \ne \mathbf{c}_j$ and $w_j < d/2$. In this case, $\chi_j(\vartheta)$ takes the value 1 (when $\vartheta > w_j$) or $\frac{1}{2}$ (when $\vartheta \le w_j$), and it never takes the value 0. So,

$$\begin{aligned} \mathsf{E}_\vartheta \{\chi_j(\vartheta)\} &= \mathsf{P}_\vartheta \{\vartheta > w_j\} + \tfrac{1}{2} \mathsf{P}_\vartheta \{\vartheta \le w_j\} = 1 - \tfrac{1}{2} \mathsf{P}_\vartheta \{\vartheta \le w_j\} \\ &= 1 - \frac{w_j}{d} = \frac{d - \mathsf{d}(\mathbf{y}_j, \hat{\mathbf{c}}_j)}{d} \le \frac{\mathsf{d}(\mathbf{y}_j, \mathbf{c}_j)}{d} , \end{aligned}$$

where the last step follows from the triangle inequality. □

We are now ready to show that there exists a threshold for which (12.2) holds.

Proposition 12.3 *If $\mathsf{d}(\mathbf{y}, \mathbf{c}) < dD/2$ then there exists a threshold $\vartheta \in \Theta(d)$ for which*

$$\tau_\vartheta + \frac{\rho_\vartheta}{2} < \frac{D}{2} .$$

Proof. Taking expected values of both sides of the equality in Lemma 12.1 we obtain

$$\mathsf{E}_\vartheta \left\{ \tau_\vartheta + \frac{\rho_\vartheta}{2} \right\} = \sum_{j=1}^{N} \mathsf{E}_\vartheta \{\chi_j(\vartheta)\} .$$

Now, by Lemma 12.2 we have

$$\sum_{j=1}^{N} \mathsf{E}_\vartheta \{\chi_j(\vartheta)\} \le \frac{1}{d} \sum_{j=1}^{N} \mathsf{d}(\mathbf{y}_j, \mathbf{c}_j) = \frac{\mathsf{d}(\mathbf{y}, \mathbf{c})}{d} < \frac{D}{2} .$$

Combining the last two equations we obtain

$$\mathsf{E}_\vartheta \left\{ \tau_\vartheta + \frac{\rho_\vartheta}{2} \right\} < \frac{D}{2} .$$

Hence, there must be at least one threshold $\vartheta \in \Theta(d)$ for which $\tau_\vartheta + \frac{1}{2}\rho_\vartheta < D/2$. □

Based on Proposition 12.3, we present in Figure 12.1 a decoding algorithm for $\mathcal{C}_{\mathrm{cont}}$. Step 1 decodes locally every sub-block $\mathbf{y}_j$ to a nearest codeword $\hat{\mathbf{c}}_j$ in $\mathcal{C}_{\mathrm{in}}$. Step 2 iterates over all thresholds $\vartheta \in \Theta(d)$ and, for each threshold, we construct the word $\mathbf{x} = \mathbf{x}(\vartheta)$ by (12.1) (in Step 2a) and apply an error–erasure decoder for $\mathcal{C}_{\mathrm{out}}$ to $\mathbf{x}$ (in Step 2b). Proposition 12.3 now guarantees that we will decode correctly for at least one threshold $\vartheta \in \Theta(d)$.

Input: received word $\mathbf{y} = (\mathbf{y}_1 \,|\, \mathbf{y}_2 \,|\, \ldots \,|\, \mathbf{y}_N) \in F^{nN}$.
Output: codeword $\mathbf{c} \in \mathcal{C}_{\mathrm{cont}}$ or a decoding-failure indicator "e".

1. For $j = 1, 2, \ldots, N$ do:
   (a) apply a nearest-codeword decoder for $\mathcal{C}_{\mathrm{in}}$ to $\mathbf{y}_j$ to produce a codeword $\hat{\mathbf{c}}_j$ of $\mathcal{C}_{\mathrm{in}}$;
   (b) let $\hat{z}_j \leftarrow \mathcal{E}_{\mathrm{in}}^{-1}(\hat{\mathbf{c}}_j)$.
2. For $\vartheta = 1, 2, \ldots, \lceil d/2 \rceil$ do:
   (a) let $\mathbf{x} = (x_1\, x_2\, \ldots\, x_N)$ be the word over $\Phi \cup \{?\}$ that is defined by
   $$x_j = \begin{cases} \hat{z}_j & \text{if } \mathsf{d}(\mathbf{y}_j, \hat{\mathbf{c}}_j) < \vartheta \\ ? & \text{otherwise} \end{cases}$$
   and let $\rho_\vartheta \leftarrow |\{j \,:\, x_j = ?\}|$; /* ρ_ϑ is the number of erasures in $\mathbf{x}$ */
   (b) apply an error–erasure decoder for $\mathcal{C}_{\mathrm{out}}$ to recover ρ_ϑ erasures and correct up to $\tau_\vartheta = \lfloor \frac{1}{2}(D-1-\rho_\vartheta) \rfloor$ errors in $\mathbf{x}$, producing either a codeword
   $$(z_1\, z_2\, \ldots\, z_N) \in \mathcal{C}_{\mathrm{out}}$$
   or a decoding-failure indicator "e";
   (c) if decoding is successful in Step 2b then do:
      i. let $\mathbf{c} \leftarrow (\mathcal{E}_{\mathrm{in}}(z_1) \,|\, \mathcal{E}_{\mathrm{in}}(z_2) \,|\, \ldots \,|\, \mathcal{E}_{\mathrm{in}}(z_N))$;
      ii. if $\mathsf{d}(\mathbf{y}, \mathbf{c}) < dD/2$ then output $\mathbf{c}$ and exit.
3. If no codeword $\mathbf{c}$ has been produced in Step 2c then return "e".

Figure 12.1. Decoding algorithm for concatenated codes (GMD decoding).

Yet, we still need to identify such a threshold among all the elements of $\Theta(d)$: we do this in Step 2c, where we test whether the Hamming distance between the computed codeword and the received word $\mathbf{y}$ is less than $dD/2$. Since we assume that the number of errors is less than $dD/2$ (and, so, less than half the minimum distance of $\mathcal{C}_{\mathrm{cont}}$), only the transmitted codeword will pass this test.

The algorithm in Figure 12.1 is known as *Forney's generalized minimum distance* (in short, GMD) *decoder.* We next analyze the complexity of this algorithm.

The decoding in Step 1a can be carried out in a brute-force manner by checking exhaustively all the codewords of $\mathcal{C}_{\mathrm{in}}$. This, in turn, requires $n \cdot |\Phi|$ comparisons of elements of F. Note, however, that when N is proportional to $|\Phi|$—e.g., when $\mathcal{C}_{\mathrm{out}}$ is taken as a primitive GRS code—then the expression $n \cdot |\Phi|$ is proportional to nN; in such a case, the complexity of Step 1 grows at most as nN^2. (Furthermore, observe that the largest value taken by the threshold ϑ in Step 2 is $\lceil d/2 \rceil$; hence, it suffices that the decoder in Step 1a attempts to correct only $\lceil d/2 \rceil - 1 = \lfloor (d-1)/2 \rfloor$ errors. This allows an efficient implementation of Step 1a if we select $\mathcal{C}_{\mathrm{in}}$ to be an alternant code with designed minimum distance d and use, for this code, the decoding algorithm of Chapter 6.)

Assuming that $\mathcal{C}_{\mathrm{cont}}$ is a linearly-concatenated code over a field F with $\mathcal{C}_{\mathrm{out}}$ taken as a GRS code over an extension field Φ, there is an efficient algorithm for implementing Step 2b using Euclid's algorithm for polynomials over Φ (see Problem 6.11). As mentioned in Section 6.6, a direct application of Euclid's algorithm, along with the computation of the syndrome and of the error values, require a number of operations in Φ that is proportional to $D \cdot N$. However, there are known methods for accelerating the GRS decoding algorithm so that its complexity becomes proportional to $N \log^2 N \log \log N$ (see the notes on Section 6.6). Translating this complexity into operations in F, it becomes proportional to at most $n^2 N \log^2 N \log \log N$. Hence, the complexity of Step 2 in Figure 12.1 grows no faster than $n^3 N \log^2 N \log \log N$, and the overall complexity of GMD decoding is therefore proportional to at most nN^2, assuming that $\mathcal{C}_{\mathrm{out}}$ is taken as a GRS code (or an extended GRS code: refer again to the notes on Section 6.6).

12.3 The Zyablov bound

In this section, we analyze the asymptotic attainable rate and relative minimum distance of linearly-concatenated codes, as the code length goes to infinity.

We recall from Section 4.5 that the q-ary entropy function $\mathsf{H}_q : [0,1] \to [0,1]$ is defined by

$$\mathsf{H}_q(x) = -x \log_q x - (1-x)\log_q(1-x) + x \log_q(q-1) ,$$

where $\mathsf{H}_q(0) = 0$ and $\mathsf{H}_q(1) = \log_q(q-1)$. Since this function is increasing in the interval $[0, 1-(1/q)]$, the inverse mapping

$$\mathsf{H}_q^{-1} : [0,1] \to [0, 1-(1/q)]$$

is well-defined, and we can use it to state the asymptotic version of the Gilbert–Varshamov bound as follows.

Theorem 12.4 *Let $F = \mathrm{GF}(q)$ and let n and rn be positive integers, where $r \in [0,1]$. There exists a linear $[n, rn, \geq \theta n]$ code over F, where*

$$\theta = \mathsf{H}_q^{-1}(1-r) .$$

This theorem is proved (see Section 4.3) by constructing an $(n(1-r)) \times n$ systematic parity-check matrix over F, column by column, such that each added column cannot be obtained as a linear combination of any $\lceil \theta n \rceil - 2$ existing columns. The number of such linear combinations, in turn, can be as large as $V_q(n-1, \lceil \theta n \rceil - 2) = q^{n(\mathsf{H}_q(\theta) - o(1))}$, where $V_q(n,t)$ is the size of a Hamming sphere with radius t in F^n, and $o(1)$ is an expression that goes to zero as n goes to infinity. Hence, an exhaustive check of all these linear combinations will require a number of operations in F that is exponential in the code length n for every fixed θ.

Suppose, however, that the code which is guaranteed by Theorem 12.4 is used as an $[n, k{=}rn]$ inner code $\mathcal{C}_{\mathrm{in}}$ in an $[n_{\mathrm{cont}}, k_{\mathrm{cont}}, d_{\mathrm{cont}}]$ linearly-concatenated code $\mathcal{C}_{\mathrm{cont}}$, with the outer code $\mathcal{C}_{\mathrm{out}}$ taken as an $[N{=}q^{rn}, K, D]$ (singly-)extended primitive GRS code over $\Phi = \mathrm{GF}(q^{rn})$, where $K = \lceil RN \rceil$ and $D = N-K+1$ $(> (1-R)N)$ for some real $R \in (0,1]$. The parameters of $\mathcal{C}_{\mathrm{cont}}$ are given by

$$n_{\mathrm{cont}} = nN = n\, q^{rn} = n\, q^{n(1-\mathsf{H}_q(\theta))} ,$$

$$k_{\mathrm{cont}} \geq rR \cdot nN = (1-\mathsf{H}_q(\theta))R \cdot nN ,$$

and

$$d_{\mathrm{cont}} > \theta(1-R) \cdot nN .$$

That is, the length of $\mathcal{C}_{\mathrm{cont}}$ can be arbitrarily large, the rate R_{cont} of $\mathcal{C}_{\mathrm{cont}}$ can be bounded from below by

$$R_{\mathrm{cont}} \geq rR = (1-\mathsf{H}_q(\theta))R , \tag{12.3}$$

and its relative minimum distance δ satisfies

$$\delta > \theta(1-R) . \tag{12.4}$$

Given a designed relative minimum distance $\delta \in (0, 1-(1/q))$, we can now maximize the right-hand side of (12.3) over all $\theta \in (0, 1-(1/q)]$ and $R \in (0, 1]$ that satisfy the constraint $\theta(1-R) \leq \delta$. This yields the *Zyablov bound*

$$R_{\mathrm{cont}} \geq R_{\mathrm{Z}}(\delta, q) ,$$

where

$$R_{\mathrm{Z}}(\delta, q) = \max_{\theta \in [\delta, 1-(1/q)]} \Big(1 - \mathsf{H}_q(\theta)\Big)\Big(1 - \frac{\delta}{\theta}\Big) . \tag{12.5}$$

(To be precise, we should restrict the maximization only to values θ that are equal to $\mathsf{H}_q^{-1}(1-(k/n))$ for some integer k; however, we are interested here in the case where n goes to infinity and so, by the continuity of the entropy function, we can take the maximum over the whole real interval $[\delta, 1-(1/q)]$.) It can be shown (Problem 12.7) that the maximum in the right-hand side of (12.5) is obtained for θ that satisfies the equation

$$\theta^2 \cdot \frac{\log_q(q-1) + \log_q((1-\theta)/\theta)}{1 + \log_q(1-\theta)} = \delta . \tag{12.6}$$

Figure 12.2 shows the Zyablov bound for $F = \mathrm{GF}(2)$, along with the Gilbert–Varshamov bound (the additional straight line that appears in the figure will be explained in Section 12.4).

While the Zyablov bound is inferior to the Gilbert–Varshamov bound, we show next that for every fixed $\delta \in (0, 1-(1/q))$, the computation of a generator matrix of $\mathcal{C}_{\mathrm{cont}}$ requires a number of operations in F that is only polynomially large in the code length n_{cont}.

Recall that a search for each column in the parity-check matrix of $\mathcal{C}_{\mathrm{in}}$ requires enumerating over (no more than) $V_q(n-1, \lceil \theta n \rceil - 2)$ linear combinations of previous columns. That number of combinations, in turn, satisfies

$$V_q(n-1, \lceil \theta n \rceil - 2) \leq q^{n \mathsf{H}_q(\theta)} = q^{(1-r)n} = N^{(1/r)-1} < n_{\mathrm{cont}}^{(1/r)-1} \tag{12.7}$$

(where we still assume that $\mathcal{C}_{\mathrm{out}}$ is an extended primitive GRS code over $\mathrm{GF}(q^{rn})$, of length $N = q^{rn}$). Now, for every fixed $\delta \in (0, 1-(1/q))$, the maximizing θ in the right-hand side of (12.5) is in the open interval $(\delta, 1-(1/q))$ and, so, $r = 1 - \mathsf{H}_q(\theta)$ is strictly positive. It follows that the power, $(1/r)-1$, of n_{cont} in (12.7) is some real constant, independent of n_{cont} (note though that this constant tends to infinity when $\delta \to 1-(1/q)$). Having found a systematic parity-check matrix of $\mathcal{C}_{\mathrm{in}}$, we effectively obtain also a generator matrix of this code.

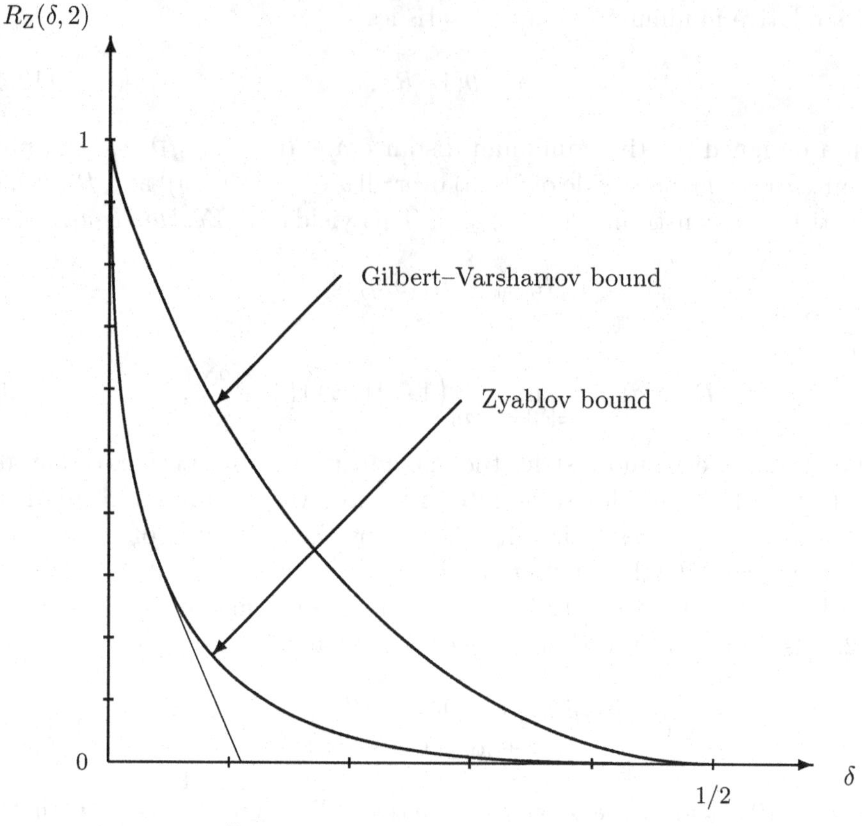

Figure 12.2. Zyablov bound for GF(2).

A generator matrix of the outer code $\mathcal{C}_{\text{out}}$ is also easy to compute, even if we need to search exhaustively for an irreducible polynomial of degree rn to represent the field $\Phi = \text{GF}(q^{rn})$: the number of monic polynomials of degree $\leq rn$ over $\text{GF}(q)$ is still smaller than $2N < 2n_{\text{cont}}$. Once we have the generator matrices of $\mathcal{C}_{\text{in}}$ and $\mathcal{C}_{\text{out}}$, we can easily obtain a generator matrix of $\mathcal{C}_{\text{cont}}$ (Problem 12.2). We thus conclude that for every fixed $\delta \in (0, 1-(1/q))$, the computation of a generator matrix of $\mathcal{C}_{\text{cont}}$ requires a number of operations in F that is only polynomially large in n_{cont}.

12.4 Justesen codes

Recall that the Zyablov bound is attained by conducting an exhaustive search for an inner code that attains the Gilbert–Varshamov bound. Such a search can be circumvented if we allow using different inner codes for distinct coordinates of the outer code. If "most" of these inner codes attain the Gilbert–Varshamov bound, then—as we show below—such a generalization of concatenation will approach the Zyablov bound.

We have shown before (Theorem 4.5) that most codes in the ensemble of linear $[n, k]$ codes over $F = \mathrm{GF}(q)$ indeed attain the Gilbert–Varshamov bound. Yet, even when we count only linear $[n, k]$ codes with systematic generator matrices, there are still at least $q^{k(n-k)}$ such codes over F. Now, if we use an extended primitive GRS code over $\mathrm{GF}(q^k)$ as an outer code, then its length is q^k—much smaller than the size of the ensemble. A sample of q^k codes from this ensemble is too small to claim that most of them attain the Gilbert–Varshamov bound. Hence, our goal is to construct an ensemble of at most q^k linear $[n, k]$ codes over F that meet this bound. We present next such an ensemble for $k < n \leq 2k$ (other ensembles are presented in Problems 12.10 and 12.11).

Let $\Phi = \mathrm{GF}(q^k)$ and fix a basis $\Omega = (\omega_1 \; \omega_2 \; \ldots \; \omega_k)$ of Φ over $F = \mathrm{GF}(q)$. Given an element $\alpha \in \Phi$, denote by $L(\alpha)$ the $k \times k$ matrix over F that represents, according to the basis Ω, the linear mapping $\varphi_\alpha : \Phi \rightarrow \Phi$ over F that is defined by

$$\varphi_\alpha : x \mapsto \alpha x \; ;$$

namely, for every column vector $\mathbf{x} \in F^k$,

$$\varphi_\alpha(\Omega \mathbf{x}) = \Omega L(\alpha) \mathbf{x} \; .$$

We now let $\mathcal{C}(\alpha)$ be the linear $[2k, k]$ code over F whose generator matrix is

$$G(\alpha) = \left(\; I \, \middle| \, (L(\alpha))^T \; \right) \; .$$

Equivalently, $\mathcal{C}(\alpha)$ consists of all vectors $(\mathbf{c}_1 \,|\, \mathbf{c}_2)$, where $\mathbf{c}_1, \mathbf{c}_2 \in F^k$ and $\mathbf{c}_2$ is related to $\mathbf{c}_1$ by

$$\Omega \mathbf{c}_2^T = \varphi_\alpha(\Omega \mathbf{c}_1^T) = \alpha \cdot \Omega \mathbf{c}_1^T \; . \qquad (12.8)$$

(The code $\mathcal{C}(\alpha)$ can also be viewed as a linearly-concatenated code over F with the outer code taken as the linear $[2, 1]$ code over Φ that is generated by $(\; 1 \quad \alpha \;)$, while the inner code is the $[k, k, 1]$ code F^k; see Problem 12.2.)

In what follows, we find it convenient to assume some ordering on the elements of Φ and we denote the jth element in Φ by α_j, where $1 \leq j \leq q^k$. Let n be an integer in the range $k < n \leq 2k$ and let G_j be the $k \times n$ matrix that consists of the first n columns of $G(\alpha_j)$. Define $\mathcal{C}_j$ to be the linear $[n, k]$ code over F whose generator matrix is G_j (equivalently, $\mathcal{C}_j$ is obtained by puncturing $\mathcal{C}(\alpha_j)$ at the last $2k{-}n$ coordinates; see Problem 2.3). The sequence

$$\mathcal{C}_1, \mathcal{C}_2, \ldots, \mathcal{C}_{q^k}$$

is called the *ensemble of* $[n, k]$ *Wozencraft codes* over F. We denote this sequence by $\mathcal{W}_F(n, k)$.

The following lemma states that a nonzero word in F^n cannot belong to too many codes in this sequence.

Lemma 12.5 *Every nonzero word* $\mathbf{c}$ *in* F^n *belongs to at most* q^{2k-n} *codes in* $\mathcal{W}_F(n,k)$.

Proof. Assume first that $n = 2k$, in which case $\mathcal{C}_j = \mathcal{C}(\alpha_j)$. Let $\mathbf{c} = (\mathbf{c}_1 \,|\, \mathbf{c}_2)$ be a nonzero word in F^n where $\mathbf{c}_1, \mathbf{c}_2 \in F^k$. From (12.8) it follows that there is at most one element $\alpha \in \Phi$ such that $\mathbf{c} \in \mathcal{C}(\alpha)$; indeed, if $\mathbf{c}_1 = \mathbf{0}$ (and $\mathbf{c}_2 \neq \mathbf{0}$) then no α can satisfy (12.8), and if $\mathbf{c}_1 \neq \mathbf{0}$ then α is given by $(\Omega \mathbf{c}_2^T)/(\Omega \mathbf{c}_1^T)$.

Next, suppose that $k < n < 2k$. A word $\mathbf{c} \in F^n$ belongs to $\mathcal{C}_j$ only if $\mathbf{c} = \mathbf{u}G_j$ for some $\mathbf{u} \in F^k$, in which case there must be an extension of $\mathbf{c}$ by $2k{-}n$ coordinates that produces a codeword $\mathbf{x} = \mathbf{x}(\mathbf{c}, \alpha_j)$ in $\mathcal{C}(\alpha_j)$: that codeword $\mathbf{x}$ is given by $\mathbf{u}G(\alpha_j)$. Given a nonzero word $\mathbf{c} \in F^n$, its q^{2k-n} possible extensions to words in F^{2k} may yield at most q^{2k-n} codewords—each belonging to at most one code in $\mathcal{W}_F(2k,k)$. Hence, $\mathbf{c}$ belongs to at most q^{2k-n} codes in $\mathcal{W}_F(n,k)$. □

We also mention that the codes in the sequence $\mathcal{W}_F(n,k)$ are all distinct. This fact is not material for the forthcoming discussion and we therefore leave the proof as an exercise (Problem 12.9).

The next proposition, which is a corollary of Lemma 12.5, provides a useful property of the distribution of the minimum distances of the codes in $\mathcal{W}_F(n,k)$.

Proposition 12.6 *The number of codes in* $\mathcal{W}_F(n,k)$ *with minimum distance less than a prescribed integer* d *is at most* $q^{2k-n} \cdot (V_q(n, d{-}1) - 1)$.

Proof. There are $V_q(n, d{-}1) - 1$ nonzero words in F^n with Hamming weight less than d, and by Lemma 12.5, each such word belongs to at most q^{2k-n} codes in $\mathcal{W}_F(n,k)$. □

In the construction that we describe next, the elements of $\mathcal{W}_F(n,k)$ will play the role of the inner code in a concatenated code. Specifically, let $F = \mathrm{GF}(q)$ and let k and n be positive integers such that $k < n \leq 2k$. For $j = 1, 2, \ldots, q^k$, we let $\mathcal{E}_j$ be a one-to-one linear mapping over F from $\Phi = \mathrm{GF}(q^k)$ onto $\mathcal{C}_j$. The outer code is taken as in Section 12.3: we fix R to be a real in $(0, 1]$ and let $\mathcal{C}_{\mathrm{out}}$ be an $[N, K, D]$ extended primitive GRS code over Φ, where $N = q^k$, $K = \lceil RN \rceil$, and $D = N{-}K{+}1$ $(> (1{-}R)N)$.

Having all these ingredients, we define a *Justesen code* as the code $\mathcal{C}_{\mathrm{Jus}}$ over F that consists of all words

$$(\mathcal{E}_1(z_1) \,|\, \mathcal{E}_2(z_2) \,|\, \ldots \,|\, \mathcal{E}_N(z_N)) \;,$$

where $(z_1 \; z_2 \; \ldots \; z_N)$ ranges over all the codewords of $\mathcal{C}_{\mathrm{out}}$.

Similarly to (proper) linearly-concatenated codes, the code $\mathcal{C}_{\mathrm{Jus}}$ is a linear $[nN, kK]$ code over F. We can obtain a lower bound on the minimum distance $\mathsf{d}(\mathcal{C}_{\mathrm{Jus}})$ by noticing that in every nonzero codeword of $\mathcal{C}_{\mathrm{Jus}}$ there are at least D nonzero sub-blocks $\mathcal{E}_j(z_j)$; so,

$$\mathsf{d}(\mathcal{C}_{\mathrm{Jus}}) \ge \min_J \sum_{j \in J} \mathsf{d}(\mathcal{C}_j) ,$$

where the minimum is taken over all subsets $J \subseteq \{1, 2, \ldots, N\}$ of size D. By Proposition 12.6 it follows that for every positive integer d,

$$\mathsf{d}(\mathcal{C}_{\mathrm{Jus}}) > d \cdot (D - q^{2k-n} V_q(n, d-1)) . \tag{12.9}$$

To obtain an asymptotic lower bound from (12.9), write $r = k/n$ and let the real θ be related to r and n by

$$\theta = \mathsf{H}_q^{-1}(1-r-\epsilon(n)) ,$$

where $n \mapsto \epsilon(n)$ is a function that satisfies both

$$\lim_{n\to\infty} \epsilon(n) = 0 \qquad \text{and} \qquad \lim_{n\to\infty} n \cdot \epsilon(n) = \infty$$

(e.g., $\epsilon(n) = (\log n)/n$). By selecting $d = \lceil \theta n \rceil$ in (12.9) we obtain

$$\begin{aligned} \mathsf{d}(\mathcal{C}_{\mathrm{Jus}}) &> \theta n \cdot \left((1-R)N - q^{(2r-1)n} \cdot q^{n\mathsf{H}_q(\theta)}\right) \\ &= \theta n N \left(1-R - q^{n(r-1+\mathsf{H}_q(\theta))}\right) \\ &= \theta n N \left(1-R - q^{-n \cdot \epsilon(n)}\right) \\ &= \theta(1-R - o(1)) \cdot nN , \end{aligned}$$

where $o(1)$ goes to zero as $n \to \infty$. We conclude that the relative minimum distance δ of $\mathcal{C}_{\mathrm{Jus}}$ is bounded from below by

$$\delta > \theta(1-R - o(1)) . \tag{12.10}$$

As for the rate R_{Jus} of $\mathcal{C}_{\mathrm{Jus}}$, we have

$$\begin{aligned} R_{\mathrm{Jus}} &\ge rR \\ &= (1-\mathsf{H}_q(\theta)-\epsilon(n))R \\ &= (1-\mathsf{H}_q(\theta)-o(1))R . \end{aligned} \tag{12.11}$$

Note that the bounds (12.10) and (12.11) are the same as (12.4) and (12.3), except for the term $o(1)$.

We can now proceed by maximizing over θ similarly to the Zyablov bound (12.5). Recall, however, that the rates of the codes in a Wozencraft ensemble must be in the interval $[\frac{1}{2}, 1)$; hence, θ is constrained to be at most $\mathsf{H}_q^{-1}(\frac{1}{2})$. We thus obtain here the lower bound

$$R_{\mathrm{Jus}} \ge R_{\mathrm{J}}(\delta, q) - o(1) ,$$

where $R_J(\delta, q)$ is given by

$$R_J(\delta, q) = \max_{\theta \in [\delta, \mathsf{H}_q^{-1}(\frac{1}{2})]} \left(1 - \mathsf{H}_q(\theta)\right)\left(1 - \frac{\delta}{\theta}\right) \qquad (12.12)$$

(note that for $\delta = \mathsf{H}_q^{-1}(\frac{1}{2})$ we get $R_J(\delta, q) = 0$). The function $\delta \mapsto R_J(\delta, q)$ coincides with $\delta \mapsto R_Z(\delta, q)$ in (12.5) whenever the maximizing θ in the latter equation is at most $\mathsf{H}_q^{-1}(\frac{1}{2})$. This occurs for values δ in the interval $[0, \delta_J(q)]$, where $\delta_J(q)$ can be computed by substituting $\theta = \mathsf{H}_q^{-1}(\frac{1}{2})$ in the left-hand side of (12.6). For instance, when $q = 2$ we get $\mathsf{H}_2^{-1}(\frac{1}{2}) \approx 0.1100$, $\delta_J(2) \approx 0.0439$, and $R_J(\delta_J(2), 2) \approx 0.3005$.

When $\delta > \delta_J(q)$, the maximum in (12.12) is obtained for $\theta = \mathsf{H}_q^{-1}(\frac{1}{2})$; hence, (12.12) becomes

$$R_J(\delta, q) = \frac{1}{2} \cdot \left(1 - \frac{\delta}{\mathsf{H}_q^{-1}(\frac{1}{2})}\right),$$

which is a straight line; this line is shown in Figure 12.2.

12.5 Concatenated codes that attain capacity

By using code concatenation, one can approach the capacity of the q-ary symmetric channel with linear codes that can be encoded and decoded in time complexity that is polynomially large in the code length. We next show how this can be done.

Let $F = \mathrm{GF}(q)$ and let p be the crossover probability of a q-ary symmetric channel (F, F, Prob) where $p < 1-(1/q)$. Also, let n and rn be positive integers such that

$$r < 1 - \mathsf{H}_q(p) .$$

We have shown (Corollary 4.18) that there always exists a linear $[n, rn]$ code $\mathcal{C}$ over F such that the decoding error probability $\mathrm{P}_{\mathrm{err}}(\mathcal{C})$ of a nearest-codeword decoder for $\mathcal{C}$ satisfies

$$\mathrm{P}_{\mathrm{err}}(\mathcal{C}) < 2q^{-n\mathrm{E}_q(p,r)}$$

for some strictly positive constant $\mathrm{E}_q(p, r)$. We can assume that $\mathcal{C}$ has a systematic generator matrix, as the value $\mathrm{P}_{\mathrm{err}}(\mathcal{C})$ remains unchanged under any permutation of the code coordinates.

A brute-force search for the code $\mathcal{C}$ requires an enumeration over all $q^{r(1-r)n^2}$ linear $[n, rn]$ codes with systematic generator matrices over F, and then testing for each possible error word $\mathbf{e}$ in F^n whether $\mathbf{e}$ is decoded correctly by a nearest-codeword decoder (yet see Problems 12.12, 12.14, and 12.15: they imply that the search can be restricted to smaller ensembles). The precise computation of $\mathrm{P_{err}}(\mathcal{C})$ requires operations in real numbers, since we need to calculate for every word $\mathbf{e} \in F^n$ the probability, $(p/(q-1))^{\mathsf{w}(\mathbf{e})}(1-p)^{n-\mathsf{w}(\mathbf{e})}$, that the error word equals $\mathbf{e}$. However, it will suffice for our purposes to find a code $\mathcal{C}$ whose value $\mathrm{P_{err}}(\mathcal{C})$ is no more than twice (say) the smallest decoding error probability, i.e.,

$$\mathrm{P_{err}}(\mathcal{C}) < 4q^{-n\mathrm{E}_q(p,r)} .$$

This, in turn, allows us to limit the precision of our computations to a number of decimal places that is linear in n. We can count the number of bit operations and operations in F that are applied while searching for $\mathcal{C}$ and we let $N_0(n, r, q)$ be an upper bound on that number.

We now use the code $\mathcal{C}$ as an inner code in a linearly-concatenated code $\mathcal{C}_{\mathrm{cont}}$, whose outer code, $\mathcal{C}_{\mathrm{out}}$, is by itself a linearly-concatenated code of length N over the field $\Phi = \mathrm{GF}(q^{rn})$, where N is taken to be at least $\max\{N_0(n, r, q), q^{rn}\}$. We further assume that $\mathcal{C}_{\mathrm{out}}$ attains the Zyablov bound and that the product of the minimum distances of its inner and outer codes is bounded from below by $\lceil \delta N \rceil$, for some real parameter $\delta \in [0, 1]$. The relationship between δ and r will be determined in the sequel. Given δ, the rate R of $\mathcal{C}_{\mathrm{out}}$ is bounded from below by

$$R \geq R_{\mathrm{Z}}(\delta, q^{rn}) .$$

The choice of δ will be such that R is close to 1.

To analyze the encoding and decoding complexity of $\mathcal{C}_{\mathrm{cont}}$, we assume that this code is defined by the pair $(\mathcal{E}, \mathcal{C}_{\mathrm{out}})$ for some one-to-one linear mapping $\mathcal{E} : \Phi \to \mathcal{C}$. The encoding consists of a multiplication by a generator matrix of $\mathcal{C}_{\mathrm{out}}$ over Φ, followed by N applications of the mapping $\mathcal{E}$. From the complexity analysis of Section 12.3 it follows that a generator matrix of (the concatenated code) $\mathcal{C}_{\mathrm{out}}$ can be obtained by a number of operations in Φ that is quadratic in N. Shifting to operations in F, this complexity becomes quadratic in nN. Hence, the overall encoding complexity of a codeword of $\mathcal{C}_{\mathrm{cont}}$ is quadratic in nN (counting also the computation of the generator matrix of $\mathcal{C}_{\mathrm{out}}$ and the N applications of the mapping $\mathcal{E}$).

Next, we turn to the decoding of $\mathcal{C}_{\mathrm{cont}}$. Let

$$\mathbf{y} = (\mathbf{y}_1 \,|\, \mathbf{y}_2 \,|\, \ldots \,|\, \mathbf{y}_N) \in F^{nN}$$

be the received word, where each $\mathbf{y}_j$ is a sub-block in F^n. Our decoder $\mathcal{D}_{\mathrm{cont}}$ of $\mathcal{C}_{\mathrm{cont}}$ consists of two decoding steps, as follows.

1. Apply a nearest-codeword decoder for $\mathcal{C}$ to each sub-block $\mathbf{y}_j$ to produce a codeword $\hat{\mathbf{c}}_j$ of $\mathcal{C}$.

2. Apply an efficient decoder (such as the GMD decoder in Figure 12.1) for the concatenated code $\mathcal{C}_{\text{out}}$ to correct up to $\lceil \delta N/2 \rceil - 1$ errors in the word
$$(\mathcal{E}^{-1}(\hat{\mathbf{c}}_1) \mid \mathcal{E}^{-1}(\hat{\mathbf{c}}_2) \mid \ldots \mid \mathcal{E}^{-1}(\hat{\mathbf{c}}_N)) \in \Phi^N$$
(note that $\lceil \delta N/2 \rceil - 1 = \lfloor (\lceil \delta N \rceil - 1)/2 \rfloor$, and recall that $\lceil \delta N \rceil$ is a lower bound on the minimum distance of $\mathcal{C}_{\text{out}}$).

Step 1 can be implemented using a number of operations in F that is proportional to $nN \cdot q^{rn}$. As for Step 2, we have shown in Section 12.2 that GMD decoding can be carried out in a number of operations in Φ that is less than quadratic in N. Translating the latter complexity to operations in F, the two decoding steps can be carried out in time complexity that is (less than) quadratic in nN.

We turn to bounding the decoding error probability, $\mathrm{P}_{\text{err}}(\mathcal{C}_{\text{cont}})$, of the decoder $\mathcal{D}_{\text{cont}}$. Clearly, decoding will fail only if $\tau = \lceil \delta N/2 \rceil$ or more of the sub-blocks $\mathbf{y}_j$ have been decoded incorrectly by a nearest-codeword decoder for $\mathcal{C}$. Now, for every given j, a sub-block $\mathbf{y}_j$ is incorrectly decoded with probability $P = \mathrm{P}_{\text{err}}(\mathcal{C})$. Furthermore, since the channel is memoryless, such incorrect decoding occurs independently of all other sub-blocks. Hence,

$$\begin{aligned}
\mathrm{P}_{\text{err}}(\mathcal{C}_{\text{cont}}) &\leq \sum_{i=\tau}^{N} \binom{N}{i} P^i (1-P)^{N-i} \\
&\leq \sum_{i=\tau}^{N} \binom{N}{i} P^i \leq P^{\tau} \sum_{i=\tau}^{N} \binom{N}{i} \\
&\leq 2^N \cdot P^{\tau} \leq 2^N \cdot P^{N\delta/2} \\
&< 4^N \cdot q^{-Nn\mathrm{E}_q(p,r)\delta/2} \leq q^{-nN(\mathrm{E}_q(p,r)\delta/2 - o(1))} , \qquad (12.13)
\end{aligned}$$

where we have recalled that $P < 4q^{-n\mathrm{E}_q(p,r)}$ and have used the notation $o(1)$ for an expression that goes to zero as $n \to \infty$. It follows from (12.13) that for every $r < 1 - \mathsf{H}_q(p)$ and $\delta > 0$ there is a sufficiently large value of n for which $\mathrm{P}_{\text{err}}(\mathcal{C}_{\text{cont}})$ decreases exponentially with the code length nN.

As for the rate, R_{cont}, of $\mathcal{C}_{\text{cont}}$, one can show (Problem 12.8) that

$$R_{\mathrm{Z}}(\delta, q^{rn}) = (1 - \sqrt{\delta})^2 - (o(1)/r)$$

and, so,

$$R_{\text{cont}} \geq rR \geq r \cdot (1 - \sqrt{\delta})^2 - o(1) . \qquad (12.14)$$

Given a designed rate $\mathcal{R} < 1 - \mathsf{H}_q(p)$, we select the rate r of the inner code so that $\mathcal{R} \leq r \leq 1 - \mathsf{H}_q(p)$ and set the value δ to

$$\delta = (1 - \sqrt{\mathcal{R}/r})^2 .$$

It follows from (12.14) that R_{cont} is bounded from below by $\mathcal{R} - o(1)$, while the error exponent in (12.13) satisfies

$$-\frac{\log_q \mathrm{P}_{\mathrm{err}}(\mathcal{C}_{\mathrm{cont}})}{nN} \geq \tfrac{1}{2}\mathrm{E}_q(p, r)(1 - \sqrt{\mathcal{R}/r})^2 - o(1) .$$

By maximizing over r we obtain

$$-\frac{\log_q \mathrm{P}_{\mathrm{err}}(\mathcal{C}_{\mathrm{cont}})}{nN} \geq \mathrm{E}_q^*(p, \mathcal{R}) - o(1) ,$$

where

$$\mathrm{E}_q^*(p, \mathcal{R}) = \max_{\mathcal{R} \leq r \leq 1 - \mathsf{H}_q(p)} \tfrac{1}{2}\mathrm{E}_q(p, r)(1 - \sqrt{\mathcal{R}/r})^2 . \tag{12.15}$$

In particular, $\mathrm{E}_q^*(p, \mathcal{R}) > 0$ whenever $\mathcal{R} < 1 - \mathsf{H}_q(p)$.

Our foregoing analysis leads to the following conclusion.

Theorem 12.7 *Let F be the field* $\mathrm{GF}(q)$ *and fix a crossover probability $p \in [0, 1-(1/q))$ of a q-ary symmetric channel. For every $\mathcal{R} < 1 - \mathsf{H}_q(p)$ there exists an infinite sequence of linearly-concatenated codes $\mathcal{C}_{\mathrm{cont}}^{(1)}, \mathcal{C}_{\mathrm{cont}}^{(2)}, \cdots, \mathcal{C}_{\mathrm{cont}}^{(i)}, \cdots$ over F such that the following conditions hold:*

(i) Each code $\mathcal{C}_{\mathrm{cont}}^{(i)}$ is a linear $[n_i, k_i]$ code over F and the values n_i and k_i can be computed from $\mathcal{R}$, q, and i in time complexity that is polynomially large in the length of the bit representations of $\mathcal{R}$, q, i, and n_i.

(ii) The code rates k_i/n_i satisfy

$$\liminf_{i \to \infty} \frac{k_i}{n_i} \geq \mathcal{R} .$$

(iii) There is an encoder for $\mathcal{C}_{\mathrm{cont}}^{(i)}$ whose time complexity is quadratic in n_i.

(iv) There is a decoder for $\mathcal{C}_{\mathrm{cont}}^{(i)}$ whose time complexity is quadratic in n_i and its decoding error probability $\mathrm{P}_{\mathrm{err}}(\mathcal{C}_{\mathrm{cont}}^{(i)})$ satisfies

$$-\liminf_{i \to \infty} \frac{1}{n_i} \log_q \mathrm{P}_{\mathrm{err}}(\mathcal{C}_{\mathrm{cont}}^{(i)}) \geq \mathrm{E}_q^*(p, \mathcal{R}) > 0 .$$

Problems

[Section 12.1]

Problem 12.1 Let the concatenated code $\mathcal{C}_{\mathrm{cont}}$ be defined over F by an (N, M, D) outer code $\mathcal{C}_{\mathrm{out}}$ over Φ and a one-to-one mapping $\mathcal{E}_{\mathrm{in}} : \Phi \to \mathcal{C}_{\mathrm{in}}$ onto an $(n, |\Phi|, d)$ inner code $\mathcal{C}_{\mathrm{in}}$ over F. Show that $\mathcal{C}_{\mathrm{cont}}$ is an $(nN, M, {\geq}dD)$ code over F.

Problem 12.2 (Generator and parity-check matrices of linearly-concatenated codes) Let $\mathcal{C}_{\text{in}}$ be a linear $[n, k]$ code over $F = \text{GF}(q)$ and let $\Omega = (\omega_1\, \omega_2\, \ldots\, \omega_k)$ be a basis of $\Phi = \text{GF}(q^k)$ over F. Fix a $k \times n$ generator matrix G_{in} of $\mathcal{C}_{\text{in}}$, and define the linearly-concatenated code $\mathcal{C}_{\text{cont}}$ over F by a linear $[N, K]$ outer code $\mathcal{C}_{\text{out}}$ over Φ and a one-to-one mapping $\mathcal{E}_{\text{in}} : \Phi \to \mathcal{C}_{\text{in}}$, where for every column vector $\mathbf{x} \in F^k$,

$$\mathcal{E}_{\text{in}}(\Omega\mathbf{x}) = \mathbf{x}^T G_{\text{in}} .$$

For each element $\alpha \in \Phi$, let $L(\alpha)$ be the $k \times k$ matrix over F that represents, according to the basis Ω, the mapping $\varphi_\alpha : \Phi \to \Phi$ defined by $\varphi_\alpha : x \mapsto \alpha x$; that is, for every $\alpha \in \Phi$ and column vector $\mathbf{x} \in F^k$,

$$\varphi_\alpha(\Omega\mathbf{x}) = \Omega L(\alpha)\mathbf{x} .$$

Also, let Q be a $k \times n$ matrix over F that satisfies $G_{\text{in}}Q^T = I$, where I is the $k \times k$ identity matrix.

1. Let
$$G_{\text{out}} = (\, g_{i,j}\,)_{i=1}^{K}\,{}_{j=1}^{N}$$
be a $K \times N$ generator matrix of $\mathcal{C}_{\text{out}}$ over Φ. Show that a $kK \times nN$ generator matrix of $\mathcal{C}_{\text{cont}}$ is given by
$$G_{\text{cont}} = \begin{pmatrix} (L(g_{1,1}))^T G_{\text{in}} & (L(g_{1,2}))^T G_{\text{in}} & \ldots & (L(g_{1,N}))^T G_{\text{in}} \\ (L(g_{2,1}))^T G_{\text{in}} & (L(g_{2,2}))^T G_{\text{in}} & \ldots & (L(g_{2,N}))^T G_{\text{in}} \\ \vdots & \vdots & \vdots & \vdots \\ (L(g_{K,1}))^T G_{\text{in}} & (L(g_{K,2}))^T G_{\text{in}} & \ldots & (L(g_{K,N}))^T G_{\text{in}} \end{pmatrix} .$$
2. Explain why a matrix Q indeed exists. Is this matrix unique?
3. Let H_{in} be an $(n-k) \times n$ parity-check matrix of $\mathcal{C}_{\text{in}}$ over F and
$$H_{\text{out}} = (\, h_{i,j}\,)_{i=1}^{N-K}\,{}_{j=1}^{N}$$
be an $(N-K) \times N$ parity-check of $\mathcal{C}_{\text{out}}$ over Φ. Show that an $(nN-kK) \times nN$ parity-check matrix of $\mathcal{C}_{\text{cont}}$ is given by
$$H_{\text{cont}} = \left(\begin{array}{cccc} H_{\text{in}} & & & 0 \\ & H_{\text{in}} & & \\ & & \ddots & \\ 0 & & & H_{\text{in}} \\ \hline L(h_{1,1})Q & L(h_{1,2})Q & \ldots & L(h_{1,N})Q \\ L(h_{2,1})Q & L(h_{2,2})Q & \ldots & L(h_{2,N})Q \\ \vdots & \vdots & \vdots & \vdots \\ L(h_{N-K,1})Q & L(h_{N-K,2})Q & \ldots & L(h_{N-K,N})Q \end{array}\right) .$$

4. Let $P(x) = P_0 + P_1 x + \ldots + P_{k-1} x^{k-1} + x^k$ be a monic primitive polynomial of degree k over F and let C_P be the $k \times k$ companion matrix of $P(x)$; i.e.,

$$C_P = \begin{pmatrix} 0 & 0 & \ldots & 0 & -P_0 \\ 1 & 0 & \ldots & 0 & -P_1 \\ 0 & 1 & \ldots & 0 & -P_2 \\ \vdots & \ddots & \ddots & 0 & \vdots \\ 0 & \ldots & 0 & 1 & -P_{k-1} \end{pmatrix}$$

(see Problem 3.9). Represent the field Φ as $F[\xi]/P(\xi)$ and take Ω as

$$(1 \;\xi\; \xi^2 \;\ldots\; \xi^{k-1}) \,.$$

Show that $L(0) = 0$ and

$$L(\xi^i) = C_P^i \,, \quad 0 \le i < q^k - 1 \,.$$

5. Let Φ be represented as $F[\xi]/P(\xi)$ where $P(x)$ is a monic primitive polynomial of degree k over F. Construct $\mathcal{C}_{\mathrm{cont}}$ with $\mathcal{C}_{\mathrm{in}} = F^k$ while $\mathcal{C}_{\mathrm{out}}$ is taken as the $[q^k - 1, K]$ narrow-sense primitive RS code over Φ with code locators $1, \xi, \xi^2, \ldots, \xi^{q^k - 2}$. Using the basis $\Omega = (1 \;\xi\; \xi^2 \;\ldots\; \xi^{k-1})$ and taking $G_{\mathrm{in}} = I$, write generator and parity-check matrices of $\mathcal{C}_{\mathrm{cont}}$ in terms of the companion matrix C_P.

Problem 12.3 (Minimum distance of dual codes of linearly-concatenated codes) Let $\mathcal{C}_{\mathrm{in}}$ be a linear $[n, k{<}n]$ code over $F = \mathrm{GF}(q)$ and $\mathcal{C}_{\mathrm{out}}$ be a linear $[N, K{<}N]$ code over $\Phi = \mathrm{GF}(q^k)$. Fix a generator matrix G_{in} of $\mathcal{C}_{\mathrm{in}}$ and a basis $\Omega = (\omega_i)_{i=1}^k$ of Φ over F, and define the linearly-concatenated code $\mathcal{C}_{\mathrm{cont}}$ over F by the outer code $\mathcal{C}_{\mathrm{out}}$ and the mapping

$$\mathcal{E}_{\mathrm{in}} : (\Omega \mathbf{x}) \mapsto \mathbf{x}^T G_{\mathrm{in}} \,,$$

where $\mathbf{x}$ ranges over all the (column) vectors in F^k.

Denote by $d^{\perp}$ and $D^{\perp}$ the minimum distances of the dual codes of $\mathcal{C}_{\mathrm{in}}$ and $\mathcal{C}_{\mathrm{out}}$, respectively, and let

$$G_{\mathrm{out}} = (\, g_{i,j} \,)_{i=1}^{K} {}_{j=1}^{N}$$

be a $K \times N$ generator matrix of $\mathcal{C}_{\mathrm{out}}$ over Φ. For each element $\alpha \in \Phi$, define the $k \times k$ matrix $L(\alpha)$ over F as in Problem 12.2.

1. Let $J = \{j_1, j_2, \ldots, j_\tau\}$ be a nonempty subset of $\{1, 2, \ldots, N\}$ of size $\tau < D^{\perp}$ and consider the $k\tau \times kK$ matrix over F that is given by

$$A_J = \begin{pmatrix} L(g_{1,j_1}) & L(g_{2,j_1}) & \ldots & L(g_{K,j_1}) \\ L(g_{1,j_2}) & L(g_{2,j_2}) & \ldots & L(g_{K,j_2}) \\ \vdots & \vdots & \vdots & \vdots \\ L(g_{1,j_\tau}) & L(g_{2,j_\tau}) & \ldots & L(g_{K,j_\tau}) \end{pmatrix} .$$

Show that

$$\mathrm{rank}(A_J) = k\tau \,.$$

Hint: Denote by G_J the $K \times \tau$ sub-matrix of G_{out} whose columns are indexed by the elements of J. Show that for every K column vectors $\mathbf{x}_1, \mathbf{x}_2, \ldots, \mathbf{x}_K$ in F^k,

$$A_J \begin{pmatrix} \mathbf{x}_1 \\ \mathbf{x}_2 \\ \vdots \\ \mathbf{x}_K \end{pmatrix} = \mathbf{0} \quad \Longleftrightarrow \quad G_J^T \begin{pmatrix} \Omega\mathbf{x}_1 \\ \Omega\mathbf{x}_2 \\ \vdots \\ \Omega\mathbf{x}_K \end{pmatrix} = \mathbf{0} .$$

Deduce that

$$\mathrm{rank}(A_J) = k \cdot \mathrm{rank}(G_J) ,$$

and then recall that every $D^{\perp} - 1$ columns in G_{out} are linearly independent.

2. Show that the minimum distance of $\mathcal{C}_{\mathrm{cont}}^{\perp}$ is at least

$$\min\{D^{\perp}, d^{\perp}\} .$$

Hint: Let G_{cont} be the generator matrix of $\mathcal{C}_{\mathrm{cont}}$ as in part 1 of Problem 12.2 and suppose that $\mathbf{c}$ is a nonzero codeword in $\mathcal{C}_{\mathrm{cont}}^{\perp}$, namely,

$$\mathbf{c}\, G_{\mathrm{cont}}^T = \mathbf{0} .$$

Write $\mathbf{c} = (\mathbf{c}_1 \,|\, \mathbf{c}_2 \,|\, \ldots \,|\, \mathbf{c}_N)$ where $\mathbf{c}_j \in F^n$, and define

$$J = \{j_1, j_2, \ldots, j_\tau\} = \Big\{ j \in \{1, 2, \ldots, N\} \;:\; \mathbf{c}_j \neq \mathbf{0} \Big\}$$

and

$$\mathbf{v}_j = \mathbf{c}_j G_{\mathrm{in}}^T , \quad j \in J .$$

Assume to the contrary that the Hamming weight of $\mathbf{c}$ is less than $\min\{D^{\perp}, d^{\perp}\}$. Argue that $\mathbf{v}_j \neq \mathbf{0}$ for every $j \in J$, while

$$(\mathbf{v}_{j_1} \,|\, \mathbf{v}_{j_2} \,|\, \ldots \,|\, \mathbf{v}_{j_\tau})\, A_J = \mathbf{0} ,$$

where A_J is as defined in part 1. Finally, apply part 1 to reach a contradiction.

3. Show that the minimum distance of $\mathcal{C}_{\mathrm{cont}}^{\perp}$ is at most $d^{\perp}$.

Hint: Consider linear combinations of the rows of the parity-check matrix H_{cont} in part 3 of Problem 12.2.

Problem 12.4 Let $F = \mathrm{GF}(q)$ and for a positive integer m, let $\mathcal{C}_m$ be an $[n_m, k_m, d_m]$ linearly-concatenated code over F obtained by taking a linear $[N, K, D]$ outer code over $\mathrm{GF}(q^m)$ and a linear $[n, m]$ inner code $\mathcal{C}_{\mathrm{in}}^{(m)}$ over F.

1. Show that

$$d_m \leq \frac{q-1}{q} \cdot \frac{nD}{1 - q^{-m}} .$$

Hint: Consider the code $\mathcal{C}'$ of length nN that consists of all the codewords in $\mathcal{C}_m$ that correspond to the q^m scalar multiples (over $\mathrm{GF}(q^m)$) of a codeword $\mathbf{z} = (z_1\ z_2\ \ldots\ z_N)$ of Hamming weight D in the outer code. Shorten the code $\mathcal{C}'$ to be of length nD by leaving only the sub-blocks of length n (over F) that are indexed by the nonzero coordinates in $\mathbf{z}$. Next, apply to $\mathcal{C}'$ the Plotkin bound from part 4 of Problem 4.23.

2. Show that the upper bound in part 1 is attained when $\mathcal{C}_{\rm in}^{(m)}$ is taken as the shortened first-order Reed–Muller code over F, which is defined as the linear $[q^m-1, m, q^{m-1}(q-1)]$ code over F with an $m \times (q^m-1)$ generator matrix whose columns range over all the nonzero vectors in F^m (see Problem 2.17).

Hereafter in this problem, let $\mathcal{C}_{\rm in}^{(m)}$ be as in part 2 and, for a fixed positive integer K, let the outer code be a $[q^m+1, K]$ doubly-extended GRS code over GF(q^m) as defined in Problem 5.2.

3. Verify that $n_m = q^{2m} - 1$.
4. Show that

$$k_m = mK = (K/2)\log_q(n_m+1)$$

and, so, $|\mathcal{C}_m| = (n_m+1)^{K/2}$.

5. Show that

$$d_m = \frac{q-1}{q} \cdot \left(n_m+1 - (K-2)\sqrt{n_m+1}\right)$$

and, so, for every fixed K,

$$\lim_{m\to\infty} \frac{d_m}{n_m} = \frac{q-1}{q} .$$

6. In comparison, show by the Plotkin bound that every infinite sequence of (n_i, M_i, d_i) codes over $F = \mathrm{GF}(q)$ with strictly increasing code sizes M_i must satisfy

$$\limsup_{i\to\infty} \frac{d_i}{n_i} \le \frac{q-1}{q} .$$

Problem 12.5 Let α be a primitive element in $\Phi = \mathrm{GF}(2^m)$ and let a and K be integers such that $a \ge 0$ and $1 \le K < 2^m$. Let $u(x)$ be a nonzero polynomial over Φ of the form

$$u(x) = u_0 x^a + u_1 x^{a+1} + \ldots + u_{K-1} x^{a+K-1} ,$$

and for $j = 0, 1, 2, \ldots, 2^m-2$ define the elements c_j by $c_j = u(\alpha^j)$.

1. Assuming that $a = 0$, obtain an upper bound, as a function of K (and independently of $u(x)$), on the number of indexes j for which $c_j = 0$. Write a polynomial $u(x)$ that attains the bound.
2. Suppose that $a = 0$ and let β be a nonzero element in Φ. Obtain an upper bound, as a function of K but independently of β or $u(x)$, on the number of indexes j for which $c_j = \beta$. Find an element β and a polynomial $u(x)$ that attain the bound.
3. Repeat parts 1 and 2 for $a = 1$.

Given $\Phi = \mathrm{GF}(2^m)$, α, a, and K as defined above, let $\mathcal{C}_{\rm out}$ be an RS code of length $N = 2^m - 1$ over Φ with a canonical generator matrix

$$G_{\rm RS} = \left(\alpha^{(a+i)j}\right)_{i=0}^{K-1}{}_{j=0}^{N-1} ,$$

and let $\mathcal{C}_{\rm cont}$ be a linearly-concatenated code over $F = \mathrm{GF}(2)$ obtained by using $\mathcal{C}_{\rm out}$ as an outer code and the set F^m as an inner code.

4. Find the dimension of $\mathcal{C}_{\text{cont}}$ as a function of K and m.

5. Assuming that $a = 0$ and $K = 1$, find the minimum distance of $\mathcal{C}_{\text{cont}}$.

6. Let t be a positive integer such that

$$K \cdot V_2(m, t) \le 2^m ,$$

where $V_2(m,t) = \sum_{i=0}^{t} \binom{m}{i}$. Show that when $a = 1$, the minimum distance of $\mathcal{C}_{\text{cont}}$ is at least

$$K \cdot \sum_{i=0}^{t} i \binom{m}{i} = K \cdot m \cdot V_2(m-1, t-1) .$$

Hint: Use part 3.

7. Using part 6, write a lower bound on the minimum distance of $\mathcal{C}_{\text{cont}}$ when $a = K = 1$.

8. Let $a = 1$, $K = 2$, and let m be odd. Using part 6, show that the minimum distance of $\mathcal{C}_{\text{cont}}$ is at least

$$m \cdot \left(2^{m-1} - \binom{m-1}{(m-1)/2} \right) .$$

[Section 12.2]

Problem 12.6 Show by example that the check in Step 2(c)ii in Figure 12.1 is necessary; that is, exhibit a case where the decoding of Step 2b is successful, yet the computed codeword $\mathbf{c}$ in Step 2(c)i does not satisfy $\mathsf{d}(\mathbf{y}, \mathbf{c}) < dD/2$.

[Section 12.3]

Problem 12.7 The purpose of this problem is to obtain several properties of the function $\delta \mapsto R_{\mathrm{Z}}(\delta, q)$, which is defined in (12.5).

1. Let θ be a fixed real in the interval $(0, 1-(1/q)]$ and consider the straight line $\delta \mapsto T_\theta(\delta)$ that is defined over the domain $(0, 1-(1/q))$ by

$$T_\theta(\delta) = \left(1 - \mathsf{H}_q(\theta)\right)\left(1 - \frac{\delta}{\theta}\right) .$$

Show that the function $\delta \mapsto T_\theta(\delta)$ lies on or below (but never above) the curve $\delta \mapsto R_{\mathrm{Z}}(\delta, q)$.

(As a side note, notice that the straight line $\delta \mapsto T_\theta(\delta)$ passes through the points $(0, 1-\mathsf{H}_q(\theta))$ and $(\theta, 0)$; these are the projections on the axes of the point $(\theta, 1-\mathsf{H}_q(\theta))$, which lies on the Gilbert–Varshamov bound.)

2. Let θ be a maximizing value of the right-hand side of (12.5) for some $\delta_0 \in (0, 1-(1/q))$. Verify that the line $\delta \mapsto T_\theta(\delta)$ passes through the point $(\delta_0, R_{\mathrm{Z}}(\delta_0, q))$.

3. Show that the maximum in the right-hand side of (12.5) is attained neither at δ nor at $1-(1/q)$.

   Hint: Compute the values $T_\delta(\delta)$ and $T_{1-(1/q)}(\delta)$.

4. Show by differentiation that the maximum in the right-hand side of (12.5) is obtained for θ that satisfies (12.6).

5. Let δ_1 and δ_2 be two reals such that $0 < \delta_1 < \delta_2 < 1-(1/q)$ and let θ_1 and θ_2 be values of θ that maximize the right-hand side of (12.5) for $\delta = \delta_1$ and $\delta = \delta_2$, respectively. Show that $\theta_1 < \theta_2$.

   Hint: First verify from part 4 that $\theta_1 \neq \theta_2$. Next, denote by λ the slope of the straight line that passes through the points $(\delta_1, R_Z(\delta_1, q))$ and $(\delta_2, R_Z(\delta_2, q))$, and show that λ is bounded from below by the slope of $\delta \mapsto T_{\theta_1}(\delta)$ and from above by the slope of $\delta \mapsto T_{\theta_2}(\delta)$; namely,

$$-\frac{1 - \mathsf{H}_q(\theta_1)}{\theta_1} \leq \lambda \leq -\frac{1 - \mathsf{H}_q(\theta_2)}{\theta_2} .$$

   Conclude from this that $\theta_1 < \theta_2$.

6. Show that for every $\delta \in (0, 1-(1/q))$ there is at most one value $\theta \in [\delta, 1-(1/q)]$ that satisfies (12.6).

   Hint: Show that the left-hand side of (12.6) is not constant on every nonempty open interval in $[\delta, 1-(1/q)]$. Then apply part 5.

7. Deduce from parts 5 and 6 that (12.6) defines a monotonically increasing differentiable function $\delta \mapsto \theta(\delta)$ from $(0, 1-(1/q))$ onto $(0, 1-(1/q))$.

8. Show that the function $\delta \mapsto R_Z(\delta, q)$ is differentiable over $(0, 1-(1/q))$.

9. Given $\delta_0 \in (0, 1-(1/q))$, let θ_0 be the value of θ that satisfies (12.6) for $\delta = \delta_0$. Show that the line $\delta \mapsto T_{\theta_0}(\delta)$ is tangent to the curve $\delta \mapsto R_Z(\delta, q)$ at the point $(\delta_0, R_Z(\delta_0, q))$.

10. Show that

$$\lim_{\delta \to 0} R_Z(\delta, q) = 1 \quad \text{and} \quad \lim_{\delta \to 1-(1/q)} R_Z(\delta, q) = 0 .$$

11. Show that the function $\delta \mapsto R_Z(\delta, q)$ is monotonically decreasing and $\cup$-convex over $(0, 1-(1/q))$.

Problem 12.8 Show that

$$(1 - \sqrt{\delta})^2 - \frac{1}{\log_2 q} \leq R_Z(\delta, q) \leq (1 - \sqrt{\delta})^2$$

and, therefore, in the limit,

$$\lim_{q \to \infty} R_Z(\delta, q) = (1 - \sqrt{\delta})^2 .$$

Hint: The Gilbert–Varshamov bound cannot exceed the Singleton bound and, so, $1 - \mathsf{H}_q(\theta) \leq 1 - \theta$.

[Section 12.4]

Problem 12.9 Let $F = \mathrm{GF}(q)$ and let n and k be positive integers such that $k < n \le 2k$. Show that the codes in the Wozencraft ensemble $\mathcal{W}_F(n, k)$ are distinct.

Hint: Using the notation of Section 12.4, consider two distinct elements α_i and α_j in $\Phi = \mathrm{GF}(q^k)$. Argue that

$$L(\alpha_i) - L(\alpha_j) = L(\alpha_i - \alpha_j)$$

and, therefore, $L(\alpha_i) - L(\alpha_j)$ is a nonsingular matrix over F. Deduce that there exists $\mathbf{u} \in F^k$ such that the vectors $\mathbf{u}(L(\alpha_i))^T$ and $\mathbf{u}(L(\alpha_j))^T$ differ on their first coordinate. Conclude that the codeword $\mathbf{u}G_i$ of $\mathcal{C}_i$ does not belong to $\mathcal{C}_j$.

Problem 12.10 The purpose of this problem is to exhibit an ensemble of linear codes that attain the Gilbert–Varshamov bound. The size of the ensemble allows it to be used instead of Wozencraft codes in the construction of Justesen codes.

Let $F = \mathrm{GF}(q)$ and let m and n be positive integers such that $m < n$. Given a monic irreducible polynomial $f(x) = f_0 + f_1 x + \ldots + f_m x^m$ of degree m over F, let $\mathcal{C}_{\mathrm{sc}}(f)$ be the linear code of length n over F with a generator matrix

$$\begin{pmatrix} f_0 & f_1 & \ldots & f_m & & & & \\ & f_0 & f_1 & \ldots & f_m & & \mathbf{0} & \\ \mathbf{0} & & \ddots & \ddots & \cdots & \ddots & & \\ & & & f_0 & f_1 & \ldots & f_m & \end{pmatrix} .$$

Denote by $X = X_F(n, m)$ the set of all codes $\mathcal{C}_{\mathrm{sc}}(f)$ over F where $f(x)$ ranges over all monic irreducible polynomials of degree m over F.

1. Show that the rate R of each code in X is $1 - (m/n)$.

2. Let $c(x)$ be a polynomial in $F_n[x]$. Show that $c(x)$ is a codeword in $\mathcal{C}_{\mathrm{sc}}(f)$ if and only if $f(x)$ divides $c(x)$. Conclude that when $f(x) \ne x$, the code $\mathcal{C}_{\mathrm{sc}}(f)$ can be obtained by shortening of a cyclic code over F (hence the subscript "sc"; see Problem 2.14).

3. Show that two codes $\mathcal{C}_{\mathrm{sc}}(f)$ and $\mathcal{C}_{\mathrm{sc}}(g)$ are equal if and only if $f(x) = g(x)$.

4. Show that
$$(q^m/m) - q^{m/2} < |X| \le q^m/m \le q^{n(1-R)} .$$
Hint: Recall from Section 7.2 the formula for the number of monic irreducible polynomials of degree m over F.

5. Show that every nonzero word in F^n belongs to at most $\lfloor (n-1)/m \rfloor$ codes in X. Are there words in F^n for which this bound is attained?

   Hint: Use part 2.

6. Show that if d is a positive integer such that
$$q^m - q^{m/2} m \ge (n-1) \cdot (V_q(n, d-1) - 1) ,$$

then there is a code in X with minimum distance at least d. Relate this result to the Gilbert–Varshamov bound.

Hint: Use part 5 to bound from above the number of codes in X whose minimum distance is less than d.

Problem 12.11 (Goppa codes) Let F and Φ be the fields $\mathrm{GF}(q)$ and $\mathrm{GF}(q^m)$, respectively, and fix $\alpha_1, \alpha_2, \ldots, \alpha_n$ to be some nonzero distinct elements in Φ. Let D be a positive integer such that $m(D-1) < n$, and suppose that $f(x)$ is a polynomial of degree $D-1$ over Φ such that $f(\alpha_j) \neq 0$ for all $1 \leq j \leq n$. Denote by $\mathcal{C}_{\mathrm{GRS}}(f)$ the $[N=n, K, D]$ GRS code over Φ with code locators $\alpha_1, \alpha_2, \ldots, \alpha_n$ and column multipliers $v_1, v_2, \ldots, v_n$, where

$$v_j = \frac{1}{f(\alpha_j)}, \quad 1 \leq j \leq n .$$

The respective alternant code

$$\mathcal{C}_{\mathrm{alt}}(f) = \mathcal{C}_{\mathrm{GRS}}(f) \cap F^n$$

is called a *Goppa code*. Assume hereafter in this problem that $D > 1$ and denote by $\Gamma = \Gamma_F(n, m, D)$ the multi-set of all Goppa codes $\mathcal{C}_{\mathrm{alt}}(f)$, where $f(x)$ ranges over all *monic irreducible* polynomials of degree $D-1$ over Φ.

1. Show that the rate of each code in Γ is at least

$$R = 1 - \frac{m(D-1)}{n} .$$

2. Are the codes $\mathcal{C}_{\mathrm{alt}}(f)$ and $\mathcal{C}_{\mathrm{alt}}(g)$ necessarily distinct for distinct monic irreducible polynomials f and g?

   Hint: Consider the case where $D = 2$, $q > n > m > 1$, the elements $\alpha_1, \alpha_2, \ldots, \alpha_n$ are all in F, and, for some $\beta \in \Phi \setminus F$,

$$f(x) = x + \beta \quad \text{and} \quad g(x) = x + \beta^q .$$

3. Show that the size of Γ (counting multiplicity) satisfies

$$|\Gamma| \leq \frac{q^{m(D-1)}}{D-1} \leq q^{n(1-R)} ,$$

   where R is the value in part 1.

4. Let $\mathbf{c}$ be a nonzero word in F^n. Show that $\mathbf{c}$ belongs to at most $\lfloor (n-1)/(D-1) \rfloor$ codes in Γ.

   Hint: Recall from part 6 of Problem 5.11 that a word $(c_1 \; c_2 \; \ldots \; c_n) \in \Phi^n$ is a codeword of $\mathcal{C}_{\mathrm{GRS}}(f)$ if and only if $f(x)$ divides the polynomial

$$\sum_{j=1}^{n} c_j \prod_{\substack{1 \leq s \leq n: \\ s \neq j}} (x - \alpha_s) .$$

5. Show that if d is a positive integer such that

$$q^{m(D-1)} - q^{m(D-1)/2}(D-1) \geq (n-1) \cdot (V_q(n, d-1) - 1) ,$$

then there is a code in Γ with minimum distance at least d. Relate this result to the Gilbert–Varshamov bound.

Hint: See part 6 of Problem 12.10.

[Section 12.5]

Problem 12.12 The purpose of this problem is to show that the Shannon Coding Theorem for the memoryless q-ary symmetric channel can be attained by the ensemble $X = X_F(n, m)$ in Problem 12.10.

The notation in Problem 12.10 is also used here. Denote by $\mathcal{S}_q(n, t)$ the set of all words in F^n whose Hamming weight is at most t.

1. For a code $\mathcal{C}$ in X and a word $\mathbf{e} \in F^n$, let $\mathrm{P_{err}}(\mathcal{C}|\mathbf{e})$ be the decoding error probability of a nearest-codeword decoder $\mathcal{D} : F^n \to \mathcal{C}$, conditioned on the error word being $\mathbf{e}$; that is,

$$\mathrm{P_{err}}(\mathcal{C}|\mathbf{e}) = \begin{cases} 1 & \text{if there is } \mathbf{c} \in \mathcal{C} \text{ such that } \mathcal{D}(\mathbf{c}+\mathbf{e}) \neq \mathbf{c} \\ 0 & \text{otherwise} \end{cases}$$

(compare with the definitions in Section 4.7). Show that for every $\mathbf{e} \in \mathcal{S}_q(n, t)$,

$$\sum_{\mathcal{C} \in X} \mathrm{P_{err}}(\mathcal{C}|\mathbf{e}) \leq \lfloor (n-1)/m \rfloor \cdot V_q(n, t) .$$

Hint: Bound from above the number of codes in X for which $\mathbf{e}$ is not a coset leader.

2. For a code $\mathcal{C}$ in X, let $\mathrm{P_{err}}(\mathcal{C}|\mathcal{S}_q(n, t))$ be the decoding error probability of a nearest-codeword decoder with respect to a given additive channel (F, F, Prob), conditioned on the error word $\mathbf{e}$ being in $\mathcal{S}_q(n, t)$. Show that

$$\frac{1}{|X|} \cdot \sum_{\mathcal{C} \in X} \mathrm{P_{err}}(\mathcal{C}|\mathcal{S}_q(n, t)) \leq \frac{\lfloor (n-1)/m \rfloor \cdot V_q(n, t)}{|X|}$$

(compare with Lemma 4.16).

3. For a code $\mathcal{C}$ in X, denote by $\mathrm{P_{err}}(\mathcal{C})$ the decoding error probability of a nearest-codeword decoder with respect to the q-ary symmetric channel with crossover probability $p \in (0, 1-(1/q))$. Suppose that the rate R of the codes in X is less than $1 - \mathsf{H}_q(p)$. Show that

$$\frac{1}{|X|} \cdot \sum_{\mathcal{C} \in X} \mathrm{P_{err}}(\mathcal{C}) \leq q^{-n(\mathrm{E}_q(p,R)-o(1))} ,$$

where

$$\mathrm{E}_q(p, R) = 1 - \mathsf{H}_q(\theta) - R$$

and

$$\theta = \frac{\log_q(1-p) + 1 - R}{\log_q(1-p) - \log_q(p/(q-1))} .$$

Here $o(1)$ stands for an expression that goes to zero as n goes to infinity (this expression may depend on q, p, or R).

Hint: Use part 2 similarly to the way Lemma 4.16 is used to prove Theorem 4.17.

Problem 12.13 The purpose of this problem is to show that the Shannon Coding Theorem for the memoryless q-ary erasure channel can be attained by the ensemble $X = X_F(n, m)$ in Problem 12.10. The notation therein is also used here, and the erasure channel is characterized by its input alphabet $F = \mathrm{GF}(q)$, output alphabet $\Phi = F \cup \{?\}$, and erasure probability p (see also Problem 4.33).

For a code $\mathcal{C}$ in X, let the decoder $\mathcal{D} : \Phi^n \to \mathcal{C} \cup \{\text{“e”}\}$ be defined by

$$\mathcal{D}(\mathbf{y}) = \begin{cases} \mathbf{c} & \text{if } \mathbf{y} \text{ agrees with a unique } \mathbf{c} \in \mathcal{C} \text{ on the non-erased locations} \\ \text{“e”} & \text{otherwise} \end{cases} .$$

For a set $J \subseteq \{1, 2, \ldots, n\}$, let $\mathrm{P}_{\mathrm{err}}(\mathcal{C}|J)$ be the decoding error probability of $\mathcal{D}$ with respect to the erasure channel, conditioned on the erasures being indexed by J.

1. Show that $\mathrm{P}_{\mathrm{err}}(\mathcal{C}|J) = 0$ if none of the supports of the nonzero codewords in $\mathcal{C}$ is contained in J.

2. Show that
$$\sum_{\mathcal{C} \in X} \mathrm{P}_{\mathrm{err}}(\mathcal{C}|J) \le \left\lfloor \frac{n-1}{m} \right\rfloor \cdot \frac{q^{|J|} - 1}{q - 1} .$$

3. For a code $\mathcal{C}$ in X, let $\mathrm{P}_{\mathrm{err}}(\mathcal{C})$ be the decoding error probability of the decoder $\mathcal{D}$ with respect to the erasure channel. Suppose that the rate R of the codes in X is less than $1-p$. Show that
$$\frac{1}{|X|} \cdot \sum_{\mathcal{C} \in X} \mathrm{P}_{\mathrm{err}}(\mathcal{C}) \le q^{-n(\mathsf{D}_q(\theta \| p) - o(1))} ,$$
where
$$\mathsf{D}_q(\theta \| p) = \theta \log_q \left(\frac{\theta}{p} \right) + (1-\theta) \log_q \left(\frac{1-\theta}{1-p} \right)$$
(which is the information divergence defined in Section 4.6) and θ is taken as the solution to
$$\theta + \mathsf{D}_q(\theta \| p) = 1 - R$$
in the open interval $(p, 1-R)$ (compare with part 3 of Problem 4.33).

Problem 12.14 Show that part 3 of Problem 12.12 and part 3 of Problem 12.13 hold when X is taken as the set of $[n, k]$ Wozencraft codes over $F = \mathrm{GF}(q)$.

Problem 12.15 Show that part 3 of Problem 12.12 and part 3 of Problem 12.13 hold when X is taken as the multi-set $\Gamma_F(n, k)$ in Problem 12.11.

Notes

[Section 12.1]

Concatenated codes were introduced by Forney in [129]. A comprehensive survey on concatenated codes is provided by Dumer in [109].

[Section 12.2]

The generalized minimum distance (GMD) decoding algorithm is due to Forney [129], [130]; see also Zyablov [405]. The decoder in Figure 12.1 is, in fact, a special case of Forney's setting where the decoder for the inner code is a nearest-codeword decoder for the Hamming metric (see also Reddy and Robinson [287] and Weldon [382]). In the more general framework of GMD decoding, the inner decoder provides for each decoded sub-block $\mathbf{y}_j$ a *reliability score* $\beta_j \in [0,1]$ of the computed codeword $\hat{\mathbf{c}}_j \in \mathcal{C}_{\mathrm{in}}$ (higher values of β_j mean that the coordinate is more reliable). Given a codeword $\mathbf{c} = (\mathbf{c}_1 \,|\, \mathbf{c}_2 \,|\, \ldots \,|\, \mathbf{c}_N) \in \mathcal{C}_{\mathrm{cont}}$, let

$$\eta_j = \eta_j(\mathbf{c}) = \begin{cases} 0 & \text{if } \hat{\mathbf{c}}_j = \mathbf{c}_j \\ 1 & \text{if } \hat{\mathbf{c}}_j \neq \mathbf{c}_j \end{cases} .$$

It is shown by Forney that there can be at most one codeword $\mathbf{c} \in \mathcal{C}_{\mathrm{out}}$ for which

$$\sum_{j=1}^{N} (-1)^{\eta_j} \beta_j > N - D ;$$

furthermore, if such a codeword $\mathbf{c}$ exists, then it will be found by iteratively changing sub-blocks $\hat{\mathbf{c}}_j$ into erasures, starting with the least reliable, and then applying an error–erasure decoder for the outer code [129, Theorems 3.1 and 3.2]. The reliability score that corresponds to the decoder in Figure 12.1 is $\beta_j = 1 - \min\{2\mathsf{d}(\mathbf{y}_j, \hat{\mathbf{c}}_j)/d, 1\}$.

[Section 12.3]

The Zyablov bound was obtained in [404]. In Section 13.8, we present a construction of codes that approach the Zyablov bound, with decoding complexity that grows *linearly* with the code length (the multiplying constant of the linear term will depend on how far we are from the Zyablov bound); see Examples 13.8 and 13.9.

[Section 12.4]

Justesen codes were presented in [199]. Improvements over Justesen codes for the low-rate range were obtained by Alon *et al.* [10], Shen [333], Sugiyama *et al.* [347], and Weldon [383], [384].

The Wozencraft ensemble is mentioned by Massey in [253, Section 2.5]. There are other known constructions of ensembles of codes—such as the ensembles of shortened cyclic codes and of Goppa codes in Problems 12.10 and 12.11, respectively—that can be used in lieu of Wozencraft codes in Justesen codes. Problem 12.10 is taken from Kasami [206] and Problem 12.11 is from MacWilliams and Sloane [249,

p. 350]. For more on Goppa codes, see Goppa [156]–[158] and MacWilliams and Sloane [249, Chapter 12].

With minor changes, the GMD decoder in Figure 12.1 is applicable also to the decoding of Justesen codes [199]. Step 1a is applied to each sub-block $\mathbf{y}_j$ using a nearest-codeword decoder for the respective code $\mathcal{C}_j \in \mathcal{W}_F(n,k)$. The value d is taken as the typical minimum distance of the codes in $\mathcal{W}_F(n,k)$ and is computed from the Gilbert–Varshamov bound.

In addition to Justesen codes, various generalizations of concatenated codes were suggested and studied by Blokh and Zyablov [57], [58], Hirasawa *et al.* [178], [179], Kasahara *et al.* [205], Sugiyama *et al.* [349], [350], Zinov'ev [397], [398], and Zinov'ev and Zyablov [400]–[402]. Some of these generalizations yield polynomial-time constructions that exceed the Zyablov bound. In particular, the generalization due to Blokh and Zyablov [58] gets arbitrarily close (yet with increasing complexity) to the *Blokh–Zyablov bound*, which is given by

$$R_{\mathrm{BZ}}(\delta, q) = 1 - \mathsf{H}_q(\delta) - \delta \int_0^{1-\mathsf{H}_q(\delta)} \frac{dx}{\mathsf{H}_q^{-1}(1-x)} .$$

Blokh and Zyablov also showed in [56] that there exists a family of binary linearly-concatenated codes with varying inner codes that attains the Gilbert–Varshamov bound. The same authors [58] and Thommesen [362], [363] then showed that the Gilbert–Varshamov bound can be attained also when the outer code is taken as a prescribed (not randomly chosen) MDS code. The case where the inner code is fixed while the outer code is randomly chosen was studied by Barg *et al.* [28].

[Section 12.5]

The exposition in this section follows Dumer's survey [109]. In his monograph on concatenated codes [129], Forney showed that linearly-concatenated codes approach the capacity of the q-ary symmetric channel and that GMD decoding yields a decoding error probability that decreases exponentially with the code length. As an outer code, Forney used an MDS code, which was much shorter than the outer code used herein. Thus, the search for the best inner code in Forney's analysis made the overall complexity super-polynomial in the code length. On the other hand, MDS outer codes yield a better exponential decay of the decoding error probability: the error exponent in (12.15) now becomes

$$\mathrm{E}_q^*(p, \mathcal{R}) = \max_{\mathcal{R} \le r \le 1-\mathsf{H}_q(p)} \tfrac{1}{2}\mathrm{E}_q(p, r)(1 - (\mathcal{R}/r)) .$$

In addition, by using MDS codes we get a denser range of code lengths for which the concatenated code construction can be realized.

While the construction herein uses an error-only decoder for the outer code, we could use an error–erasure decoder instead, where the erasures are flagged in locations where the decoder of the inner code detects a number of errors that exceeds a certain threshold (see Problem 4.31). Forney's analysis shows that such a GMD decoder yields a better error exponent. See also Blokh and Zyablov [56], [58] and Thommesen [363].

Instead of searching for the best inner code among all linear codes, one can try to identify much smaller ensembles with average decoding error probability that behaves like the ensemble of all linear codes. This approach was investigated by Delsarte and Piret in [99]. Indeed, the Wozencraft ensemble $\mathcal{W}_F(n,k)$, as well as the ensembles in Problems 12.10 and 12.11, approach the capacity of the q-ary symmetric channel with decoding error probability that decays exponentially with the code length; see Problems 12.12, 12.14, and 12.15 (our definition of the ensemble $\mathcal{W}_F(n,k)$ restricts the rate k/n to be at least $1/2$, but, as shown by Weldon in [383], it is rather easy to generalize the definition also to lower rates). When using the best code in such ensembles as an inner code, we can use an MDS code as an outer code, while still keeping the encoding complexity only polynomially large in the code length. As was the case in Section 12.3, the resulting encoding complexity is some power c of the code length, where c goes to infinity when the rate goes to zero. Such a complexity increase is circumvented by Delsarte and Piret in [99] at the expense of a poorer error exponent. In fact, if the error exponent can be compromised, then the time complexity in Theorem 12.7 can be improved by inserting another level of concatenation: the construction of Theorem 12.7 is used as a linear $[n_i, k_i]$ inner code over $F = \mathrm{GF}(q)$, with an $[N_i, K_i]$ outer GRS code over $\mathrm{GF}(q^{k_i})$. The complexity of encoding and decoding will then be dictated by the GRS code and will amount to $O(N_i \log^2 N_i \log\log N_i)$ field operations in $\mathrm{GF}(q^{k_i})$, with each such operation being implemented using $O(n_i \log n_i \log\log n_i)$ field operations in F (see the notes on Sections 3.3 and 6.6).

In Example 13.10 in Section 13.8, we present an improvement on Theorem 12.7, whereby the expression for the decoding complexity grows only linearly with the code length (the multiplying constant in that expression grows as the rate approaches the capacity of the q-ary symmetric channel).

We point out that Theorem 12.7 is known to hold with a better lower bound on the error exponent in part (iv) of that theorem (see [99]).

A counterpart of Theorem 12.7 (with respective capacity and error exponent values) can be stated also for the memoryless q-ary erasure channel. An attaining construction for this channel can be obtained by using an $[N, K]$ GRS code as an outer MDS code and the best code in $\mathcal{W}_F(n,k)$ (say) as an inner code (see Problems 12.13–12.15). In the first decoding step, the inner decoder attempts to recover the erasures in each sub-block of length n through solving linear equations. Sub-blocks in which the solution is not unique are then flagged as erasures to the outer decoder, which, in turn, can recover them using Euclid's algorithm.

For constructions of concatenated codes for arbitrary discrete memoryless channels, see Uyematsu and Okamoto [369].

Chapter 13

Graph Codes

Concatenated codes are examples of compound constructions, as they are obtained by combining two codes—an inner code and an outer code—with a certain relationship between their parameters. This chapter presents another compound construction, now combining an (inner) code $\mathcal{C}$ over some alphabet F with an undirected graph $\mathcal{G} = (V, E)$. In the resulting construction, which we refer to as a graph code and denote by $(\mathcal{G}, \mathcal{C})$, the degrees of all the vertices in $\mathcal{G}$ need to be equal to the length of $\mathcal{C}$, and the code $(\mathcal{G}, \mathcal{C})$ consists of all the words of length $|E|$ over F in which certain subwords, whose locations are defined by $\mathcal{G}$, belong to $\mathcal{C}$. The main result to be obtained in this chapter is that there exist explicit constructions of graph codes that can be decoded in linear-time complexity, such that the code rate is bounded away from zero, and so is the fraction of symbols that are allowed to be in error.

We start this chapter by reviewing several concepts from graph theory. We then focus on regular graphs, i.e., graphs in which all vertices have the same degree. We will be interested in the expansion properties of such graphs; namely, how the number of outgoing edges from a given set of vertices depends on the size of this set. We present a lower bound on this number in terms of the second largest eigenvalue of the adjacency matrix of the graph: a small value of this eigenvalue implies a large expansion.

Two families of regular graphs with good expansion properties will be described. The first construction is based on binary linear codes whose minimum distance is close to half the code length; while the analysis of the resulting graphs is fairly elementary, the number of vertices in these graphs for fixed degree cannot grow arbitrarily. A second construction to be shown—due to Lubotzky, Phillips, Sarnak (LPS), and Margulis—is essentially optimal in that it attains an asymptotic lower bound on the second largest eigenvalue; furthermore, for a given degree, this construction yields infinitely many graphs. A complete analysis of the LPS construction is beyond our scope herein; still, we will make use of these graphs to construct good families of graph codes.

We then turn to analyzing the parameters of graph codes $(\mathcal{G}, \mathcal{C})$, making use of the tools that we will have developed for bounding the expansion of regular graphs. The same tools will be used also in analyzing the performance of an iterative decoding algorithm which we present for a special class of graph codes. In particular, we show that the algorithm is capable of correcting any error pattern whose Hamming weight does not exceed approximately one quarter of the lower bound on the minimum distance of $(\mathcal{G}, \mathcal{C})$. A complexity analysis of this algorithm reveals that its running time is proportional to the number of vertices in $\mathcal{G}$, under the assumption that the code $\mathcal{C}$ is fixed. We end this chapter by viewing graph codes through the lens of concatenated codes. Following this approach, we show how incorporating GMD decoding into the iterative decoder enhances the latter so that it can correct twice as many errors. Furthermore, we show that the Singleton bound can be approached by the outer code of concatenated codes that are derived from certain generalizations of graph codes.

13.1 Basic concepts from graph theory

This section and Sections 13.2 and 13.3 provide some background material that will be used throughout the chapter.

An *(undirected simple finite) graph* is a pair (V, E), where V is a nonempty finite set of *vertices* and E is a (possibly empty) set of *edges*, where by an edge we mean a subset of V of size (exactly) 2. (The adjective "simple" indicates that the graph has neither parallel edges nor self-loops; that is, E is a proper set—rather than a multi-set—and none of its elements has size 1.)

Let u be a vertex in a graph $\mathcal{G} = (V, E)$. A vertex $v \in V$ is said to be *adjacent* to u if $\{u, v\} \in E$. The set of all vertices in V that are adjacent to u is called the *neighborhood* of u in $\mathcal{G}$ and is denoted by $\mathcal{N}(u)$. The *degree* of u in $\mathcal{G}$, denoted by $\deg_{\mathcal{G}}(u)$, is defined as the size of $\mathcal{N}(u)$. We extend the term neighborhood also to any subset $S \subseteq V$ by

$$\mathcal{N}(S) = \bigcup_{u \in S} \mathcal{N}(u) .$$

Example 13.1 Figure 13.1 depicts a graph $\mathcal{G} = (V, E)$ where

$$V = \{00, 01, 10, 11\} \tag{13.1}$$

and

$$E = \Big\{\{00, 01\}, \{00, 10\}, \{01, 10\}, \{01, 11\}, \{10, 11\}\Big\} . \tag{13.2}$$

Two vertices—namely, 01 and 10—have degree 3, while 00 and 11 have degree 2. □

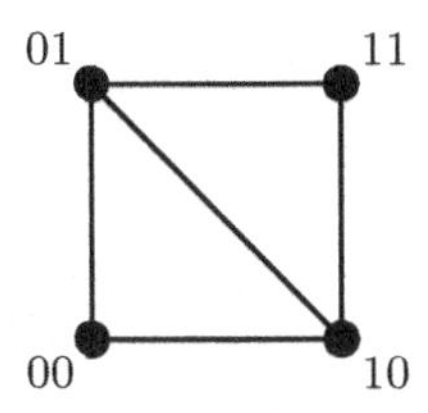

Figure 13.1. Graph with four vertices.

Given a graph $\mathcal{G} = (V, E)$, an edge $e \in E$ is *incident* with a vertex $u \in V$ if $e = \{u, v\}$ for some $v \in V$. The set of edges that are incident with u will be denoted by $E(u)$, and we have $|E(u)| = |\mathcal{N}(u)| = \deg_{\mathcal{G}}(u)$. If an edge e is incident with a vertex u then u is called an *endpoint* of e. For two subsets $S, T \subseteq V$ (not necessarily disjoint), we denote by $E_{S,T}$ the set of all edges that have an endpoint in S and an endpoint in T, that is,

$$E_{S,T} = \Big\{ e \in E \; : \; |e \cap S| > 0 \text{ and } |e \cap T| > 0 \Big\} .$$

The *edge cut* associated with a subset $S \subseteq V$, denoted by $\partial(S)$, is defined as

$$\partial(S) = E_{S,(V \setminus S)} \cdot$$

A *subgraph* of a graph $\mathcal{G} = (V, E)$ is a graph $\mathcal{G}' = (V', E')$ where $V' \subseteq V$ and $E' \subseteq E$. Given a nonempty subset S of V, the *induced* subgraph of $\mathcal{G}$ on S is the subgraph $\mathcal{G}_S = (S, E_{S,S})$ of $\mathcal{G}$.

Let u and v be vertices in a graph $\mathcal{G} = (V, E)$. A *path* of length $\ell > 0$ from u to v in $\mathcal{G}$ is a finite sequence of edges

$$\{u, u_1\}\{u_1, u_2\}\{u_2, u_3\} \ldots \{u_{\ell-2}, u_{\ell-1}\}\{u_{\ell-1}, v\} , \tag{13.3}$$

where $u_1, u_2, \ldots, u_{\ell-1} \in V$ (a path of length 1 from u to v is just an edge $\{u, v\}$, and a path of length 0 is defined formally for the case $u = v$ as consisting of one vertex—u—with no edges). A path (13.3) is called a *cycle* if $u = v$. A graph $\mathcal{G}$ is *connected* if for every two distinct vertices u and v in $\mathcal{G}$ there is a path from u to v in $\mathcal{G}$. The *distance* between two vertices u and v in $\mathcal{G}$ is the smallest length of any path from u to v in $\mathcal{G}$. We denote the distance by $\mathsf{d}_{\mathcal{G}}(u, v)$ and define it to be zero if $u = v$ and infinity if $\mathcal{G}$ contains no path from u to v. The *diameter* of $\mathcal{G} = (V, E)$, denoted by $\operatorname{diam}(\mathcal{G})$, is defined as

$$\operatorname{diam}(\mathcal{G}) = \max_{u,v \in V} \mathsf{d}_{\mathcal{G}}(u, v) .$$

Clearly, $\mathcal{G}$ is connected if and only if $\operatorname{diam}(\mathcal{G}) < \infty$.

A graph $\mathcal{G} = (V, E)$ is *bipartite* if V can be partitioned into two subsets, V' and V'', such that every edge $e \in E$ has one endpoint in V' and one endpoint in V''. We will denote a bipartite graph by $(V' : V'', E)$.

Lemma 13.1 *A graph $\mathcal{G}$ is bipartite if and only if it contains no cycles of odd length.*

The proof is left as an exercise (Problem 13.5).

Example 13.2 The graph in Figure 13.2 is bipartite, with

$$V' = \{000, 011, 101, 110\} \quad \text{and} \quad V'' = \{001, 010, 100, 111\} \tag{13.4}$$

(the set V' consists of the even-weight binary triples, and these vertices are marked in the figure distinctively from the elements of V''). On the other hand, the graph in Figure 13.1 is non-bipartite. □

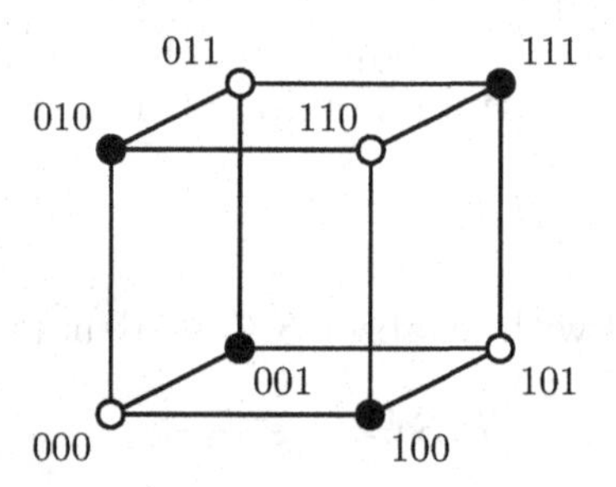

Figure 13.2. Bipartite graph.

The *adjacency matrix* of a graph $\mathcal{G} = (V, E)$, denoted by $A_\mathcal{G}$, is a $|V| \times |V|$ integer matrix over $\{0, 1\}$ whose rows and columns are indexed by the set V, and for every $u, v \in V$, the entry in $A_\mathcal{G}$ that is indexed by (u, v) is given by

$$(A_\mathcal{G})_{u,v} = \begin{cases} 1 & \text{if } \{u, v\} \in E \\ 0 & \text{otherwise} \end{cases} .$$

(Notice that our definition of $A_\mathcal{G}$ does not assume any specific ordering on V, even though we may sometimes prefer certain orderings for the sake of convenience. All of the properties that we will be interested in will be independent of the choice of such an ordering—as long as the same order is applied to both the rows and the columns of $A_\mathcal{G}$.)

We remark that $A_\mathcal{G}$ is a symmetric matrix; as such, its eigenvalues are real and its set of eigenvectors spans $\mathbb{R}^{|V|}$. Furthermore, eigenvectors that are associated with distinct eigenvalues are orthogonal, which means that we can find an orthonormal basis of $\mathbb{R}^{|V|}$ that consists of eigenvectors of $A_\mathcal{G}$. We will make use of these facts in the sequel.

When $\mathcal{G}$ is a bipartite graph $(V' : V'', E)$, the adjacency matrix takes the form

$$A_{\mathcal{G}} = \left(\begin{array}{c|c} 0 & X_{\mathcal{G}} \\ \hline X_{\mathcal{G}}^T & 0 \end{array} \right) , \tag{13.5}$$

where $X_{\mathcal{G}}$ is a $|V'| \times |V''|$ *transfer matrix*, whose rows and columns are indexed by V' and V'', respectively, and $(X_{\mathcal{G}})_{u,v} = 1$ if and only if $\{u, v\} \in E$ (when writing (13.5), we are assuming an ordering on $V' \cup V''$ where the vertices in V' precede those in V'').

The *incidence matrix* of $\mathcal{G} = (V, E)$, denoted by $C_{\mathcal{G}}$, is defined as the $|E| \times |V|$ integer matrix over $\{0, 1\}$ whose rows and columns are indexed by E and V, respectively, and for every $e \in E$ and $v \in V$,

$$(C_{\mathcal{G}})_{e,v} = \begin{cases} 1 & \text{if } v \in e \\ 0 & \text{otherwise} \end{cases} .$$

We also define the matrices $L_{\mathcal{G}}^+$ and $L_{\mathcal{G}}^-$ by

$$L_{\mathcal{G}}^{\pm} = D_{\mathcal{G}} \pm A_{\mathcal{G}} ,$$

where $D_{\mathcal{G}}$ is a $|V| \times |V|$ diagonal matrix whose entries along the diagonal are

$$(D_{\mathcal{G}})_{u,u} = \deg_{\mathcal{G}}(u) \qquad \text{for every } u \in V .$$

The matrix $L_{\mathcal{G}}^-$ is known as the *Laplace matrix* of $\mathcal{G}$, and we will sometimes use the notation $L_{\mathcal{G}}$ (with the superscript "$-$" omitted) when referring to this matrix.

Example 13.3 Let $\mathcal{G} = (V, E)$ be the graph in Example 13.1. The adjacency matrix and incidence matrix of $\mathcal{G}$ are given by

$$A_{\mathcal{G}} = \begin{pmatrix} 0 & 1 & 1 & 0 \\ 1 & 0 & 1 & 1 \\ 1 & 1 & 0 & 1 \\ 0 & 1 & 1 & 0 \end{pmatrix} \qquad \text{and} \qquad C_{\mathcal{G}} = \begin{pmatrix} 1 & 1 & 0 & 0 \\ 1 & 0 & 1 & 0 \\ 0 & 1 & 1 & 0 \\ 0 & 1 & 0 & 1 \\ 0 & 0 & 1 & 1 \end{pmatrix} ,$$

respectively, where we have arranged the rows and columns in each matrix according to the order in which the vertices and edges are written in (13.1) and (13.2). □

Example 13.4 The adjacency matrix of the bipartite graph $\mathcal{G} = (V' : V'', E)$ in Example 13.2 is given by (13.5), where

$$X_{\mathcal{G}} = \begin{pmatrix} 1 & 1 & 1 & 0 \\ 1 & 1 & 0 & 1 \\ 1 & 0 & 1 & 1 \\ 0 & 1 & 1 & 1 \end{pmatrix} ,$$

assuming the ordering in (13.4). □

Let $\mathcal{G} = (V, E)$ be a graph. An *orientation* on $\mathcal{G}$ is an ordering on the two endpoints of each edge $e \in E$, thus making the graph effectively *directed* (this ordering can be set independently of the ordering on V mentioned earlier in connection with the indexing of the rows and columns of $A_{\mathcal{G}}$). A graph $\mathcal{G}$ with a particular orientation will be denoted by $\vec{\mathcal{G}}$.

The incidence matrix of a directed (i.e., oriented) graph $\vec{\mathcal{G}}$, denoted by $C_{\vec{\mathcal{G}}}$, is the $|E| \times |V|$ integer matrix over $\{0, 1, -1\}$ whose entries are given by

$$(C_{\vec{\mathcal{G}}})_{e,v} = \begin{cases} 1 & \text{if } v \text{ is the smallest endpoint of } e \\ -1 & \text{if } v \text{ is the largest endpoint of } e \\ 0 & \text{if } v \notin e \end{cases} .$$

The following lemma can be easily verified (Problem 13.8).

Lemma 13.2 *For every graph $\mathcal{G} = (V, E)$ and every orientation on $\mathcal{G}$,*

$$L_{\mathcal{G}}^{-} = C_{\vec{\mathcal{G}}}^{T} C_{\vec{\mathcal{G}}} \quad \textit{and} \quad L_{\mathcal{G}}^{+} = C_{\mathcal{G}}^{T} C_{\mathcal{G}} .$$

(Lemma 13.2 holds also for the degenerate case $|E| = 0$, if we define the product of two "matrices" of respective orders $|V| \times 0$ and $0 \times |V|$ to be the $|V| \times |V|$ all-zero matrix.)

For real column vectors $\mathbf{x}, \mathbf{y} \in \mathbb{R}^m$, we will use the notation $\langle \mathbf{x}, \mathbf{y} \rangle$ for the scalar product $\mathbf{x}^T \mathbf{y}$; the norm $\sqrt{\langle \mathbf{x}, \mathbf{x} \rangle}$ will be denoted by $\|\mathbf{x}\|$. In many cases, the entries of a vector $\mathbf{x}$ will be indexed by the set of vertices V of a graph; we will then write $\mathbf{x} = (x_u)_{u \in V}$ if we want to specify these entries.

Corollary 13.3 *For every graph $\mathcal{G} = (V, E)$ and every real vector $\mathbf{x} = (x_u)_{u \in V}$,*

$$\langle \mathbf{x}, L_{\mathcal{G}}^{\pm} \mathbf{x} \rangle = \sum_{\{u,v\} \in E} (x_u \pm x_v)^2 \geq 0 .$$

Proof. Fix some orientation on $\mathcal{G}$. By Lemma 13.2 we have,

$$\langle \mathbf{x}, L_{\mathcal{G}}^{-} \mathbf{x} \rangle = \mathbf{x}^T C_{\vec{\mathcal{G}}}^{T} C_{\vec{\mathcal{G}}} \mathbf{x} = \|C_{\vec{\mathcal{G}}} \mathbf{x}\|^2 = \sum_{\{u,v\} \in E} (x_u - x_v)^2 .$$

Similarly,

$$\langle \mathbf{x}, L_{\mathcal{G}}^{+} \mathbf{x} \rangle = \mathbf{x}^T C_{\mathcal{G}}^{T} C_{\mathcal{G}} \mathbf{x} = \|C_{\mathcal{G}} \mathbf{x}\|^2 = \sum_{\{u,v\} \in E} (x_u + x_v)^2 .$$

□

13.2 Regular graphs

A graph is called *n-regular* if all of its vertices have the same degree $n > 0$.

The next proposition presents properties of the eigenvalues of the adjacency matrix of a (connected) n-regular graph.

Proposition 13.4 *Let $\mathcal{G} = (V, E)$ be an n-regular graph and let $\lambda_1 \geq \lambda_2 \geq \ldots \geq \lambda_{|V|}$ be the eigenvalues of $A_{\mathcal{G}}$. Then the following conditions hold:*

(i) $\lambda_1 = n$ and the all-one vector $\mathbf{1}$ is an eigenvector associated with λ_1.

(ii) If $\mathcal{G}$ is connected and bipartite then $\lambda_{|V|} = -n$ and $|\lambda_i| < n$ for $1 < i < |V|$.

(iii) If $\mathcal{G}$ is connected and non-bipartite then $|\lambda_i| < n$ for $1 < i \leq |V|$.

Proof. The sum of entries along each row in $A_{\mathcal{G}}$ is n; hence, $A_{\mathcal{G}}\mathbf{1} = n \cdot \mathbf{1}$, i.e., $\mathbf{1}$ is an eigenvector of $A_{\mathcal{G}}$ associated with the eigenvalue n.

For an n-regular graph $\mathcal{G}$ we have

$$L_{\mathcal{G}} = L_{\mathcal{G}}^{-} = n \cdot I - A_{\mathcal{G}} ,$$

where I is the $|V| \times |V|$ identity matrix; therefore, the eigenvalues $\mu_1 \leq \mu_2 \leq \ldots \leq \mu_{|V|}$ of $L_{\mathcal{G}}$ are related to those of $A_{\mathcal{G}}$ by

$$\mu_i = n - \lambda_i , \quad 1 \leq i \leq |V| .$$

Thus, part (i) will be proved once we show that $\mu \geq 0$ for every eigenvalue μ of $L_{\mathcal{G}}$.

Let $\mathbf{x} = (x_u)_{u \in V}$ be a real eigenvector of $L_{\mathcal{G}}$ associated with an eigenvalue μ ($\mathbf{x}$ is also an eigenvector of $A_{\mathcal{G}}$ associated with the eigenvalue $n-\mu$). By Corollary 13.3 we have,

$$\mu\|\mathbf{x}\|^2 = \langle \mathbf{x}, \mu\mathbf{x} \rangle = \langle \mathbf{x}, L_{\mathcal{G}}\mathbf{x} \rangle = \sum_{\{u,v\} \in E} (x_u - x_v)^2 \geq 0 , \tag{13.6}$$

thereby proving part (i).

In the remaining part of the proof we assume that $\mathcal{G}$ is connected. This, in turn, implies that the inequality holds in (13.6) with equality if and only if all the entries of $\mathbf{x}$ are equal.

As our next step, we show that $\lambda_i \geq -n$ for every $1 \leq i \leq |V|$, with equality holding if and only if $\mathcal{G}$ is bipartite and $i = |V|$. To this end, we proceed similarly to what we have done in the proof of part (i), except that now we replace the matrix $L_{\mathcal{G}}$ by

$$L_{\mathcal{G}}^{+} = n \cdot I + A_{\mathcal{G}} ,$$

whose eigenvalues are given by $\mu_i = n + \lambda_i$, $1 \le i \le |V|$. Letting $\mathbf{x} = (x_u)_{u \in V}$ be a real eigenvector of $L_\mathcal{G}^+$ associated with an eigenvalue μ of $L_\mathcal{G}^+$, by Corollary 13.3 we have

$$\mu \|\mathbf{x}\|^2 = \langle \mathbf{x}, \mu\mathbf{x} \rangle = \langle \mathbf{x}, L_\mathcal{G}^+ \mathbf{x} \rangle = \sum_{\{u,v\} \in E} (x_u + x_v)^2 \ge 0 , \qquad (13.7)$$

with equality holding if and only if $x_u + x_v = 0$ for every edge $\{u, v\} \in E$. We now distinguish between two cases.

Case 1: $\mathcal{G}$ is bipartite. Here the set V can be partitioned into V' and V'', and equality in (13.7) is attained if and only if $\mathbf{x} = (x_u)_{u \in V}$ is such that x_u is equal to some nonzero constant c for every $u \in V'$ and is equal to $-c$ for every $u \in V''$: it can be easily verified that such a vector is, indeed, an eigenvector of $A_\mathcal{G}$ associated with the eigenvalue $-n$. This proves part (ii).

Case 2: $\mathcal{G}$ is non-bipartite. By Lemma 13.1, $\mathcal{G}$ contains a cycle of odd length ℓ, say,

$$\{u, u_1\}\{u_1, u_2\}\{u_2, u_3\} \ldots \{u_{\ell-2}, u_{\ell-1}\}\{u_{\ell-1}, u_\ell\} ,$$

where $u_\ell = u$; furthermore, by connectivity we can assume that the cycle passes through each vertex of $\mathcal{G}$ at least once. Now, equality in (13.7) can hold only if $x_{u_i} = (-1)^i x_u$ for all $1 \le i \le \ell$. But then $x_u = x_{u_\ell} = (-1)^\ell x_u = -x_u$, thereby implying that $\mathbf{x} = \mathbf{0}$. Hence, no eigenvector of $A_\mathcal{G}$ can satisfy (13.7) with equality. This completes the proof of part (iii). □

Corollary 13.5 *Let $\mathcal{G} = (V, E)$ be an n-regular graph (not necessarily connected). Then*

$$\max_\lambda |\lambda| = n ,$$

where λ ranges over all eigenvalues of $A_\mathcal{G}$.

The proof is left as an exercise (Problem 13.14).

The second largest eigenvalue of the adjacency matrix of an n-regular graph will turn out to be rather useful in our analysis in the sequel. We will denote hereafter the ratio between this eigenvalue and n by $\gamma_\mathcal{G}$; namely, if $\lambda_1 \ge \lambda_2 \ge \ldots \ge \lambda_{|V|}$ are the eigenvalues of $A_\mathcal{G}$, then

$$\gamma_\mathcal{G} = \frac{\lambda_2}{n} .$$

13.3 Graph expansion

Theorem 13.7 below will show that an n-regular graph $\mathcal{G} = (V, E)$ for which $\gamma_\mathcal{G}$ is small has a large *expansion*; namely, every nonempty subset S of V has a "large" edge cut associated with it. The proof of the theorem makes use of the following lemma.

Lemma 13.6 *Let A be an $m \times m$ real symmetric matrix (where $m \geq 2$) with eigenvalues $\lambda_1 \geq \lambda_2 \geq \ldots \geq \lambda_m$ and let $\mathbf{x}_1$ be a real eigenvector associated with λ_1. Suppose that $\mathbf{y} \in \mathbb{R}^m$ is such that $\langle \mathbf{y}, \mathbf{x}_1 \rangle = 0$. Then*

$$\langle \mathbf{y}, A\mathbf{y} \rangle \leq \lambda_2 \|\mathbf{y}\|^2 .$$

Proof. Without loss of generality we can assume that $\|\mathbf{x}_1\| = 1$. For $i = 2, 3, \ldots, m$, let $\mathbf{x}_i$ be a real eigenvector of A associated with λ_i such that the set $\{\mathbf{x}_1, \mathbf{x}_2, \ldots, \mathbf{x}_m\}$ forms an orthonormal basis of $\mathbb{R}^m$. Write

$$\mathbf{y} = \sum_{i=1}^{m} \beta_i \mathbf{x}_i ,$$

where

$$\beta_i = \langle \mathbf{y}, \mathbf{x}_i \rangle , \quad 1 \leq i \leq m .$$

In particular,

$$\beta_1 = \langle \mathbf{y}, \mathbf{x}_1 \rangle = 0 .$$

Hence,

$$\mathbf{y} = \sum_{i=2}^{m} \beta_i \mathbf{x}_i$$

and

$$A\mathbf{y} = \sum_{i=2}^{m} \beta_i A\mathbf{x}_i = \sum_{i=2}^{m} \lambda_i \beta_i \mathbf{x}_i ,$$

from which we obtain

$$\langle \mathbf{y}, A\mathbf{y} \rangle = \Big\langle \sum_{i=2}^{m} \beta_i \mathbf{x}_i, \sum_{i=2}^{m} \lambda_i \beta_i \mathbf{x}_i \Big\rangle = \sum_{i=2}^{m} \lambda_i \beta_i^2 \leq \lambda_2 \sum_{i=2}^{m} \beta_i^2 = \lambda_2 \|\mathbf{y}\|^2 ,$$

as claimed. □

Theorem 13.7 *Let $\mathcal{G} = (V, E)$ be an n-regular graph and S be a subset of V. Then,*

$$|\partial(S)| \geq (1-\gamma_{\mathcal{G}})n \cdot |S| \left(1 - \frac{|S|}{|V|}\right) .$$

Proof. Write $\sigma = |S|/|V|$ and define the vector $\mathbf{y} = (y_u)_{u \in V}$ by

$$y_u = \begin{cases} 1-\sigma & \text{if } u \in S \\ -\sigma & \text{otherwise} \end{cases} , \quad u \in V .$$

It is easy to see that $\langle \mathbf{y}, \mathbf{1} \rangle = 0$ and that $\|\mathbf{y}\|^2 = \sigma(1-\sigma)|V|$. Hence, by Lemma 13.6 we have

$$\langle \mathbf{y}, A_{\mathcal{G}}\mathbf{y} \rangle \leq \gamma_{\mathcal{G}} n \|\mathbf{y}\|^2 = \gamma_{\mathcal{G}} \sigma(1-\sigma) n |V| .$$

Recalling that $L_{\mathcal{G}} = L_{\mathcal{G}}^{-} = n \cdot I - A_{\mathcal{G}}$ we obtain

$$\langle \mathbf{y}, L_{\mathcal{G}}\mathbf{y} \rangle = n\|\mathbf{y}\|^2 - \langle \mathbf{y}, A_{\mathcal{G}}\mathbf{y} \rangle \geq (1-\gamma_{\mathcal{G}})\sigma(1-\sigma) n |V| .$$

On the other hand, by Corollary 13.3 we also have

$$\langle \mathbf{y}, L_{\mathcal{G}}\mathbf{y} \rangle = \sum_{\{u,v\} \in \partial(S)} \underbrace{(y_u - y_v)}_{\pm 1}{}^2 + \sum_{\{u,v\} \in E \setminus \partial(S)} \underbrace{(y_u - y_v)}_{0}{}^2 = |\partial(S)| .$$

The result follows. □

As a counterpart of Theorem 13.7 we have the simple upper bound

$$|\partial(S)| \leq n \cdot |S| .$$

Therefore, when $\gamma_{\mathcal{G}}$ is close to zero and $|S|$ is much smaller than $|V|$, then Theorem 13.7 is essentially tight.

Let $\mathcal{G} = (V, E)$ be a graph and ξ be a nonnegative real. We say that $\mathcal{G}$ is an *(n, ξ)-expander* if it is n-regular and for every $S \subseteq V$,

$$|\partial(S)| \geq \xi \cdot n \cdot |S| \left(1 - \frac{|S|}{|V|}\right) . \tag{13.8}$$

The next corollary follows from this definition and Theorem 13.7.

Corollary 13.8 *Every n-regular graph $\mathcal{G}$ is an (n, ξ)-expander for every $\xi \in [0, 1-\gamma_{\mathcal{G}}]$.*

The following lemma provides an upper bound on the average degree within the induced subgraph of an (n, ξ)-expander on a given nonempty subset of vertices.

Lemma 13.9 *Let $\mathcal{G} = (V, E)$ be an (n, ξ)-expander and S be a nonempty subset of V of size $\sigma|V|$. The average degree of the vertices within the induced subgraph $\mathcal{G}_S$ satisfies*

$$\frac{1}{|S|} \sum_{u \in S} \deg_{\mathcal{G}_S}(u) = \frac{2|E_{S,S}|}{|S|} \leq (\xi \cdot \sigma + 1 - \xi) n .$$

Proof. The (first) equality follows from Problem 13.2. To show the inequality, consider the sum, $\sum_{u \in S} \deg_{\mathcal{G}}(u)$, of the degrees of the vertices of S in $\mathcal{G}$. On the one hand, this sum equals $n \cdot |S|$; on the other hand, it equals $2|E_{S,S}| + |\partial(S)|$. Now combine the equality

$$2|E_{S,S}| + |\partial(S)| = n \cdot |S|$$

with (13.8). □

The last two results yield the next corollary.

Corollary 13.10 *Let $\mathcal{G} = (V, E)$ be an n-regular graph and S be a nonempty subset of V of size $\sigma|V|$. The average degree of the vertices within the induced subgraph $\mathcal{G}_S$ is at most*

$$((1-\gamma_{\mathcal{G}})\sigma + \gamma_{\mathcal{G}})n \; .$$

While small values of $\gamma_{\mathcal{G}}$ imply a greater expansion, $\gamma_{\mathcal{G}}$ cannot be too small, as seen from the following lower bound.

Proposition 13.11 *Let $\mathcal{G} = (V, E)$ be an n-regular graph such that* $\mathrm{diam}(\mathcal{G}) \geq 4$. *Then*

$$\gamma_{\mathcal{G}} \geq \frac{1}{\sqrt{n}} \; .$$

Proof. Let s^+ and s^- be two vertices in $\mathcal{G}$ at distance at least 4 apart. Define the vector $\mathbf{y} = (y_u)_{u \in V}$ by

$$y_u = \begin{cases} \pm 1 & \text{if } u = s^\pm \\ \pm 1/\sqrt{n} & \text{if } u \in \mathcal{N}(s^\pm) \\ 0 & \text{otherwise} \end{cases} , \quad u \in V \; .$$

Since $\mathsf{d}_{\mathcal{G}}(s^+, s^-) \geq 4$, the sets $\{s^+\} \cup \mathcal{N}(s^+)$ and $\{s^-\} \cup \mathcal{N}(s^-)$ are disjoint; so, $\mathbf{y}$ is well-defined and, in addition, $\langle \mathbf{y}, \mathbf{1} \rangle = 0$. Therefore, by Lemma 13.6 we obtain

$$\langle \mathbf{y}, L_{\mathcal{G}} \mathbf{y} \rangle = n\|\mathbf{y}\|^2 - \langle \mathbf{y}, A_{\mathcal{G}} \mathbf{y} \rangle \geq (1-\gamma_{\mathcal{G}}) n \|\mathbf{y}\|^2 = 4(1-\gamma_{\mathcal{G}}) n \; .$$

On the other hand, by Corollary 13.3 we also have

$$\begin{aligned} \langle \mathbf{y}, L_{\mathcal{G}} \mathbf{y} \rangle &= \sum_{\{u,v\} \in E} (y_u - y_v)^2 \\ &= \sum_{s \in \{s^+, s^-\}} \sum_{u \in \mathcal{N}(s)} \Bigl((y_u - y_s)^2 + \sum_{v \in \mathcal{N}(u) \setminus \{s\}} (y_u - y_v)^2 \Bigr) \\ &\leq \sum_{s \in \{s^+, s^-\}} \sum_{u \in \mathcal{N}(s)} \Bigl(\underbrace{(y_u - y_s)^2}_{(1-(1/\sqrt{n}))^2} + \underbrace{(n-1) y_u^2}_{1-(1/n)} \Bigr) \\ &= 4n \left(1 - (1/\sqrt{n})\right) , \end{aligned}$$

where the second equality and the inequality follow from observing that when $u \in \mathcal{N}(s^{\pm})$, the sets $\mathcal{N}(u)$ and $\mathcal{N}(s^{\mp}) \cup \{s^{\mp}\}$ are disjoint. We conclude that

$$4(1-\gamma_{\mathcal{G}})n \le \langle \mathbf{y}, L_{\mathcal{G}}\mathbf{y} \rangle \le 4n\left(1 - (1/\sqrt{n})\right) ,$$

thus completing the proof. □

The proof of Proposition 13.11 can be extended to also prove the lower bound

$$\gamma_{\mathcal{G}} \ge \frac{2\sqrt{n-1}}{n}\left(1 - \frac{1}{\kappa}\right) + \frac{1}{\kappa n} , \tag{13.9}$$

whenever there exist two edges e^+ and e^- in $\mathcal{G}$ such that the distance between each endpoint of e^+ and each endpoint of e^- is at least twice the integer κ. In particular, κ can be taken as $\lfloor(\mathrm{diam}(\mathcal{G}))/2\rfloor - 1$. Now, the diameter of $\mathcal{G}$ can be bounded from below by an expression that tends to infinity as $|V|$ grows (see Problem 13.4). The next result follows.

Theorem 13.12 *For every fixed positive integer n,*

$$\liminf_{\mathcal{G}} \gamma_{\mathcal{G}} \ge \frac{2\sqrt{n-1}}{n} ,$$

where the limit is taken over any infinite sequence of distinct n-regular graphs $\mathcal{G}$.

13.4 Expanders from codes

Let Q be a finite group and B be a subset of $Q \setminus \{1\}$ that is closed under inversion, namely, $\alpha \in B \Longrightarrow \alpha^{-1} \in B$. The *Cayley graph* $\mathcal{G}(Q, B)$ is defined as the graph (Q, E) where

$$E = \Big\{ \{\alpha, \beta\} \;:\; \alpha\beta^{-1} \in B \Big\} .$$

It is easily seen that $\mathcal{G}(Q, B)$ is a $|B|$-regular (undirected) graph.

Let $F = \mathrm{GF}(2)$ and let k be a positive integer. Regarding F^k as a group whose operation is the addition of vectors, we will consider in this section Cayley graphs $\mathcal{G}(F^k, B)$ where $B \subseteq F^k \setminus \{\mathbf{0}\}$. In these graphs, the edges are the subsets $\{\mathbf{u}, \mathbf{u}'\} \subseteq F^k$ (of size 2) such that $\mathbf{u} + \mathbf{u}' \in B$.

The next proposition characterizes the eigenvalues of the adjacency matrix of the Cayley graph $\mathcal{G}(F^k, B)$. For two column vectors $\mathbf{u}$ and $\mathbf{v}$ of the same length over a finite field of prime size, denote by $\langle \mathbf{u}, \mathbf{v} \rangle$ the smallest nonnegative integer such that $\mathbf{u}^T\mathbf{v} = \langle \mathbf{u}, \mathbf{v} \rangle \cdot 1$, where 1 is the field unity. Recall from Problem 2.23 that the $2^k \times 2^k$ *Sylvester-type Hadamard matrix*, denoted by $\mathcal{H}_k$, is the real matrix whose rows and columns are indexed by the elements of F^k, and

$$(\mathcal{H}_k)_{\mathbf{u},\mathbf{v}} = (-1)^{\langle \mathbf{u},\mathbf{v} \rangle} , \quad \mathbf{u}, \mathbf{v} \in F^k .$$

It is known that

$$\mathcal{H}_k \mathcal{H}_k^T = 2^k \cdot I ;$$

equivalently, the rows of $\mathcal{H}_k$ are orthogonal, and so are the columns.

Proposition 13.13 *Let $F = \mathrm{GF}(2)$ and let B be a subset of $F^k \setminus \{\mathbf{0}\}$. The eigenvalues of the adjacency matrix of the Cayley graph $\mathcal{G}(F^k, B)$ are given by*

$$\lambda_{\mathbf{v}} = \sum_{\mathbf{u} \in B} (-1)^{\langle \mathbf{u},\mathbf{v} \rangle} , \quad \mathbf{v} \in F^k .$$

Proof. Denote by A the adjacency matrix of $\mathcal{G}(F^k, B)$ and consider the matrix product $A\mathcal{H}_k$. For every $\mathbf{u}, \mathbf{v} \in F^k$ we have

$$\begin{aligned} (A\mathcal{H}_k)_{\mathbf{u},\mathbf{v}} &= \sum_{\substack{\mathbf{u}' \in F^k: \\ \mathbf{u}+\mathbf{u}' \in B}} (\mathcal{H}_k)_{\mathbf{u}',\mathbf{v}} = \sum_{\mathbf{u}' \in B} (\mathcal{H}_k)_{\mathbf{u}+\mathbf{u}',\mathbf{v}} \\ &= (\mathcal{H}_k)_{\mathbf{u},\mathbf{v}} \sum_{\mathbf{u}' \in B} (\mathcal{H}_k)_{\mathbf{u}',\mathbf{v}} = (\mathcal{H}_k)_{\mathbf{u},\mathbf{v}} \lambda_{\mathbf{v}} . \end{aligned}$$

It follows that for every $\mathbf{v} \in F^k$, the column of $\mathcal{H}_k$ that is indexed by $\mathbf{v}$ is an eigenvector of A associated with the eigenvalue $\lambda_{\mathbf{v}}$. □

Corollary 13.14 *Let $\mathcal{C}$ be a linear $[n, k, d]$ code over $F = \mathrm{GF}(2)$ whose dual code, $\mathcal{C}^\perp$, has minimum distance $d^\perp \geq 3$. Let B be the set of columns of a given generator matrix of $\mathcal{C}$. The Cayley graph $\mathcal{G}(F^k, B)$ is a connected n-regular graph, and the eigenvalues of the adjacency matrix of $\mathcal{G}(F^k, B)$ are given by*

$$n - 2\,\mathsf{w}(\mathbf{c}) , \quad \mathbf{c} \in \mathcal{C} .$$

In particular,

$$\gamma_{\mathcal{G}(F^k,B)} = 1 - \frac{2d}{n} .$$

Proof. Let B be the set of columns of a given generator matrix $G = (\mathbf{g}_1 \ \mathbf{g}_2 \ \ldots \ \mathbf{g}_n)$ of $\mathcal{C}$. From $\mathrm{rank}(G) = k$ we get that $\mathcal{G}(F^k, B)$ is connected; furthermore, since $d^\perp \geq 3$, all the columns of G are distinct and, so, $|B| = n$ and $\mathcal{G}(F^k, B)$ is n-regular. Fix some column vector $\mathbf{v} \in F^k$ and let $\mathbf{c} = (c_1\ c_2\ \ldots\ c_n)$ be the codeword $\mathbf{v}^T G$ in $\mathcal{C}$. By Proposition 13.13, the eigenvalue $\lambda_{\mathbf{v}}$ of $A_{\mathcal{G}(F^k,B)}$ is given by

$$\begin{aligned}\lambda_{\mathbf{v}} &= \sum_{\mathbf{u}\in B}(-1)^{\langle \mathbf{u},\mathbf{v}\rangle} = \sum_{j=1}^{n}(-1)^{\langle \mathbf{v},\mathbf{g}_j\rangle} \\ &= \sum_{j=1}^{n}(-1)^{c_j} = \Bigl(\sum_{j\,:\,c_j=0} 1\Bigr) - \Bigl(\sum_{j\,:\,c_j=1} 1\Bigr) \\ &= n - 2\,\mathsf{w}(\mathbf{c})\end{aligned}$$

(where we have regarded the elements of F also as the integers 0 and 1). As $\mathbf{v}$ ranges over the elements of F^k, the vector $\mathbf{c} = \mathbf{v}^T G$ ranges over all codewords of $\mathcal{C}$. □

Example 13.5 The graph in Example 13.2, with the vertex names transposed, is the Cayley graph $\mathcal{G}(F^3, B)$, where B consists of the columns of the 3×3 identity matrix over $F = \mathrm{GF}(2)$. Clearly, this matrix is a generator matrix of the $[3, 3, 1]$ code $\mathcal{C} = F^3$. Therefore, by Corollary 13.14, the eigenvalues of $A_{\mathcal{G}(F^3,B)}$ are given by

$$3,\ 1,\ 1,\ 1, -1, -1, -1, -3$$

(the requirement on $d^\perp$ holds vacuously in this case, since $\mathcal{C}^\perp$ consists of one codeword only; see also Problem 13.28). □

Example 13.6 Recall from Problem 2.17 that the first-order Reed–Muller code over $F = \mathrm{GF}(2)$ is a linear $[2^m, m{+}1]$ code $\mathcal{C}_0$ over F with an $(m{+}1) \times 2^m$ generator matrix whose columns range over all the vectors in F^{m+1} with a first entry equaling 1. The code $\mathcal{C}_0$ contains the all-one word (in F^{2^m}) as a codeword, and its minimum distance equals 2^{m-1}.

Consider the $[n, k, d]$ linearly-concatenated code $\mathcal{C}_{\mathrm{cont}}$ over F with $\mathcal{C}_0$ taken as the inner code, while the outer code is a $[2^{m+1}, t, 2^{m+1}{-}t{+}1]$ singly-extended normalized GRS code $\mathcal{C}_{\mathrm{GRS}}$ over $\Phi = \mathrm{GF}(2^{m+1})$. As shown in Problem 5.1, the dual code of $\mathcal{C}_{\mathrm{GRS}}$ is also a singly-extended normalized GRS code and, so, the all-one word (in $\Phi^{2^{m+1}}$), being a row in a canonical parity-check matrix of $\mathcal{C}_{\mathrm{GRS}}^\perp$, is a codeword of $\mathcal{C}_{\mathrm{GRS}}$.

The parameters of $\mathcal{C}_{\mathrm{cont}}$ are given by $n = 2^{2m+1}$, $k = t(m{+}1)$, and

$$d \ge (2^{m+1}{-}t{+}1) \cdot 2^{m-1} = \frac{1}{2}\left(n - (t{-}1)\sqrt{n/2}\right) .$$

In addition, it can be verified by Problem 12.3 that when $t > 1$, the minimum distance of $\mathcal{C}_{\mathrm{cont}}^\perp$ is at least 3. Letting B be the set of columns of a generator matrix of $\mathcal{C}_{\mathrm{cont}}$, we obtain from Corollary 13.14 that the Cayley graph $\mathcal{G}(F^k, B)$ is an n-regular graph with $2^k = 2^{t(m+1)} = (2n)^{t/2}$ vertices, and

$$\gamma_{\mathcal{G}(F^k,B)} = 1 - \frac{2d}{n} \le \frac{(t-1)\sqrt{n/2}}{n} = \frac{t-1}{\sqrt{2n}} .$$

Hence, $\gamma_{\mathcal{G}(F^k,B)}$ approaches zero at a rate of $1/\sqrt{n}$ for every fixed t, while the number of vertices grows polynomially with n.

Observe that since the word $(\beta\ \beta\ \ldots\ \beta)$ is a codeword of C_{GRS} for every $\beta \in \Phi$, the word $(\mathbf{u}\,|\,\mathbf{u}\,|\,\ldots\,|\,\mathbf{u})$ is a codeword of $\mathcal{C}_{\mathrm{cont}}$ for every $\mathbf{u} \in \mathcal{C}_0$. In particular, the all-one word is a codeword of $\mathcal{C}_{\mathrm{cont}}$. Therefore, by Corollary 13.14 we get that $-n$ is an eigenvalue of the adjacency matrix of $\mathcal{G}(F^k, B)$, and from Proposition 13.4 we can conclude that this graph is bipartite. □

13.5 Ramanujan graphs

A graph $\mathcal{G} = (V, E)$ is called a *Ramanujan graph* if it is a connected n-regular graph such that every eigenvalue $\lambda \notin \{n, -n\}$ of $A_{\mathcal{G}}$ satisfies

$$|\lambda| \le 2\sqrt{n-1} .$$

In particular,

$$\gamma_{\mathcal{G}} \le \frac{2\sqrt{n-1}}{n} .$$

In view of Theorem 13.12, a Ramanujan graph $\mathcal{G}$ has essentially the smallest possible value for $\gamma_{\mathcal{G}}$.

We next describe a construction of Ramanujan graphs due to Lubotzky, Phillips, and Sarnak (in short, LPS), and Margulis: for every integer n such that $n-1$ is a prime congruent to 1 modulo 4, their construction yields an infinite sequence of n-regular Ramanujan graphs.

Given $F = \mathrm{GF}(q)$ and an integer k, consider the set of all $k \times k$ matrices over F whose determinant equals 1. This set forms a subgroup of all $k \times k$ nonsingular matrices over F under matrix multiplication and is called the *special linear group* $\mathrm{SL}_k(q)$. The subset of the $k \times k$ diagonal matrices

$$T_k(q) = \{a \cdot I \;:\; a \in F^*,\ a^k = 1\}$$

forms a normal subgroup of $\mathrm{SL}_k(q)$ and the factor group of $\mathrm{SL}_k(q)$ by $T_k(q)$ is called the *projective special linear group* $\mathrm{PSL}_k(q)$ (see Problem A.15). For the case $k = 2$ (which we will be interested in) we have

$$\mathrm{PSL}_2(q) = \mathrm{SL}_2(q)/\{I, -I\} ;$$

namely, the elements of $\mathrm{PSL}_2(q)$ are all the 2×2 matrices over F whose determinant is 1, where the matrices A and $-A$ are regarded as the same element $\pm A$. The unity element of $\mathrm{PSL}_2(q)$ is $\pm I$, and one can easily show (see Problem 13.31) that $|\mathrm{SL}_2(q)| = q(q^2-1)$ and

$$|\mathrm{PSL}_2(q)| = \begin{cases} q(q^2-1)/2 & \text{if } q \text{ is odd} \\ q(q^2-1) & \text{if } q \text{ is even} \end{cases} .$$

Fix q to be a prime congruent to 1 modulo 4 and let $F = \mathrm{GF}(q)$. Given an integer a (which may be negative), we will use the notation $\overline{a}$ to stand for the element in F such that $\overline{a} = a \cdot 1$, where 1 is the multiplicative unity in F. An integer a is said to be a *quadratic residue* modulo q if there is a nonzero element $\eta \in F$ such that $\eta^2 = \overline{a}$ in F. Since $q \equiv 1 \pmod 4$, the integer -1 is a quadratic residue modulo q (see Problem 3.23).

Let p be a prime other than q such that $p \equiv 1 \pmod 4$ and p is a quadratic residue modulo q. The graphs we describe are Cayley graphs $\mathcal{G}(\mathrm{PSL}_2(q), B)$ for sets B which we define next.

Denote by $\Upsilon(p)$ the set of all integer quadruples $\mathbf{z} = (z_0\ z_1\ z_2\ z_3)$, where z_0 is an odd positive integer, z_1, z_2, and z_3 are even integers (which may be negative), and $z_0^2 + z_1^2 + z_2^2 + z_3^2 = p$. Given $\mathbf{z} \in \Upsilon(p)$, consider the following 2×2 matrix over F,

$$M_{\mathbf{z}} = \pm \frac{1}{\eta} \begin{pmatrix} \overline{z}_0 + \imath \overline{z}_1 & \overline{z}_2 + \imath \overline{z}_3 \\ -\overline{z}_2 + \imath \overline{z}_3 & \overline{z}_0 - \imath \overline{z}_1 \end{pmatrix} ,$$

where $\imath$ and η are elements of F that satisfy

$$\imath^2 = -1 \quad \text{and} \quad \eta^2 = \overline{p}$$

(all operations in F). It can be verified (Problem 13.32) that $\det(M_{\mathbf{z}}) = 1$ and that $M_{\mathbf{z}}^{-1} = M_{\mathbf{z}^*}$, where $\mathbf{z}^* = (z_0\ {-z_1}\ {-z_2}\ {-z_3})$. Next, define the subset B of $\mathrm{PSL}_2(q) \setminus \{\pm I\}$ by

$$B = \{\pm M_{\mathbf{z}} \ : \ \mathbf{z} \in \Upsilon(p)\} .$$

This set is closed under inversion, and we denote the resulting $|\Upsilon(p)|$-regular Cayley graph $\mathcal{G}(\mathrm{PSL}_2(q), B)$ by $\mathcal{G}_{\mathrm{LPS}}(p, q)$.

The size of $\Upsilon(p)$ is determined by the following theorem in number theory, which is quoted here without proof.

Theorem 13.15 (Jacobi's Four Square Theorem) *Let m be a positive integer. The number of ordered integer quadruples $(z_0\ z_1\ z_2\ z_3)$ such that $z_0^2 + z_1^2 + z_2^2 + z_3^2 = m$ is equal to*

$$8 \cdot \sum_{\substack{d \mid m: \\ d>0,\ 4 \nmid d}} d .$$

In particular, if m is a prime then this number equals $8(m+1)$.

Corollary 13.16 *For every prime* $p \equiv 1 \pmod 4$,

$$|\Upsilon(p)| = p+1 .$$

Proof. By Theorem 13.15, the number of ordered integer quadruples $(z_0\ z_1\ z_2\ z_3)$ such that $z_0^2 + z_1^2 + z_2^2 + z_3^2 = p$ is equal to $8(p+1)$. Since $p \equiv 1 \pmod 4$, there is precisely one odd integer in every such quadruple. Restricting this integer to be the first entry in the quadruple reduces the number of quadruples by a factor of 4, while requiring that this odd integer be also positive yields a further reduction by a factor of 2. Hence, $|\Upsilon(p)| = p+1$. □

The next theorem was proved by Lubotzky, Phillips, and Sarnak, and independently by Margulis, and is quoted here without proof.

Theorem 13.17 *Let* p *and* q *be two distinct primes such that* $p \equiv q \equiv 1 \pmod 4$ *and* p *is a quadratic residue modulo* q. *Then* $\mathcal{G}_{\mathrm{LPS}}(p, q)$ *is a non-bipartite* $(p+1)$*-regular Ramanujan graph.*

Given a prime $p \equiv 1 \pmod 4$, it follows from the Law of Quadratic Reciprocity that an odd prime $q \neq p$ is a quadratic residue modulo p if and only if p is a quadratic residue modulo q (see the notes on this section at the end of the chapter). We thus obtain from Theorem 13.17 that for any given prime $p \equiv 1 \pmod 4$, the graph $\mathcal{G}_{\mathrm{LPS}}(p, q)$ is a $(p+1)$-regular Ramanujan graph for all primes $q \neq p$ such that $q \equiv 1 \pmod 4$ and q is a quadratic residue modulo p. We can therefore take q to be any prime that satisfies the congruence

$$q \equiv (1-p)s + p \pmod{4p} , \tag{13.10}$$

where s is any quadratic residue modulo p (indeed, such values of q yield the right remainders when divided by 4 and by p). From the extension of the Prime Number Theorem to arithmetic progressions we get that primes q which satisfy (13.10) are frequent among the prime numbers.

The graph $\mathcal{G}_{\mathrm{LPS}}(p, q)$ can be easily transformed into a bipartite $(p+1)$-regular Ramanujan graph with $q(q^2-1)$ vertices through the construction in Problem 13.13.

13.6 Codes from expanders

We next describe a construction of codes based on graphs. The construction uses the following ingredients:

- An n-regular graph $\mathcal{G} = (V, E)$.
- A code of length n over an alphabet F.

For every vertex $u \in V$, we assume an ordering on the set, $E(u)$, of the edges that are incident with u. The size of E will be denoted by N, and we have $|V| = 2N/n$ (see Problem 13.2).

For a word $\mathbf{x} = (x_e)_{e \in E}$ (whose entries are indexed by E) in F^N, denote by $(\mathbf{x})_{E(u)}$ the sub-word of $\mathbf{x}$ that is indexed by $E(u)$, that is, $(\mathbf{x})_{E(u)} = (x_e)_{e \in E(u)}$.

The *graph code* $\mathsf{C} = (\mathcal{G}, \mathcal{C})$ is defined by

$$\mathsf{C} = \left\{ \mathbf{c} \in F^N \;:\; (\mathbf{c})_{E(u)} \in \mathcal{C} \text{ for every } u \in V \right\} .$$

The next two propositions provide lower bounds on the rate and the relative minimum distance of the code C, for the case where $\mathcal{C}$ is linear.

Proposition 13.18 *Let $\mathcal{G} = (V, E)$ be an n-regular graph with $|E| = N$ edges and let $\mathcal{C}$ be a linear $[n, k{=}rn]$ code over a field F. Then $\mathsf{C} = (\mathcal{G}, \mathcal{C})$ is a linear $[N, K{=}RN]$ code over F, where*

$$R \geq 2r - 1 .$$

Proof. Let H be a $((1{-}r)n) \times n$ parity-check matrix of $\mathcal{C}$ over F. Then C can be characterized through a set of $|V|(1{-}r)n$ linear constraints over F, namely,

$$\mathsf{C} = \left\{ \mathbf{c} \in F^N \;:\; H\left((\mathbf{c})_{E(u)}\right)^T = \mathbf{0} \text{ for every } u \in V \right\} .$$

Hence, C is linear and

$$N - K \leq n|V|(1{-}r) = 2N(1{-}r)$$

or

$$\frac{K}{N} \geq 2r - 1 ,$$

as claimed. □

Proposition 13.19 *Let $\mathcal{G} = (V, E)$ be an $(n, \xi{>}0)$-expander with $|E| = N$ edges and let $\mathcal{C}$ be a linear $[n, k, d{=}\theta n]$ code over F. The minimum distance $D = \delta N$ of $\mathsf{C} = (\mathcal{G}, \mathcal{C})$ satisfies*

$$\delta \geq \frac{\theta(\theta + \xi - 1)}{\xi}$$

(whenever the rate of C is positive). In particular,

$$\delta \geq \frac{\theta(\theta - \gamma_{\mathcal{G}})}{1 - \gamma_{\mathcal{G}}} .$$

Proof. We assume that $\xi > 1 - \theta$ (≥ 0), or else the result trivially holds. Let $\mathbf{c} = (c_e)_{e \in E}$ be a nonzero codeword of C and denote by $Y \subseteq E$ the support of $\mathbf{c}$, namely,

$$Y = \left\{ e \in E \,:\, c_e \neq 0 \right\} .$$

Let S be the set of all vertices in $\mathcal{G}$ that are endpoints of edges in Y and consider the subgraph $\mathcal{G}(Y) = (S, Y)$ of $\mathcal{G}$. Since the minimum distance of $\mathcal{C}$ is d, the degree of each vertex in $\mathcal{G}(Y)$ must be at least d. Therefore, the average degree in $\mathcal{G}(Y)$ satisfies

$$\frac{2\,\mathsf{w}(\mathbf{c})}{|S|} = \frac{2|Y|}{|S|} \geq d . \tag{13.11}$$

On the other hand, by Lemma 13.9 we have

$$\frac{2|Y|}{|S|} \leq \frac{2|E_{S,S}|}{|S|} \leq (\xi \cdot \sigma + 1 - \xi)n ,$$

where $\sigma = |S|/|V|$. Combining with (13.11) yields

$$(\xi \cdot \sigma + 1 - \xi)n \geq d$$

or

$$\sigma \geq \frac{\theta + \xi - 1}{\xi} .$$

Using (13.11) again, we obtain

$$\mathsf{w}(\mathbf{c}) \geq \frac{d}{2} \cdot |S| = \frac{d}{2} \cdot \sigma \cdot |V| \geq \frac{\theta n}{2} \cdot \frac{\theta + \xi - 1}{\xi} \cdot \frac{2N}{n} = \frac{\theta(\theta + \xi - 1)N}{\xi} ,$$

and the result follows by letting $\mathbf{c}$ be a codeword of Hamming weight δN. □

Fix $\mathcal{C}$ to be a linear $[n, k{=}rn, d{=}\theta n]$ code over $F = \mathrm{GF}(q)$ that attains the Gilbert–Varshamov bound in Theorem 4.10, and let the graph $\mathcal{G} = (V, E)$ be taken from an infinite sequence of n-regular Ramanujan graphs. It follows from Propositions 13.18 and 13.19 that $\mathsf{C} = (\mathcal{G}, \mathcal{C})$ is a linear $[N, RN, \delta N]$ code over F, where $R \geq 2r - 1 \geq 1 - 2\mathsf{H}_q(\theta)$ and $\delta \geq \theta^2 - o(1)$, with $o(1)$ standing for an expression that goes to zero as $n \to \infty$; thus,

$$R \geq 1 - 2\,\mathsf{H}_q(\sqrt{\delta}) - o(1) . \tag{13.12}$$

Note that since the sequence of graphs is infinite, the code length, $N = |E|$, can take arbitrarily large values.

The bound (13.12) is inferior to the Gilbert–Varshamov bound or even to the Zyablov bound (see Section 12.3): for $q = 2$, the lower bound $1 -$

$2\,\mathsf{H}_2(\sqrt{\delta})$ vanishes already for $\delta = (\mathsf{H}_2^{-1}(\frac{1}{2}))^2 \approx 0.012$. Yet, the advantage of the codes $(\mathcal{G}, \mathcal{C})$ lies in their decoding complexity. For the case where $\mathcal{G}$ is bipartite, we show in Section 13.7 a decoder for $(\mathcal{G}, \mathcal{C})$ that can correct up to (approximately) $(1/4) \cdot \delta N$ errors in time complexity that grows *linearly* with the code length N. Furthermore, in Section 13.8, we present a generalization of the construction $(\mathcal{G}, \mathcal{C})$ that attains the Zyablov bound, with a linear-time decoding algorithm that corrects any number of errors up to (approximately) half the (designed) minimum distance of the code.

13.7 Iterative decoding of graph codes

Let $\mathsf{C} = (\mathcal{G}, \mathcal{C})$ be a linear graph code of length N over $F = \mathrm{GF}(q)$, where $\mathcal{G} = (V' : V'', E)$ is a bipartite n-regular graph with $|E| = n|V'| = n|V''| = N$ and $\mathcal{C}$ is a linear $[n, k, d{=}\theta n]$ code over F. In this section, we analyze the iterative algorithm in Figure 13.3 as a decoder for the code C. The algorithm assumes a decoder $\mathcal{D} : F^n \rightarrow \mathcal{C}$ that recovers correctly any pattern of less than $d/2$ errors (e.g., a nearest-codeword decoder). The number of iterations, ν, in Figure 13.3 is proportional to $\log |V'|$; a concrete value for ν will be given later on.

Input: received word $\mathbf{y} \in F^N$.
Output: word $\mathbf{z} \in F^N$ or a decoding-failure indicator "e".

1. $\mathbf{z} \leftarrow \mathbf{y}$.
2. For $i = 1, 2, \ldots, \nu$ do:
   (a) If i is odd then $U \equiv V'$, else $U \equiv V''$.
   (b) For every $u \in U$ do: $(\mathbf{z})_{E(u)} \leftarrow \mathcal{D}\left((\mathbf{z})_{E(u)}\right)$.
3. Return $\mathbf{z}$ if $\mathbf{z} \in \mathsf{C}$ (and "e" otherwise).

Figure 13.3. Iterative decoder for a graph code $\mathsf{C} = (\mathcal{G}, \mathcal{C})$.

Let $\mathbf{y} = (y_e)_{e \in E}$ denote the received word over F. The algorithm assigns to each edge $e \in E$ a label $z_e \in F$, which is initially set to y_e. The labels are represented in Figure 13.3 as a word $\mathbf{z} = (z_e)_{e \in E}$. The algorithm then performs ν iterations, where at iteration i, the decoder $\mathcal{D}$ of $\mathcal{C}$ is applied to each word $(\mathbf{z})_{E(u)} = (z_e)_{e \in E(u)}$, with u ranging over the set V' (if i is odd) or V'' (if i is even).

As part of our analysis, we show (in Proposition 13.23 below) that the algorithm in Figure 13.3 can correct any error word whose Hamming weight does not exceed

$$\frac{N\theta}{2} \cdot \sigma ,$$

where σ is any prescribed positive real such that

$$\sigma < \frac{(\theta/2) - \gamma_{\mathcal{G}}}{1 - \gamma_{\mathcal{G}}}$$

(the particular choice of σ will affect the value of ν). Assuming that $\gamma_{\mathcal{G}}$ is much smaller than θ, the guaranteed number of correctable errors, $(N\theta/2)\cdot\sigma$, can reach approximately half the number of errors that we expect to be able to correct based on Proposition 13.19. As pointed out earlier, the advantage of the proposed algorithm is manifested through its complexity: we show that the algorithm can be implemented using $O(N)$ operations in F, assuming that the code $\mathcal{C}$ is fixed (in particular, this assumption implies that $\mathcal{D}$ can be implemented in constant time). In Section 13.8, we present another linear-time decoding algorithm, which can correct twice as many errors (yet the multiplicative constant in the linear expression for the decoding complexity is larger).

Our analysis of the algorithm in Figure 13.3 makes use of several lemmas. The following lemma is an improvement on Corollary 13.10 for the case of bipartite graphs.

Lemma 13.20 *Let $\mathcal{G} = (V' : V'', E)$ be a bipartite n-regular graph and let $S \subseteq V'$ and $T \subseteq V''$ be subsets of sizes $|S| = \sigma|V'|$ and $|T| = \tau|V''|$, respectively, where $\sigma + \tau > 0$. The average degree of the vertices within the induced subgraph $\mathcal{G}_{S \cup T}$ satisfies*

$$\begin{aligned}
\frac{1}{|S \cup T|} \sum_{u \in S \cup T} \deg_{\mathcal{G}_{S\cup T}}(u) &= \frac{2|E_{S,T}|}{|S| + |T|} \\
&\leq \frac{2}{\sigma+\tau}\left(\sigma\tau + \gamma_{\mathcal{G}}\sqrt{\sigma(1-\sigma)\tau(1-\tau)}\right) n \\
&\leq \left((1-\gamma_{\mathcal{G}})\frac{2\sigma\tau}{\sigma+\tau} + \gamma_{\mathcal{G}} \cdot \frac{2\sqrt{\sigma\tau}}{\sigma+\tau}\right) n \\
&\leq \left((1-\gamma_{\mathcal{G}})\frac{2\sigma\tau}{\sigma+\tau} + \gamma_{\mathcal{G}}\right) n . \qquad (13.13)
\end{aligned}$$

The proof of Lemma 13.20 is left as a guided exercise (Problem 13.20). Note that the expression (13.13) looks like the upper bound of Corollary 13.10, except that σ therein is replaced by the harmonic mean of σ and τ.

Lemma 13.21 *Let $\mathcal{G} = (V' : V'', E)$ be a bipartite n-regular graph with $\gamma_{\mathcal{G}} > 0$ and let θ be a positive real. Suppose that there exist nonempty subsets*

$S \subseteq V'$ and $T \subseteq V''$ of sizes $|S| = \sigma|V'|$ and $|T| = \tau|V''|$, respectively, such that

$$u \in T \implies |\mathcal{N}(u) \cap S| \geq \frac{\theta n}{2} .$$

Then

$$\sqrt{\frac{\sigma}{\tau}} \geq \frac{(\theta/2) - (1-\gamma_{\mathcal{G}})\sigma}{\gamma_{\mathcal{G}}} .$$

Proof. By the condition of the lemma we can bound $|E_{S,T}|$ from below by

$$|E_{S,T}| = \sum_{u \in T} |\mathcal{N}(u) \cap S| \geq \frac{\theta n}{2} \cdot |T| = \frac{\theta n}{2} \cdot \tau |V'| .$$

On the other hand, by Lemma 13.20 we have the upper bound

$$|E_{S,T}| \leq \left((1-\gamma_{\mathcal{G}})\sigma\tau + \gamma_{\mathcal{G}}\sqrt{\sigma\tau}\right) n|V'| .$$

Combining these two bounds on $|E_{S,T}|$ we obtain

$$\frac{\theta\tau}{2} \cdot n|V'| \leq |E_{S,T}| \leq \left((1-\gamma_{\mathcal{G}})\sigma\tau + \gamma_{\mathcal{G}}\sqrt{\sigma\tau}\right) n|V'| ,$$

and dividing by $\gamma_{\mathcal{G}}\tau n|V'|$ yields the desired result. □

Let $\mathsf{C} = (\mathcal{G}, \mathcal{C})$ be a graph code over F, where $\mathcal{G} = (V' : V'', E)$ is a bipartite n-regular graph and $\mathcal{C}$ is a linear $[n, k, d]$ code over F. Suppose that $\mathbf{c} = (c_e)_{e \in E}$ is the transmitted codeword and let $\mathbf{y} = (y_e)_{e \in E}$ be the received word to which the algorithm in Figure 13.3 is applied. For $i = 1, 2, \ldots, \nu$, denote by $\mathbf{z}_i = (z_{i,e})_{e \in E}$ and U_i the values of the word $\mathbf{z}$ and the set U, respectively, at the end of iteration i in Figure 13.3; that is,

$$U_i = \begin{cases} V' & \text{if } i \text{ is odd} \\ V'' & \text{if } i \text{ is even} \end{cases} .$$

Also define the sets of edges

$$Y_i = \left\{ e \in E \,:\, z_{i,e} \neq c_e \right\} , \quad i = 1, 2, \ldots, \nu ,$$

and the sets of vertices

$$S_i = \left\{ u \in U_i \,:\, |E(u) \cap Y_i| > 0 \right\} , \quad i = 1, 2, \ldots, \nu ; \tag{13.14}$$

that is, an edge e belongs to Y_i if it corresponds to an erroneous coordinate in $\mathbf{z}_i$, and a vertex $u \in U_i$ belongs to S_i if it is an endpoint of at least one edge in Y_i. We denote the ratio $|S_i|/|V'|$ by σ_i.

The next lemma provides a useful property of the evolution of the values σ_i.

Lemma 13.22 *Let $\mathcal{G} = (V' : V'', E)$ be a bipartite n-regular graph with $|E| = N$ edges and let $\mathcal{C}$ be a linear $[n, k, d{=}\theta n]$ code over F such that $\theta > 2\gamma_\mathcal{G} > 0$. Denote by β the value*

$$\beta = \beta(\theta, \gamma_\mathcal{G}) = \frac{(\theta/2) - \gamma_\mathcal{G}}{1 - \gamma_\mathcal{G}} ,$$

and let σ be a prescribed real in the range $0 < \sigma < \beta$. Suppose that a codeword $\mathbf{c} = (c_e)_{e \in E}$ of $\mathsf{C} = (\mathcal{G}, \mathcal{C})$ is transmitted and a word $\mathbf{y} = (y_e)_{e \in E}$ over F is received, where

$$\mathsf{d}(\mathbf{y}, \mathbf{c}) \le \frac{N\theta}{2} \cdot \sigma .$$

Then, for $i = 1, 2, \ldots, \nu$,

$$\begin{aligned} |S_i| = \sigma_i |V'| \;&\le\; \left(\left(1 - \frac{\sigma}{\beta}\right) \left(\frac{\theta}{2\gamma_\mathcal{G}}\right)^{i-1} + \frac{\sigma}{\beta} \right)^{-2} \cdot \sigma |V'| \\ &\le\; \frac{\sigma |V'|}{(1 - (\sigma/\beta))^2} \cdot \left(\frac{2\gamma_\mathcal{G}}{\theta}\right)^{2i-2} , \end{aligned}$$

namely, σ_i decreases exponentially with i.

Proof. Denote by Y_0 the set of erroneous coordinates in $\mathbf{y}$, that is,

$$Y_0 = \left\{ e \in E \;:\; y_e \neq c_e \right\} .$$

Since the decoder $\mathcal{D} : F^N \to \mathcal{C}$ errs only when it attempts to correct $\lceil d/2 \rceil$ or more errors, we have,

$$u \in S_i \quad \Longrightarrow \quad |E(u) \cap Y_{i-1}| \ge \frac{d}{2} , \quad i = 1, 2, \ldots, \nu . \tag{13.15}$$

In particular, for $i = 1$ we get

$$\mathsf{d}(\mathbf{y}, \mathbf{c}) = |Y_0| \ge \sum_{u \in S_1} |E(u) \cap Y_0| \ge \frac{d}{2} \cdot |S_1| ;$$

therefore,

$$\sigma_1 |V'| = |S_1| \le \frac{2}{d} \cdot \mathsf{d}(\mathbf{y}, \mathbf{c}) \le \frac{2}{\theta n} \cdot \frac{N\theta}{2} \cdot \sigma = \sigma |V'| ,$$

thereby proving the claim for $i = 1$.

Let ℓ be the smallest positive integer (possibly ∞) for which $\sigma_\ell = 0$. It is clear from the algorithm in Figure 13.3 that $\sigma_i = 0$ for every $i \ge \ell$; hence, we focus hereafter in the proof on the range $1 < i < \ell$. The proof for this range is carried out by applying Lemma 13.21 to the graph $(U_{i-1} : U_i, E)$, with S and T in that lemma being taken as S_{i-1} and S_i, respectively. From (13.14) and (13.15) we obtain that

$$u \in S_i \quad \Longrightarrow \quad |\mathcal{N}(u) \cap S_{i-1}| \geq \frac{\theta n}{2} ,$$

and by Lemma 13.21 we conclude that for every $1 < i < \ell$,

$$\sqrt{\frac{\sigma_{i-1}}{\sigma_i}} \geq \frac{\theta}{2\gamma_{\mathcal{G}}} - \frac{1-\gamma_{\mathcal{G}}}{\gamma_{\mathcal{G}}} \sigma_{i-1} . \tag{13.16}$$

Since $\sigma_1 \leq \sigma < \beta$, we get that the right-hand side of (13.16) is greater than 1 for $i = 2$; therefore, $\sigma_1/\sigma_2 > 1$ and, so, $\sigma_2 < \sigma_1 < \beta$. Continuing by induction on i, it follows that $\sigma_{i-1}/\sigma_i > 1$ for all $1 < i < \ell$, thus implying that $\sigma_i \leq \sigma$ for all $i > 0$. This, in turn, allows us to replace σ_{i-1} by $\sqrt{\sigma_{i-1}\sigma}$ in the right-hand side of (13.16). When we do so and divide by $\sqrt{\sigma_{i-1}}$, we obtain

$$\frac{1}{\sqrt{\sigma_i}} \geq \frac{\theta/(2\gamma_{\mathcal{G}})}{\sqrt{\sigma_{i-1}}} - \frac{(1-\gamma_{\mathcal{G}})\sqrt{\sigma}}{\gamma_{\mathcal{G}}} .$$

Finally, by changing the inequality into an equality (thereby we may only overestimate σ_i), we get a first-order non-homogeneous linear recurrence in $1/\sqrt{\sigma_i}$, the solution of which yields

$$\frac{1}{\sqrt{\sigma_i}} = \left(\left(1 - \frac{\sigma}{\beta}\right) \cdot \left(\frac{\theta}{2\gamma_{\mathcal{G}}}\right)^{i-1} + \frac{\sigma}{\beta} \right) \frac{1}{\sqrt{\sigma}} .$$

The result follows. □

Based on the previous lemma, we can now prove the next proposition, which states that by properly selecting the number of iterations ν, the algorithm in Figure 13.3 is indeed a graph code decoder.

Proposition 13.23 *Let $\mathcal{G} = (V' : V'', E)$ be a bipartite n-regular graph with $|E| = N$ edges and let $\mathcal{C}$ be a linear $[n, k, d{=}\theta n]$ code over F such that $\theta > 2\gamma_{\mathcal{G}} > 0$. Fix σ to be a positive real such that*

$$\sigma < \beta = \frac{(\theta/2) - \gamma_{\mathcal{G}}}{1 - \gamma_{\mathcal{G}}} ,$$

and suppose that a codeword $\mathbf{c}$ of $\mathsf{C} = (\mathcal{G}, \mathcal{C})$ is transmitted and a word $\mathbf{y} \in F^N$ is received, where

$$\mathsf{d}(\mathbf{y}, \mathbf{c}) \leq \frac{N\theta}{2} \cdot \sigma .$$

If the algorithm in Figure 13.3 is applied to $\mathbf{y}$ *with*

$$\nu = \left\lfloor \log_{\theta/(2\gamma_{\mathcal{G}})} \left(\frac{\beta\sqrt{\sigma|V'|} - \sigma}{\beta - \sigma} \right) \right\rfloor + 2$$

(taking $\nu = 1$ *when* $|V'| \le \sigma/\beta^2$*), then the value of* $\mathbf{z}$ *upon termination of the algorithm equals* $\mathbf{c}$.

Proof. By Lemma 13.22 we get that $|S_\nu| = \sigma_\nu|V'| < 1$; namely, S_ν is empty (this applies also to the case where $|V'| \le \sigma/\beta^2$: here we have $|S_1| \le \sigma|V'| \le (\sigma/\beta)^2 < 1$). □

Observe that the dependence of ν on $|V'|$ is logarithmic, with an additive term that grows (arbitrarily) as σ gets closer to β. While Proposition 13.23 provides a number of iterations ν that guarantees successful decoding, the word $\mathbf{z}$ in Figure 13.3 will already contain the correct codeword once it remains unchanged during a full execution of Step 2b. This condition, in turn, can serve as an early stopping rule for the algorithm.

Next, we turn to a complexity analysis of the algorithm. We assume here that the code $\mathcal{C}$ is fixed and, so, the decoder $\mathcal{D} : F^n \to \mathcal{C}$ can be implemented in constant time. It is rather easy to see that the decoder $\mathcal{D}$ is applied in Figure 13.3 at most $|V'|\nu = O(|V'| \log |V'|)$ times (where we absorb into the "O" notation additive and multiplicative terms which depend on θ, $\gamma_{\mathcal{G}}$, and σ). However, a finer analysis reveals that this bound can be reduced to $O(|V'|)$. We demonstrate this next.

First, observe that during each iteration $i \ge 3$, we need to apply the decoder $\mathcal{D}$ to $(\mathbf{z})_{E(u)}$ in Step 2b for a given vertex $u \in U_i$, only if at least one entry in $(\mathbf{z})_{E(u)}$—say, the entry indexed by the edge $\{u, v\} \in E(u)$—has been altered during iteration $i-1$. Yet, such an alteration occurs only if v is a vertex in U_{i-1} such that $|E(v) \cap Y_{i-2}| > 0$; the latter inequality, in turn, holds only if $v \in \mathcal{N}(S_{i-2})$ (see Figure 13.4, where the dotted line represents an edge in Y_{i-2} and u' is its endpoint in S_{i-2}). We thus conclude that $\mathcal{D}$ needs to be applied to $(\mathbf{z})_{E(u)}$ in iteration i only if $u \in \mathcal{N}(\mathcal{N}(S_{i-2}))$. Summing

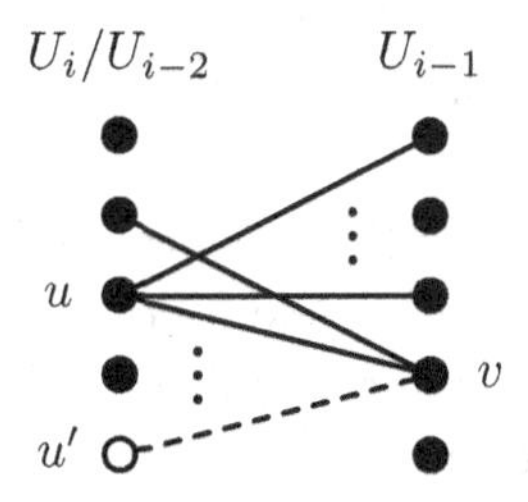

Figure 13.4. The dotted edge belongs to Y_{i-2} and u' is its endpoint in S_{i-2}.

over all iterations (including the first two), we obtain that the number of applications of the decoder $\mathcal{D}$ can be bounded from above by

$$\begin{aligned}
2|V'| + \sum_{i=3}^{\nu} |\mathcal{N}(\mathcal{N}(S_{i-2}))| &\le 2|V'| + n^2 \sum_{i=3}^{\nu} |S_{i-2}| \\
&< 2|V'| + \frac{\sigma n^2 |V'|}{(1-(\sigma/\beta))^2} \cdot \sum_{i=3}^{\infty} \left(\frac{2\gamma_{\mathcal{G}}}{\theta}\right)^{2(i-3)} \\
&= \left(2 + \frac{\sigma n^2}{(1-(\sigma/\beta))^2} \cdot \frac{1}{1-(2\gamma_{\mathcal{G}}/\theta)^2}\right) |V'| \\
&= O(|V'|) ,
\end{aligned}$$

where the second inequality follows from Lemma 13.22.

(When assessing the overall decoding complexity, we also need to take into account the complexity of finding the neighbors of any vertex in $\mathcal{G}$; yet, we can assume that the neighborhoods of each vertex are pre-computed and stored, say, as a table in memory or hard-wired into the decoding hardware.)

While the decoding complexity of C is linear in $|V'|$ (or N), the *encoding* (which can be carried out by multiplying the information word with a generator matrix of C) may still have time complexity that is quadratic in N.

13.8 Graph codes in concatenated schemes

The bipartite graph codes $(\mathcal{G}, \mathcal{C})$, which we considered in Section 13.7, can be generalized in several ways. For instance, we can insert the flexibility of associating different codes $\mathcal{C}$ to the two partition elements of the vertex set of $\mathcal{G}$. We do this next.

Let $\mathcal{G} = (V' : V'', E)$ be a bipartite n-regular graph with $|E| = n|V'| = n|V''| = N$ and let $\mathcal{C}'$ be a linear $[n, k{=}rn, \theta n]$ code over $F = \mathrm{GF}(q)$. Also, let $\mathcal{C}''$ be a (second) linear $[n, Rn, \delta n]$ code over F. We define the code $\mathsf{C} = (\mathcal{G}, \mathcal{C}' : \mathcal{C}'')$ over F by

$$\mathsf{C} = \left\{ \mathbf{c} \in F^N \; : \; \begin{array}{l} (\mathbf{c})_{E(u)} \in \mathcal{C}' \text{ for every } u \in V' \text{ and} \\ (\mathbf{c})_{E(u)} \in \mathcal{C}'' \text{ for every } u \in V'' \end{array} \right\} . \qquad (13.17)$$

The construction $(\mathcal{G}, \mathcal{C}' : \mathcal{C}'')$ will be referred to hereafter as a *generalized* graph code.

The construction $\mathsf{C} = (\mathcal{G}, \mathcal{C}' : \mathcal{C}'')$ can be further generalized if we regard C as a concatenated code over F in the following manner. Let Φ denote the alphabet F^k. Fix some one-to-one (and onto) mapping $\mathcal{E} : \Phi \to \mathcal{C}'$ that is linear over F, and define the one-to-one mapping $\psi_{\mathcal{E}} : \mathsf{C} \to \Phi^{|V'|}$ by

$$\psi_{\mathcal{E}}(\mathbf{c}) = \left(\mathcal{E}^{-1}((\mathbf{c})_{E(u)})\right)_{u \in V'} , \quad \mathbf{c} \in \mathsf{C} ;$$

i.e., the entry of $\psi_{\mathcal{E}}(\mathbf{c})$ that is indexed by a vertex $u \in V'$ equals the unique element in Φ that is mapped by $\mathcal{E}$ to the sub-word $(\mathbf{c})_{E(u)}$ (by the definition of $(\mathcal{G}, \mathcal{C}', \mathcal{C}'')$, this sub-word is indeed a codeword of $\mathcal{C}'$ and, thus, an image of $\mathcal{E}$). Next, define the code C_Φ over Φ by

$$\mathsf{C}_\Phi = \{\psi_{\mathcal{E}}(\mathbf{c}) \;:\; \mathbf{c} \in \mathsf{C}\} \;. \tag{13.18}$$

Through this code construction, we can represent the generalized graph code C as a concatenated code $(\mathcal{E}, \mathsf{C}_\Phi)$, with an inner code $\mathcal{C}'$ over F and an outer code C_Φ over Φ. While such a characterization of C as a concatenated code may seem to be somewhat artificial, it does introduce the possibility of using C_Φ as an outer code with inner codes other than $\mathcal{C}'$. We explore this potential further by first stating several properties of the code C_Φ.

Proposition 13.24 *Let $\mathsf{C} = (\mathcal{G}, \mathcal{C}' : \mathcal{C}'')$ be a generalized graph code over $F = \mathrm{GF}(q)$, where $\mathcal{G} = (V' : V'', E)$ is a bipartite n-regular graph, $\mathcal{C}'$ is a linear $[n, k{=}rn, \theta n]$ code over F, and $\mathcal{C}''$ is a linear $[n, Rn, \delta n]$ code over F. Let Φ denote the alphabet F^k, and define the code C_Φ over Φ by (13.18). Then the following conditions hold:*

(i) The code C_Φ is a linear space over F.

(ii) The rate of C_Φ is bounded from below by

$$1 - \frac{1}{r} + \frac{R}{r} \,.$$

(iii) The relative minimum distance of C_Φ is bounded from below by

$$\frac{\delta - \gamma_{\mathcal{G}}\sqrt{\delta/\theta}}{1 - \gamma_{\mathcal{G}}} \,.$$

The proof of Proposition 13.24 is given as a guided exercise (Problem 13.36). It follows from part (ii) of the proposition that the rate of C_Φ approaches R when $r \to 1$, while from part (iii) we get that its relative minimum distance approaches δ when $\gamma_{\mathcal{G}}/\sqrt{\theta} \to 0$.

Example 13.7 Consider the case where $\mathcal{C}'$ and $\mathcal{C}''$ are taken as GRS codes over $F = \mathrm{GF}(q)$ (which is possible if $n < q$). We show that the respective code C_Φ can get arbitrarily close to the asymptotic version of the Singleton bound (as stated in Section 4.5), provided that n (and q) are sufficiently large. Specifically, we fix $\theta = \epsilon$ for some small $\epsilon \in (0, 1]$ (in which case $r > 1 - \epsilon$), and then select q and n so that $q > n \geq 4/\epsilon^3$. Assuming that $\mathcal{G}$ is a bipartite n-regular Ramanujan graph, we have $\gamma_{\mathcal{G}} \leq (2\sqrt{n{-}1})/n <$

$\epsilon^{3/2}$, or $\gamma_{\mathcal{G}}/\sqrt{\theta} < \epsilon$. By Proposition 13.24, the rate and relative minimum distance of the resulting code C_Φ are bounded from below by

$$1 - \frac{1}{r} + \frac{R}{r} > 1 - \frac{1}{1-\epsilon} + \frac{R}{1-\epsilon} > R - \epsilon$$

and

$$\frac{\delta - \gamma_{\mathcal{G}}\sqrt{\delta/\theta}}{1-\gamma_{\mathcal{G}}} > \frac{\delta - \epsilon\sqrt{\delta}}{1-\epsilon^{3/2}} > \delta - \epsilon > 1 - R - \epsilon ,$$

respectively. In addition, if q and n are selected to be (no larger than) $O(1/\epsilon^3)$, then the alphabet size of C_Φ is

$$|\Phi| = q^{rn} = 2^{O((\log(1/\epsilon))/\epsilon^3)} ; \tag{13.19}$$

namely, it does not grow with the length, $|V'|$, of C_Φ. □

Example 13.8 Suppose that F (= GF(q)) is an extension field of a field K with extension degree $[F : K] = m$, and let the parameters of C_Φ be selected as in Example 13.7. Construct a concatenated code $\mathcal{C}_{\mathrm{cont}} = (\mathcal{E}_{\mathrm{in}}, \mathsf{C}_\Phi)$ over K, where $\mathcal{E}_{\mathrm{in}}$ is a one-to-one linear mapping over K from Φ onto a linear $[\ell, r_{\mathrm{in}}\ell, \delta_{\mathrm{in}}\ell]$ code $\mathcal{C}_{\mathrm{in}}$ over K that attains the Gilbert–Varshamov bound, with $\ell = (r/r_{\mathrm{in}})mn$. The concatenated code $\mathcal{C}_{\mathrm{cont}}$ is thus a linear code of length $\ell|V'|$ over K, whose rate and relative minimum distance are bounded from below by

$$r_{\mathrm{in}}(R - \epsilon) \geq \left(1 - \mathsf{H}_{|K|}(\delta_{\mathrm{in}})\right)(R - \epsilon)$$

and

$$\delta_{\mathrm{in}}(\delta - \epsilon) > \delta_{\mathrm{in}}(1 - R - \epsilon) ,$$

respectively. By comparing these lower bounds with Equations (12.3) and (12.4) in Section 12.3, we conclude that the code $\mathcal{C}_{\mathrm{cont}}$ can get arbitrarily close to the Zyablov bound when $\epsilon \to 0$. □

When using C_Φ as an outer code in a concatenated code, one would also prefer having an efficient decoder for C_Φ that can recover both errors and erasures; this, in turn, would imply an efficient generalized minimum distance (in short, GMD) decoder for the entire concatenated code (see Section 12.2).

A combined error–erasure decoder for C_Φ is presented in Figure 13.5. This decoder is similar to the decoder for C, which was presented in Figure 13.3. The received word $\mathbf{y}$ is now over the alphabet $\Phi \cup \{?\}$ and its entries, $\mathbf{y}_u$, are indexed by $u \in V'$. Step 1 transforms the received word into a word $\mathbf{z} \in F^{|E|}$ by encoding each non-erased entry of $\mathbf{y}$ to a codeword of $\mathcal{C}'$ and mapping each erased entry to an erased sub-block of length n. The inverse mapping is then applied in Step 3. The main loop in Step 2 is

---

Input: received word $\mathbf{y} = (\mathbf{y}_u)_{u \in V'}$ in $(\Phi \cup \{?\})^{|V'|}$.
Output: word in $\Phi^{|V'|}$ or a decoding-failure indicator "e".

1. For $u \in V'$ do: $\quad (\mathbf{z})_{E(u)} \leftarrow \begin{cases} \mathcal{E}(\mathbf{y}_u) & \text{if } \mathbf{y}_u \in \Phi \\ ??\ldots? & \text{if } \mathbf{y}_u = ? \end{cases}$.
2. For $i = 2, 3, \ldots, \nu$ do:
   (a) If i is odd then $U \equiv V'$ and $\mathcal{D} \equiv \mathcal{D}'$, else $U \equiv V''$ and $\mathcal{D} \equiv \mathcal{D}''$.
   (b) For every $u \in U$ do: $(\mathbf{z})_{E(u)} \leftarrow \mathcal{D}\left((\mathbf{z})_{E(u)}\right)$.
3. Return $\psi_{\mathcal{E}}(\mathbf{z})$ if $\mathbf{z} \in \mathsf{C}$ (and "e" otherwise).

Figure 13.5. Iterative decoder for C_Φ.

essentially the same as its counterpart in Figure 13.3, except that now we have two decoders, $\mathcal{D}' : F^n \to \mathcal{C}'$ and $\mathcal{D}'' : (F \cup \{?\})^n \to \mathcal{C}''$: the former recovers correctly any pattern with less than $\theta n/2$ errors over F and the latter recovers correctly any pattern of a errors and b erasures, provided that $2a + b < \delta n$. Observe that iteration $i = 1$ was skipped in Step 2 in Figure 13.5 (since the non-erased sub-blocks of $\mathbf{z}$ are already initialized to be codewords of $\mathcal{C}'$) and that erasure decoding may occur only when $i = 2$.

The next proposition states that the algorithm in Figure 13.5 is indeed a decoder for C_Φ.

Proposition 13.25 *Let $\mathsf{C} = (\mathcal{G}, \mathcal{C}' : \mathcal{C}'')$ be a generalized graph code over $F = \mathrm{GF}(q)$, where $\mathcal{G} = (V' : V'', E)$ is a bipartite n-regular graph, $\mathcal{C}'$ is a linear $[n, k{=}rn, \theta n]$ code over F, and $\mathcal{C}''$ is a linear $[n, Rn, \delta n]$ code over F such that $\sqrt{\theta\delta} > 2\gamma_\mathcal{G} > 0$. Let Φ denote the alphabet F^k, and define the code C_Φ over Φ by (13.18). Fix σ to be a positive real such that*

$$\sigma < \beta = \frac{(\delta/2) - \gamma_\mathcal{G}\sqrt{\delta/\theta}}{1 - \gamma_\mathcal{G}} \,,$$

and suppose that a transmission of a codeword $\mathbf{c} \in \mathsf{C}_\Phi$ results in a word $\mathbf{y} \in (\Phi \cup \{?\})^{|V'|}$ with t errors and ρ erasures, where

$$t + \frac{\rho}{2} \leq \sigma |V'| \,.$$

Apply the algorithm in Figure 13.5 to $\mathbf{y}$ with

$$\nu = 2\left\lfloor \log\left(\frac{\beta\sqrt{\sigma|V'|} - \sigma}{\beta - \sigma}\right)\right\rfloor + 3 \,, \tag{13.20}$$

where the base of the logarithm equals $\theta\delta/(4\gamma_{\mathcal{G}}^2)$ *(take* $\nu = 1$ *when* $|V'| \le \sigma/\beta^2$*). Then the value of* $\psi_{\mathcal{E}}(\mathbf{z})$ *upon termination of the algorithm equals* $\mathbf{c}$.

The proof of Proposition 13.25 is given as an exercise (Problems 13.37–13.39). As was the case with the decoder of Figure 13.3, the decoder herein needs to apply the decoders $\mathcal{D}'$ and $\mathcal{D}''$ only $O(|V'|)$ times.

The following examples present several applications of the algorithm in Figure 13.5.

Example 13.9 We analyze the performance of a GMD decoder for the concatenated code $\mathcal{C}_{\mathrm{cont}}$ in Example 13.8, while using the algorithm in Figure 13.5 to decode the outer code C_Φ. We assume that K, m, and n are fixed and, so, a nearest-codeword decoder for $\mathcal{C}_{\mathrm{in}}$ can be implemented in constant time. Thus, a GMD decoder for $\mathcal{C}_{\mathrm{cont}}$ has time complexity that is linear in $|V'|$ and is capable of correcting any error pattern in which the ratio between the number of errors and the code length $\ell|V'|$ is less than

$$\delta_{\mathrm{in}} \cdot \frac{(\delta/2) - \gamma_{\mathcal{G}}\sqrt{\delta/\theta}}{1 - \gamma_{\mathcal{G}}} > \delta_{\mathrm{in}} \cdot \frac{(\delta/2) - \epsilon\sqrt{\delta}}{1 - \epsilon^{3/2}} \ge \tfrac{1}{2}\delta_{\mathrm{in}}(\delta - 2\epsilon)$$

(refer to the analysis of GMD decoding in Section 12.2 and combine it with Proposition 13.25). The *construction* of $\mathcal{C}_{\mathrm{cont}}$ (e.g., computing one of its generator matrices) has time complexity that is polynomially large in the code length. □

Example 13.10 Let F and K be as in Example 13.8, and consider a $|K|$-ary symmetric channel (K, K, Prob) with crossover probability $p < 1-(1/|K|)$. We have shown in Section 12.5 that one can approach the capacity of this channel with concatenated codes (over K) that can be encoded and decoded in time complexity that is polynomially large in the code length. We next verify that this can be achieved with the code $\mathcal{C}_{\mathrm{cont}}$ of Example 13.8, where we now select $\mathcal{C}_{\mathrm{in}}$ so that it has a nearest-codeword decoder whose decoding error probability, $P = \mathrm{P}_{\mathrm{err}}(\mathcal{C}_{\mathrm{in}})$, decreases exponentially with ℓ $(= (r/r_{\mathrm{in}})mn)$, whenever $r_{\mathrm{in}} < 1 - \mathsf{H}_{|K|}(p)$; i.e.,

$$P < 4 \cdot |K|^{-\mathrm{E}\cdot\ell}$$

for some constant $\mathrm{E} = \mathrm{E}_{|K|}(p, r_{\mathrm{in}}) > 0$ (the multiplier 4 is inserted here to make the notation consistent with that in Section 12.5).

The code $\mathcal{C}_{\mathrm{cont}}$ can be decoded by first applying a nearest-codeword decoder for $\mathcal{C}_{\mathrm{in}}$ to each sub-block of length ℓ within the received word, followed by an application of the decoder of Figure 13.5 (without attempting to correct erasures). As was the case in Example 13.9, this decoding process has time complexity that is linear in the code length.

Recalling that the rate of the outer code C_Φ is greater than $R - \epsilon$, we can repeat the analysis of Section 12.5 and replace (12.14) therein by the following lower bound on the overall rate R_{cont} of $\mathcal{C}_{\mathrm{cont}}$,

$$R_{\mathrm{cont}} > r_{\mathrm{in}}(R - \epsilon) \geq r_{\mathrm{in}} \cdot (1 - \delta - o(1)) ,$$

where $o(1)$ stands for an expression that goes to zero as n goes to infinity (note that when K is fixed, then, by the construction in Example 13.7, if n goes to infinity, then so must $|F|$ and, consequently, m and ℓ). Thus, given a designed rate $\mathcal{R} < 1 - \mathsf{H}_{|K|}(p)$, we select the rate r_{in} of the inner code so that $\mathcal{R} \leq r_{\mathrm{in}} < 1 - \mathsf{H}_{|K|}(p)$ and take $\delta = 1 - (\mathcal{R}/r_{\mathrm{in}})$. By Proposition 13.25, the decoder in Figure 13.5 can recover a codeword of C_Φ provided that the erroneous fraction is smaller than

$$\frac{(\delta/2) - \gamma_{\mathcal{G}}\sqrt{\delta/\theta}}{1 - \gamma_{\mathcal{G}}} > \frac{(\delta/2) - \epsilon\sqrt{\delta}}{1 - \epsilon^{3/2}} \geq \tfrac{1}{2}\delta - \epsilon .$$

With our choice of parameters we therefore have

$$R_{\mathrm{cont}} \geq \mathcal{R} - o(1) ,$$

and the attainable decoding error probability, $\mathrm{P}_{\mathrm{err}}(\mathcal{C}_{\mathrm{cont}})$, is bounded from above—similarly to (12.13)—by

$$\mathrm{P}_{\mathrm{err}}(\mathcal{C}_{\mathrm{cont}}) \leq \left(2P^{(\delta/2)-\epsilon}\right)^{|V'|} = |K|^{-\ell|V'|(\mathrm{E}\cdot\delta/2 - o(1))} .$$

We readily reach the following lower bound on the error exponent:

$$\begin{aligned} -\frac{\log_{|K|} \mathrm{P}_{\mathrm{err}}(\mathcal{C}_{\mathrm{cont}})}{\ell|V'|} &\geq \tfrac{1}{2}\mathrm{E} \cdot \delta - o(1) \\ &= \tfrac{1}{2}\mathrm{E}_{|K|}(p, r_{\mathrm{in}})\,(1 - (\mathcal{R}/r_{\mathrm{in}})) - o(1) . \qquad (13.21) \end{aligned}$$

Finally, we maximize the expression (13.21) over r_{in} within the range $\mathcal{R} \leq r_{\mathrm{in}} \leq 1 - \mathsf{H}_{|K|}(p)$. (Since

$$(1 - \sqrt{\mathcal{R}/r_{\mathrm{in}}})^2 < 1 - (\mathcal{R}/r_{\mathrm{in}})$$

for $r_{\mathrm{in}} > \mathcal{R} > 0$, this maximization results in a lower bound which, in fact, is better than (12.15).) □

Example 13.11 Let $\mathsf{C} = (\mathcal{G}, \mathcal{C})$ be a graph code over F where $\mathcal{G} = (V' : V'', E)$ is a bipartite n-regular graph with $|E| = N$ edges and $\mathcal{C}$ is a linear $[n, k, d{=}\theta n]$ code over F. Denoting $\Phi = F^k$, we regard C as a concatenated code over F, with the outer code being C_Φ and the inner code being $\mathcal{C}$. It follows from Proposition 13.25 that by applying GMD decoding to this concatenated code (whose length is $n|V'| = N$), we can correct any pattern with up to

$$\theta n \cdot \sigma |V'| = N\theta \cdot \sigma$$

errors (over F), where σ is any positive real such that

$$\sigma < \frac{(\theta/2) - \gamma_{\mathcal{G}}}{1 - \gamma_{\mathcal{G}}} .$$

This is an improvement by a factor of 2 compared to Proposition 13.23. Note, however, that the GMD decoder requires $\lceil d/2 \rceil$ applications of the algorithm in Figure 13.5. Since d is fixed, the resulting complexity is still linear in the code length, yet the multiplicative constant is bigger compared to the algorithm in Figure 13.3. □

Problems

[Section 13.1]

Problem 13.1 Let $\mathcal{G} = (V, E)$ be a connected graph. Show that $|E| \geq |V| - 1$.

Problem 13.2 Show that in every graph $\mathcal{G} = (V, E)$,

$$\sum_{u \in V} \deg_{\mathcal{G}}(u) = 2|E| .$$

Problem 13.3 Let $\mathcal{G} = (V, E)$ be a graph. Show that $\mathsf{d}_{\mathcal{G}} : V \times V \to \mathbb{R}$ is a metric.

Problem 13.4 Let $\mathcal{G} = (V, E)$ be a graph and let n be the maximum degree of any vertex in V. Show that when $n > 1$,

$$\operatorname{diam}(\mathcal{G}) \geq \begin{cases} \frac{1}{2}(|V|-1) & \text{if } n = 2 \\ \log_{n-1}\left(\frac{(n-2)|V| + 2}{n}\right) & \text{if } n > 2 \end{cases} .$$

Hint: Show that when $n > 2$, the number of vertices at distance ℓ or less from a given vertex in $\mathcal{G}$ is bounded from above by

$$1 + n \sum_{i=0}^{\ell-1} (n-1)^i = \frac{n(n-1)^\ell - 2}{n - 2} .$$

Then claim that this bound must be at least $|V|$ for $\ell = \operatorname{diam}(\mathcal{G})$.

Problem 13.5 Show that a graph $\mathcal{G}$ is bipartite if and only if it contains no cycles of odd length.

Problem 13.6 Let $\mathcal{G} = (V, E)$ be a connected bipartite graph. Show that V can be uniquely partitioned into subsets V' and V'' such that $\mathcal{G} = (V' : V'', E)$.

Problem 13.7 Let F be an alphabet of size q. The *k-dimensional Hamming graph* over F is the graph $\mathcal{G}_{k,q} = (V, E)$, where $V = F^k$ and

$$E = \Big\{\{\mathbf{u}, \mathbf{u}'\} \; : \; \mathsf{d}(\mathbf{u}, \mathbf{u}') = 1\Big\}$$

(with $\mathsf{d}(\cdot, \cdot)$ standing for Hamming distance). Figure 13.2 depicts the three-dimensional Hamming graph over $F = \{0, 1\}$.

1. Show that for every two vertices $\mathbf{u}, \mathbf{u}' \in F^k$,
$$\mathsf{d}_{\mathcal{G}_{k,q}}(\mathbf{u}, \mathbf{u}') = \mathsf{d}(\mathbf{u}, \mathbf{u}') \; .$$
2. Show that
$$\mathrm{diam}(\mathcal{G}_{k,q}) = k \; .$$
3. Show that for every vertex $\mathbf{u} \in F^k$,
$$\deg_{\mathcal{G}_{k,q}}(\mathbf{u}) = k(q-1) \; .$$
4. Show that
$$|E| = \tfrac{1}{2}kq^k(q-1) \; .$$
5. Show that $\mathcal{G}_{k,q}$ is bipartite if and only if $q = 2$.

Problem 13.8 Show that for every graph $\mathcal{G} = (V, E)$ and every orientation on $\mathcal{G}$,

$$L_{\mathcal{G}}^{-} = C_{\vec{\mathcal{G}}}^T C_{\vec{\mathcal{G}}}$$

and

$$L_{\mathcal{G}}^{+} = C_{\mathcal{G}}^T C_{\mathcal{G}} \; .$$

Problem 13.9 Let $\mathcal{G} = (V, E)$ be a connected graph. Show that the rank of the incidence matrix $C_{\mathcal{G}}$ of $\mathcal{G}$ is given by

$$\mathrm{rank}(C_{\mathcal{G}}) = \begin{cases} |V| - 1 & \text{if } \mathcal{G} \text{ is bipartite} \\ |V| & \text{otherwise} \end{cases} \; .$$

Hint: A vector $\mathbf{x} = (x_u)_{u \in V}$ belongs to the right kernel of $C_{\mathcal{G}}$ if and only if $x_u = -x_v$ for every $\{u, v\} \in E$. What is the dimension of this kernel?

Problem 13.10 Let $\mathcal{G} = (V' : V'', E)$ be a bipartite graph. Show that λ is an eigenvalue of $A_{\mathcal{G}}$ if and only if so is $-\lambda$ and that both λ and $-\lambda$ have the same (algebraic and geometric) multiplicity. (The algebraic multiplicity of an eigenvalue is its multiplicity as a root of the characteristic polynomial of the matrix, and the geometric multiplicity is the dimension of the linear space that is formed by the associated eigenvectors. In the case of a symmetric matrix, these two multiplicities are equal.)

Hint: Let $\mathbf{x}$ be an eigenvector of $A_{\mathcal{G}}$ associated with the eigenvalue λ. Consider the vector $\mathbf{x}'$ obtained by negating the entries in $\mathbf{x}$ that are indexed by V'.

Problem 13.11 Let A and B be matrices over a field F with orders $m \times n$ and $n \times m$, respectively.

1. Show that the sets of nonzero eigenvalues of AB and BA (in any extension field of F) are the same, and every nonzero eigenvalue has the same geometric multiplicity in both matrices.

   Hint: Suppose that
   $$AB\mathbf{x} = \lambda \mathbf{x}$$
   for $\lambda \neq 0$ and $\mathbf{x} \neq \mathbf{0}$. By left-multiplying both sides of the equation by B, deduce that $B\mathbf{x}$ is an eigenvector of BA associated with the eigenvalue λ (note that this requires also showing that $B\mathbf{x} \neq \mathbf{0}$). Then verify that if $\mathbf{x}_1$ and $\mathbf{x}_2$ are two linearly independent eigenvectors of AB associated with the same eigenvalue $\lambda \neq 0$, then $B\mathbf{x}_1$ and $B\mathbf{x}_2$ are linearly independent as well.

2. Show that when $B = A^T$, then part 1 holds also with respect to the algebraic multiplicity.

   Hint: AA^T and $A^T A$ are symmetric.

(By using Jordan canonical forms, it can be shown that part 1 holds also with respect to the algebraic multiplicity, even when $B \neq A^T$.)

Problem 13.12 Let $A = (A_{i,j})_{i=1}^{m}{}_{j=1}^{n}$ and $B = (B_{i,j})_{i=1}^{r}{}_{j=1}^{s}$ be matrices of orders $m \times n$ and $r \times s$, respectively, over a field F. Recall from Problem 2.21 that the Kronecker product (or direct product) of A and B is defined as the $mr \times ns$ matrix $A \otimes B$ whose entries are given by

$$(A \otimes B)_{r(i-1)+i', s(j-1)+j'} = A_{i,j} B_{i',j'} , \quad 1 \le i \le m,\ 1 \le j \le n,\ 1 \le i' \le r,\ 1 \le j' \le s .$$

1. Let A, B, C, and D be matrices over F such that the number of columns in A (respectively, B) equals the number of rows in C (respectively, D). Show that
   $$(A \otimes B)(C \otimes D) = (AC) \otimes (BD) .$$

2. Show that if A and B are nonsingular square matrices over F then so is $A \otimes B$, and
   $$(A \otimes B)^{-1} = A^{-1} \otimes B^{-1} .$$

3. Let A and B be square matrices over F that can be decomposed into the diagonal forms
   $$A = P \Lambda P^{-1} \quad \text{and} \quad B = Q M Q^{-1} ,$$
   where Λ (respectively, M) is a diagonal matrix whose diagonal consists of the eigenvalues of A (respectively, B), and P and Q are nonsingular matrices. Show that
   $$A \otimes B = (P \otimes Q)(\Lambda \otimes M)(P \otimes Q)^{-1} .$$

4. Show that the eigenvalues of the matrix $A \otimes B$ in part 3 are given by $\lambda\mu$, where λ (respectively, μ) ranges over the eigenvalues of A (respectively, B).

(By using Jordan canonical forms, one can show that the characterization of the eigenvalues of $A \otimes B$ in part 4 holds for any two square matrices A and B.)

Problem 13.13 Let $\mathcal{G} = (V, E)$ be a graph. Define the bipartite graph $\mathcal{G}' = (U' : U'', E')$ by

$$U' = \{u' : u \in V\} , \quad U'' = \{u'' : u \in V\} ,$$

and

$$E' = \Big\{\{u', v''\} : \{u, v\} \in E\Big\} .$$

1. Show that

$$A_{\mathcal{G}'} = \left(\begin{array}{c|c} 0 & A_{\mathcal{G}} \\ \hline A_{\mathcal{G}} & 0 \end{array}\right) .$$

2. Show that the eigenvalues of $A_{\mathcal{G}'}$ are given by $\pm\lambda$, where λ ranges over all the eigenvalues of $A_{\mathcal{G}}$.

   Hint: Apply part 4 of Problem 13.12 to the matrix

$$\begin{pmatrix} 0 & 1 \\ 1 & 0 \end{pmatrix} \otimes A_{\mathcal{G}} .$$

3. Under what conditions on $\mathcal{G}$ is the graph $\mathcal{G}'$ connected?

[Section 13.2]

Problem 13.14 Let $\mathcal{G}$ be an n-regular graph (not necessarily connected) and let $\lambda_1 \geq \lambda_2 \geq \cdots$ be the eigenvalues of $A_{\mathcal{G}}$.

1. Show that if $\mathcal{G}$ is not connected then $\lambda_1 = \lambda_2 = n$.

   Hint: If $\mathcal{G}$ is not connected, then it consists of two isolated n-regular subgraphs, $\mathcal{G}'$ and $\mathcal{G}''$. After re-ordering of vertices, the adjacency matrix of $\mathcal{G}$ can be written in the block-diagonal form

$$A_{\mathcal{G}} = \left(\begin{array}{c|c} A_{\mathcal{G}'} & 0 \\ \hline 0 & A_{\mathcal{G}''} \end{array}\right) .$$

2. Show that $|\lambda_i| \leq n$ for every eigenvalue λ_i of $A_{\mathcal{G}}$ (whether $\mathcal{G}$ is connected or not).

Problem 13.15 Let $\mathcal{G} = (V, E)$ be a graph with $|E| > 0$. The *edge graph* of $\mathcal{G}$ is the graph $\widehat{\mathcal{G}} = (E, \widehat{E})$, where

$$\widehat{E} = \Big\{\{e, e'\} : e, e' \in E \text{ and } |e \cap e'| = 1\Big\} .$$

1. Show that

$$A_{\widehat{\mathcal{G}}} = C_{\mathcal{G}} C_{\mathcal{G}}^T - 2 \cdot I ,$$

   where $C_{\mathcal{G}}$ is the $|E| \times |V|$ incidence matrix of $\mathcal{G}$ and I is the $|E| \times |E|$ identity matrix.

2. Show that if $\mathcal{G}$ is n-regular then $\widehat{\mathcal{G}}$ is $2(n-1)$-regular.

3. Suppose that $\mathcal{G}$ is n-regular. Show that the eigenvalues of $A_{\widehat{\mathcal{G}}}$, other than -2, are given by $n + \lambda - 2$, where λ ranges over all the eigenvalues of $A_{\mathcal{G}}$ other than $-n$. Furthermore, show that for every eigenvalue $\lambda \neq -n$ of $A_{\mathcal{G}}$, the (algebraic and geometric) multiplicity of the eigenvalue $n+\lambda-2$ in $A_{\widehat{\mathcal{G}}}$ is the same as that of λ in $A_{\mathcal{G}}$.

   Hint: Use Lemma 13.2 and Problem 13.11.

4. Let m be the multiplicity of the eigenvalue $-n$ in $A_{\mathcal{G}}$, where m is defined to be zero if $-n$ is not an eigenvalue. Show that -2 is an eigenvalue of $A_{\widehat{\mathcal{G}}}$ if and only if $|E| > |V| - m$; furthermore, show that whenever -2 is an eigenvalue of $A_{\widehat{\mathcal{G}}}$, its multiplicity is $|E| - |V| + m$.

Problem 13.16 Let $\mathcal{G} = (V, E)$ be a connected graph with $|V| > 1$. Consider the bipartite graph $\widetilde{\mathcal{G}} = (E : V, \widetilde{E})$, where

$$\widetilde{E} = \Big\{ \{e, v\} \; : \; e \in E \text{ and } v \in e \Big\} .$$

1. Show that $\widetilde{\mathcal{G}}$ is connected.

2. Show that

$$A_{\widetilde{\mathcal{G}}} = \left(\begin{array}{c|c} 0 & C_{\mathcal{G}} \\ \hline C_{\mathcal{G}}^T & 0 \end{array} \right) ,$$

   where $C_{\mathcal{G}}$ is the $|E| \times |V|$ incidence matrix of $\mathcal{G}$.

3. Show that

$$A_{\widetilde{\mathcal{G}}}^2 = \left(\begin{array}{c|c} C_{\mathcal{G}} C_{\mathcal{G}}^T & 0 \\ \hline 0 & C_{\mathcal{G}}^T C_{\mathcal{G}} \end{array} \right) .$$

4. Show that the (algebraic and geometric) multiplicity of the zero eigenvalue in $A_{\widetilde{\mathcal{G}}}$ equals

$$\begin{cases} |E| - |V| + 2 & \text{if } \mathcal{G} \text{ is bipartite} \\ |E| - |V| & \text{otherwise} \end{cases} .$$

   Hint: Use Problem 13.9.

5. Suppose that $\mathcal{G}$ is n-regular (as well as connected). Show that the nonzero eigenvalues of $A_{\widetilde{\mathcal{G}}}$ are given by $\pm\sqrt{n+\lambda}$, where λ ranges over all the eigenvalues of $A_{\mathcal{G}}$ other than $-n$. Furthermore, show that the multiplicity of each nonzero eigenvalue $\pm\sqrt{n+\lambda}$ in $A_{\widetilde{\mathcal{G}}}$ is the same as that of λ in $A_{\mathcal{G}}$.

   Hint: Use Lemma 13.2 and Problems 13.10 and 13.11.

6. Let $\mathcal{G}$ be the graph in Figure 13.2. Verify that $\widetilde{\mathcal{G}}$ is given by the graph in Figure 13.6. What is the rule for the naming of the vertices of E in that figure?

Problem 13.17 Let $\mathcal{G} = (V' : V'', E)$ be a bipartite n-regular graph and let $X_{\mathcal{G}}$ be the $|V'| \times |V''|$ transfer matrix of $\mathcal{G}$ (where $|V'| = |V''|$).

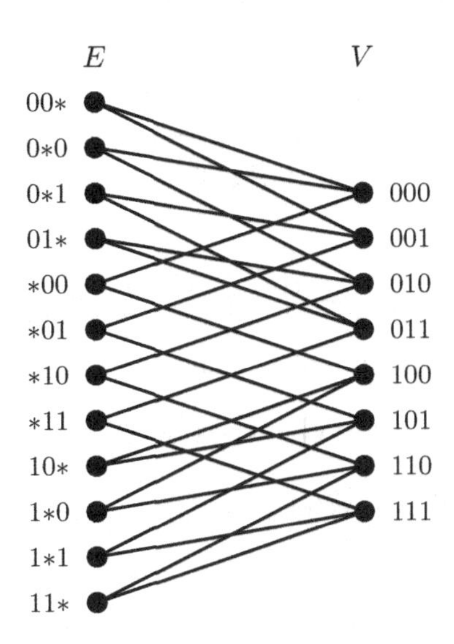

Figure 13.6. Graph $\widetilde{\mathcal{G}}$ for Problem 13.16.

1. Show that

$$A_{\widetilde{\mathcal{G}}}^2 = \left(\begin{array}{c|c} X_{\mathcal{G}} X_{\mathcal{G}}^T & 0 \\ \hline 0 & X_{\mathcal{G}}^T X_{\mathcal{G}} \end{array} \right) .$$

2. Show that the sum of entries along every row in each of the (symmetric) matrices $X_{\mathcal{G}} X_{\mathcal{G}}^T$ and $X_{\mathcal{G}}^T X_{\mathcal{G}}$ equals n^2. Deduce that n^2 is the largest eigenvalue of each of these matrices.

3. Show that $\gamma_{\mathcal{G}}^2 n^2$ is the second largest eigenvalue of $X_{\mathcal{G}} X_{\mathcal{G}}^T$ and of $X_{\mathcal{G}}^T X_{\mathcal{G}}$, whenever $|V'| > 1$.

   Hint: Use Problems 13.10 and 13.11.

[Section 13.3]

Problem 13.18 Let $\mathcal{G} = (V, E)$ be an n-regular graph and let $\chi : V \to \mathbb{R}$ be a function on the vertices of $\mathcal{G}$. Define the function $\omega : E \to \mathbb{R}$ for every edge $e = \{u, v\}$ in $\mathcal{G}$ by

$$\omega(e) = \chi(u)\chi(v) .$$

Denote by $\mathsf{E}_{\mathcal{G}}\{\omega\}$ the average of the values of the function ω over the edges of $\mathcal{G}$, namely,

$$\mathsf{E}_{\mathcal{G}}\{\omega\} = \frac{2}{n|V|} \sum_{e \in E} \omega(e) .$$

(The notation $\mathsf{E}_{\mathcal{G}}\{\cdot\}$ is interpreted here also as an expected value of a random variable over the set of edges or the set of vertices of $\mathcal{G}$. To this end, assume a uniform distribution over the edges of $\mathcal{G}$; this, in turn, induces a uniform distribution over the vertices of $\mathcal{G}$, if the selection of a vertex is carried out by first randomly selecting an edge e with probability $2/(n|V|)$, and then choosing each of the endpoints of e with probability $\frac{1}{2}$.)

1. Show that

$$\begin{aligned}\mathsf{E}_{\mathcal{G}}\{\omega\} &\le (1-\gamma_{\mathcal{G}})\left(\mathsf{E}_{\mathcal{G}}\{\chi\}\right)^2 + \gamma_{\mathcal{G}}\mathsf{E}_{\mathcal{G}}\{\chi^2\} \\ &= \left(\mathsf{E}_{\mathcal{G}}\{\chi\}\right)^2 + \gamma_{\mathcal{G}}\mathsf{Var}_{\mathcal{G}}\{\chi\}\,,\end{aligned}$$

where

$$\mathsf{E}_{\mathcal{G}}\{\chi\} = \frac{1}{|V|}\sum_{u\in V}\chi(u)\,, \quad \mathsf{E}_{\mathcal{G}}\{\chi^2\} = \frac{1}{|V|}\sum_{u\in V}(\chi(u))^2\,,$$

and $\mathsf{Var}_{\mathcal{G}}\{\chi\}$ is the variance of χ, namely,

$$\mathsf{Var}_{\mathcal{G}}\{\chi\} = \mathsf{E}_{\mathcal{G}}\{\chi^2\} - \left(\mathsf{E}_{\mathcal{G}}\{\chi\}\right)^2\,.$$

Hint: Define the column vector $\mathbf{y} = (y_u)_{u\in V}$ by

$$y_u = \chi(u) - \mathsf{E}_{\mathcal{G}}\{\chi\}\,, \quad u \in V\,.$$

First verify that

$$\langle \mathbf{y}, \mathbf{1}\rangle = 0 \qquad \text{and} \qquad \|\mathbf{y}\|^2 = \mathsf{Var}_{\mathcal{G}}\{\chi\}\cdot|V|\,.$$

Then consider the column vector $\mathbf{x} = (\chi(u))_{u\in V}$ and observe that

$$\mathsf{E}_{\mathcal{G}}\{\omega\} = \frac{\langle \mathbf{x}, A_{\mathcal{G}}\mathbf{x}\rangle}{n|V|}\,.$$

Using the relationship $\mathbf{x} = \mathbf{y} + \mathsf{E}_{\mathcal{G}}\{\chi\}\cdot\mathbf{1}$ and Lemma 13.6, show that

$$\langle \mathbf{x}, A_{\mathcal{G}}\mathbf{x}\rangle = \langle \mathbf{y}, A_{\mathcal{G}}\mathbf{y}\rangle + \left(\mathsf{E}_{\mathcal{G}}\{\chi\}\right)^2\cdot n|V| \le \left(\gamma_{\mathcal{G}}\mathsf{Var}_{\mathcal{G}}\{\chi\} + \left(\mathsf{E}_{\mathcal{G}}\{\chi\}\right)^2\right)\cdot n|V|\,.$$

2. Derive Corollary 13.10 from part 1 by selecting, for a given subset $S \subseteq V$, the function χ to be

$$\chi(u) = \begin{cases} 1 & \text{if } u \in S \\ 0 & \text{otherwise}\end{cases}\,.$$

Problem 13.19 Let $\mathcal{G} = (V' : V'', E)$ be a bipartite n-regular graph with $|V'| > 1$ and let $\mathbf{s} = (s_u)_{u\in V'}$ and $\mathbf{t} = (t_u)_{u\in V''}$ be two column vectors in $\mathbb{R}^{|V'|}$. Denote by σ and τ the averages

$$\sigma = \frac{1}{|V'|}\sum_{u\in V'} s_u \qquad \text{and} \qquad \tau = \frac{1}{|V'|}\sum_{u\in V''} t_u\,,$$

and let the column vectors $\mathbf{y}$ and $\mathbf{z}$ in $\mathbb{R}^{|V'|}$ be defined by

$$\mathbf{y} = \mathbf{s} - \sigma\cdot\mathbf{1} \qquad \text{and} \qquad \mathbf{z} = \mathbf{t} - \tau\cdot\mathbf{1}\,.$$

1. Show that

$$\langle \mathbf{y}, X_{\mathcal{G}}\mathbf{z}\rangle = \langle \mathbf{s}, X_{\mathcal{G}}\mathbf{t}\rangle - \sigma\tau n|V'|\,,$$

where $X_{\mathcal{G}}$ is the transfer matrix of $\mathcal{G}$.

Hint: $X_{\mathcal{G}}\mathbf{1} = X_{\mathcal{G}}^T\mathbf{1} = n\cdot\mathbf{1}$.

2. Show that

$$\|X_{\mathcal{G}}\mathbf{z}\|^2 = \langle \mathbf{z}, X_{\mathcal{G}}^T X_{\mathcal{G}}\mathbf{z}\rangle \le \gamma_{\mathcal{G}}^2 n^2 \|\mathbf{z}\|^2 .$$

Hint: Apply Lemma 13.6 to the matrix $X_{\mathcal{G}}^T X_{\mathcal{G}}$ while making use of Problem 13.17.

3. Show that

$$\Big|\langle \mathbf{s}, X_{\mathcal{G}}\mathbf{t}\rangle - \sigma\tau n|V'|\Big| = |\langle \mathbf{y}, X_{\mathcal{G}}\mathbf{z}\rangle| \le \|\mathbf{y}\| \cdot \|X_{\mathcal{G}}\mathbf{z}\| \le \gamma_{\mathcal{G}} n \|\mathbf{y}\|\cdot\|\mathbf{z}\| .$$

Hint: Use the Cauchy–Schwartz inequality.

4. Let the vector $\mathbf{x} \in \mathbb{R}^{2|V'|}$ be defined by

$$\mathbf{x} = \begin{pmatrix} \mathbf{s} \\ \mathbf{t} \end{pmatrix} .$$

Show that

$$\langle \mathbf{x}, A_{\mathcal{G}}\mathbf{x}\rangle = 2\langle \mathbf{s}, X_{\mathcal{G}}\mathbf{t}\rangle$$

and deduce that

$$\Big|\langle \mathbf{x}, A_{\mathcal{G}}\mathbf{x}\rangle - 2\sigma\tau n|V'|\Big| \le 2\gamma_{\mathcal{G}} n \|\mathbf{y}\|\cdot\|\mathbf{z}\| .$$

Problem 13.20 The purpose of this problem is to improve on Corollary 13.10 for the case of bipartite regular graphs.

Let $\mathcal{G} = (V' : V'', E)$ be a bipartite n-regular graph with $|V'| > 1$ and let $\chi : (V' \cup V'') \to \mathbb{R}$ be a function on the vertices of $\mathcal{G}$. Define the function $\omega : E \to \mathbb{R}$ and the average $\mathsf{E}_{\mathcal{G}}\{\omega\}$ as in Problem 13.18, namely,

$$\omega(e) = \chi(u)\chi(v) \quad \text{for every edge } e = \{u, v\} \text{ in } \mathcal{G}$$

and

$$\mathsf{E}_{\mathcal{G}}\{\omega\} = \frac{1}{n|V'|}\sum_{e\in E} \omega(e) .$$

1. Show that

$$\Big|\mathsf{E}_{\mathcal{G}}\{\omega\} - \mathsf{E}'_{\mathcal{G}}\{\chi\}\cdot\mathsf{E}''_{\mathcal{G}}\{\chi\}\Big| \le \gamma_{\mathcal{G}}\sqrt{\mathsf{Var}'_{\mathcal{G}}\{\chi\}\cdot\mathsf{Var}''_{\mathcal{G}}\{\chi\}} ,$$

where

$$\mathsf{E}'_{\mathcal{G}}\{\chi^i\} = \frac{1}{|V'|}\sum_{u\in V'} (\chi(u))^i , \qquad \mathsf{E}''_{\mathcal{G}}\{\chi^i\} = \frac{1}{|V''|}\sum_{u\in V''} (\chi(u))^i ,$$

$$\mathsf{Var}'_{\mathcal{G}}\{\chi\} = \mathsf{E}'_{\mathcal{G}}\{\chi^2\} - \left(\mathsf{E}'_{\mathcal{G}}\{\chi\}\right)^2 , \quad \text{and} \quad \mathsf{Var}''_{\mathcal{G}}\{\chi\} = \mathsf{E}''_{\mathcal{G}}\{\chi^2\} - \left(\mathsf{E}''_{\mathcal{G}}\{\chi\}\right)^2 .$$

Hint: As in Problem 13.18, define the column vector $\mathbf{x} = (\chi(u))_{u\in V'\cup V''}$ and notice that

$$\mathsf{E}_{\mathcal{G}}\{\omega\} = \frac{\langle \mathbf{x}, A_{\mathcal{G}}\mathbf{x}\rangle}{2n|V'|} .$$

Sub-divide $\mathbf{x}$ into the sub-vectors $\mathbf{s} = (\chi(u))_{u\in V'}$ and $\mathbf{t} = (\chi(u))_{u\in V''}$, and apply part 4 of Problem 13.19.

2. Let $S \subseteq V'$ and $T \subseteq V''$ be subsets of sizes $|S| = \sigma|V'|$ and $|T| = \tau|V''|$, respectively. Show that

$$\left| \frac{|E_{S,T}|}{n|V'|} - \sigma\tau \right| \le \gamma_{\mathcal{G}} \sqrt{\sigma(1-\sigma)\tau(1-\tau)} ,$$

and obtain the following upper bound on the sum of degrees of the vertices within the induced subgraph $\mathcal{G}_{S \cup T}$:

$$\sum_{u \in S \cup T} \deg_{\mathcal{G}_{S\cup T}}(u) = 2|E_{S,T}| \le 2\left(\sigma\tau + \gamma_{\mathcal{G}} \sqrt{\sigma(1-\sigma)\tau(1-\tau)}\right) n|V'| .$$

Hint: Select the function χ in part 1 to be

$$\chi(u) = \begin{cases} 1 & \text{if } u \in S \cup T \\ 0 & \text{otherwise} \end{cases} .$$

3. Deduce from part 2 the following upper bound on the average degree in $\mathcal{G}_{S\cup T}$, whenever $\sigma + \tau > 0$:

$$\begin{aligned} \frac{1}{|S \cup T|} \sum_{u \in S \cup T} \deg_{\mathcal{G}_{S\cup T}}(u) &\le \left((1-\gamma_{\mathcal{G}}) \frac{2\sigma\tau}{\sigma+\tau} + \gamma_{\mathcal{G}} \cdot \frac{2\sqrt{\sigma\tau}}{\sigma+\tau} \right) n \\ &\le \left((1-\gamma_{\mathcal{G}}) \frac{2\sigma\tau}{\sigma+\tau} + \gamma_{\mathcal{G}} \right) n . \end{aligned}$$

Hint:

$$\sqrt{\sigma\tau} + \sqrt{(1-\sigma)(1-\tau)} \le 1 .$$

(It follows from part 2 that for every two subsets $S \subseteq V'$ and $T \subseteq V''$, the fraction of edges of $\mathcal{G}$ that belong to $\mathcal{G}_{S\cup T}$ is concentrated around $\sigma\tau$: the smaller $\gamma_{\mathcal{G}}$ is, the sharper the bound is on this concentration.)

Problem 13.21 Let $\mathcal{G} = (V' : V'', E)$ be a bipartite n-regular graph and let S be a nonempty subset of V'' of size $\sigma|V''|$. The purpose of this problem is to obtain the lower bound

$$|\mathcal{N}(S)| \ge \frac{|S|}{(1-\gamma_{\mathcal{G}}^2)\sigma + \gamma_{\mathcal{G}}^2} .$$

Denote by $\mathbf{s} = (s_u)_{u \in V''}$ the vector

$$s_u = \begin{cases} 1 & \text{if } u \in S \\ 0 & \text{if } u \in V'' \setminus S \end{cases} , \quad u \in V'' .$$

1. Show that

$$|\mathcal{N}(S)| \ge \frac{n^2|S|^2}{\|X_{\mathcal{G}}\mathbf{s}\|^2} ,$$

where $X_{\mathcal{G}}$ is the $|V'| \times |V''|$ transfer matrix of $\mathcal{G}$.

Hint: Observing that $\mathcal{N}(S)$ is the support of the vector $X_{\mathcal{G}}\mathbf{s}$, use the $\cup$-convexity of the function $z \mapsto z^2$ to argue that

$$\frac{\|X_{\mathcal{G}}\mathbf{s}\|^2}{|\mathcal{N}(S)|} = \frac{\sum_{u \in V'} (X_{\mathcal{G}}\mathbf{s})_u^2}{|\mathcal{N}(S)|} \ge \left(\frac{\sum_{u \in V'} (X_{\mathcal{G}}\mathbf{s})_u}{|\mathcal{N}(S)|} \right)^2 = \frac{n^2|S|^2}{|\mathcal{N}(S)|^2}$$

(see Jensen's inequality in the notes on Section 1.4).

2. Let $\mathbf{y}$ denote the vector $\mathbf{s} - \sigma \cdot \mathbf{1}$. Justify the following chain of equalities:

$$\langle X_{\mathcal{G}}\mathbf{y}, X_{\mathcal{G}}\mathbf{1}\rangle = \langle \mathbf{y}, X_{\mathcal{G}}^T X_{\mathcal{G}}\mathbf{1}\rangle = n^2\langle \mathbf{y}, \mathbf{1}\rangle = 0 .$$

3. Justify the following steps:

$$\begin{aligned} \|X_{\mathcal{G}}\mathbf{s}\|^2 &= \|X_{\mathcal{G}}(\mathbf{y} + \sigma \cdot \mathbf{1})\|^2 \\ &= \|X_{\mathcal{G}}\mathbf{y}\|^2 + 2\sigma\langle X_{\mathcal{G}}\mathbf{y}, X_{\mathcal{G}}\mathbf{1}\rangle + \sigma^2\|X_{\mathcal{G}}\mathbf{1}\|^2 \\ &= \|X_{\mathcal{G}}\mathbf{y}\|^2 + \sigma^2 n^2 |V'| \\ &\le \gamma_{\mathcal{G}}^2\sigma(1-\sigma)n^2|V'| + \sigma^2 n^2|V'| \\ &= \left((1-\gamma_{\mathcal{G}}^2)\sigma + \gamma_{\mathcal{G}}^2\right) n^2 |S| . \end{aligned}$$

Hint: See part 2 of Problem 13.19.

4. Deduce from parts 1 and 3 that

$$|\mathcal{N}(S)| \ge \frac{|S|}{(1-\gamma_{\mathcal{G}}^2)\sigma + \gamma_{\mathcal{G}}^2} .$$

(The latter inequality can alternatively be obtained by applying part 2 of Problem 13.20 to the set $T = \mathcal{N}(S)$, in which case $|E_{S,T}| = n|S|$. A lower bound on $|\mathcal{N}(S)|$ can be obtained also when S is a subset of $V' \cup V''$ that is not wholly contained in V'' (or V'). To this end, partition S into the disjoint sets $S \cap V'$ and $S \cap V''$. The neighborhoods $\mathcal{N}(S \cap V')$ and $\mathcal{N}(S \cap V'')$ are also disjoint and, so,

$$|\mathcal{N}(S)| = |\mathcal{N}(S \cap V')| + |\mathcal{N}(S \cap V'')| .$$

Now apply part 4 to obtain lower bounds on $|\mathcal{N}(S \cap V')|$ and $|\mathcal{N}(S \cap V'')|$.)

Problem 13.22 Let $\mathcal{G} = (V' : V'', E)$ be a bipartite n-regular graph and S be a subset of V''. Denote $\sigma = |S|/|V''|$ and $J_S(u) = |\mathcal{N}(u) \cap S|$. The purpose of this problem is to show that

$$\frac{1}{|V'|}\sum_{u \in V'} (J_S(u) - \sigma n)^2 \le \gamma_{\mathcal{G}}^2 n^2 \sigma(1-\sigma) .$$

1. Using the notation in Problem 13.21, justify the following chain of equalities:

$$(X_{\mathcal{G}}\mathbf{y})_u = (X_{\mathcal{G}}\mathbf{s})_u - \sigma \cdot (X_{\mathcal{G}}\mathbf{1})_u = |\mathcal{N}(u) \cap S| - \sigma n = J_S(u) - \sigma n , \quad u \in V .$$

2. Conclude from part 2 of Problem 13.19 that

$$\frac{1}{|V'|}\sum_{u \in V'} (J_S(u) - \sigma n)^2 = \frac{\|X_{\mathcal{G}}\mathbf{y}\|^2}{|V'|} \le \gamma_{\mathcal{G}}^2 n^2 \sigma(1-\sigma) .$$

(Note that for every subset $S \subseteq V''$ of size $\sigma|V''|$,

$$\frac{1}{|V'|}\sum_{u \in V'} J_S(u) = \sigma n .$$

It follows from this problem that $\gamma_{\mathcal{G}}^2 n^2 \sigma(1-\sigma)$ is an upper bound on the variance of $J_S(u)$, when the latter is regarded as a random variable with u being uniformly distributed over V'.)

Problem 13.23 Let $\mathcal{G} = (V, E)$ be a non-bipartite n-regular graph. Denote by $\lambda_1 \geq \lambda_2 \geq \cdots$ the eigenvalues of $A_{\mathcal{G}}$ and define

$$\varrho = \frac{1}{n} \max_{2 \leq i \leq |V|} |\lambda_i| .$$

Let S be a subset of V of size $\sigma|V|$. Show that

$$|\mathcal{N}(S)| \geq \frac{|S|}{(1-\varrho^2)\sigma + \varrho^2} .$$

Hint: Apply the bound in part 4 of Problem 13.21 to the graph $\mathcal{G}'$ obtained by the construction in Problem 13.13.

Problem 13.24 Let $\mathcal{G} = (V, E)$ be a non-bipartite n-regular graph and S be a subset of V. Denote $\sigma = |S|/|V|$ and $J_S(u) = |\mathcal{N}(u) \cap S|$. Show that

$$\frac{1}{|V|} \sum_{u \in V} \left(J_S(u) - \sigma n\right)^2 \leq \varrho^2 n^2 \sigma(1-\sigma) ,$$

where ϱ is as in Problem 13.23.

Hint: See Problem 13.22.

[Section 13.4]

Problem 13.25 Let $\mathcal{H}_k$ be the $2^k \times 2^k$ Sylvester-type Hadamard matrix. Show that for every $k \geq 0$,

$$\mathcal{H}_{k+1} = \begin{pmatrix} 1 & 1 \\ 1 & -1 \end{pmatrix} \otimes \mathcal{H}_k ,$$

where $\otimes$ stands for the Kronecker product of matrices (see Problem 13.12).

Problem 13.26 (Generalization of Proposition 13.13) Let $F = \mathrm{GF}(p)$ where p is a prime and let B be a subset of $F^k \setminus \{\mathbf{0}\}$ that is closed under negation, that is $\mathbf{u} \in B \Longrightarrow -\mathbf{u} \in B$. Denote by ω a root of order p of unity in the complex field $\mathbb{C}$. Show that the eigenvalues of the adjacency matrix of the Cayley graph $\mathcal{G}(F^k, B)$ are given by

$$\lambda_{\mathbf{v}} = \sum_{\mathbf{u} \in B} \omega^{\langle \mathbf{u}, \mathbf{v} \rangle} = \sum_{\mathbf{u} \in B} \cos\left(\frac{2\pi}{p} \langle \mathbf{u}, \mathbf{v} \rangle\right) , \quad \mathbf{v} \in F^k ,$$

where $\pi = 3.14159 \cdots$.

Hint: Consider the $p^k \times p^k$ matrix $\mathcal{H}_{k,p}$ whose rows and columns are indexed by F^k and

$$(\mathcal{H}_{k,p})_{\mathbf{u},\mathbf{v}} = \omega^{\langle \mathbf{u}, \mathbf{v} \rangle} , \quad \mathbf{u}, \mathbf{v} \in F^k .$$

Show that the column of $\mathcal{H}_{k,p}$ that is indexed by $\mathbf{v}$ is an eigenvector of $A_{\mathcal{G}(F^k,B)}$ associated with the eigenvalue $\lambda_{\mathbf{v}}$. Also, verify that the columns of $\mathcal{H}_{k,p}$ are linearly independent over $\mathbb{C}$ by computing the product of $\mathcal{H}_{k,p}$ with its conjugate transpose $\mathcal{H}_{k,p}^*$.

Problem 13.27 Let $\mathcal{G}$ be the graph in Figure 13.7.

1. Show that $\mathcal{G}$ is bipartite and compute its adjacency matrix $A_{\mathcal{G}}$.

2. Compute the eigenvalues of $A_{\mathcal{G}}$.

   Hint: The eigenvalues can be found by direct inspection of the matrix $A_{\mathcal{G}}$. Alternatively, verify that $\mathcal{G}$ is the Cayley graph $\mathcal{G}(F^3, B)$, where $F = \mathrm{GF}(2)$ and B is the set of columns of the matrix

$$G = \begin{pmatrix} 1 & 0 & 0 & 1 \\ 0 & 1 & 0 & 1 \\ 0 & 0 & 1 & 1 \end{pmatrix} .$$

   Then use Corollary 13.14.

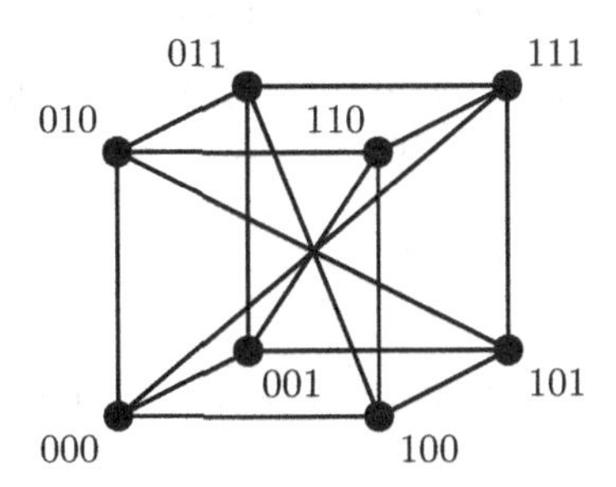

Figure 13.7. Graph $\mathcal{G}$ for Problem 13.27.

Problem 13.28 Let $\mathcal{G}_{k,2}$ be the Hamming graph over $F = \mathrm{GF}(2)$ as defined in Problem 13.7. Compute the eigenvalues of $A_{\mathcal{G}_{k,2}}$.

Hint: Verify that $\mathcal{G}_{k,2}$ can be represented as a Cayley graph $\mathcal{G}(F^k, B)$, where B is the set of columns of the $k \times k$ identity matrix over F. Then use Corollary 13.14.

Problem 13.29 (Generalization of Corollary 13.14) Let p be a prime and let $\mathcal{C}$ be a linear $[n, k, d]$ code over $F = \mathrm{GF}(p)$ whose dual code, $\mathcal{C}^{\perp}$, has minimum distance $d^{\perp} \geq 3$. Let B be the set of all the nonzero scalar multiples of the columns of a given generator matrix of $\mathcal{C}$.

1. Show that the Cayley graph $\mathcal{G}(F^k, B)$ is a connected $(p-1)n$-regular graph.

2. Show that the eigenvalues of the adjacency matrix of $\mathcal{G}(F^k, B)$ are given by

$$(p-1)n - p \cdot \mathsf{w}(\mathbf{c}) \, , \quad \mathbf{c} \in \mathcal{C} \, .$$

   Hint: See Problem 13.26.

3. Show that

$$\gamma_{\mathcal{G}(F^k,B)} = 1 - \frac{pd}{(p-1)n} \, .$$

Problem 13.30 Let $F = \mathrm{GF}(2)$ and let $x \mapsto \mathsf{H}_2(x)$ be the binary entropy function.

1. Let δ be a real in the interval $[0, \frac{1}{2})$ and R be a positive real smaller than $1 - \mathsf{H}_2(\delta)$. Show that for sufficiently large integer n, there exists a linear $[n, k, d]$ code $\mathcal{C}$ over F such that $k \geq nR$, $d \geq \delta n$, the all-one word is a codeword of $\mathcal{C}$, and the minimum distance of $\mathcal{C}^\perp$ is at least 3.

   Hint: Assume a uniform distribution over all $k \times n$ matrices over F whose first row is all-one. Show that the probability that such a matrix has two identical columns is at most $\binom{n}{2} 2^{1-k}$, and proceed along the lines of the proof of Theorem 4.5, while making use of Lemma 4.7.

2. Show that for $\varepsilon \in (0, 1]$,

$$1 - \mathsf{H}_2\left(\tfrac{1}{2}(1-\varepsilon)\right) > \tfrac{1}{2}(\log_2 \mathsf{e}) \cdot \varepsilon^2 ,$$

   where $\mathsf{e} = 2.71828\cdots$ is the base of natural logarithms.

3. Fix ε to be a real in the interval $(0, 1]$. Show that for every sufficiently large integer n there exists a bipartite n-regular Cayley graph $\mathcal{G}(F^k, B) = (V, E)$ such that $|V| = 2^k \geq \mathsf{e}^{\varepsilon^2 n/2}$ and

$$\gamma_{\mathcal{G}(F^k,B)} \leq \varepsilon .$$

4. Recall from Section 12.3 that the Zyablov bound for the binary field takes the form

$$R_{\mathrm{Z}}(\delta, 2) = \max_{\theta \in [\delta, \frac{1}{2}]} \Bigl(1 - \mathsf{H}_2(\theta)\Bigr)\left(1 - \frac{\delta}{\theta}\right), \qquad \delta \in (0, \tfrac{1}{2}) .$$

   Show that for every $\varepsilon \in (0, 1)$,

$$R_{\mathrm{Z}}\left(\tfrac{1}{2}(1-\varepsilon), 2\right) > \tfrac{2}{27}(\log_2 \mathsf{e}) \cdot \varepsilon^3 .$$

   Hint: Substitute $\delta = \frac{1}{2}(1-\varepsilon)$ and $\theta = \frac{1}{2} - \frac{1}{3}\varepsilon$ in $(1 - \mathsf{H}_2(\theta))(1 - (\delta/\theta))$.

5. Show that by relaxing the lower bound on $|V|$ in part 3 to $\mathsf{e}^{2\varepsilon^3 n/27}$, the set B therein can be found in time complexity $O(n^c)$, where c depends on ε.

[Section 13.5]

Problem 13.31 Let q be a power of a prime and k be a positive integer.

1. Show that

$$|\mathrm{SL}_k(q)| = q^{k-1} \prod_{i=0}^{k-2} (q^k - q^i) .$$

   Hint: The number of $k \times k$ nonsingular matrices over $\mathrm{GF}(q)$ is equal to $\prod_{i=0}^{k-1} (q^k - q^i)$.

2. Show that

$$|\mathrm{PSL}_k(q)| = \frac{|\mathrm{SL}_k(q)|}{\gcd(k, q-1)} .$$

Problem 13.32 Let $F = \mathrm{GF}(q)$ such that $q \equiv 1 \pmod 4$ and let $\imath$ be a root of $x^2 + 1 = 0$ in F. Fix $\alpha_0, \alpha_1, \alpha_2, \alpha_3$, and η to be elements in F such that $\alpha_0^2 + \alpha_1^2 + \alpha_2^2 + \alpha_3^2 = \eta^2$, and consider the 2×2 matrix

$$M = \frac{1}{\eta} \begin{pmatrix} \alpha_0 + \imath\alpha_1 & \alpha_2 + \imath\alpha_3 \\ -\alpha_2 + \imath\alpha_3 & \alpha_0 - \imath\alpha_1 \end{pmatrix}$$

over F.

1. Show that $\det(M) = 1$.
2. Show that
$$M^{-1} = \frac{1}{\eta} \begin{pmatrix} \alpha_0 - \imath\alpha_1 & -\alpha_2 - \imath\alpha_3 \\ \alpha_2 - \imath\alpha_3 & \alpha_0 + \imath\alpha_1 \end{pmatrix} .$$

[Section 13.6]

Problem 13.33 Let $\mathcal{G} = (V, E)$ be a connected n-regular graph and $\mathcal{C}$ be the $[n, n-1, 2]$ parity code over $F = \mathrm{GF}(q)$.

1. Show that the $|V| \times |E|$ matrix $C_{\mathcal{G}}^T$, when viewed as a matrix over F, is a parity-check matrix of the graph code $(\mathcal{G}, \mathcal{C})$.
2. Show that the dimension of $(\mathcal{G}, \mathcal{C})$ is given by
$$\begin{cases} |E| - |V| + 1 & \text{if } q \text{ is even or } \mathcal{G} \text{ is bipartite} \\ |E| - |V| & \text{otherwise} \end{cases} .$$

   Hint: Characterize the *left* kernel of $C_{\mathcal{G}}^T$ (i.e., the right kernel of $C_{\mathcal{G}}$) over F (see also Problem 13.9).

Problem 13.34 Let $F = \mathrm{GF}(q)$ and let $\mathcal{G} = (V, E)$ be a non-bipartite n-regular Ramanujan graph. For every vertex $u \in V$, assume an ordering on the set, $\mathcal{N}(u)$, of neighbors of u. Write $N = |V|$, and for a word $\mathbf{x} = (x_u)_{u \in V}$, let $(\mathbf{x})_{\mathcal{N}(u)}$ be the sub-word of $\mathbf{x}$ that is indexed by the n elements of $\mathcal{N}(u)$.

Denote by Φ the alphabet F^n and define the mapping $\varphi_{\mathcal{G}} : F^N \to \Phi^N$ by

$$\varphi_{\mathcal{G}}(\mathbf{x}) = \left((\mathbf{x})_{\mathcal{N}(u)}\right)_{u \in V} , \quad \mathbf{x} \in F^N ;$$

namely, the entry of $\varphi_{\mathcal{G}}(\mathbf{x})$ that is indexed by a vertex u equals the sub-word $(\mathbf{x})_{\mathcal{N}(u)}$ (in Φ).

Let $\mathcal{C}$ be a linear $[N, K{=}rN, D{=}\theta N]$ code over F and define the code $\varphi_{\mathcal{G}}(\mathcal{C})$ over Φ by

$$\varphi_{\mathcal{G}}(\mathcal{C}) = \{\varphi_{\mathcal{G}}(\mathbf{c}) \,:\, \mathbf{c} \in \mathcal{C}\} .$$

1. Show that $\varphi_{\mathcal{G}}(\mathcal{C})$ is a linear space over F.
2. Let the real δ_n be defined by
$$\delta_n = \frac{\theta n}{\theta n + 4(1-\theta)} .$$

   Show that the relative minimum distance of $\varphi_{\mathcal{G}}(\mathcal{C})$ is greater than δ_n.

Hint: First verify that the minimum distance of the code $\varphi_{\mathcal{G}}(\mathcal{C})$ equals the minimum Hamming weight (over Φ) of any nonzero codeword of this code (a codeword is nonzero if at least one of its coordinates is not the zero vector of Φ). Then apply Problem 13.23 to the graph $\mathcal{G}$, with S taken as the support of a nonzero codeword $\mathbf{c} \in \mathcal{C}$.

3. Show that the rate R_n of $\varphi_{\mathcal{G}}(\mathcal{C})$ is related to δ_n by

$$R_n = \frac{r}{n} = \kappa \cdot \left(\frac{1}{\delta_n} - 1 \right) ,$$

where

$$\kappa = \frac{r\theta}{4(1-\theta)}$$

(note that κ is a constant that depends on r and θ but not on n).

4. Recall from Problem 12.8 that the Zyablov bound satisfies

$$R_{\mathrm{Z}}(\delta, q^n) \leq (1 - \sqrt{\delta})^2 .$$

Show that

$$R_n > (1 - \sqrt{\delta_n})^2 \geq R_{\mathrm{Z}}(\delta_n, q^n) \quad \text{whenever} \quad \delta_n > \left(\frac{1-\kappa}{1+\kappa} \right)^2 .$$

(It follows from this problem that the Zyablov bound can be exceeded by the codes $\varphi_{\mathcal{G}}(\mathcal{C})$ for a certain range of alphabet sizes and relative minimum distances. The code length N can be made arbitrarily large, by taking $\mathcal{G}$ as the LPS construction and $\mathcal{C}$ as the Justesen construction of Section 12.4.)

[Section 13.7]

Problem 13.35 Let $\mathsf{C} = (\mathcal{G}, \mathcal{C})$ be a linear $[N, K, D]$ graph code over $F = \mathrm{GF}(q)$, where $\mathcal{G} = (V, E)$ is an $(n, \xi{>}0)$-expander with $|E| = n|V|/2 = N$ and $\mathcal{C}$ is a linear $[n, k{=}rn, d{=}\theta n{>}1]$ code over F. A codeword $\mathbf{c} \in \mathsf{C}$ is transmitted through an additive channel (F, F, Prob) and a word $\mathbf{y} \in F^N$ is received, where hereafter the entries of words in F^N are indexed by the elements of E. The purpose of this problem is to bound from below the fraction, $\zeta = \zeta_{\mathsf{C}}(\mathbf{y}, \mathbf{c})$, of vertices $u \in V$ for which $(\mathbf{y})_{E(u)} \notin \mathcal{C}$, by an expression that is a function of $\mathsf{d}(\mathbf{y}, \mathbf{c})$ and the parameters of C.

(As an application of such a lower bound, consider an error-detection algorithm that performs t statistically independent trials, where each trial consists of selecting a vertex $u \in V$ uniformly at random and testing whether $(\mathbf{y})_{E(u)} \in \mathcal{C}$. The complexity of each such test depends on the parameters of $\mathcal{C}$, yet not on N, and the probability of misdetecting an error event after t statistically independent trials is $(1 - \zeta)^t$.)

Denote by $Y \subseteq E$ the support of the error word $\mathbf{y} - \mathbf{c}$, let S be the set of all vertices in $\mathcal{G}$ that are endpoints of edges in Y, and let T be the following subset of S:

$$T = \{u \in S \, : \, (\mathbf{y})_{E(u)} \in \mathcal{C}\} .$$

Define the quantities w, σ, and τ by

$$w = \frac{\mathsf{d}(\mathbf{y}, \mathbf{c})}{N}\,, \qquad \sigma = \frac{|S|}{|V|}\,, \qquad \text{and} \qquad \tau = \frac{|T|}{|V|}\,,$$

respectively. Note that $\zeta = \sigma - \tau$.

1. Show that

$$w \le \sigma \cdot (\xi \cdot \sigma + 1 - \xi)\,.$$

Hint: Justify the following chain of equalities and inequalities:

$$\frac{wn}{\sigma} = \frac{2\,\mathsf{d}(\mathbf{y}, \mathbf{c})}{\sigma |V|} = \frac{2|Y|}{|S|} \le \frac{2|E_{S,S}|}{|S|} \le (\xi \cdot \sigma + 1 - \xi)n\,.$$

2. Show that

$$w \ge \theta \cdot \tau + \frac{1}{n}(\sigma - \tau)\,.$$

Hint: Argue that the degree of each vertex of T in the subgraph (S, Y) of $\mathcal{G}$ is at least $d = \theta n$, while the degree of each of the other vertices in that subgraph is at least 1; so,

$$2|Y| \ge d|T| + |S| - |T|\,.$$

3. Deduce from parts 1 and 2 that ζ can be bounded from below by

$$\zeta = \sigma - \tau \ge \frac{1}{1 - (1/d)} \left(\Delta + \sqrt{\xi^{-1} w + \Delta^2} - \theta^{-1} w\right)\,,$$

where $\Delta = \frac{1}{2}(1 - \xi^{-1})$. (Note that when $(\theta n =)\, d \to \infty$ and $\xi \to 1$, this lower bound becomes $\sqrt{w} - (w/\theta)$.)

4. Show that the lower bound in part 3 takes its maximum value when

$$w = \frac{\theta^2 - (1 - \xi)^2}{4\xi}\,,$$

and that it is strictly positive whenever

$$0 < w < \frac{\theta(\theta + \xi - 1)}{\xi}$$

(compare with Proposition 13.19).

[Section 13.8]

Problem 13.36 Let $F = \mathrm{GF}(q)$ and define the code $\mathsf{C} = (\mathcal{G}, \mathcal{C}' : \mathcal{C}'')$ by (13.17), where $\mathcal{G} = (V' : V'', E)$ is a bipartite n-regular graph, $\mathcal{C}'$ is a linear $[n, k{=}rn, \theta n]$ code over F, and $\mathcal{C}''$ is a (second) linear $[n, Rn, \delta n]$ code over F. Let Φ denote the alphabet F^k, and define the code C_Φ over Φ by (13.18).

1. Show that C_Φ is a linear space over F.

2. Show that the rate of C_Φ is bounded from below by
$$1 - \frac{1}{r} + \frac{R}{r}$$
and, so, that rate approaches R when $r \to 1$.

3. Show that the relative minimum distance of C_Φ is bounded from below by
$$\frac{\delta - \gamma_\mathcal{G}\sqrt{\delta/\theta}}{1 - \gamma_\mathcal{G}}$$
and, so, it approaches δ when $\gamma_\mathcal{G}/\sqrt{\theta} \to 0$.

Hint: Verify first that the minimum distance of C_Φ equals the minimum Hamming weight (over Φ) of any nonzero codeword of C_Φ. Then follow the steps of the proof of Proposition 13.19, with the set S therein taken as the union $S \cup T$, where $S \subseteq V'$ and $T \subseteq V''$. Letting $\sigma = |S|/|V'|$ and $\tau = |T|/|V''|$, replace (13.11) by the inequality
$$|Y| \ge \max\{\theta \cdot \sigma, \delta \cdot \tau\} \cdot n|V'|$$
and, using Lemma 13.20, bound $|Y|$ from above by
$$|Y| \le \left((1-\gamma_\mathcal{G})\sigma\tau + \gamma_\mathcal{G}\sqrt{\sigma\tau}\right) n|V'| \,.$$
Finally, combine these two bounds on $|Y|$ by distinguishing between the following two cases: (i) $\sigma/\tau \le \delta/\theta$ and (ii) $\sigma/\tau \ge \delta/\theta$.

Problem 13.37 Let C_Φ be the code in Problem 13.36, where $\sqrt{\theta\delta} > 2\gamma_\mathcal{G} > 0$, and fix σ to be a positive real such that
$$\sigma < \beta = \frac{(\delta/2) - \gamma_\mathcal{G}\sqrt{\delta/\theta}}{1 - \gamma_\mathcal{G}} \,.$$
The purpose of this problem is to show that the decoder in Figure 13.5 recovers correctly any error word in $\Phi^{|V'|}$ with at most $\sigma|V'|$ errors (and no erasures).

Let $\psi_\mathcal{E}(\mathbf{c})$ be the transmitted codeword where $\mathbf{c} = (c_e)_{e\in E}$ is in C, and let $\mathbf{y} = (\mathbf{y}_u)_{u\in V'}$ be the received word (over Φ) to which the algorithm in Figure 13.5 is applied. For $i = 2, 3, \ldots, \nu$, denote by $\mathbf{z}_i = (z_{i,e})_{e\in E}$ and U_i the values of the word $\mathbf{z}$ and the set U, respectively, at the end of iteration i in Step 2 in Figure 13.5; extend this notation also to $i = 1$ by letting U_1 be V' and $\mathbf{z}_1$ be the value of $\mathbf{z}$ at the end of Step 1. Define the sets $S_1, S_2, \ldots, S_\nu$ by (13.14); namely, S_i consists of all vertices $u \in U_i$ such that $(\mathbf{z}_i)_{E(u)} \ne (\mathbf{c})_{E(u)}$; in particular, S_1 stands for the set of error locations. Denote by σ_i the ratio $|S_i|/|V'|$ and assume that $\sigma_1 \le \sigma$. Let ℓ be the smallest positive integer (possibly ∞) such that $\sigma_\ell = 0$.

1. Show that
$$\sqrt{\frac{\sigma_{i-1}}{\sigma_i}} \ge \begin{cases} \dfrac{\delta}{2\gamma_\mathcal{G}} - \dfrac{1-\gamma_\mathcal{G}}{\gamma_\mathcal{G}}\sigma_{i-1} & \text{for even } 0 < i < \ell \\[2ex] \dfrac{\theta}{2\gamma_\mathcal{G}} - \dfrac{1-\gamma_\mathcal{G}}{\gamma_\mathcal{G}}\sigma_{i-1} & \text{for odd } 1 < i < \ell \end{cases} .$$
Hint: See how Equation (13.16) is derived.

2. Show by induction on i that

$$\frac{\sigma_{i-1}}{\sigma_i} \geq \begin{cases} \delta/\theta & \text{for even } 0 < i < \ell \\ \theta/\delta & \text{for odd } 1 < i < \ell \end{cases} .$$

3. Show that

$$\frac{1}{\sqrt{\sigma_i}} \geq \begin{cases} \dfrac{1}{\sqrt{\sigma}} & \text{for } i = 1 \\ \dfrac{\delta/(2\gamma_{\mathcal{G}})}{\sqrt{\sigma_{i-1}}} - \dfrac{(1-\gamma_{\mathcal{G}})\sqrt{\sigma}}{\gamma_{\mathcal{G}}} & \text{for even } 0 < i < \ell \\ \dfrac{\theta/(2\gamma_{\mathcal{G}})}{\sqrt{\sigma_{i-1}}} - \dfrac{(1-\gamma_{\mathcal{G}})\sqrt{\sigma_2}}{\gamma_{\mathcal{G}}} & \text{for odd } 1 < i < \ell \end{cases} .$$

4. Show that for even $0 < i < \ell$,

$$\frac{2\gamma_{\mathcal{G}}/\theta}{\sqrt{\sigma_{i+1}}} + \frac{2(1-\gamma_{\mathcal{G}})\sqrt{\sigma_2}}{\theta} \geq \frac{1}{\sqrt{\sigma_i}} \geq \frac{\delta/(2\gamma_{\mathcal{G}})}{\sqrt{\sigma_{i-1}}} - \frac{(1-\gamma_{\mathcal{G}})\sqrt{\sigma}}{\gamma_{\mathcal{G}}} .$$

5. Verify that for even $0 < i < \ell$,

$$\begin{aligned} \frac{1}{\sqrt{\sigma_{i+1}}} &\geq \frac{\theta\delta/(4\gamma_{\mathcal{G}}^2)}{\sqrt{\sigma_{i-1}}} - \frac{1-\gamma_{\mathcal{G}}}{\gamma_{\mathcal{G}}} \left(\frac{\theta\sqrt{\sigma}}{2\gamma_{\mathcal{G}}} + \sqrt{\sigma_2} \right) \\ &\geq \frac{\theta\delta/(4\gamma_{\mathcal{G}}^2)}{\sqrt{\sigma_{i-1}}} - \frac{1-\gamma_{\mathcal{G}}}{\gamma_{\mathcal{G}}} \left(\frac{\theta}{2\gamma_{\mathcal{G}}} + \sqrt{\frac{\theta}{\delta}} \right) \sqrt{\sigma} \\ &= \frac{\theta\delta}{4\gamma_{\mathcal{G}}^2} \left(\frac{1}{\sqrt{\sigma_{i-1}}} - \frac{\sqrt{\sigma}}{\beta} \right) + \frac{\sqrt{\sigma}}{\beta} . \end{aligned}$$

6. By solving the linear recurrence in $1/\sqrt{\sigma_{i+1}}$, conclude that for even $0 \leq i < \ell-1$,

$$\frac{1}{\sqrt{\sigma_{i+1}}} \geq \left(\left(1 - \frac{\sigma}{\beta}\right) \cdot \left(\frac{\theta\delta}{4\gamma_{\mathcal{G}}^2} \right)^{i/2} + \frac{\sigma}{\beta} \right) \frac{1}{\sqrt{\sigma}} .$$

7. Show that when the number ν in Figure 13.5 equals (13.20), then the return value of the decoder in Figure 13.5 is the correct codeword $\psi_{\mathcal{E}}(\mathbf{c})$.

Problem 13.38 Let $\mathcal{G} = (V' : V'', E)$ be a bipartite n-regular graph and let $\chi : (V' \cup V'') \to [0,1]$ be a function on the vertices of $\mathcal{G}$. Write

$$\sigma = \frac{1}{|V'|} \sum_{u \in V'} \chi(u) \qquad \text{and} \qquad \tau = \frac{1}{|V''|} \sum_{u \in V''} \chi(u) .$$

1. Show that

$$\begin{aligned} \frac{1}{n|V'|} \sum \chi(u)\chi(v) &\leq \sigma\tau + \gamma_{\mathcal{G}} \sqrt{\sigma(1-\sigma)\tau(1-\tau)} \\ &\leq (1-\gamma_{\mathcal{G}})\sigma\tau + \gamma_{\mathcal{G}}\sqrt{\sigma\tau} , \end{aligned}$$

where the summation is taken over all ordered pairs $(u, v) \in V' \times V''$ such that $\{u, v\} \in E$.

Hint: Using the notation of Problem 13.20, show that when the images of χ are in $[0, 1]$, then

$$\mathsf{Var}'_{\mathcal{G}}\{\chi\} \leq \sigma(1-\sigma) \qquad \text{and} \qquad \mathsf{Var}''_{\mathcal{G}}\{\chi\} \leq \tau(1-\tau) .$$

2. Suppose that the restriction of χ to V'' is not identically zero and that $\gamma_{\mathcal{G}} > 0$, and let δ be a real number for which the following condition is satisfied for every $u \in V''$:

$$\chi(u) > 0 \quad \Longrightarrow \quad \sum_{v \in \mathcal{N}(u)} \chi(v) \geq \frac{\delta n}{2} .$$

Show that

$$\sqrt{\frac{\sigma}{\tau}} \geq \frac{(\delta/2) - (1-\gamma_{\mathcal{G}})\sigma}{\gamma_{\mathcal{G}}} .$$

Hint: Follow the proof of Lemma 13.21.

Problem 13.39 Let C_Φ be the code in Problem 13.36, where $\sqrt{\theta\delta} > 2\gamma_{\mathcal{G}} > 0$, and fix σ to be a positive real such that

$$\sigma < \beta = \frac{(\delta/2) - \gamma_{\mathcal{G}}\sqrt{\delta/\theta}}{1 - \gamma_{\mathcal{G}}} .$$

Show that the decoder in Figure 13.5 recovers correctly any pattern that consists of t errors (over Φ) and ρ erasures, provided that

$$t + \frac{\rho}{2} \leq \sigma |V'| .$$

Hint: For $i \geq 2$, let U_i be the value of the set U at the end of iteration i in Figure 13.5, and let S_i be the set of all vertices $u \in U_i$ such that $(\mathbf{z})_{E(u)}$ is in error at the end of that iteration. Let $\chi_1 : (V' \cup V'') \to \{0, \frac{1}{2}, 1\}$ be the function

$$\chi_1(u) = \begin{cases} 1 & \text{if } u \in V' \text{ and } \mathbf{y}_u \text{ is in error} \\ \frac{1}{2} & \text{if } u \in V' \text{ and } \mathbf{y}_u \text{ is an erasure} \\ 0 & \text{otherwise} \end{cases}$$

and, for $i \geq 2$, define the function $\chi_i : (V' \cup V'') \to \{0, \frac{1}{2}, 1\}$ recursively by

$$\chi_i(u) = \begin{cases} 1 & \text{if } u \in S_i \\ 0 & \text{if } u \in U_i \setminus S_i \\ \chi_{i-1}(u) & \text{if } u \in U_{i-1} \end{cases} ,$$

where $U_1 = V'$. Denoting $\sigma_i = (1/|V'|) \sum_{u \in U_i} \chi_i(u)$, first show that

$$\sigma_1 |V'| = t + \frac{\rho}{2} .$$

Then, using part 2 of Problem 13.38, derive the inequalities in part 1 of Problem 13.37.

Notes

[Section 13.1]

There are quite a few textbooks on graph theory and related topics; see, for example, Bollobás [60], Brualdi and Ryser [67], Diestel [104], and West [385].

The following generalization of the notion of graphs will be used below.

A *hyper-graph* is a pair (V, E), where V is a nonempty finite set of vertices and E is a (possibly empty) set of *hyper-edges*, where by a hyper-edge we mean a nonempty subset of V of size *at least* 2 (see Berge [33]). A hyper-graph is *t-uniform* if each hyper-edge contains t vertices. Thus, an ordinary graph is a 2-uniform hyper-graph. A hyper-graph is called n-regular if each vertex is contained in n hyper-edges.

[Section 13.2]

Proposition 13.4 is, in effect, a special case of the *Perron–Frobenius Theorem*, which we quote next.

A real matrix is called nonnegative if all of its entries are nonnegative. A nonnegative real square matrix A is called *irreducible* if for every row index u and column index v there exists a nonnegative integer $\ell_{u,v}$ such that $(A^{\ell_{u,v}})_{u,v} > 0$ (note that we do not assume symmetry). The 1×1 matrix $A = (\,0\,)$ is referred to as the *trivial* irreducible matrix.

Thus, for every graph $\mathcal{G}$, the adjacency matrix $A_{\mathcal{G}}$ is irreducible if and only if $\mathcal{G}$ is connected.

Given a nontrivial irreducible matrix A, its *period* is defined by

$$\mathsf{p}(A) = \gcd\left\{\ell \in \mathbb{Z}^+ \;:\; (A^{\ell})_{u,u} > 0 \text{ for at least one index } u\right\} .$$

If the period equals 1 then the matrix is said to be *aperiodic* or *primitive*.

In particular, if $\mathcal{G}$ is a connected graph with more than one vertex, then the diagonal entries of $A_{\mathcal{G}}^{\ell}$ are nonzero for every even ℓ; so, $\mathsf{p}(A_{\mathcal{G}}) \,|\, 2$. Thus, by Lemma 13.1,

$$\mathsf{p}(A_{\mathcal{G}}) = \begin{cases} 2 & \text{if } \mathcal{G} \text{ is bipartite} \\ 1 & \text{otherwise} \end{cases} .$$

Theorem 13.26 (Perron–Frobenius Theorem for irreducible matrices) *Let A be a nontrivial irreducible matrix. There exists an eigenvalue λ of A such that the following conditions hold:*

1. *λ is real and $\lambda > 0$.*
2. *There are right and left eigenvectors associated with λ that are strictly positive; that is, each of their components is strictly positive.*
3. *$\lambda \ge |\mu|$ for any other eigenvalue μ of A.*
4. *The algebraic (and geometric) multiplicity of λ is 1.*
5. *There are exactly $\mathsf{p}(A)$ eigenvalues μ of A for which $|\mu| = \lambda$: those eigenvalues have the form $\lambda\omega^i$, where ω is a complex root of order $\mathsf{p}(A)$ of unity, and each of those eigenvalues has algebraic multiplicity 1.*

The proof of Theorem 13.26 can be found in Gantmacher [141, Chapter 13], Minc [263, Chapter 1], Seneta [328, Chapter 1], or Varga [371, Chapter 2].

The eigenvalue λ in Theorem 13.26 is sometimes called the *Perron eigenvalue* of the irreducible matrix A, and is denoted here by $\lambda(A)$ (if A is the trivial irreducible matrix define $\lambda(A) = 0$).

One consequence of Theorem 13.26 is the following result.

Proposition 13.27 *Let $A = (a_{u,v})_{u,v}$ be an irreducible matrix. Then,*

$$\min_u \sum_v a_{u,v} \leq \lambda(A) \leq \max_u \sum_v a_{u,v} ,$$

where equality in one side implies equality in the other.

Proof. Let $(y_v)_v$ be a strictly positive left eigenvector associated with $\lambda(A)$. Then $\sum_u y_u a_{u,v} = \lambda(A) y_v$ for every index v. Summing over v, we obtain,

$$\sum_u y_u \sum_v a_{u,v} = \lambda(A) \sum_v y_v$$

or

$$\lambda(A) = \frac{\sum_u y_u \sum_v a_{u,v}}{\sum_v y_v} .$$

That is, $\lambda(A)$ is a weighted average (over u) of the values $\sum_v a_{u,v}$. □

As a special case of the last proposition we get that if A is the adjacency matrix of a connected n-regular graph, then $\lambda(A) = n$.

When a nonnegative square matrix A is not irreducible (i.e., when A is *reducible*), then, by applying a permutation matrix P to its rows and to its columns, one can reach a block-triangular form

$$P^{-1}AP = \begin{pmatrix} A_1 & B_{1,2} & B_{1,3} & \cdots & B_{1,k} \\ & A_2 & B_{2,3} & \cdots & B_{2,k} \\ & & A_3 & \ddots & \vdots \\ & 0 & & \ddots & B_{k-1,k} \\ & & & & A_k \end{pmatrix} ,$$

where $A_1, A_2, \ldots, A_k$ are irreducible. Theorem 13.26 applies to each nontrivial block A_i, and the set of eigenvalues of A is the union of the sets of eigenvalues of the blocks A_i. Thus, $\lambda(A) = \max_i \lambda(A_i)$ is an eigenvalue of A, and $|\mu| \leq \lambda(A)$ for every eigenvalue μ of A.

[Section 13.3]

The relationship between graph expansion and the second largest eigenvalue was found independently by Alon and Milman [13], [14] and by Tanner [357]; see also Alon [8], [9], Alon and Spencer [16, Section 9.2], and West [385, Section 8.6].

Theorem 13.7 and Corollary 13.10 are from Alon and Milman [13] and Alon and Chung [11], respectively. Theorem 13.12 is reported by Alon in [8], and a proof of

Equation (13.9) can be found in [271]. A simplified version of that proof was given here to yield Proposition 13.11.

Problem 13.20 is a modification of Corollary 9.2.5 in Alon and Spencer [16]. Problem 13.21 is from Tanner [357], and Problems 13.22 and 13.24 are from Alon *et al.* [10] and Alon and Spencer [16, Theorem 9.2.4].

[Section 13.4]

The columns of a Sylvester-type Hadamard matrix $\mathcal{H}_k$—and, more generally, the columns of the matrix $\mathcal{H}_{k,p}$ in Problem 13.26—contain the values of an (additive) character of $(\mathrm{GF}(p))^k$ (or, rather, of $\mathrm{GF}(p^k)$); see Problem 3.36. As pointed out in the notes on Section 3.6, the notion of characters can be generalized to every finite Abelian group Q: a nonzero mapping $\chi : Q \to \mathbb{C}$ is a character of Q if

$$\chi(\alpha + \beta) = \chi(\alpha)\chi(\beta) \qquad \text{for every } \alpha, \beta \in Q .$$

Accordingly, Proposition 13.13 and Problem 13.26 can be generalized to Cayley graphs $\mathcal{G}(Q, B)$ for every finite Abelian group Q; namely, the eigenvalues of $A_{\mathcal{G}(Q,B)}$ are given by

$$\lambda_\chi = \sum_{\alpha \in B} \chi(\alpha) ,$$

where χ ranges over all characters of Q (see Babai [26] and Lovász [239]).

Corollary 13.14 is taken from Alon and Roichman [15].

The analysis in Problem 13.30 yields for fixed (small) $\varepsilon \in (0, 1]$ a polynomial-time construction of a sequence of linear $[n, k, d]$ codes $\mathcal{C}$ over $F = \mathrm{GF}(2)$ with the following properties: the code dimension k is proportional to $\varepsilon^m n$ for some positive constant m, the all-one word is a codeword of $\mathcal{C}$, and $d \geq \frac{1}{2}(1-\varepsilon)n$ (the columns of the generator matrix of $\mathcal{C}$ then form the set B in part 5 of Problem 13.30). It follows from the properties of $\mathcal{C}$ that this code has a linear $[n, k-1]$ sub-code in which every nonzero codeword has Hamming weight within the range $\frac{1}{2}(1 \pm \varepsilon)n$. In addition to obtaining expanders, codes with these properties have other applications as well; see Naor and Naor [267].

[Section 13.5]

The construction of Ramanujan graphs due to Lubotzky *et al.* and Margulis is from [240] and [252], where a proof of Theorem 13.17 can be found.

For a proof of Theorem 13.15, see Hardy and Wright [171, Sections 20.11 and 20.12].

We summarize next several properties of quadratic residues (some of these properties were already covered in Problems 3.23 and 3.26).

Let q be an odd prime and a be an integer. Recall from Problem 3.26 that the Legendre symbol of a modulo q is defined by

$$\left(\frac{a}{q}\right) = \begin{cases} 1 & \text{if } a \text{ is a quadratic residue modulo } q \\ 0 & \text{if } q \mid a \\ -1 & \text{otherwise} \end{cases} .$$

The proof of the following two lemmas is given as an exercise (parts 1 and 2 of Problem 3.26).

Lemma 13.28 *Let q be an odd prime and let a and b be integers. Then*

$$\left(\frac{ab}{q}\right) = \left(\frac{a}{q}\right) \cdot \left(\frac{b}{q}\right) .$$

Lemma 13.29 (Euler's criterion) *Let q be an odd prime and a be an integer not divisible by q. Then,*

$$\left(\frac{a}{q}\right) \equiv a^{(q-1)/2} \pmod{q} .$$

Corollary 13.30 *For every odd prime q,*

$$\left(\frac{-1}{q}\right) = (-1)^{(q-1)/2} = \begin{cases} 1 & q \equiv 1 \pmod{4} \\ -1 & \textit{otherwise} \end{cases} .$$

(See also part 4 of Problem 3.23.)

Let $F = \mathrm{GF}(q)$ where q is an odd prime, and recall the following definitions from Chapter 10. An element $\alpha \in F$ is called "negative" if α belongs to the set $\{\frac{1}{2}(q+1), \frac{1}{2}(q+3), \ldots, q-1\}$. For an element $\alpha \in F$, denote by $\langle\alpha\rangle$ the smallest nonnegative integer such that $\alpha = \langle\alpha\rangle \cdot 1$. The *Lee weight* of an element $\alpha \in F$, denoted as $|\alpha|$, takes nonnegative integer values and is defined by

$$|\alpha| = \begin{cases} \langle\alpha\rangle & \text{if } \alpha \text{ is nonnegative} \\ q - \langle\alpha\rangle & \text{otherwise} \end{cases} .$$

Also recall that for an integer a, the notation $\overline{a}$ stands for the element in $\mathrm{GF}(q)$ such that $\overline{a} = a \cdot 1$.

Lemma 13.31 (Gauss' criterion) *Let q be an odd prime and a be an integer not divisible by q. Denote by μ the number of negative elements within the following subset*

$$\left\{\overline{m \cdot a} \;:\; m = 1, 2, \ldots, \tfrac{1}{2}(q-1)\right\}$$

of $\mathrm{GF}(q)$. Then

$$\left(\frac{a}{q}\right) = (-1)^{\mu} .$$

Proof. Let $F = \mathrm{GF}(q)$ and for $m = 1, 2, \ldots, \frac{1}{2}(q-1)$, denote by α_m the element $\overline{m \cdot a}$. It is easy to see that $\alpha_m \neq \pm\alpha_{m'}$ for any two distinct integers m and m' in the range $1 \leq m, m' \leq \frac{1}{2}(q-1)$ and, hence, $|\alpha_m| \neq |\alpha_{m'}|$. This means that for $m = 1, 2, \ldots, \frac{1}{2}(q-1)$, the value $|\alpha_m|$ ranges over the whole set $\{1, 2, \ldots, \frac{1}{2}(q-1)\}$. Letting F^- denote the negative elements of F, we obtain

$$\begin{aligned} &\left(\tfrac{1}{2}(q-1)\right)! \cdot (\overline{a})^{(q-1)/2} \\ &\quad = \prod_{m=1}^{(q-1)/2} \alpha_m = \Bigl(\prod_{\substack{m: \\ \alpha_m \in F^-}} (-|\alpha_m|)\Bigr)\Bigl(\prod_{\substack{m: \\ \alpha_m \notin F^-}} |\alpha_m|\Bigr) = (-1)^{\mu} \underbrace{\prod_{m=1}^{(q-1)/2} |\alpha_m|}_{(\frac{1}{2}(q-1))!} . \end{aligned}$$

The result now follows from Lemma 13.29. $\square$

Corollary 13.32 *For every odd prime* q,

$$\left(\frac{2}{q}\right) = (-1)^{(q^2-1)/8} = \begin{cases} 1 & q \equiv \pm 1 \pmod 8 \\ -1 & \textit{otherwise} \end{cases} .$$

Proof. For $1 \le m \le \frac{1}{2}(q-1)$, the element $\overline{2m}$ is negative if and only if $\frac{1}{4}(q-1) < m \le \frac{1}{2}(q-1)$. Using the notation of Lemma 13.31, we have $\mu = \frac{1}{2}(q-1) - \lfloor \frac{1}{4}(q-1) \rfloor$, and a simple check reveals that μ is even if and only if $q \equiv \pm 1 \pmod 8$. □

Theorem 13.33 (Law of Quadratic Reciprocity) *For any distinct odd primes* p *and* q,

$$\left(\frac{p}{q}\right) \cdot \left(\frac{q}{p}\right) = (-1)^{(p-1)(q-1)/4} .$$

The proof, which makes use of Lemma 13.31, can be found in Hardy and Wright [171, Sections 6.11–6.13]. See also Ireland and Rosen [195, Chapter 5].

We mentioned in Section 13.5 the Prime Number Theorem for arithmetic progressions. We next quote this theorem; for a proof, see Davenport [90] or Huxley [190, Chapter 17].

Theorem 13.34 *Let* a *and* b *be integers such that* $a > 0$ *and* $\gcd(a, b) = 1$ *and denote by* $\pi_{a,b}(x)$ *the size of the set*

$$\{p \le x \;:\; p \textit{ prime}, \; p \equiv b \pmod a\} .$$

Then,

$$\lim_{x \to \infty} \frac{\pi_{a,b}(x)}{\mathrm{Li}(x)} = \frac{1}{\phi(a)} ,$$

where

$$\mathrm{Li}(x) = \int_2^x \frac{dt}{\ln t}$$

and $\phi(\cdot)$ *is the Euler function.*

[Section 13.6]

Graph codes were first introduced by Tanner [356]. In Tanner's description of a graph code $(\mathcal{G}, \mathcal{C})$, the n-regular graph $\mathcal{G} = (V, E)$ is represented by the respective bipartite form $\widetilde{\mathcal{G}}$ defined in Problem 13.16 and seen in the example in Figure 13.6. Thus, in the graph $\widetilde{\mathcal{G}}$, the vertices to the left (the set E) stand for the coordinates in a codeword and are commonly referred to as *message nodes* or *variable nodes*; the vertices to the right (the set V) represent the code constraints and are commonly called *check nodes*.

The construction proposed by Tanner was more general, allowing $\mathcal{G}$ to be a uniform regular hyper-graph rather than an (ordinary) regular graph (see the notes on Section 13.1). Specifically, Tanner's construction is defined through a t-uniform n-regular hyper-graph $\mathcal{G} = (V, E)$ and a code $\mathcal{C}$ of length n over an alphabet F. As was the case with ordinary graphs, we denote by $E(u)$ the set of hyper-edges incident with $u \in V$ and assume some ordering on $E(u)$. The hyper-graph code $\mathsf{C} = (\mathcal{G}, \mathcal{C})$ is defined by

$$\mathsf{C} = \left\{ \mathbf{c} \in F^N \ : \ (\mathbf{c})_{E(u)} \in \mathcal{C} \ \text{ for every } u \in V \right\} ,$$

where $N = |E| = n|V|/t$. When shifting to a bipartite representation akin to that in Problem 13.16, each of the N variable nodes has degree n while each of the $|V|$ check nodes now has degree t. It can be easily seen that if $\mathcal{C}$ is taken as a linear $[n, rn, d]$ code over $F = \mathrm{GF}(q)$, then the rate R of C satisfies

$$R \geq 1 - (1 - r)t$$

(Proposition 13.18 is a special case of this bound for $t = 2$).

As noted by Tanner, one possible generalization of hyper-graph codes is allowing them to have a different code $\mathcal{C}_u$ (instead of the same code $\mathcal{C}$) for each vertex $u \in V$, and we in fact consider an instance of such a generalization in Section 13.8. When $\mathcal{G} = (V, E)$ is a t-uniform n-regular hyper-graph and the codes $\mathcal{C}_u$ are taken to be linear $[n, n{-}1, 2]$ codes over $F = \mathrm{GF}(q)$ (e.g., the parity code), the resulting construction is called a *(t, n)-doubly-regular low-density parity-check* (in short, LDPC) *code* (see Problem 13.33). Such a code has a parity-check matrix in which each column has Hamming weight t and each row has Hamming weight n; the code rate is then bounded from below by $1 - (|V|/|E|) = 1 - (t/n)$. When only the columns (respectively, rows) of the parity-check matrix are constrained to have a given Hamming weight w, the LDPC code is said to be *left* (respectively, *right*) *w-regular.*

The notion of LDPC codes was introduced by Gallager [138], [139]. In particular, he showed that doubly-regular LDPC codes over $\mathrm{GF}(q)$ attain the Gilbert–Varshamov bound: given positive reals $\delta < 1 - (1/q)$ and $R < 1 - \mathsf{H}_q(\delta)$, for every sufficiently large n there exists an infinite family of $(t{=}\lceil n(1{-}R) \rceil, n)$-doubly-regular LDPC codes over $\mathrm{GF}(q)$ with relative minimum distance at least δ. Litsyn and Shevelev [235] analyzed the asymptotic average weight distribution of (left, right, and doubly) regular LDPC codes over $\mathrm{GF}(2)$. When their results are incorporated into Gallager's analysis in [139], one gets that doubly-regular LDPC codes attain the capacity of the binary symmetric channel: given a crossover probability $p \in [0, \frac{1}{2})$, for every fixed positive $R < 1 - \mathsf{H}_2(p)$ and sufficiently large n there exists an infinite family of $(t{=}\lceil n(1{-}R) \rceil, n)$-doubly-regular LDPC codes over $\mathrm{GF}(2)$ such that the decoding error probability of a nearest-codeword decoder goes to zero as the code length increases (similar results, for related families of codes, were obtained by MacKay [245] and Miller and Burshtein [261]). On the other hand, it was demonstrated by Gallager that if we fix the right degree n, the rate of any right n-regular LDPC code over $\mathrm{GF}(2)$ must be (below and) bounded away from $1 - \mathsf{H}_2(p)$ in order to guarantee a decoding error probability that vanishes with the code length; see also Burshtein *et al.* [69] and Sason and Urbanke [318].

Proposition 13.19 is a slight improvement on a result originally obtained by Sipser and Spielman in [340].

Problem 13.34, which presents another way of constructing codes based on expanders, is due to Alon *et al.* [10]. More references to work on expander-based constructions of codes are mentioned in the notes on Section 13.8 below.

[Section 13.7]

The iterative decoder of Figure 13.3 was first analyzed by Sipser and Spielman in [340]. Their analysis applied also to non-bipartite graphs, but their guaranteed number of correctable errors was approximately 12 times smaller than what one gets from Proposition 13.23. The presentation here follows along the lines of Zémor in [394].

[Section 13.8]

The results in this section are based mostly on the work of Barg and Zémor [29], [30] and Skachek and Roth [308], [342].

Expanders and graph codes were used as building blocks by Spielman in [343] to obtain the first known construction of codes which can be decoded and *encoded* in linear time, where both the code rate and the fraction of allowable erroneous entries are bounded away from zero. Guruswami and Indyk then constructed in [165]–[167] linear-time encodable and decodable codes that approach the Singleton bound. Their codes are based on a combination of Spielman codes with two generalized graph codes, $(\mathcal{G}_1, \mathcal{C}_1', \mathcal{C}_1'')$ at rate $1-\epsilon$ (for small $\epsilon > 0$) and $(\mathcal{G}_2, \mathcal{C}_2', \mathcal{C}_2'')$ at the designed rate R. The codes $\mathcal{C}_1''$ and $\mathcal{C}_2''$ are selected to be the whole space, thus allowing linear-time encoding of the two graph codes. The resulting construction has relative minimum distance at least $1-R-\epsilon$ and alphabet size $2^{O\left((\log(1/\epsilon))/(\epsilon^4 R)\right)}$. Roth and Skachek have suggested in [308] an improved linear-time encodable and decodable construction, where the alphabet size is reduced to the expression (13.19) in Example 13.7. See also Alon and Luby [12].

Starting with the work of Gallager [138], [139], the iterative decoding methods that have been mostly studied in relation to LDPC codes go under the collective name *message-passing algorithm* (in short, MPA). We will not discuss this algorithm here, except for citing several references. Richardson and Urbanke [294] analyzed the performance of the MPA on the ensemble of doubly-regular LDPC codes over the binary symmetric channel with crossover probability p. They showed that there exists a threshold (called the "LDPC MPA capacity"), which depends on p and is smaller than the channel capacity $1 - \mathsf{H}_2(p)$, such that for a random LDPC code of length N over GF(2) at rate smaller than that threshold, the MPA reduces the number of errors below εN for any prescribed $\varepsilon > 0$ (the error reduction has a probability of failure that decreases exponentially with N, and the number of iterations of the algorithm depends on p and ε). Luby *et al.* [241] and Richardson *et al.* [293] then showed that higher rates can be attained if the LDPC codes are not regular. Further analysis on the performance of the MPA, for both regular and non-regular LDPC codes, was done by Burshtein and Miller [71].

Luby *et al.* [241] used graph codes as auxiliary codes in an LDPC construction to take care of the final "clean-up" of residual errors left by the iterative decoder for the LDPC code (Burshtein and Miller showed in [70] that, in fact, it is unnecessary to insert an auxiliary graph code, since the primary LDPC code turns out to induce sufficient graph expansion for handling the residual errors).

Chapter 14

Trellis and Convolutional Codes

In Chapter 1, we introduced the concept of a block code with a certain application in mind: the codewords in the code serve as the set of images of the channel encoder. The encoder maps a message into a codeword which, in turn, is transmitted through the channel, and the receiver then decodes that message (possibly incorrectly) from the word that is read at the output of the channel. In this model, the encoding of a message is independent of any previous or future transmissions—and so is the decoding.

In this chapter, we consider a more general coding model, where the encoding and the decoding are context-dependent. The encoder may now be in one of finitely many states, which contain information about the history of the transmission. Such a finite-state encoder still maps messages to codewords, yet the mapping depends on the state which the encoder is currently in, and that state is updated during each message transmission. Finite-state encoders will be specified through directed graphs, where the vertices stand for the states and the edges define the allowed transitions between states. The mapping from messages to codewords will be determined by the edge names and by labels that we assign to the edges.

The chapter is organized as follows. We first review several concepts from the theory of directed graphs. We then introduce the notion of trellis codes, which can be viewed as the state-dependent counterpart of block codes: the elements of a trellis code form the set of images of a finite-state encoder. We next turn to describing how trellis codes can be encoded and decoded. In particular, given an encoder for a trellis code, we present an algorithm for implementing a maximum-likelihood decoder for such an encoder.

The remaining part of the chapter is devoted to convolutional codes, which play the role of linear codes among trellis codes. Convolutional codes are defined through a special class of labeled directed graphs, referred to as linear finite-state machines. We present methods for constructing encoders for convolutional codes and then develop tools for analyzing the decoding error probability (per message) of maximum-likelihood decoders for these encoders.

14.1 Labeled directed graphs

A *labeled directed finite graph* (in short, a labeled digraph) is a quintuple $\mathcal{G} = (V, E, \iota, \tau, L)$, where V is a nonempty finite set of *states* and E is a (possibly empty) finite set of *edges*, with the following three functions that are defined on E,

$$\iota : E \to V \ , \quad \tau : E \to V \ , \quad \text{and} \quad L : E \to \Sigma \ ,$$

for some finite alphabet Σ. For each edge $e \in E$, we call $\iota(e)$, $\tau(e)$, and $L(e)$ the *initial state*, *terminal state*, and *label*, respectively, of e. A *self-loop* is an edge $e \in E$ for which $\iota(e) = \tau(e)$, and two edges $e, e' \in E$ are parallel if $\iota(e) = \iota(e')$ and $\tau(e) = \tau(e')$. Unlike our definition of (unlabeled) undirected graphs in Section 13.1, we allow both self-loops and parallel edges in labeled digraphs; still, in all circumstances that we will be interested in, parallel edges will never carry the same label. We will sometimes denote an edge e with initial state s, terminal state $\tilde{s}$, and label c by

$$s \xrightarrow{c} \tilde{s} \ .$$

For the sake of simplicity, we will abbreviate the notation (V, E, ι, τ, L) into just a triple (V, E, L).

Example 14.1 Figure 14.1 shows a labeled digraph $\mathcal{G}$ with a set of states

$$V = \{\alpha, \beta, \gamma, \delta\}$$

and eight edges, out of which two are self-loops. The labels, taking values in the alphabet $\Sigma = \{a, b, c, d\}$, are written next to the edges. □

Borrowing similar terms from undirected graphs, we say that a labeled digraph $\mathcal{G}' = (V', E', L')$ is a subgraph of $\mathcal{G} = (V, E, L)$ if $V' \subseteq V$ and $E' \subseteq E$ (with the initial and terminal states of each edge in $\mathcal{G}'$ being the same as in $\mathcal{G}$), and L' is the restriction of L to the domain E'. Given a nonempty subset V' of V, the induced subgraph of $\mathcal{G}$ on V' is the subgraph (V', E', L') of $\mathcal{G}$ where

$$E' = \{e \in E \ : \ \iota(e), \tau(e) \in V'\} \ .$$

A *path* of length $\ell > 0$ in a labeled digraph $\mathcal{G} = (V, E, L)$ is a sequence of ℓ edges

$$\pi = e_0 e_1 \ldots e_{\ell-1} \ ,$$

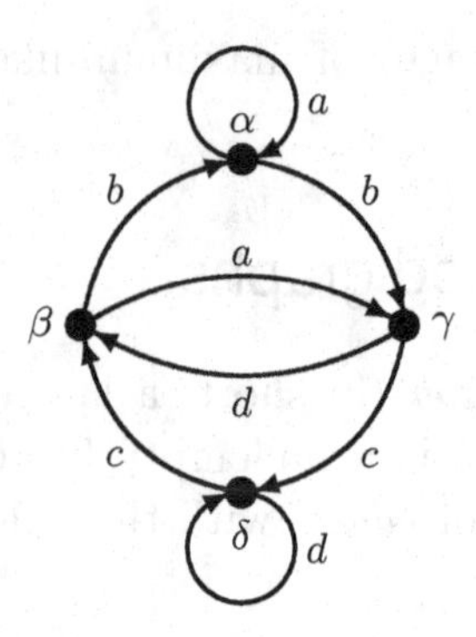

Figure 14.1. Labeled digraph $\mathcal{G}$ for Example 14.1.

where $\iota(e_{t+1}) = \tau(e_t)$ for $0 \le t < \ell-1$. The states $s = \iota(e_0)$ and $\tilde{s} = \tau(e_{\ell-1})$ are called the initial and terminal states, respectively, of the path π, and will be denoted by $\iota(\pi)$ and $\tau(\pi)$; we then say that the path is from state s to state $\tilde{s}$. We formally define zero-length paths as consisting of one state and no edges (the only state in the path then serves as both the initial state and the terminal state of that path). We will also use the notion of infinite paths where $\ell = \infty$ (in which case the terminal state is undefined). A finite path π is called a *cycle* if $\iota(\pi) = \tau(\pi)$. Clearly, $\mathcal{G}$ contains an infinite path if and only if it contains a cycle.

For each (possibly infinite) path $\pi = e_0 e_1 \ldots e_{\ell-1}$ in $\mathcal{G}$, we can associate the word

$$L(\pi) = L(e_0)L(e_1)\ldots L(e_{\ell-1})$$

of length ℓ over Σ; we will say that the path π *generates* the word $L(\pi)$. For example, the cycle

$$\alpha \xrightarrow{a} \alpha \xrightarrow{b} \gamma \xrightarrow{d} \beta \xrightarrow{b} \alpha$$

in Figure 14.1 generates the word *abdb*.

14.1.1 Irreducible digraphs

The following definition classifies digraphs according to their connectivity.

A (labeled) digraph $\mathcal{G} = (V, E, L)$ is *irreducible* or *strongly-connected* or *controllable* if for every *ordered* pair of states $(s, \tilde{s}) \in V \times V$ there is a path from s to $\tilde{s}$ in $\mathcal{G}$ (as zero-length paths are allowed, there is always a path from a state to itself). Note that irreducibility does not depend on the labeling L. A digraph that is not irreducible is called *reducible*. For example, the digraph $\mathcal{G}$ in Figure 14.2 is irreducible, while $\hat{\mathcal{G}}$ is reducible.

Let $\mathcal{G} = (V, E, L)$ be a digraph (which may be either irreducible or reducible), and define the following relation on the set V: two states s and $\tilde{s}$ in V are called *bi-connected* if there is a (possibly zero-length) path from s to $\tilde{s}$

Figure 14.2. Irreducible and reducible digraphs.

and a path from $\tilde{s}$ to s. It can be shown (Problem 14.2) that bi-connection is an equivalence relation; namely, it satisfies reflexivity, symmetry, and transitivity. The induced subgraphs of $\mathcal{G}$ on the equivalence classes of this relation are called the *irreducible components* of $\mathcal{G}$. Each irreducible component of $\mathcal{G}$ is an irreducible digraph, which is maximal in the following sense: no irreducible component of $\mathcal{G}$ is a proper subgraph of any irreducible subgraph of $\mathcal{G}$. At least one of the irreducible components of $\mathcal{G}$ must be an *irreducible sink*: no state in that component has an outgoing edge in $\mathcal{G}$ to any other irreducible component of $\mathcal{G}$.

We next present a useful property of irreducible digraphs.

Let $\mathcal{G} = (V, E, L)$ be an irreducible digraph and assume that $\mathcal{G}$ contains at least one edge (and, hence, it contains a cycle; the trivial digraph, which consists of one state and no edges, is the only irreducible digraph that contains no cycles). Let s be a state in V and ℓ be a nonnegative integer, and consider the set $V(s, \ell)$ of all terminal states of paths of length ℓ in $\mathcal{G}$ that start at state s. It turns out that there are positive integers Δ and p, which depend only on $\mathcal{G}$, such that from every state in $V(s, \ell)$ there is a path of length exactly

$$\Delta - (\ell \text{ MOD } \mathsf{p})$$

back to state s, where ℓ MOD p denotes the remainder of ℓ when divided by p. The constant p is called the *period* of $\mathcal{G}$ and it equals the greatest common divisor of the lengths of cycles in $\mathcal{G}$ (see Problems 14.3 and 14.4). When the period equals 1 then the digraph is said to be *aperiodic* or *primitive*. For example, the digraph $\mathcal{G}'$ in Figure 14.3 is aperiodic, while $\mathcal{G}''$ has period 2.

Figure 14.3. Digraphs with periods 1 and 2.

We will refer to Δ as a *back-length* of $\mathcal{G}$. Clearly, there are infinitely many back-lengths for a given digraph; in fact, every sufficiently large multiple of p is a back-length (the condition that $\mathsf{p} \,|\, \Delta$ is necessary—see Problem 14.4). For our purposes any finite back-length will suffice, although smaller values will be preferable. Turning again to Figure 14.3, we can take $\Delta = 1$ for $\mathcal{G}'$ and $\Delta = 2$ for $\mathcal{G}''$.

14.1.2 Lossless digraphs

We now turn to classifying labeled digraphs according to their labeling.

A labeled digraph $\mathcal{G} = (V, E, L)$ is called *deterministic* if for every state in V, no two outgoing edges from that state have the same label. The labeled digraph in Example 14.1 is deterministic, and so are the labeled digraphs $\mathcal{G}_1$ and $\mathcal{G}_2$ in Figure 14.4. The labeled digraphs $\mathcal{G}_3$ and $\mathcal{G}_4$, on the other hand, are not deterministic.

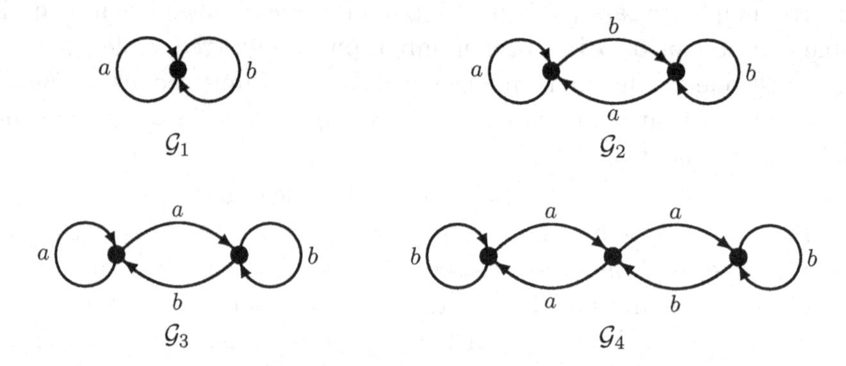

Figure 14.4. Lossless digraphs.

When a labeled digraph $\mathcal{G}$ is deterministic, then the initial state of a path and the word that it generates uniquely identify the path. We next generalize this property and introduce a wider family of labeled digraphs, which contains deterministic digraphs as a subset.

A labeled digraph $\mathcal{G}$ is called *lossless* if distinct paths in $\mathcal{G}$ with the same initial state and the same terminal state always generate distinct words; equivalently, for every two paths π and π' in $\mathcal{G}$,

$$\text{if } \iota(\pi) = \iota(\pi') \text{ and } \tau(\pi) = \tau(\pi') \text{ and } L(\pi) = L(\pi'), \text{ then } \pi = \pi' .$$

It is straightforward to see that every deterministic digraph is lossless. The converse is not true: all the labeled digraphs in Figure 14.4 are lossless, although two of them—namely, $\mathcal{G}_3$ and $\mathcal{G}_4$—are not deterministic. Not every labeled digraph is lossless; for example, if we changed the label of the right self-loop in $\mathcal{G}_2$ into a, then the resulting labeled digraph would not be lossless.

14.1.3 Sequence sets of labeled digraphs

Labeled digraphs will be used in this chapter to define sets of sequences (these sets, in turn, will serve as our codes). To this end, we need several basic terms, which we introduce next.

Let $\mathcal{G} = (V, E, L)$ be a labeled digraph. For the definitions that follow we need one state in V to be designated as the *start state* of $\mathcal{G}$; e.g., if we assume some ordering on the elements of V, then the start state can be the smallest state in V. We let $\iota(\mathcal{G})$ denote the start state of $\mathcal{G}$ and define $\mathbf{C}(\mathcal{G})$ to be the set of all (distinct) infinite words that are generated in $\mathcal{G}$ by infinite paths that start at state $\iota(\mathcal{G})$. We refer to the set $\mathbf{C}(\mathcal{G})$ as the *sequence set* of $\mathcal{G}$.

For example, the sequence set of the labeled digraph $\mathcal{G}_1$ in Figure 14.4 is the set of all infinite words over the alphabet $\Sigma = \{a, b\}$. The same holds for the labeled digraph $\mathcal{G}_2$ in that figure, regardless of the choice of $\iota(\mathcal{G}_2)$.

We will find it convenient to deal with labeled digraphs that are irreducible. Therefore, we will adopt the convention that the start state of a labeled digraph $\mathcal{G}$ will always belong to some irreducible sink $\mathcal{G}_0$ of $\mathcal{G}$, and the start state of $\mathcal{G}_0$ will be the same as that of $\mathcal{G}$. Based on this convention we have

$$\mathbf{C}(\mathcal{G}) = \mathbf{C}(\mathcal{G}_0) ,$$

which effectively means that it suffices to consider only sequence sets of irreducible labeled digraphs.

For practical reasons, we will also need to consider words that are generated by finite paths. Given a positive integer ℓ, two sets of words of length ℓ will be of interest: the first set, which we denote by $\mathbf{C}_\ell(\mathcal{G})$, consists of all distinct words of length ℓ that are generated in $\mathcal{G}$ by paths of length ℓ that start at state $\iota(\mathcal{G})$. The second set, denoted by $\mathbf{C}_\ell^\circ(\mathcal{G})$, is defined similarly, except that we require that the generating paths be *cycles*, namely, they also *terminate* in $\iota(\mathcal{G})$.

Thus, for the labeled digraph $\mathcal{G}_1$ in Figure 14.4 we have

$$\mathbf{C}_\ell(\mathcal{G}_1) = \mathbf{C}_\ell^\circ(\mathcal{G}_1) = \{a, b\}^\ell$$

for every $\ell \geq 1$. As for the labeled digraph $\mathcal{G}_2$ in that figure, we still have

$$\mathbf{C}_\ell(\mathcal{G}_2) = \{a, b\}^\ell$$

(regardless of the choice of $\iota(\mathcal{G}_2)$), yet the set $\mathbf{C}_\ell^\circ(\mathcal{G}_2)$ is now a proper subset of $\mathbf{C}_\ell(\mathcal{G}_2)$: it consists of the words in $\mathbf{C}_\ell(\mathcal{G}_2)$ that end with a (if $\iota(\mathcal{G}_2)$ is chosen to be the left state in $\mathcal{G}_2$) or b (otherwise).

When a labeled digraph $\mathcal{G}$ is deterministic, then every word in $\mathbf{C}_\ell(\mathcal{G})$ is generated in $\mathcal{G}$ by exactly one path from $\iota(\mathcal{G})$. Similarly, when $\mathcal{G}$ is lossless, then every word in $\mathbf{C}_\ell^\circ(\mathcal{G})$ is generated in $\mathcal{G}$ by exactly one cycle from $\iota(\mathcal{G})$.

As mentioned in Section 14.1.1, when a labeled digraph $\mathcal{G}$ is irreducible (with at least one edge), then every path of length h can be extended to a cycle by adding $\Delta - (h \text{ MOD } \mathsf{p})$ edges, where p and Δ are the period and back-length of $\mathcal{G}$, respectively. This, in turn, implies that every word in $\mathbf{C}_h(\mathcal{G})$ is a prefix of some word in $\mathbf{C}^\circ_\ell(\mathcal{G})$, where

$$\ell = \Delta + h - (h \text{ MOD } \mathsf{p}) . \tag{14.1}$$

14.1.4 Trellis diagram of labeled digraphs

Let $\mathcal{G} = (V, E, L)$ be a labeled digraph. In the sequel, we will find it helpful on occasions to describe the paths in $\mathcal{G}$ through the *trellis diagram* of $\mathcal{G}$. The trellis diagram, which we denote by $\mathsf{T}(\mathcal{G})$, is an infinite labeled directed graph whose set of states is given by the infinite set

$$V^{(0)} \cup V^{(1)} \cup V^{(2)} \cup \cdots ,$$

where

$$V^{(t)} = \{s^{(t)} \;:\; s \in V\} , \quad \text{for every } t \geq 0 .$$

The subset $V^{(t)}$ is called *layer* t of $\mathsf{T}(\mathcal{G})$. The set of edges of $\mathsf{T}(\mathcal{G})$ is defined as

$$E^{(0)} \cup E^{(1)} \cup E^{(2)} \cup \cdots , \tag{14.2}$$

where

$$E^{(t)} = \{e^{(t)} \;:\; e \in E\} ,$$

and the initial state, terminal state, and label of each edge $e^{(t)}$ in $\mathsf{T}(\mathcal{G})$ are given by

$$(\iota(e))^{(t)} , \quad (\tau(e))^{(t+1)} , \quad \text{and} \quad L(e) ,$$

respectively; equivalently, for each edge

$$s \xrightarrow{c} \tilde{s}$$

in $\mathcal{G}$, we endow $\mathsf{T}(\mathcal{G})$ with the edges

$$s^{(t)} \xrightarrow{c} \tilde{s}^{(t+1)} , \quad \text{for every } t \geq 0 .$$

Thus, edges in $\mathsf{T}(\mathcal{G})$ whose initial states are in layer t have their terminal states in layer $t+1$.

Example 14.2 Figure 14.5 shows the trellis diagram of the labeled digraph in Figure 14.1. □

There is a straightforward one-to-one correspondence which we can define between the paths of length ℓ in $\mathcal{G}$ and paths from layer 0 to layer ℓ in $\mathsf{T}(\mathcal{G})$; specifically, we associate the path

$$s_0 \xrightarrow{c_0} s_1 \xrightarrow{c_1} s_2 \xrightarrow{c_2} \cdots \xrightarrow{c_{\ell-1}} s_\ell$$

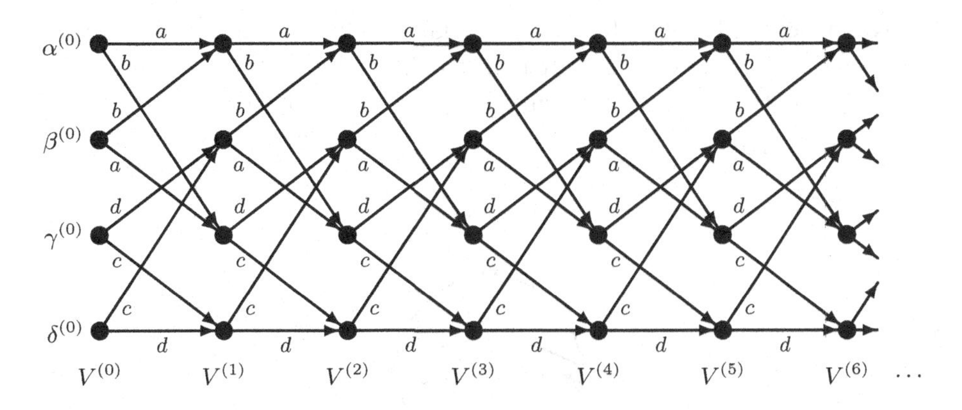

Figure 14.5. Trellis diagram for the labeled digraph $\mathcal{G}$ in Figure 14.1.

in $\mathcal{G}$ with the path

$$s_0^{(0)} \xrightarrow{c_0} s_1^{(1)} \xrightarrow{c_1} s_2^{(2)} \xrightarrow{c_2} \ldots \xrightarrow{c_{\ell-1}} s_\ell^{(\ell)}$$

in $\mathsf{T}(\mathcal{G})$. Obviously, both paths generate the same word.

A similar correspondence exists between infinite paths from state $\iota(\mathcal{G})$ in $\mathcal{G}$ and paths from state $(\iota(\mathcal{G}))^{(0)}$ in $\mathsf{T}(\mathcal{G})$. Thus, the elements of the sequence set $\mathbf{C}(\mathcal{G})$ of $\mathcal{G}$ are the infinite words that are generated in $\mathsf{T}(\mathcal{G})$ by paths from state $(\iota(\mathcal{G}))^{(0)}$.

14.1.5 Regular digraphs

A labeled digraph $\mathcal{G} = (V, E, L)$ is called *M-regular* if each state in V has exactly M (> 0) outgoing edges. The labeled digraph $\mathcal{G}$ in Figure 14.1 is 2-regular, and so is every labeled digraph in Figure 14.4. In the case of M-regular digraphs, we find it convenient to name edges as pairs $[s; u]$, where s is the initial state of the edge and u is an element of a prescribed set Υ of size M. Thus,

$$E = V \times \Upsilon = \{[s; u] \ : \ s \in V \ , \ \ u \in \Upsilon\} \ . \tag{14.3}$$

We refer to the component u in the name of an edge as the *tag* of the edge. Any such naming of edges will be considered valid as long as it assigns distinct tags to edges with the same initial state. Note also the difference between a tag and a label: the tag is considered to be part of the identity of an edge (in a regular digraph), while a label is the value of the function L at that edge. The outgoing edges from each state must all have distinct tags, but not necessarily distinct labels.

Example 14.3 Let $\mathcal{G}$ be the 2-regular labeled digraph in Figure 14.1. We select $\Upsilon = \{0, 1\}$ and assign tags to the edges of $\mathcal{G}$, as shown in Figure 14.6. For the sake of simplicity and clarity, we have abbreviated each edge name $[s; u]$ in the figure into just $[u]$, where the edge tag u is italicized. □

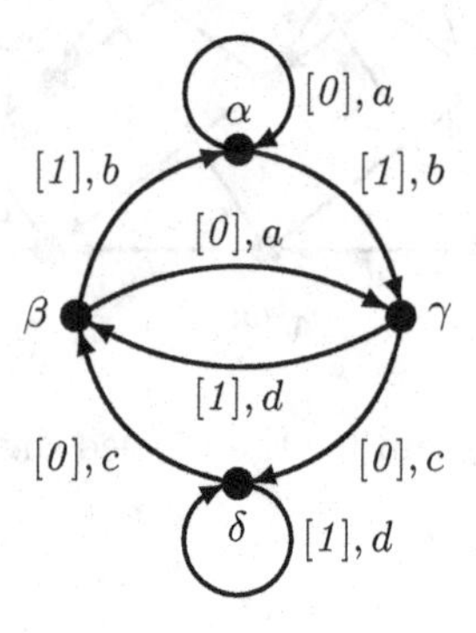

Figure 14.6. Labeled digraph $\mathcal{G}$ for Example 14.1.

A regular labeled digraph $\mathcal{G}$ whose edges are named as in (14.3) is sometimes called a *finite-state machine* (in short, FSM). A FSM $\mathcal{G}$ induces for every $\ell > 0$ a mapping

$$\mathcal{E}_\ell : \Upsilon^\ell \rightarrow \mathbf{C}_\ell(\mathcal{G})$$

which is defined by

$$\mathcal{E}_\ell(u_0 u_1 \ldots u_{\ell-1}) = L(\pi) , \tag{14.4}$$

where π is the unique path from $\iota(\mathcal{G})$ in $\mathcal{G}$ whose edges are tagged by $u_0 u_1 \ldots u_{\ell-1}$, i.e.,

$$\pi = [s_0; u_0][s_1; u_1] \ldots [s_{\ell-1}; u_{\ell-1}] , \tag{14.5}$$

where $s_0 = \iota(\mathcal{G})$ and $s_{t+1} = \tau\left([s_t; u_t]\right)$ for $0 \le t < \ell-1$. The mapping $\mathcal{E}_\ell$ will be referred to in the sequel as a *(finite-state) path encoder* that is associated with $\mathcal{G}$.

14.2 Trellis codes

We now turn to defining codes that are based on labeled digraphs.

Let n and M be positive integers and F be a finite alphabet. An (n, M) *trellis code* over F is a set of infinite sequences over F^n that equals $\mathbf{C}(\mathcal{G})$ for some M-regular lossless digraph $\mathcal{G} = (V, E, L)$ with labeling $L : E \rightarrow F^n$. We then say that $\mathcal{G}$ *presents* the trellis code and refer to the set of images of the mapping L as *codewords* of $\mathcal{G}$ and to the elements of the trellis code as *codeword sequences*.

We will denote a given (n, M) trellis code over F by $\mathbf{C}(\mathcal{G})$ (or, in short, just $\mathbf{C}$), where $\mathcal{G}$ is one of its presenting M-regular lossless digraphs. Note that the presenting digraph of a given trellis code $\mathbf{C}$ is not unique. For example, the deterministic digraphs $\mathcal{G}_1$ and $\mathcal{G}_2$ in Figure 14.4 both present the same $(1, 2)$ trellis code over $F = \{a, b\}$: this code consists of all infinite sequences over F.

There is no loss of generality in assuming that the presenting lossless digraph $\mathcal{G}$ is irreducible. Indeed, if it is reducible, then—according to our convention—we select $\iota(\mathcal{G})$ to be a state in an irreducible sink $\mathcal{G}_0$ of $\mathcal{G}$, in which case $\mathbf{C} = \mathbf{C}(\mathcal{G}) = \mathbf{C}(\mathcal{G}_0)$. The labeled digraph $\mathcal{G}_0$, in turn, is lossless, irreducible, and M-regular.

The sets $\mathbf{C}_\ell(\mathcal{G})$ and $\mathbf{C}_\ell^\circ(\mathcal{G})$ are defined for a given trellis code $\mathbf{C} = \mathbf{C}(\mathcal{G})$ as in Section 14.1.3. While the set $\mathbf{C}_\ell(\mathcal{G})$ does not depend on the particular lossless digraph $\mathcal{G}$ that is selected to present $\mathbf{C}$, the set $\mathbf{C}_\ell^\circ(\mathcal{G})$ does (e.g., the deterministic digraphs $\mathcal{G}_1$ and $\mathcal{G}_2$ in Figure 14.4 both present the same trellis code, yet $\mathbf{C}_\ell^\circ(\mathcal{G}_1) \neq \mathbf{C}_\ell^\circ(\mathcal{G}_2)$). We point out, however, that regardless of the choice of the (lossless) digraph $\mathcal{G}$, each codeword sequence in $\mathbf{C}_\ell^\circ(\mathcal{G})$ is generated by exactly one cycle in $\mathcal{G}$ that starts at $\iota(\mathcal{G})$.

Both sets $\mathbf{C}_\ell(\mathcal{G})$ and $\mathbf{C}_\ell^\circ(\mathcal{G})$ will be treated as block codes of length ℓ over the alphabet F^n, and their elements will be referred to as (finite) codeword sequences.

Example 14.4 A $(2, 2)$ trellis code $\mathbf{C}$ over $F = \{0, 1\}$ is presented by the lossless digraph $\mathcal{G}'$ in Figure 14.7, where we let $\iota(\mathcal{G}') = \alpha$. This digraph is the same as the one in Figure 14.1, except that the labels have been changed according to the rule

$$a \to 00\,, \quad b \to 11\,, \quad c \to 01\,, \quad \text{and} \quad d \to 10\,.$$

The trellis code $\mathbf{C}$ can be described also by the trellis diagram $\mathsf{T}(\mathcal{G}')$ of $\mathcal{G}'$, which is shown in Figure 14.8 (compare with Figure 14.5). The elements of

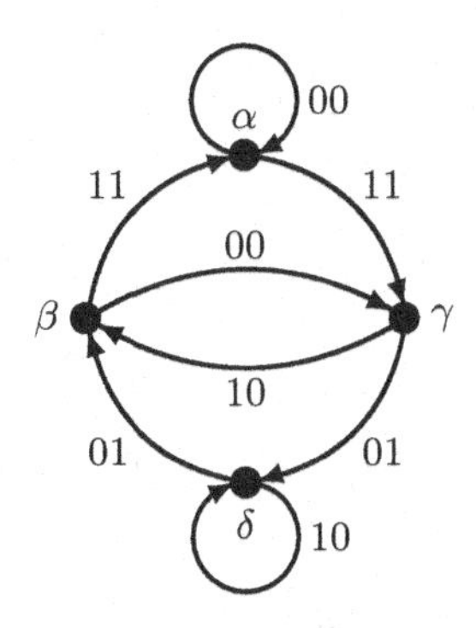

Figure 14.7. Labeled digraph $\mathcal{G}'$ that presents the trellis code in Example 14.4.

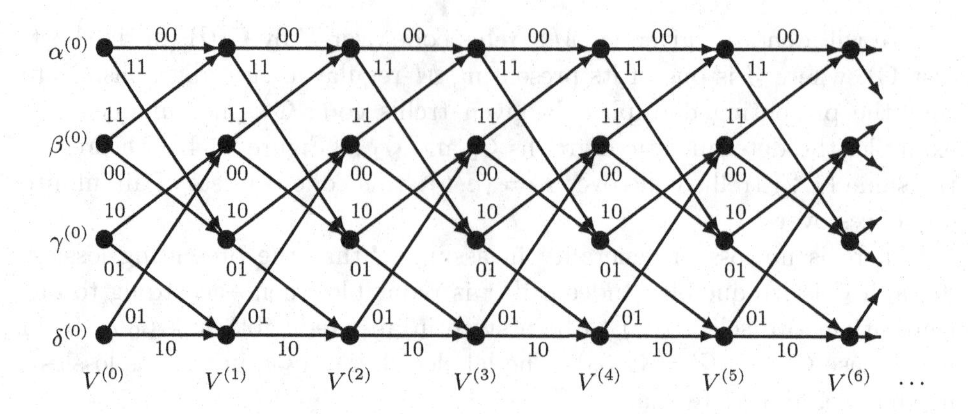

Figure 14.8. Trellis diagram $\mathsf{T}(\mathcal{G}')$ for Example 14.4.

$\mathbf{C}_\ell(\mathcal{G}')$ are the codeword sequences over F^2 that are generated by paths of length ℓ in $\mathsf{T}(\mathcal{G}')$ that start at state $\alpha^{(0)}$. The same holds for the set $\mathbf{C}^\circ_\ell(\mathcal{G}')$, except that now the paths also terminate in state $\alpha^{(\ell)}$. □

A special instance of trellis codes is obtained when one of the presenting digraphs $\mathcal{G}$ has one state only. Here $\mathbf{C}_1(\mathcal{G}) = \mathbf{C}^\circ_1(\mathcal{G})$ and

$$\mathbf{C}(\mathcal{G}) = \left\{ \mathbf{c}_0\mathbf{c}_1\mathbf{c}_2\cdots \;:\; \mathbf{c}_t \in \mathbf{C}_1(\mathcal{G}) \;\text{ for every }\; t \geq 0 \right\} ,$$

that is, the codewords of $\mathcal{G}$ are *freely-concatenable*.

14.2.1 Rate and free distance of trellis codes

Let $\mathbf{C}$ be an (n, M) trellis code over F. The *rate* of $\mathbf{C}$ is defined as

$$R = \frac{\log_{|F|} M}{n} .$$

For instance, the trellis code in Example 14.4 has rate $\frac{1}{2}$.

The following result relates the rate of a trellis code $\mathbf{C}(\mathcal{G})$ with the rates of the block codes $\mathbf{C}_\ell(\mathcal{G})$ and $\mathbf{C}^\circ_\ell(\mathcal{G})$.

Proposition 14.1 *Let $\mathbf{C} = \mathbf{C}(\mathcal{G})$ be an (n, M) trellis code over an alphabet of size q. Then*

$$\lim_{\ell\to\infty} \frac{\log_{q^n} |\mathbf{C}_\ell(\mathcal{G})|}{\ell} = \limsup_{\ell\to\infty} \frac{\log_{q^n} |\mathbf{C}^\circ_\ell(\mathcal{G})|}{\ell} = \frac{\log_q M}{n} .$$

In other words, the rates of the block codes $\mathbf{C}_\ell(\mathcal{G})$ and $\mathbf{C}^\circ_\ell(\mathcal{G})$ can get arbitrarily close to the rate of the trellis code $\mathbf{C}$.

The proof of Proposition 14.1 is given as a guided exercise in Problem 14.6. We remark that the "lim sup" in Proposition 14.1 can be changed into "lim" only when the (irreducible) presenting digraph $\mathcal{G}$ is aperiodic. If the period p is greater than 1 then the set $\mathbf{C}_\ell^\circ(\mathcal{G})$ is empty unless p divides ℓ.

The next term that we present can be regarded as an extension of the notion of minimum distance to trellis codes. Our definition makes use of the following notation. Let

$$\mathbf{x}_0\mathbf{x}_1\ldots\mathbf{x}_{\ell-1} \quad \text{and} \quad \mathbf{y}_0\mathbf{y}_1\ldots\mathbf{y}_{\ell-1}$$

be two (possibly infinite) sequences over F^n. The Hamming distance over F between these sequences is defined by

$$\mathsf{d}_F(\mathbf{x}_0\mathbf{x}_1\ldots\mathbf{x}_{\ell-1}, \mathbf{y}_0\mathbf{y}_1\ldots\mathbf{y}_{\ell-1}) = \sum_{t=0}^{\ell-1} \mathsf{d}(\mathbf{x}_t, \mathbf{y}_t) .$$

Equivalently, the Hamming distance over F is the (ordinary) Hamming distance between the two words

$$(\mathbf{x}_0 \,|\, \mathbf{x}_1 \,|\, \ldots \,|\, \mathbf{x}_{\ell-1}) \quad \text{and} \quad (\mathbf{y}_0 \,|\, \mathbf{y}_1 \,|\, \ldots \,|\, \mathbf{y}_{\ell-1}) ,$$

which are of length ℓn over F.

Let $\mathbf{C}$ be an $(n, M{>}1)$ trellis code over F. We define the *free distance* of $\mathbf{C}$ by

$$\begin{aligned} \mathsf{d}_{\text{free}}(\mathbf{C}) &= \min \mathsf{d}_F(\mathbf{c}_0\mathbf{c}_1\mathbf{c}_2\cdots, \mathbf{c}_0'\mathbf{c}_1'\mathbf{c}_2'\cdots) \\ &= \min \sum_{t=0}^{\infty} \mathsf{d}(\mathbf{c}_t, \mathbf{c}_t') , \end{aligned}$$

where the minimum is taken over all pairs $(\mathbf{c}_0\mathbf{c}_1\mathbf{c}_2\cdots, \mathbf{c}_0'\mathbf{c}_1'\mathbf{c}_2'\cdots)$ of distinct codeword sequences in $\mathbf{C}$.

Example 14.5 Let $\mathbf{C}$ be the $(2, 2)$ trellis code over $F = \{0, 1\}$ as in Example 14.4 and let $\mathcal{G}' = (V, E, L)$ be the presenting digraph that is shown in Figure 14.7. The codeword sequences

$$11\ 10\ 11\ 00\ 00\ 00\ \cdots \quad \text{and} \quad 11\ 01\ 01\ 11\ 00\ 00\ \cdots$$

can be generated in $\mathcal{G}'$ by paths that start at state $\iota(\mathcal{G}') = \alpha$; the respective paths in $\mathsf{T}(\mathcal{G}')$ are shown in the partial trellis diagram in Figure 14.9. Comparing these two codeword sequences, we get that $\mathsf{d}_{\text{free}}(\mathbf{C}) \leq 5$. We will show in Example 14.11 below that $\mathsf{d}_{\text{free}}(\mathbf{C})$ is exactly 5 in this case. $\square$

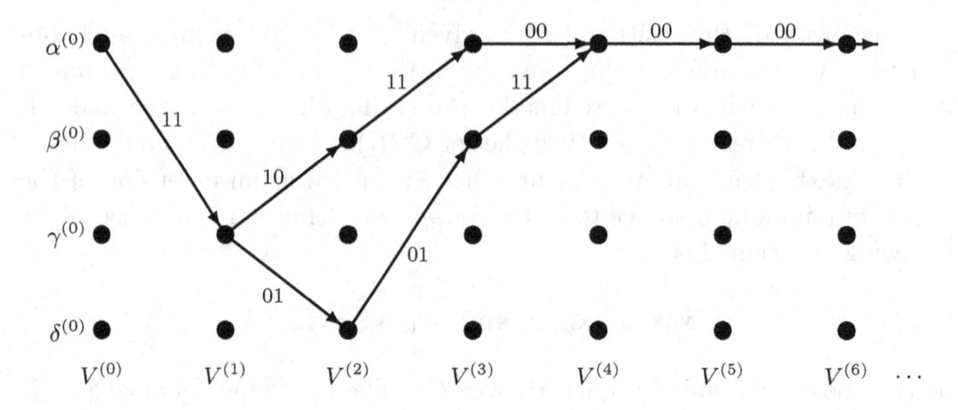

Figure 14.9. Partial trellis diagram for Example 14.5.

The free distance of a trellis code $\mathbf{C}(\mathcal{G})$ can be related to the minimum distance, <u>over F</u>, of the block codes $\mathbf{C}_\ell^\circ(\mathcal{G})$; assuming that $|\mathbf{C}_\ell^\circ(\mathcal{G})| > 1$, this minimum distance is defined by

$$\mathsf{d}_F(\mathbf{C}_\ell^\circ(\mathcal{G})) = \min \mathsf{d}_F(\mathbf{c}_0\mathbf{c}_1 \ldots \mathbf{c}_{\ell-1}, \mathbf{c}'_0\mathbf{c}'_1 \ldots \mathbf{c}'_{\ell-1}) ,$$

where $(\mathbf{c}_0\mathbf{c}_1 \ldots \mathbf{c}_{\ell-1}, \mathbf{c}'_0\mathbf{c}'_1 \ldots \mathbf{c}'_{\ell-1})$ ranges over all pairs of distinct codeword sequences in $\mathbf{C}_\ell^\circ(\mathcal{G})$. It is easy to verify that

$$\mathsf{d}_F(\mathbf{C}_\ell^\circ(\mathcal{G})) \geq \mathsf{d}_{\text{free}} . \tag{14.6}$$

14.2.2 Encoding of trellis codes

Let $\mathbf{C} = \mathbf{C}(\mathcal{G})$ be an (n, M) trellis code over an alphabet F of size q. We are going to use $\mathbf{C}$ for transmission by encoding sequences of messages into codeword sequences of $\mathbf{C}_\ell^\circ(\mathcal{G})$. Recall that the latter is a block code of length ℓ over F^n, with the property that each of its codeword sequences is generated by a unique cycle from state $\iota(\mathcal{G})$; this uniqueness, in turn, will allow a one-to-one mapping from sequences of messages to codeword sequences of $\mathbf{C}_\ell^\circ(\mathcal{G})$.

To construct the encoder, we fix $\mathcal{G} = (V, E, L)$ to be one of the presenting M-regular lossless digraphs of $\mathbf{C}$; we further assume (without loss of generality) that $\mathcal{G}$ is irreducible and denote by p and Δ its period and back-length, respectively. We name the edges of $\mathcal{G}$ as in (14.3), where Υ is taken as the set of messages. The value of ℓ will be typically large—in particular, at least Δ. We should also select it to be a multiple of p, or else $\mathbf{C}_\ell^\circ(\mathcal{G})$ would be empty.

The mapping from message sequences to codeword sequences will essentially be carried out by the path encoder

$$\mathcal{E}_\ell : \Upsilon^\ell \to \mathbf{C}_\ell(\mathcal{G}) ,$$

which was defined in (14.4) and (14.5): this encoder maps a sequence

$$u_0 u_1 \ldots u_{\ell-1} \in \Upsilon^\ell \tag{14.7}$$

through a path

$$\pi = [s_0; u_0][s_1; u_1] \ldots [s_{\ell-1}; u_{\ell-1}]$$

(from $s_0 = \iota(\mathcal{G})$) to the codeword sequence

$$\mathbf{c}_0 \mathbf{c}_1 \ldots \mathbf{c}_{\ell-1} = L([s_0; u_0]) L([s_1; u_1]) \ldots L([s_{\ell-1}; u_{\ell-1}]) \,. \tag{14.8}$$

However, the mapping $\mathcal{E}_\ell$ does require a modification, since we need π to be a cycle in $\mathcal{G}$. Therefore, we proceed as follows: the first

$$h = \ell - \Delta$$

elements in the sequence (14.7) are set to the messages that are to be transmitted, while the remaining $\ell - h = \Delta$ elements are set so that the sub-path

$$\pi' = [s_h; u_h][s_{h+1}; u_{h+1}] \ldots [s_{\ell-1}; u_{\ell-1}]$$

terminates in $\iota(\mathcal{G})$. Observe that (14.1) is satisfied for the selected value of h, since both ℓ and Δ are divisible by p; hence, such a path π' always exists. The modification to $\mathcal{E}_\ell$ that we have just described defines a mapping

$$\mathcal{E}_\ell^\circ : \Upsilon^h \to \mathbf{C}_\ell^\circ(\mathcal{G}) \,,$$

where

$$\mathcal{E}_\ell^\circ(u_0 u_1 \ldots u_{h-1}) = \mathcal{E}_\ell(u_0 u_1 \ldots u_{h-1} u_h u_{h+1} \ldots u_{\ell-1}) \,. \tag{14.9}$$

We refer to $\mathcal{E}_\ell^\circ$ as a *(finite-state) cycle encoder* that is associated with $\mathcal{G}$. The trailing Δ elements

$$u_h u_{h+1} \ldots u_{\ell-1}$$

will be called *dummy messages*: these are added only to guarantee that the path π is a cycle (and that the generated codeword sequence is an element of $\mathbf{C}_\ell^\circ(\mathcal{G})$).

The actual encoding is now carried out by the cycle encoder $\mathcal{E}_\ell^\circ$. If no errors have occurred, then knowledge of the initial (and terminal) state s_0 and the codeword sequence (14.8) allows the receiving end to reconstruct the cycle π and, hence, the message sequence $u_0 u_1 \ldots u_{h-1}$; in other words, the mapping $\mathcal{E}_\ell^\circ$ is one-to-one into $\mathbf{C}_\ell^\circ(\mathcal{G})$.

While the mapping $\mathcal{E}_\ell^\circ$ is not necessarily onto $\mathbf{C}_\ell^\circ(\mathcal{G})$, its set of images does have a rate that approaches the rate of the trellis code $\mathbf{C}$ when ℓ goes to infinity. Indeed, the number of images is $M^h = M^{\ell-\Delta}$ and, so,

$$\frac{\log_{q^n} M^{\ell-\Delta}}{\ell} = \left(1 - \frac{\Delta}{\ell}\right) \cdot \frac{\log_q M}{n} \,.$$

14.3 Decoding of trellis codes

We next turn to the problem of decoding trellis codes. Let $\mathbf{C}$ be an (n, M) trellis code over an alphabet F and let $\mathcal{G} = (V, E, L)$ be an irreducible M-regular lossless digraph that presents $\mathbf{C}$. We assume that a codeword sequence

$$\mathbf{c}_0\mathbf{c}_1 \ldots \mathbf{c}_{\ell-1} \in \mathbf{C}_\ell^\circ(\mathcal{G})$$

is generated by a cycle encoder $\mathcal{E}_\ell^\circ$ that is associated with $\mathcal{G}$. This sequence is regarded as a word of length ℓn over F and transmitted as such through a probabilistic channel $S = (F, \Phi, \mathsf{Prob})$, whose input alphabet, F, is the same as the alphabet of the codewords of $\mathbf{C}$. We denote by

$$\mathbf{y}_0\mathbf{y}_1 \ldots \mathbf{y}_{\ell-1} \qquad (14.10)$$

the respective received sequence, where each word $\mathbf{y}_t$ belongs to Φ^n. Since we treat the set $\mathbf{C}_\ell^\circ(\mathcal{G})$ as a block code of length ℓ over F^n, we will adapt the description of the channel S accordingly and regard S as having an input alphabet F^n and output alphabet Φ^n, simply by grouping together non-overlapping sub-blocks of n symbols (both in the input and the output) into one new symbol.

14.3.1 Maximum-likelihood decoder for trellis codes

A maximum-likelihood decoder (MLD) for $\mathbf{C}_\ell^\circ(\mathcal{G})$ with respect to the channel S maps the sequence (14.10) into the codeword sequence

$$\hat{\mathbf{c}}_0\hat{\mathbf{c}}_1 \ldots \hat{\mathbf{c}}_{\ell-1} \in \mathbf{C}_\ell^\circ(\mathcal{G}) \qquad (14.11)$$

that maximizes the conditional probability distribution

$$\mathsf{Prob}\{\, \mathbf{y}_0\mathbf{y}_1 \ldots \mathbf{y}_{\ell-1} \text{ received} \mid \hat{\mathbf{c}}_0\hat{\mathbf{c}}_1 \ldots \hat{\mathbf{c}}_{\ell-1} \text{ transmitted} \,\} .$$

We assume hereafter in the discussion that S is a memoryless channel (see Problem 1.8); in this case,

$$\mathsf{Prob}\{\, \mathbf{y}_0\mathbf{y}_1 \ldots \mathbf{y}_{\ell-1} \text{ received} \mid \hat{\mathbf{c}}_0\hat{\mathbf{c}}_1 \ldots \hat{\mathbf{c}}_{\ell-1} \text{ transmitted} \,\} = \prod_{t=0}^{\ell-1} \mathsf{Prob}\{\, \mathbf{y}_t \text{ received} \mid \hat{\mathbf{c}}_t \text{ transmitted} \,\} .$$

Taking logarithms, we thus find that an MLD for $\mathbf{C}_\ell^\circ(\mathcal{G})$ with respect to S decodes (14.10) into the codeword sequence (14.11) that minimizes the sum

$$\sum_{t=0}^{\ell-1} \chi(\mathbf{y}_t, \hat{\mathbf{c}}_t) , \qquad (14.12)$$

where

$$\chi(\mathbf{y}_t, \hat{\mathbf{c}}_t) = -\log \mathsf{Prob}\{\, \mathbf{y}_t \text{ received} \mid \hat{\mathbf{c}}_t \text{ transmitted} \,\} \,, \quad 0 \le t < \ell$$

(the base of the logarithm can be chosen arbitrarily, as long as it is greater than 1). We refer to the value $\chi(\mathbf{y}_t, \hat{\mathbf{c}}_t)$ as the *cost* of the codeword $\hat{\mathbf{c}}_t$ with respect to the received word $\mathbf{y}_t$, and to the sum (14.12) as the cost of the codeword sequence $(\hat{\mathbf{c}}_t)_{t=0}^{\ell-1}$ with respect to the received sequence $(\mathbf{y}_t)_{t=0}^{\ell-1}$.

Example 14.6 Let S be the memoryless q-ary symmetric channel with positive crossover probability $p < 1 - (1/q)$. In this case we have for every $\mathbf{x}, \mathbf{y} \in F^n$,

$$\begin{aligned} &\mathsf{Prob}\{\, \mathbf{y} \text{ received} \mid \mathbf{x} \text{ transmitted} \,\} \\ &\quad = (p/(q{-}1))^{\mathsf{d}(\mathbf{y},\mathbf{x})} (1{-}p)^{n-\mathsf{d}(\mathbf{y},\mathbf{x})} \\ &\quad = (1{-}p)^n \left(\frac{p}{(1{-}p)(q{-}1)} \right)^{\mathsf{d}(\mathbf{y},\mathbf{x})} . \end{aligned}$$

Hence here,

$$\chi(\mathbf{y}_t, \hat{\mathbf{c}}_t) = a + b \cdot \mathsf{d}(\mathbf{y}_t, \hat{\mathbf{c}}_t) \,,$$

where

$$a = -n \log{(1{-}p)} \quad \text{and} \quad b = -\log \left(\frac{p}{(1{-}p)(q{-}1)} \right) .$$

Since $b > 0$, we get the (familiar) result that an MLD for $\mathbf{C}_\ell^\circ(\mathcal{G})$ with respect to S returns a codeword sequence (14.11) that minimizes the Hamming distance

$$\sum_{t=0}^{\ell-1} \mathsf{d}(\mathbf{y}_t, \hat{\mathbf{c}}_t) = \mathsf{d}_F(\mathbf{y}_0 \mathbf{y}_1 \ldots \mathbf{y}_{\ell-1}, \hat{\mathbf{c}}_0 \hat{\mathbf{c}}_1 \ldots \hat{\mathbf{c}}_{\ell-1})$$

(see Problem 1.7). □

14.3.2 Viterbi's algorithm

We next turn to the problem of effectively computing the codeword sequence $(\hat{\mathbf{c}}_t)_{t=0}^{\ell-1}$ whose cost (14.12) with respect to the received sequence is the smallest. The problem of minimizing (14.12) is equivalent to finding a cycle π of length ℓ in $\mathcal{G}$ that starts at $\iota(\mathcal{G})$ and generates a codeword sequence with a minimal cost.

Given the sequence (14.10) of received words over Φ^n, we associate with each edge $e \in E$ and integer $t \in \{0, 1, \ldots, \ell{-}1\}$ the following *edge cost* of e at time t:

$$\chi_t(e) = \chi(\mathbf{y}_t, L(e)) \,.$$

The cost of a path

$$\pi = e_0 e_1 \ldots e_{t-1}$$

with initial state $\iota(\mathcal{G})$ and length $t \le \ell$ is defined as the sum of the edge costs along that path, taking the "time" of each edge as its location along the path, namely,

$$\chi(\pi) = \chi_0(e_0) + \chi_1(e_1) + \ldots + \chi_{t-1}(e_{t-1}) . \tag{14.13}$$

We readily obtain that $\chi(\pi)$ equals the cost of the codeword sequence, $L(\pi)$, that π generates. Finally, we associate with each state $s \in V$ its *state cost* at time t,

$$\chi_t(s) = \min_{\pi} \chi(\pi) , \tag{14.14}$$

where the minimum is taken over all paths of length t in $\mathcal{G}$ from state $\iota(\mathcal{G})$ to s; if no such paths exist, we set $\chi_t(s) = \infty$. The cost of $\iota(\mathcal{G})$ at time 0 is defined as 0, and the path that minimizes (14.14) will be denoted by $\pi_t(s)$.

Our problem of minimizing (14.12) will be solved once we find a minimizing cycle $\pi_\ell(\iota(\mathcal{G}))$.

We can accomplish this by the algorithm that is shown in Figure 14.10. This algorithm, known as *Viterbi's algorithm*, computes the state costs

---

Input: edge costs $\chi_t(e)$ for $e \in E$ and $0 \le t < \ell$.
Output: path $\pi_{\min}$.

1. For every $s \in V$ do:

$$\chi_0(s) \leftarrow \begin{cases} 0 & \text{if } s = \iota(\mathcal{G}) \\ \infty & \text{otherwise} \end{cases} .$$

2. Let $\pi_0(\iota(\mathcal{G})) \leftarrow$ path of length 0.
3. For $t = 0, 1, \ldots, \ell-1$ do:

   For every state $s \in V$ do:

   (a) find an incoming edge e to s in $\mathcal{G}$ that minimizes the sum

   $$\chi_t(e) + \chi_t(\iota(e)) ;$$

   (b) let $\chi_{t+1}(s) \leftarrow \chi_t(e) + \chi_t(\iota(e))$;
   (c) let $\pi_{t+1}(s) \leftarrow \pi_t(\iota(e))\, e$.
4. Return $\pi_{\min} = \pi_\ell(\iota(\mathcal{G}))$.

Figure 14.10. Viterbi's algorithm for a labeled digraph $\mathcal{G} = (V, E, L)$.

$(\chi_t(s))_{s \in V}$ iteratively for times $t = 0, 1, \ldots, \ell$, along with respective minimizing paths $(\pi_t(s))_{s \in V}$. The algorithm is based on the following observation, which is easily proved by induction on t: for each state $s \in V$,

$$\chi_{t+1}(s) = \min_{\substack{e \in E: \\ \tau(e)=s}} \{\chi_t(e) + \chi_t(\iota(e))\}$$

(that is, the minimum is taken over all the incoming edges e to state s in $\mathcal{G}$).

Viterbi's algorithm can be visualized also through the trellis diagram $\mathsf{T}(\mathcal{G})$. With each edge $e^{(t)}$ in

$$E^{(0)} \cup E^{(1)} \cup \ldots \cup E^{(\ell-1)}$$

(see (14.2)), we associate the edge cost $\chi_t(e)$. Then, for $t = 0, 1, \ldots, \ell$, the algorithm finds a path with a smallest cost from state $(\iota(\mathcal{G}))^{(0)}$ (in layer 0) to each state in layer t. The returned value of the algorithm is a path with the smallest cost from $(\iota(\mathcal{G}))^{(0)}$ to $(\iota(\mathcal{G}))^{(\ell)}$.

Example 14.7 Let $\mathbf{C} = \mathbf{C}(\mathcal{G}')$ be the $(2,2)$ trellis code over $F = \{0,1\}$ as in Example 14.4, and suppose that a codeword sequence of $\mathbf{C}_6^{\circ}(\mathcal{G}')$ is transmitted through a binary symmetric channel with crossover probability $p < \frac{1}{2}$ and that the received sequence is

$$\mathbf{y}_0\mathbf{y}_1 \cdots \mathbf{y}_5 \;=\; 11\ 00\ 00\ 01\ 00\ 10\ .$$

Figure 14.11 shows layers 0 through 6 of the trellis diagram $\mathsf{T}(\mathcal{G}')$ in Figure 14.8, except that in each layer t, we have replaced every edge label $\mathbf{c}$ by the Hamming distance $\mathsf{d}(\mathbf{y}_t, \mathbf{c})$ (these distances appear as italic numbers next to each edge in Figure 14.11).

We now apply Viterbi's algorithm to the labeled digraph $\mathcal{G}'$ in Figure 14.7, using for simplicity the integer values $\mathsf{d}(\mathbf{y}_t, L(e))$ as our edge costs

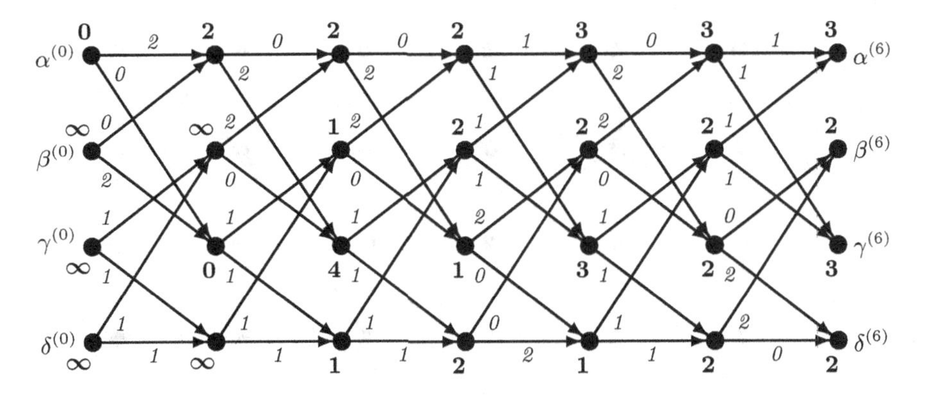

Figure 14.11. Edge and state costs in the trellis diagram $\mathsf{T}(\mathcal{G}')$ for Example 14.7.

$\chi_t(e)$ (while this deviates from our original definition of edge costs, it follows from Example 14.6 that this modification does not affect the minimization). The resulting state costs are shown as boldface numbers in Figure 14.11, where each value $\chi_t(s)$ is written next to the state $s^{(t)}$.

The (unique) minimum-cost path $\pi_{\min}$ is given by

$$\alpha \xrightarrow{11} \gamma \xrightarrow{10} \beta \xrightarrow{00} \gamma \xrightarrow{01} \delta \xrightarrow{01} \beta \xrightarrow{11} \alpha ,$$

and the respective path in the trellis diagram $\mathsf{T}(\mathcal{G}')$ is shown in Figure 14.12. The decoded codeword sequence is therefore

$$\hat{\mathbf{c}}_0\hat{\mathbf{c}}_1\hat{\mathbf{c}}_2\hat{\mathbf{c}}_3\hat{\mathbf{c}}_4\hat{\mathbf{c}}_5 = 11\ 10\ 00\ 01\ 01\ 11 ,$$

and its Hamming distance from the received sequence is

$$\sum_{t=0}^{5} \mathsf{d}(\mathbf{y}_t, \hat{\mathbf{c}}_t) = 3 .$$

The sequence $(\hat{\mathbf{c}}_t)_{t=0}^5$ is the closest in $\mathbf{C}_6^\circ(\mathcal{G})$ (with respect to the Hamming distance) to the received sequence $(\mathbf{y}_t)_{t=0}^5$. □

By looking at the two nested loops in Step 3 in Figure 14.10, we easily see that the algorithm can be implemented using a number of real operations that is proportional to $\ell \cdot |E|$. For each iteration over t, the algorithm needs to keep track of $|V|$ paths of length t, thereby requiring space whose size (in bits) is proportional to $\ell \cdot |V| \log M$. Additional space is then required to record the $|V|$ real values of the state costs. (Note that the computations at iteration $t+1$ of Step 3 only require the results of the computations at iteration t. Therefore, there is no need to record the whole history of values of the state costs.)

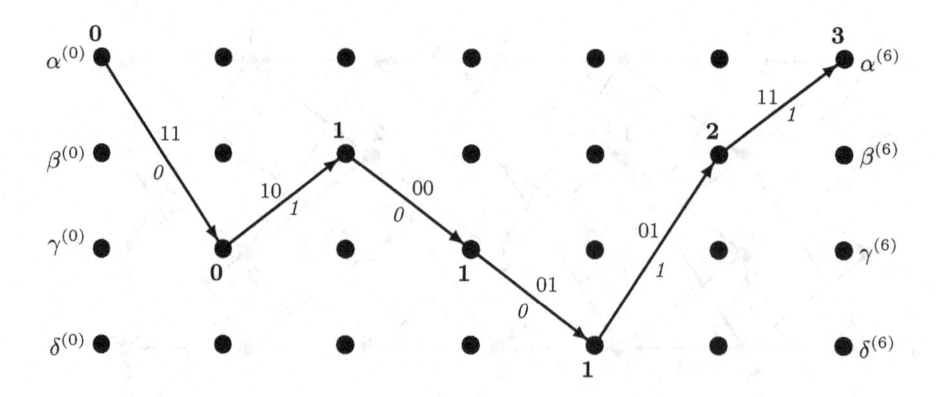

Figure 14.12. Minimum-cost path in $\mathsf{T}(\mathcal{G}')$ for Example 14.7.

14.3.3 Decoding error probability

It should be noted that while Viterbi's algorithm implements an MLD for $\mathbf{C}_\ell^\circ(\mathcal{G})$, the decoding error probability of this decoder will typically approach 1 as ℓ becomes large. This can be seen already for the simple case where the presenting digraph $\mathcal{G}$ has one state only. Here, the transmitted codeword sequence

$$\mathbf{c}_0\mathbf{c}_1 \ldots \mathbf{c}_{\ell-1}$$

consists of elements of an (n, M) block code $\mathbf{C}_1(\mathcal{G})$ over F, which are freely-concatenable. An application of Viterbi's algorithm to the respective received sequence

$$\mathbf{y}_0\mathbf{y}_1 \cdots \mathbf{y}_{\ell-1}$$

becomes a sequence of applications of an MLD $\mathcal{D}$ for $\mathbf{C}_1(\mathcal{G})$ to each word $\mathbf{y}_t$, independently of words at other times t. This results in a decoded codeword sequence

$$\hat{\mathbf{c}}_0\hat{\mathbf{c}}_1 \ldots \hat{\mathbf{c}}_{\ell-1} ,$$

where $\hat{\mathbf{c}}_t = \mathcal{D}(\mathbf{y}_t)$ for $0 \leq t < \ell$. Now, except for very special codes or channels, the decoding error probability of $\mathcal{D}$ will be bounded from below by some strictly positive real θ. It follows that there is a codeword sequence $\mathbf{c}_0\mathbf{c}_1 \ldots \mathbf{c}_{\ell-1}$ for which

$$\mathsf{Prob}\left\{\hat{\mathbf{c}}_0\hat{\mathbf{c}}_1 \ldots \hat{\mathbf{c}}_{\ell-1} = \mathbf{c}_0\mathbf{c}_1 \ldots \mathbf{c}_{\ell-1} \mid \mathbf{c}_0\mathbf{c}_1 \ldots \mathbf{c}_{\ell-1} \text{ transmitted}\right\} \leq (1-\theta)^\ell ,$$

and this upper bound becomes zero as $\ell \to \infty$.

This seemingly absurd phenomenon is easily settled by observing that in each transmission, we actually send ℓ messages rather than just one message; so, accordingly, we need to normalize the decoding error probability per each sent message. And we will indeed do that in Section 14.7.2, for a certain family of trellis codes over the q-ary symmetric channel.

14.4 Linear finite-state machines

In this section, we introduce a family of regular labeled digraphs with some underlying linear properties. The trellis codes that these digraphs present will then be the focus of our discussion in the upcoming sections of this chapter.

Let F be the finite field $\mathrm{GF}(q)$ and let m, n, and k be positive integers. Fix P, B, Q, and D to be matrices over F with orders

$$m \times m , \quad k \times m , \quad m \times n , \quad \text{and} \quad k \times n ,$$

respectively, and consider the labeled digraph $\mathcal{G} = (V, E, L)$ which has the following form:

- $V = F^m$; we denote a typical state in $\mathcal{G}$ as a row vector $\mathbf{s} \in F^m$.
- $E = V \times F^k$; we denote a typical edge in $\mathcal{G}$ by $[\mathbf{s}; \mathbf{u}]$, where $\mathbf{s} \in V$ and $\mathbf{u}$ is a row vector in F^k that serves as the edge tag.
- For every edge $e = [\mathbf{s}; \mathbf{u}]$ in E,
$$\iota(e) = \mathbf{s} \quad \text{and} \quad \tau(e) = \mathbf{s}P + \mathbf{u}B \,. \tag{14.15}$$
- The range of the labeling L is $\Sigma = F^n$, and for every edge $e = [\mathbf{s}; \mathbf{u}]$ in E,
$$L(e) = \mathbf{s}Q + \mathbf{u}D \,. \tag{14.16}$$

A labeled digraph $\mathcal{G}$ that satisfies these properties for some P, B, Q, and D, is called a *k-to-n linear finite-state machine* (in short, a $k : n$ LFSM) over F. (We can formally extend the definition of LFSMs to include also the case $m = 0$: here $|V| = 1$, making (14.15) vacuous and reducing (14.16) to just $L(e) = \mathbf{u}D$.)

It is easy to see that $\mathcal{G}$ is a q^k-regular digraph. While an LFSM $\mathcal{G}$ is completely determined by the quadruple (P, B, Q, D), different quadruples may define isomorphic LFSMs (i.e., LFSMs that differ only in their state names or edge names). We illustrate this in the next example (see also Problem 14.8).

Example 14.8 Let $F = \mathrm{GF}(2)$ and let the matrices P, B, Q, and D (over F) be given by

$$P = \begin{pmatrix} 0 & 1 \\ 1 & 1 \end{pmatrix}, \quad B = \begin{pmatrix} 0 & 1 \end{pmatrix}, \quad Q = \begin{pmatrix} 0 & 0 \\ 0 & 1 \end{pmatrix}, \quad \text{and} \quad D = \begin{pmatrix} 1 & 1 \end{pmatrix}.$$

These matrices define a $1 : 2$ LFSM $\mathcal{G}$ over F, which is shown in Figure 14.13 (the state names are underlined in the figure and the edge names are abbreviated to indicate only the edge tags, which are italicized). The LFSM $\mathcal{G}$ is identical to the labeled digraph in Figure 14.7, except that the edges have been assigned tags and the names of the states have been changed as follows:

$$\alpha \to \underline{00}\,, \quad \beta \to \underline{10}\,, \quad \gamma \to \underline{01}\,, \quad \text{and} \quad \delta \to \underline{11}\,.$$

Figure 14.13 shows a second $1 : 2$ LFSM, $\tilde{\mathcal{G}}$, over F that is defined by the matrices

$$\tilde{P} = \begin{pmatrix} 0 & 0 \\ 1 & 0 \end{pmatrix}, \quad \tilde{B} = \begin{pmatrix} 0 & 1 \end{pmatrix}, \quad \tilde{Q} = \begin{pmatrix} 1 & 1 \\ 1 & 0 \end{pmatrix}, \quad \text{and} \quad \tilde{D} = \begin{pmatrix} 1 & 1 \end{pmatrix}.$$

Both $\mathcal{G}$ and $\tilde{\mathcal{G}}$ are identical, except for the edge tags. □

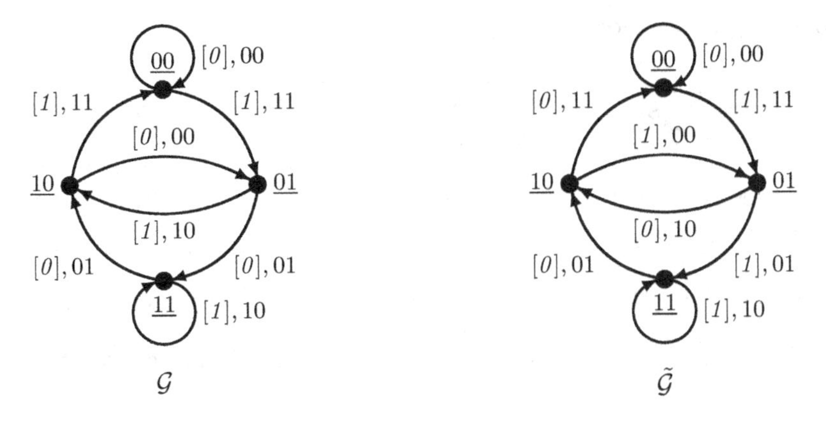

Figure 14.13. LFSMs $\mathcal{G}$ and $\tilde{\mathcal{G}}$ for Example 14.8.

We select the start state of an LFSM to be the all-zero vector $\mathbf{0}_m$. (As we will be referring in the sequel to all-zero vectors in various vector spaces over F, we will use subscripts—such as m here—to identify the dimensions of these spaces.) An LFSM $\mathcal{G}$ is not necessarily irreducible, but the all-zero state $\mathbf{0}_m$ belongs to an irreducible sink $\mathcal{G}_0$ of $\mathcal{G}$. The irreducible sink $\mathcal{G}_0$ is also an LFSM: it is aperiodic and every integer $\ell \geq m$ serves as a back-length of $\mathcal{G}_0$ (see Problems 14.10 and 14.11).

14.4.1 Response matrix of LFSMs

The next step in our study of LFSMs is obtaining an algebraic characterization of the sequence set of a given LFSM.

Let $\mathcal{G} = (V{=}F^m, E, L)$ be a $k : n$ LFSM over F that is defined by the quadruple (P, B, Q, D). Consider an infinite path in $\mathcal{G}$,

$$[\mathbf{s}_0; \mathbf{u}_0][\mathbf{s}_1; \mathbf{u}_1][\mathbf{s}_2; \mathbf{u}_2] \cdots ,$$

and let

$$\mathbf{c}_0\mathbf{c}_1\mathbf{c}_2 \cdots$$

be the infinite sequence that it generates (each label $\mathbf{c}_t$ is an element of $\Sigma = F^n$). By (14.15) and (14.16) we get that for every $t \geq 0$,

$$\mathbf{s}_{t+1} = \mathbf{s}_t P + \mathbf{u}_t B \tag{14.17}$$

and

$$\mathbf{c}_t = \mathbf{s}_t Q + \mathbf{u}_t D \,. \tag{14.18}$$

Henceforth, we associate with the infinite sequences

$$\mathbf{s}_0\mathbf{s}_1\mathbf{s}_2 \cdots , \quad \mathbf{u}_0\mathbf{u}_1\mathbf{u}_2 \cdots , \quad \text{and} \quad \mathbf{c}_0\mathbf{c}_1\mathbf{c}_2 \cdots$$

the following expressions in the indeterminate x:

$$\mathbf{s}(x) = \sum_{t=0}^{\infty} \mathbf{s}_t x^t , \quad \mathbf{u}(x) = \sum_{t=0}^{\infty} \mathbf{u}_t x^t , \quad \text{and} \quad \mathbf{c}(x) = \sum_{t=0}^{\infty} \mathbf{c}_t x^t .$$

Each of these expressions forms a vector over the ring, $F[[x]]$, of (infinite) formal power series over F (see Section 6.3.2); namely,

$$\mathbf{s}(x) \in (F[[x]])^m , \quad \mathbf{u}(x) \in (F[[x]])^k , \quad \text{and} \quad \mathbf{c}(x) \in (F[[x]])^n .$$

Multiplying both sides of (14.17) by x^{t+1} and summing over $t \geq 0$, we obtain

$$\mathbf{s}(x) - \mathbf{s}_0 = x\,(\mathbf{s}(x)P + \mathbf{u}(x)B)$$

or

$$\mathbf{s}(x)(I - xP) = \mathbf{u}(x) \cdot xB + \mathbf{s}_0 , \tag{14.19}$$

where I stands for the $m \times m$ identity matrix. Now, $I - xP$ is an $m \times m$ invertible matrix over $F[[x]]$; indeed,

$$(I - xP)\Big(\sum_{t=0}^{\infty} x^t P^t\Big) = \Big(\sum_{t=0}^{\infty} x^t P^t\Big) - \Big(\sum_{t=0}^{\infty} x^{t+1} P^{t+1}\Big) = I .$$

We thus conclude from (14.19) that, over $F[[x]]$,

$$\mathbf{s}(x) = (\mathbf{u}(x) \cdot xB + \mathbf{s}_0)(I - xP)^{-1} . \tag{14.20}$$

Equation (14.20) expresses the state sequence $(\mathbf{s}_t)_{t=0}^{\infty}$ along an infinite path, in terms of the tag sequence $(\mathbf{u}_t)_{t=0}^{\infty}$ and the initial state $\mathbf{s}_0$ of the path. In a similar manner, we can obtain from (14.18) an expression that relates $\mathbf{c}(x)$ to $\mathbf{u}(x)$ and $\mathbf{s}_0$. Specifically, if we multiply both sides of (14.18) by x^t and sum over $t \geq 0$, we get

$$\mathbf{c}(x) = \mathbf{s}(x)Q + \mathbf{u}(x)D ,$$

and plugging (14.20) into the latter equality yields

$$\mathbf{c}(x) = \mathbf{u}(x)G(x) + \mathbf{s}_0(I - xP)^{-1}Q , \tag{14.21}$$

where $G(x)$ is the following $k \times n$ matrix over $F[[x]]$:

$$G(x) = \big(\, g_{i,j}(x) \,\big)_{i=1}^{k}\,_{j=1}^{n} \;=\; xB(I - xP)^{-1}Q + D . \tag{14.22}$$

We will refer to $G(x)$ as the *response matrix* of the LFSM $\mathcal{G}$.

Of particular interest is the case where $\mathbf{s}_0 = \mathbf{0}_m$ $(= \iota(\mathcal{G}))$, for which we obtain from (14.21) a characterization of the sequence set $\mathbf{C}(\mathcal{G})$ of $\mathcal{G}$. We summarize this case as follows.

Proposition 14.2 *Let $\mathcal{G}$ be a $k : n$ LFSM over $F = \mathrm{GF}(q)$ and $G(x)$ be the response matrix of $\mathcal{G}$. Denote by $\mathrm{span}(G(x))$ the linear span over $F[[x]]$ of the rows of $G(x)$, namely,*

$$\mathrm{span}(G(x)) = \Big\{ \mathbf{c}(x) \in (F[[x]])^n \; : \; \mathbf{c}(x) = \mathbf{u}(x)G(x) \;\; \textit{for some} \;\; \mathbf{u}(x) \in (F[[x]])^k \Big\} .$$

Then

$$\mathbf{C}(\mathcal{G}) = \Big\{ \mathbf{c}_0\mathbf{c}_1\mathbf{c}_2 \cdots \; : \; \sum_{t=0}^{\infty} \mathbf{c}_t x^t \in \mathrm{span}(G(x)) \Big\} .$$

Based on Proposition 14.2, we will not distinguish hereafter between the sets $\mathbf{C}(\mathcal{G})$ and $\mathrm{span}(G(x))$. Along similar lines, we will regard each set $\mathbf{C}_\ell(\mathcal{G})$ or $\mathbf{C}_\ell^\circ(\mathcal{G})$ as a subset of $(F_\ell[x])^n$, with each sequence $\mathbf{c}_0\mathbf{c}_1 \ldots \mathbf{c}_{\ell-1}$ being identified with the vector $\sum_{t=0}^{\ell-1} \mathbf{c}_t x^t$ in $(F_\ell[x])^n$.

From (14.22) we can obtain some properties of the entries of the response matrix $G(x)$. To this end, we first recall from linear algebra that the inverse of an invertible square matrix U over any given commutative ring can be written as

$$U^{-1} = \frac{1}{\det(U)} \cdot \mathrm{Adj}(U) ,$$

where $\mathrm{Adj}(U)$ is the adjoint matrix of U. We now apply this formula to the matrix $I - xP$ (as a matrix over $F[[x]]$) and obtain from (14.22) that

$$G(x) = \frac{1}{\det(I - xP)} \Big(xB \cdot \mathrm{Adj}(I - xP) \cdot Q + \det(I - xP) \cdot D \Big) .$$

The determinant $\det(I - xP)$ is a polynomial in $F_{m+1}[x]$, which takes the value 1 at $x = 0$; as such, it is an invertible element in $F[[x]]$. Also, each entry of $\mathrm{Adj}(I - xP)$ is a cofactor of $I - xP$ (namely, it is a minor of $I - xP$, up to a sign change); as such, it is a polynomial in $F_m[x]$. It follows that each entry $g_{i,j}(x)$ of $G(x)$ can be expressed as a ratio

$$g_{i,j}(x) = \frac{\omega_{i,j}(x)}{\sigma_{i,j}(x)} , \quad 1 \le i \le k , \quad 1 \le j \le n ,$$

where $\omega_{i,j}(x)$ and $\sigma_{i,j}(x)$ are polynomials in $F[x]$ such that

$$\sigma_{i,j}(0) = 1 \quad \text{and} \quad \max\{\deg \sigma_{i,j}, \deg \omega_{i,j}\} \le m .$$

Furthermore, we can assume without loss of generality that each ratio $\omega_{i,j}(x)/\sigma_{i,j}(x)$ is reduced, i.e.,

$$\gcd(\sigma_{i,j}(x), \omega_{i,j}(x)) = 1 .$$

Hence, in addition to being a matrix over $F[[x]]$, the response matrix $G(x)$ is also a matrix over the field of rational functions over F; we denote this field by $F(x)$ (see the proof of Lemma 9.1). We mention that there exists a field, denoted by $F((x))$, which contains both $F[[x]]$ as a subring and $F(x)$ as a subfield. The field $F((x))$ consists of all the *Laurent series* over F: these series are similar to infinite formal power series, except that they may also contain finitely many terms with negative powers of x. We define the field $F((x))$ formally in Problem 14.14.

Example 14.9 Let $F = \mathrm{GF}(2)$ and let the matrices P, B, Q, and D be as in Example 14.8. Here

$$I - xP = \begin{pmatrix} 1 & x \\ x & 1+x \end{pmatrix}$$

and, so,

$$\begin{aligned} (I - xP)^{-1} &= \frac{1}{\det(I - xP)} \cdot \mathrm{Adj}(I - xP) \\ &= \frac{1}{1+x+x^2} \begin{pmatrix} 1+x & x \\ x & 1 \end{pmatrix} . \end{aligned}$$

Therefore,

$$\begin{aligned} G(x) &= xB(I - xP)^{-1}Q + D \\ &= \begin{pmatrix} 1 & \dfrac{1+x^2}{1+x+x^2} \end{pmatrix} . \end{aligned}$$

We could do a similar computation also with the matrices $\tilde{P}$, $\tilde{B}$, $\tilde{Q}$, and $\tilde{D}$ in Example 14.8, in which case we would get the response matrix

$$\begin{aligned} \tilde{G}(x) &= x\tilde{B}(I - x\tilde{P})^{-1}\tilde{Q} + \tilde{D} \\ &= \begin{pmatrix} 1+x+x^2 & 1+x^2 \end{pmatrix} . \end{aligned}$$

□

14.4.2 Lossless LFSMs

As mentioned at the beginning of Section 14.4, our goal in the sequel is to study trellis codes whose presenting labeled digraphs are LFSMs. Such digraphs need to be lossless, and the next result presents a necessary and sufficient condition for them to be so.

Proposition 14.3 *Let $\mathcal{G}$ be a $k : n$ LFSM over $F = \mathrm{GF}(q)$ and $G(x)$ be the response matrix of $\mathcal{G}$. Then $\mathcal{G}$ is lossless if and only if*

$$\mathrm{rank}(G(x)) = k$$

(over $F(x)$).

The proof of Proposition 14.3 is given as a guided exercise in Problem 14.15.

Let $\mathcal{G}$ be a lossless LFSM and $G(x)$ be the response matrix of $\mathcal{G}$. It follows from Proposition 14.3 that $G(x)$ has rank k in any field that contains $F(x)$ as a subfield. In particular, this holds for the field of Laurent series $F((x))$, which also contains $F[[x]]$ as a subring (see Problem 14.14; recall that the entries of $G(x)$ are in the intersection $F(x) \cap F[[x]]$). Hence, the rows of $G(x)$ are linearly independent over $F[[x]]$ and, so, when $\mathcal{G}$ is lossless we have that

$$\text{span}(G(x)) = \left\{ \mathbf{u}(x)G(x) \ : \ \mathbf{u}(x) \in (F[[x]])^k \right\}$$

(where distinct elements $\mathbf{u}(x)$ in $(F[[x]])^k$ yield distinct elements $\mathbf{u}(x)G(x)$ in $\text{span}(G(x))$).

14.5 Convolutional codes

Let $F = \text{GF}(q)$ and let k and n be positive integers such that $k \leq n$. An $[n, k]$ *convolutional code* over F is an (n, q^k) trellis code $\mathbf{C}$ over F that is presented by a lossless $k : n$ LFSM over F. We will specify a convolutional code through one of its presenting LFSMs $\mathcal{G}$ and use the notation $\mathbf{C}(\mathcal{G})$ (or simply $\mathbf{C}$). The rate of an $[n, k]$ convolutional code over F is given by

$$R = \frac{k}{n} .$$

If $G(x)$ is the response matrix of a presenting lossless LFSM $\mathcal{G}$, then from Propositions 14.2 and 14.3 we get that

$$\mathbf{C} = \mathbf{C}(\mathcal{G}) = \text{span}(G(x)) \qquad \text{and} \qquad \text{rank}(G(x)) = k$$

(following the convention set in Section 14.4.1, we do not distinguish here between an infinite codeword sequence $\mathbf{c}_0\mathbf{c}_1\mathbf{c}_2 \cdots$ over F^n and its representation as the element $\sum_{t=0}^{\infty} \mathbf{c}_t x^t$ of $(F[[x]])^n$). Borrowing the term from ordinary linear codes, we then say that $G(x)$ is a *generator matrix* of the convolutional code $\mathbf{C}$.

Example 14.10 We have shown in Example 14.8 that the labeled digraph in Figure 14.7 is a (lossless) $1 : 2$ LFSM over $F = \text{GF}(2)$. Thus, the trellis code $\mathbf{C}$ in Example 14.4 is a $[2, 1]$ convolutional code of rate $\frac{1}{2}$ over F. From the computation in Example 14.9 we get that

$$G(x) = \begin{pmatrix} 1 & \frac{1+x^2}{1+x+x^2} \end{pmatrix}$$

is a generator matrix of $\mathbf{C}$. This generator matrix is *systematic*, as its first k $(= 1)$ columns form the identity matrix. □

We can state an alternate (equivalent) definition of convolutional codes that is based directly on their generator matrices (without going through the presenting LFSMs). Specifically, let $F = \mathrm{GF}(q)$ and let $G(x)$ be a $k \times n$ matrix of rank k over $F(x) \cap F[[x]]$; that is, the entries of $G(x)$ are of the form

$$g_{i,j}(x) = \frac{\omega_{i,j}(x)}{\sigma_{i,j}(x)}, \quad 1 \le i \le k, \quad 1 \le j \le n, \tag{14.23}$$

where $\omega_{i,j}(x)$ and $\sigma_{i,j}(x)$ are polynomials in $F[x]$ such that $\sigma_{i,j}(0) = 1$. Then $\mathrm{span}(G(x))$ is an $[n, k]$ convolutional code over F.

To see why this definition coincides with our earlier one, it suffices to show that $G(x)$ is the response matrix of some $k : n$ LFSM $\mathcal{G}$ over F (since $\mathrm{rank}(G(x)) = k$, we are guaranteed by Proposition 14.3 that $\mathcal{G}$ will then be lossless). We leave it as an exercise (Problems 14.17 and 14.18) to show that such an LFSM $\mathcal{G}$ can indeed be realized.

Based on the latter definition, we can view convolutional codes as an extension of the notion of linear codes to alphabets that are infinite rings—$F[[x]]$ in our case. In fact, the underlying ring $F[[x]]$ is "almost" a field: it would have become one had we allowed its elements to contain also finitely many negative powers of x (Problem 14.14).

We emphasize that when defining a convolutional code, the generator matrix $G(x)$ is restricted to be over the intersection of $F(x)$ and $F[[x]]$. A convolutional code $\mathbf{C}$ has infinitely many generator matrices: given any generator matrix $G(x)$, we can obtain other generator matrices of $\mathbf{C}$ by applying invertible linear operations over $F(x) \cap F[[x]]$ to the rows of $G(x)$ (see Problem 14.21). In particular, if the entries of $G(x)$ are given by (14.23), then for $i = 1, 2, \ldots, k$, we can multiply row i in $G(x)$ by the least common multiplier of the denominators $\sigma_{i,1}(x), \sigma_{i,2}(x), \ldots, \sigma_{i,n}(x)$. This, in turn, will produce a generator matrix whose entries are all polynomials in $F[x]$ (e.g., see the generator matrix $\tilde{G}(x)$ in Example 14.9).

Every $[n, k]$ convolutional code $\mathbf{C}$ over F is a linear space over F. Thus, if $\mathbf{c}_1(x)$ and $\mathbf{c}_2(x)$ are elements of $\mathbf{C}$, then so is their difference $\mathbf{c}_1(x) - \mathbf{c}_2(x)$. It follows that the free distance of $\mathbf{C}$ equals

$$\mathsf{d}_{\mathrm{free}}(\mathbf{C}) = \min \sum_{t=0}^{\infty} \mathsf{w}(\mathbf{c}_t),$$

where the minimum is taken over all nonzero codeword sequences $\mathbf{c}_0\mathbf{c}_1\mathbf{c}_2\cdots$ in $\mathbf{C}$ (compare with Proposition 2.1).

Given a (possibly finite) sequence $\mathbf{c}(x) = \sum_{t=0}^{\infty} \mathbf{c}_t x^t$ in $(F[[x]])^n$, we will use the notation $\mathsf{w}_F(\mathbf{c}(x))$ for the sum

$$\sum_{t=0}^{\infty} \mathsf{w}(\mathbf{c}_t);$$

i.e., $\mathsf{w}_F(\mathbf{c}(x))$ is the Hamming weight of $\mathbf{c}(x)$ over F. Using this notation, the free distance of $\mathbf{C}$ can be written as

$$\mathsf{d}_{\text{free}}(\mathbf{C}) = \min_{\mathbf{c}(x) \in \mathbf{C} \setminus \{\mathbf{0}\}} \mathsf{w}_F(\mathbf{c}(x)) .$$

Example 14.11 Let $\mathbf{C}$ be the $[2, 1]$ convolutional code over $F = \mathrm{GF}(2)$ as in Example 14.10. We will use the presenting lossless digraph $\mathcal{G}$ in Figure 14.13 to identify the minimum-weight nonzero codeword sequences in $\mathbf{C}$.

We first observe that an infinite run of zeros can be generated in $\mathcal{G}$ only by repeating the self-loop at state $\underline{00}$. Therefore, an infinite codeword sequence with finite Hamming weight can be generated in $\mathcal{G}$ only by paths from state $\underline{00}$ that eventually return to state $\underline{00}$ and then repeat the self-loop at that state forever. It follows that infinite codeword sequences with nonzero finite weight are generated by paths that take the form

$$\underline{00} \xrightarrow{00} \underline{00} \xrightarrow{00} \ldots \xrightarrow{00} \underline{00} \xrightarrow{11} \underline{01} \longrightarrow \ldots \longrightarrow \underline{10} \xrightarrow{11} \underline{00} \xrightarrow{00} \underline{00} \xrightarrow{00} \cdots .$$

Now, the smallest weight that we can gain by any path from $\underline{01}$ to $\underline{10}$ is 1 (through the edge from $\underline{01}$ to $\underline{10}$). We thus conclude that the minimum-weight nonzero codeword sequences in $\mathbf{C}$ take the form

$$00\ 00\ \ldots\ 00\ 11\ 10\ 11\ 00\ 00\ \cdots$$

and, so, $\mathsf{d}_{\text{free}}(\mathbf{C}) = 5$ (compare with Example 14.5). □

14.6 Encoding of convolutional codes

Since convolutional codes are trellis codes, the encoding method that was described in Section 14.2.2 applies to them as well. Specifically, given an $[n, k]$ convolutional code $\mathbf{C}$ over F, we can select an (irreducible) lossless $k : n$ LFSM $\mathcal{G} = (V{=}F^m, E, L)$ over F that presents $\mathbf{C}$ and define for every $\ell \geq 1$ the path encoder

$$\mathcal{E}_\ell : (F^k)^\ell \rightarrow \mathbf{C}_\ell(\mathcal{G})$$

as in (14.4) and (14.5). The cycle encoder

$$\mathcal{E}_\ell^\circ : (F^k)^h \rightarrow \mathbf{C}_\ell^\circ(\mathcal{G})$$

is then defined by (14.9), where $h = \ell - \Delta$ for some back-length Δ of $\mathcal{G}$; e.g., by Problem 14.10 we can take $\Delta = m$.

Alternatively, we can obtain the mapping $\mathcal{E}_\ell$ directly from a generator matrix $G(x)$ of $\mathbf{C}$: such a mapping will then be a path encoder that is associated with any LFSM whose response matrix is $G(x)$. We will use the following notation. For an element $a(x) = \sum_{t=0}^{\infty} a_t x^t$ in $F[[x]]$, denote by

$$a(x) \text{ MOD } x^{\ell}$$

the polynomial $\sum_{t=0}^{\ell-1} a_t x^t$; the operation MOD extends in a straightforward manner to vectors in $(F[[x]])^n$. Using this notation, we now define $\mathcal{E}_\ell$ to be the function from $(F_\ell[x])^k$ to $\mathbf{C}_\ell(\mathcal{G}) \subseteq (F_\ell[x])^n$ that is given by

$$\mathcal{E}_\ell(\mathbf{u}(x)) = \mathbf{u}(x)G(x) \text{ MOD } x^{\ell} . \tag{14.24}$$

Suppose first that $k = n = 1$, i.e., $G(x)$ is the scalar $\omega(x)/\sigma(x)$ in $F[[x]]$, where $\sigma(0) = 1$. In this case, we can implement (14.24) by a circuit that multiplies a formal power series $u(x) \in F[[x]]$—in particular, a polynomial $u(x) \in F_\ell[x]$—by the fixed multiplier $\omega(x)/\sigma(x)$ in $F(x)$. Such a circuit is shown in Figure 14.14 for the polynomials

$$\sigma(x) = 1 + \lambda_1 x + \lambda_2 x^2 + \ldots + \lambda_m x^m$$

and

$$\omega(x) = \gamma_0 + \gamma_1 x + \gamma_2 x^2 + \ldots + \gamma_m x^m ,$$

for some nonnegative integer m. (Without loss of generality we can assume that $m = \max\{\deg \sigma, \deg \omega\}$; in fact, the same value of m can be taken in the LFSM realization of $G(x)$ in Problem 14.17.)

The circuit in Figure 14.14 is, in a way, a montage of Figure 5.1 in Section 5.3 and Figure 6.4 in Section 6.7.1. It consists of m cascaded delay units, which are controlled by a clock and are initially reset to zero. The input to the circuit is a sequence of coefficients $u_0 u_1 u_2 \cdots$ of the multiplicand $u(x)$. At every clock tick $t \geq 0$, the coefficient u_t is fed into the circuit, and for each $r = 1, 2, \ldots, m$, the contents of the rth delay unit is multiplied by λ_r. The sum of the m resulting products is then subtracted from the input u_t to produce a value named b_t (compare with Figure 6.4: the multiplicand $u(x)$ plays the role of the input therein). At the same time, the contents of the rth delay unit is multiplied also by γ_r and the resulting products for all r are summed with $\gamma_0 b_t$ to produce a value named c_t (note that the same computation is carried out in Figure 5.1, except that the order of coefficients is reversed both in the input sequence and in the multiplying polynomial). Clock tick t ends by shifting the contents of the delay units to the right and feeding the newly computed b_t into the leftmost delay unit.

Suppose that we run this circuit infinitely, and let $b(x)$ and $c(x)$ denote the formal power series $\sum_{t=0}^{\infty} b_t x^t$ and $\sum_{t=0}^{\infty} c_t x^t$, respectively. The values b_t are computed for every $t \geq 0$ by the formula

$$b_t = u_t - \sum_{r=1}^{m} \lambda_r b_{t-r} , \tag{14.25}$$

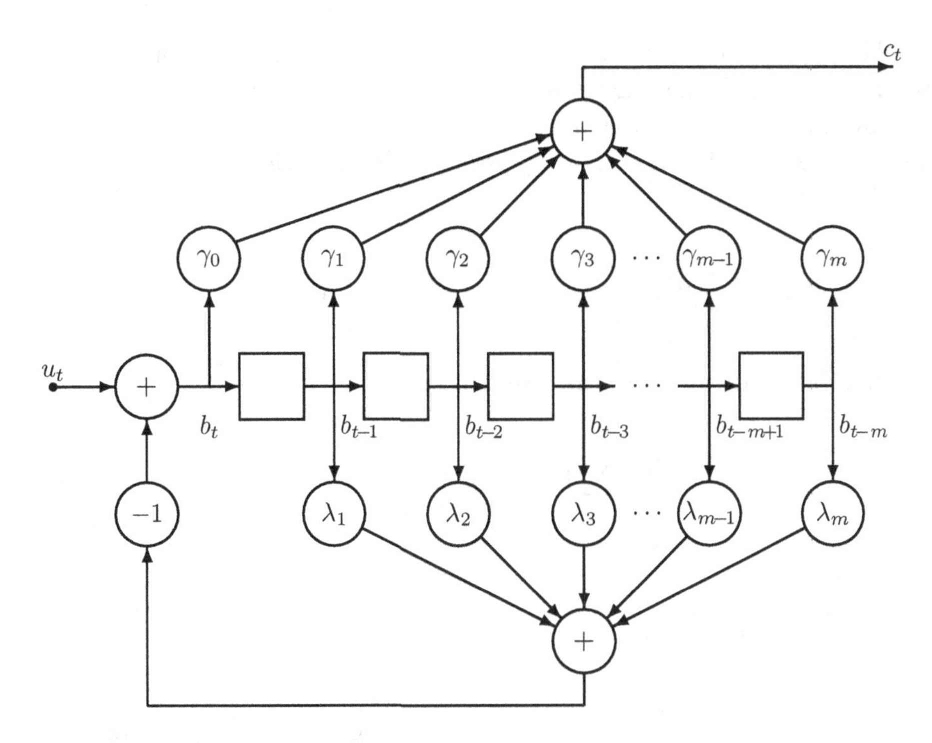

Figure 14.14. Multiplication circuit by $\omega(x)/\sigma(x)$.

where we assume that $b_r = 0$ for $r < 0$. Therefore,

$$b_t + \sum_{r=1}^{m} \lambda_r b_{t-r} = u_t \ ,$$

and when we multiply both sides by x^t and sum over t, we obtain

$$\sigma(x)b(x) = u(x) \ .$$

Hence,

$$b(x) = \frac{u(x)}{\sigma(x)} \ . \tag{14.26}$$

In a similar manner, we see that the values c_t are computed for every $t \geq 0$ by

$$c_t = \sum_{r=0}^{m} \gamma_r b_{t-r}$$

and, so,

$$c(x) = \omega(x)b(x) \ .$$

Together with (14.26) we thus conclude that

$$c(x) = \frac{\omega(x)}{\sigma(x)} \cdot u(x) \ .$$

We can use the circuit in Figure 14.14 as a building block for implementing the mapping $\mathcal{E}_\ell$ in (14.24) also for general k and n. For $i = 1, 2, \ldots, k$, denote by $(\mathbf{u}(x))_i$ the ith entry of the argument $\mathbf{u}(x)$ of $\mathcal{E}_\ell$ (each such entry is an element of $F_\ell[x]$). From (14.23) we get that the n entries of $\mathbf{c}(x) = \mathbf{u}(x)G(x)$ are given by

$$(\mathbf{c}(x))_j = \sum_{i=1}^{k} (\mathbf{u}(x))_i g_{i,j}(x) = \sum_{i=1}^{k} \frac{\omega_{i,j}(x)}{\sigma_{i,j}(x)} \cdot (\mathbf{u}(x))_i \, , \quad 1 \le j \le n \, .$$

The number of delay units that we will need then equals

$$\sum_{i=1}^{k} \sum_{j=1}^{n} \max\{\deg \sigma_{i,j}, \deg \omega_{i,j}\} \, . \tag{14.27}$$

This number may be reduced if we bring all entries in every given row of $G(x)$ to have a common denominator; the entries of $G(x)$ then take the form $\hat{\omega}_{i,j}(x)/\sigma_i(x)$, where $\sigma_i(0) = 1$,

$$\deg \sigma_i \le \sum_{j=1}^{n} \deg \sigma_{i,j} \, , \quad \text{and} \quad \deg \hat{\omega}_{i,j} = \deg \omega_{i,j} - \deg \sigma_{i,j} + \deg \sigma_i \, .$$

With this structure of $G(x)$, the multiplication circuits that correspond to the entries of row i can share both the delay units and the division portion in Figure 14.14 (i.e., the part of the figure below the delay units). Thus, we will need

$$m = \sum_{i=1}^{k} \max\{\deg \sigma_i, \deg \hat{\omega}_{i,1}, \deg \hat{\omega}_{i,2}, \ldots, \deg \hat{\omega}_{i,n}\} \tag{14.28}$$

delay units, and this number is never greater than (14.27).

If we now multiply row i of $G(x)$ by $\sigma_i(x)$ for each i, we get a generator matrix whose entries are all polynomials in $F[x]$. In this case, the division portions can be eliminated altogether, and Figure 14.14 reduces to just a multiplication by the (numerator) polynomial $\omega(x) = \omega_{i,j}(x)$; multiplication by a fixed polynomial, in turn, is sometimes called convolution, which is where the codes got their name from. We therefore conclude that if the entries $g_{i,j}(x)$ of $G(x)$ are all polynomials in $F[x]$, then we can implement the respective path encoder—which is commonly referred to in this case as a *convolutional encoder*—by using at most

$$\sum_{i=1}^{k} \max_{j=1}^{n} \deg g_{i,j} \tag{14.29}$$

delay units, and the circuits contain no feedback. The value in (14.29) is sometimes called the *constraint length* of the convolutional encoder.

Having shown how the mapping $\mathcal{E}_\ell$ can be obtained from $G(x)$, we next discuss how we construct the cycle encoder $\mathcal{E}_\ell^\circ$; namely, we determine the sequence of dummy messages

$$\mathbf{u}_{\ell-\Delta}\mathbf{u}_{\ell-\Delta+1}\ldots\mathbf{u}_{\ell-1}$$

at the end of the transmission so that the encoding circuit is led into the all-zero state. We assume that all entries in row i of $G(x)$ share the same denominator $\sigma_i(x)$, in which case Δ can be taken as the value m in (14.28) (see Problem 14.18). Write

$$\sigma_i(x) = 1 + \sum_{r=1}^{m} \lambda_{r,i}x^r \quad \text{and} \quad (\mathbf{u}(x))_i = \sum_{t=0}^{\infty} u_{t,i}x^t ,$$

and let $b_{t,i}$ denote the value that is fed at clock tick t to the leftmost delay unit of the circuit in Figure 14.14 that corresponds to row i. We readily get from (14.25) that when the input $u_{t,i}$ at clock ticks $\ell-m \le t < \ell$ is set to

$$u_{t,i} = \sum_{r=1}^{m} \lambda_{r,i}b_{t-r,i} ,$$

then we end up with

$$b_{\ell-m,i} = b_{\ell-m+1,i} = \ldots = b_{\ell-1,i} = 0 .$$

In particular, when all the entries in $G(x)$ are polynomials in $F[x]$, then we reach the all-zero state $\mathbf{0}_m$ simply by taking

$$\mathbf{u}_{\ell-m} = \mathbf{u}_{\ell-m+1} = \ldots = \mathbf{u}_{\ell-1} = \mathbf{0}_k .$$

Example 14.12 Let $\mathbf{C}$ be the $[2,1]$ convolutional code over $F = \mathrm{GF}(2)$ as in Example 14.10 and select the generator matrix

$$G(x) = \begin{pmatrix} 1 & \frac{1+x^2}{1+x+x^2} \end{pmatrix} .$$

Figure 14.15 shows a circuit that implements the path encoder that is associated with $G(x)$. The message to be encoded at time t comprises one bit, denoted by u_t, and the generated codeword at time t is the two-bit vector

$$\mathbf{c}_t = (c_{t,1}\ c_{t,2}) .$$

Since the generator matrix $G(x)$ is systematic, the message u_t appears as part of the generated codeword $\mathbf{c}_t$. □

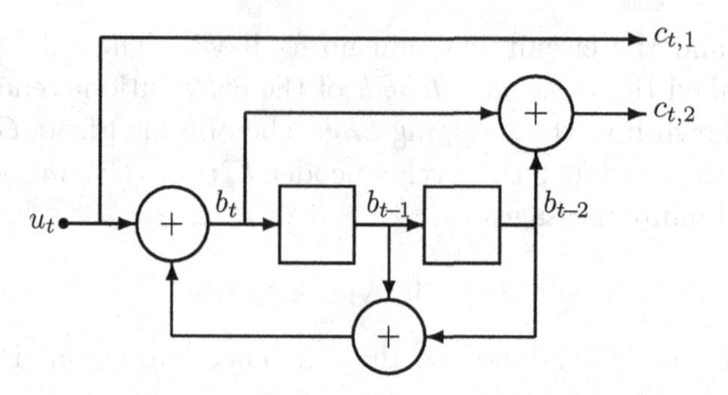

Figure 14.15. Encoding circuit for Example 14.12.

Example 14.13 Let **C** and $G(x)$ be as in Example 14.12. By looking at the circuit in Figure 14.15, we can see that

$$\begin{aligned} c_{t,2} &= b_t + b_{t-2} \\ &= \underbrace{(u_t + b_{t-1} + b_{t-2})}_{b_t} + b_{t-2} = u_t + b_{t-1} \,. \end{aligned}$$

Therefore, the path encoder of Example 14.12 can also be implemented by the circuit in Figure 14.16.

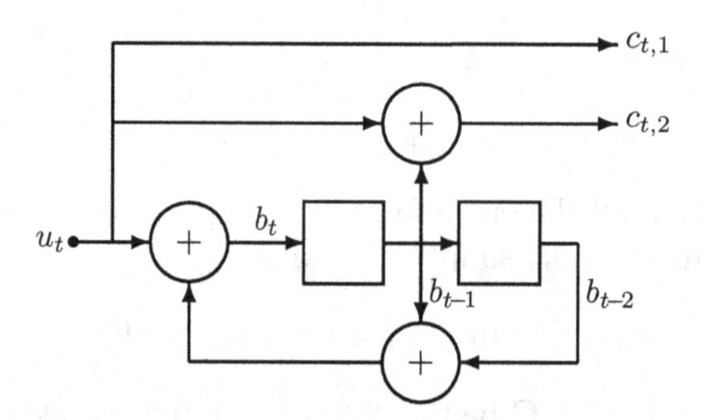

Figure 14.16. Encoding circuit for Example 14.13.

Let

$$\mathbf{s}_t = (b_{t-2}\ b_{t-1})$$

denote the contents of the two delay units in Figure 14.16 at time t. It is easily seen from the figure that

$$\mathbf{s}_{t+1} = \mathbf{s}_t P + u_t B \,,$$

where

$$P = \begin{pmatrix} 0 & 1 \\ 1 & 1 \end{pmatrix} \quad \text{and} \quad B = \begin{pmatrix} 0 & 1 \end{pmatrix} .$$

Also, we see that the generated codeword $\mathbf{c}_t$ is obtained from u_t and $\mathbf{s}_t$ by

$$\mathbf{c}_t = \mathbf{s}_t Q + u_t D ,$$

where

$$Q = \begin{pmatrix} 0 & 0 \\ 0 & 1 \end{pmatrix} \quad \text{and} \quad D = \begin{pmatrix} 1 & 1 \end{pmatrix} .$$

The matrix quadruple (P, B, Q, D) defines the $1:2$ LFSM $\mathcal{G}$ in Figure 14.13, and Figure 14.16 implements the path encoder that is associated with $\mathcal{G}$ by effectively simulating the state sequence $(\mathbf{s}_t)_t$ along the unique path from $\mathbf{s}_0 = \mathbf{0}_m$ whose edges are tagged by the input $(u_t)_t$. Recall that we have shown in Example 14.9 that $G(x)$ is indeed the response matrix of $\mathcal{G}$. □

Example 14.14 Let $\mathbf{C}$ be again as in Example 14.12. A third encoding circuit for $\mathbf{C}$ is shown in Figure 14.17. This circuit implements a path encoder that is associated with the generator matrix

$$\tilde{G}(x) = \begin{pmatrix} 1+x+x^2 & 1+x^2 \end{pmatrix}$$

(the same path encoder is also associated with the $1:2$ LFSM $\tilde{\mathcal{G}}$ in Figure 14.13). The matrix $\tilde{G}(x)$ is non-systematic; on the other hand, all the entries in $\tilde{G}(x)$ are polynomials and, so, the circuit in Figure 14.17 contains no feedback, as it does not perform any divisions. □

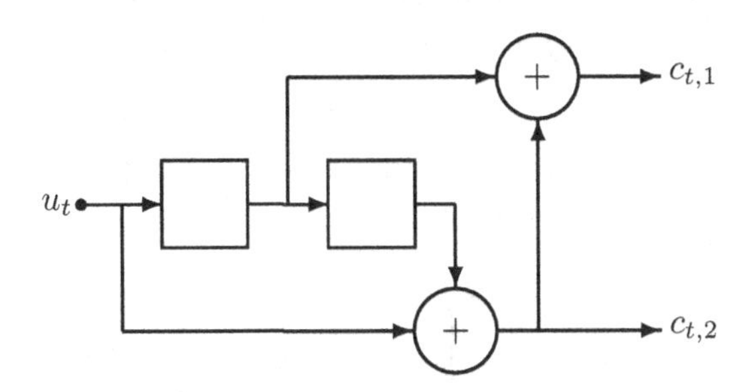

Figure 14.17. Encoding circuit for Example 14.14.

14.7 Decoding of convolutional codes

As with all other trellis codes, MLDs for convolutional codes can be implemented by Viterbi's algorithm from Section 14.3.2. In this section, we analyze the decoding error probability of such an MLD, while focusing on the q-ary symmetric channel. Following the discussion in Section 14.3.3, our interest will be in bounding from above the probability of erring on a *given* message—as opposed to the probability of erring on *at least one* message—within the transmitted sequence.

14.7.1 Length–weight enumerator of LFSMs

In our analysis of the decoding error probability, we will need to know the Hamming weight distribution of certain codeword sequences that are generated by a given LFSM presentation of the convolutional code. In this section, we characterize those codeword sequences and show how their weight distribution can be computed.

Let F be a finite Abelian group and n be a positive integer, and let $\mathcal{G} = (V, E, L)$ be a labeled digraph with labeling $L : E \rightarrow F^n$. The *generalized adjacency matrix* of $\mathcal{G}$, denoted by $A_{\mathcal{G}}(z)$, is the following $|V| \times |V|$ matrix over the ring $\mathbb{R}[z]$ (of real polynomials in the indeterminate z): the rows and columns of $A_{\mathcal{G}}$ are indexed by the states of V, and for every $s, \tilde{s} \in V$, the entry of $A_{\mathcal{G}}(z)$ that is indexed by $(s, \tilde{s})$ is given by

$$(A_{\mathcal{G}}(z))_{s,\tilde{s}} = \sum_{\substack{e \in E: \\ \iota(e)=s,\, \tau(e)=\tilde{s}}} z^{\mathsf{w}(L(e))} .$$

In other words, entry $(s, \tilde{s})$ is the *(Hamming) weight enumerator* of the block code in F^n whose codewords are the labels of the edges from state s to state $\tilde{s}$ in $\mathcal{G}$ (recall that we used the term weight enumerator also in Section 4.4, for the special case of block codes that are linear). We adopt the convention that the first row and the first column in $A_{\mathcal{G}}$ correspond to the start state $\iota(\mathcal{G})$.

When we substitute $z = 1$ in $A_{\mathcal{G}}(z)$, we get the (ordinary) adjacency matrix of $\mathcal{G}$: the value $(A_{\mathcal{G}}(1))_{s,\tilde{s}}$ equals the number of edges in $\mathcal{G}$ from s to $\tilde{s}$.

Example 14.15 The generalized adjacency matrix of the labeled digraph $\mathcal{G}'$ in Figure 14.7 is given by

$$A_{\mathcal{G}'}(z) = \begin{pmatrix} 1 & 0 & z^2 & 0 \\ z^2 & 0 & 1 & 0 \\ 0 & z & 0 & z \\ 0 & z & 0 & z \end{pmatrix} ,$$

where the rows and columns have been arranged assuming the ordering $\alpha < \beta < \gamma < \delta$ on the states of $\mathcal{G}'$. This is also the generalized adjacency matrix of the LFSMs $\mathcal{G}$ and $\tilde{\mathcal{G}}$ in Figure 14.13. □

We specialize from now on to labeled digraphs $\mathcal{G}$ that are LFSMs.

Let $\mathcal{G} = (V{=}F^m, E, L)$ be a $k : n$ LFSM over $F = \mathrm{GF}(q)$. The self-loop $[\mathbf{0}_m; \mathbf{0}_k]$ at state $\mathbf{0}_m$ will hereafter be referred to as the *trivial cycle*. A *fundamental cycle* in $\mathcal{G}$ is a positive-length nontrivial cycle whose initial (and terminal) state is $\mathbf{0}_m$, yet otherwise $\mathbf{0}_m$ is not visited along the cycle. Namely, a path

$$\psi = e_0 e_1 \ldots e_{\ell-1}$$

in $\mathcal{G}$ is called a fundamental cycle if $\ell > 0$, $\psi \neq [\mathbf{0}_m; \mathbf{0}_k]$, and

$$\iota(e_0) = \tau(e_{\ell-1}) = \mathbf{0}_m ,$$

yet

$$\iota(e_t) \neq \mathbf{0}_m \quad \text{for every } 0 < t < \ell .$$

The set of all fundamental cycles of length ℓ in $\mathcal{G}$ will be denoted by Ψ_ℓ. Clearly, every positive-length cycle that starts at $\mathbf{0}_m$ can be expressed as a concatenation of fundamental cycles and trivial cycles.

We now define the *length–weight enumerator* of $\mathcal{G}$ by

$$W_{\mathcal{G}}(x, z) = \sum_{\ell=1}^{\infty} \sum_{i=0}^{\infty} W_{\ell,i} x^\ell z^i ,$$

where $W_{\ell,i}$ is the number of fundamental cycles in $\mathcal{G}$ of length ℓ that generate sequences whose Hamming weight over F is i; equivalently,

$$W_{\mathcal{G}}(x, z) = \sum_{\ell=1}^{\infty} x^\ell \sum_{\psi \in \Psi_\ell} z^{\mathsf{w}_F(L(\psi))} . \tag{14.30}$$

At this point, we regard $W_{\mathcal{G}}(x, z)$ as an element of the set of *bivariate formal power series* over the real field $\mathbb{R}$. This set is given by

$$\mathbb{R}[[x, z]] = \left\{ b(x, z) = \textstyle\sum_{\ell,i=0}^{\infty} b_{\ell,i} x^\ell z^i \; : \; b_{\ell,i} \in \mathbb{R} \right\}$$

and, by a straightforward generalization from the univariate case, it can be shown that $\mathbb{R}[[x, z]]$ forms a ring. Later on, we will substitute real values for x and z in $W_{\mathcal{G}}(x, z)$, in which case the range of convergence will be of interest.

The length–weight enumerator $W_{\mathcal{G}}(x, z)$ can be computed from the generalized adjacency matrix $A_{\mathcal{G}}(z)$ in a manner that we now describe. We assume here that $|V| > 1$ (leaving the case $|V| = 1$ to Problem 14.25), and define $\mathcal{G}^*$ as the subgraph of $\mathcal{G}$ that is induced on $V \setminus \{\mathbf{0}_m\}$. The generalized adjacency matrix of $\mathcal{G}^*$ can be related to that of $\mathcal{G}$ by

$$A_{\mathcal{G}}(z) = \left(\begin{array}{c|c} a(z) & \mathbf{b}(z) \\ \hline \mathbf{v}(z) & A_{\mathcal{G}^*}(z) \end{array} \right) ,$$

where

$$a(z) = (A_{\mathcal{G}}(z))_{\mathbf{0}_m, \mathbf{0}_m} ,$$

and $\mathbf{b}(z)$ and $\mathbf{v}(z)$ are row and column vectors, respectively, in $(\mathbb{R}[z])^{|V|-1}$.

Next, we calculate the contribution of the cycles in Ψ_ℓ to the sum (14.30), for each length value ℓ. For $\ell = 1$ we have

$$\sum_{\psi \in \Psi_1} z^{\mathsf{w}_F(\psi)} = a(z) - 1 \,. \tag{14.31}$$

As for larger values of ℓ, any cycle in Ψ_ℓ takes the form

$$e_0 \pi e_{\ell-1} \,,$$

where $\iota(e_0) = \tau(e_{\ell-1}) = \mathbf{0}_m$ while the sub-path π is entirely contained in $\mathcal{G}^*$. It readily follows from the rules of matrix multiplication that

$$\sum_{\psi \in \Psi_\ell} z^{\mathsf{w}_F(\psi)} = \mathbf{b}(z) \left(A_{\mathcal{G}^*}(z)\right)^{\ell-2} \mathbf{v}(z) \,. \tag{14.32}$$

From (14.31) and (14.32) we conclude that, over $\mathbb{R}[[x, z]]$,

$$\begin{aligned} W_{\mathcal{G}}(x, z) &= x \cdot (a(z)-1) + \mathbf{b}(z) \Big(\sum_{\ell=2}^{\infty} x^\ell \left(A_{\mathcal{G}^*}(z)\right)^{\ell-2} \Big) \mathbf{v}(z) & (14.33) \\ &= x \cdot (a(z)-1) + x^2 \cdot \mathbf{b}(z) \left(I - x A_{\mathcal{G}^*}(z)\right)^{-1} \mathbf{v}(z) \,. & (14.34) \end{aligned}$$

Example 14.16 For the LFSM $\mathcal{G}$ in Figure 14.13 we have

$$a(z) = 1 \,, \quad \mathbf{b}(z) = (0 \;\; z^2 \;\; 0) \,, \quad \mathbf{v}(z) = \begin{pmatrix} z^2 \\ 0 \\ 0 \end{pmatrix} ,$$

and

$$A_{\mathcal{G}^*}(z) = \begin{pmatrix} 0 & 1 & 0 \\ z & 0 & z \\ z & 0 & z \end{pmatrix} .$$

The inverse of $I - x A_{\mathcal{G}^*}(z)$ is given by

$$\left(I - x A_{\mathcal{G}^*}(z)\right)^{-1} = \frac{1}{1-xz-x^2z} \begin{pmatrix} 1-xz & x-x^2z & x^2z \\ xz & 1-xz & xz \\ xz & x^2z & 1-x^2z \end{pmatrix}$$

and, so, we get the length–weight enumerator

$$\begin{aligned} W_{\mathcal{G}}(x, z) &= x \cdot (a(z)-1) + x^2 \cdot \mathbf{b}(z) \left(I - x A_{\mathcal{G}^*}(z)\right)^{-1} \mathbf{v}(z) \\ &= \frac{x^3 z^5}{1 - xz - x^2 z} \,. \end{aligned}$$

□

We can extend the notation of (partial) derivatives from $\mathbb{R}[[x]]$ or $\mathbb{R}[x, z]$ also to $\mathbb{R}[[x, z]]$: the partial derivative of $W_\mathcal{G}(x, z)$ with respect to x can be calculated from (14.33) to yield

$$W'_\mathcal{G}(x, z) = \frac{\partial W_\mathcal{G}(x, z)}{\partial x} = a(z) - 1 + \mathbf{b}(z)\Big(\sum_{\ell=2}^{\infty} \ell \cdot x^{\ell-1} \left(A_{\mathcal{G}^*}(z)\right)^{\ell-2}\Big)\mathbf{v}(z) . \tag{14.35}$$

In Section 14.7.2, we will compute upper bounds on the decoding error probability and, for this purpose, we will treat the derivative of the length–weight enumerator as a function

$$(x, z) \mapsto W'_\mathcal{G}(x, z) ,$$

where x and z range over the nonnegative reals. We next identify a range of values of (x, z) for which the sum (14.35) converges.

For every nonnegative real z, let

$$\lambda(z) = \lambda(A_{\mathcal{G}^*}(z))$$

be the spectral radius of $A_{\mathcal{G}^*}(z)$, i.e., $\lambda(z)$ is the largest absolute value of any eigenvalue of $A_{\mathcal{G}^*}(z)$. By transforming the matrix $A_{\mathcal{G}^*}(z)$ into its Jordan canonical form, we readily get that the entries of $(A_{\mathcal{G}^*}(z))^\ell$ are bounded from above by $f(\ell, z) \cdot (\lambda(z))^\ell$, where $f(\ell, z)$ is a function that depends on $\mathcal{G}$ and grows polynomially with ℓ. It follows that the sum (14.35) converges whenever x and z are nonnegative reals such that

$$x \cdot \lambda(z) < 1 .$$

In Section 14.7.2, we will be particularly interested in the values of $W'_\mathcal{G}(x, z)$ when $x = 1$, and from our analysis here we conclude that $W'_\mathcal{G}(1, z)$ converges whenever $\lambda(z) < 1$. It turns out that when $\mathcal{G}$ is lossless, the latter inequality holds for some $z \geq 0$, if and only if $\mathcal{G}$ satisfies the following condition (see Problem 14.26):

None of the nontrivial cycles in $\mathcal{G}$ generates an all-zero sequence.

We will elaborate more on this condition in Section 14.8.

Example 14.17 The characteristic polynomial of the matrix $A_{\mathcal{G}^*}(z)$ in Example 14.16 is given by

$$\det(\xi I - A_{\mathcal{G}^*}(z)) = \det \begin{pmatrix} \xi & -1 & 0 \\ -z & \xi & -z \\ -z & 0 & \xi - z \end{pmatrix} = \xi(\xi^2 - \xi z - z) .$$

Therefore, for $z \geq 0$,

$$\lambda(z) = \frac{1}{2}\left(z + \sqrt{z^2+4z}\right)$$

and

$$\lambda(z) < 1 \quad \Longleftrightarrow \quad z < \tfrac{1}{2} .$$

□

14.7.2 Bounds on the decoding error probability

Let $\mathbf{C}$ be an $[n,k]$ convolutional code over $F = \mathrm{GF}(q)$ and let $\mathcal{G} = (V{=}F^m, E, L)$ be a lossless $k:n$ LFSM that presents $\mathbf{C}$. Our goal in this section is to obtain a computable upper bound on the decoding error probability *per message* of an MLD, given that we use a cycle encoder that is associated with $\mathcal{G}$. We obtain the upper bound by using the tools of Section 14.7.1, but first we need to define what we mean by the per-message decoding error probability.

We assume that a codeword sequence

$$\mathbf{c}(x) = \sum_{t=0}^{\ell-1} \mathbf{c}_t x^t$$

of $\mathbf{C}_\ell^\circ(\mathcal{G})$ is transmitted through a q-ary symmetric channel with crossover probability p. The codeword sequence $\mathbf{c}(x)$ defines a unique cycle

$$\pi = [\mathbf{s}_0; \mathbf{u}_0][\mathbf{s}_1; \mathbf{u}_1]\ldots[\mathbf{s}_{\ell-1}; \mathbf{u}_{\ell-1}]$$

in $\mathcal{G}$ from state $\mathbf{s}_0 = \mathbf{0}_m$, where $\mathbf{c}_t = L([\mathbf{s}_t; \mathbf{u}_t])$ for $0 \leq t < \ell$; we refer to π as the *correct cycle*. The sequence of edge tags along π,

$$\mathbf{u}(x) = \sum_{t=0}^{\ell-1} \mathbf{u}_t x^t ,$$

consists of the transmitted messages—including the dummy messages that guarantee that π is indeed a cycle. The received sequence at the output of the channel will be denoted by

$$\mathbf{y}(x) = \sum_{t=0}^{\ell-1} \mathbf{y}_t x^t ,$$

where $\mathbf{y}_t \in F^n$.

An MLD (as in Figure 14.10) is applied to $\mathbf{y}(x)$ and produces a *decoded cycle*

$$\hat{\pi} = [\hat{\mathbf{s}}_0; \hat{\mathbf{u}}_0][\hat{\mathbf{s}}_1; \hat{\mathbf{u}}_1]\ldots[\hat{\mathbf{s}}_{\ell-1}; \hat{\mathbf{u}}_{\ell-1}]$$

in $\mathcal{G}$ from state $\hat{\mathbf{s}}_0 = \mathbf{0}_m$, with the property that the codeword sequence

$$\hat{\mathbf{c}}(x) = \sum_{t=0}^{\ell-1} \hat{\mathbf{c}}_t x^t$$

(in $\mathbf{C}^\circ_\ell(\mathcal{G})$) that $\hat{\pi}$ generates maximizes the conditional probability

$$\mathsf{Prob}(\, \mathbf{y}(x) \mid \hat{\mathbf{c}}(x)\,) = \mathsf{Prob}\{\, \mathbf{y}(x) \text{ received} \mid \hat{\mathbf{c}}(x) \text{ transmitted}\,\} \qquad (14.36)$$

(when $p < 1 - (1/q)$, the sequence $\hat{\mathbf{c}}(x)$ is a closest codeword sequence in $\mathbf{C}^\circ_\ell(\mathcal{G})$ to $\mathbf{y}(x)$ with respect to the Hamming distance over F). The decoded message sequence is then given by

$$\hat{\mathbf{u}}(x) = \sum_{t=0}^{\ell-1} \hat{\mathbf{u}}_t x^t \,,$$

which is the sequence of edge tags along $\hat{\pi}$.

We now define the *decoding error probability per message* of the MLD by

$$\mathrm{P}_{\mathrm{err}}(\mathcal{G}) = \sup_{1 \le \ell < \infty} \; \max_{\mathbf{c}(x) \in \mathbf{C}^\circ_\ell(\mathcal{G})} \; \max_{0 \le t < \ell} \mathrm{P}_{\mathrm{err}}(\mathbf{c}(x), t; \ell) \,, \qquad (14.37)$$

where $\mathrm{P}_{\mathrm{err}}(\mathbf{c}(x), t; \ell)$ is the probability that the MLD decodes the message at time t incorrectly, conditioned on transmitting the codeword sequence $\mathbf{c}(x)$; that is,

$$\mathrm{P}_{\mathrm{err}}(\mathbf{c}(x), t; \ell) = \mathsf{Prob}\{\, \hat{\mathbf{u}}_t \neq \mathbf{u}_t \mid \mathbf{c}(x) \text{ transmitted}\,\} \,. \qquad (14.38)$$

(When maximizing over $\mathbf{c}(x)$ in (14.37), we assume that every codeword sequence of $\mathbf{C}^\circ_\ell(\mathcal{G})$ can in effect be generated by the cycle encoder. This only means that our definition of $\mathrm{P}_{\mathrm{err}}(\mathcal{G})$ is somewhat conservative: a cycle encoder is typically not onto $\mathbf{C}^\circ_\ell(\mathcal{G})$.)

To facilitate the analysis of $\mathrm{P}_{\mathrm{err}}(\mathcal{G})$, we first reduce to the case where the correct cycle π consists of ℓ repetitions of the trivial cycle. This reduction is based on the following three observations:

1. By the linearity of $\mathbf{C}^\circ_\ell(\mathcal{G})$ over F, the difference $\hat{\mathbf{c}}(x) - \mathbf{c}(x)$ is a codeword sequence in $\mathbf{C}^\circ_\ell(\mathcal{G})$.

2. The cycle from $\mathbf{0}_m$ in $\mathcal{G}$ that generates $\hat{\mathbf{c}}(x) - \mathbf{c}(x)$ is tagged by $\hat{\mathbf{u}}(x) - \mathbf{u}(x)$.

3. By the definition of the q-ary symmetric channel, the probability distribution of $\mathbf{y}(x) - \mathbf{c}(x)$ does not depend on the transmitted codeword sequence $\mathbf{c}(x)$.

It follows from these properties that if $\hat{\mathbf{c}}(x)$ maximizes the expression (14.36) for a given $\mathbf{y}(x)$, then $\hat{\mathbf{c}}(x) - \mathbf{c}(x)$ will maximize that expression for $\mathbf{y}(x) - \mathbf{c}(x)$. Therefore, we subtract $\mathbf{c}(x)$ from $\mathbf{y}(x)$ and $\hat{\mathbf{c}}(x)$ and, accordingly, subtract $\mathbf{u}(x)$ from $\hat{\mathbf{u}}(x)$, thereby allowing us to assume from now on in the analysis that both $\mathbf{u}(x)$ and $\mathbf{c}(x)$ are all-zero.

Figure 14.18 shows the correct cycle and an example of a decoded cycle for the LFSM $\mathcal{G}$ in Figure 14.13, the way these cycles are seen in the trellis diagram of $\mathcal{G}$. Within the time frame that is covered in the figure, the decoded cycle contains two sub-paths that are fundamental cycles: one of length 3 which starts at time 2, and the other of length 4 which starts at time 8. Thus, the decoded cycle lies off the correct cycle at times 2–4 and 8–11.

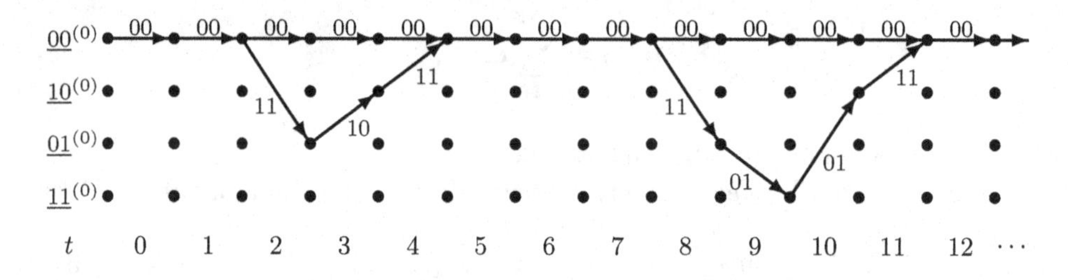

Figure 14.18. Correct cycle and decoded cycle.

In our next step of the analysis, we bound from above the conditional probability (14.38), which can now be re-written in the form

$$\mathsf{Prob}\{ \hat{\mathbf{u}}_t \neq \mathbf{0}_k \mid \mathbf{0}_n^\ell \} ,$$

where the conditioning is on the event of transmitting the all-zero codeword sequence (of length ℓ over F^n). We start with the simple inequality

$$\mathsf{Prob}\{ \hat{\mathbf{u}}_t \neq \mathbf{0}_k \mid \mathbf{0}_n^\ell \} \leq \mathsf{Prob}\Big\{ [\hat{\mathbf{s}}_t; \hat{\mathbf{u}}_t] \neq [\mathbf{0}_m; \mathbf{0}_k] \;\Big|\; \mathbf{0}_n^\ell \Big\} . \tag{14.39}$$

Notice that the right-hand side of (14.39) is the conditional probability of the event that the decoded cycle $\hat{\pi}$ is off the correct cycle at time t.

To compute the right-hand side of (14.39), we make use of the following definition: given a fundamental cycle $\psi \in \Psi_h$, denote by $\mathcal{X}(\psi, t)$ the event that the decoded cycle $\hat{\pi}$ contains ψ as a sub-path starting at time t, namely, $\mathcal{X}(\psi, t)$ is the event

$$[\hat{\mathbf{s}}_t; \hat{\mathbf{u}}_t][\hat{\mathbf{s}}_{t+1}; \hat{\mathbf{u}}_{t+1}] \ldots [\hat{\mathbf{s}}_{t+h-1}; \hat{\mathbf{u}}_{t+h-1}] = \psi$$

(for the sake of simplicity, we allow the time index t to be negative or greater than $\ell - h$; in these cases, the probability of the event $\mathcal{X}(\psi, t)$ will be defined to be zero). Now, $\hat{\pi}$ will be off the correct cycle at time t, if and only if time

t belongs to a sub-path of $\hat{\pi}$ that forms a fundamental cycle; for example, the decoded cycle in Figure 14.18 at time 9 is off the correct cycle, since the decoded cycle enters a fundamental cycle of length 4 at time 8. Based on this reasoning, we conclude that the event

$$[\hat{\mathbf{s}}_t; \hat{\mathbf{u}}_t] \neq [\mathbf{0}_m; \mathbf{0}_k]$$

(of being off the correct cycle at time t) is identical to the following union of (disjoint) events:

$$\bigcup_{h=1}^{\infty} \bigcup_{\psi \in \Psi_h} \bigcup_{j=0}^{h-1} \mathcal{X}(\psi, t-j) .$$

Hence, we reach the following expression for the right-hand side of (14.39):

$$\mathsf{Prob}\Big\{ [\hat{\mathbf{s}}_t; \hat{\mathbf{u}}_t] \neq [\mathbf{0}_m; \mathbf{0}_k] \;\Big|\; \mathbf{0}_n^\ell \Big\} = \sum_{h=1}^{\infty} \sum_{j=0}^{h-1} \sum_{\psi \in \Psi_h} \mathsf{Prob}\{\mathcal{X}(\psi, t-j) \mid \mathbf{0}_n^\ell\} . \tag{14.40}$$

Next, we bound from above the conditional probability of the event $\mathcal{X}(\psi, t)$ for $\psi \in \Psi_h$. By the definition of an MLD it follows that the event $\mathcal{X}(\psi, t)$ occurs only if the following inequality is satisfied by the sub-sequence $\mathbf{y}_t \mathbf{y}_{t+1} \ldots \mathbf{y}_{t+h-1}$:

$$\mathsf{Prob}\left(\mathbf{y}_t \mathbf{y}_{t+1} \ldots \mathbf{y}_{t+h-1} \mid L(\psi) \right) \geq \mathsf{Prob}\left(\mathbf{y}_t \mathbf{y}_{t+1} \ldots \mathbf{y}_{t+h-1} \mid \mathbf{0}_n^h \right) \tag{14.41}$$

(we use here the abbreviated notation $\mathsf{Prob}(\cdot|\cdot)$ as in (14.36): the first argument in this notation stands for a received sequence and the second argument is the transmitted sequence on which we condition). Indeed, (14.41) is a necessary condition that an MLD would prefer the cycle ψ over h repetitions of the trivial cycle.

We proceed with the next lemma, which will be used to bound from above the conditional probability that the inequality (14.41) is satisfied. This lemma is general in that it applies to every memoryless channel.

Lemma 14.4 (The Bhattacharyya bound) *Let* (F, Φ, Prob) *be a discrete memoryless channel and let*

$$\mathbf{w} = w_1 w_2 \ldots w_N \quad \textit{and} \quad \mathbf{w}' = w_1' w_2' \ldots w_N'$$

be two words over F. *Denote by* $\mathrm{P}(\mathbf{w}, \mathbf{w}')$ *the probability that the received word* $\mathbf{y} \in \Phi^N$ *satisfies the inequality*

$$\mathsf{Prob}\left(\mathbf{y}|\mathbf{w}'\right) \geq \mathsf{Prob}\left(\mathbf{y}|\mathbf{w}\right) ,$$

conditioned on the event that $\mathbf{w}$ *was transmitted. Then*

$$\mathrm{P}(\mathbf{w}, \mathbf{w}') \leq \prod_{j \,:\, w_j \neq w_j'} \sum_{y \in \Phi} \sqrt{\mathsf{Prob}(y|w_j) \cdot \mathsf{Prob}(y|w_j')} . \tag{14.42}$$

Lemma 14.4 is a special case of part 3 of Problem 1.9: simply take the code $\mathcal{C}$ therein to be $\{\mathbf{w}, \mathbf{w}'\}$. It also follows from that problem that for the q-ary symmetric channel, the right-hand side of (14.42) becomes

$$\gamma^{\mathsf{d}(\mathbf{w},\mathbf{w}')} ,$$

where

$$\boxed{\gamma = 2\sqrt{\frac{p(1-p)}{q-1}} + \frac{p(q-2)}{q-1}} .$$

If we now substitute

$$\mathbf{w} \leftarrow \mathbf{0}_n^h , \quad \mathbf{w}' \leftarrow L(\psi) , \quad \text{and} \quad \mathbf{y} \leftarrow \mathbf{y}_t\mathbf{y}_{t+1}\cdots\mathbf{y}_{t+h-1}$$

in Lemma 14.42, we get that the conditional probability of the event (14.41) is bounded from above by

$$\gamma^{\mathsf{w}_F(L(\psi))} .$$

This is also an upper bound on $\mathsf{Prob}\{\mathcal{X}(\psi,t) \mid \mathbf{0}_n^\ell\}$, since the event (14.41) contains $\mathcal{X}(\psi,t)$. Therefore, we get from (14.40) the following upper bound on the conditional probability of being off the correct cycle at time t:

$$\begin{aligned} \mathsf{Prob}\Big\{ [\hat{\mathbf{s}}_t; \hat{\mathbf{u}}_t] \neq [\mathbf{0}_m; \mathbf{0}_k] \;\Big|\; \mathbf{0}_n^\ell \Big\} & \\ \leq \sum_{h=1}^{\infty} h \sum_{\psi\in\Psi_h} \gamma^{\mathsf{w}_F(L(\psi))} & \\ = \sum_{h=1}^{\infty} h \sum_{i=0}^{\infty} W_{h,i}\gamma^i & \\ = \left.\frac{\partial W_{\mathcal{G}}(x,z)}{\partial x}\right|_{(x,z)=(1,\gamma)} = W'_{\mathcal{G}}(1,\gamma) . & \end{aligned}$$

Combining this with (14.37)–(14.39) finally yields the upper bound

$$\boxed{\mathrm{P_{err}}(\mathcal{G}) \leq W'_{\mathcal{G}}(1,\gamma)} , \tag{14.43}$$

under the assumption that $W'_{\mathcal{G}}(1,\gamma)$ converges.

Example 14.18 Let $\mathcal{G}$ be the LFSM as in Figure 14.13. We have shown in Example 14.16 that

$$W_{\mathcal{G}}(x,z) = \frac{x^3z^5}{1 - xz - x^2z} .$$

The partial derivative of $W_{\mathcal{G}}(x,z)$ with respect to x is given by

$$W'_{\mathcal{G}}(x,z) = \frac{3x^2z^5 - x^3z^6(2+x)}{(1-xz-x^2z)^2}$$

and, so, for the binary symmetric channel with crossover probability p we obtain

$$\mathrm{P_{err}}(\mathcal{G}) \le W'_{\mathcal{G}}(1,\gamma) = \frac{3\gamma^5(1-\gamma)}{(1-2\gamma)^2} ,$$

where

$$\gamma = 2\sqrt{p(1-p)} .$$

Now, we have seen in Example 14.17 that $W'_{\mathcal{G}}(1,\gamma)$ converges whenever $0 \le \gamma < \frac{1}{2}$. This range corresponds to values of p that satisfy either $p < \frac{1}{2} - \frac{1}{4}\sqrt{3}$ or $p > \frac{1}{2} + \frac{1}{4}\sqrt{3}$. □

The behavior of the upper bound (14.43) is strongly dictated by the free distance of $\mathbf{C}$: as exhibited in Problem 14.29, this bound becomes proportional to

$$\gamma^{\mathsf{d_{free}}(\mathbf{C})}$$

for small values of γ. See also Problem 14.30 for an improvement on the bound (14.43).

14.8 Non-catastrophic generator matrices

Let $\mathbf{C}$ be an $[n,k]$ convolutional code that is presented by an irreducible lossless $k:n$ LFSM $\mathcal{G} = (V{=}F^m, E, L)$ over $F = \mathrm{GF}(q)$, and let $G(x)$ be the response matrix of $\mathcal{G}$ ($G(x)$ is therefore a generator matrix of $\mathbf{C}$). In Section 14.7.1, we mentioned a certain condition, which guarantees convergence of $W'_{\mathcal{G}}(1,z)$ for at least one nonnegative real z. That condition required that no nontrivial cycle in $\mathcal{G}$ generate an all-zero codeword sequence. In this section, we discuss this condition in more detail.

Suppose that $\mathcal{G}$ contains a cycle $\psi \ne [\mathbf{0}_m; \mathbf{0}_k]$ that generates an all-zero codeword sequence, and let $\mathbf{s}_0$ denote the initial (and terminal) state of ψ. Obviously, $\mathbf{s}_0$ cannot be $\mathbf{0}_m$, or else $\mathcal{G}$ would not be lossless; in fact, since any cyclic shift of ψ is also a cycle, we find for the same reason that $\mathbf{0}_m$ is not visited at all along ψ.

Consider first the case where the edge tags along the cycle ψ are all equal to $\mathbf{0}_k$. It turns out that in this case the state $\mathbf{s}_0$ is *indistinguishable* from state $\mathbf{0}_m$ in the following sense: two paths from $\mathbf{0}_m$ and $\mathbf{s}_0$, respectively, that have the same tag sequence, also generate the same codeword sequence. This means that $\mathcal{G}$ is an inefficient LFSM realization of the generator matrix $G(x)$, as the state $\mathbf{s}_0$ is actually redundant. In fact, there is a way by which $\mathcal{G}$ can be transformed into another LFSM where such redundant states are eliminated while the response matrix $G(x)$ remains the same (see Problem 14.13). A lossless LFSM whose states are pairwise distinguishable is called *observable* or *reduced*.

Next, assume that there is at least one edge tag along ψ that does not equal $\mathbf{0}_k$. If we happen to choose $\mathcal{G}$ for the encoding, then a respective MLD may run into a severe problem, as we now demonstrate. Denote by π_0 the shortest path from $\mathbf{0}_m$ to $\mathbf{s}_0$ and by π_1 the shortest path back from $\mathbf{s}_0$ to $\mathbf{0}_m$. Consider the path

$$\psi' = \pi_0 \psi^r \pi_1$$

in $\mathcal{G}$, where r is a large integer and ψ^r denotes r repetitions of ψ. The path ψ' is a cycle that starts at $\mathbf{0}_m$, and the codeword sequence that it generates is almost all-zero: any nonzero codeword in that sequence is generated only by the prefix π_0 or the suffix π_1 of ψ'. As r becomes large, the lengths of π_0 and π_1 become negligible compared to r and, thus, to the overall length of ψ'. Assuming that the correct cycle consists of repetitions of the trivial cycle (and that the crossover probability of the channel is less than $1-(1/q)$), we reach the conclusion that there are error patterns with a relatively small number of channel errors—e.g., the error pattern that equals $L(\psi')$—which will cause an MLD to incorrectly decide upon ψ' instead of the correct cycle. Yet, with the exception of the initial and terminal states, ψ' never even touches the correct cycle; furthermore, each of the r repetitions of ψ within ψ' produces at least one nonzero—and therefore erroneous—decoded message.

The event which we have just described is called *catastrophic error propagation*: a bounded number of channel errors may cause an arbitrarily large number of decoding errors in the message sequence. Accordingly, we say that the irreducible lossless LFSM $\mathcal{G}$ is *catastrophic* if it contains a cycle ψ such that the edge tags along ψ are not all zero yet the edge labels are.

Example 14.19 All three LFSMs in Figure 14.19 present the following simple $[3, 1]$ convolutional code over $F = \mathrm{GF}(2)$:

$$\mathbf{C} = \left\{ \sum_{t=0}^{\infty} \mathbf{c}_t x^t \;:\; \mathbf{c}_t \in \{000, 111\} \text{ for all } t \geq 0 \right\} .$$

The single-state LFSM $\mathcal{G}_1$ is defined by a matrix quadruple (P_1, B_1, Q_1, D_1), where P_1, B_1, and Q_1 are vacuous (having no rows or no columns) and

$$D_1 = \begin{pmatrix} 1 & 1 & 1 \end{pmatrix} .$$

Clearly, $\mathcal{G}_1$ is both observable and non-catastrophic.

The two-state LFSM $\mathcal{G}_2$ is defined by

$$P_2 = \begin{pmatrix} 0 \end{pmatrix}, \quad B_2 = \begin{pmatrix} 1 \end{pmatrix}, \quad Q_2 = \begin{pmatrix} 0 & 0 & 0 \end{pmatrix}, \quad \text{and} \quad D_2 = \begin{pmatrix} 1 & 1 & 1 \end{pmatrix}.$$

This LFSM is not observable yet still non-catastrophic.

The matrix quadruple (P_3, B_3, Q_3, D_3) that defines $\mathcal{G}_3$ is almost the same as that of $\mathcal{G}_2$; the only difference is that here

$$Q_3 = \begin{pmatrix} 1 & 1 & 1 \end{pmatrix}.$$

The LFSM $\mathcal{G}_3$, however, is catastrophic. □

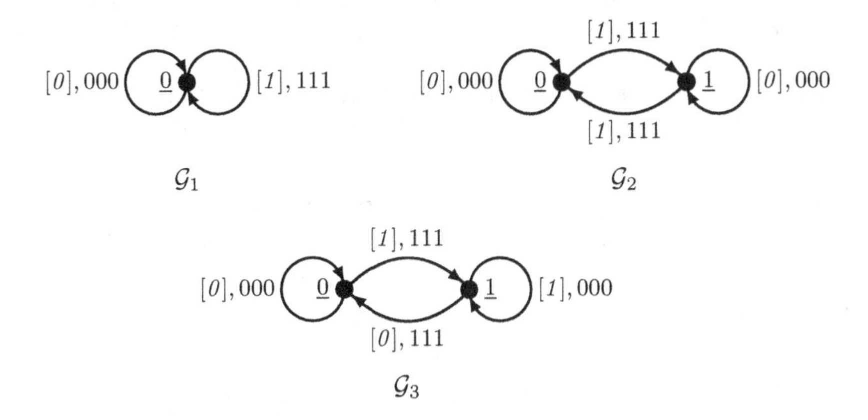

Figure 14.19. LFSMs for Example 14.19.

We can determine whether an irreducible lossless LFSM $\mathcal{G}$ is catastrophic also from the response matrix $G(x)$ of $\mathcal{G}$: it can be shown (Problem 14.31) that $\mathcal{G}$ is catastrophic if and only if there exists a formal power series $\mathbf{u}(x) \in F[[x]]$ such that

$$\mathsf{w}_F(\mathbf{u}(x)) = \infty \quad \text{and} \quad \mathsf{w}_F(\mathbf{u}(x)G(x)) < \infty. \tag{14.44}$$

Thus, the catastrophic property can be seen as a characteristic of the generator matrix which we select for the encoding: the LFSM realizations of a given generator matrix $G(x)$ are either all catastrophic or all non-catastrophic, depending on whether (14.44) holds for some $\mathbf{u}(x) \in F[[x]]$. This, in turn, motivates the following definition: given a convolutional code $\mathbf{C}$ over F, we say that a generator matrix $G(x)$ of $\mathbf{C}$ is catastrophic if there exists $\mathbf{u}(x) \in F[[x]]$ that satisfies (14.44).

Turning to the three LFSMs in Example 14.19, their respective response matrices are

$$G_1(x) = G_2(x) = \begin{pmatrix} 1 & 1 & 1 \end{pmatrix} \qquad \text{and} \qquad G_3(x) = \begin{pmatrix} 1+x & 1+x & 1+x \end{pmatrix}.$$

We can see that (14.44) holds for $G_3(x)$ when we take

$$\mathbf{u}(x) = \sum_{t=0}^{\infty} x^t .$$

The following result presents a necessary and sufficient algebraic condition that a given generator matrix $G(x)$ is non-catastrophic, provided that the entries of $G(x)$ are all polynomials.

Proposition 14.5 *Let $G(x)$ be a $k \times n$ generator matrix over $F[x]$ of an $[n, k]$ convolutional code over F. Then $G(x)$ is non-catastrophic if and only if the greatest common divisor of the $\binom{n}{k}$ determinants of the $k \times k$ sub-matrices of $G(x)$ equals some power of x.*

Proof. We show the "only if" direction; the "if" part is given as a guided exercise in Problem 14.32. Suppose that all the determinants of the $k \times k$ sub-matrices of $G(x)$ are divisible (in $F[x]$) by a monic irreducible polynomial $a(x) \neq x$ over F. Denote by K the extension field $F[\xi]/a(\xi)$ of F and consider the $k \times n$ matrix $G(\xi)$ over K, which is obtained by substituting $x = \xi$ in $G(x)$. All the $k \times k$ sub-matrices of $G(\xi)$ are singular and, so, over K,

$$\text{rank}(G(\xi)) < k .$$

It follows that by applying a sequence of elementary linear operations to the rows of $G(\xi)$, we can always reach a $k \times n$ matrix $\tilde{G}(\xi)$ over K whose last row is all-zero. Each of these linear operations belongs to one of the following two types: interchanging two rows or adding a scalar multiple (over K) of one row to another row. Elementary operations of the first type are represented by $k \times k$ permutation matrices over F with entries that are 0's and 1's; elementary operations of the second type are triangular matrices having 1's on the main diagonal and only one nonzero element off the main diagonal. In either case, the representing matrix of the elementary operation can be obtained by substituting $x = \xi$ in a $k \times k$ matrix over $F[x]$ whose determinant is ± 1. Thus, the sequence of elementary operations can be expressed as the matrix product

$$\tilde{G}(\xi) = T_r(\xi) T_{r-1}(\xi) \ldots T_1(\xi) G(\xi) ,$$

where $\det(T_i(x)) = \pm 1$ over $F[x]$ for $i = 1, 2, \ldots, r$. In particular, each matrix $T_i(x)$ has an inverse, $(T_i(x))^{-1}$, over $F[x]$. We conclude that the matrix

$$T(x) = T_r(x) T_{r-1}(x) \ldots T_1(x)$$

and its inverse (over $F(x)$) are both over $F[x]$.

Consider now the $k \times n$ matrix

$$\hat{G}(x) = T(x) G(x)$$

over $F[x]$. Clearly, $\hat{G}(\xi) = \tilde{G}(\xi)$ over K; hence, the last row in $\hat{G}(x)$ is a multiple of $a(x)$. Define the vectors

$$\hat{\mathbf{u}}(x) = (0 \;\; 0 \;\; \ldots \;\; 0 \;\; 1/a(x))$$

and

$$\mathbf{u}(x) = \hat{\mathbf{u}}(x)T(x)$$

in $(F[[x]])^k$. On the one hand, $\hat{\mathbf{u}}(x)\hat{G}(x)$ is a vector over $F[x]$ and, so,

$$\mathsf{w}_F(\mathbf{u}(x)G(x)) = \mathsf{w}_F(\hat{\mathbf{u}}(x)\hat{G}(x)) < \infty .$$

On the other hand,

$$\mathsf{w}_F\left(\mathbf{u}(x)(T(x))^{-1}\right) = \mathsf{w}_F\left(\hat{\mathbf{u}}(x)\right) = \mathsf{w}_F\left(1/a(x)\right) = \infty ,$$

which readily implies that

$$\mathsf{w}_F(\mathbf{u}(x)) = \infty .$$

This completes the proof that $G(x)$ is catastrophic. □

The next result guarantees that every convolutional code over F has a non-catastrophic generator matrix; moreover, there is such a matrix whose entries are all polynomials over F.

Proposition 14.6 *Every $[n, k]$ convolutional code over F has a $k \times n$ generator matrix over $F[x]$ that is non-catastrophic.*

Proof. Given an $[n, k]$ convolutional code $\mathbf{C}$ over F, let $G(x)$ be any $k \times n$ generator matrix of $\mathbf{C}$ with entries in $F[x]$ and let $f(x)$ be the greatest common divisor of the determinants of the $k \times k$ sub-matrices of $G(x)$. Suppose that $f(x)$ is divisible by a monic irreducible polynomial $a(x) \neq x$ over F. As we did in the proof of Proposition 14.5, we apply to the rows of $G(x)$ elementary linear operations, resulting in a $k \times n$ matrix

$$\hat{G}(x) = T(x)G(x)$$

over $F[x]$ whose last row is a multiple of $a(x)$. Since $T(x)$ is invertible over $F[[x]]$ with $\det(T(x)) = \pm 1$, we get that $\hat{G}(x)$ is a generator matrix of $\mathbf{C}$ and also that $f(x)$ is the greatest common divisor of the determinants of the $k \times k$ sub-matrices of $\hat{G}(x)$.

Next, we reduce that greatest common divisor to $f(x)/a(x)$, by multiplying the last row of $\hat{G}(x)$ by the inverse, $1/a(x)$, of $a(x)$ in $F[[x]]$. This results in a new generator matrix of $\mathbf{C}$ over $F[x]$, and we can now repeat the process, with this new matrix playing the role of $G(x)$, until we reach a stage where the only irreducible factor (if any) of the greatest common divisor is x. By Proposition 14.5, the generator matrix at that point is non-catastrophic. □

The proof of Proposition 14.6 can be made into a rather efficient algorithm for transforming a (possibly catastrophic) generator matrix $G(x)$ (over $F[x]$) into a non-catastrophic one. In such an algorithm, we will need to find an irreducible factor $a(x)$ of the (as yet unknown) greatest common divisor $f(x)$, and we can accomplish this as follows. By applying elementary operations (over $F(x)$) to the rows of $G(x)$, we identify a $k \times k$ nonsingular sub-matrix $G_0(x)$ of $G(x)$ and bring it into a triangular form over $F(x)$. The diagonal of the resulting matrix already provides a factorization of $\det(G_0(x))$ over F, although the factors are not necessarily irreducible. Then, using available factoring algorithms (like those mentioned in the notes on Section 3.2), we compute the full irreducible factorization of $\det(G_0(x))$ over F; clearly, each irreducible factor of $f(x)$ also divides $\det(G_0(x))$. Finally, for each irreducible factor $a(x)$ of $\det(G_0(x))$, we compute the rank of $G(\xi)$ in the field $F[\xi]/a(\xi)$: the rank will be smaller than k if and only if $a(x)$ divides $f(x)$. Once we know an irreducible factor $a(x)$ of $f(x)$, we proceed as outlined in the proof of Proposition 14.6: we find a new generator matrix over $F[x]$ where now the greatest common divisor is $f(x)/a(x)$, and then continue recursively, until the greatest common divisor becomes a power of x.

Having found a non-catastrophic generator matrix of $\mathbf{C}$, it follows from Problem 14.19 that we can always realize it by an LFSM that is lossless, irreducible, observable, and of course non-catastrophic.

The next theorem lists several properties of irreducible LFSM realizations of a given matrix $G(x)$ whose entries are polynomials over F.

Theorem 14.7 *Let F be a finite field and $G(x)$ be a $k \times n$ matrix over $F[x]$. The following properties hold for every irreducible LFSM over F whose response matrix is $G(x)$:*

(i) $\mathcal{G}$ is lossless if and only if $\mathrm{rank}(G(x)) = k$ *(over $F(x)$).*

(ii) $\mathcal{G}$ is deterministic if and only if $\mathrm{rank}(G(0)) = k$ *(over F).*

(iii) $\mathcal{G}$ is (lossless and) non-catastrophic if and only if $\mathrm{rank}(G(\alpha)) = k$ *over every finite extension field K of F, for every nonzero element α in K.*

Part (i) of the theorem is already covered in Proposition 14.3, and the other two parts are given as exercises in Problems 14.33 and 14.34.

Problems

[Section 14.1]

Problem 14.1 Let $\mathcal{G} = (V, E, L)$ be a (labeled) digraph in which each state has at least $M > 1$ outgoing edges. Show that for every state $s \in V$ and every integer $\ell > \log_M |V|$ there exist at least two distinct paths π and π' of length ℓ in $\mathcal{G}$ such that

$$\iota(\pi) = \iota(\pi') = s \quad \text{and} \quad \tau(\pi) = \tau(\pi') .$$

Problem 14.2 Let $\mathcal{G} = (V, E, L)$ be a digraph and define the bi-connection relation on the set V as in Section 14.1.1.

1. Show that bi-connection is an equivalence relation.

Let $V_1, V_2, \ldots, V_k$ be the equivalence classes of the bi-connection relation and for $i = 1, 2, \ldots, k$, let $\mathcal{G}_i$ be the induced subgraph of $\mathcal{G}$ on V_i. The subgraphs $\mathcal{G}_1, \mathcal{G}_2, \ldots, \mathcal{G}_k$ are the irreducible components of $\mathcal{G}$.

2. Show that each irreducible component of $\mathcal{G}$ is an irreducible digraph.
3. Show that no irreducible component of $\mathcal{G}$ is a proper subgraph of any irreducible subgraph of $\mathcal{G}$.
4. Show that if there is a path in $\mathcal{G}$ from a state in an irreducible component $\mathcal{G}_i$ to a state in another irreducible component $\mathcal{G}_j$, then there can be no path in $\mathcal{G}$ from any state in $\mathcal{G}_j$ to any state in $\mathcal{G}_i$.
5. Show that $\mathcal{G}$ must contain at least one irreducible sink; namely, there is at least one equivalence class V_i such that any edge in E with initial state in V_i must also have its terminal state in V_i.

   Hint: Use part 4.

Problem 14.3 Let $\ell_1, \ell_2, \ldots, \ell_k$ be positive integers and let p denote their greatest common divisor, namely,

$$\mathsf{p} = \gcd(\ell_1, \ell_2, \ldots, \ell_k) .$$

Show that every sufficiently large multiple of p can be written as a *nonnegative* integer linear combination of $\ell_1, \ell_2, \ldots, \ell_k$; i.e., there exists an integer t_0 (which depends on $\ell_1, \ell_2, \ldots, \ell_k$) such that for every integer $t \geq t_0$ one can find nonnegative integers $a_1, a_2, \ldots, a_k$ such that

$$\mathsf{p}t = \sum_{i=1}^{k} a_i \ell_i .$$

Hint: First argue that it suffices to consider the case where $\mathsf{p} = 1$. Then apply Euclid's algorithm in Problem A.3 to show that there exist integers $b_1, b_2, \ldots, b_k$ (some of which may be negative) such that

$$\sum_{i=1}^{k} b_i \ell_i = \gcd(\ell_1, \ell_2, \ldots, \ell_k) = 1 .$$

Define

$$\alpha = \sum_{i\,:\,b_i>0} b_i \ell_i \qquad \text{and} \qquad \beta = -\sum_{i\,:\,b_i<0} b_i \ell_i$$

and select $t_0 = \beta(\beta-1)$. Given $t \geq t_0$, write it in the form

$$t = \beta q + r \; ,$$

where $q \geq \beta-1$ and $0 \leq r \leq \beta-1$. Show that

$$t = \alpha r + \beta(q-r) \; ,$$

and conclude that the coefficients a_i can be taken as follows:

$$a_i = \begin{cases} b_i r & \text{if } b_i \geq 0 \\ b_i(r-q) & \text{if } b_i < 0 \end{cases} \; , \quad i = 1, 2, \ldots, k \; .$$

Problem 14.4 (Period of irreducible digraphs) Let $\mathcal{G} = (V, E, L)$ be an irreducible (labeled) digraph and assume that $\mathcal{G}$ contains at least one edge. Define the period p of $\mathcal{G}$ as the greatest common divisor of all the cycle lengths in $\mathcal{G}$.

1. Let s and $\tilde{s}$ be two states in V. Show that the lengths of any two paths from s to $\tilde{s}$ in $\mathcal{G}$ are congruent modulo the period p.

   Hint: Let π_1 and π_2 be two distinct paths from s to $\tilde{s}$ and let π_3 be a path from $\tilde{s}$ back to s. Consider the lengths of the cycles $\pi_1\pi_3$ and $\pi_2\pi_3$.

2. Two states s and $\tilde{s}$ in V are said to be *congruent* if there is a path from s to $\tilde{s}$ whose length is divisible by the period p. Show that congruence is an equivalence relation.

3. Show that the congruence relation in part 2 partitions V into p equivalence classes; furthermore, these classes can be labeled $V_0, V_1, \ldots, V_{\mathsf{p}-1}$ such that for every $i = 0, 1, \ldots, \mathsf{p}-1$, edges outgoing from states in V_i terminate in V_{i+1} (where V_{p} is read as V_0).

4. Show that there exists an integer N (which depends on the digraph $\mathcal{G}$) such that for every $t \geq N$ and every two congruent states s and $\tilde{s}$ in V, there is a path of length (exactly) $\mathsf{p}t$ in $\mathcal{G}$ from s to $\tilde{s}$.

   Hint: Let $\ell_1, \ell_2, \ldots, \ell_k$ be lengths of cycles in $\mathcal{G}$ such that $\gcd(\ell_1, \ell_2, \ldots, \ell_k) = \mathsf{p}$ and let s_i be the initial (and terminal) state of a cycle of length ℓ_i. Assume in addition that the states s_1 and s_k are congruent (explain why this assumption is allowed). Based on Problem 14.3, show that for every sufficiently large t, there is a path ψ_t from s_1 to s_k in $\mathcal{G}$ of length $\mathsf{p}t$. Given any two congruent states s and $\tilde{s}$ in V, denote by π the shortest path from s to s_1 and by $\tilde{\pi}$ the shortest path from s_k to $\tilde{s}$. Consider the paths

   $$\pi \psi_t \tilde{\pi}$$

   for sufficiently large t.

5. Let N be as in part 4. Show that for every $t > N$, the integers $\mathsf{p}t$ are back-lengths of $\mathcal{G}$; that is, every path π of length ℓ in $\mathcal{G}$ can be extended by
$$(\mathsf{p}t) - (\ell \text{ MOD } \mathsf{p})$$
edges so that the resulting path is a cycle (that terminates in $\iota(\pi)$).

   Hint: Given a state $s \in V$, let V_i be the congruence equivalence class that contains s and $V(s, \ell)$ be the set of terminal states of the paths of length ℓ from state s. Show that any path of length $\mathsf{p} - (\ell \text{ MOD } \mathsf{p})$ from any state in $V(s, \ell)$ terminates in V_i. Then use part 4.

6. (Converse to part 5) Show that every back-length of $\mathcal{G}$ must be divisible by p.

Problem 14.5 Let $\mathcal{G}$ be a labeled digraph, and suppose that there is a nonnegative integer N such that the following property holds: for any two paths
$$\pi = e_0 e_1 \ldots e_N \quad \text{and} \quad \pi' = e_0' e_1' \ldots e_N'$$
of length $N+1$ in $\mathcal{G}$,
$$\text{if } \iota(\pi) = \iota(\pi') \text{ and } L(\pi) = L(\pi'), \text{ then } e_0 = e_0' \ ;$$
that is, if π and π' have the same initial state and generate the same word (of length $N+1$), then they also agree on their first edge. The smallest such N, if any, for which this holds is called the *anticipation* of $\mathcal{G}$. If there is no such N then the anticipation is defined as ∞.

1. What is the anticipation of deterministic digraphs?
2. Show that the labeled digraph $\mathcal{G}_3$ in Figure 14.4 has anticipation 1 while $\mathcal{G}_4$ has infinite anticipation.
3. Let $\mathcal{G}$ be a labeled digraph with anticipation $\mathcal{A} < \infty$ and let
$$\mathbf{w} = w_0 w_1 \ldots w_{\ell-1}$$
be a word of length $\ell > \mathcal{A}$ that is generated in $\mathcal{G}$ by at least one path that starts at state s_0. Show that all paths from s_0 in $\mathcal{G}$ that generate $\mathbf{w}$ agree on their first $\ell - \mathcal{A}$ edges.
4. Let $\mathcal{G}$ be a labeled digraph in which each state has at least one outgoing edge. Show that if $\mathcal{G}$ has finite anticipation then $\mathcal{G}$ is lossless.

[Section 14.2]

Problem 14.6 Let F be an alphabet of size q and let $\mathcal{G} = (V, E, L)$ be an M-regular lossless digraph with labeling $L : E \to F^n$. The purpose of this problem is to show that
$$\lim_{\ell\to\infty} \frac{\log_{q^n} |\mathbf{C}_\ell(\mathcal{G})|}{\ell} = \limsup_{\ell\to\infty} \frac{\log_{q^n} |\mathbf{C}_\ell^\circ(\mathcal{G})|}{\ell} = \frac{\log_q M}{n} \ .$$
By looking at the irreducible sink in $\mathcal{G}$ that contains $\iota(\mathcal{G})$, assume hereafter in this problem that $\mathcal{G}$ is irreducible, and let p and Δ be the period and back-length of $\mathcal{G}$, respectively.

1. Let h be a positive integer and define the integer $\ell(h)$ by

$$\ell(h) = \Delta + h - (h \text{ MOD } \mathsf{p}) .$$

Show that

$$|\mathbf{C}_{h+\Delta}(\mathcal{G})| \ge |\mathbf{C}_{\ell(h)}(\mathcal{G})| \ge |\mathbf{C}^{\circ}_{\ell(h)}(\mathcal{G})| \ge M^h .$$

2. Deduce from part 1 that

$$\liminf_{\ell\to\infty} \frac{\log_{q^n} |\mathbf{C}_\ell(\mathcal{G})|}{\ell} \ge \frac{\log_q M}{n}$$

and that

$$\limsup_{\ell\to\infty} \frac{\log_{q^n} |\mathbf{C}^{\circ}_\ell(\mathcal{G})|}{\ell} \ge \frac{\log_q M}{n} .$$

3. Based on part 2, show that

$$\lim_{\ell\to\infty} \frac{\log_{q^n} |\mathbf{C}_\ell(\mathcal{G})|}{\ell} = \limsup_{\ell\to\infty} \frac{\log_{q^n} |\mathbf{C}^{\circ}_\ell(\mathcal{G})|}{\ell} = \frac{\log_q M}{n} .$$

Hint:

$$|\mathbf{C}^{\circ}_\ell(\mathcal{G})| \le |\mathbf{C}_\ell(\mathcal{G})| \le M^\ell .$$

[Section 14.3]

Problem 14.7 Let $\mathcal{G} = (V, E, L)$ be an M-regular lossless graph with labeling $L : E \to F^n$ and let

$$\mathcal{E}^{\circ}_\ell : \Upsilon^h \to \mathbf{C}^{\circ}_\ell(\mathcal{G})$$

be a cycle encoder that is associated with $\mathcal{G}$, as defined in Section 14.2.2 (note that $\mathcal{E}^{\circ}_\ell$ is typically not onto $\mathbf{C}^{\circ}_\ell(\mathcal{G})$). Explain how Viterbi's algorithm in Figure 14.10 can be modified so that it minimizes (14.12) only over the codeword sequences $\hat{\mathbf{c}}_0\hat{\mathbf{c}}_1\ldots\hat{\mathbf{c}}_{\ell-1}$ that are images of $\mathcal{E}^{\circ}_\ell$ (rather than over all the codeword sequences in $\mathbf{C}^{\circ}_\ell(\mathcal{G})$).

[Section 14.4]

Problem 14.8 Let $F = \mathrm{GF}(q)$ and let $\mathcal{G} = (V{=}F^m, E, L)$ be a $k : n$ LFSM over F that is defined by the matrix quadruple (P, B, Q, D). Show that by renaming of states and tags in $\mathcal{G}$, it can also be defined by the quadruple $(XPX^{-1}, YBX^{-1}, XQ, YD)$, for every $m \times m$ (respectively, $k \times k$) nonsingular matrix X (respectively, Y) over F.

(Observe that no such matrices X and Y can yield the quadruple $(\tilde{P}, \tilde{B}, \tilde{Q}, \tilde{D})$ in Example 14.8 from the quadruple (P, B, Q, D) therein. Nevertheless, the LFSMs $\mathcal{G}$ and $\tilde{\mathcal{G}}$ that these two quadruples define are isomorphic.)

Problem 14.9 (Irreducible LFSMs) Let $F = \text{GF}(q)$ and let $\mathcal{G} = (V{=}F^m, E, L)$ be a $k : n$ LFSM over F that is defined by the matrix quadruple (P, B, Q, D). For any positive integer ℓ, denote by Γ_ℓ the $(\ell k) \times m$ matrix over F that is given by

$$\Gamma_\ell = \begin{pmatrix} B \\ BP \\ BP^2 \\ \vdots \\ BP^{\ell-1} \end{pmatrix} .$$

1. Let $\mathbf{s}$ and $\tilde{\mathbf{s}}$ be two states in $\mathcal{G}$. Show that there is a path of length ℓ from $\mathbf{s}$ to $\tilde{\mathbf{s}}$ in $\mathcal{G}$, if and only if there exist tags $\mathbf{u}_0, \mathbf{u}_1, \ldots, \mathbf{u}_{\ell-1} \in F^k$ such that

$$\tilde{\mathbf{s}} = \mathbf{s}P^\ell + \sum_{t=0}^{\ell-1} \mathbf{u}_t BP^t .$$

2. Let $\mathbf{s}$ be a state in $\mathcal{G}$ and let ℓ be a positive integer. Show that there are paths of length ℓ from $\mathbf{s}$ to each state in $\mathcal{G}$, if and only if

$$\text{rank}(\Gamma_\ell) = m .$$

3. Show that for every $\ell > m$, the rows of Γ_ℓ are spanned by the rows of Γ_m.

   Hint: Let

$$a(z) = \det(zI - P) = z^m + \sum_{t=0}^{m-1} a_t z^t$$

   be the characteristic polynomial of P over F. By the Cayley–Hamilton Theorem,

$$P^m = -\sum_{t=0}^{m-1} a_t P^t .$$

4. Show that $\mathcal{G}$ is irreducible if and only if

$$\text{rank}(\Gamma_m) = m .$$

Problem 14.10 Let $\mathcal{G} = (V{=}F^m, E, L)$ be an irreducible $k : n$ LFSM over F. Using Problem 14.9, show that $\mathcal{G}$ is aperiodic and that every integer $\ell \geq m$ is a back-length of $\mathcal{G}$.

Problem 14.11 Let $F = \text{GF}(q)$ and let $\mathcal{G} = (V{=}F^m, E, L)$ be a $k : n$ LFSM over F that is defined by the matrix quadruple (P, B, Q, D). Define the matrix Γ_ℓ as in Problem 14.9, and denote by $V(\mathbf{0}_m)$ the set of terminal states of all finite paths in $\mathcal{G}$ that start at state $\mathbf{0}_m$.

1. Show that $V(\mathbf{0}_m)$ equals the linear span of the rows of Γ_m.
2. Show that for every $\ell \geq m$ there is a path in $\mathcal{G}$ of length (exactly) ℓ from state $\mathbf{0}_m$ to each state in $V(\mathbf{0}_m)$.

3. Show that for every $\ell \geq m$ and every state $\mathbf{s} \in V(\mathbf{0}_m)$ there is a path of length ℓ in $\mathcal{G}$ from $\mathbf{s}$ back to state $\mathbf{0}_m$.

   Hint: Show that for every m vectors
$$\mathbf{u}_0, \mathbf{u}_1, \ldots, \mathbf{u}_{m-1} \in F^k$$
   there are ℓ vectors
$$\mathbf{u}'_0, \mathbf{u}'_1, \ldots, \mathbf{u}'_{\ell-1} \in F^k$$
   such that
$$\Big(\sum_{t=0}^{m-1} \mathbf{u}_t B P^t\Big) P^\ell + \sum_{t=0}^{\ell-1} \mathbf{u}'_t B P^t = \mathbf{0}_m \ .$$

4. Conclude from part 3 that state $\mathbf{0}_m$ belongs to an irreducible sink of $\mathcal{G}$.

5. Show that the irreducible sink of $\mathcal{G}$ that contains state $\mathbf{0}_m$ is itself a $k:n$ LFSM over F.

   Hint: Write F^m as a direct sum of two linear spaces, one of which is $V(\mathbf{0}_m)$.

Problem 14.12 Let $\mathcal{G}$ be a lossless $k:n$ LFSM over $F = \mathrm{GF}(q)$.

1. Show that for every $\ell \geq 1$, the sets $\mathbf{C}_\ell(\mathcal{G})$ and $\mathbf{C}^\circ_\ell(\mathcal{G})$ are linear spaces over F.

2. Show that for every $\ell > m$, the dimensions of $\mathbf{C}_\ell(\mathcal{G})$ and $\mathbf{C}^\circ_\ell(\mathcal{G})$ (as linear spaces over F) are at least $(\ell - m)k$.

   Hint: Use Problem 14.10.

Problem 14.13 (Observable LFSMs) Let $F = \mathrm{GF}(q)$ and let $\mathcal{G} = (V{=}F^m, E, L)$ be a $k:n$ LFSM over F that is defined by the matrix quadruple (P, B, Q, D). For any positive integer ℓ, denote by Ω_ℓ the following $m \times (\ell n)$ matrix over F:
$$\Omega_\ell = (\ Q \mid PQ \mid P^2 Q \mid \ldots \mid P^{\ell-1} Q\ ) \ .$$

1. Show that for every $\ell > m$, the columns of Ω_ℓ are spanned by the columns of Ω_m.

   Hint: Use arguments similar to those in part 3 of Problem 14.9.

Two states $\mathbf{s}$ and $\tilde{\mathbf{s}}$ in V are said to be *distinguishable* in $\mathcal{G}$ if there exist paths π and $\tilde{\pi}$ that start at $\mathbf{s}$ and $\tilde{\mathbf{s}}$, respectively, such that the sequences of edge tags along π and $\tilde{\pi}$ are the same, yet $L(\pi) \neq L(\tilde{\pi})$.

2. Show that the following conditions are equivalent:

   (i) $\mathbf{s}$ and $\tilde{\mathbf{s}}$ are distinguishable in $\mathcal{G}$.

   (ii) $(\tilde{\mathbf{s}} - \mathbf{s})(I - xP)^{-1} Q \neq \mathbf{0}_n$.

   (iii) There exists a positive integer ℓ such that $(\tilde{\mathbf{s}} - \mathbf{s})\Omega_\ell \neq \mathbf{0}_{\ell n}$.

   (iv) $(\tilde{\mathbf{s}} - \mathbf{s})\Omega_m \neq \mathbf{0}_{mn}$.

   (v) The following property holds for *every* two paths π and $\tilde{\pi}$ of length m that start at $\mathbf{s}$ and $\tilde{\mathbf{s}}$, respectively: if π and $\tilde{\pi}$ have the same tag sequence, then $L(\pi) \neq L(\tilde{\pi})$.

3. The LFSM $\mathcal{G}$ is called *observable* or *reduced* if every two distinct states in V are distinguishable in $\mathcal{G}$. Show that $\mathcal{G}$ is observable if and only if

$$\text{rank}(\Omega_m) = m .$$

4. Let U be the set of states in V that are *indistinguishable* from state $\mathbf{0}_m$ in $\mathcal{G}$. Show that U forms a linear subspace of V of dimension $m - \text{rank}(\Omega_m)$ over F and that

$$\mathbf{s} \in U \implies \mathbf{s}P \in U .$$

5. Show that—regardless of whether $\mathcal{G}$ is observable or not—there always exists an observable $k : n$ LFSM $\hat{\mathcal{G}} = (\hat{V}{=}F^h, \hat{E}, \hat{L})$ whose response matrix is the same as that of $\mathcal{G}$ and

$$h = \text{rank}(\Omega_m) .$$

Hint: Write F^m as a direct sum of two linear spaces, one of which is U.

Problem 14.14 Let F be a field. A *Laurent series* over F (in the indeterminate x) is an expression of the form

$$\sum_{t=\mu}^{\infty} a_t x^t ,$$

where $a_t \in F$ and μ is a (possibly negative) integer. That is, a Laurent series can include also terms with negative powers of x, as long as the number of such terms is finite. The set of all Laurent series over F will be denoted by $F((x))$.

Given two Laurent series

$$a(x) = \sum_{t=\mu}^{\infty} a_t x^t \quad \text{and} \quad b(x) = \sum_{t=\nu}^{\infty} b_t x^t ,$$

their sum and product in $F((x))$ are defined, respectively, by

$$a(x) + b(x) = \sum_{t=-\infty}^{\infty} (a_t + b_t) x^t = \sum_{t=\min\{\mu,\nu\}}^{\infty} (a_t + b_t) x^t$$

and

$$a(x)b(x) = \sum_{t=-\infty}^{\infty} \Big(\sum_{r=-\infty}^{\infty} a_r b_{t-r} \Big) x^t = \sum_{t=\mu+\nu}^{\infty} \Big(\sum_{r=\mu}^{t-\nu} a_r b_{t-r} \Big) x^t ,$$

where a_t (respectively, b_t) is defined to be zero when $t < \mu$ (respectively, $t < \nu$).

1. Show that $F((x))$ is a commutative ring with unity under these definitions of addition and multiplication.

2. Show that the ring $F[[x]]$ of formal power series over F is a subring of $F((x))$.

3. Show that every nonzero element in $F((x))$ has a multiplicative inverse and, so, $F((x))$ is a field.

Hint: Let $a(x) = \sum_{t=\mu}^{\infty} a_t x^t$ be an element in $F((x))$ such that $a_\mu \neq 0$. Show that its multiplicative inverse in $F((x))$ is given by $b(x) = \sum_{t=-\mu}^{\infty} b_t x^t$, where

$$b_{-\mu} = \frac{1}{a_\mu} \qquad \text{and} \qquad b_t = -\frac{1}{a_\mu} \sum_{r=1}^{t+\mu} a_{r+\mu} b_{t-r} \; , \quad t > -\mu \; .$$

4. Show that the field $F(x)$ of rational functions over F is a subfield of $F((x))$.

Problem 14.15 Let $\mathcal{G}$ be a $k:n$ LFSM over $F = \mathrm{GF}(q)$ that is defined by the matrix quadruple (P, B, Q, D) and let $G(x)$ be the response matrix of $\mathcal{G}$. The purpose of this problem is to show that $\mathcal{G}$ is lossless if and only if $\mathrm{rank}(G(x)) = k$.

Suppose first that $\mathrm{rank}(G(x)) < k$ (over $F(x)$); i.e., there exists a nonzero vector $\mathbf{u}(x) \in (F(x))^k$ such that

$$\mathbf{u}(x) G(x) = \mathbf{0}_n \; .$$

1. Show that without loss of generality, one can assume that the vector $\mathbf{u}(x)$ satisfies the following properties:

   (i) $\mathbf{u}(x) \in (F_\ell[x])^k$ for some (finite) integer ℓ, i.e.,

   $$\mathbf{u}(x) = \mathbf{u}_0 + \mathbf{u}_1 x + \ldots + \mathbf{u}_{\ell-1} x^{\ell-1}$$

   for some ℓ vectors $\mathbf{u}_0, \mathbf{u}_1, \ldots, \mathbf{u}_{\ell-1} \in F^k$.

   (ii) $\mathbf{u}_0 \neq \mathbf{0}_k$.

   (iii) Each entry in $\mathbf{u}(x)$ is divisible (in $F[x]$) by the polynomial $\det(I - xP)$.

2. Under the assumptions on $\mathbf{u}(x)$ in part 1, let the infinite state sequence

   $$\mathbf{s}_0 \mathbf{s}_1 \mathbf{s}_2 \cdots$$

   be defined inductively by $\mathbf{s}_0 = \mathbf{0}_m$ and

   $$\mathbf{s}_{t+1} = \mathbf{s}_t P + \mathbf{u}_t B \; , \quad t \geq 0 \; ,$$

   where $\mathbf{u}_t = \mathbf{0}_k$ for every $t \geq \ell$. Show that there is an index $r > 0$ such that $\mathbf{s}_r = \mathbf{s}_0$.

   Hint: Show that for the given $\mathbf{s}_0$ and $\mathbf{u}(x)$, the right-hand side of (14.20) is a vector over $F[x]$.

3. Let the infinite sequence $\mathbf{s}_0 \mathbf{s}_1 \mathbf{s}_2 \cdots$ and the index r be as in part 2. Show that the cycle

   $$[\mathbf{s}_0; \mathbf{u}_0][\mathbf{s}_1; \mathbf{u}_1] \ldots [\mathbf{s}_{r-1}; \mathbf{u}_{r-1}]$$

   generates the all-zero sequence. Deduce that $\mathcal{G}$ is not lossless.

Turning to proving the converse result, assume now that $\mathrm{rank}(G(x)) = k$.

4. Let π and π' be two distinct infinite paths in $\mathcal{G}$ that start at the same state. Show that π and π' generate distinct sequences.
5. Conclude from part 4 that $\mathcal{G}$ is lossless.

Problem 14.16 Let $\mathcal{G}$ be a lossless $k : n$ LFSM over $F = \mathrm{GF}(q)$ that is defined by the matrix quadruple (P, B, Q, D) and let $G(x)$ be the response matrix of $\mathcal{G}$. The purpose of this problem is to show that $\mathcal{G}$ has finite anticipation (see Problem 14.5).

Let the subset J of $\{1, 2, \ldots, n\}$ index k columns in $G(x)$ that are linearly independent over $F(x)$ (why do such columns exist?) and let $G_J(x)$ be the $k \times k$ sub-matrix of $G(x)$ that is formed by these columns. The determinant of $G_J(x)$, which is a nonzero element of $F(x)$, can be written as

$$\det(G_J(x)) = \frac{\omega(x)}{\sigma(x)} ,$$

where both $\sigma(x)$ and $\omega(x)$ are nonzero polynomials in $F[x]$. Denote by r the largest integer such that x^r divides $\omega(x)$ in $F[x]$.

Fix $\mathbf{s}_0$ to be a state in $\mathcal{G}$ and let

$$\mathbf{c}(x) = \sum_{t=0}^{\infty} \mathbf{c}_t x^t$$

be an infinite sequence that is generated by some path from state $\mathbf{s}_0$ in $\mathcal{G}$; by part 4 of Problem 14.15 this path is unique. Denote the tag sequence along this path by

$$\mathbf{u}(x) = \sum_{t=0}^{\infty} \mathbf{u}_t x^t .$$

1. Define the vector $\mathbf{v}(x) \in (F[[x]])^n$ by
$$\mathbf{v}(x) = \mathbf{c}(x) - \mathbf{s}_0 (I - xP)^{-1} Q$$
and let $\mathbf{v}_J(x)$ be the sub-vector of $\mathbf{v}(x)$ that is indexed by J. Show that
$$\sigma(x) \cdot \mathbf{v}_J(x) \cdot \mathrm{Adj}(G_J(x)) = \mathbf{u}(x) \cdot \omega(x) .$$

2. Show that $\mathbf{u}_0$ can be effectively computed only from $\mathbf{s}_0$ and $\mathbf{c}_0, \mathbf{c}_1, \ldots, \mathbf{c}_r$ (i.e., it is unnecessary to know the values $\mathbf{c}_t$ for $t > r$ in order to determine $\mathbf{u}_0$).

[Section 14.5]

Problem 14.17 Let

$$\sigma(x) = 1 + \lambda_1 x + \lambda_2 x^2 + \ldots + \lambda_m x^m$$

and

$$\omega(x) = \gamma_0 + \gamma_1 x + \gamma_2 x^2 + \ldots + \gamma_m x^m$$

be polynomials over a field F, where

$$m \geq \max\{\deg \sigma, \deg \omega\} .$$

Define the column vector $\mathbf{q} = \mathbf{q}(\sigma, \omega)$ in F^m by

$$\mathbf{q} = \begin{pmatrix} \gamma_m \\ \gamma_{m-1} \\ \vdots \\ \gamma_1 \end{pmatrix} - \gamma_0 \begin{pmatrix} \lambda_m \\ \lambda_{m-1} \\ \vdots \\ \lambda_1 \end{pmatrix}$$

and let P be the $m \times m$ companion matrix of the reverse polynomial

$$x^m \cdot \sigma(x^{-1}) = x^m + \lambda_1 x^{m-1} + \ldots + \lambda_m ;$$

that is,

$$P = \begin{pmatrix} 0 & 0 & \ldots & 0 & -\lambda_m \\ 1 & 0 & \ldots & 0 & -\lambda_{m-1} \\ 0 & 1 & \ldots & 0 & -\lambda_{m-2} \\ \vdots & \ddots & \ddots & 0 & \vdots \\ 0 & \ldots & 0 & 1 & -\lambda_1 \end{pmatrix}$$

(see Problem 3.9). Show that

$$\frac{\omega(x)}{\sigma(x)} = x\,\mathbf{e}(I - xP)^{-1}\mathbf{q} + \omega(0) ,$$

where I is the $m \times m$ identity matrix and $\mathbf{e}$ is the row vector

$$(0\,0\,\ldots\,0\,1)$$

in F^m.

Hint: Using Problem 3.9, identify the characteristic polynomial $a(z)$ of P and then show that the last row of $(zI - P)^{-1}$ equals

$$\frac{1}{a(z)} \cdot (1 \;\; z \;\; z^2 \;\; \ldots \;\; z^{m-1}) .$$

Problem 14.18 (LFSM realizations of generator matrices) Let

$$G(x) = \left(\omega_{i,j}(x)/\sigma_i(x) \right)_{i=1}^{k} {}_{j=1}^{n}$$

be a $k \times n$ matrix over $F(x)$, where $\sigma_i(x)$ and $\omega_{i,j}(x)$ are polynomials over F such that $\sigma_i(0) = 1$ (thus, it is assumed here that the entries in each row of $G(x)$ are brought to a common denominator).

For $1 \leq i \leq k$, define

$$m_i = \max\{\deg \sigma_i, \deg \omega_{i,1}, \deg \omega_{i,2}, \ldots, \deg \omega_{i,n}\} ,$$

and let $\mathbf{q}_{i,j}$ be the column vector $\mathbf{q}(\sigma_i, \omega_{i,j}) \in F^{m_i}$ as defined in Problem 14.17. Also, denote by P_i the $m_i \times m_i$ companion matrix of the reverse polynomial

$$x^{m_i} \cdot \sigma_i(x^{-1}) ,$$

and by $\mathbf{e}_i$ the row vector $(0\,0\,\ldots\,0\,1)$ in F^{m_i}. Define m to be the sum $\sum_{i=1}^{k} m_i$.

Show that $G(x)$ can be written in the form

$$G(x) = xB(I - xP)^{-1}Q + D ,$$

where P, B, Q, and D are matrices over F of orders $m \times m$, $k \times m$, $m \times n$, and $k \times n$, respectively, which are given by

$$P = \begin{pmatrix} \boxed{P_1} & & & \\ & \boxed{P_2} & & \\ & & \ddots & \\ & & & \boxed{P_k} \end{pmatrix},$$

$$B = \begin{pmatrix} \boxed{\mathbf{e}_1} & & & \\ & \boxed{\mathbf{e}_2} & & \\ & & \ddots & \\ & & & \boxed{\mathbf{e}_k} \end{pmatrix},$$

$$Q = \begin{pmatrix} \boxed{\mathbf{q}_{1,1}} & \boxed{\mathbf{q}_{1,2}} & \cdots & \boxed{\mathbf{q}_{1,n}} \\ \boxed{\mathbf{q}_{2,1}} & \boxed{\mathbf{q}_{2,2}} & \cdots & \boxed{\mathbf{q}_{2,n}} \\ \vdots & \vdots & \vdots & \vdots \\ \boxed{\mathbf{q}_{k,1}} & \boxed{\mathbf{q}_{k,2}} & \cdots & \boxed{\mathbf{q}_{k,n}} \end{pmatrix},$$

and

$$D = \left(\omega_{i,j}(0) \right)_{i=1}^{k} {}_{j=1}^{n}$$

(the entries outside the marked blocks in P and B are all zero).

Problem 14.19 Let $F = \mathrm{GF}(q)$ and let $G(x)$ be a $k \times n$ matrix of rank k over $F(x) \cap F[[x]]$. Show that $G(x)$ is a response matrix of some $k : n$ LFSM $\mathcal{G}$ over F that is lossless, irreducible, and observable.

Hint: See Problems 14.11, 14.13, and 14.18.

Problem 14.20 Let $\mathbf{C} = \mathbf{C}(\mathcal{G})$ be an $[n, k]$ convolutional code over $F = \mathrm{GF}(q)$ and suppose that $\mathbf{C}$ has a $k \times n$ generator matrix $G(x)$ whose entries are all in the ground field F.

1. What is the smallest number of states in any lossless LFSM that presents $\mathbf{C}$?
2. Show that the set of codewords in $\mathbf{C}_1(\mathcal{G})$ forms a linear $[n, k]$ block code over F whose generator matrix is $G(x)$.

Problem 14.21 Let $\mathbf{C}$ be an $[n, k]$ convolutional code over $F = \mathrm{GF}(q)$ and let $G(x)$ be a generator matrix of $\mathbf{C}$ over $F(x) \cap F[[x]]$. Show that a $k \times n$ matrix

$\hat{G}(x)$ over $F(x) \cap F[[x]]$ is a generator matrix of $\mathbf{C}$, if and only if it can be written as

$$\hat{G}(x) = T(x)G(x) ,$$

where $T(x)$ is a $k \times k$ matrix over $F(x) \cap F[[x]]$ such that $\det(T(0)) \neq 0$ (over F).

Hint: Show that a $k \times k$ matrix $T(x)$ over $F(x) \cap F[[x]]$ has an inverse over $F(x) \cap F[[x]]$, if and only if $\det(T(0)) \neq 0$.

Problem 14.22 (Wyner–Ash codes) Let $F = \mathrm{GF}(q)$ and let $n = q^m$ for some positive integer m. Consider the $[n, n-1]$ convolutional code $\mathbf{C}$ over F with a generator matrix

$$G(x) = \left(\begin{array}{c|c} & 1 + x \cdot g_1(x) \\ I & 1 + x \cdot g_2(x) \\ & \vdots \\ & 1 + x \cdot g_{n-1}(x) \end{array} \right) ,$$

where I is the $(n-1) \times (n-1)$ identity matrix and $g_1(x), g_2(x), \ldots, g_{n-1}(x)$ range over all the nonzero elements in $F_m[x]$. For example, when $F = \mathrm{GF}(2)$ and $m = 2$,

$$G(x) = \left(\begin{array}{c|c} & 1+x \\ I & 1+x^2 \\ & 1+x+x^2 \end{array} \right) .$$

Show that

$$\mathsf{d}_{\mathsf{free}}(\mathbf{C}) = 3 .$$

Hint: Let $\mathbf{c}(x) = \mathbf{u}(x)G(x)$ be a nonzero codeword sequence in $\mathbf{C}$. Assume first that $\mathbf{u}(x)$, as a vector in $(F[[x]])^{n-1}$, has just one nonzero entry (in $F[[x]]$), say $u_i(x)$, and show that it suffices to rule out only the case where $u_i(x) = 1 + ax^r$ for some $a \in F$. Then assume that $\mathbf{u}(x)$ has two nonzero entries, $u_i(x)$ and $u_j(x)$, and reduce the problem to the case where $u_i(x) = 1$ and $u_j(x) = ax^r$.

Problem 14.23 (Upper bound on the free distance) Let $\mathcal{G} = (V{=}F^m, E, L)$ be a lossless $k : n$ LFSM over $F = \mathrm{GF}(q)$. Denote by $\mathsf{d}_{\mathsf{max}}(\nu, \kappa, q)$ the largest minimum distance of any linear $[\nu, \kappa{>}0]$ block code over F. Show that

$$\mathsf{d}_{\mathsf{free}}(\mathbf{C}(\mathcal{G})) \leq \min_{\ell > m} \mathsf{d}_{\mathsf{max}}(\ell n, (\ell - m)k, q) .$$

Hint: See Problem 14.12.

[Section 14.6]

Problem 14.24 Let

$$\sigma(x) = 1 + \lambda_1 x + \lambda_2 x^2 + \ldots + \lambda_m x^m$$

and

$$\omega(x) = \gamma_0 + \gamma_1 x + \gamma_2 x^2 + \ldots + \gamma_m x^m$$

be polynomials over a field F, and consider the circuit in Figure 14.20. The m delay units are initially reset to zero and at each clock tick $t \geq 0$, the circuit is fed with the coefficient of x^t in the formal power series

$$u(x) = \sum_{t=0}^{\infty} u_t x^t$$

and produces the respective coefficient in

$$c(x) = \sum_{t=0}^{\infty} c_t x^t .$$

Show that the circuit implements a multiplication by $\omega(x)/\sigma(x)$ in $F[[x]]$; i.e., $c(x)$ is related to $u(x)$ by

$$c(x) = \frac{\omega(x)}{\sigma(x)} \cdot u(x) .$$

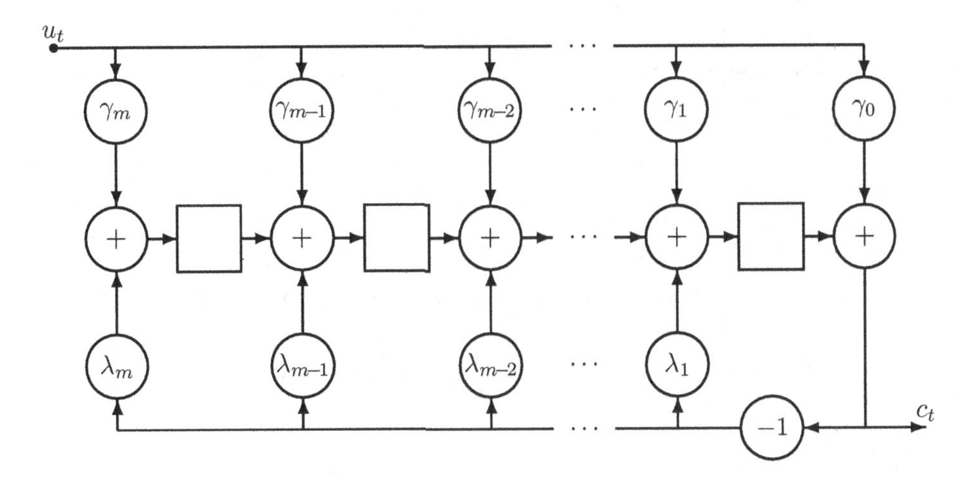

Figure 14.20. Multiplication circuit for Problem 14.24.

[Section 14.7]

Problem 14.25 Let $\mathcal{G}$ be a single-state LFSM over F and let $A_{\mathcal{G}}(z)$ be its 1×1 generalized adjacency matrix. What is the length–weight enumerator of $\mathcal{G}$?

Problem 14.26 Let $\mathcal{G} = (V{=}F^m, E, L)$ be a lossless $k : n$ LFSM over F with $|V| > 1$ and let $\mathcal{G}^*$ be the induced subgraph of $\mathcal{G}$ on $V \setminus \{\mathbf{0}_m\}$. Denote by $\lambda(z)$ the spectral radius of the generalized adjacency matrix $A_{\mathcal{G}^*}(z)$.

1. Suppose that no nontrivial cycle in $\mathcal{G}$ generates an all-zero sequence. Show that $\lambda(0) = 0$. (Since $\lambda(z)$ is continuous in z, it follows that there is a positive-length interval of nonnegative values of z for which $\lambda(z) < 1$.)

   Hint: Show that $(A_{\mathcal{G}^*}(0))^{|V|-1} = 0$ (i.e., $A_{\mathcal{G}^*}(0)$ is nilpotent).

2. Suppose now that there is a nontrivial cycle in $\mathcal{G}$ that generates an all-zero sequence (since $\mathcal{G}$ is lossless, that cycle must be wholly contained in $\mathcal{G}^*$). Show that $\lambda(z) \geq 1$ for every $z \geq 0$.

   Hint: Show that there exists $\mathbf{s} \in V \setminus \{\mathbf{0}_m\}$ such that $\left((A_{\mathcal{G}^*}(z))^\ell\right)_{\mathbf{s},\mathbf{s}} \geq 1$ for infinitely many integers ℓ and for every $z \geq 0$. Then use the fact that the trace of $(A_{\mathcal{G}^*}(z))^\ell$ (which is the sum of entries on its main diagonal) is bounded from above by $(|V|-1) \cdot (\lambda(z))^\ell$.

Problem 14.27 Let $F = \mathrm{GF}(q)$ and consider transmission of codeword sequences of a convolutional code $\mathbf{C}$ over F through a memoryless q-ary erasure channel $(F, F \cup \{?\}, \mathsf{Prob})$ with erasure probability p. Verify that the bound (14.43) on the per-message decoding error probability holds also for this channel, if one substitutes

$$\gamma = p\,.$$

Hint: Compute the right-hand side of (14.42) for this channel.

Problem 14.28 Let $\mathcal{G}$ be an irreducible lossless $k : n$ LFSM over F and for every $i \geq 0$, denote by W_i the number of fundamental cycles in $\mathcal{G}$ that generate sequences whose Hamming weight (over F) is i. The values W_i are related to the coefficients of the length–weight enumerator $W_{\mathcal{G}}(x, z)$ by

$$W_i = \sum_{\ell=1}^{\infty} W_{\ell,i}\,.$$

1. What is the value of W_0?
2. Show that $W_i < \infty$ for all $i \geq 0$, if and only if no nontrivial cycle in $\mathcal{G}$ generates an all-zero sequence.

Assume hereafter in this problem that no nontrivial cycle in $\mathcal{G}$ generates an all-zero sequence.

3. Show that
$$W_{\mathcal{G}}(1, z) = \sum_{i=0}^{\infty} W_i z^i\,.$$
4. Show that $\mathsf{d}_{\mathsf{free}}(\mathbf{C}(\mathcal{G}))$ equals the smallest i such that $W_i > 0$.
5. Show that the inequality (14.6) is attained for some ℓ, that is,
$$\mathsf{d}_{\mathsf{free}}(\mathbf{C}(\mathcal{G})) = \min_{\ell} \mathsf{d}_F(\mathbf{C}^\circ_\ell(\mathcal{G}))\,,$$
   where the minimum is taken over all ℓ such that $|\mathbf{C}^\circ_\ell(\mathcal{G})| > 1$.

Problem 14.29 Let $\mathcal{G}$ be a $k : n$ LFSM over F and assume that no nontrivial cycle in $\mathcal{G}$ generates an all-zero sequence (by parts 1–3 of Problem 14.15, this condition implies that the response matrix of $\mathcal{G}$ has rank k and, therefore, $\mathcal{G}$ is lossless). Show that for $d = \mathsf{d}_{\mathsf{free}}(\mathbf{C}(\mathcal{G}))$,

$$0 < \lim_{z \rightarrow 0} \frac{W'_{\mathcal{G}}(1, z)}{z^d} = \sum_{\ell=1}^{\infty} \ell \cdot W_{\ell,d} < \infty\,.$$

Problem 14.30 (Improvements on the bound (14.43)) Let $\mathcal{G} = (V{=}F^m, E, L)$ be a $k : n$ LFSM over $F = \mathrm{GF}(q)$ and assume that no nontrivial cycle in $\mathcal{G}$ generates an all-zero sequence. For a path

$$\pi = [\mathbf{s}_0; \mathbf{u}_0][\mathbf{s}_1; \mathbf{u}_1] \ldots [\mathbf{s}_{\ell-1}; \mathbf{u}_{\ell-1}]$$

in $\mathcal{G}$, define the *tag support* $\mathcal{T}(\pi)$ as the set of indexes t for which $\mathbf{u}_t \neq \mathbf{0}_k$. The *tag weight* of π is the size of its tag support; namely, the tag weight is the Hamming weight over F^k of the sequence of edge tags along π.

Define the *tag-weight enumerator* of $\mathcal{G}$ by

$$U_{\mathcal{G}}(x, z) = \sum_{h=0}^{\infty} \sum_{i=0}^{\infty} U_{h,i} x^h z^i ,$$

where $U_{h,i}$ is the number of fundamental cycles in $\mathcal{G}$ with tag weight h (over F^k) that generate sequences with Hamming weight i (over F).

1. Let the $|V| \times |V|$ matrix $A_{\mathcal{G}}(x, z)$ be defined by

$$(A_{\mathcal{G}}(x, z))_{\mathbf{s},\tilde{\mathbf{s}}} = \sum_{\substack{e \in E: \\ \iota(e)=\mathbf{s},\, \tau(e)=\tilde{\mathbf{s}}}} x^{\delta(e)} z^{\mathsf{w}(L(e))} ,$$

   where

$$\delta(e) = \begin{cases} 0 & \text{if the edge tag of } e \text{ equals } \mathbf{0}_k \\ 1 & \text{otherwise} \end{cases} .$$

   Show that

$$U_{\mathcal{G}}(x, z) = x \cdot (a(z) - 1) + x \cdot \mathbf{b}(z) \left(I - A_{\mathcal{G}^*}(x, z) \right)^{-1} \mathbf{v}(x, z) ,$$

   where

$$a(z) = (A_{\mathcal{G}}(x, z))_{\mathbf{0}_m, \mathbf{0}_m} ,$$

   and $\mathbf{b}(x)$, $\mathbf{v}(x, z)$, and $A_{\mathcal{G}^*}(x, z)$ are obtained from $A_{\mathcal{G}}(x, z)$ as follows:

$$A_{\mathcal{G}}(x, z) = \left(\begin{array}{c|c} a(z) & x \cdot \mathbf{b}(z) \\ \hline \mathbf{v}(x, z) & A_{\mathcal{G}^*}(x, z) \end{array} \right) .$$

2. Show that for the LFSMs $\mathcal{G}$ and $\tilde{\mathcal{G}}$ in Figure 14.13,

$$U_{\mathcal{G}}(x, z) = \frac{x^2 z^5 (x - x^2 z + z)}{1 - 2xz + x^2 z^2 - z^2} \qquad \text{and} \qquad U_{\tilde{\mathcal{G}}}(x, z) = \frac{x z^5}{1 - 2xz} .$$

3. Show that the upper bound (14.43) can be improved to

$$\mathrm{P_{err}}(\mathcal{G}) \leq U'_{\mathcal{G}}(1, \gamma) ,$$

   where $U'_{\mathcal{G}}(x, z)$ is the partial derivative of $U_{\mathcal{G}}(x, z)$ with respect to x.

Hint: Using the notation of Section 14.7.2, show that the event

$$\hat{\mathbf{u}}_t \neq \mathbf{0}_k$$

(of decoding into the wrong message at time t) is identical to the union of events

$$\bigcup_{h=1}^{\infty} \bigcup_{\psi \in \Theta_h} \bigcup_{j \in \mathcal{T}(\psi)} \mathcal{X}(\psi, t-j) ,$$

where Θ_h stands for the set of fundamental cycles in $\mathcal{G}$ with tag weight h.

4. Verify that for the LFSMs $\mathcal{G}$ and $\tilde{\mathcal{G}}$ in Figure 14.13,

$$U'_{\mathcal{G}}(1, \gamma) = \frac{\gamma^5(3 - 6\gamma + 2\gamma^2)}{(1 - 2\gamma)^2} \qquad \text{and} \qquad U'_{\tilde{\mathcal{G}}}(1, \gamma) = \frac{\gamma^5}{(1 - 2\gamma)^2} ,$$

whenever $\gamma \in [0, \frac{1}{2})$.

[Section 14.8]

Problem 14.31 Let $\mathcal{G}$ be an irreducible lossless $k : n$ LFSM over $F = \mathrm{GF}(q)$ and let $G(x)$ be the response matrix of $\mathcal{G}$. Show that the following conditions are equivalent:

(i) $\mathcal{G}$ is non-catastrophic.

(ii) For every nonnegative integer w there exists a (finite) integer $N(w)$ such that for every $\mathbf{u}(x) \in F[[x]]$,

$$\mathsf{w}_F(\mathbf{u}(x)G(x)) \leq w \quad \Longrightarrow \quad \mathsf{w}_F(\mathbf{u}(x)) \leq N(w) .$$

(iii) $G(x)$ is non-catastrophic, i.e., for every $\mathbf{u}(x) \in (F[[x]])^k$,

$$\mathsf{w}_F(\mathbf{u}(x)G(x)) < \infty \quad \Longrightarrow \quad \mathsf{w}_F(\mathbf{u}(x)) < \infty .$$

Problem 14.32 Let $G(x)$ be a $k \times n$ catastrophic generator matrix of an $[n, k]$ convolutional code over F, and suppose that the entries of $G(x)$ are all in $F[x]$. The purpose of this problem is to show that there is a monic irreducible polynomial $a(x) \neq x$ over F that divides all $\binom{n}{k}$ determinants of the $k \times k$ sub-matrices of $G(x)$.

Let $\mathcal{G} = (V{=}F^m, E, L)$ be a lossless LFSM that realizes $G(x)$ as in Problem 14.18, and denote by (P, B, Q, D) the defining quadruple of $\mathcal{G}$. Based on that problem, assume further that the nonzero entries in P are all below the main diagonal.

1. Show that there are nonzero vectors $\mathbf{s}_0 \in V$ and $\mathbf{u}(x) \in (F_\ell[x])^k$ for some $\ell < \infty$ such that

$$\Big(\mathbf{u}(x) \sum_{i=0}^{\infty} x^{\ell i}\Big) G(x) = -\mathbf{s}_0 (I - xP)^{-1} Q .$$

Hint: By Problem 14.31, there is an irreducible sink in $\mathcal{G}$ that is catastrophic. Therefore, there exists a cycle ψ in $\mathcal{G}$ from a state $\mathbf{s}_0 \neq \mathbf{0}_m$ such that the edge tags along ψ are not all zero yet the labels are. Select the coefficients of $\mathbf{u}(x)$ to be the edge tags along ψ.

Hereafter, let $\mathbf{s}_0$, $\mathbf{u}(x)$, and ℓ be as in part 1.

2. Show that
$$\mathbf{u}(x)G(x) = (x^\ell - 1)\mathbf{s}_0(I - xP)^{-1}Q \, .$$

3. Let $G_0(x)$ be a $k \times k$ sub-matrix of $G(x)$. Show that
$$\mathbf{u}(x)\det(G_0(x)) = (x^\ell - 1)\mathbf{s}_0(I - xP)^{-1}Q_0 \cdot \mathrm{Adj}(G_0(x)) \, ,$$
where Q_0 is an $m \times k$ sub-matrix of Q formed by the columns of Q with the same indexes as the columns of $G_0(x)$ within $G(x)$.

   Hint: $G_0(x) \cdot \mathrm{Adj}(G_0(x)) = \det(G_0(x)) \cdot I$.

4. Show that the vector equality in part 3 is over $F[x]$ (i.e., the entries of the vectors are all polynomials).

   Hint: Under the assumption on the matrix P, what is the determinant of $(I - xP)$?

5. Let $u_j(x)$ denote any nonzero entry in $\mathbf{u}(x)$. Show that for every $k \times k$ sub-matrix $G_0(x)$ of $G(x)$,
$$x^\ell - 1 \mid u_j(x)\det(G_0(x)) \, .$$

6. Show that there is a monic irreducible factor $a(x)$ $(\neq x)$ of $x^\ell - 1$ over F that divides $\det(G_0(x))$ for all $k \times k$ sub-matrices $G_0(x)$ of $G(x)$.

Problem 14.33 Let $\mathbf{C}$ be an $[n, k]$ convolutional code over F and let $G(x)$ be a generator matrix of $\mathbf{C}$.

1. Show that the following conditions are equivalent:

   (i) $\mathrm{rank}(G(0)) = k$ (over F).

   (ii) $\mathbf{C}$ has a generator matrix that contains the $k \times k$ identity matrix as a sub-matrix; i.e., the generator matrix is systematic—or can made into one by permutation of columns.

   (iii) Every LFSM presentation of $\mathbf{C}$ is deterministic.

Suppose now that the entries of $G(x)$ are all in $F[x]$, and let r be the largest integer such that x^r divides all the determinants of the $k \times k$ sub-matrices of $G(x)$.

2. Show that there exists an $[n, k]$ convolutional code $\mathbf{C}'$ over F that satisfies the following properties:

   (i) $\mathbf{C}'$ has a (possibly column-permuted) systematic generator matrix.

   (ii) $\mathbf{C} \subseteq \mathbf{C}'$.

   (iii) If $\mathbf{c}(x) \in \mathbf{C}'$ then $x^r \cdot \mathbf{c}(x) \in \mathbf{C}$.

   Hint: Follow the steps of the proof of Proposition 14.6 with $a(x) = x$. Notice that multiplying a row of $\hat{G}(x)$ therein by $1/x$ results now in a generator matrix of another convolutional code.

Problem 14.34 Let $G(x)$ be a $k \times n$ matrix over $F[x]$ where F is a finite field and let $\mathcal{G}$ be an irreducible LFSM realization of $G(x)$. Show that $\mathcal{G}$ is lossless and non-catastrophic if and only if $\text{rank}(G(\alpha)) = k$ over every finite extension field K of F, for every nonzero element $\alpha \in K$.

Hint: Consider first the case where $\text{rank}(G(x)) < k$ (over $F(x)$). Next, assume that $\text{rank}(G(x)) = k$ and deduce from Proposition 14.5 that $G(x)$ is catastrophic if and only if there is a nonzero element α in a finite extension field of F such that the minimal polynomial of α (with respect to F) divides $\det(G_0(x))$ for every $k \times k$ sub-matrix $G_0(x)$ of $G(x)$.

Notes

There is a vast body of literature on convolutional codes and trellis codes, and this chapter only attempts to be an introductory exposition to this subject. More information on these codes can be found in Blahut [46, Chapter 12], Forney [131]–[133], Johannesson and Zigangirov [197], Lin and Costello [230, Chapters 10–12], McEliece [258], [259, Chapter 10], and Viterbi and Omura [374, Chapters 4–6].

[Section 14.1]

Labeled digraphs serve as a descriptive tool in many areas. Their use in this chapter is very similar to the role they play in automata theory, symbolic dynamics, and constrained coding theory. For example, our notion of a *sequence set* is almost the same as a *regular language* in automata theory, a *sofic system* in symbolic dynamics, and a *constrained system* in constrained coding. See Hopcroft *et al.* [188, Chapters 2–4], Kohavi [217], Lind and Marcus [232], and Marcus *et al.* [251].

[Section 14.2]

In our definition of a trellis code, we require that the presenting lossless digraph be regular. The next result implies that there is no loss of generality in this requirement.

Proposition 14.8 *Let $\mathcal{G}$ be an irreducible lossless digraph and λ be the spectral radius of the adjacency matrix of $\mathcal{G}$. There exists an irreducible M-regular lossless digraph $\mathcal{G}'$ such that $\mathbf{C}(\mathcal{G}') \subseteq \mathbf{C}(\mathcal{G})$, if and only if $M \leq \lambda$.*

Proposition 14.8 follows from a key theorem in constrained coding theory, by Adler *et al.* [2]. Their theorem provides an algorithm, known as the *state-splitting algorithm*, for obtaining the regular digraph $\mathcal{G}'$ out of $\mathcal{G}$. This algorithm serves as a primary tool for constructing encoders for constrained systems; see Marcus *et al.* [251, Section 4]. The proof of Adler *et al.* in [2] makes use of Perron–Frobenius theory for nonnegative matrices. (See the notes on Section 13.2 at the end of Chapter 13. It follows from this theory that the spectral radius λ in Proposition 14.8 is, in fact, an eigenvalue of the adjacency matrix of $\mathcal{G}$. The same can be said about the spectral radius $\lambda(z)$ of the generalized adjacency matrix $A_{\mathcal{G}^*}(z)$ in Section 14.7.1.)

[Section 14.3]

Viterbi's algorithm was suggested in [373] for the decoding of convolutional codes. Viterbi's algorithm is an instance of an optimization algorithm which is known in computer science and operations research as dynamic programming. See, for example, Hillier and Lieberman [177, Chapter 11].

Other decoding algorithms are discussed in the notes on Section 14.7 below.

[Section 14.4]

Linear finite-state machines are found in other disciplines as well, such as signal processing and linear dynamical systems. In these applications, the underlying field is typically the real or the complex field, rather than a finite field. The terms "controllable" and "observable" are borrowed from these disciplines, and so is "response matrix," which has been adapted from the term "impulse response" used in linear systems. See Kailath [202] and Kwakernaak and Sivan [221, Chapter 5] (for the linear systems approach to LFSMs) and Kohavi [217, Chapter 15] (for the finite automata approach).

[Section 14.5]

Convolutional codes were first introduced by Elias in [115]. Many of the convolutional codes with the largest-known free distance were found by computer search, and they are available through tables. See Blahut [46, Section 12.5] and Lin and Costello [230, Section 11.3].

The codes in Problem 14.22 are due to Wyner and Ash [389].

Convolutional codes can be used as ingredients for constructing other codes. One important example of the latter is the class of *turbo codes*. A turbo code is defined through two systematic generator matrices,

$$G_1(x) = \left(\; I \;\middle|\; A_1(x) \; \right) \quad \text{and} \quad G_2(x) = \left(\; I \;\middle|\; A_2(x) \; \right) \,,$$

of two respective convolutional codes, an $[n_1, k]$ code $\mathbf{C}_1$ and an $[n_2, k]$ code $\mathbf{C}_2$, both over the same field F. The sequence of messages $\mathbf{u}(x)$ to be sent is encoded into a codeword sequence $\mathbf{c}_1(x) = \mathbf{u}(x)G_1(x)$, and a *permuted* (interleaved) copy of $\mathbf{u}(x)$, denoted by $\hat{\mathbf{u}}(x)$, is mapped into the sequence $\hat{\mathbf{c}}_2(x) = \hat{\mathbf{u}}(x)A_2(x)$. The transmitted sequence is then given by $(\, \mathbf{c}_1(x) \,|\, \hat{\mathbf{c}}_2(x) \,)$, which is a vector of length $n_1{+}n_2{-}k$ over $F[[x]]$. Turbo codes were introduced by Berrou *et al.* in [41], [42], along with an iterative method for decoding them. While the theory of turbo codes is not yet fully understood, they have exhibited remarkable performance in empirical results.

[Section 14.7]

The time complexity of Viterbi's algorithm, when applied to a $k : n$ LFSM $\mathcal{G} = (V{=}F^m, E, L)$ over $F = \mathrm{GF}(q)$, is proportional to $\ell \cdot q^{m+k}$, where ℓ is the length of the received sequence. In practical applications, the parameters k and n are usually small; the value of m, however, may need to be large so that we meet the desired decoding error probability. This, in turn, poses a limitation on the use of Viterbi's algorithm (such a limitation applies also to the computation of the formula (14.34) for the length–weight enumerator of $\mathcal{G}$, since the number of rows and columns in $A_{\mathcal{G}^*}(z)$ equals q^m-1).

Alternate decoding algorithms exist with a considerably smaller time complexity, yet with some sacrifice in the decoding error probability. These algorithms usually go under the collective term *sequential decoding*. Early sequential decoders were suggested by Wozencraft [387] and Wozencraft and Reiffen [388], and the predominant algorithms currently known are *Fano's algorithm* [120] and the *stack algorithm* due to Jelinek [196] and Zigangirov [395].

We next provide a brief description of these two algorithms. Both algorithms operate on the trellis diagram $\mathsf{T}(\mathcal{G})$, where it is assumed that states from which there are no paths to state $(\iota(\mathcal{G}))^{(\ell)}$ have been pruned. In addition, since the algorithms compare paths of different lengths, a certain adjustment for the length is inserted into the definition (14.13) of a path cost.

Fano's algorithm constructs a path by starting at state $(\iota(\mathcal{G}))^{(0)}$ in layer 0 and then making its way greedily towards subsequent layers, always selecting an edge that yields the lowest (adjusted) cost to the traversed path so far. This process continues until one of the following occurs: (a) we reach state $(\iota(\mathcal{G}))^{(\ell)}$, in which case the algorithm terminates, or (b) the path cost exceeds a certain threshold T. In the latter case, the algorithm backtracks one layer and tries another outgoing edge, if any. The threshold T is updated throughout the execution of the algorithm: backtracking loosens it while progress tightens it.

The stack algorithm does not maintain a running threshold; instead, it keeps a list ("stack") of all the paths from $(\iota(\mathcal{G}))^{(0)}$ that have been traversed so far, sorted according to their cost. In each step of the algorithm, the path π with the lowest cost in the list is replaced by all q^k paths in $\mathsf{T}(\mathcal{G})$ that are obtained by extending π by one edge. The list is re-sorted and the process continues until we reach state $(\iota(\mathcal{G}))^{(\ell)}$. Practical space constraints may cause the list to be overflowed, in which case the worst paths are removed from the list during the re-sorting.

A detailed description of Fano's algorithm and the stack algorithm can be found in several references: see Blahut [46, Section 12.9], Forney [133], Lin and Costello [230, Chapter 12], McEliece [259, Section 10.4], and Viterbi and Omura [374, Chapter 6].

Our decoding error analysis for MLDs in Section 14.7.2 was done for a *given* convolutional code. It is known that convolutional codes attain the capacity of the q-ary symmetric channel, with a per-message decoding error probability that decays exponentially with the constraint length of the encoder. See Shulman and Feder [337], Viterbi and Omura [374, Chapter 5], and Zigangirov [396].

[Section 14.8]

Proposition 14.5 is due to Massey and Sain [256].

Appendix

Basics in Modern Algebra

We summarize below several concepts—mainly from modern algebra—which are used throughout the book. Some of the properties of these concepts are included in the problems that follow.

A *group* is a nonempty set G with a binary operation "$\cdot$" that satisfies the following properties:

- Closure: $a \cdot b \in G$ for every $a, b \in G$.
- Associativity: $(a \cdot b) \cdot c = a \cdot (b \cdot c)$ for every $a, b, c \in G$.
- Unity element: there exists an element $1 \in G$ such that $1 \cdot a = a \cdot 1 = a$ for every $a \in G$.
- Inverse element: for every element $a \in G$ there is an element $a^{-1} \in G$ such that $a \cdot a^{-1} = a^{-1} \cdot a = 1$.

A group G is called *commutative* or *Abelian* if $a \cdot b = b \cdot a$ for every $a, b \in G$.

For an element a in a group G and a positive integer n, the notation a^n will stand for

$$\underbrace{a \cdot a \cdot \ldots \cdot a}_{n \text{ times}},$$

and a^{-n} will denote the element $(a^{-1})^n$ in G. Also, define $a^0 = 1$.

A group G is called *cyclic* if there is an element $a \in G$, called a *generator*, such that each element in G has the form a^n for some integer n.

An element a in a group G has *finite order* if there is a positive integer n such that $a^n = 1$. The smallest n for which this holds is called the *order* of a and is denoted by $\mathcal{O}(a)$.

A nonempty subset H of a group G is a *subgroup* of G if H is a group with respect to the operation of G.

Let H be a subgroup of a group G. Two elements $a, b \in G$ are said to be *congruent modulo* H if and only if $a \cdot b^{-1} \in H$.

Let H be a subgroup of a group G and let a be an element of G. The *(right) coset* Ha (of H in G) is the set $\{ h \cdot a \, : \, h \in H \}$.

A subgroup H of a group G is called *normal* if for every $a \in G$ and $h \in H$, the element aha^{-1} belongs to H.

A *ring* is a nonempty set R with two binary operations "+" and "·" that satisfy the following properties:

- R is a commutative group with respect to "+".
- Associativity of "·": $(a \cdot b) \cdot c = a \cdot (b \cdot c)$ for every $a, b, c \in R$.
- Distributivity: $a \cdot (b+c) = (a \cdot b) + (a \cdot c)$ and $(b+c) \cdot a = (b \cdot a) + (c \cdot a)$ for every $a, b, c \in R$.

The unity element of "+" is called the *zero element* and is denoted by 0. The inverse of an element $a \in R$ with respect to "+" is denoted by $-a$ and the notation $a - b$ stands for $a + (-b)$. When writing expressions, the operation "·" takes precedence over "+" and is usually omitted when no confusion arises.

A *ring with unity* is a ring R in which the operation "·" has a unity element; namely, there is an element $1 \in R$ such that $1 \cdot a = a \cdot 1 = a$ for every $a \in R$.

A *commutative ring* is a ring in which the operation "·" is commutative.

An *integral domain* is a commutative ring with unity in which

$$ab = 0 \quad \Longrightarrow \quad a = 0 \text{ or } b = 0 \, ;$$

namely, there are no *zero divisors*.

A nonempty subset S of a ring R is a *subring* of R if S is a ring with respect to the operations "+" and "·" of R.

A subring I of R is called an *ideal* if for every $r \in R$ and $a \in I$, both ar and ra are in I.

A *field* is a commutative ring in which the nonzero elements form a group with respect to the operation "·".

Problems

Problem A.1 Let $\mathbb{Z}^+$ denote the set of positive integers (excluding zero) and let $\phi : \mathbb{Z}^+ \to \mathbb{Z}^+$ be the Euler function; namely, for every positive integer n,

$$\phi(n) = \left| \left\{ i \in \{1, 2, \ldots, n\} \, : \, \gcd(i, n) = 1 \right\} \right|$$

(where $\gcd(\cdot, \cdot)$ denotes the greatest common divisor and $|\cdot|$ denotes the size of a set).

1. Show that $\phi(p^e) = p^{e-1}(p-1)$ for every prime p and positive integer e.
2. Let $\prod_{j=1}^{s} p_j^{e_j}$ be the factorization of n into distinct primes $p_1, p_2, \ldots, p_s$. Show that

$$\phi(n) = n \cdot \prod_{j=1}^{s} \left(1 - \frac{1}{p_j}\right) .$$

3. Let m and n be positive integers such that $\gcd(m, n) = 1$. Show that $\phi(mn) = \phi(m)\phi(n)$.
4. Show that for every positive divisor m of n,

$$\phi(n/m) = \left|\left\{i \in \{1, 2, \ldots, n\} \ : \ \gcd(i, n) = m\right\}\right| .$$

5. Show that for every positive integer n,

$$\sum_{m \mid n} \phi(m) = \sum_{m \mid n} \phi(n/m) = n ,$$

where the summation is taken over all positive integers m that divide n.

Problem A.2 Define the *Möbius function* $\mu : \mathbb{Z}^+ \to \{-1, 0, 1\}$ as follows. Given a positive integer n, let $\prod_{j=1}^{s} p_j^{e_j}$ be the factorization of n into distinct primes $p_1, p_2, \ldots, p_s$. Then,

$$\mu(n) = \begin{cases} 1 & \text{if } n = 1 \\ (-1)^s & \text{if } e_j = 1 \text{ for } 1 \le j \le s \\ 0 & \text{otherwise} \end{cases} .$$

1. (The Möbius inversion formula) Let $h : \mathbb{Z}^+ \to \mathbb{R}$ and $H : \mathbb{Z}^+ \to \mathbb{R}$ be two real-valued functions defined over the domain of positive integers. Show that the following two conditions are equivalent:

   (i) For every $n \in \mathbb{Z}^+$,

$$H(n) = \sum_{m \mid n} h(m) .$$

   (ii) For every $n \in \mathbb{Z}^+$,

$$h(n) = \sum_{m \mid n} \mu(m) H(n/m) .$$

   Hint: Show that

$$\sum_{m \mid n} \mu(m) = \begin{cases} 1 & \text{if } n = 1 \\ 0 & \text{if } n > 1 \end{cases} .$$

2. Show that

$$\frac{\phi(n)}{n} = \sum_{m \mid n} \frac{\mu(m)}{m} .$$

   Hint: Use part 5 of Problem A.1.

```
r_{-1} ← a; r_0 ← b;
s_{-1} ← 1; s_0 ← 0;
t_{-1} ← 0; t_0 ← 1;
for (i ← 1; r_{i-1} ≠ 0; i++) {
    q_i ← ⌊r_{i-2}/r_{i-1}⌋;
    r_i ← r_{i-2} − q_i r_{i-1};
    s_i ← s_{i-2} − q_i s_{i-1};
    t_i ← t_{i-2} − q_i t_{i-1};
}
```

Figure A.1. Euclid's algorithm for integers.

Problem A.3 (Extended Euclid's algorithm for integers) Let a and b be nonnegative integers, not both zero, and consider the algorithm in Figure A.1 for computing remainders r_i, quotients q_i, and auxiliary values s_i and t_i.

Let ν denote the largest index i for which $r_i \neq 0$. Prove the following four properties (apply induction in properties 1–3):

1. $s_i a + t_i b = r_i$ for $i = -1, 0, \ldots, \nu{+}1$.
2. If c divides both a and b then c divides r_i for $i = -1, 0, \ldots, \nu{+}1$.
3. r_ν divides r_i for $i = \nu{-}1, \nu{-}2, \ldots, -1$.
4. $r_\nu = \gcd(a, b)$.

Problem A.4 Let n be a positive integer. Show that the set $\{0, 1, 2, \ldots, n{-}1\}$, with the operation of integer addition modulo n, is a cyclic group (the sum of two integers modulo n is the remainder obtained when their ordinary sum is divided by n).

Problem A.5 Show that the unity element in a group is unique.

Problem A.6 Show that the inverse of a given element in a group is unique.

Problem A.7 Show that $(a^n)^{-1} = (a^{-1})^n$ for every element a in a group and every positive integer n.

Problem A.8 Show that every element in a finite group has finite order.

Problem A.9 Let a be an element of finite order in a group G.

1. Show that for every positive integer ℓ,

$$a^\ell = 1 \quad \text{if and only if} \quad \mathcal{O}(a) \,|\, \ell\,.$$

Hint: $a^\ell = a^r$, where r is the remainder of ℓ when divided by $\mathcal{O}(a)$.

2. Show that for every positive integer n,

$$\mathcal{O}(a^n) = \frac{\mathcal{O}(a)}{\gcd(\mathcal{O}(a), n)} .$$

Problem A.10 Let a and b be elements with finite orders m and n, respectively, in a commutative group and suppose that $\gcd(m, n) = 1$. Show that $\mathcal{O}(a \cdot b) = mn$.

Hint: First verify that $e = \mathcal{O}(a \cdot b)$ divides mn. Then suppose to the contrary that $e < mn$. Argue that this implies the existence of a prime divisor p of n (say) such that $e \,|\, (mn/p)$. Compute $(ab)^{mn/p}$ and show that it is equal both to 1 and to $b^{mn/p}$.

Problem A.11 Let G be a cyclic group of size n with a generator a and let m be a positive divisor of n. Show that the elements of G of order m are given by a^i, where i ranges over the set

$$\Big\{ i \in \{1, 2, \ldots, n\} \ : \ \gcd(i, n) = n/m \Big\} .$$

Conclude that there are $\phi(m)$ elements of order m in G; in particular, there are $\phi(n)$ generators in G.

Problem A.12 Let G be a group. Show that a nonempty subset H of G is a subgroup of G if and only if every two (possibly equal) elements $a, b \in H$ satisfy $a \cdot b^{-1} \in H$.

Problem A.13 Let H be a nonempty finite subset of a group G and suppose that H is closed under the operation of G. Show that H is a subgroup of G.

Problem A.14 Let H be a subgroup of a group G.

1. Show that congruence modulo H is an equivalence relation; namely, it is reflexive, symmetric, and transitive.
2. Show that the equivalence classes in G of the relation of congruence modulo H are the cosets of H in G.
3. Show that if H is finite then all the cosets of H in G have the same size.
4. (Lagrange's Theorem) Show that if G is finite then $|H|$ divides $|G|$.

Problem A.15 Let H be a normal subgroup of a group G. Show that the cosets of H in G form a group with respect to the operation $(Ha)\cdot(Hb) = (Hab)$. (This group is called the *factor group* or *quotient group* of G by H and is denoted by G/H.)

Problem A.16 Let a be an element of finite order in a group G.

1. Show that the set $\{a^i\}_{i=0}^{\mathcal{O}(a)-1}$ forms a subgroup of G.
2. Show that if G is finite then $\mathcal{O}(a)$ divides $|G|$.

   Hint: Apply Lagrange's Theorem (part 4 of Problem A.14).

Problem A.17 Let n be a positive integer.

1. Show that the set $\{b \in \{1, 2, \ldots, n\} : \gcd(b, n) = 1\}$ with the operation of integer multiplication modulo n is a commutative group.

   Hint: Use part 1 of Problem A.3 to show that every element in the set has an inverse.

2. (The Euler–Fermat Theorem) Let a be an integer such that $\gcd(a, n) = 1$. Show that n divides $a^{\phi(n)} - 1$.

3. (Fermat's Little Theorem) Suppose that n is a prime and let a be an integer. Show that n divides $a^n - a$.

Problem A.18 Show that $0 \cdot a = a \cdot 0 = 0$ for every element a in a ring R.

Problem A.19 Let $\mathbb{Z}$ denote the set of integers with "+" and "·" standing, respectively, for ordinary addition and multiplication of integers.

1. Verify that $\mathbb{Z}$ is an integral domain.

2. Show that for every positive integer m, the integer multiples of m form an ideal in $\mathbb{Z}$. Is this ideal a ring with unity?

Problem A.20 Let m be a positive integer. Show that the $m \times m$ matrices over $\mathbb{R}$ form a ring with unity, where "+" and "·" stand, respectively, for addition and multiplication of matrices.

Problem A.21 For a positive integer n, denote by $\mathbb{Z}_n$ the set $\{0, 1, 2, \ldots, n-1\}$ with the operations "+" and "·" standing, respectively, for addition and multiplication of integers modulo n.

1. Show that $\mathbb{Z}_n$ is a commutative ring with unity.

2. Show that when n is a prime then $\mathbb{Z}_n$ is a field.

Bibliography

(Citing chapters of a reference are listed next to it in bracketed italics.)

[1] L.M. ADLEMAN, *The function field sieve, Proc. 1st Int'l Symp. Algorithmic Number Theory (ANTS-I),* Ithaca, New York (1994), L.M. Adleman, M.-D. A. Huang (Editors), Lecture Notes in Computer Science, Volume 877, Springer, Berlin, 1994, pp. 108–121 *[3]*.

[2] R.L. ADLER, D. COPPERSMITH, M. HASSNER, *Algorithms for sliding block codes—an application of symbolic dynamics to information theory, IEEE Trans. Inform. Theory,* 29 (1983), 5–22 *[14]*.

[3] S.S. AGAIAN, *Hadamard Matrices and Their Applications,* Lecture Notes in Mathematics, Volume 1168, Springer, Berlin, 1985 *[2]*.

[4] M. AGRAWAL, S. BISWAS, *Primality and identity testing via Chinese remaindering, J. ACM,* 50 (2003), 429–443 *[3]*.

[5] M. AGRAWAL, N. KAYAL, N. SAXENA, *PRIMES is in P, Ann. Math.,* 160 (2004), 781–793 *[3]*.

[6] A.V. AHO, J.E. HOPCROFT, J.D. ULLMAN, *The Design and Analysis of Computer Algorithms,* Addison-Wesley, Reading, Massachusetts, 1974 *[3, 5, 6]*.

[7] M. ALEKHNOVICH, *Linear Diophantine equations over polynomials and soft decoding of Reed–Solomon codes, IEEE Trans. Inform. Theory,* 51 (2005), 2257–2265 *[9]*.

[8] N. ALON, *Eigenvalues and expanders, Combinatorica,* 6 (1986), 83–96 *[13]*.

[9] N. ALON, *Eigenvalues, geometric expanders, sorting in rounds, and Ramsey theory, Combinatorica,* 6 (1986), 207–219 *[13]*.

[10] N. ALON, J. BRUCK, J. NAOR, M. NAOR, R.M. ROTH, *Construction of asymptotically good low-rate error-correcting codes through pseudo-random graphs, IEEE Trans. Inform. Theory,* 38 (1992), 509–516 *[12, 13]*.

[11] N. ALON, F.R.K. CHUNG, *Explicit construction of linear sized tolerant networks, Discrete Math.,* 72 (1988), 15–19 *[13]*.

[12] N. ALON, M. LUBY, *A linear time erasure-resilient code with nearly optimal recovery, IEEE Trans. Inform. Theory,* 42 (1996), 1732–1736 *[13]*.

[13] N. ALON, V.D. MILMAN, *Eigenvalues, expanders, and superconcentrators, Proc. 25th Annual IEEE Symp. Foundations of Computer Science (FOCS'1984),* Singer Island, Florida (1984), IEEE Computer Society Press, Los Alamitos, California, 1984, pp. 320–322 *[13]*.

[14] N. ALON, V.D. MILMAN, λ_1*, isoperimetric inequalities for graphs, and superconcentrators, J. Comb. Theory B,* 38 (1985), 73–88 *[13]*.

[15] N. ALON, Y. ROICHMAN, *Random Cayley graphs and expanders, Random Struct. Algorithms,* 5 (1994), 271–284 *[13]*.

[16] N. ALON, J.H. SPENCER, *The Probabilistic Method,* Second Edition, Wiley, New York, 2000 *[13]*.

[17] S. ARORA, L. BABAI, J. STERN, Z. SWEEDYK, *The hardness of approximate optima in lattices, codes, and systems of linear equations, J. Comput. Syst. Sci.,* 54 (1997), 317–331 *[2]*.

[18] E.F. ASSMUS, JR., J.D. KEY, *Designs and their Codes,* Cambridge Tracts in Mathematics, Volume 103, Cambridge University Press, Cambridge, 1992 *[2, 5, 8]*.

[19] J. ASTOLA, *On the non-existence of certain perfect Lee-error-correcting codes, Ann. Univ. Turku A 1,* 167 (1975), 1–13 *[10]*.

[20] J. ASTOLA, *On perfect codes in the Lee metric, Ann. Univ. Turku A 1,* 176 (1978), 1–56 *[10]*.

[21] J. ASTOLA, *A note on perfect Lee-codes over small alphabets, Discrete Appl. Math.,* 4 (1982), 227–228 *[10]*.

[22] J.T. ASTOLA, *An Elias-type bound for Lee-codes over large alphabets and its application to perfect codes, IEEE Trans. Inform. Theory,* 28 (1982), 111–113 *[10]*.

[23] J.T. ASTOLA, *Concatenated codes for the Lee metric, IEEE Trans. Inform. Theory,* 28 (1982), 778–779 *[10]*.

[24] J. ASTOLA, *On the asymptotic behaviour of Lee-codes, Discrete Appl. Math.,* 8 (1984), 13–23 *[10]*.

[25] D. AUGOT, L. PECQUET, *A Hensel lifting to replace factorization in list-decoding of algebraic–geometric and Reed–Solomon codes, IEEE Trans. Inform. Theory,* 46 (2000), 2605–2614 *[9]*.

[26] L. BABAI, *Spectra of Cayley graphs, J. Comb. Theory B,* 27 (1979), 180–189 *[13]*.

[26'] L. BABAI, H. ORAL, K.T. PHELPS, *Eulerian self-dual codes, SIAM J. Discrete Math.,* 7 (1994), 325–330 *[2]*.

[27] A. Barg, G.D. Forney, Jr., *Random codes: minimum distances and error exponents, IEEE Trans. Inform. Theory,* 48 (2002), 2568–2573 *[4].*

[28] A. Barg, J. Justesen, C. Thommesen, *Concatenated codes with fixed inner code and random outer code, IEEE Trans. Inform. Theory,* 47 (2001), 361–365 *[12].*

[29] A. Barg, G. Zémor, *Error exponents of expander codes, IEEE Trans. Inform. Theory,* 48 (2002), 1725–1729 *[13].*

[30] A. Barg, G. Zémor, *Concatenated codes: serial and parallel, IEEE Trans. Inform. Theory,* 51 (2005), 1625–1634 *[13].*

[31] L.A. Bassalygo, *A necessary condition for the existence of perfect codes in the Lee metric, Math. Notes,* 15 (1974), 178–181 *[10].*

[32] M. Ben-Or, *Probabilistic algorithms in finite fields, Proc. 22nd Annual IEEE Symp. Foundations of Computer Science (FOCS'1981),* Nashville, Tennessee (1981), IEEE Computer Society Press, Los Alamitos, California, 1981, pp. 394–398 *[3].*

[33] C. Berge, *Hypergraphs: Combinatorics of Finite Sets,* North-Holland, Amsterdam, 1989 *[13].*

[34] E.R. Berlekamp, *Factoring polynomials over large finite fields, Math. Comput.,* 24 (1970), 713–735 *[3].*

[35] E.R. Berlekamp, *Long primitive binary BCH codes have distance* $d \sim 2n \ln R^{-1}/\log n \cdots$, *IEEE Trans. Inform. Theory,* 18 (1972), 415–426 *[5, 8].*

[36] E.R. Berlekamp, *Algebraic Coding Theory,* Revised Edition, Aegean Park Press, Laguna Hills, California, 1984 *[Prf., 2, 3, 5, 6, 8, 10].*

[37] E.R. Berlekamp, *Bounded distance +1 soft-decision Reed–Solomon decoding, IEEE Trans. Inform. Theory,* 42 (1996), 704–720 *[6, 9].*

[38] E.R. Berlekamp, R.J. McEliece, H.C.A. van Tilborg, *On the inherent intractability of certain coding problems, IEEE Trans. Inform. Theory,* 24 (1978), 384–386 *[2].*

[39] E.R. Berlekamp, H. Rumsey, G. Solomon, *On the solution of algebraic equations over finite fields, Inform. Control,* 10 (1967), 553–564 *[3].*

[40] S.D. Berman, *On the theory of group codes, Cybernetics,* 3 (1967), 25–31 *[8].*

[41] C. Berrou, A. Glavieux, *Near optimum error correcting coding and decoding: turbo-codes, IEEE Trans. Commun.,* 44 (1996), 1261–1271 *[14].*

[42] C. Berrou, A. Glavieux, P. Thitimajshima, *Near Shannon limit error-correcting coding and decoding: turbo-codes, Conf. Record 1993 IEEE Int'l Conf. Communications (ICC'1993),* Geneva, Switzerland (1993), pp. 1064–1070 *[14].*

[43] A. BHATTACHARYYA, *On a measure of divergence between two statistical populations defined by their probability distributions, Bull. Calcutta Math. Soc.,* 35 (1943), 99–110 *[1].*

[44] N.L. BIGGS, *Discrete Mathematics,* Second Edition, Oxford University Press, Oxford, 2002 *[11].*

[45] S.R. BLACKBURN, *Fast rational interpolation, Reed–Solomon decoding, and the linear complexity profile of sequences, IEEE Trans. Inform. Theory,* 43 (1997), 537–548 *[9].*

[46] R.E. BLAHUT, *Theory and Practice of Error Control Codes,* Addison-Wesley, Reading, Massachusetts, 1983 *[Prf., 3, 6, 14].*

[47] R.E. BLAHUT, *A universal Reed–Solomon decoder, IBM J. Res. Develop.,* 28 (1984), 150–158 *[3, 6].*

[48] R.E. BLAHUT, *Algebraic Methods for Signal Processing and Communications Coding,* Springer, New York, 1992 *[3].*

[49] I.F. BLAKE, R.C. MULLIN, *The Mathematical Theory of Coding,* Academic Press, New York, 1975 *[Prf.].*

[50] M. BLAUM, J. BRADY, J. BRUCK, J. MENON, *EVENODD: an efficient scheme for tolerating double disk failures in RAID architectures, IEEE Trans. Comput.,* 445 (1995), 192–202 *[11].*

[51] M. BLAUM, J. BRUCK, *MDS array codes for correcting a single criss-cross error, IEEE Trans. Inform. Theory,* 46 (2000), 1068–1077 *[11].*

[52] M. BLAUM, J. BRUCK, A. VARDY, *MDS array codes with independent parity symbols, IEEE Trans. Inform. Theory,* 42 (1996), 529–542 *[11].*

[53] M. BLAUM, R.M. ROTH, *New array codes for multiple phased burst correction, IEEE Trans. Inform. Theory,* 39 (1993), 66–77 *[11].*

[54] M. BLAUM, R.M. ROTH, *On lowest density MDS codes, IEEE Trans. Inform. Theory,* 45 (1999), 46–59 *[11].*

[55] V.M. BLINOVSKII, *Covering the Hamming space with sets translated by linear code vectors, Probl. Inform. Transm.,* 26 (1990), 196–201 *[4].*

[56] E.L. BLOKH, V.V. ZYABLOV, *Existence of linear concatenated binary codes with optimal correcting properties, Probl. Inform. Transm.,* 9 (1973), 271–276 *[12].*

[57] E.L. BLOKH, V.V. ZYABLOV, *Coding of generalized concatenated codes, Probl. Inform. Transm.,* 10 (1974), 218–222 *[5, 12].*

[58] E.L. BLOKH, V.V. ZYABLOV, *Linear Concatenated Codes,* Nauka, Moscow, 1982 (in Russian) *[5, 12].*

[59] M.A. de Boer, *Almost MDS codes, Designs Codes Cryptogr.*, 9 (1996), 143–155 *[11]*.

[60] B. Bollobás, *Modern Graph Theory,* Springer, New York, 1998 *[13]*.

[61] R.C. Bose, D.K. Ray-Chaudhuri, *On a class of error correcting binary group codes, Inform. Control,* 3 (1960), 68–79 *[5]*.

[62] R.C. Bose, D.K. Ray-Chaudhuri, *Further results on error correcting binary group codes, Inform. Control,* 3 (1960), 279–290 *[5]*.

[63] P.A.H. Bours, *Construction of fixed-length insertion/deletion correcting runlength-limited codes, IEEE Trans. Inform. Theory,* 40 (1994), 1841–1856 *[10]*.

[64] D.W. Boyd, *On a problem of Byrnes concerning polynomials with restricted coefficients, Math. Comput.*, 66 (1997), 1697–1703 *[10]*.

[65] D.W. Boyd, *On a problem of Byrnes concerning polynomials with restricted coefficients, II, Math. Comput.*, 71 (2002), 1205–1217 *[10]*.

[66] D. Le Brigand, *On computational complexity of some algebraic curves over finite fields, Proc. 3rd Int'l Conf. Algebraic Algorithms and Error Correcting Codes (AAECC-3),* Grenoble, France (1985), J. Calmet (Editor), Lecture Notes in Computer Science, Volume 229, Springer, Berlin, 1986, pp. 223–227 *[4]*.

[67] R.A. Brualdi, H.J. Ryser, *Combinatorial Matrix Theory,* Cambridge University Press, Cambridge, 1991 *[11, 13]*.

[68] J. Bruck, M. Naor, *The hardness of decoding linear codes with preprocessing, IEEE Trans. Inform. Theory,* 36 (1990), 381–385 *[2]*.

[69] D. Burshtein, M. Krivelevich, S. Litsyn, G. Miller, *Upper bounds on the rate of LDPC codes, IEEE Trans. Inform. Theory,* 48 (2002), 2437–2449 *[13]*.

[70] D. Burshtein, G. Miller, *Expander graph arguments for message-passing algorithms, IEEE Trans. Inform. Theory,* 47 (2001), 782–790 *[13]*.

[71] D. Burshtein, G. Miller, *Bounds on the performance of belief propagation decoding, IEEE Trans. Inform. Theory,* 48 (2002), 112–122 *[13]*.

[72] K.A. Bush, *Orthogonal arrays of index unity, Ann. Math. Stat.*, 23 (1952) 426–434 *[5]*.

[73] L. Carlitz, S. Uchiyama, *Bounds for exponential sums, Duke Math. J.*, 24 (1957), 37–41 *[5]*.

[74] L.R.A. Casse, *A solution to Beniamino Segre's "Problem $I_{r,q}$" for q even, Atti Accad. Naz. Lincei Rend.*, 46 (1969), 13–20 *[11]*.

[75] L.R.A. CASSE, D.G. GLYNN, *The solution to Beniamino Segre's problem* $I_{r,q}$, $r = 3$, $q = 2^h$, *Geom. Dedic.*, 13 (1982), 157–163 *[11]*.

[76] G. CASTAGNOLI, J.L. MASSEY, P.A. SCHOELLER, N. VON SEEMANN, *On repeated-root cyclic codes, IEEE Trans. Inform. Theory,* 37 (1991), 337–342 *[8]*.

[77] U. CHENG, *On the continued fraction and Berlekamp's algorithm, IEEE Trans. Inform. Theory,* 30 (1984), 541–544 *[6]*.

[78] J.C.-Y. CHIANG, J.K. WOLF, *On channels and codes for the Lee metric, Inform. Control,* 19 (1971), 159–173 *[10]*.

[79] R.T. CHIEN, *Cyclic decoding procedures for Bose–Chaudhuri–Hocquenghem codes, IEEE Trans. Inform. Theory,* 10 (1964), 357–363 *[6]*.

[80] D.V. CHUDNOVSKY, G.V. CHUDNOVSKY, *Algebraic complexity and algebraic curves over finite fields, J. Complexity,* 4 (1988), 285–316 *[3]*.

[81] G. COHEN, I. HONKALA, S. LITSYN, A. LOBSTEIN, *Covering Codes,* North-Holland, Amsterdam, 1997 *[4]*.

[82] G.D. COHEN, M.G. KARPOVSKY, H.F. MATTSON, JR., J.R. SCHATZ, *Covering radius—survey and recent results, IEEE Trans. Inform. Theory,* 31 (1985), 328–343 *[4]*.

[83] G.D. COHEN, A.C. LOBSTEIN, N.J.A. SLOANE, *Further results on the covering radius of codes, IEEE Trans. Inform. Theory,* 32 (1986), 680–694 *[4]*.

[84] C.J. COLBOURN, J.H. DINITZ (EDITORS), *The CRC Handbook of Combinatorial Designs,* CRC Press, New York, 1996 *[11]*.

[85] D. COPPERSMITH, *Fast evaluation of logarithms in fields of characteristic two, IEEE Trans. Inform. Theory,* 30 (1984), 587–594 *[3]*.

[86] D. COPPERSMITH, A.M. ODLYZKO, R. SCHROEPPEL, *Discrete logarithms in* GF(p), *Algorithmica,* 1 (1986), 1–15 *[3]*.

[87] T.M. COVER, J.A. THOMAS, *Elements of Information Theory,* Wiley, New York, 1991 *[1, 4]*.

[88] I. CSISZÁR, J. KÖRNER, *Information Theory: Coding Theorems for Discrete Memoryless Systems,* Second Edition, Akadémiai Kiadó, Budapest, 1997 *[1, 4]*.

[89] D. DABIRI, I.F. BLAKE, *Fast parallel algorithms for decoding Reed–Solomon codes based on remainder polynomials, IEEE Trans. Inform. Theory,* 41 (1995), 873–885 *[9]*.

[90] H. DAVENPORT, *Multiplicative Number Theory,* Third Edition, revised by H.L. Montgomery, Springer, New York, 2000 *[13]*.

[91] P.J. DAVIS, *Circulant Matrices,* Second Edition, Chelsea, New York, 1994 *[10].*

[92] V.A. DAVYDOV, *Codes correcting errors in the modulus metric, Lee metric, and operator errors, Probl. Inform. Transm.,* 29 (1993), 208–216 *[10].*

[93] P. DELSARTE, *Bounds for unrestricted codes, by linear programming, Philips Res. Rep.,* 27 (1972), 272–289 *[4].*

[94] P. DELSARTE, *An algebraic approach to the association schemes of coding theory, Philips Res. Rep. Suppl.,* 10 (1973) *[4].*

[95] P. DELSARTE, *Four fundamental parameters of a code and their combinatorial significance, Inform. Control,* 23 (1973), 407–438 *[4].*

[96] P. DELSARTE, *On subfield subcodes of modified Reed–Solomon codes, IEEE Trans. Inform. Theory,* 21 (1975), 575–576 *[5].*

[97] P. DELSARTE, *Bilinear forms over a finite field, with applications to coding theory, J. Comb. Theory A,* 25 (1978), 226–241 *[11].*

[98] P. DELSARTE, J.-M. GOETHALS, *Alternating bilinear forms over* GF(q), *J. Comb. Theory A,* 19 (1975), 26–50 *[10].*

[99] P. DELSARTE, P. PIRET, *Algebraic constructions of Shannon codes for regular channels, IEEE Trans. Inform. Theory,* 28 (1982), 593–599 *[12].*

[100] P. DELSARTE, P. PIRET, *Do most binary linear codes achieve the Goblick bound on the covering radius?, IEEE Trans. Inform. Theory,* 32 (1986), 826–828 *[4].*

[101] J. DÉNES, A.D. KEEDWELL, *Latin Squares and their Applications,* Academic Press, New York, 1974 *[11].*

[102] J. DÉNES, A.D. KEEDWELL, *Latin Squares—New Developments in the Theory and Applications,* Annals of Discrete Mathematics, Volume 46, North-Holland, Amsterdam, 1991 *[11].*

[103] O. DEUTSCH, *Decoding methods for Reed–Solomon codes over polynomial rings,* M.Sc. dissertation, Computer Science Department, Technion, Haifa, Israel, 1994 (in Hebrew) *[6, 11].*

[104] R. DIESTEL, *Graph Theory,* Second Edition, Springer, New York, 2000 *[13].*

[105] C. DING, T. HELLESETH, W. SHAN, *On the linear complexity of Legendre sequences, IEEE Trans. Inform. Theory,* 44 (1998), 1276–1278 *[7].*

[106] S.M. DODUNEKOV, I.N. LANDJEV, *On near-MDS codes, J. Geom.,* 54 (1995), 30–43 *[11].*

[107] S.M. DODUNEKOV, I.N. LANDJEV, *Near-MDS codes over some small fields, Discrete Math.,* 213 (2000), 55–65 *[11].*

[108] J.L. DORNSTETTER, *On the equivalence between Berlekamp's and Euclid's algorithms, IEEE Trans. Inform. Theory,* 33 (1987), 428–431 *[6].*

[109] I. DUMER, *Concatenated codes and their multilevel generalizations,* in *Handbook of Coding Theory, Volume II,* V.S. Pless, W.C. Huffman (Editors), North-Holland, Amsterdam, 1998, pp. 1911–1988 *[12].*

[110] I. DUMER, D. MICCIANCIO, M. SUDAN, *Hardness of approximating the minimum distance of a linear code, IEEE Trans. Inform. Theory,* 49 (2003), 22–37 *[2].*

[111] A. DÜR, *The automorphism groups of Reed–Solomon codes, J. Comb. Theory A,* 44 (1987), 69–82 *[5].*

[112] A. DÜR, *On linear MDS codes of length* $q+1$ *over* GF(q) *for even* q, *J. Comb. Theory A,* 49 (1988), 172–174 *[8].*

[113] E. ELEFTHERIOU, R. CIDECIYAN, *On codes satisfying* M*th-order running digital sum constraints, IEEE Trans. Inform. Theory,* 37 (1991), 1294–1313 *[10].*

[114] P. ELIAS, *Error-free coding, IRE Trans. Inform. Theory,* 4 (1954), 29–37 *[5].*

[115] P. ELIAS, *Coding for noisy channels, IRE Int'l Conv. Rec.,* 1955, 37–46 *[14].*

[116] P. ELIAS, *Error-correcting codes for list decoding, IEEE Trans. Inform. Theory,* 37 (1991), 5–12 *[9].*

[117] T. ETZION, A. VARDY, *Perfect binary codes: constructions, properties, and enumeration, IEEE Trans. Inform. Theory,* 40 (1994), 754–763 *[4].*

[118] A. FALDUM, W. WILLEMS, *Codes of small defect, Designs Codes Cryptogr.,* 10 (1997), 341–350 *[11].*

[119] G. FALKNER, W. HEISE, B. KOWOL, E. ZEHENDNER, *On the existence of cyclic optimal codes, Atti. Sem. Mat. Fis. Univ. Modena,* 28 (1979), 326–341 *[8].*

[120] R.M. FANO, *A heuristic discussion of probabilistic decoding, IEEE Trans. Inform. Theory,* 9 (1963), 64–74 *[14].*

[121] P.G. FARRELL, *A survey of array error control codes, Eur. Trans. Telecommun. Relat. Technol.,* 3 (1992), 441–454 *[5].*

[122] U. FEIGE, D. MICCIANCIO, *The inapproximability of lattice and coding problems with preprocessing, Proc. 17th Annual IEEE Conf. Computational Complexity (CCC'2002),* Montréal, Québec (2002), IEEE Computer Society Press, Los Alamitos, California, 2002, pp. 44–52 *[2].*

[123] G.-L. FENG, *Two fast algorithms in the Sudan decoding procedure, 37th Annual Allerton Conf. Communication, Control, and Computing,* Urbana-Champaign, Illinois (1999), pp. 545–554 *[9].*

[124] G.-L. FENG, K.K. TZENG, *A generalized Euclidean algorithm for multisequence shift-register synthesis, IEEE Trans. Inform. Theory,* 35 (1989) 584–594 *[6].*

[125] G.-L. FENG, K.K. TZENG, *A generalization of the Berlekamp–Massey algorithm for multisequence shift-register synthesis with applications to decoding cyclic codes, IEEE Trans. Inform. Theory,* 37 (1991) 1274–1287 *[6].*

[126] P. FITZPATRICK, G.H. NORTON, *Finding a basis for the characteristic ideal of an n-dimensional linear recurring sequence, IEEE Trans. Inform. Theory,* 36 (1990), 1480–1487 *[6].*

[127] P. FITZPATRICK, G.H. NORTON, *The Berlekamp–Massey algorithm and linear recurring sequences over a factorial domain, Appl. Algebra Eng. Commun. Comput.,* 6 (1995) 309–323 *[6].*

[128] G.D. FORNEY, JR., *On decoding BCH codes, IEEE Trans. Inform. Theory,* 11 (1965) 549–557 *[6].*

[129] G.D. FORNEY, JR., *Concatenated Codes,* MIT Press, Cambridge, Massachusetts, 1966 *[4, 5, 12].*

[130] G.D. FORNEY, JR., *Generalized minimum distance decoding, IEEE Trans. Inform. Theory,* 12 (1966) 125–131 *[12].*

[131] G.D. FORNEY, JR., *Convolutional codes I: algebraic structure, IEEE Trans. Inform. Theory,* 16 (1970), 720–738 (see the correction in *IEEE Trans. Inform. Theory,* 17 (1971), 360) *[14].*

[132] G.D. FORNEY, JR., *Convolutional codes II: maximum-likelihood decoding, Inform. Control,* 25 (1974), 222–266 *[14].*

[133] G.D. FORNEY, JR., *Convolutional codes III: sequential decoding, Inform. Control,* 25 (1974), 267–297 *[14].*

[134] G. FREIMAN, S. LITSYN, *Asymptotically exact bounds on the size of high-order spectral-null codes, IEEE Trans. Inform. Theory,* 45 (1999), 1798–1807 *[10].*

[135] E.M. GABIDULIN, *Theory of codes with maximum rank distance, Probl. Inform. Transm.,* 21 (1985), 1–12 *[11].*

[136] E.M. GABIDULIN, *Optimum codes correcting lattice errors, Problemy Peredachi Informatsii,* 21 No. 2 (April–June 1985), 103–108 (in Russian) *[11].*

[137] E.M. GABIDULIN, V.I. KORZHIK, *Codes correcting lattice-pattern errors, Izvestiya VUZ Radioelektronika,* 15 (1972), 492–498 (in Russian) *[11].*

[138] R.G. GALLAGER, *Low-density parity-check codes, IRE Trans. Inform. Theory,* 8 (1962), 21–28 *[13].*

[139] R.G. GALLAGER, *Low-Density Parity-Check Codes,* MIT Press, Cambridge, Massachusetts, 1963 *[13]*.

[140] R.G. GALLAGER, *Information Theory and Reliable Communication,* Wiley, New York, 1968 *[1, 4]*.

[141] F.R. GANTMACHER, *Matrix Theory, Volume II,* Chelsea, New York, 1960 *[13]*.

[142] S. GAO, M.A. SHOKROLLAHI, *Computing roots of polynomials over function fields of curves,* in *Coding Theory and Cryptography: from Enigma and Geheimschreiber to Quantum Theory,* D. Joyner (Editor), Springer, Berlin, 2000, pp. 214–228 *[9]*.

[143] M.R. GAREY, D.S. JOHNSON, *Computers and Intractability: a Guide to the Theory of NP-Completeness,* Freeman, New York, 1979 *[2]*.

[144] J. VON ZUR GATHEN, J. GERHARD, *Modern Computer Algebra,* Cambridge University Press, Cambridge, 1999 *[3, 5, 6]*.

[145] J. GEORGIADES, *Cyclic* $(q+1, k)$*-codes of odd order* q *and even dimension* k *are not optimal, Atti. Sem. Mat. Fis. Univ. Modena,* 30 (1982), 284–285 *[8]*.

[146] E.N. GILBERT, *A comparison of signalling alphabets, Bell Syst. Tech. J.,* 31 (1952), 504–522 *[4]*.

[147] D.G. GLYNN, *The non-classical* 10*-arc of* PG$(4, 9)$, *Discrete Math.,* 59 (1986), 43–51 *[11]*.

[148] T.J. GOBLICK, JR., *Coding for a discrete information source with a distortion measure,* Ph.D. dissertation, Department of Electrical Engineering, Massachusetts Institute of Technology, Cambridge, Massachusetts, 1962 *[4]*.

[149] M.J.E. GOLAY, *Notes on digital coding, Proc. IEEE,* 37 (1949), 657 *[2, 4]*.

[150] M.J.E. GOLAY, *Anent codes, priorities, patents, etc., Proc. IEEE,* 64 (1976), 572 *[2]*.

[151] O. GOLDREICH, R. RUBINFELD, M. SUDAN, *Learning polynomials with queries: the highly noisy case, SIAM J. Discrete Math.,* 13 (2000), 535–570 *[4, 9]*.

[152] S.W. GOLOMB, *Shift Register Sequences,* Revised Edition, Aegean Park Press, Laguna Hills, California, 1982 *[3, 6, 7]*.

[153] S.W. GOLOMB, E.C. POSNER, *Rook domains, Latin squares, affine planes, and error-distribution codes, IEEE Trans. Inform. Theory,* 10 (1964), 196–208 *[11]*.

[154] S.W. GOLOMB, L.R. WELCH, *Algebraic coding and the Lee metric,* in *Error Correcting Codes,* H.B. Mann (Editor), Wiley, New York, 1968, pp. 175–194 *[10]*.

[155] S.W. GOLOMB, L.R. WELCH, *Perfect codes in the Lee metric and the packing of polyominoes*, *SIAM J. Appl. Math.*, 18 (1970), 302–317 *[10]*.

[156] V.D. GOPPA, *A new class of linear correcting codes*, *Probl. Inform. Transm.*, 6 (1970), 207–212 *[5, 12]*.

[157] V.D. GOPPA, *A rational representation of codes and (L, g)-codes*, *Probl. Inform. Transm.*, 7 (1971), 223–229 *[5, 12]*.

[158] V.D. GOPPA, *Binary symmetric channel capacity is attained with irreducible codes*, *Probl. Inform. Transm.*, 10 (1974), 89–90 *[5, 12]*.

[159] D.M. GORDON, *Discrete logarithms in* GF(p) *using the number field sieve*, *SIAM J. Discrete Math.*, 6 (1993), 124–138 *[3]*.

[160] D.C. GORENSTEIN, N. ZIERLER, *A class of error-correcting codes in p^m symbols*, *J. Soc. Ind. Appl. Math.*, 9 (1961), 207–214 *[5, 6]*.

[161] R.L. GRAHAM, N.J.A. SLOANE, *On the covering radius of codes*, *IEEE Trans. Inform. Theory*, 31 (1985), 385–401 *[4]*.

[162] S. GRAVIER, M. MOLLARD, C. PAYAN, *On the non-existence of 3-dimensional tiling in the Lee metric*, *Europ. J. Combinatorics*, 19 (1998), 567–572 *[10]*.

[163] J.H. GRIESMER, *A bound for error-correcting codes*, *IBM J. Res. Develop.*, 4 (1960), 532–542 *[4]*.

[164] B.R. GULATI, E.G. KOUNIAS, *On bounds useful in the theory of symmetrical factorial designs*, *J. Roy. Statist. Soc. B*, 32 (1970), 123–133 *[11]*.

[165] V. GURUSWAMI, P. INDYK, *Expander-based constructions of efficiently decodable codes*, *Proc. 42nd Annual IEEE Symp. Foundations of Computer Science (FOCS'2001)*, Las Vegas, Nevada (2001), IEEE Computer Society Press, Los Alamitos, California, 2001, pp. 658–667 *[13]*.

[166] V. GURUSWAMI, P. INDYK, *Linear-time codes to correct a maximum possible fraction of errors*, *39th Annual Allerton Conf. Communication, Control, and Computing*, Urbana-Champaign, Illinois (2001) *[13]*.

[167] V. GURUSWAMI, P. INDYK, *Near-optimal linear-time codes for unique decoding and new list-decodable codes over smaller alphabets*, *Proc. 34th Annual ACM Symp. Theory of Computing (STOC'2002)*, Montréal, Québec (2002), ACM, New York, 2002, pp. 812–821 *[13]*.

[168] V. GURUSWAMI, M. SUDAN, *Improved decoding of Reed–Solomon and algebraic–geometry codes*, *IEEE Trans. Inform. Theory*, 45 (1999), 1757–1767 *[9]*.

[169] R.W. HAMMING, *Error detecting and error correcting codes*, *Bell Syst. Tech. J.*, 26 (1950), 147–160 *[2, 4]*.

[170] A.R. HAMMONS, JR., P.V. KUMAR, A.R. CALDERBANK, N.J.A. SLOANE, P. SOLÉ, *The $\mathbb{Z}_4$-linearity of Kerdock, Preparata, Goethals, and related codes, IEEE Trans. Inform. Theory,* 40 (1994), 301–319 *[10].*

[171] G.H. HARDY, E.M. WRIGHT, *An Introduction to the Theory of Numbers,* Fifth Edition, Oxford University Press, Oxford, 1979 *[10, 13].*

[172] C.R.P. HARTMANN, K.K. TZENG, *Generalizations of the BCH bound, Inform. Control,* 20 (1972), 489–498 *[8].*

[173] A. HASAN, V.K. BHARGAVA, T. LE-NGOC, *Algorithms and architectures for the design of a VLSI Reed–Solomon codec,* in *Reed–Solomon Codes and their Applications,* S.B. Wicker, V.K. Bhargava (Editors), IEEE Press, New York, 1994, 60–107 *[6].*

[174] H.J. HELGERT, *Alternant codes, Inform. Control,* 26 (1974), 369–380 *[5].*

[175] A.E. HEYDTMANN, J.M. JENSEN, *On the equivalence of the Berlekamp–Massey and the Euclidean algorithms for decoding, IEEE Trans. Inform. Theory,* 46 (2000), 2614–2624 *[6].*

[176] H.M. HILDEN, D.G. HOWE, E.J. WELDON, JR., *Shift error correcting modulation codes, IEEE Trans. Magn.,* 27 (1991), 4600–4605 *[10].*

[177] F.S. HILLIER, G.J. LIEBERMAN, *Introduction to Operations Research,* Seventh Edition, McGraw-Hill, Boston, Massachusetts, 2001 *[14].*

[178] S. HIRASAWA, M. KASAHARA, Y. SUGIYAMA, T. NAMEKAWA, *Certain generalizations of concatenated codes—exponential error bounds and decoding complexity, IEEE Trans. Inform. Theory,* 26 (1980), 527–534 *[5, 12].*

[179] S. HIRASAWA, M. KASAHARA, Y. SUGIYAMA, T. NAMEKAWA, *An improvement of error exponents at low rates for the generalized version of concatenated codes, IEEE Trans. Inform. Theory,* 27 (1981), 350–352 *[5, 12].*

[180] S. HIRASAWA, M. KASAHARA, Y. SUGIYAMA, T. NAMEKAWA, *Modified product codes, IEEE Trans. Inform. Theory,* 30 (1984), 299–306 *[5].*

[181] J.W.P. HIRSCHFELD, *Projective Geometries over Finite Fields,* Second Edition, Oxford University Press, Oxford, 1998 *[11].*

[182] J.W.P. HIRSCHFELD, G. KORCHMÁROS, *On the embedding of an arc into a conic in a finite plane, Finite Fields Appl.,* 2 (1996), 274–292 *[11].*

[183] J.W.P. HIRSCHFELD, G. KORCHMÁROS, *On the number of rational points on an algebraic curve over a finite field, Bull. Belg. Math. Soc. Simon Stevin,* 5 (1998), 313–340 *[11].*

[184] J.W.P. HIRSCHFELD, L. STORME, *The packing problem in statistics, coding theory, and finite projective spaces, J. Statist. Planning Infer.,* 72 (1998), 355–380 *[11].*

[185] J.W.P. HIRSCHFELD, L. STORME, *The packing problem in statistics, coding theory and finite projective spaces: update 2001,* in *Finite Geometries: Proc. 4th Isle of Thorns Conference,* Chelwood Gate, UK (2000), A. Blokhuis, J.W.P. Hirschfeld, D. Jungnickel, J.A. Thas (Editors), Developments in Mathematics, Volume 3, Kluwer, Dordrecht, 2001, pp. 201–246 *[11]*.

[186] J.W.P. HIRSCHFELD, J.A. THAS, *General Galois Geometries,* Oxford University Press, Oxford, 1991 *[2, 11]*.

[187] A. HOCQUENGHEM, *Codes correcteurs d'erreurs, Chiffres,* 2 (1959), 147–156 *[5]*.

[188] J.E. HOPCROFT, R. MOTWANI, J.D. ULLMAN, *Introduction to Automata Theory, Languages, and Computation,* Second Edition, Addison–Wesley, Boston, Massachusetts, 2001 *[14]*.

[189] L.-K. HUA, *Introduction to Number Theory,* Springer, Berlin, 1982 *[10]*.

[190] M.N. HUXLEY, *The Distribution of Prime Numbers,* Oxford University Press, London, 1972 *[13]*.

[191] I. IIZUKA, M. KASAHARA, T. NAMEKAWA, *Block codes capable of correcting both additive and timing errors, IEEE Trans. Inform. Theory,* 26 (1980), 393–400 *[10]*.

[192] K.A.S. IMMINK, *Coding Techniques for Digital Recorders,* Prentice-Hall, New York, 1991 *[5, 10]*.

[193] K.A.S. IMMINK, *Codes for Mass Data Storage Systems,* Second Edition, Shannon Foundation Publishers, Eindhoven, The Netherlands, 2004 *[10]*.

[194] K.A.S. IMMINK, G.F.M. BEENKER, *Binary transmission codes with higher order spectral zeros at zero frequency, IEEE Trans. Inform. Theory,* 33 (1987), 452–454 *[10]*.

[195] K. IRELAND, M. ROSEN, *A Classical Introduction to Modern Number Theory,* Second Edition, Springer, New York, 1990 *[13]*.

[196] F. JELINEK, *Fast sequential decoding algorithm using a stack, IBM J. Res. Develop.,* 13 (1969), 675–685 *[14]*.

[197] R. JOHANNESSON, K.S. ZIGANGIROV, *Fundamentals of Convolutional Coding,* IEEE Press, New York, 1999 *[14]*.

[198] S.M. JOHNSON, *A new upper bound for error-correcting codes, IRE Trans. Inform. Theory,* 8 (1962), 203–207 *[4]*.

[199] J. JUSTESEN, *A class of constructive asymptotically good algebraic codes, IEEE Trans. Inform. Theory,* 18 (1972), 652–656 *[12]*.

[200] J. JUSTESEN, *On the complexity of decoding Reed–Solomon codes, IEEE Trans. Inform. Theory,* 22 (1976), 237–238 *[6]*.

[201] J. JUSTESEN, T. HØHOLDT, *Bounds on list decoding of MDS codes, IEEE Trans. Inform. Theory,* 47 (2001), 1604–1609 *[9]*.

[202] T. KAILATH, *Linear Systems,* Prentice-Hall, Englewood Cliffs, New Jersey, 1980 *[14]*.

[203] M. KAMINSKI, D.G. KIRKPATRICK, N.H. BSHOUTY, *Addition requirements for matrix and transposed matrix products, J. Algorithms,* 9 (1988), 354–364 *[6]*.

[204] R. KARABED, P.H. SIEGEL, *Matched spectral-null codes for partial-response channels, IEEE Trans. Inform. Theory,* 37 (1991), 818–855 *[10]*.

[205] M. KASAHARA, Y. SUGIYAMA, S. HIRASAWA, T. NAMEKAWA, *New classes of binary codes constructed on the basis of concatenated codes and product codes, IEEE Trans. Inform. Theory,* 22 (1976), 462–468 *[5, 12]*.

[206] T. KASAMI, *An upper bound on k/n for affine-invariant codes with fixed d/n, IEEE Trans. Inform. Theory,* 15 (1969), 174–176 *[12]*.

[207] T. KASAMI, S. LIN, W.W. PETERSON, *New generalizations of the Reed–Muller codes—part I: primitive codes, IEEE Trans. Inform. Theory,* 14 (1968), 189-199 *[8]*.

[208] G.L. KATSMAN, M.A. TSFASMAN, S.G. VLĂDUT, *Modular curves and codes with a polynomial construction, IEEE Trans. Inform. Theory,* 30 (1984), 353–355 *[4]*.

[209] A.M. KERDOCK, *A class of low-rate nonlinear codes, Inform. Control,* 20 (1972), 182–187 *[10]*.

[210] O. KEREN, S.N. LITSYN, *A class of array codes correcting multiple column erasures, IEEE Trans. Inform. Theory,* 43 (1997), 1843–1851 *[11]*.

[211] O. KEREN, S.N. LITSYN, *Codes correcting phased burst erasures, IEEE Trans. Inform. Theory,* 44 (1998), 416–420 *[11]*.

[212] L. KHACHIYAN, *On the complexity of approximating extremal determinants in matrices, J. Complexity,* 11 (1995), 138–153 *[2, 5]*.

[213] J.-H. KIM, H.-Y. SONG, *Trace representation of Legendre sequences, Designs Codes Cryptogr.,* 24 (2001), 343–348 *[7]*.

[214] J.F.C. KINGMAN, *The exponential decay of Markov transition probabilities, Proc. Lond. Math. Soc.,* 13 (1963), 337–358 *[10]*.

[215] D.E. KNUTH, *The Art of Computer Programming, Volume 2: Seminumerical Algorithms,* Third Edition, Addison-Wesley, Reading, Massachusetts, 1998 *[3]*.

[216] R. KOETTER, A. VARDY, *Algebraic soft-decision decoding of Reed–Solomon codes, IEEE Trans. Inform. Theory,* 49 (2003), 2809–2825 *[9]*.

[217] Z. KOHAVI, *Switching and Finite Automata Theory,* Second Edition, McGraw-Hill, New York, 1978 *[14]*.

[218] V.D. KOLESNIK, E.T. MIRONCHIKOV, *Cyclic Reed–Muller codes and their decoding, Probl. Inform. Transm.,* 4 (1968), 15–19 *[8]*.

[219] M. KRAWTCHOUK, *Sur une généralisation des polynomes d'Hermite, Comptes Rendus,* 189 (1929), 620–622 *[4]*.

[220] A.V. KUZNETSOV, A.J.H. VINCK, *A coding scheme for single peak-shift correction in (d, k)-constrained channels, IEEE Trans. Inform. Theory,* 39 (1993), 1444–1450 *[10]*.

[221] H. KWAKERNAAK, R. SIVAN, *Modern Signals and Systems,* Prentice-Hall, Englewood Cliffs, New Jersey, 1991 *[14]*.

[222] C.Y. LEE, *Some properties of nonbinary error-correcting codes, IRE Trans. Inform. Theory,* 4 (1958), 77–82 *[10]*.

[223] T. LEPISTÖ, *A modification of Elias-bound and nonexistence theorem for perfect codes in the Lee-metric, Inform. Control,* 49 (1981), 109–124 *[10]*.

[224] T. LEPISTÖ, *A note on perfect Lee-codes over small alphabets, Discrete Appl. Math.,* 3 (1981), 73–74 *[10]*.

[225] V.I. LEVENSHTEIN, *Binary codes capable of correcting deletions, insertions, and reversals, Soviet Physics—Doklady,* 10 (1966), 707–710 *[10]*.

[226] V.I. LEVENSHTEIN, *One method of constructing quasilinear codes providing synchronization in the presence of errors, Probl. Inform. Transm.,* 7 (1971), 215–222 *[10]*.

[227] V.I. LEVENSHTEIN, *On perfect codes in deletion and insertion metric, Discrete Math. Appl.,* 2 (1992), 241–258 *[10]*.

[228] V.I. LEVENSHTEIN, A.J.H. VINCK, *Perfect (d, k)-codes capable of correcting single peak-shifts, IEEE Trans. Inform. Theory,* 39 (1993), 656–662 *[10]*.

[229] R. LIDL, H. NIEDERREITER, *Finite Fields,* Second Edition, Cambridge University Press, Cambridge, 1997 *[3, 5, 7, 10]*.

[230] S. LIN, D.J. COSTELLO, JR., *Error Control Coding: Fundamentals and Applications,* Prentice-Hall, Englewood Cliffs, New Jersey, 1983 *[Prf., 4, 5, 14]*.

[231] S. LIN, E.J. WELDON, JR., *Long BCH codes are bad, Inform. Control,* 11 (1967), 445–451 *[5, 8]*.

[232] D. LIND, B. MARCUS, *An Introduction to Symbolic Dynamics and Coding,* Cambridge University Press, Cambridge, 1995 *[14]*.

[233] J.H. VAN LINT, *A survey of perfect codes, Rocky Mountain J. Math.,* 5 (1975), 199–224 *[4]*.

[234] J.H. VAN LINT, *Repeated-root cyclic codes, IEEE Trans. Inform. Theory,* 37 (1991), 343–345 *[8].*

[235] S. LITSYN, V. SHEVELEV, *On ensembles of low-density parity-check codes: asymptotic distance distributions, IEEE Trans. Inform. Theory,* 48 (2002), 887–908 *[13].*

[236] A. LOBSTEIN, *The hardness of solving subset sum with preprocessing, IEEE Trans. Inform. Theory,* 36 (1990), 943–946 *[2].*

[237] E. LOUIDOR, *Lowest-density MDS codes over super-alphabets,* M.Sc. dissertation, Computer Science Department, Technion, Haifa, Israel, 2004 *[11].*

[238] E. LOUIDOR, R.M. ROTH, *Lowest density MDS codes over extension alphabets, IEEE Trans. Inform. Theory,* 52 (2006), 3186–3197 *[11].*

[239] L. LOVÁSZ, *Spectra of graphs with transitive groups, Period. Math. Hungar.,* 6 (1975), 191–195 *[13].*

[240] A. LUBOTZKY, R. PHILLIPS, P. SARNAK, *Ramanujan graphs, Combinatorica,* 8 (1988), 261–277 *[13].*

[241] M.G. LUBY, M. MITZENMACHER, M.A. SHOKROLLAHI, D.A. SPIELMAN, *Improved low-density parity-check codes using irregular graphs, IEEE Trans. Inform. Theory,* 47 (2001), 585–598 *[13].*

[242] D.G. LUENBERGER, *Linear and Nonlinear Programming,* Second Edition, Addison-Wesley, Reading, Massachusetts, 1984 *[4].*

[243] H. LÜNEBURG, *Translation Planes,* Springer, Berlin, 1980 *[11].*

[244] X. MA, X.-M. WANG, *On the minimal interpolation problem and decoding RS codes, IEEE Trans. Inform. Theory,* 46 (2000), 1573–1580 *[9].*

[245] D.J.C. MACKAY, *Good error-correcting codes based on very sparse matrices, IEEE Trans. Inform. Theory,* 45 (1999), 399–431 *[13].*

[246] S. MACLANE, G. BIRKHOFF, *Algebra,* Third Edition, Chelsea, New York, 1967 *[3].*

[247] F.J. MACWILLIAMS, *Combinatorial problems of elementary group theory,* Ph.D. dissertation, Department of Mathematics, Harvard University, Cambridge, Massachusetts, 1962 *[4].*

[248] F.J. MACWILLIAMS, *A theorem on the distribution of weights in a systematic code, Bell Syst. Tech. J.,* 42 (1963), 79–94 *[4].*

[249] F.J. MACWILLIAMS, N.J.A. SLOANE, *The Theory of Error-Correcting Codes,* North-Holland, Amsterdam, 1977 *[Prf., 2, 3, 4, 5, 8, 11, 12].*

[250] S. MARCUGINI, A. MILANI, F. PAMBIANCO, *NMDS codes of maximal length over* F_q, $8 \le q \le 11$, *IEEE Trans. Inform. Theory,* 48 (2002), 963–966 *[11].*

[251] B.H. MARCUS, R.M. ROTH, P.H. SIEGEL, *Constrained systems and coding for recording channels*, in *Handbook of Coding Theory, Volume II*, V.S. Pless, W.C. Huffman (Editors), North-Holland, Amsterdam, 1998, pp. 1635–1764 *[14]*.

[252] G.A. MARGULIS, *Explicit group-theoretical constructions of combinatorial schemes and their application to the design of expanders and concentrators*, *Probl. Inform. Transm.*, 24 (1988), 39–46 *[13]*.

[253] J.L. MASSEY, *Threshold Decoding*, MIT Press, Cambridge, Massachusetts, 1963 *[12]*.

[254] J.L. MASSEY, *Shift-register synthesis and BCH decoding*, *IEEE Trans. Inform. Theory*, 15 (1969), 122–127 *[6]*.

[255] J.L. MASSEY, D.J. COSTELLO, JR., J. JUSTESEN, *Polynomial weights and code constructions*, *IEEE Trans. Inform. Theory*, 19 (1973), 101–110 *[8]*.

[256] J.L. MASSEY, M.K. SAIN, *Inverses of linear sequential circuits*, *IEEE Trans. Comput.*, 17 (1968), 330-337 *[14]*.

[257] L.E. MAZUR, *Codes correcting errors of large weight in Lee metric*, *Probl. Inform. Transm.*, 9 (1973), 277–281 *[10]*.

[258] R.J. MCELIECE, *The algebraic theory of convolutional codes*, in *Handbook of Coding Theory, Volume I*, V.S. Pless, W.C. Huffman (Editors), North-Holland, Amsterdam, 1998, pp. 1067–1138 *[14]*.

[259] R.J. MCELIECE, *The Theory of Information and Coding*, Second Edition, Cambridge University Press, Cambridge, 2002 *[Prf., 1, 4, 14]*.

[260] R.J. MCELIECE, E.R. RODEMICH, H. RUMSEY, JR., L.R. WELCH, *New upper bounds on the rate of a code via the Delsarte–MacWilliams inequalities*, *IEEE Trans. Inform. Theory*, 23 (1977), 157–166 *[4]*.

[261] G. MILLER, D. BURSHTEIN, *Bounds on the maximum-likelihood decoding error probability of low-density parity-check codes*, *IEEE Trans. Inform. Theory*, 47 (2001), 2696–2710 *[13]*.

[262] W. MILLS, *Continued fractions and linear recurrences*, *Math. Comput.*, 29 (1975), 173–180 *[6]*.

[263] H. MINC, *Nonnegative Matrices*, Wiley, New York, 1988 *[13]*.

[264] C.M. MONTI, G.L. PIEROBON, *Codes with a multiple spectral null at zero frequency*, *IEEE Trans. Inform. Theory*, 35 (1989), 463–472 *[10]*.

[265] D.E. MULLER, *Application of Boolean algebra to switching circuit design and to error detection*, *IRE Trans. Comput.*, 3 (1954), 6–12 *[2]*.

[266] K. NAKAMURA, *A class of error-correcting codes for DPSK channels*, *Conf. Record 1979 IEEE Int'l Conf. Communications (ICC'1979)*, Boston, Massachusetts (1979), pp. 45.4.1–45.4.5 *[10]*.

[267] J. NAOR, M. NAOR, *Small-bias probability spaces: efficient constructions and applications, SIAM J. Comput.*, 22 (1993), 838–856 *[13]*.

[268] G.L. NEMHAUSER, L.A. WOLSEY, *Integer and Combinatorial Optimization,* Wiley, New York, 1988 *[4]*.

[269] R.R. NIELSEN, T. HØHOLDT, *Decoding Reed–Solomon codes beyond half the minimum distance,* in *Coding Theory, Cryptography and Related Areas,* J. Buchmann, T. Høholdt, H. Stichtenoth, H. Tapia-Recillas (Editors), Springer, Berlin, 2000, pp. 221–236 *[9]*.

[270] A.F. NIKIFOROV, S.K. SUSLOV, V.B. UVAROV, *Classical Orthogonal Polynomials of a Discrete Variable,* Springer, Berlin, 1991 *[4]*.

[271] A. NILLI, *On the second eigenvalue of a graph, Discrete Math.*, 91 (1991), 207–210 *[13]*.

[272] J.-S. NO, H.-K. LEE, H. CHUNG, H.-Y. SONG, K. YANG, *Trace representation of Legendre sequences of Mersenne prime period, IEEE Trans. Inform. Theory,* 42 (1996), 2254–2255 *[7]*.

[273] A. ODLYZKO, *Discrete logarithms: the past and the future, Designs Codes Cryptogr.*, 19 (2000), 129–145 *[3]*.

[274] H. O'KEEFFE, P. FITZPATRICK, *Gröbner basis solution of constrained interpolation problems, Linear Algebra Appl.*, 351–352 (2002), 533–551 *[9]*.

[275] V. OLSHEVSKY, M.A. SHOKROLLAHI, *A displacement approach to efficient decoding of algebraic-geometric codes, Proc. 31st ACM Symp. Theory of Computing (STOC'1999),* Atlanta, Georgia (1999), ACM, New York, 1999, pp. 235–244 *[9]*.

[276] A. ORLITSKY, *Interactive communication of balanced distributions and of correlated files, SIAM J. Discrete Math.*, 6 (1993), 548–564 *[10]*.

[277] W.W. PETERSON, *Encoding and error-correction procedures for the Bose–Chaudhuri codes, IRE Trans. Inform. Theory,* 6 (1960), 459–470 *[6]*.

[278] W.W. PETERSON, E.J. WELDON, JR., *Error-Correcting Codes,* Second Edition, MIT Press, Cambridge, Massachusetts, 1972 *[Prf., 2, 4]*.

[279] E. PETRANK, R.M. ROTH, *Is code equivalence easy to decide?, IEEE Trans. Inform. Theory,* 43 (1997), 1602–1604 *[2]*.

[280] V. PLESS, *Introduction to the Theory of Error Correcting Codes,* Third Edition, Wiley, New York, 1998 *[Prf., 4]*.

[281] V.S. PLESS, W.C. HUFFMAN (EDITORS), *Handbook of Coding Theory, Volumes I–II,* North-Holland, Amsterdam, 1998 *[Prf.]*.

[282] M. PLOTKIN, *Binary codes with specified minimum distances, IRE Trans. Inform. Theory,* 6 (1960), 445–450 *[4]*.

[283] K.C. POHLMANN, *The Compact Disc Handbook,* Second Edition, A–R Editions, Madison, Wisconsin, 1992 *[5].*

[284] K.A. POST, *Nonexistence theorems on perfect Lee codes over large alphabets, Inform. Control,* 29 (1975), 369–380 *[10].*

[285] F.P. PREPARATA, *A class of optimum nonlinear double-error-correcting codes, Inform. Control,* 13 (1968), 378-400 *[10].*

[286] M.O. RABIN, *Probabilistic algorithms in finite fields, SIAM J. Comput.,* 9 (1980), 273–280 *[3].*

[287] S.M. REDDY, J.P. ROBINSON, *Random error and burst correction by iterated codes, IEEE Trans. Inform. Theory,* 18 (1972), 182-185 *[12].*

[288] I.S. REED, *A class of multiple-error-correcting codes and the decoding scheme, IRE Trans. Inform. Theory,* 4 (1954), 38–49 *[2].*

[289] I.S. REED, G. SOLOMON, *Polynomial codes over certain finite fields, J. SIAM,* 8 (1960), 300-304 *[5].*

[290] J.A. REEDS, N.J.A. SLOANE, *Shift-register synthesis (modulo m), SIAM J. Comput.,* 14 (1985), 505–513 *[6].*

[291] O. REGEV, *Improved inapproximability of lattice and coding problems with preprocessing, Proc. 18th Annual IEEE Conf. Computational Complexity (CCC'2003),* Århus, Denmark (2003), IEEE Computer Society Press, Los Alamitos, California, 2003, pp. 363–370 *[2].*

[292] S.H. REIGER, *Codes for the correction of "clustered" errors, IRE Trans. Inform. Theory,* 6 (1960), 16–21 *[4].*

[293] T.J. RICHARDSON, M.A. SHOKROLLAHI, R.L. URBANKE, *Design of capacity-approaching irregular low-density parity-check codes, IEEE Trans. Inform. Theory,* 47 (2001), 619–637 *[13].*

[294] T.J. RICHARDSON, R.L. URBANKE, *The capacity of low-density parity-check codes under message-passing decoding, IEEE Trans. Inform. Theory,* 47 (2001), 599–618 *[13].*

[295] A.I. RIIHONEN, *A note on perfect Lee-code, Discrete Appl. Math.,* 2 (1980), 259–260 *[10].*

[296] C. ROOS, *A new lower bound for the minimum distance of a cyclic code, IEEE Trans. Inform. Theory,* 29 (1983), 330–332 *[8].*

[297] R.M. ROTH, *Maximum-rank array codes and their application to crisscross error correction, IEEE Trans. Inform. Theory,* 37 (1991), 328–336 *[11].*

[298] R.M. ROTH, *Spectral-null codes and null spaces of Hadamard submatrices, Designs Codes Cryptogr.,* 9 (1996), 177–191 *[10].*

[299] R.M. Roth, *Tensor codes for the rank metric, IEEE Trans. Inform. Theory,* 42 (1996), 2146–2157 *[11].*

[300] R.M. Roth, A. Lempel, *A construction of non-Reed–Solomon type MDS codes, IEEE Trans. Inform. Theory,* 35 (1989), 655–657 *[5].*

[301] R.M. Roth, A. Lempel, *On MDS codes via Cauchy matrices, IEEE Trans. Inform. Theory,* 35 (1989), 1314–1319 *[5, 11].*

[302] R.M. Roth, G. Ruckenstein, *Efficient decoding of Reed–Solomon codes beyond half the minimum distance, IEEE Trans. Inform. Theory,* 46 (2000), 246–257 *[6, 9].*

[303] R.M. Roth, G. Seroussi, *On generator matrices of MDS codes, IEEE Trans. Inform. Theory,* 31 (1985), 826–830 *[5].*

[304] R.M. Roth, G. Seroussi, *On cyclic MDS codes of length q over GF(q), IEEE Trans. Inform. Theory,* 32 (1986), 284–285 *[8].*

[305] R.M. Roth, G. Seroussi, *Reduced-redundancy product codes for burst error correction, IEEE Trans. Inform. Theory,* 44 (1998), 1395–1406 *[5].*

[306] R.M. Roth, P.H. Siegel, *Lee-metric BCH codes and their application to constrained and partial-response channels, IEEE Trans. Inform. Theory,* 40 (1994), 1083–1096 *[10].*

[307] R.M. Roth, P.H. Siegel, A. Vardy, *High-order spectral-null codes—constructions and bounds, IEEE Trans. Inform. Theory,* 40 (1994), 1826–1840 *[10].*

[308] R.M. Roth, V. Skachek, *Improved nearly-MDS expander codes, IEEE Trans. Inform. Theory,* 52 (2006), 3650–3661 *[13].*

[309] G. Ruckenstein, *Error decoding strategies for algebraic codes,* Ph.D. dissertation, Computer Science Department, Technion, Haifa, Israel, 2002 *[9].*

[310] G. Ruckenstein, R.M. Roth, *Bounds on the list-decoding radius of Reed–Solomon codes, SIAM J. Discrete Math.,* 17 (2003), 171–195 *[9].*

[311] G.E. Sacks, *Multiple error correction by means of parity checks, IRE Trans. Inform. Theory,* 4 (1958), 145–147 *[4].*

[312] Y. Saitoh, *Theory and design of error-control codes for byte-organized/input-restricted storage devices where unidirectional/peak-shift errors are predominant,* Ph.D. dissertation, Division of Electrical and Computer Engineering, Yokohama National University, Yokohama, Japan, 1993 *[10].*

[313] Y. Saitoh, T. Ohno, H. Imai, *Construction techniques for error-control runlength-limited block codes, IEICE Trans. Fundamentals,* E76-A (1993), 453–458 *[10].*

[314] S. SAKATA, *Finding a minimal set of linear recurring relations capable of generating a given finite two-dimensional array, J. Symb. Comput.,* 5 (1988), 321–337 *[6].*

[315] S. SAKATA, *Extension of the Berlekamp–Massey algorithm to N dimensions, Inform. Comput.,* 1990 (84), 207–239 *[6].*

[316] S. SAKATA, Y. NUMAKAMI, M. FUJISAWA, *A fast interpolation method for list decoding of RS and algebraic–geometric codes, Proc. 2000 IEEE Int'l Symp. Inform. Theory (ISIT'2000),* Sorrento, Italy (2000), p. 479 *[9].*

[317] D.V. SARWATE, *On the complexity of decoding Goppa codes, IEEE Trans. Inform. Theory,* 23 (1977), 515–516 *[6].*

[318] I. SASON, R. URBANKE, *Parity-check density versus performance of binary linear block codes over memoryless symmetric channels, IEEE Trans. Inform. Theory,* 49 (2003), 1611–1635 *[13].*

[319] C. SATYANARAYANA, *Lee metric codes over integer residue rings, IEEE Trans. Inform. Theory,* 25 (1979), 250–254 *[10].*

[320] O. SCHIROKAUER, *Discrete logarithms and local units, Philos. Trans. R. Soc. Lond. A,* 345 (1993), 409–423 *[3].*

[321] O. SCHIROKAUER, *Using number fields to compute logarithms in finite fields, Math. Comput.,* 69 (2000), 1267–1283 *[3].*

[322] A. SCHÖNHAGE, *Schnelle Multiplikation von Polynomen über Körpern der Charakteristik 2, Acta Informatica,* 7 (1977), 395–398 *[3].*

[323] A. SCHÖNHAGE, V. STRASSEN, *Schnelle Multiplikation grosser Zahlen, Computing,* 7 (1971), 281–292 *[3].*

[324] A. SCHRIJVER, *Theory of Linear and Integer Programming,* Wiley, New York, 1986 *[4].*

[325] B. SEGRE, *Ovals in a finite projective plane, Canad. J. Math.,* 7 (1955), 414–416 *[11].*

[326] B. SEGRE, *Introduction to Galois Geometries, Atti Accad. Naz. Lincei Mem.,* 8 (1967), 133–236 *[11].*

[327] N. SENDRIER, *Finding the permutation between equivalent linear codes: the support splitting algorithm, IEEE Trans. Inform. Theory,* 46 (2000), 1193–1203 *[2].*

[328] E. SENETA, *Non-negative Matrices and Markov Chains,* Second Edition, Springer, New York, 1980 *[10, 13].*

[329] G. SEROUSSI, R.M. ROTH, *On MDS extensions of generalized Reed–Solomon codes, IEEE Trans. Inform. Theory,* 32 (1986), 349–354 *[11].*

[330] C.E. SHANNON, *A mathematical theory of communication*, *Bell Syst. Tech. J.*, 27 (1948), 379–423 and 623–656 *[1, 4]*.

[331] C.E. SHANNON, R.G. GALLAGER, E.R. BERLEKAMP, *Lower bounds to error probability for coding on discrete memoryless channels*, *Inform. Control*, 10 (1967), 65–103 and 522–552 *[4]*.

[332] B.D. SHARMA, R.K. KHANNA, *On m-ary Gray codes*, *Inform. Sci.*, 15 (1978), 31–43 *[10]*.

[333] B.-Z. SHEN, *A Justesen construction of binary concatenated codes that asymptotically meet the Zyablov bound for low rate*, *IEEE Trans. Inform. Theory*, 39 (1993), 239–242 *[12]*.

[334] V. SHOUP, *New algorithms for finding irreducible polynomials over finite fields*, *Math. Comput.*, 54 (1990), 435–447 *[3]*.

[335] V. SHOUP, *Fast construction of irreducible polynomials over finite fields*, *J. Symb. Comput.*, 17 (1994), 371–391 *[3]*.

[336] V. SHOUP, *A new polynomial factorization algorithm and its implementation*, *J. Symb. Comput.*, 20 (1995), 363–397 *[3]*.

[337] N. SHULMAN, M. FEDER, *Improved error exponent for time-invariant and periodically time-variant convolutional codes*, *IEEE Trans. Inform. Theory*, 46 (2000), 97–103 *[14]*.

[338] K.W. SHUM, I. ALESHNIKOV, P.V. KUMAR, H. STICHTENOTH, V. DEOLALIKAR, *A low-complexity algorithm for the construction of algebraic-geometric codes better than the Gilbert–Varshamov bound*, *IEEE Trans. Inform. Theory*, 47 (2001), 2225-2241 *[4]*.

[339] R.C. SINGLETON, *Maximum distance q-nary codes*, *IEEE Trans. Inform. Theory*, 10 (1964), 116–118 *[4]*.

[340] M. SIPSER, D.A. SPIELMAN, *Expander codes*, *IEEE Trans. Inform. Theory*, 42 (1996), 1710–1722 *[13]*.

[341] V. SKACHEK, T. ETZION, R.M. ROTH, *Efficient encoding algorithm for third-order spectral-null codes*, *IEEE Trans. Inform. Theory*, 44 (1998), 846–851 *[10]*.

[342] V. SKACHEK, R.M. ROTH, *Generalized minimum distance iterative decoding of expander codes*, *Proc. IEEE Information Theory Workshop*, Paris, France (April 2003), pp. 245–248 *[13]*.

[343] D.A. SPIELMAN, *Linear-time encodable and decodable error-correcting codes*, *IEEE Trans. Inform. Theory*, 42 (1996), 1723–1731 *[13]*.

[344] L. STORME, J.A. THAS, *Generalized Reed–Solomon codes and normal rational curves: an improvement of results by Seroussi and Roth*, in *Advances in*

Finite Geometries and Designs: Proc. 3rd Isle of Thorns Conference, Chelwood Gate, UK (1990), J.W.P. Hirschfeld, D.R. Hughes, J.A. Thas (Editors), Oxford University Press, Oxford, 1991, pp. 369–389 *[11]*.

[345] L. STORME, J.A. THAS, *M.D.S. codes and arcs in* PG(n, q) *with* q *even: an improvement of the bounds of Bruen, Thas, and Blokhuis, J. Comb. Theory A,* 62 (1993), 139–154 *[11]*.

[346] M. SUDAN, *Decoding of Reed–Solomon codes beyond the error-correction bound, J. Complexity,* 13 (1997), 180–193 *[9]*.

[347] Y. SUGIYAMA, M. KASAHARA, S. HIRASAWA, T. NAMEKAWA, *A modification of the constructive asymptotically good codes of Justesen for low rates, Inform. Control,* 25 (1974), 341–350 *[12]*.

[348] Y. SUGIYAMA, M. KASAHARA, S. HIRASAWA, T. NAMEKAWA, *A method for solving key equation for decoding Goppa codes, Inform. Control,* 27 (1975), 87–99 *[6]*.

[349] Y. SUGIYAMA, M. KASAHARA, S. HIRASAWA, T. NAMEKAWA, *A new class of asymptotically good codes beyond the Zyablov bound, IEEE Trans. Inform. Theory,* 24 (1978), 198–204 *[12]*.

[350] Y. SUGIYAMA, M. KASAHARA, S. HIRASAWA, T. NAMEKAWA, *Superimposed concatenated codes, IEEE Trans. Inform. Theory,* 26 (1980), 735–736 *[5, 12]*.

[351] G. SZEGŐ, *Orthogonal Polynomials,* Fourth Edition, Amer. Math. Soc. Colloq. Publ., Volume 23, AMS, Providence, Rhode Island, 1975 *[4]*.

[352] I. TAL, *List decoding of Lee metric codes,* M.Sc. dissertation, Computer Science Department, Technion, Haifa, Israel, 2003 *[10]*.

[353] I. TAL, R.M. ROTH, *On list decoding of alternant codes in the Hamming and Lee metrics, Proc. 2003 IEEE Int'l Symp. Inform. Theory (ISIT'2003),* Yokohama, Japan (2003), p. 364 *[9, 10]*.

[354] L.G. TALLINI, B. BOSE, *On efficient high-order spectral-null codes, IEEE Trans. Inform. Theory,* 45 (1999), 2594–2601 *[10]*.

[355] E. TANAKA, T. KASAI, *Synchronization and substitution error-correcting codes for the Levenshtein metric, IEEE Trans. Inform. Theory,* 22 (1976), 156–162 *[10]*.

[356] R.M. TANNER, *A recursive approach to low-complexity codes, IEEE Trans. Inform. Theory,* 27 (1981), 533–547 *[13]*.

[357] R.M. TANNER, *Explicit concentrators from generalized* N*-gons, SIAM J. Algebr. Discrete Methods,* 5 (1984), 287–293 *[13]*.

[358] G.M. TENENGOLTS, *Class of codes correcting bit loss and errors in the preceding bit, Autom. Remote Control,* 37 (1976), 797–802 *[10]*.

[359] G. TENENGOLTS, *Nonbinary codes, correcting single deletion or insertion, IEEE Trans. Inform. Theory,* 30 (1984), 766–769 *[10].*

[360] J.A. THAS, *Complete arcs and algebraic curves in* PG$(2,q)$, *J. Algebra,* 106 (1987), 451–464 *[11].*

[361] J.A. THAS, *Projective geometry over a finite field,* in *Handbook of Incidence Geometry: Buildings and Foundations,* F. Buekenhout (Editor), Elsevier Science, Amsterdam, 1995, 295–347 *[2, 11].*

[362] C. THOMMESEN, *The existence of binary linear concatenated codes with Reed–Solomon outer codes which asymptotically meet the Gilbert–Varshamov bound, IEEE Trans. Inform. Theory,* 29 (1983), 850–853 *[12].*

[363] C. THOMMESEN, *Error-correcting capabilities of concatenated codes with MDS outer codes on memoryless channels with maximum-likelihood decoding, IEEE Trans. Inform. Theory,* 33 (1987), 632–640 *[12].*

[364] A. TIETÄVÄINEN, *On the nonexistence of perfect codes over finite fields, SIAM J. Appl. Math.,* 24 (1973), 88–96 *[4].*

[365] M.A. TSFASMAN, S.G. VLĂDUT, T. ZINK, *Modular curves, Shimura curves, and Goppa codes, better than Varshamov–Gilbert bound, Math. Nachr.,* 109 (1982), 21–28 *[4].*

[366] J.D. ULLMAN, *Near-optimal, single-synchronization-error-correcting code, IEEE Trans. Inform. Theory,* 12 (1966), 418–424 *[10].*

[367] J.D. ULLMAN, *On the capabilities of codes to correct synchronization errors, IEEE Trans. Inform. Theory,* 13 (1967), 95–105 *[10].*

[368] W. ULRICH, *Non-binary error correction codes, Bell Syst. Tech. J.,* 36 (1957), 1341–1387 *[10].*

[369] T. UYEMATSU, E. OKAMOTO, *A construction of codes with exponential error bounds on arbitrary discrete memoryless channels, IEEE Trans. Inform. Theory,* 43 (1997), 992–996 *[12].*

[370] A. VARDY, *The intractability of computing the minimum distance of a code, IEEE Trans. Inform. Theory,* 43 (1997), 1757–1766 *[2, 5].*

[371] R.S. VARGA, *Matrix Iterative Analysis,* Prentice-Hall, Englewood Cliffs, New Jersey, 1962 *[13].*

[372] R.R. VARSHAMOV, *Estimate of the number of signals in error correcting codes, Dokl. Akad. Nauk SSSR,* 117 (1957), 739–741 *[4].*

[373] A.J. VITERBI, *Error bounds for convolutional codes and an asymptotically optimum decoding algorithm, IEEE Trans. Inform. Theory,* 13 (1967), 260–269 *[14].*

[374] A.J. VITERBI, J.K. OMURA, *Principles of Digital Communication and Coding,* McGraw-Hill, New York, 1979 *[4, 14]*.

[375] J.F. VOLOCH, *Arcs in projective planes over prime fields, J. Geom.,* 38 (1990), 198–200 *[11]*.

[376] J.F. VOLOCH, *Complete arcs in Galois planes of non-square order,* in *Advances in Finite Geometries and Designs: Proc. 3rd Isle of Thorns Conference,* Chelwood Gate, UK (1990), J.W.P. Hirschfeld, D.R. Hughes, J.A. Thas (Editors), Oxford University Press, Oxford, 1991, pp. 401–406 *[11]*.

[377] B.L. VAN DER WAERDEN, *Algebra, Volume 1,* Seventh Edition, Ungar, New York, 1970, and Springer, New York, 1991 *[10]*.

[378] W.D. WALLIS, A.P. STREET, J.S. WALLIS, *Combinatorics: Room Squares, Sum-Free Sets, Hadamard Matrices,* Lecture Notes in Mathematics, Volume 292, Springer, Berlin, 1972 *[2]*.

[379] A. WEIL, *On some exponential sums, Proc. Nat. Acad. Sci. USA,* 34 (1948), 204–207 *[5]*.

[380] L.R. WELCH, E.R. BERLEKAMP, *Error correction for algebraic block codes,* US Patent 4,633,470 (1986) *[6, 9]*.

[381] L.R. WELCH, R.A. SCHOLTZ, *Continued fractions and Berlekamp's algorithm, IEEE Trans. Inform. Theory,* 25 (1979), 19–27 *[6]*.

[382] E.J. WELDON, JR., *Decoding binary block codes on Q-ary output channels, IEEE Trans. Inform. Theory,* 17 (1971), 713–718 *[12]*.

[383] E.J. WELDON, JR., *Justesen's construction—the low-rate case, IEEE Trans. Inform. Theory,* 19 (1973), 711–713 *[12]*.

[384] E.J. WELDON, JR., *Some results on the problem of constructing asymptotically good error-correcting codes, IEEE Trans. Inform. Theory,* 21 (1975), 412–417 *[12]*.

[385] D.B. WEST, *Introduction to Graph Theory,* Second Edition, Prentice-Hall, Upper Saddle River, New Jersey, 2001 *[13]*.

[386] S. WINOGRAD, *Arithmetic Complexity of Computations,* CBMS-NSF Regional Conference Series in Applied Mathematics, Volume 33, SIAM, Philadelphia, Pennsylvania, 1980 *[3]*.

[387] J.M. WOZENCRAFT, *Sequential decoding for reliable communication, IRE Nat. Conv. Rec.,* 5 (1957), 11–25 *[14]*.

[388] J.M. WOZENCRAFT, B. REIFFEN, *Sequential Decoding,* MIT Press and Wiley, New York, 1961 *[14]*.

[389] A.D. WYNER, R.B. ASH, *Analysis of recurrent codes, IEEE Trans. Inform. Theory,* 9 (1963), 143–156 *[14]*.

[390] A.D. WYNER, R.L. GRAHAM, *An upper bound on minimum distance for a k-ary code, Inform. Control,* 13 (1968), 46–52 *[10]*.

[391] R.K. YARLAGADDA, J.E. HERSHEY, *Hadamard Matrix Analysis and Synthesis, with Applications to Communications and Signal/Image Processing,* Kluwer, Boston, Massachusetts, 1996 *[2]*.

[392] G.V. ZAITSEV, V.A. ZINOV'EV, N.V. SEMAKOV, *Minimum-check-density codes for correcting bytes of errors, erasures, or defects, Probl. Inform. Transm.,* 19 (1983), 197–204 *[11]*.

[393] E. ZEHENDNER, *A non-existence theorem for cyclic MDS-codes, Atti. Sem. Mat. Fis. Univ. Modena,* 32 (1983), 203–205 *[8]*.

[394] G. ZÉMOR, *On expander codes, IEEE Trans. Inform. Theory,* 47 (2001), 835–837 *[13]*.

[395] K.S. ZIGANGIROV, *Some sequential decoding procedures, Probl. Inform. Transm.,* 2 (1966), 1–10 *[14]*.

[396] K.S. ZIGANGIROV, *On the error probability of sequential decoding on the BSC, IEEE Trans. Inform. Theory,* 18 (1972), 199–202 *[14]*.

[397] V.A. ZINOV'EV, *Generalized cascade codes, Probl. Inform. Transm.,* 12 (1976), 2–9 *[5, 12]*.

[398] V.A. ZINOV'EV, *Generalized concatenated codes for channels with error bursts and independent errors, Probl. Inform. Transm.,* 17 (1981), 254–260 *[5, 12]*.

[399] V.A. ZINOV'EV, V.K. LEONT'EV, *The nonexistence of perfect codes over Galois fields, Probl. Control Inform. Theory,* 2 No. 2 (1973), English supplement, 16–24 *[4]*.

[400] V.A. ZINOV'EV, V.V. ZYABLOV, *Decoding of nonlinear generalized cascade codes, Probl. Inform. Transm.,* 14 (1978), 110–114 *[5, 12]*.

[401] V.A. ZINOV'EV, V.V. ZYABLOV, *Correction of error bursts and independent errors using generalized cascaded codes, Probl. Inform. Transm.,* 15 (1979), 125–134 *[5, 12]*.

[402] V.A. ZINOV'EV, V.V. ZYABLOV, *Codes with unequal protection of information symbols, Probl. Inform. Transm.,* 15 (1979), 197–205 *[5, 12]*.

[403] R. ZIPPEL, *Effective Polynomial Computation,* Kluwer, Boston, Massachusetts, 1993 *[3, 9]*.

[404] V.V. ZYABLOV, *An estimate of the complexity of constructing binary linear cascade codes, Probl. Inform. Transm.,* 7 (1971), 3–10 *[12]*.

[405] V.V. ZYABLOV, *Optimization of concatenated decoding algorithms, Probl. Inform. Transm.,* 9 (1973), 19–24 *[12]*.

List of Symbols

(Most of the symbol descriptions include a reference to the page where the symbol is defined.)

$\mathbb{N}$	set of natural numbers (including 0)
$\mathbb{Z}$	integer ring
$\mathbb{Z}^+$	set of positive integers (excluding 0)
$\mathbb{Z}_q$	ring of integer residues modulo q
$\mathbb{Q}$	rational field
$\mathbb{R}$	real field
$\mathbb{C}$	complex field
$\mathrm{GF}(q)$	finite field of size q, 50
F, Φ, K	common notation for a field
$\lvert S \rvert$	size of a set S
$\binom{n}{i}$	binomial coefficient
$\phi(n)$	Euler function, 522
$\mu(n)$	Möbius function, 523
$\left(\frac{a}{q}\right)$	Legendre symbol of a modulo q, 80
χ	additive character, 85
ψ	multiplicative character, 78
$\lfloor x \rfloor$	largest integer not greater than the real x
$\lceil x \rceil$	smallest integer not smaller than the real x
$\mathrm{Re}\{\cdot\}$	real part of a complex number
$O(f)$	expression that grows at most linearly with f
$o(1)$	expression that goes to zero as the parameters go to infinity
$\mathbf{x}, \mathbf{y}, \mathbf{z}$	common notation for a word or a vector
$\mathbf{0}$	all-zero word (or vector)
$\mathbf{1}$	all-one word (or vector)
F^n	set of all words (vectors) of length n over an alphabet (field) F, 3
$F^{m \times n}$	set of all $m \times n$ arrays (matrices) over an alphabet (field) F, 353
$(A)_{i,j}$	entry that is indexed by (i, j) in a matrix A
$\mathrm{rank}(A)$	rank of a matrix A
$\ker(A)$	(right) kernel of a matrix A

$\det(A)$	determinant of a matrix A
$\mathrm{Adj}(A)$	adjoint of a matrix A
$\dim W$	dimension of a linear space W
$A^T, \mathbf{x}^T$	transpose of a matrix A or a vector $\mathbf{x}$
$A^*, \mathbf{x}^*$	conjugate transpose of a complex matrix A or a vector $\mathbf{x}$
A^{c}	matrix consisting of the inverses of the entries of a matrix A, 169
$\langle \mathbf{x}, \mathbf{y} \rangle$	scalar product of real vectors $\mathbf{x}$ and $\mathbf{y}$, 400
$\\|\mathbf{x}\\|$	norm of a real vector $\mathbf{x}$, 400
$A \otimes B$	Kronecker product of matrices A and B, 45
I	identity matrix
$\mathcal{H}_k$	Sylvester-type Hadamard matrix of order $2^k \times 2^k$, 45
$\mathrm{SL}_n(q)$	n-dimensional special linear group over $\mathrm{GF}(q)$, 409
$\mathrm{PSL}_n(q)$	n-dimensional projective special linear group over $\mathrm{GF}(q)$, 409
$a(x)$	polynomial or a formal power series in the indeterminate x, 189
$a'(x)$	formal derivative of a polynomial (or a formal power series) $a(x)$, 65
$a^{[\ell]}(x)$	ℓth Hasse derivative (hyper-derivative) of a polynomial $a(x)$, 87
$\deg a(x), \deg a$	degree of a polynomial $a(x)$, 51
$F[x]$	set of all polynomials over a field F, 51
$F_n[x]$	set of all polynomials with degree $< n$ over a field F, 52
$F[x]/a(x)$	ring of polynomial residues modulo a polynomial $a(x)$, 56
$F[[x]]$	ring of formal power series over a field F, 189
$F(x)$	field of rational functions over a field F, 268
$F((x))$	field of Laurent series over a field F, 476
$b(x, z)$	bivariate polynomial in the indeterminates x and z, 268
$\deg_{\mu,\nu} b(x, z)$	(μ, ν)-degree of a bivariate polynomial $b(x, z)$, 268
$F[x, z]$	set of all bivariate polynomials over a field F, 268
$b^{[s,t]}(x, z)$	(s, t)th Hasse derivative of a bivariate polynomial $b(x, z)$, 276
$a \mid b$	a divides b (relation), 52
$a \equiv b \pmod{c}$	a is congruent to b modulo c (relation), 52
$a \text{ MOD } b$	remainder of a when divided by b (binary operation), 242
$\gcd(a, b)$	greatest common divisor of a and b, 52
$\mathcal{O}(a)$	(multiplicative) order of an element a in a group or a field, 51
F^*	multiplicative group of a field F, 51
$\langle \alpha \rangle$	smallest nonnegative integer m such that $\alpha = m \cdot 1$, for an element α in $\mathbb{Z}_q$, 298
$\mathsf{c}(F)$	characteristic of a field F, 62
$[\Phi : F]$	extension degree of a field Φ with respect to a subfield F, 57
Ω	basis of an extension field, 58
$\mathrm{T}_{\Phi:F}(\alpha)$	trace of an element α in a field Φ (with respect to a subfield F), 83
C_α	conjugacy class of a field element α (with respect to a subfield), 219
$M_\alpha(x)$	minimal polynomial of a field element α (with respect to a subfield), 219

$\mathcal{I}(n,q)$	number of monic irreducible polynomials of degree n over $\mathrm{GF}(q)$, 225
$\mathcal{P}(n,q)$	number of monic primitive polynomials of degree n over $\mathrm{GF}(q)$, 229
$\mathsf{d}_{(\mathcal{L})}(\mathbf{x},\mathbf{y})$	Hamming (Lee) distance between words $\mathbf{x}$ and $\mathbf{y}$, 6
$\mathsf{w}_{(\mathcal{L})}(\mathbf{x})$	Hamming (Lee) weight of a word $\mathbf{x}$, 6
$\lvert\alpha\rvert$	Lee weight of an element α in $\mathbb{Z}_q$, 299
$\chi_{\mathcal{L}}(q)$	average Lee weight of the elements of $\mathbb{Z}_q$, 326
$\mathcal{S}$	sphere in a Hamming or Lee metric, 98
$V_q(n,t)$	size of a Hamming sphere with radius t in F^n, where $\lvert F\rvert = q$, 95
$V_{\mathcal{L}\mid q}(n,t)$	size of a Lee sphere with radius t in $\mathbb{Z}_q^n$, 317
$\mathcal{C}$	code, 5
$\mathbf{C}$	trellis code or a convolutional code, 461
n	length of a code, 5
M	size of a code, 5
$d, \mathsf{d}_{(\mathcal{L})}(\mathcal{C})$	minimum Hamming (Lee) distance of a code $\mathcal{C}$, 6
$\mathsf{d}_{\mathrm{free}}(\mathbf{C})$	free distance of a trellis code or a convolutional code $\mathbf{C}$, 463
(n,M,d)	concise notation for the parameters of a code, 6
k	dimension of a (linear) code, 5
$[n,k,d], [n,k]$	concise notation for the parameters of a linear code, 26
R	rate of a code, 3
δ	relative minimum distance, 104
r	covering radius, 123
$(W_i)_{i=0}^n$	(Hamming) weight distribution, 99
$W_{\mathcal{C}}(z)$	(Hamming) weight enumerator of a code $\mathcal{C}$, 99
$W_{\mathcal{C}}^{\mathrm{h}}(x,z)$	homogeneous (Hamming) weight enumerator of a code $\mathcal{C}$, 99
$\mathcal{C}^{\perp}$	dual code of a linear code, 30
G	generator matrix of a linear code, 27
H	parity-check matrix of a linear code, 29
$G(x)$	generator matrix of a convolutional code, 477
$g(x)$	generator polynomial of a cyclic code, 245
$h(x)$	check polynomial of a cyclic code, 246
$\mathcal{C}_{\mathrm{in}}, \mathcal{C}_{\mathrm{out}}$	inner code, outer code in a concatenated code, 154
$\mathcal{C}_1 * \mathcal{C}_2$	product code of $\mathcal{C}_1$ and $\mathcal{C}_2$, 44
$\mathcal{C}_{\mathrm{GRS}}$	generalized Reed–Solomon (GRS) code, 148
$\mathcal{C}_{\mathrm{RS}}$	Reed–Solomon code, 151
$\mathcal{C}_{\mathrm{alt}}$	alternant code, 157
$\mathcal{C}_{\mathrm{BCH}}$	BCH code, 162
$G_{\mathrm{GRS}}, H_{\mathrm{GRS}}$	canonical generator matrix and parity-check matrix of a GRS code, 148
$\mathcal{C}_{\mathrm{RM}}$	Reed–Muller code, 41
$\mathcal{W}_F(n,k)$	ensemble of $[n,k]$ Wozencraft codes over F, 375

S	channel, 3
$\mathsf{cap}(S)$	capacity of a channel S, 10
F, Φ	input alphabet and output alphabet of a channel, 3
?	erasure symbol, 15
$\mathsf{Prob}\{\mathcal{A}\}$	probability of an event $\mathcal{A}$
$\mathsf{E}\{X\}$	expected value of a random variable X
$\mathsf{Var}\{X\}$	variance of a random variable X
$\mathbf{c}, \mathbf{y}, \mathbf{e}$	common notation for codeword, received word, and error word (respectively)
$\mathcal{D}, \mathcal{D}_{\mathrm{MLD}}$	decoder, maximum-likelihood decoder, 7
"e"	error-detection indicator, 14
$\mathrm{P}_{\mathrm{err}}, \mathrm{P}_{\mathrm{mis}}$	decoding error probability, decoding misdetection probability, 7
$\Lambda(x)$	error locator polynomial, 185
$\Gamma(x)$	error evaluator polynomial, 186
$\Lambda(x) : \mathrm{V}(x)$	error locator ratio, 307
$S(x)$	syndrome polynomial, 185
$\mathrm{ord}(\sigma, \omega)$	recurrence order of a polynomial pair $(\sigma(x), \omega(x))$, 197
$\mathsf{H}_q(x)$	q-ary entropy function, 105
$\mathsf{D}_q(\theta \| p)$	information divergence (Kullback–Leibler distance), 111
$\mathrm{E}_q(p, R)$	decoding error exponent, 117
$\mathcal{J}(M, \theta, q)$	Johnson bound, 129
$\mathcal{K}_\ell(y; n, q)$	Krawtchouk polynomial, 103
$\Delta_\ell(\mathcal{C})$	largest decoding radius of any list-ℓ decoder for a code $\mathcal{C}$, 289
$\Theta_\ell(R')$	bound on the relative decoding radius of the Guruswami–Sudan algorithm, 278
$\Theta_\ell(R', q)$	bound on the relative decoding radius of the Koetter–Vardy algorithm, 283
$L_q(k)$	largest length of any linear MDS code of dimension k over $\mathrm{GF}(q)$, 338
$\Gamma_q(k)$	smallest n such that $[n, k]$ MDS codes over $\mathrm{GF}(q)$ are all extended GRS codes, 342
$R_{\mathrm{Z}}(\delta, q)$	Zyablov bound, 373
$\mathcal{G} = (V, E)$	graph (or hyper-graph) with vertex set V and (hyper-)edge set E, 396
$\mathcal{G} = (V, E, L)$	labeled digraph with state set V, edge set E, and labeling L, 453
u, v	common notation for a vertex in a graph
s	common notation for a state in a digraph
e	common notation for an edge in a graph or a digraph
π	common notation for a path in a digraph
$\iota(e), \iota(\pi)$	initial state of an edge e or a path π in a digraph, 453
$\tau(e), \tau(\pi)$	terminal state of an edge e or a path π in a digraph, 453
p	period of an irreducible digraph, 455

$(V' : V'', E)$	bipartite graph with partition elements V' and V'' of the vertex set, 398
$\mathcal{G}_S$	induced subgraph of $\mathcal{G}$ on a set of vertices S, 397
$\mathcal{G}(Q, S)$	Cayley graph for a finite group Q and a subset $S \subseteq Q$, 406
$\mathcal{G}_{\mathrm{LPS}}(p, q)$	Lubotzky-Phillips-Sarnak (and also Margulis) Ramanujan graph, 410
$\deg_{\mathcal{G}}(u)$	degree of a vertex u in a graph $\mathcal{G}$, 396
$\mathcal{N}(u), \mathcal{N}(S)$	neighborhood of a vertex u or a set of vertices S in a graph, 396
$E(u)$	set of edges incident with a vertex u in a graph, 397
$E_{S,T}$	set of all edges with one endpoint in S and one endpoint in T, 397
$\partial(S)$	edge cut associated with a set of vertices S, 397
$\mathsf{d}_{\mathcal{G}}(u, v)$	distance between vertices u and v in a graph $\mathcal{G}$, 397
$\mathrm{diam}(\mathcal{G})$	diameter of a graph $\mathcal{G}$, 397
$A_{\mathcal{G}}$	adjacency matrix of a graph or a digraph $\mathcal{G}$, 398
$A_{\mathcal{G}}(z)$	generalized adjacency matrix of a labeled digraph $\mathcal{G}$, 486
$\gamma_{\mathcal{G}}$	second largest eigenvalue divided by the degree of a regular graph $\mathcal{G}$, 402
$X_{\mathcal{G}}$	transfer matrix of a bipartite graph $\mathcal{G}$, 399
$C_{\mathcal{G}}$	incidence matrix of a graph $\mathcal{G}$, 399
$L_{\mathcal{G}}^{-}, L_{\mathcal{G}}$	Laplace matrix of a graph $\mathcal{G}$, 399
$\vec{\mathcal{G}}$	orientation on a graph $\mathcal{G}$, 400
$(\mathcal{G}, \mathcal{C})$	graph code defined by a graph $\mathcal{G}$ and a code $\mathcal{C}$, 412
$\mathsf{C}(\mathcal{G})$	sequence set of a labeled digraph $\mathcal{G}$, 457
$\mathsf{T}(\mathcal{G})$	trellis diagram of a labeled digraph $\mathcal{G}$, 458
(P, B, Q, D)	matrix quadruple that defines a linear finite-state machine, 471
$W_{\mathcal{G}}(x, z)$	length-weight enumerator of a linear finite-state machine $\mathcal{G}$, 487

Index

adjacency matrix
  of digraphs, 486
  of graphs, 398
  Lee, 325
alphabet, 3
alternant code, 70, 157, 179
  decoding of, 197, 204
    Lee-metric, 306
    list, 280, 328
  designed minimum distance of, 157, 250
  dual code of, 175, 180
  Lee-metric, 302
  list decoding of, 280, 328
  over $\mathbb{Z}$, 328
aperiodic irreducible digraph, 455
aperiodic irreducible matrix, 445
arc (in projective geometry), 361
  complete, 363
autocorrelation
  of Legendre sequences, 80
  of maximal-length sequences, 87
AWGN channel, 17

basis
  complementary, 85
  dual, 85
  normal, 240
BCH bound, 253
BCH code, 162, 181, 244, 250
  consecutive root sequence of, 163
  decoding of, *see* alternant code, decoding of
  designed minimum distance of, 163, 250
  excess root of, 251
  root of, 163, 250
Berlekamp code, 314, 330
Berlekamp–Massey algorithm, 200, 217
Bhattacharyya bound, 21, 25, 493
bi-connection, 454
binary erasure channel, 15
binary symmetric channel (BSC), 4, 450
bipartite graph, 362, 398
bit-shift error, 327
Blahut's algorithm, 217
block code, 5
Blokh–Zyablov bound, 393
bound
  BCH, 253
  Bhattacharyya, 21, 25, 493
  Blokh–Zyablov, 393
  Carlitz–Uchiyama, 179
  Chernoff, 139
  decoding-radius, 290
  Elias, 108
    Lee-metric, 332
  Gilbert–Varshamov, 97, 137, 176, 181, 393
    asymptotic, 107, 372
    Lee-metric, 320, 330
  Griesmer, 120, 136
  Hamming, *see* bound, sphere-packing
  Hartmann–Tzeng, 265
  Johnson, 107, 128, 139, 289
    Lee-metric, 330
  linear programming, 103, 110, 138
  MDS code length, 338
  MRRW, 110
  Plotkin, 37, 127, 131, 139, 294
    Lee-metric, 326, 330
  Reiger, 122
  Roos, 265
  Singleton, *see* Singleton bound
  sphere-covering, 123
  sphere-packing, 95, 122, 136
    asymptotic, 107
    Lee-metric, 318, 330
  union, 137
  Zyablov, 373, 392, 413, 422, 438, 440
burst, 45, 122, 137, 257

cap (in projective geometry), 47
capacity, 10, 16, 24, 110
Carlitz–Uchiyama bound, 179
catastrophic error propagation, 496
Cauchy matrix, 168, 336, 356, 362
Cayley graph, 406, 447
channel, 1
  additive, 5
  AWGN, 17
  binary symmetric (BSC), 4, 450

- discrete memoryless (DMC), 19, 23, 46, 142, 466
- erasure, 15, 25, 126, 134, 391, 514
- Gaussian (AWGN), 17
- non-symmetric, 20, 46
- probabilistic, 3
- symmetric, 4, 19, 24, 32, 110, 113, 117, 125, 133, 145, 378, 390, 394, 424, 450, 467, 490

character
- of Abelian groups, 91, 447
- additive, 85, 99, 447
- multiplicative, 78
- quadratic, 80
- trivial, 79

characteristic (of fields), 62
check node, 449
check polynomial, 246
Chernoff bound, 139
Chien search, 186, 215, 285
circulant matrix, 325, 330, 356, 362
code, 5
- algebraic-geometry, 138
- almost-MDS (AMDS), 363
- alternant, *see* alternant code
- array, 353
- BCH, *see* BCH code
- Berlekamp, 314, 330
- block, 5
- concatenated, *see* concatenated code
- constant-weight, 121, 139, 352
- convolutional, *see* convolutional code
- cyclic, *see* cyclic code
- Delsarte–Goethals, 328
- dimension, 5
- double-error-correcting, 70, 161
- dual, *see* dual code
- equidistant, 128
- equivalence, 29, 47
- generalized Reed–Solomon, *see* GRS code
- Golay, 96, 136, 255
- Goppa, 182, 389
- graph, *see* graph code
- Gray, 321, 328
- group, 37, 299
- Hamming, *see* Hamming code
- inner (in concatenated codes), 154, 366
- Justesen, 376
- Kerdock, 328
- LDPC, 362, 450
- length, 5
- lengthening of, 123
- linear, *see* linear code
- low-density parity-check, 362, 450
- maximal, 123
- maximum distance separable (MDS), *see* MDS code
- minimum distance, 6
- near-MDS (NMDS), 363
- negacyclic, 323, 330
- outer (in concatenated codes), 154, 366
- parity, 27
- perfect, 96, 137, 256
  - Lee-metric, 319, 330
- Preparata, 328
- product, 44, 178
- punctured, 36
- rate, 5
- redundancy, 27
- Reed–Muller, *see* Reed–Muller code
- Reed–Solomon, *see* RS code
- repetition, 28
- self-dual, 31
- shortened, 40
- simplex, 41, 120
  - weight distribution of, 99
- size, 5
- spectral-null, 329
- Spielman, 451
- trellis, 460
  - decoding of, 466
  - encoding of, 464
  - free distance of, 463
- turbo, 519
- Wozencraft, 375
- Wyner–Ash, 512

codeword, 3, 460
communication system, 1
companion matrix, 73, 383, 510
complementary basis, 85
concatenated code, 154, 172, 178, 408, 420
- decoding of, 178, 371, 396, 422
- dual code of, 383
- linearly-, 154, 367

concave function, 9, 23
conjugate element
- in the complex field, 325
- in cyclotomic extension fields, 241
- in finite extension fields, 218

conjugate transpose (of complex matrices), 326
convex function, 9
convolutional code, 477
- constraint length of encoders of, 483
- decoding of, 485, 519
- encoding of, 479
- free distance of, 478, 512
- generator matrix of, 477

coset
- leader, 34
- of linear codes, 34
- of subgroups, 522

cover (of arrays), 354, 362
covering radius, 123, 137
- of GRS codes, 166
- of MDS codes, 166

crisscross error, 362
cycle (in graphs), 397
- in digraphs, 454

cyclic code, 242
    dual code of, 247
    encoding of, 245
    Hamming, 244, 254, 264
    of length q over GF(q), 257, 265
    of length q+1 over GF(q), 262, 265
    repeated-root, 265
    root of, 248
    shortened, 388
cyclotomic coset, 230
cyclotomic extension field, 240

dB, 18
decoding, 7
    of alternant codes, *see* alternant code, decoding of
    of BCH codes, *see* alternant code, decoding of
    complexity, 48
    of concatenated codes, 178, 371, 396, 422
    of convolutional codes, 485, 519
    error probability, 7, 110, 132, 133, 471, 485, 491, 519
    generalized minimum distance (GMD), 178, 371, 396, 422
    of graph codes, 414
    of GRS codes, *see* GRS decoding
    of Hamming codes, 35
    hard-decision, 18
    iterative, 414, 451
    of linear codes, 33
    list, *see* list decoding
        of GRS codes, *see* GRS list decoding
    maximum *a posteriori*, 8
    maximum-likelihood, 8, 32, 140, 466
    misdetection probability, 22, 125, 132, 140, 440
    nearest-codeword, 9, 33
    sequential, 520
    soft-decision, 18
    standard array, 33
    syndrome, 34
    of trellis codes, 466
degree (of extension fields), 57
degree (of polynomials), 51
    of bivariate polynomials, 268
degree (of vertices in graphs), 396
derivative, 65, 87, 194, 300, 322
    Hasse (or hyper-), 87, 276, 310, 329
designed minimum distance, 157, 163, 250
diameter (of graphs), 397
digraph (and labeled digraph), 400, 453
    adjacency matrix of, 486
    anticipation of, 503
    aperiodic irreducible, 455
    controllable, 454, 519
    deterministic, 456
    induced, 453
    irreducible, 454, 519
    irreducible component of, 455, 501
    irreducible sink of, 455, 501
    lossless, 456
    period of, 455, 502
    primitive irreducible, 455
    regular, 459
    sequence set of, 457
    strongly-connected, 454, 519
    tag in, 459
    trellis diagram of, 458
dimension (of codes), 5, 26
direct product (of matrices), 45, 428
directed graph, *see* digraph (and labeled digraph)
discrete logarithm, 59, 81, 91
discrete memoryless channel (DMC), 19, 23, 46, 142, 466
distance
    cover, 354, 362
    free, 463, 478, 512
    in graphs, 397
    Hamming, 6
    Kullback–Leibler, 111
    Lee, 299
    rank, 19, 353, 361
divergence, 111
dual basis, 85
dual code, 30
    of alternant codes, 175, 180
    of concatenated codes, 383
    of cyclic codes, 247
    of extended GRS codes, 163
    of GRS codes, 148
    of Hamming codes, 41
    of MDS codes, 119
    of RS codes, 257
    self-, 31
    of subfield sub-codes, 175
dual linear programming problem, 138
dynamic programming, 519

edge (in graphs), 396
    cut, 397
    in digraphs, 453
Elias bound, 108
    Lee-metric, 332
encoder, 3
entropy function
    binary, 9
    q-ary, 24, 105
erasure, 15
    burst, 45, 257
    channel, 15, 25, 126, 134, 391, 514
error, 12
    bit-shift, 327
    burst, 45, 122, 137, 257
    correction, 12
    crisscross, 362
    detection, 13
    evaluator polynomial, 186

exponent, 143, 381, 393, 425
location, 5
locator polynomial, 185
Lee-metric, 307
locator ratio, 307
peak-shift, 327
synchronization, 327
value, 5
word, 5
error-evaluator polynomial, 186
error-locator polynomial, 185
Lee-metric, 307
error-locator ratio, 307
Euclid's algorithm
for integers, 50, 501, 524
for polynomials, 52, 71, 90, 191, 215, 309
Euler–Fermat Theorem, 526
Euler function, 62, 229, 449, 522
expander (graph), 404
exponent (of polynomials), 206, 227, 247
extension field, 57, 218
arithmetic in, 59, 74, 90
conjugate element in, 218, 241
cyclotomic, 240

factorization of polynomials, 56, 90
Fano's algorithm, 520
Fermat's Little Theorem, 526
field, 522
characteristic of, 62
extension, *see* extension field
finite, *see* finite field
Galois, *see* finite field
prime, 50
of rational functions, 268, 476
splitting, 65
finite field, 50, 218, 240
characteristic of, 62
extension field of, *see* extension field
isomorphism, 227
prime, 50
product of elements in, 77
sum of powers of elements in, 77, 150
finite-state machine (FSM), 460
linear, *see* linear finite-state machine (LFSM)
formal derivative, *see* derivative
formal power series, 189, 237, 300, 474
bivariate, 487
Forney's algorithm, 195, 215
Fourier transform, 81, 92, 217
free distance, 463, 478, 512
Frobenius mapping, 64

Galois field, *see* finite field
Gaussian (AWGN) channel, 17
Gaussian elimination, 189, 274, 296
Gaussian noise, 17
generalized minimum distance (GMD) decoder, 178, 371, 396, 422
generalized Reed–Solomon code, *see* GRS code
generator matrix (of convolutional codes), 477
catastrophic, 497
LFSM realization of, 510
systematic, 477, 517
generator matrix (of linear codes), 27
systematic, 29
generator polynomial
of cyclic codes, 245
of negacyclic codes, 323
of RS codes, 152
Gilbert–Varshamov bound, 97, 137, 176, 181, 393
asymptotic, 107, 372
Lee-metric, 320, 330
GMD decoder, 178, 371, 396, 422
Golay code, 96, 136, 255
Goppa code, 182, 389
graph
adjacency matrix of, 398
bipartite, 362, 398
transfer matrix of, 399
Cayley, 406, 447
code, *see* graph code
connected, 397
directed, *see* digraph (and labeled digraph)
edge cut in, 397
expander, 404
Hamming, 427
hyper-, 445
incidence matrix of, 399
induced, 397
isomorphism, 47
labeled directed, *see* digraph (and labeled digraph)
Laplace matrix of, 399
matching in, 362
oriented, 400
Ramanujan, 409, 447
regular, 401
undirected simple, 396
graph code, 412
approaching the Singleton bound, 421
decoding of, 414
error detection with, 440
generalized, 420
Gray code, 321, 328
Griesmer bound, 120, 136
group, 521
Abelian, 521
code, 37, 299
commutative, 521
cyclic, 521
factor, 525
projective special linear, 409

quotient, 525
special linear, 409
GRS code, 148
canonical generator matrix of, 167
canonical parity-check matrix of, 148
code locator of, 148
decoding of, *see* GRS decoding
doubly-extended, 163, 335
dual code of, 148
encoding of, 152, 177, 216
extended, 150, 336
dual code of, 163
Lee-metric, 304
list decoding of, *see* GRS list decoding
narrow-sense, 150
normalized, 149
primitive, 149
singly-extended, 150, 335
systematic generator matrix of, 167
triply-extended, 165, 337
GRS decoding, 184
Berlekamp–Massey, 200, 217
bivariate interpolation, 269
Blahut's time-domain, 217
Chien search in, 186, 215
erasure, 207, 216
with Euclid's algorithm, 191, 215
extended, 210, 216
Forney's algorithm for, 195, 215
key equation of, 186, 215
Lee-metric, 308
Lee-metric, 312
list, *see* GRS list decoding
Peterson–Gorenstein–Zierler, 189, 215
singly-extended, 210, 216
Sugiyama *et al.* (SKHN), 215
Welch–Berlekamp, 215, 217
equations of, 211, 215, 271, 291, 295
GRS list decoding, 271
Guruswami–Sudan, 278, 296
Koetter–Vardy, 282, 296
radius, 271
Sudan's, 274, 296
Guruswami–Sudan algorithm, 278, 296

Hadamard matrix, 45, 49, 406, 436, 447
Hamming bound, *see* sphere-packing bound
Hamming code, 32, 47
binary, 32
cyclic, 244, 254, 264
decoding of, 35
list, 291
dual code of, 41
extended binary, 32
list decoding of, 291
weight distribution of, 101, 121
Hamming distance, 6
Hamming graph, 427
Hartmann–Tzeng bound, 265
Hasse derivative, 87, 310, 329
of bivariate polynomials, 276
hyper-derivative, *see* Hasse derivative
hyper-graph, 445

ideal, 522
incidence matrix, 399
infinite formal power series, 189, 237, 300, 474
bivariate, 487
integer (of fields), 63
integer programming, 103, 137
integral domain, 522
interleaver, 44
interpolation, 76
bivariate, 270
noisy, 151, 177
rational, 211
invertible element (in a ring), 190
irreducible matrix, 445
irreducible polynomial, 54, 90
enumeration of, 225

Jacobi's Four Square Theorem, 410
Jensen's inequality, 23
Johnson bound, 107, 128, 139, 289
Lee-metric, 330
Justesen code, 376

key equation, 186, 215
Lee-metric, 308
Koetter–Vardy algorithm, 282, 296
König's Theorem, 362
Krawtchouk polynomial, 103, 124, 137
Kronecker product (of matrices), 45, 428
Kullback–Leibler distance, 111

labeled directed graph, *see* digraph (and labeled digraph)
Lagrange's Theorem, 525
Laplace matrix, 399
Latin square, 351, 361
Laurent series, 476, 507
Law of Large Numbers, 11
Law of Quadratic Reciprocity, 411, 449
LDPC code, 362, 450
Lee
adjacency matrix, 325
distance, 299
weight, 299, 448
Legendre sequence, 80
autocorrelation of, 80
linear recurrence of, 239, 241
Legendre symbol, 80, 447
length (of codes), 5
likelihood ratio, 20
linear code, 26
decoding of, 33
encoding of, 28
over rings, 43, 299, 328, 478
linear-feedback shift register (LFSR), 198, 217

maximal-length, 237
linear finite-state machine (LFSM), 472
catastrophic, 496, 516
deterministic, 517
irreducible, 473, 505
length–weight enumerator of, 487
lossless, 476, 508
observable, 496, 506, 519
reduced, 496, 506
response matrix of, 474, 519
linear programming, 137
bound, 103, 110, 138
linearized polynomial, 83, 353
list decoding, 267
of alternant codes, 280, 328
error probability, 267
of GRS codes, *see* GRS list decoding
of Hamming codes, 291
radius, 267
bound on, 290
log likelihood ratio, 20
low-density parity-check code, 362, 450

MacWilliams' identities, 104, 124
matching (in graphs), 362
matrix
adjacency, 398, 486
aperiodic irreducible, 445
Cauchy, 168, 336, 356, 362
circulant, 325, 330, 356, 362
companion, 73, 383, 510
complex conjugate transpose of, 326
direct product of, 45, 428
generator, 27, 477
incidence, 399
irreducible, 445
Kronecker product of, 45, 428
Laplace, 399
Lee adjacency, 325
parity-check, 29
primitive irreducible, 445
super-regular, 335
Sylvester-type Hadamard, 45, 49, 406, 436, 447
transfer, 399
Vandermonde, 75
inverse of, 166
maximum-likelihood decoder, 8, 32, 140, 466
MDS code, 94, 119, 148, 164, 334
almost (AMDS), 363
bound on the length of, 338
conjecture, 342, 363
dual code of, 119
near (NMDS), 363
over polynomial rings, 355
uniqueness of, 347
weight distribution of, 104
MDS conjecture, 342, 363
Mersenne prime, 77
message node, 449
message-passing algorithm (MPA), 451
metric, 6
cover, 354, 362
Hamming, 6
Lee, 299
rank, 19, 353, 361
minimal polynomial, 219, 241
minimum distance, 6
computation complexity of, 48, 177
designed, 157, 163, 250
Lee, 299
relative, 104
Möbius function, 224, 523
MRRW bound, 110
multiplication circuit, 153
multiplicative order
in fields, 51
in groups, 521
in rings, 72, 206, 228
mutual information, 23

negacyclic code, 323, 330
Newton's identities, 301, 328
node
check, 449
message, 449
variable, 449
normal basis, 240
normal rational curve, 362

order, *see* multiplicative order
orthogonal array, 334, 361

parity-check matrix, 29
parity code, 27
path (in graphs), 397
in digraphs, 453
peak-shift error, 327
perfect code, 96, 137, 256
Lee-metric, 319, 330
period (of irreducible digraphs), 455, 502
period (of sequences), 206, 237, 305
Perron–Frobenius Theorem, 445, 518
Peterson–Gorenstein–Zierler algorithm, 189, 215
phase-shift keying (PSK), 327
Plotkin bound, 37, 127, 131, 139, 294
Lee-metric, 326, 330
polynomial, 51
bivariate, 268
characteristic, 73
check, 246
degree, 51
error-evaluator, 186
error-locator, 185
Lee-metric, 307
evaluation, 177, 216
exponent of, 206, 227, 247
factorization, 56, 90
generator, 152, 245, 323
interpolation, 76, 151, 177, 211, 270

irreducible, 54, 90
enumeration of, 225
Krawtchouk, 103, 124, 137
linearized, 83, 353
minimal, 219, 241
monic, 52
multivariate, 177, 295
primitive, 228
enumeration of, 229
quadratic, 78, 88
root of, 59
syndrome, 185
power series, 189, 237, 300, 474
bivariate, 487
Prime Number Theorem, 411, 449
primitive element, 51, 61, 91
primitive irreducible digraph, 455
primitive irreducible matrix, 445
primitive polynomial, 228
enumeration of, 229
product code, 44, 178
projective geometry, 47, 361
Prouhet–Tarry problem, 329

q-ary erasure channel, 15, 25, 126, 134, 391, 514
q-ary symmetric channel, 4, 19, 24, 32, 110, 113, 117, 125, 133, 145, 378, 390, 394, 424, 467, 490
quadratic polynomial, 78, 88
Quadratic Reciprocity Law, 411, 449
quadratic residue, 77, 239, 410, 447
Euler's criterion for, 80, 448
Gauss' criterion for, 448

random coding, 137
Gallager's error exponent of, 143
rate, 5
of convolutional codes, 477
of trellis codes, 462
recurrence order, 197, 214
redundancy, 27
Reed–Muller code, 41, 48, 177, 260, 265
first-order, 41, 46, 120, 139, 327, 385, 408
Reed–Solomon code, *see* RS code
Reiger bound, 122
remaindering circuit, 73, 153
repetition code, 28
ring, 522
Roos bound, 265
root, 59
of BCH codes, 163, 250
of bivariate polynomials, 268
computation, 90, 284, 296
of cyclic codes, 248
multiplicity, 60, 76
of RS codes, 152, 163, 250
RS code
conventional, 151, 243
dual code of, 257
encoding of, 152, 178
extended, 152
generalized, *see* GRS code
generator polynomial of, 152
root of, 152, 163, 250

self-dual code, 31
sequence, 189
Legendre, 80
autocorrelation of, 80
linear recurrence of, 239, 241
linear-recurring, 198, 217, 237
M-, 237, 241
maximal-length, 237, 241
autocorrelation of, 87
periodic, 206, 305
trace, 236, 240
Shannon Coding Theorem, 139
for BSC, 10
with error detection, 132
for q-ary erasure channel, 135
for q-ary symmetric channel, 117
Shannon Converse Coding Theorem, 139
for BSC, 10
for q-ary erasure channel, 134
for q-ary symmetric channel, 113
shift register, 198, 217, 237
signal-to-noise ratio (SNR), 18
simplex code, 41, 120
weight distribution of, 99
Singleton bound, 94, 136, 332, 334
asymptotic, 105, 421, 451
cover-metric, 354, 362
rank-metric, 353, 361
size (of codes), 5
spectral-null code, 329
sphere, 12
Lee, 317
volume, 95
sphere-covering bound, 123
sphere-packing bound, 95, 122, 136
asymptotic, 107
Lee-metric, 318, 330
splitting field, 65
stack algorithm, 520
standard array decoding, 33
state (in digraphs), 453
subfield sub-code, 174
subgroup, 521
normal, 522
subring, 522
Sudan's algorithm, 274, 296
Sugiyama *et al.* (SKHN) algorithm, 215
super-multiplicative sequence, 331
super-regular matrix, 335
Sylvester-type Hadamard matrix, 45, 49, 406, 436, 447
synchronization error, 327
syndrome, 34

polynomial, 185

tag (of edges), 459
term rank (of arrays), 362
trace
of elements in a finite field, 83
of matrices, 331, 514
sequence, 236, 240
transfer matrix (of bipartite graphs), 399
trellis code, 460
decoding of, 466
encoding of, 464
free distance of, 463
triangle inequality, 6
turbo code, 519

union bound, 137
unit element (in a ring), 190
unity element, 521

variable node, 449
vertex (in graphs), 396
Viterbi's algorithm, 468, 519
volume
of Hamming spheres, 95
of Lee spheres, 317

weight, 6
cover, 354, 362
distribution, *see* weight distribution
enumerator, 99, 486
Lee, 299, 448
weight distribution, 99
of cycles in LFSMs, 487
of Hamming codes, 101, 121
of MDS codes, 104
of simplex codes, 99
Weil's Theorem, 179
Welch–Berlekamp
algorithm, 215, 217
equations, 211, 215, 271, 291, 295
Wyner–Ash code, 512

Zyablov bound, 373, 392, 413, 422, 438, 440

Printed in the United States
By Bookmasters